MATHEMATICAL FOUNDATIONS FOR LINEAR CIRCUITS AND SYSTEMS IN ENGINEERING

MATHEMATICAL FOUNDATIONS FOR LINEAR CIRCUITS AND SYSTEMS IN ENGINEERING

JOHN J. SHYNK
Department of Electrical and Computer Engineering
University of California, Santa Barbara

WILEY

Published by John Wiley & Sons, Inc., Hoboken, New Jersey
Published simultaneously in Canada

Library of Congress Cataloging-in-Publication Data applied for:

ISBN: 9781119073475

Set in 10/12pt, TimesLTStd by SPi Global, Chennai, India.

10 9 8 7 6 5 4 3 2 1

1 2016

To Tokie

In memory of N. and A.

CONTENTS

PREFACE

The main goal of this book is to provide the mathematical background needed for the study of linear circuits and systems in engineering. It is more rigorous than the material found in most circuit theory books, and it is appropriate for upper-division undergraduate students and first-year graduate students. The book has the following features:

- A comparison of linear circuits and mechanical systems that are modeled by similar ordinary differential equations. This provides a greater understanding of the behavior of different types of linear time-invariant circuits and systems.
- Numerous tables and figures summarize several mathematical techniques and provide example results. Although the focus of the book is on the equations used in engineering models, it includes over 250 figures and plots generated using MATLAB that reinforce the material and illustrate subtle points.
- Several appendices provide background material on set theory, series expansions, various identities, and the Lambert W-function. An extensive summary of important functions and their transforms encountered in the study of linear systems is included in Appendix A.
- A brief introduction to the theory of generalized functions, which are defined by their properties under an integral. This theory is connected to the Laplace and Fourier transforms covered later, which are specific types of integral transforms of time-domain functions.

After the overview in Chapter 1, which includes a brief review of functions and calculus, the book is divided into two parts:

- **Part I**: Circuits and Mechanical Systems; Linear Equations and Matrices; Complex Numbers and Functions (Chapters 2–4).
- **Part II**: Signals, Generalized Functions, and Fourier Series; Differential Equation Models for Linear Systems; Laplace Transforms and Linear Systems; Fourier Transforms and Frequency Responses (Chapters 5–8).

Chapter 2 describes circuits consisting of resistors (R), capacitors (C), and inductors (L), as well as Kirchoff's circuit laws and mesh and nodal analysis techniques. There is a brief study of nonlinear diode circuits and then a discussion of some mechanical systems that have the same time-domain properties as RL, RC, and RLC circuits. Linear algebra and systems of linear equations are covered in Chapter 3, along with the matrix determinant, matrix subspaces, LU and LDU decompositions, and eigendecompositions. Equations that model the voltages and currents in a resistive circuit are represented using matrices, and the solutions are derived using either Cramer's rule or Gaussian elimination. Chapter 4 contains a thorough discussion of complex numbers, with material not covered in most books on linear circuits and systems. It includes matrix representations of complex quantities, exponential rotations on the complex plane, the constant angular velocity of time-varying complex functions, and a brief discussion of quaternions.

Chapter 5 gives definitions of several signals that describe the dynamic behavior of linear circuits and systems, including ordinary functions such as the exponential function and singular functions like the Dirac delta function. A brief introduction to the theory of generalized functions is provided, which illustrates several of their properties and in particular how their derivatives are found. This chapter also includes Fourier series representations of periodic signals and a view of their coefficients as cross-correlations between the original signal and sinusoidal signals with increasing frequency. First- and second-order ordinary differential equations used to model RL, RC, and RLC circuits are then covered in Chapter 6. The solutions are derived entirely in the time domain, and it is demonstrated that second-order linear systems can have three types of responses depending on their parameter values. Phasor notation and impedance for circuits with sinusoidal source signals are also discussed.

The final two chapters describe transform techniques for solving the ODEs developed in Chapter 6 and for illustrating their frequency characteristics. Chapter 7 defines the unilateral and bilateral Laplace transforms, focusing on causal systems with initial conditions. Several Laplace transform properties are proved, and these are used to solve linear ODEs as well as linear circuits directly in the s-domain. Partial fraction expansions for different pole configurations are discussed in detail, and the significance of pole locations relative to the imaginary axis on the complex plane is described. Finally, Chapter 8 covers the Fourier transform and describes how it is related to the Laplace transform. Various first- and second-order filters are discussed, including the four basic types: low-pass, high-pass, band-pass, and band-reject. These low-order filters are extended using high-order Butterworth filters to generate sharper frequency responses. Amplitude modulation with and without suppressed carrier is also briefly discussed.

Several appendices provide background material for the topics covered in this book:

- Extensive summaries of 21 functions that include their Fourier and Laplace transforms and various signal properties.
- Two tables of inverse Laplace transforms where the Laplace transform is given first, so that the time-domain function is found without computing a partial fraction expansion.
- Trigonometric identities, summation formulas, quadratic and cubic formulas, derivatives, several integrals, and their properties.
- Set theory, set operations, Venn diagrams, and partitions.
- Series expansions, including Taylor, Maclaurin, and Laurent series, and the different types of singularities.
- The Lambert W-function, which is useful for finding an explicit expression for some nonlinear equations that cannot be solved using ordinary functions.

The book is designed for a two-quarter sequence or a single semester covering continuous-time linear systems and related signals, represented by ordinary and singular functions. For a two-quarter sequence, the chapters might be covered as follows:

- Fall quarter: Chapters 1–5. Winter quarter: Chapters 6–8.

Depending on prerequisite courses, it may not be necessary to cover all the topics on matrices in Chapter 3, in which case some of that chapter would serve as reference material for related courses. This is probably the case for the semester system where some material would be emphasized less in order to complete the chapters on the Laplace and Fourier transforms. Since this book is mainly mathematical in nature, it does not cover all circuit theory techniques as is usually done at the sophomore level in electrical engineering programs. Instead, just enough material on circuits has been included, as well as some mechanical systems, in order for the reader to learn how systems of linear equations and ODEs are developed. It is the goal of this book to provide a more comprehensive mathematical study of the various topics than is usually done in circuits courses and to explain subtle points with examples and figures throughout the chapters.

I would like to thank S. Chandrasekaran, R. Pauplis, and A. Nguyen-Le for discussions of some of the material in this book. I am indebted to my students in the ECE 2 series on circuits and systems whose questions have provided the motivation to write a book that focuses on mathematical models for signals and systems. Their comments have led directly to some of the discussions and illustrative examples. Finally, thanks to my editors at Wiley, B. Kurzman and A. Castro, for supporting this project, and to R. Roberts and N. Swaminathan for their assistance during the final stages of production.

J.J.S.
Santa Barbara, CA
October 2015

NOTATION AND BIBLIOGRAPHY

We provide an overview of the notation used in this book.

- Lowercase i and v represent currents and voltages that may or may not be time-varying. If they are time-varying, we may explicitly write $i(t)$ and $v(t)$.
- Uppercase I and V represent constant currents and voltages.
- Bold lowercase $\mathbf{b}$ denotes a column vector, and bold uppercase $\mathbf{A}$ denotes a matrix.
- Notation such as $C(\mathbf{A})$ refers to the column space of matrix $\mathbf{A}$ and should not be confused with the usual notation for a function such as $f(t)$.
- Roman letters s are units (seconds), and italic letters s are variables ($s = \sigma + j\omega$). Similarly for F (farads), F (force), and so on.
- The notation $\langle f, \phi \rangle$ represents the generalized function f with test function ϕ. It should not be confused with the inner product notation $\langle \mathbf{x}, \mathbf{y} \rangle$, which is written in this book using transpose $\mathbf{x}^T\mathbf{y}$.
- In order to be concise, in many equations, expressions like $1/2\pi j$ are equivalent to $1/(2\pi j)$.

In the glossary near the end of the book, there is a summary of the notation and symbols used throughout the book: (i) general symbols and notation, (ii) Greek symbols, (iii) calligraphic symbols (for different sets of numbers and transforms), (iv) mathematical notation (including relational and arrow symbols), (v) physical parameter values (for circuits and mechanical systems), and (vi) abbreviations (acronyms).

At the end of the book, the bibliography contains many references that the reader might find useful for further study of the topics in each of the chapters and the appendices. References are not cited in the text except in cases for material that might be less familiar and is not covered in most books on linear circuits and systems.

ABOUT THE COMPANION WEBSITE

This book is accompanied by a companion website:
http://www.wiley.com/go/linearcircuitsandsystems

The website includes:

- Solutions Manual available for instructors
- MATLAB files for some problems
- Updated errata

1

OVERVIEW AND BACKGROUND

1.1 INTRODUCTION

In this book, we develop and examine several *mathematical models* consisting of one or more equations that are used in engineering to represent various *physical systems*. Usually, the goal is to solve these equations for the unknown dependent variables, and if that is not possible, the equations can be used to *simulate* the behavior of a system using computer software such as MATLAB.[1] In most engineering courses, the equations are usually *linear* or can be linearized as an approximation, but sometimes they are nonlinear and may be difficult to solve. From such models, it is possible to design and analyze components of a proposed system in order to achieve required performance specifications before developing a prototype and actually implementing the physical system.

Definition: System A *system* is a collection of interacting elements or devices that together result in a more complicated structure than the individual components alone, for the purpose of generating a specific type of signal or realizing a particular process.

The term system, as used in this book, also describes several interrelated equations called a *system of equations*, which are usually linear and can be represented by a

[1]MATLAB® is a registered trademark of The Mathworks, Inc., 3 Apple Hill Drive, Natick, MA.

Mathematical Foundations for Linear Circuits and Systems in Engineering, First Edition. John J. Shynk.

Companion Website: http://www.wiley.com/go/linearcircuitsandsystems

matrix equation. The distinction between a physical system and a system of linear equations will be evident from the specific application.

Definition: Mathematical Model A *mathematical model* is an equation or set of equations used to represent a physical system, from which it is possible to predict the properties of the system and its output response to an input, given known parameters, certain variables, and initial conditions.

Generally, we are interested in the *dynamic behavior* of a system over time as it responds to one or more time-varying input signals. A block diagram of a system with single input $x(t)$ and single output $y(t)$ (single-input single-output (SISO)) is shown in Figure 1.1(a), where t is continuous time. The time variable can be defined for the entire real line $\mathcal{R}$: $-\infty < t < \infty$, but often we assume nonnegative $\mathcal{R}^+$: $0 \leq t < \infty$. In this scenario, a mathematical model provides the means to observe how $y(t)$ varies with $x(t)$ over t, assuming known initial conditions (usually at $t = 0$), so that we can predict the future behavior of the system. For the electric circuits described in Chapter 2, the inputs and outputs are *currents* through or *voltages* across the circuit components. For convenience, Table 1.1 summarizes the notation for different sets of numbers used in this book (though quaternions are only briefly discussed in Chapter 4).

Figure 1.1(b) shows a linear SISO system with sinusoidal input $\cos(2\pi f_o t)$ where f_o is ordinary frequency in hertz (Hz). As discussed in Chapter 7, a sinusoidal signal is an *eigenfunction* of a linear system, which means that the output is also sinusoidal with the same frequency f_o. For such a signal, the output differs from the input by having a different magnitude, which is A in the figure, and possibly a phase shift ϕ. This is an important characteristic of linear systems that allows us to investigate them in the so-called *frequency domain*, which provides information about their properties beyond those observed in the time domain.

In order to more easily solve for the unknown variables of a mathematical model, the techniques usually require knowledge of *matrices* and *complex numbers*. The matrices covered in Chapter 3 are useful for describing a system of linear equations

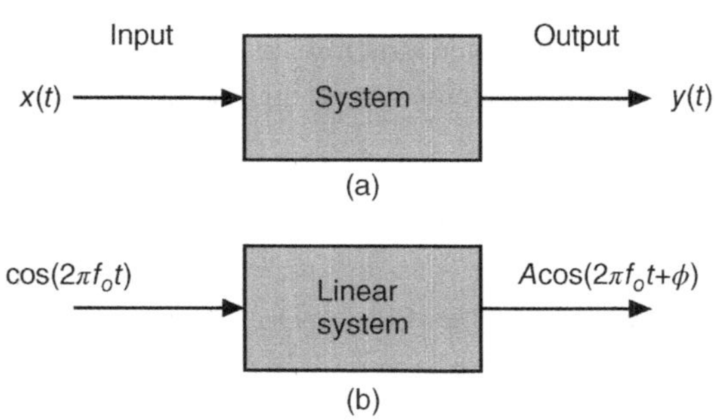

Figure 1.1 Systems with a single input and a single output (SISO). (a) General system with input $x(t)$ and output $y(t)$. (b) Linear system with sinusoidal input and output.

TABLE 1.1 Symbols for Sets of Numbers

Symbol	Domain x	Set
$\mathcal{R}$	$x \in (-\infty, \infty)$	Real numbers
$\mathcal{R}^+$	$x \in [0, \infty)$	Nonnegative real numbers
$\mathcal{Z}$	$x \in \{\ldots, -2, -1, 0, 1, 2, \ldots\}$	Integers
$\mathcal{Z}^+$	$x \in \{0, 1, 2, \ldots\}$	Nonnegative integers
$\mathcal{N}$	$x \in \{1, 2, \ldots\}$	Natural numbers
$\mathcal{Q}$	$x = a/b$ with $a, b \in \mathcal{Z}$ and $b \neq 0$	Rational numbers
$\mathcal{I}$	$x = jb$ with $j = \sqrt{-1}$ and $b \in \mathcal{R}$	Imaginary numbers
$\mathcal{C}$	$x = a + jb$ with $j = \sqrt{-1}$ and $a, b \in \mathcal{R}$	Complex numbers
$\mathcal{H}$	$x = a + ib_1 + jb_2 + kb_3$ with $i = j = k = \sqrt{-1}$ and $a, b_1, b_2, b_3 \in \mathcal{R}$	Quaternions

with constant coefficients. Chapter 4 provides the motivation for complex numbers and summarizes many of their properties. Chapter 5 introduces several different waveforms that are used to represent the signals of a system: inputs, outputs, as well as internal waveforms. These include the well-known sinusoidal and exponential signals, as well as the *unit step function* and the *Dirac delta function*. The theory of *generalized functions* and some of their properties are briefly introduced. Systems represented by linear ordinary differential equation*s* (ODEs) are then covered in Chapter 6, where they are solved using conventional *time-domain* techniques. The reader will find that such techniques are straightforward for first- and second-order ODEs, especially for the linear circuits covered in this book, but are more difficult to use for higher order systems.

Chapter 7 describes methods based on the *Laplace transform* that are widely used in engineering to solve linear ODEs with constant coefficients. The Laplace transform converts an ODE into an *algebraic equation* that is more easily solved using matrix techniques. Finally, Chapter 8 introduces methods for analyzing a system in the frequency domain, which provides a characterization of its frequency response to different input waveforms. In particular, we can view linear circuits and systems as *filters* that modify the frequency content of their input signals.

We focus on *continuous-time* systems, which means $\{x(t), y(t)\}$ are defined with *support* $t \in \mathcal{R}$ or $t \in \mathcal{R}^+$ where the functions are nonzero. *Discrete-time* systems and signals are defined for a countable set of time instants such as $\mathcal{Z}$, $\mathcal{Z}^+$, or $\mathcal{N}$. Different but related techniques are used to examine discrete-time systems, though these are beyond the scope of this book.

1.2 MATHEMATICAL MODELS

Consider again the system in Figure 1.1(a) and assume that we have access only to its input $x(t)$ and output $y(t)$ as implied by the block diagram. There is no direct information about the internal structure of the system, and the only way we can learn about its properties is by providing input signals and observing the output signals. Such an unknown system is called a "black box" (because we cannot see inside), and the procedure of examining its input/output characteristics is a type of *reverse engineering*. We mention this because the mathematical models used to represent physical devices and systems are typically verified and even derived from experiments with various types of input/output signals. Such an approach yields the *transfer characteristic* of the system, and for linear and time-invariant (LTI) systems, we can write a specific *transfer function* as described in Chapter 7.

Example 1.1 Suppose input x of an unknown system is varied over $\mathcal{R}$ and we observe the output y shown in Figure 1.2. This characteristic does not change with time, and so we have suppressed the time argument for the input and output. The plot of y is flat for three intervals: $-\infty < x \le -2$, $-1 < x \le 2$, and $3 < x < \infty$, and it is linearly increasing for two intervals: $-2 < x \le -1$ and $2 < x \le 3$. For this *piecewise linear function*, the equation for each interval has the form $y = ax + b$ where $a = \Delta y/\Delta x$ is the *slope* and b is the *ordinate*, which is the point where the line crosses the y-axis if it were extended to $x = 0$. For the first linearly increasing region, the slope is obviously $a = (1-0)/[-1-(-2)] = 1$. When $x = 0$, the extended line crosses the y-axis at $y = 2$, which gives $b = 2$. Similarly, for the second linearly increasing region, $a = (3-1)/(3-2) = 2$ and $b = -3$. The remaining three regions have zero slope but different ordinates (these equations are of the form $y = b$), and so the overall transfer characteristic for this system is

$$y = \begin{cases} 0, & x \le -2 \\ x+2, & -2 < x \le -1 \\ 1, & -1 < x \le 2 \\ 2x-3, & 2 < x \le 3 \\ 3, & x > 3. \end{cases} \tag{1.1}$$

The values of y match at the boundaries for each interval of x as shown in the figure. The *mapping* in (1.1) is a mathematical model for a particular system that can be used to study its behavior even if it is included as part of a larger system. Note that this

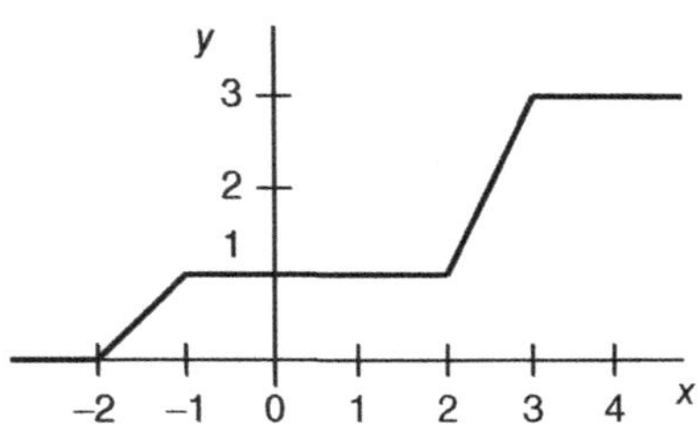

Figure 1.2 Input/output characteristic for the nonlinear system in Example 1.1.

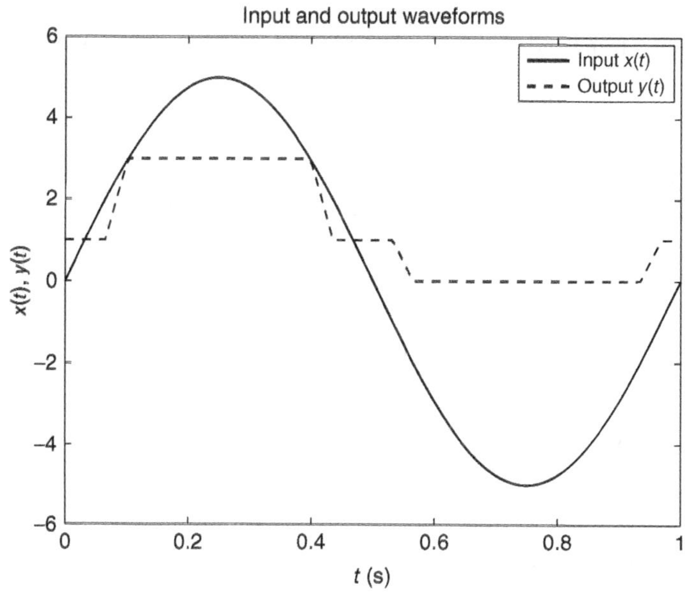

Figure 1.3 Output $y(t)$ for the transfer characteristic in (1.1) in Example 1.1 with input $x(t) = 5\sin(2\pi t)$ for $t \in [0, 1]$.

input/output characteristic does not provide any direct information about the individual components or the internal dynamics of the system. When the input $x(t)$ is a function of time, the output $y(t)$ is also time varying. For example, suppose that $x(t) = 5\sin(2\pi t)$ as illustrated in Figure 1.3 for one period of the sine function with frequency $f_o = 1$ Hz. The output $y(t)$ is computed using (1.1) at each time instant on the closed interval $t \in [0, 1]$ in seconds (s). Observe that $y(t)$ is *truncated* relative to the input waveform due to this particular input/output mapping. Similar results for $y(t)$ can be derived for any input function $x(t)$ by using the model in (1.1).

The output $y(t)$ is *not* sinusoidal because the function in Figure 1.2 is piecewise linear, and so, overall it is *nonlinear*. Sinusoidal signals are not eigenfunctions for nonlinear systems as demonstrated in this example. Eigenfunctions and their defining properties are covered later in Chapter 7. The *fundamental frequency* of the output in Figure 1.3 is $f_o = 1$ Hz because the waveform for all $t \in \mathcal{R}$ consists of repetitions of the 1 s segment dashed curve. The waveform within this segment also has variations, which result in *harmonics* of f_o. This means that sinusoidal components with integer multiples of f_o are also present in $y(t)$. It is possible to determine these harmonics using a *Fourier series* representation of $y(t)$ as discussed in Chapter 5.

Example 1.2 Consider the following mapping:

$$y = 2x - 3, \quad x \in \mathcal{R}, \tag{1.2}$$

which is one component of (1.1) with support extended to the entire real line, and so, the input is not truncated. For $x(t) = 5\sin(2\pi t)$, the output of this system is

$$y(t) = 10\sin(2\pi t) - 3, \tag{1.3}$$

which has the same frequency $f_o = 1$ Hz as the input; there are no harmonics of f_o. However, this system is not linear because it introduces a DC ("direct current") component at $f = 0$ Hz, which causes the output to be shifted downward, as illustrated in Figure 1.4 (the dashed line). The function in (1.2) is actually *affine* because of the nonzero ordinate $b = -3$. A linear function is obtained by dropping the ordinate:

$$y = 2x, \qquad x \in \mathcal{R}, \tag{1.4}$$

which has the output in Figure 1.4 (the dotted line). This is a trivial system because the peak amplitude 10 of the output is unchanged for any input frequency f_o, and the phase shift ϕ is always zero.

A linear system that is modeled by an ODE has a more complicated representation than the simple scaling in (1.4), and the amplitude and phase of its output generally change with frequency f_o. By varying the frequency of the input and observing the output of a linear system, we can derive its frequency response. This representation of a system indicates which frequency components of a signal are attenuated or

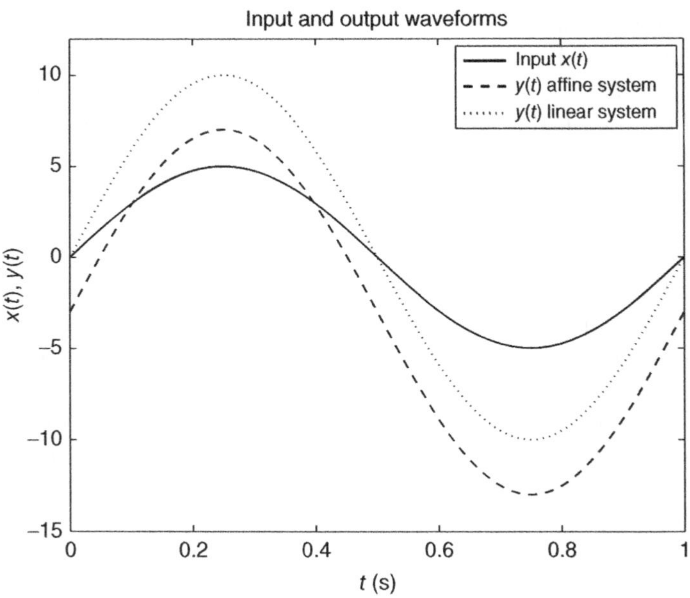

Figure 1.4 Output $y(t)$ for the transfer characteristics in (1.2) and (1.4) in Example 1.2 with input $x(t) = 5\sin(2\pi t)$ for $t \in [0, 1]$.

amplified and whether they are shifted in time. Using this approach, the system can be viewed as a type of filter that modifies the frequency characteristics of the input signal. For example, a low-pass filter retains only low-frequency components while attenuating or blocking high frequencies. It is useful in many applications such as noise reduction in communication systems. The frequency response of a system is investigated further in Chapter 8 where we cover the *Fourier transform.*

Example 1.3 An example of a system of linear equations is

$$a_{11}y_1(t) + a_{12}y_2(t) = x_1(t), \tag{1.5}$$

$$a_{21}y_1(t) + a_{22}y_2(t) = x_2(t), \tag{1.6}$$

where $\{y_1(t), y_2(t)\}$ are unknown outputs, $\{x_1(t), x_2(t)\}$ are known inputs, and $\{a_{mn}\}$ are constant coefficients. (Many books on linear algebra have x and y interchanged. We use the form in (1.5) and (1.6) for notational consistency throughout the book, where known x is the input and unknown y is the output.) These equations can be viewed as a multiple-input multiple-output (MIMO) system as depicted in Figure 1.5. It is straightforward to solve for the unknown variables $\{y_1(t), y_2(t)\}$ by first rearranging (1.6) as

$$y_2(t) = x_2(t)/a_{22} - a_{21}y_1(t)/a_{22}, \tag{1.7}$$

and then substituting (1.7) into (1.5):

$$a_{11}y_1(t) + a_{12}x_2(t)/a_{22} - a_{12}a_{21}y_1(t)/a_{22} = x_1(t), \tag{1.8}$$

which gives

$$y_1(t) = \frac{x_1(t) - a_{12}x_2(t)/a_{22}}{a_{11} - a_{12}a_{21}/a_{22}} = \frac{a_{22}x_1(t) - a_{12}x_2(t)}{a_{11}a_{22} - a_{12}a_{21}}, \tag{1.9}$$

and likewise for the other output:

$$\begin{aligned} y_2(t) &= x_2(t)/a_{22} - (a_{21}/a_{22})\frac{a_{22}x_1(t) - a_{12}x_2(t)}{a_{11}a_{22} - a_{12}a_{21}} \\ &= \frac{a_{11}x_2(t) - a_{21}x_1(t)}{a_{11}a_{22} - a_{12}a_{21}}. \end{aligned} \tag{1.10}$$

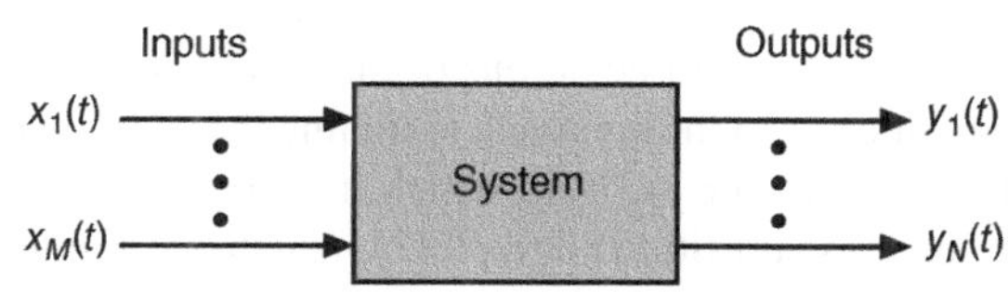

Figure 1.5 Multiple-input and multiple-output (MIMO) system.

The reader may recognize that if (1.5) and (1.6) are written in matrix form as described in Chapter 3, then the denominator in (1.9) and (1.10) is the *determinant* $\det(\mathbf{A}) = a_{11}a_{22} - a_{12}a_{21}$ of the matrix

$$\mathbf{A} \triangleq \begin{bmatrix} a_{11} & a_{12} \\ a_{21} & a_{22} \end{bmatrix}. \tag{1.11}$$

It is usually convenient to write such systems of equations in matrix form, because it is then straightforward to examine their properties based on the structure and elements of $\mathbf{A}$. Moreover, we can write the solution of the linear equations $\mathbf{A}\mathbf{y}(t) = \mathbf{x}(t)$ via the matrix inverse as $\mathbf{y}(t) = \mathbf{A}^{-1}\mathbf{x}(t)$, where for this two-dimensional matrix, the column vectors are

$$\mathbf{x}(t) \triangleq \begin{bmatrix} x_1(t) \\ x_2(t) \end{bmatrix}, \quad \mathbf{y}(t) \triangleq \begin{bmatrix} y_1(t) \\ y_2(t) \end{bmatrix}. \tag{1.12}$$

For a numerical example, let the matrix elements be $a_{11} = a_{21} = a_{22} = 1$ and $a_{12} = -0.1$, and assume the inputs are constant: $x_1(t) = 0$ and $x_2(t) = 1$. Then from (1.9) and (1.10), we have the explicit solution $y_1(t) = 1/11 \approx 0.0909$ and $y_2(t) = 10/11 \approx 0.9091$.

Example 1.4 In this example, we examine a *nonlinear* system to illustrate the difficulty of solving for the output variables of such models. A MIMO system is described by two equations, the first of which is nonlinear:

$$a_{11}y_1(t) + a_{12}\exp(\alpha y_2(t)) = x_1(t), \tag{1.13}$$

$$a_{21}y_1(t) + a_{22}y_2(t) = x_2(t), \tag{1.14}$$

where α and the coefficients $\{a_{mn}\}$ are constant parameters. This system is similar to the one in Example 1.3, except that a_{12} multiplies the *exponential function*

$$\exp(\alpha y_2(t)) \triangleq e^{\alpha y_2(t)}, \tag{1.15}$$

where e is Napier's constant which is reviewed later in this chapter. The inputs are again $\{x_1(t), x_2(t)\}$, and we would like to find a solution for $\{y_1(t), y_2(t)\}$. Unlike the linear system of equations in the previous example, eliminating one variable by substituting one equation into the other does not yield a closed-form solution because of the exponential function. Figure 1.6(a) shows examples of these two equations, obtained by plotting y_1 versus y_2 for the parameters used at the end of Example 1.3 and with $\alpha = 4$. Since $\{y_n\}$ must simultaneously satisfy both equations, it is clear that the solution for this system of equations occurs where the two curves (solid and dashed) in the figure intersect. One approach to finding the solution is *iterative*, where an initial estimate is chosen for y_2, from which it is possible to solve for y_1 using (1.14). This value for y_1 is substituted into (1.13), which is rewritten as follows:

$$y_2 = (1/\alpha)\ln((x_1 - a_{11}y_1)/a_{12}), \tag{1.16}$$

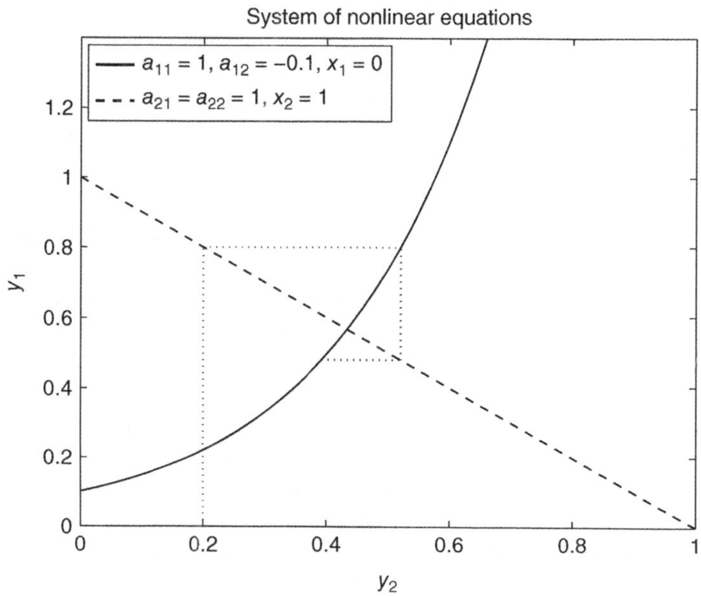

(a)

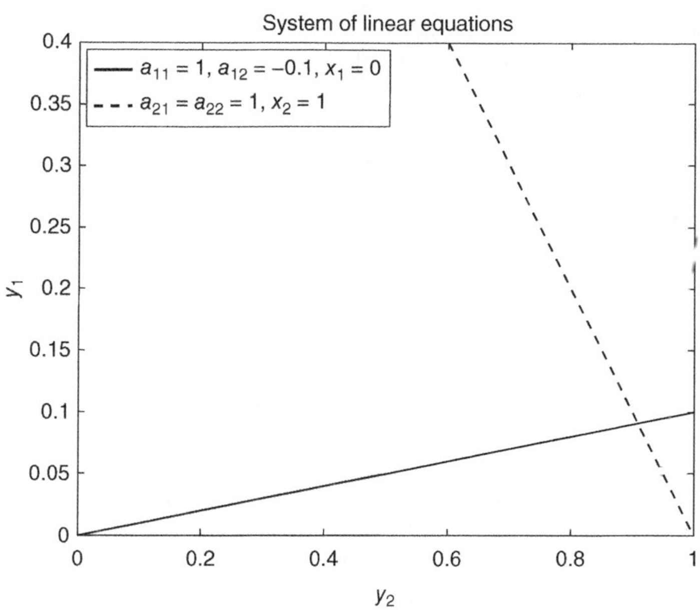

(b)

Figure 1.6 Systems of equations. (a) Nonlinear system in (1.13) and (1.14) in Example 1.4 with $\alpha = 4$. (b) Linear system in (1.5) and (1.6) in Example 1.3.

where ln (·) is the *natural logarithm*. This equation yields a new value for y_2, which is used in (1.14) to compute y_1, and the procedure is repeated several times until $\{y_n\}$ no longer change (up to some desired numerical precision), and so they have *converged* to a solution. For the aforementioned parameters and initial value $y_2 = 0.2$, we find using MATLAB that the solution is $y_1 \approx 0.5664$ and $y_2 \approx 0.4336$, which is verified by the intersecting curves in Figure 1.6(a). The first four iterations are denoted by the dotted lines in the figure, which we see approach the solution. For comparison purposes, Figure 1.6(b) shows the two lines for the linear system in Example 1.3 using the same coefficient values. The solution is located where the two lines intersect: $y_1 \approx 0.0909$ and $y_2 \approx 0.9091$. Since this system of equations is linear, we can solve for y_1 and y_2 explicitly as was done in (1.9) and (1.10) (there is no need to perform iterations).

We mention that it is possible to find a type of explicit solution for the system of nonlinear equations in the previous example by using the Lambert W-function described in Appendix F, which includes some examples. Nonlinear circuit equations for the diode are briefly discussed in Chapter 2, and an explicit solution using the Lambert W-function for a simple diode circuit is derived in Appendix F. Although an explicit solution is obtained, it turns out that the Lambert W-function cannot be written in terms of ordinary functions, and so, it must be solved numerically.

The transfer characteristic in Example 1.1 is *static* because it describes the output $y(t)$ for a given input $x(t)$ independently of the time variable t. For many physical systems, the transfer characteristic also depends on other factors, such as the *rate* at which $x(t)$ changes over time. This type of system is modeled by an ODE. In subsequent chapters, we describe techniques used to evaluate and solve linear ODEs for systems in general as in Figure 1.1 and for linear circuits in particular.

Example 1.5 An example of a linear ODE is

$$\frac{d^2}{dt^2}y(t) + a_1\frac{d}{dt}y(t) + a_0 y(t) = x(t), \quad t \in \mathcal{R}, \tag{1.17}$$

where time t is the *independent variable*, $y(t)$ is the *unknown* dependent variable, and $x(t)$ is the *known* dependent variable. For the system in Figure 1.1(a), $x(t)$ is the input and $y(t)$ is the output. The coefficients $\{a_0, a_1\}$ are fixed, and the goal is to find a solution for $y(t)$ given these parameters as well as the initial conditions $y(0)$ and $y'(0)$. The superscript denotes the ordinary derivative of $y(t)$ with respect to t, which is then evaluated at $t = 0$:

$$y'(0) \triangleq \left.\frac{d}{dt}y(t)\right|_{t=0.} \tag{1.18}$$

Equation 1.17 is a *second-order* ODE because it contains the second derivative of $y(t)$; higher order derivatives are considered in Chapter 7. An implementation based on *integrators* is illustrated in Figure 1.7. This configuration is preferred in practice because differentiators enhance additive noise in a system (Kailath, 1980), which can overwhelm the signals of interest. Integrators, on the other hand, average out

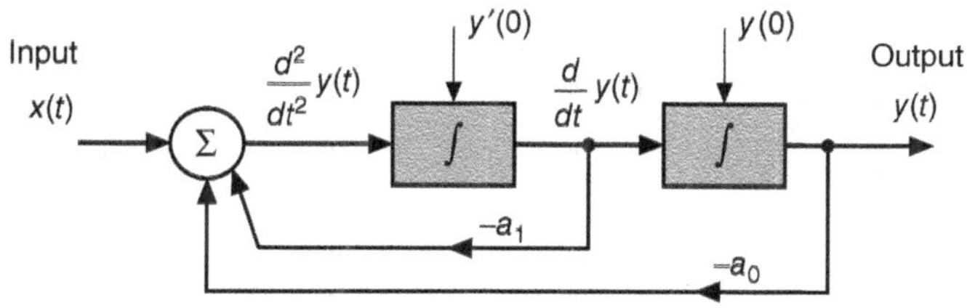

Figure 1.7 Integrator implementation of a second-order linear ODE.

additive noise, which often has a zero average value. This implementation is obtained by bringing the $\{a_0, a_1\}$ terms of (1.17) to the right-hand side of the equation such that the output of the summing element in the figure is

$$\frac{d^2}{dt^2}y(t) = x(t) - a_1\frac{d}{dt}y(t) - a_0y(t). \tag{1.19}$$

The cascaded integrators sequentially yield $dy(t)/dt$ and $y(t)$. The solution to (1.17) when $x(t) = 0$ and $a_0 = a_1 = 2$ is

$$y(t) = \exp(-t)[2\sin(t) + \cos(t)], \quad t \in \mathcal{R}^+, \tag{1.20}$$

where the nonzero initial conditions $y(0) = y'(0) = 1$ have been assumed. This waveform is plotted in Figure 1.8 (the solid line) from which we can easily verify the initial conditions. It is straightforward to show that (1.20) is the solution of (1.19) by differentiating $y(t)$:

$$\begin{aligned}\frac{d}{dt}y(t) &= -\exp(-t)[2\sin(t) + \cos(t)] + \exp(-t)[2\cos(t) - \sin(t)] \\ &= \exp(-t)[\cos(t) - 3\sin(t)],\end{aligned} \tag{1.21}$$

$$\begin{aligned}\frac{d^2}{dt^2}y(t) &= -\exp(-t)[\cos(t) - 3\sin(t)] + \exp(-t)[-\sin(t) - 3\cos(t)] \\ &= \exp(-t)[2\sin(t) - 4\cos(t)].\end{aligned} \tag{1.22}$$

Substituting these expressions into (1.17) with $a_0 = a_1 = 2$, we find that all terms cancel to give 0. By changing the coefficients, a different output response is obtained. For example, when $a_0 = 2$ and $a_1 = 3$, the solution is purely exponential:

$$y(t) = 3\exp(-t) - 2\exp(-2t), \quad t \in \mathcal{R}^+. \tag{1.23}$$

This is also plotted in Figure 1.8 for the same initial conditions and input $x(t) = 0$ (the dashed line). The solutions in (1.20) and (1.23) are known as *underdamped* and *overdamped*, respectively. It turns out that there is a third type of solution for a second-order ODE called *critically damped*, which is obtained by changing the coefficient values. All three solutions are discussed in greater detail in Chapters 6 and 7.

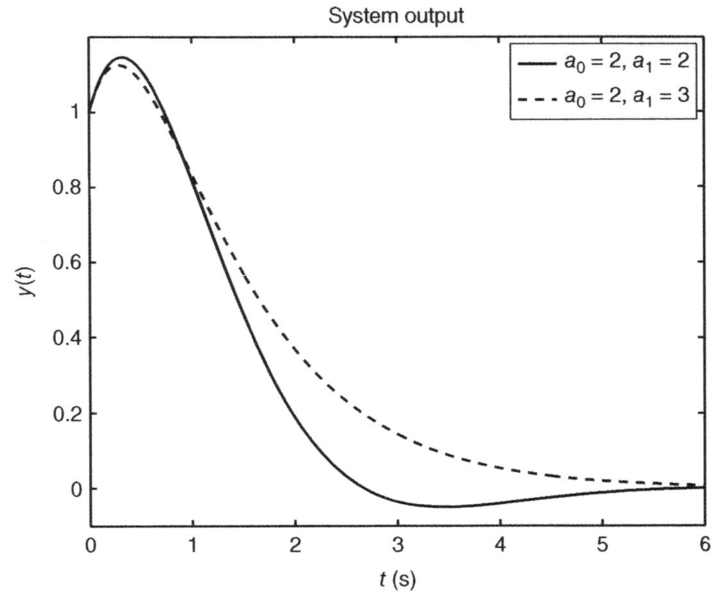

Figure 1.8 Solutions for the second-order ODE in Example 1.5 with constant coefficients. The input is $x(t) = 0$ and the initial conditions are nonzero: $y(0) = y'(0) = 1$.

1.3 FREQUENCY CONTENT

As mentioned earlier, the main goal of this book is to develop mathematical models for circuits and systems, and to describe techniques for finding expressions (solutions) for the dependent variables of interest. In addition, we are interested in the *frequency content* of signals and the *frequency response* of different types of systems. This frequency information illustrates various properties of signals and systems beyond that observed from their time-domain representations.

The most basic signal is sinusoidal with angular frequency $\omega_o = 2\pi f_o$ in radians/second (rad/s) and ordinary frequency f_o in Hz. It turns out that all *periodic* signals can be represented by a sum of sinusoidal signals with fundamental frequency f_o and integer multiples nf_o for $n \in \mathcal{Z}$ called harmonics. For example, the periodic rectangular waveform in Figure 1.9(a) has the frequency *spectrum* shown in Figure 1.9(b), with the magnitude of each frequency component indicated on the vertical axis. Lower harmonics have greater magnitudes, demonstrating that this waveform is dominated by low frequencies. This frequency representation for periodic signals is known as the Fourier series and is covered in Chapter 5. Aperiodic signals, which do not repeat, have a frequency representation known as the Fourier transform. Whereas the Fourier series consists of integer multiples of a fundamental frequency, the Fourier transform is a *continuum of frequencies* as illustrated in Figure 1.10 for triangular and rectangular waveforms. Both of these signals are dominated by low-frequency content.

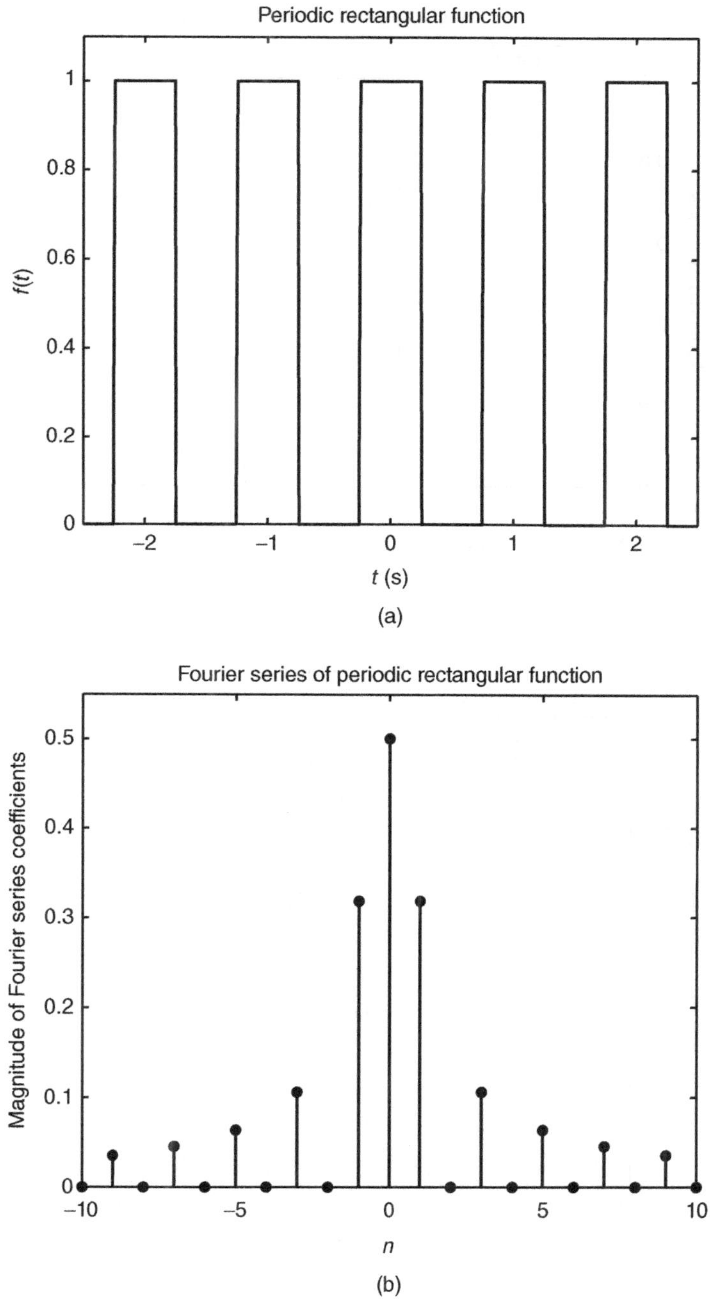

Figure 1.9 Periodic rectangular waveform. (a) Time-domain representation. (b) Magnitude of frequency spectrum: Fourier series with harmonics $n f_o$ and $f_o = 1$ Hz.

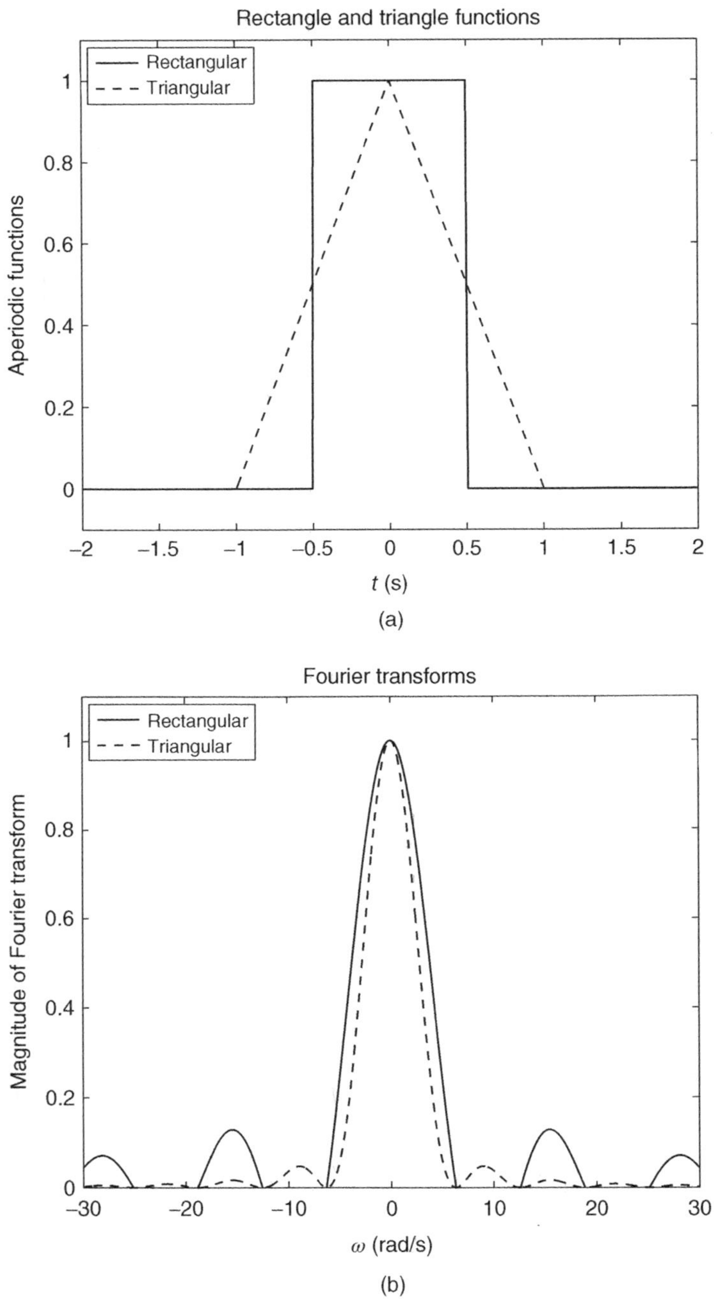

Figure 1.10 Aperiodic waveforms. (a) Time-domain representation. (b) Magnitude of frequency spectrum: Fourier transform.

(a)

(b)

Figure 1.11 Two-dimensional image and spectrum. (a) Spatial representation. (b) Magnitude of frequency spectrum in two dimensions. White denotes a greater magnitude. (The vertical and horizontal white lines are the frequency axes where $\omega_1 = 0$ and $\omega_2 = 0$. A log scale is used to better visualize variations in the spectrum.)

Although we focus on one-dimensional signals in this book, which are generally a function of the independent variable time t, Figure 1.11 shows a two-dimensional *image* and its frequency representation. The two independent variables in Figure 1.11(a) are given by the horizontal (width) and vertical (height) axes, and the information contained in the image is indicated by a gray scale from white to black. Similarly, the magnitude of the spectrum in Figure 1.11(b) is represented by a gray scale, with white denoting a greater magnitude for particular frequencies. Two frequency variables $\{\omega_1, \omega_2\}$ are used in the Fourier transform of a two-dimensional image (Bracewell, 1978); these are the horizontal and vertical axes in Figure 1.11(b). Low frequencies are located around the center of the plot, and high frequencies (positive and negative) extend outward to the edges of the spectrum plot. Once again, we have a signal with mostly low-frequency content; in fact, the spectrum is dominated by the white "star" located about the center where $\omega_1 = \omega_2 = 0$. This occurs because there is not much spatial variation across the image in Figure 1.11(a). In general, greater variations in the time/spatial domain correspond to higher frequencies with greater magnitudes in the Fourier/frequency domain.

Systems are often designed to have a particular frequency response where some frequencies are emphasized and others are attenuated. For example, a system that passes low frequencies and attenuates high frequencies is called a *low-pass filter*. Likewise, systems can be designed to have a high-pass, band-pass, or band-reject frequency response. Conventional amplitude modulated (AM) radio is an example of a system that incorporates band-pass filters to select a transmitted signal located in a specific radio frequency *channel*. Such a channel is defined by a *center frequency* and a *bandwidth* over which the signal can be transmitted without interfering with other signals in nearby channels. The Fourier transform and different types of filters are covered in Chapter 8.

In the rest of this chapter, we provide a review of some basic topics that the reader has probably studied to some extent, and which form the basis of the material covered throughout this book.

1.4 FUNCTIONS AND PROPERTIES

We begin with a summary of basic definitions for functions of a single independent variable.

Definition: Function The *function* $y = f(x)$ is a unique mapping from input x to output y.

Although x yields a single value y, more than one value of x could map to the same y. (Note, however, that it is possible to define multiple-output functions; an example of this is the Lambert W-function discussed in Appendix F.)

Definition: Domain and Range The *domain* of function $f(x)$ consists of those values of x for which the function is defined. The *range* of a function is the set of values $y = f(x)$ generated when x is varied over the domain.

Example 1.6 For $f(x) = x^2$, the *natural* domain is $\mathcal{R}$ (although it is possible to restrict the domain to some finite interval), and the corresponding range is $\mathcal{R}^+$. The domain for $f(x) = \log(x)$ is $\mathcal{R}^+$ and its range is $\mathcal{R}$.

Definition: Support The *support* of a function is the set of x values for which $f(x)$ is nonzero.

Example 1.7 The domain of the *unit step* function is $\mathcal{R}$:

$$u(x) \triangleq \begin{cases} 1, & x \geq 0 \\ 0, & x < 0, \end{cases} \tag{1.24}$$

but its support is $\mathcal{R}^+$. Similarly, the domain of the truncated sine function $\sin(\omega_o t)u(t)$ is $\mathcal{R}$ and its support is $\mathcal{R}^+$. Even though sine is 0 for integer multiples of π, the support is still $\mathcal{R}^+$ because sine is a continuous function and those points (which form a *countable* set) are not excluded from the support.

Definition: Inverse Image and Inverse Function The *inverse image* $x = f^{-1}(y)$ is the set of all values x that map to y. The inverse image of a function may not yield a unique x. If a single $x = f^{-1}(y)$ is generated for each y, then $f(x)$ is *one-to-one* and the inverse image is equivalent to the *inverse function* $x = f^{-1}(y) \triangleq g(y)$.

Example 1.8 For the quadratic function $y = x^2$, it is obvious that each $x \in \mathcal{R}$ gives a single y. Solving for x yields $x = \pm\sqrt{y}$. Since x is not unique for each y, the square root is *not* the inverse function. An inverse function does not exist for $y = x^2$. However, $x = f^{-1}(y) = \pm\sqrt{y}$ describes the inverse image; for example, the inverse image of $y = 9$ is the *set* of values $x = \{-3, 3\}$. The one-to-one function $y = 2x + 1$ has inverse function $x = g(y) = (y - 1)/2$. The natural logarithm $y = \ln(x)$ is also one-to-one with inverse function $x = g(y) = \exp(y)$.

Definition: Linear Function A *linear function* $f(x)$ has the following two properties:

$$f(x_1 + x_2) = f(x_1) + f(x_2), \quad f(\alpha x) = \alpha f(x), \tag{1.25}$$

where $\alpha \in \mathcal{R}$ is any constant.

The line representing a linear function necessarily passes through the origin: $y(x) = 0$ when $x = 0$.

Example 1.9 The circuit model shown in Figure 1.12(a) for a resistor with resistance R has the form $v = Ri$ known as *Ohm's law*. It is a linear function.

$$v_1 = Ri_1, \quad v_2 = Ri_2 \Rightarrow v_1 + v_2 = R(i_1 + i_2) = Ri_1 + Ri_2, \tag{1.26}$$

$$v = Ri \Rightarrow \alpha v = R(\alpha i) = \alpha Ri, \tag{1.27}$$

where v is a voltage and i is a current (both are defined in Chapter 2). An example of a nonlinear function is the piecewise linear circuit model for a diode that is in series with resistor R:

$$i = \begin{cases} (v - v_c)/R, & v \geq v_c \\ 0, & v < v_c, \end{cases} \tag{1.28}$$

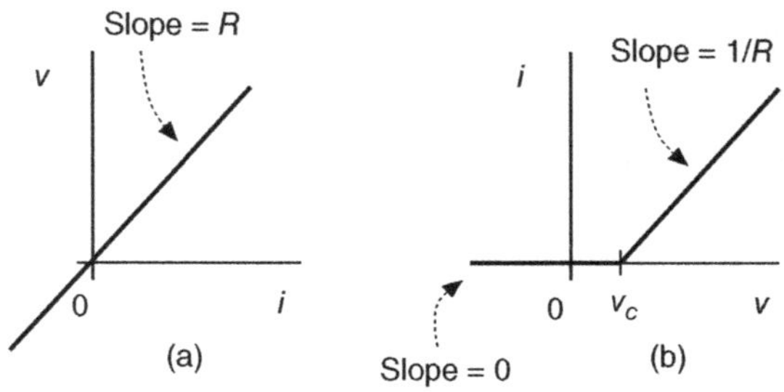

Figure 1.12 Device models used in Example 1.9. (a) Linear model for resistor R. (b) Nonlinear model for diode D with resistance R.

where v_c is a cutoff voltage; typically $v_c \approx 0.7$ V (volt). Although this equation has straight-line components (it is piecewise linear), overall it is nonlinear as depicted in Figure 1.12(b) because it does not satisfy (1.25). Suppose $v_1 = -2$ V such that $i_1 = 0$ A (ampere), and $v_2 = 1.7$ V such that $i_2 = (1/R)$ A. Then $v_1 + v_2 = -0.3$ V $\Rightarrow\; i = 0$ A, which is not equal to $i_1 + i_2 = (1/R)$ A.

The general equation $y = ax + b$ for a line is not linear even though it is straight and is used to describe the different parts of a piecewise linear function (as in Example 1.1). A linear function based on the properties in (1.25) must pass through the origin.

Definition: Affine Function *Affine function* $g(x)$ is a linear function $f(x)$ with additive scalar b such that the ordinate is nonzero:

$$g(x) = f(x) + b. \tag{1.29}$$

An affine function does not satisfy either requirement in (1.25) for a linear function:

$$g(x_1 + x_2) = f(x_1 + x_2) + b \neq g(x_1) + g(x_2) = f(x_1) + b + f(x_2) + b, \tag{1.30}$$

$$g(\alpha x) = f(\alpha x) + b \neq \alpha g(x) = \alpha f(x) + \alpha b, \tag{1.31}$$

where $\alpha \in \mathcal{R}$ is any nonzero constant.

Definition: Continuous Function Function $f(x)$ is *continuous* at x_o if there exists $\epsilon > 0$ for every $\delta > 0$ such that

$$|x - x_o| < \epsilon \;\Rightarrow\; |f(x) - f(x_o)| < \delta. \tag{1.32}$$

More simply we can write

$$\lim_{\epsilon \to 0} |f(x_o + \epsilon) - f(x_o)| = 0, \tag{1.33}$$

where ϵ is either positive or negative such that $f(x_o + \epsilon)$ approaches x_o from the right or the left, respectively.

All the functions shown in Figures 1.8 and 1.12 are continuous. An example of a function that is continuous only *from the right* is shown in Figure 1.13. Approaching x_o from the left, the function jumps to the higher value b. A solid circle indicates that the function is continuous approaching from the right, meaning that the function is b at x_o. A function that is continuous at x_o from the left is similarly defined with the solid and open circles in Figure 1.13 interchanged. If a function is left- and right-continuous at x_o, then it is strictly continuous at that point as defined in (1.32) and (1.33).

Functions of a real variable can have different types of discontinuities. The plot in Figure 1.13 shows a function with a *jump discontinuity*. Another example is the unit step function $u(t)$ in (1.24), which is used extensively throughout this book. Similar to the example in Figure 1.13, $u(t)$ is continuous from the right but not from the left. A function that is *nowhere continuous* is the Dirichlet function, given by

$$f(x) = \begin{cases} 1, & x \in \mathcal{Q} \\ 0, & x \in \mathcal{R} - \mathcal{Q}, \end{cases} \tag{1.34}$$

where $\mathcal{Q}$ is the set of rational numbers. It is not possible to accurately plot this function using MATLAB (or any other mathematics software). Another type of discontinuity is an *infinite discontinuity*, also called an *asymptotic discontinuity*. Examples include

$$f(x) = 1/x, \quad f(x) = 1/(x-1)(x-2), \tag{1.35}$$

where, in the first case, the discontinuity is at $x = 0$, and in the second case, there are discontinuities at $x = \{1, 2\}$. The second function is plotted in Figure 1.14(a), which we see is continuous except at the two points indicated by the vertical dotted lines. With the terminology of functions of complex variables considered later in this book (see Chapter 5 and Appendix E), these *singularities* are called *poles*.

Consider the function

$$f(x) = \frac{\sin(x)}{x}, \tag{1.36}$$

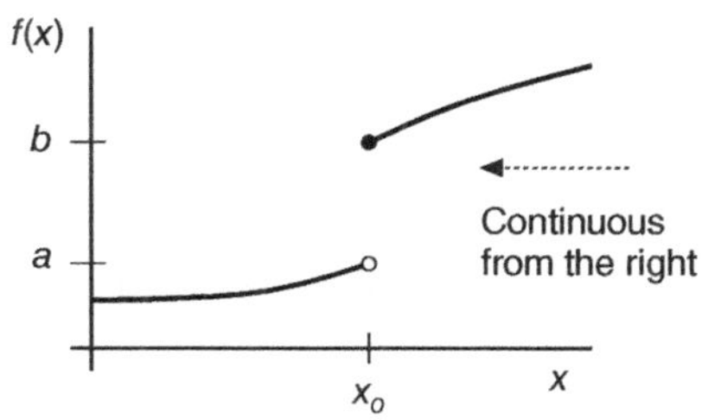

Figure 1.13 Example of a function with a discontinuity at x_o.

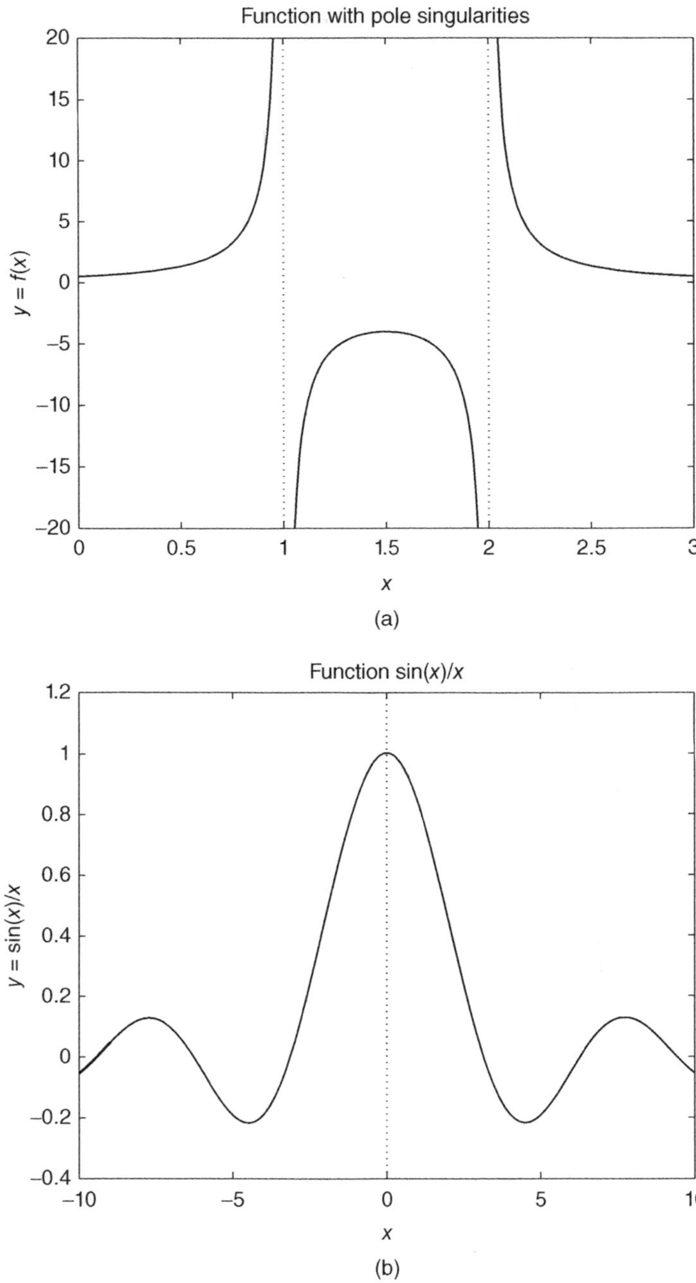

Figure 1.14 (a) Function with two pole singularities at $x = \{1, 2\}$. (b) Function with a removable pole singularity at $x = 0$.

which appears to have a pole at $x = 0$. It turns out, however, that this pole is cancelled by the numerator such that $f(0) = 1$. This can be seen using L'Hôpital's rule

$$\frac{d\sin(x)/dx|_{x=0}}{dx/dx|_{x=0}} = \cos(0) = 1. \tag{1.37}$$

Such singularities are called *removable*. This function, which is plotted in Figure 1.14(b), is known as the unnormalized sinc function, and should not be confused with the usual sinc function $\text{sinc}(x) \triangleq \sin(\pi x)/\pi x$ discussed in subsequent chapters. Another example of a removable singularity is the following rational function, where a factor in the numerator cancels the denominator:

$$f(x) = \frac{x^2 - 1}{x + 1} = x - 1, \tag{1.38}$$

and so $f(-1) = -2$. A function with a singularity for which there is no limit is called an *essential singularity*. The classic example is

$$f(x) = \sin(1/x), \tag{1.39}$$

which is plotted in Figure 1.15. Observe that as x approaches 0, there is no single finite value for the function.

Finally, ordinary functions can be divided into two basic types.

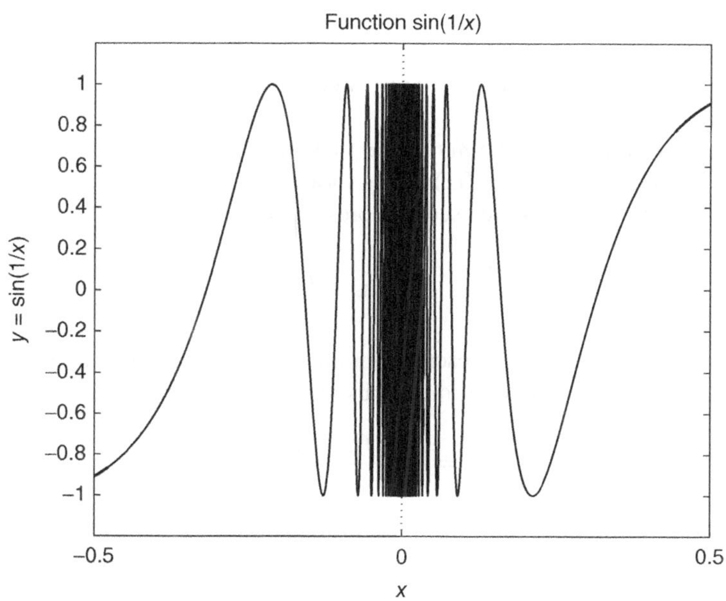

Figure 1.15 Function with an essential singularity at $x = 0$.

Definition: Algebraic Functions and Transcendental Equations An *algebraic function* $f(x)$ satisfies the following polynomial equation:

$$p_n(x)f^n(x) + \ldots + p_1(x)f(x) + p_0(x) = 0, \tag{1.40}$$

where $\{p_m(x)\}$ are polynomials in x and $n \in \mathcal{N}$ (the natural numbers $\{1, 2, \ldots\}$). All other equations are *transcendental equations*, such as those containing exponential, logarithmic, and trigonometric functions.

Example 1.10 Examples of algebraic functions are

$$f(x) = x^4 + x_2 - x + 1, \quad f(x) = \sqrt{x}, \quad f(x) = 1/x^2, \tag{1.41}$$

and examples of transcendental functions are

$$f(x) = \log(x), \quad f(x) = \tan^{-1}(x), \quad f(x) = \cos(x)\tan(x). \tag{1.42}$$

Both types of functions/equations are considered in this book. In Chapter 4, we find that the solutions to some algebraic equations require *complex numbers*. The class of ordinary functions is extended in Chapter 5 to *generalized functions*, which include the Dirac delta function $\delta(x)$ and its derivatives $\delta^{(n)}(x)$.

1.5 DERIVATIVES AND INTEGRALS

In this section, definitions for the ordinary derivative of a function of one independent variable and its Riemann integral are reviewed.

Definition: Derivative The *derivative* of function $f(x)$ is another function that gives the rate of change of $y = f(x)$ as x is varied.

The following notations are used to represent the derivative of $y = f(x)$:

$$\frac{dy}{dx}, \quad \frac{d}{dx}f(x), \quad f'(x), \quad \dot{y}, \tag{1.43}$$

though the last form is usually reserved for the derivative of $y(t)$ with respect to time t: $\dot{y} = dy/dt$. The derivative of a continuous function is generated from the following limit:

$$\frac{d}{dx}f(x) = \lim_{\Delta x \to 0} \frac{f(x + \Delta x) - f(x)}{\Delta x}, \tag{1.44}$$

where a *secant line* connects the points $\{x, f(x)\}$ and $\{x + \Delta x, f(x + \Delta x)\}$. As $\Delta x \to 0$, the *family* of secant lines approach the *tangent line* at x as shown for the function in Figure 1.16. The next example demonstrates how to use this definition of the derivative for two of the functions in Example 1.8.

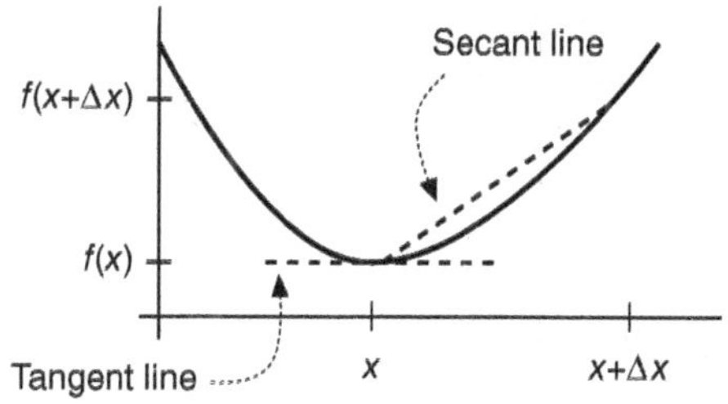

Figure 1.16 Finite approximation of the derivative of $f(x)$ at x.

Example 1.11 For $y = x^2$:

$$\frac{dy}{dx} = \lim_{\Delta x \to 0} \frac{(x+\Delta x)^2 - x^2}{\Delta x} = \lim_{\Delta x \to 0} \frac{2x\Delta x + \Delta x^2}{\Delta x} = 2x, \tag{1.45}$$

and for $y = 2x + 1$:

$$\frac{dy}{dx} = \lim_{\Delta x \to 0} \frac{[2(x+\Delta x)+1] - (2x+1)}{\Delta x} = \lim_{\Delta x \to 0} \frac{2\Delta x}{\Delta x} = 2. \tag{1.46}$$

For the latter affine function, the derivative is a constant equal to the slope. In general, the derivative varies with x, as it does for the quadratic function $f(x) = x^2$.

Example 1.12 Consider the derivative of the absolute value function $y = |x|$:

$$\begin{aligned}\frac{dy}{dx} &= \lim_{\Delta x \to 0} \frac{|x+\Delta x| - |x|}{\Delta x} \\ &= \lim_{\Delta x \to 0} \begin{cases} (x+\Delta x - x)/\Delta x, & x > 0 \\ (-x - \Delta x + x)/\Delta x, & x < 0. \end{cases}\end{aligned} \tag{1.47}$$

Thus, as $\Delta x \to 0$:

$$\frac{dy}{dx} = \begin{cases} 1, & x > 0 \\ -1, & x < 0. \end{cases} \tag{1.48}$$

Although the absolute value function is continuous at all points, its derivative does not exist at $x = 0$ because the ratio in (1.47) is not defined there in the limit. However, since this is usually not an issue in practice, $d|x|/dx = \text{sgn}(x)$ is often used where $\text{sgn}(x)$ is the *signum function*:

$$\text{sgn}(x) \triangleq \begin{cases} 1, & x > 0 \\ 0, & x = 0 \\ -1, & x < 0. \end{cases} \tag{1.49}$$

The derivative of function $y(x)$ can be extended to include points where $dy(x)/dx$ is not defined by using the theory of generalized functions, as discussed in Chapter 5. This is even more evident for the derivative of the signum function:

$$\frac{d}{dx}\text{sgn}(x) = 2\delta(x), \tag{1.50}$$

where $\delta(x)$ is the Dirac delta function. This result cannot be derived using the difference approach in (1.44)

It is not necessary to use the limit in (1.44) to find derivatives because many results have already been established for a wide range of functions. For convenience, Appendix C summarizes the derivatives of several ordinary functions. The following important properties of the derivative are provided without proof, which can be used to derive results for more complicated functions.

- *Addition and scalar multiplication*:

$$\frac{d}{dx}[\alpha f(x) + \beta g(x)] = \alpha\frac{d}{dx}f(x) + \beta\frac{d}{dx}g(x), \tag{1.51}$$

  with $\alpha, \beta \in \mathcal{R}$.
- *Product rule*:

$$\frac{d}{dx}[f(x)g(x)] = g(x)\frac{d}{dx}f(x) + f(x)\frac{d}{dx}g(x). \tag{1.52}$$

- *Quotient rule*:

$$\frac{d}{dx}\frac{f(x)}{g(x)} = [1/g^2(x)]\left[g(x)\frac{d}{dx}f(x) - f(x)\frac{d}{dx}g(x)\right]. \tag{1.53}$$

- *Chain rule*:

$$\frac{d}{dx}f(g(x)) = \frac{d}{dg(x)}f(g(x))\frac{d}{dx}g(x). \tag{1.54}$$

As shown earlier, the *independent variable* of the function in a derivative is often suppressed for notational convenience. For example, we usually just write dy/dx, which is the same as $dy(x)/dx$; we also use $y'(x)$ as was done for the initial condition $y'(0)$ in Figure 1.7. For the nth-order derivative, a superscript is used: $y^{(n)}(x)$, or multiple primes $y''(x)$, or multiple dots for time derivatives $\ddot{y}(t)$.

Example 1.13 The chain rule is useful for finding the derivative of a *composite function* where the variable of one equation depends on another variable. Let the two functions be

$$f(y) = 4y^2 - y + 3, \quad y = g(x) = x^2 + 1. \tag{1.55}$$

The derivatives are

$$\frac{d}{dy}f(y) = 8y - 1 = 8g(x) - 1, \quad \frac{d}{dx}g(x) = 2x, \tag{1.56}$$

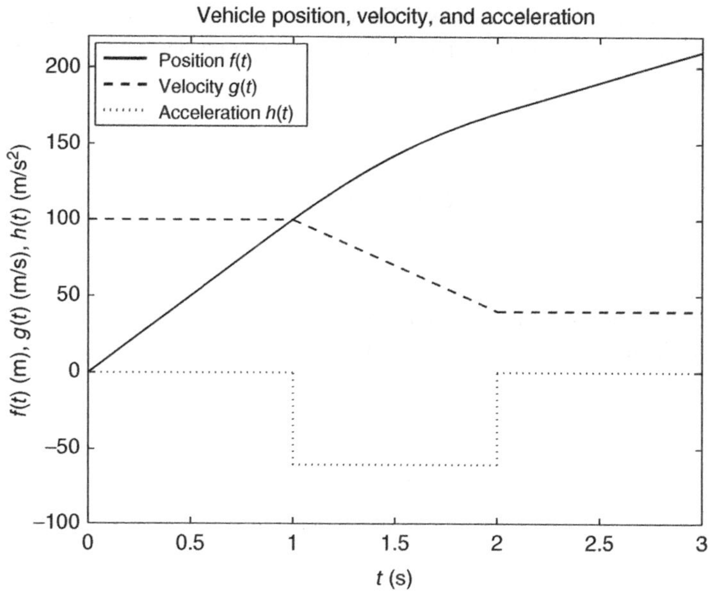

Figure 1.17 Vehicle position, velocity, and acceleration waveforms used in Example 1.14.

and the chain rule yields

$$\frac{d}{dx}f(g(x)) = [8g(x) - 1]2x = 16x^3 + 14x. \tag{1.57}$$

This is verified by substituting $g(x)$ into $f(y)$ and differentiating once with respect to x:

$$f(g(x)) = 4x^4 + 7x^2 + 6 \Rightarrow \frac{d}{dx}f(g(x)) = 16x^3 + 14x. \tag{1.58}$$

Substituting one equation into the other is usually a tedious process, which is a step the chain rule eliminates. The product and quotient formulas also simplify finding derivatives because it is not necessary to multiply or divide the functions, respectively, before computing derivatives.

Example 1.14 Suppose the position of a vehicle in meters (m) along one Cartesian coordinate over time t is described by the following piecewise linear function:

$$f(t) = \begin{cases} 100t, & 0 \le t \le 1 \\ -30t^2 + 160t - 30, & 1 < t \le 2 \\ 40t + 90, & 2 < t \le 3, \end{cases} \tag{1.59}$$

where the units of t are seconds (s). The distance traveled versus time is illustrated in Figure 1.17 (the solid line). The velocity is the time derivative of this function, denoted by $g(t) = \dot{f}(t)$ with units m/s:

$$g(t) = \begin{cases} 100, & 0 \leq t \leq 1 \\ -60t + 160, & 1 < t \leq 2 \\ 40, & 2 < t \leq 3. \end{cases} \tag{1.60}$$

In Figure 1.17, we see that the velocity is initially 100 m/s and then it decreases linearly to 40 m/s (the dashed line). The acceleration is the time derivative of the velocity $h(t) = \dot{g}(t) = \ddot{f}(t)$, which has units m/s^2:

$$h(t) = \begin{cases} 0, & 0 \leq t < 1 \\ -60, & 1 \leq t < 2 \\ 0, & 2 \leq t \leq 3. \end{cases} \tag{1.61}$$

The vehicle has nonzero acceleration only when its velocity is decreasing from 100 to 40 m/s (it is actually a *deceleration* because $h(t)$ is negative). Unlike the first two functions, $h(t)$ is not continuous.

Example 1.15 The derivative can be used to find *saddle points*, the *minimum*, or the *maximum* of a function (if they exist). Consider the cubic function $f(x) = (x-1)^3$ plotted in Figure 1.18. The first derivative is $f'(x) = 3(x-1)^2$ and the second derivative is $f''(x) = 6(x-1)$, both of which are also plotted in Figure 1.18. Observe that $f(x)$ has a saddle point at $x = 1$: the derivative $f'(x)$ (the dashed line) is 0 there, but $x = 1$ is neither a maximum nor a minimum of $f(x)$. The second derivative $f''(x)$ is also 0 at $x = 1$, which means that the quadratic function $f'(x)$ has a minimum there. It is a minimum (and not a maximum) because the second derivative of $f'(x)$, given by $f^{(3)}(x) = 6$, is positive.

Definition: Indefinite Integral The *indefinite integral* of $f(x)$ is another function $g(x)$ such that $dg(x)/dx = f(x)$.

The indefinite integral $g(x)$ is also called the *antiderivative*, and so, integration is the inverse operation of differentiation. It is represented by

$$g(x) = \int f(x)dx = F(x) + c, \tag{1.62}$$

where c is a constant independent of x, and $F(x)$ is the antiderivative when $c = 0$. Thus, the antiderivative is not unique; instead, we say it is unique *up to a constant*. The value of c is determined by *boundary conditions*.

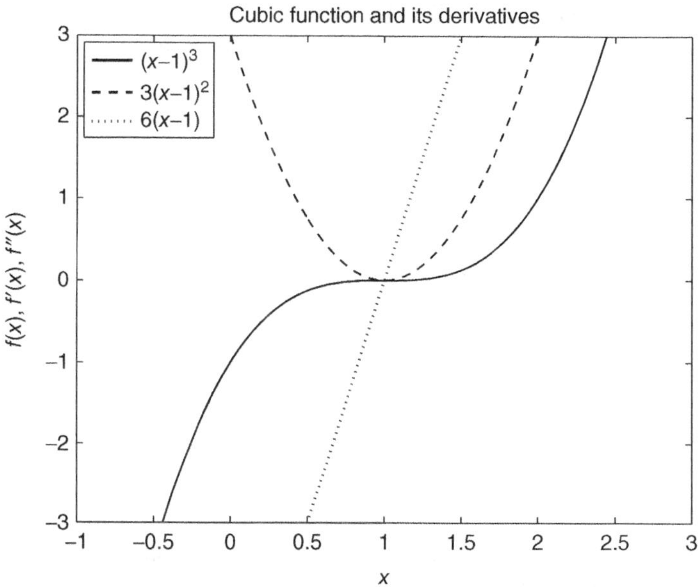

Figure 1.18 Cubic function $f(x) = (x-1)^3$ in Example 1.15 and its derivatives.

Example 1.16 For the scenario in Example 1.14, the velocity is the indefinite integral of acceleration:

$$g(t) = \int h(t)dt = \begin{cases} c_1, & 0 \le t \le 1 \\ -60t + c_2, & 1 < t \le 2 \\ c_3, & 2 < t \le 3, \end{cases} \tag{1.63}$$

where $\{c_n\}$ are constants that are determined by the boundary conditions for the subintervals $[0, 1]$, $(1, 2]$, and $(2, 3]$. In order to continue, we need the initial velocity, which in this case is 100 m/s, yielding $c_1 = 100$. Similarly, the final velocity gives $c_3 = 40$ m/s. The middle coefficient is derived by assuming that the velocity does not change instantaneously. Thus, at $t = 1$:

$$-60 \times 1 + c_2 = 100 \implies c_2 = 160 \text{ m/s}, \tag{1.64}$$

which can also be derived at $t = 2$:

$$-60 \times 2 + c_2 = 40 \implies c_2 = 160 \text{ m/s}. \tag{1.65}$$

Combining these terms gives the expression for $g(t)$ in (1.60).

Definition: Definite Integral The *definite integral* of function $f(x)$ is a real number derived from the indefinite integral with specific limits of integration:

$$g(b) - g(a) = \int_a^b f(x)dx. \tag{1.66}$$

It gives the *area* under $f(x)$ on the interval $[a, b]$.

For a definite integral, the constant c appearing in (1.62) is of no concern because it cancels when evaluated at the limits:

$$g(b) - g(a) = [F(b) + c] - [F(a) + c] = F(b) - F(a). \tag{1.67}$$

Note that seemingly simple integrals require special attention. For example, it is not clear how to evaluate

$$\int_a^b f(x)dx = \int_{-1}^1 (1/x)dx, \tag{1.68}$$

because $f(x) = 1/x$ has a singularity at $x = 0$. Such functions are sometimes called *pseudofunctions* and the integral is improper.

Definition: Improper Integral The following integral is *improper* if $f(x)$ is infinite for some x in $[a, b]$:

$$\int_a^b f(x)dx, \tag{1.69}$$

or if $a = -\infty$, $b = \infty$, or both.

In both situations, we must carefully evaluate the integral as demonstrated in the next example.

Example 1.17 The following integral is improper because the function is unbounded at $x = 1$:

$$\int_1^2 \frac{dx}{x-1}. \tag{1.70}$$

This expression is examined by changing the lower limit to ϵ and letting $\epsilon \to 1$:

$$\lim_{\epsilon \to 1} \int_\epsilon^2 \frac{dx}{x-1} = \lim_{\epsilon \to 1} [\ln(|2-1|) - \ln(|\epsilon - 1|)] = \infty. \tag{1.71}$$

Similarly, for

$$\int_2^\infty \frac{dx}{x-1}, \tag{1.72}$$

we have

$$\lim_{\epsilon \to \infty} \int_2^\epsilon \frac{dx}{x-1} = \lim_{\epsilon \to \infty} [\ln(|\epsilon - 1|) - \ln(|2-1|)] = \infty. \tag{1.73}$$

Both of these integrals are *divergent*. Suppose the denominator in (1.70) is squared:

$$\int_1^2 \frac{dx}{(x-1)^2}. \tag{1.74}$$

Then

$$\lim_{\epsilon \to 1} \int_\epsilon^2 \frac{dx}{(x-1)^2} = \lim_{\epsilon \to 1} \left. \frac{-1}{x-1} \right|_\epsilon^2 = \infty, \tag{1.75}$$

which is also divergent. However, the integral

$$\int_2^\infty \frac{dx}{(x-1)^2} \tag{1.76}$$

is *convergent*:

$$\lim_{\epsilon \to \infty} \int_2^\epsilon \frac{dx}{(x-1)^2} = \lim_{\epsilon \to \infty} \left. \frac{-1}{x-1} \right|_2^\epsilon = 1. \tag{1.77}$$

Although $f(x) = 1/\sqrt{x-1}$ is undefined at $x = 1$, the following integral is convergent:

$$\int_1^2 \frac{dx}{\sqrt{x-1}} = 2\sqrt{x-1}\Big|_1^2 = 2. \tag{1.78}$$

The three functions in this example all have a singularity at $x = 1$ as shown in Figure 1.19. Since $1/\sqrt{x-1}$ is *imaginary* for $x < 1$, it is plotted only for $x > 1$. Imaginary and complex numbers are covered in Chapter 4.

The definite integral in (1.66) is known as a *Riemann integral* in order to distinguish it from other types of integrals (such as the Lebesgue integral, which is beyond the scope of this book). It can be defined in terms of the following *Riemann sum*:

$$\int_a^b f(x)dx = \lim_{\Delta x_n \to 0} \sum_{n=0}^{N-1} f(x_n)\Delta x_n, \tag{1.79}$$

such that $N \to \infty$ with $\Delta x_n \triangleq x_{n+1} - x_n$, $x_0 = a$, and $x_N = b$. In a Riemann sum, the interval $[a, b]$ on the x-axis is divided into nonoverlapping subintervals, which together cover the entire interval. This collection of subintervals is called a *partition* of $[a, b]$. Observe that we have used the smaller value x_n of Δ_n for the argument of $f(x_n)$, in which case the sum is known as a *lower Riemann sum*. If instead x_{n+1} is used, then it is called an *upper Riemann sum*. In the limit as $\Delta x_n \to 0$, both sums converge to the same quantity for a continuous function, giving the definite integral of $f(x)$ on $[a, b]$. Examples of the lower and upper Riemann sums are indicated by the shaded regions in Figure 1.20.

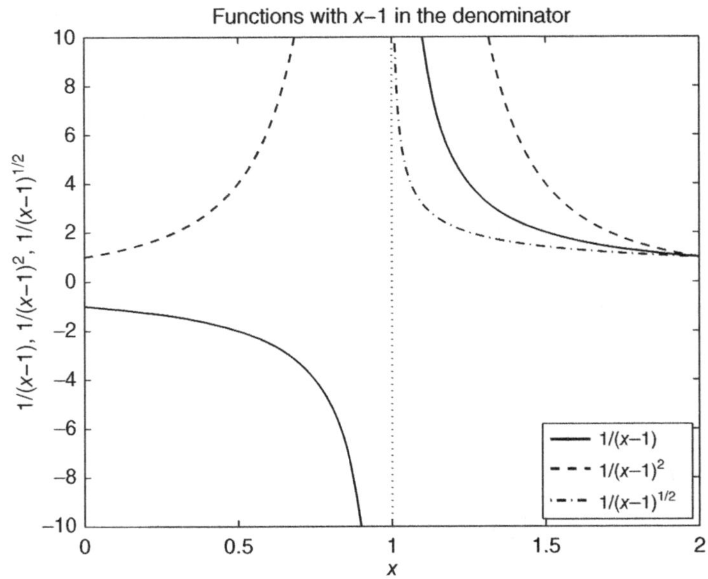

Figure 1.19 Three functions in Example 1.17 with singularities at $x = 1$.

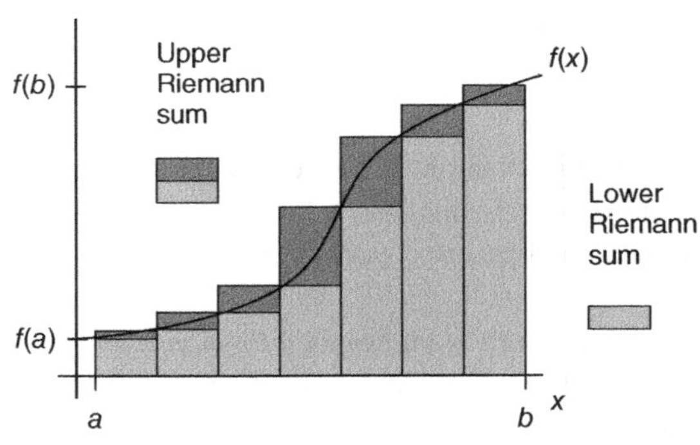

Figure 1.20 Lower and upper Riemann sums approximating the integral of $f(x)$ on $[a, b]$.

Although it is not necessary for the subintervals to have the same width, it is usually convenient to do so with $\Delta x_n = (b - a)/N \triangleq \Delta$ for all n such that $x_n = a + n\Delta$ and (1.79) becomes

$$\int_a^b f(x)dx = \frac{b-a}{N} \lim_{N\to\infty} \sum_{n=0}^{N-1} f(a + n\Delta). \tag{1.80}$$

Example 1.18 Consider again the functions in Example 1.8. The area under $f(x) = x^2$ on $[0, 2]$ is

$$\int_0^2 x^2 dx = \frac{2}{N} \lim_{N\to\infty} \sum_{n=0}^{N-1} (n\Delta)^2 = \frac{8}{N^3} \sum_{n=0}^{N-1} n^2, \tag{1.81}$$

where we have assumed equal-length subintervals and substituted $\Delta = 2/N$. A closed-form expression in Appendix C for the last sum in (1.81) yields

$$\int_0^2 x^2 dx = \lim_{N\to\infty} (8/N^3)(1/6)[(N-1)N][2(N-1)+1]$$
$$= (8/6) \lim_{N\to\infty} \frac{2N^3 - 3N^2 + N}{N^3} = 8/3. \tag{1.82}$$

Since the indefinite integral of $f(x) = x^2$ is $g(x) = x^3/3 + c$, we confirm that the area of $f(x)$ on $[0, 2]$ is $8/3$. For $f(x) = 2x + 1$ on $[-1, 2]$:

$$\int_{-1}^2 (2x+1)dx = \lim_{N\to\infty} \frac{3}{N} \sum_{n=0}^{N-1} [2(-1+n\Delta)+1]$$
$$= \lim_{N\to\infty} \left[(18/N^2) \sum_{n=0}^{N-1} n - (3/N) \sum_{n=0}^{N-1} 1 \right], \tag{1.83}$$

where $\Delta = 3/N$. The last sum is N, and using another closed-form expression from Appendix C for the first sum in (1.83), the area is

$$\int_{-1}^2 (2x+1)dx = \lim_{N\to\infty} (18/N^2)[(N-1)N/2] - 3 = 6. \tag{1.84}$$

The indefinite integral of $f(x) = 2x + 1$ is $g(x) = x^2 + x + c$, and from this we verify that the definite integral on $[-1, 2]$ is 6.

It is not necessary that the sum in (1.79) be used to derive integrals because many results have already been established for a wide range of functions. Appendix C includes some indefinite integrals as well as a few definite integrals. The following important properties of integration are provided without proof.

- *Integration by parts*:

$$\int w(x)\frac{dv(x)}{dx}dx = w(x)v(x) - \int \frac{dw(x)}{dx}v(x)dx. \tag{1.85}$$

- *Leibniz's integral rule*:

$$\frac{d}{dx}\int_{a(x)}^{b(x)} f(v)dv = f(b(x))\frac{d}{dx}b(x) - f(a(x))\frac{d}{dx}a(x). \tag{1.86}$$

The following expressions are special cases that are used often in engineering problems:

$$\frac{d}{dx}\int_a^x f(v)dv = f(x), \quad \frac{d}{dx}\int_x^b f(v)dv = -f(x). \tag{1.87}$$

Example 1.19 Consider the indefinite integral

$$f(x) = \int x\exp(\alpha x)dx, \tag{1.88}$$

where α is a constant. In order to use integration by parts, we equate the following: $w(x) = x$ and $dv(x)/dx = \exp(\alpha x)$ which yield $dw(x)/dx = 1$ and $v(x) = (1/\alpha)\exp(\alpha x)$. The expression in (1.85) gives

$$f(x) = (x/\alpha)\exp(\alpha x) - (1/\alpha)\int \exp(\alpha x)dx, \tag{1.89}$$

whose integral is straightforward to evaluate:

$$\begin{aligned} f(x) &= (x/\alpha)\exp(\alpha x) - (1/\alpha^2)\exp(\alpha x) + c \\ &= [(\alpha x - 1)/\alpha^2]\exp(\alpha x) + c, \end{aligned} \tag{1.90}$$

where c is the constant of integration. For an example of Leibniz's integral rule, consider

$$g(x) = \int_x^{x^2} \exp(\alpha u)du, \tag{1.91}$$

which has derivative

$$\begin{aligned} \frac{d}{dx}g(x) &= \exp(\alpha x^2)\frac{d}{dx}x^2 - \exp(\alpha x)\frac{d}{dx}x \\ &= 2x\exp(\alpha x^2) - \exp(\alpha x). \end{aligned} \tag{1.92}$$

This is verified by performing the integration:

$$g(x) = (1/\alpha)[\exp(\alpha x^2) - \exp(\alpha x)], \tag{1.93}$$

and then differentiating (using the chain rule):

$$\frac{d}{dx}g(x) = (1/\alpha)[2\alpha x\exp(\alpha x^2) - \alpha\exp(\alpha x)], \tag{1.94}$$

which simplifies to (1.92). Leibniz's integral rule allows us to find the derivative in (1.92) without first computing the integral in (1.91).

Derivatives and integrals appear in the linear ODEs and *integro-differential* equations discussed in Chapters 6 and 7.

1.6 SINE, COSINE, AND π

Next, we discuss some properties of sinusoidal functions and indicate how they arise in practice. Consider the circle shown in Figure 1.21, which has unit radius $r = 1$ and is called the *unit circle*, and so, its circumference is 2π. The famous constant $\pi = 3.141592653589\ldots$ is the ratio of the circumference of any circle and its diameter. Since it cannot be expressed as the ratio of two integers, π is an *irrational number* (of course, this also means that if the circumference of a circle is an integer, then its diameter is not). The circumference in the figure can be divided into 360 equal lengths (arcs), and each "pie slice" projected back to the origin is defined to have an angle of 1°. The distance along the unit circle yields the corresponding angle in radians. The example in Figure 1.21 illustrates that an angle of $\pi/2$ in the first quadrant relative to the positive horizontal axis is actually one-quarter distance along the circle circumference of 2π: $2\pi/4 = \pi/2$.

It is well known from trigonometry that sine of an angle formed by a right triangle is defined as the ratio of the lengths of the distant side y and the hypotenuse r: $\sin(\theta) \triangleq y/r$. Similarly, cosine of θ is defined as the ratio of the lengths of the adjacent side of a right triangle and its hypotenuse: $\cos(\theta) \triangleq x/r$. Since $x^2 + y^2 = r^2$, we immediately have that for any angle θ:

$$\sin^2(\theta) + \cos^2(\theta) = 1, \tag{1.95}$$

where either $\theta \in [0, 360°]$ or $\theta \in [0, 2\pi]$ radians. It is also clear from Figure 1.21 and the connection between sine and cosine that

$$\sin(\theta \pm \pi/2) = \pm\cos(\theta), \tag{1.96}$$

$$\cos(\theta \pm \pi/2) = \mp\sin(\theta). \tag{1.97}$$

Plotting sine and cosine as functions of θ, we find that sine lags cosine by $\pi/2$ radians (90°).

Suppose now that the angle is written as $\theta(t) = \omega_o t$ so that it varies with time, where ω_o is *angular frequency* with units of rad/s. Thus, with a fixed ω_o, any point on the radial line from the origin to the circle with radius r has the same *constant*

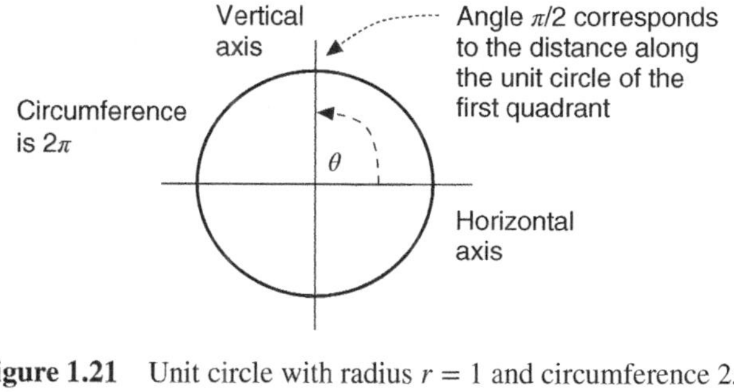

Figure 1.21 Unit circle with radius $r = 1$ and circumference 2π.

angular velocity as it sweeps counterclockwise with increasing t. This result follows because the derivative is a constant $d\theta(t)/dt = \omega_o$. From Figure 1.22, observe how the *functions* $\sin(\omega_o t)$ and $\cos(\omega_o t)$ are generated. As time increases, $\cos(\omega_o t)$ is the *projection* of the end of the radial line onto the horizontal axis; the cosine function is the length of this projection as it varies over $[-1, 1]$ (for $r = 1$). Likewise, $\sin(\omega_o t)$ is the projection of the end of the radial line onto the vertical axis. A projection is defined to be negative for cosine when it is located to the left of the origin on the horizontal axis, and it is negative for sine when it is below the origin on the vertical axis.

Summarizing, the time-varying functions $\sin(\omega_o t)$ and $\cos(\omega_o t)$ follow from the usual definitions of the sine and cosine of an angle, except that the angle varies as $\theta(t) = \omega_o t$. By convention, the angle is defined with respect to the positive horizontal axis as depicted in Figure 1.22 for four different time instants (snapshots). These plots illustrate why the sine and cosine functions are 90° out of phase with respect to each other: as $\sin(\omega_o t)$ increases, $\cos(\omega_o t)$ decreases and vice versa. They are *orthogonal* functions:

$$\int_a^b \sin(\omega_o t)\cos(\omega_o t)dt = 0, \tag{1.98}$$

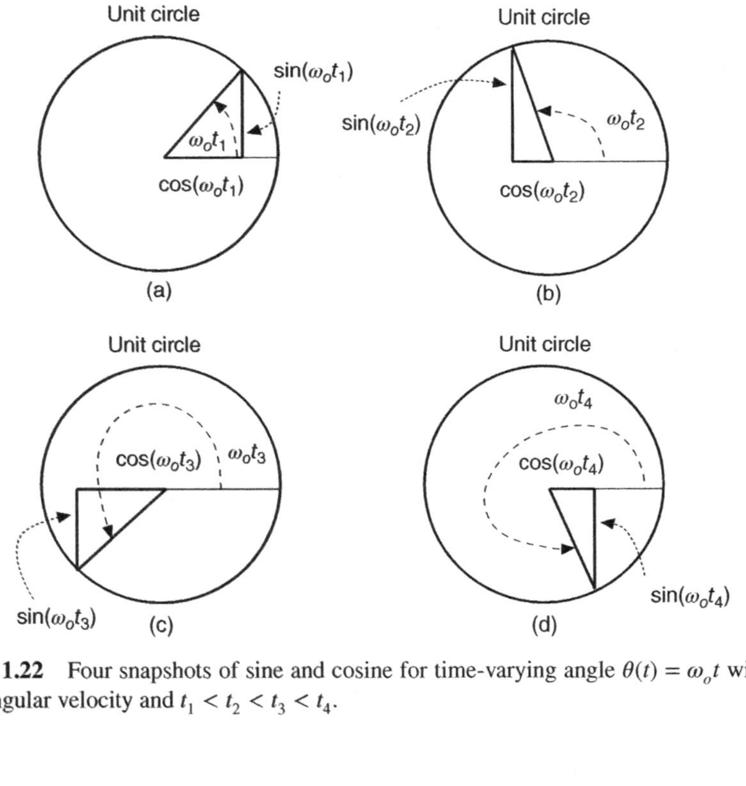

Figure 1.22 Four snapshots of sine and cosine for time-varying angle $\theta(t) = \omega_o t$ with constant angular velocity and $t_1 < t_2 < t_3 < t_4$.

when $(b-a)\omega_o$ is an integer multiple of π. This result is verified by using a trigonometric identity from Appendix C:

$$\begin{aligned}\int_a^b \sin(\omega_o t)\cos(\omega_o t)dt &= (1/2)\int_a^b [\sin(2\omega_o t)+\sin(0)]dt \\ &= (-1/4\omega_o)\cos(2\omega_o t)|_a^b \\ &= [\cos(2\omega_o a)-\cos(2\omega_o b)]/4\omega_o, \end{aligned} \tag{1.99}$$

which is 0 when $\cos(2\omega_o b)=\cos(2\omega_o a)$. Since cosine is periodic with period 2π, we require $2\omega_o b = 2\omega_o a + n2\pi$ for $n \in \mathcal{Z}$, which means $(b-a)\omega_o = n\pi$. Figure 1.23 shows a plot of (1.99) for $a=0$ and $\omega_o = 1$ rad/s as b is varied from 0 to 5π. The integral is 0 for $b=\{0,\pi,2\pi,3\pi,4\pi,5\pi\}$, and the maximum area is $1/2$ for this value of ω_o. The orthogonality property is also evident from a geometric viewpoint because the vertical and horizontal dashed lines in Figure 1.22 are orthogonal: they form the previously mentioned right triangle. The fact that the radial line sweeps along a *circle* gives rise to the specific *smooth* shapes of the sine and cosine waveforms, derived as projections on the two axes.

Figure 1.24(a) shows the sine waveform in Figure 1.22 with $\omega_o = 1$ rad/s. The function approaches its maximum with a decreasing derivative, which is the cosine

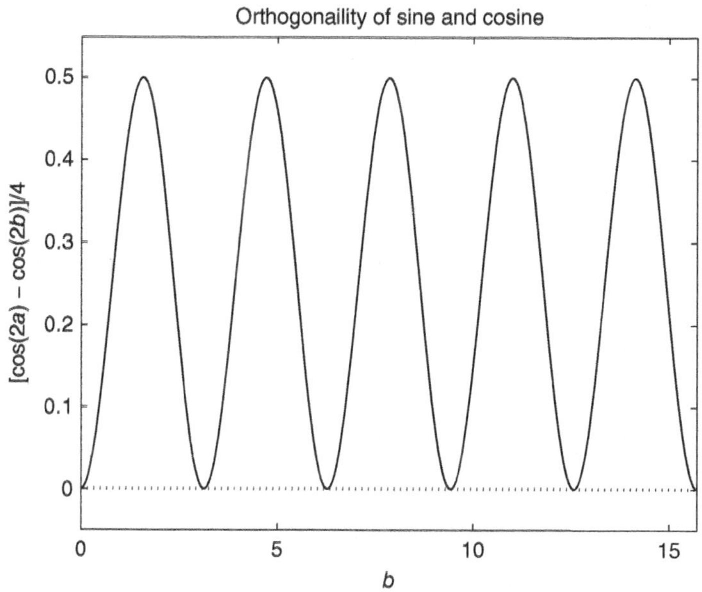

Figure 1.23 Orthogonality of sine and cosine for $a=0$ and $\omega_o = 1$ rad/s.

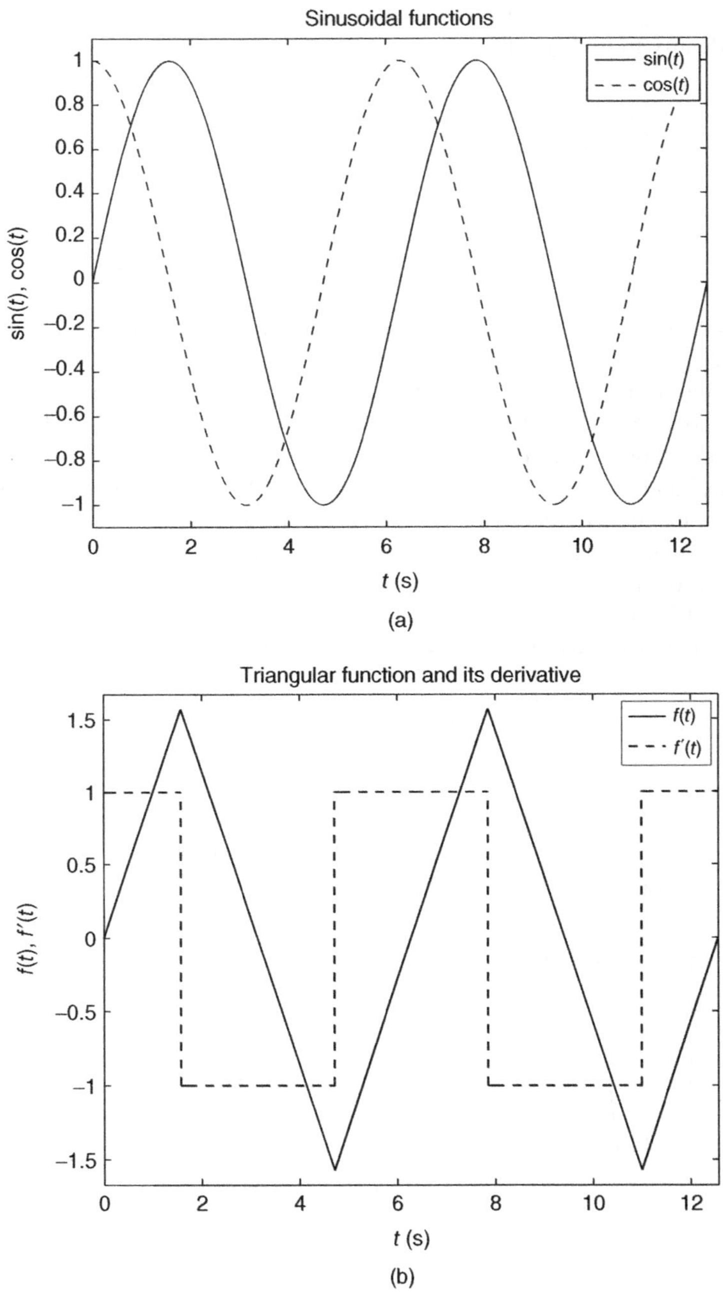

Figure 1.24 Periodic waveforms. (a) Sine waveform sin(t) and its derivative cos(t). (b) Triangular waveform and its rectangular derivative.

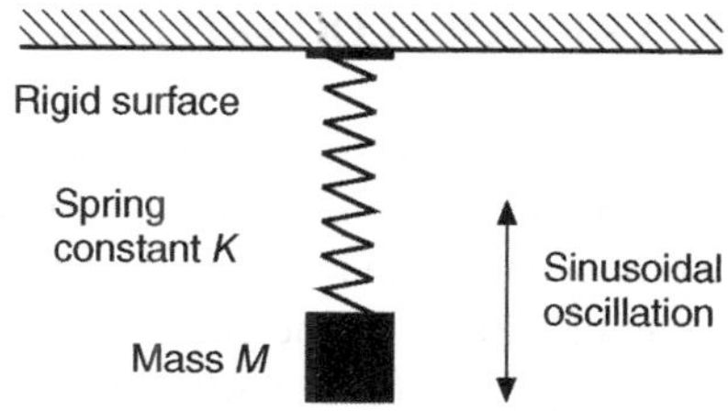

Figure 1.25 Mass on a spring influenced by gravity.

waveform also shown in the figure. (The orthogonality of these two waveforms is also apparent from this figure.) This smooth behavior of its derivative is unlike that of the triangular waveform in Figure 1.24(b) whose derivative is a constant until the function reaches its maximum, at which point the derivative abruptly changes sign. It turns out that many physical phenomena are modeled accurately using sinusoidal functions. Apparently, many physical systems behave in a sinusoidal manner because the underlying physics yield gradual variations rather than abrupt changes. This also means that the physical mechanisms of many systems have the dynamic of constant angular velocity along a circle on the plane as in Figure 1.22.

An example of a mechanical process is an object (mass) attached to a spring as depicted in Figure 1.25. If the object is extended downward and released, its up-and-down trajectory is sinusoidal. As the spring is stretched, its linear velocity gradually decreases and it becomes exactly 0 at its maximum distance, just like a sinusoidal waveform. This behavior is due to the physical properties of the spring and the force of gravity. The object does not have constant *linear* velocity, and it does not abruptly change direction at its minimum and maximum distance from the rigid surface. The amplitude and frequency of the waveform depend on the mass M of the object, the *spring constant* K, and the initial position of the object, which are discussed further in Chapter 2.

We demonstrate in Chapter 4 that the sine and cosine axes as depicted in Figure 1.21 can be represented on the *complex plane*, where the horizontal axis (associated with cosine) is the *real axis* and the vertical axis (associated with sine) is the *imaginary axis*. It turns out that both sine and cosine can be written together using complex notation as follows:

$$\exp(j\omega_o t) = \cos(\omega_o t) + j\sin(\omega_o t), \tag{1.100}$$

where $j \triangleq \sqrt{-1}$ and $\exp(1) = e$ is Napier's constant. This two-dimensional formulation called *Euler's formula* is widely used in engineering to represent signals and waveforms, and $\exp(j\omega_o t)$ is an *eigenfunction* of a linear system as discussed in Chapter 7.

1.7 NAPIER'S CONSTANT *e* AND LOGARITHMS

Napier's constant e is another important irrational number used in mathematics and engineering. It is motivated by the following *compound interest* problem. Suppose one has an initial monetary amount x_o called the *principal*, which accumulates interest at an annual percentage rate of $100r\%$. At the end of 1 year when a single interest payment is made, the new principal is $x_o(1 + r)$, where for now we assume $0 < r \leq 1$. Suppose instead that an interest payment is made after 6 months, and the total amount available then accumulates interest until the end of the year. The amount after one-half year is $x_o(1 + r/2)$. Since this is the principal for the second half of the year, we have a total amount of $x_o(1 + r/2)(1 + r/2) = x_o(1 + r/2)^2$ at the end of the year. Similarly, by dividing the year into thirds, the amount at the end of the year is $x_o(1 + r/3)^3$, and in general, for n interest payments, the principal is $x_o(1 + r/n)^n$ at the end of 1 year.

It can be shown that for $x_o = 1$ and $r = 1$ (corresponding to a 100% interest rate), the limit is Napier's constant:

$$\lim_{n\to\infty} (1 + 1/n)^n = e = 2.718281828459\ldots \tag{1.101}$$

This convergence to e is demonstrated in Figure 1.26. It is an interesting result that the total monetary amount after 1 year of essentially *continuous* interest payments

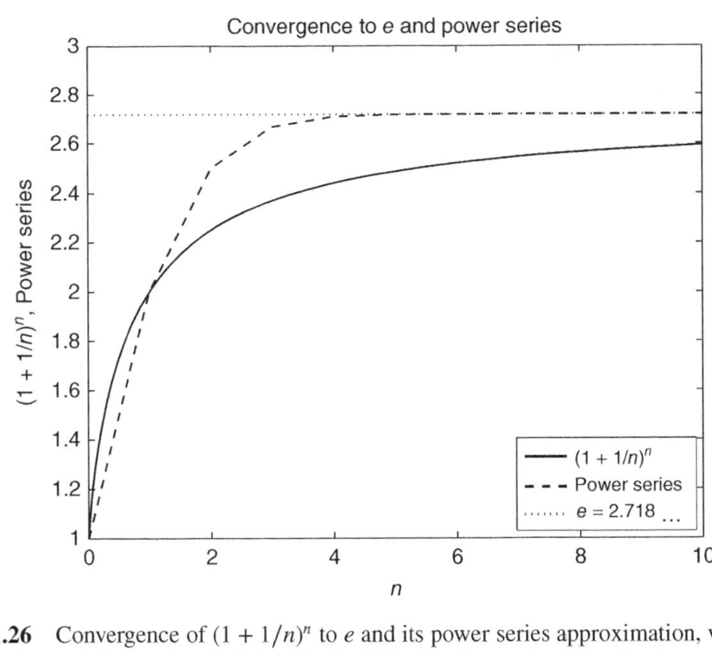

Figure 1.26 Convergence of $(1 + 1/n)^n$ to e and its power series approximation, where n is the upper limit of the sum in (1.104). (The individual points at integer n for the power series have been connected by lines for ease of viewing.)

(because $n \to \infty$) is finite and given exactly by e. For general x_o and r, the limit is

$$\lim_{n\to\infty} x_o(1 + r/n)^n = x_o e^r, \tag{1.102}$$

such that $r > 0$ results in a gain on the original principal x_o, and $r < 0$ yields a loss. These correspond to *exponential growth* and *exponential decay*, respectively.

The constant e has the following alternative representations.

- *Limits*:

$$e = \lim_{n\to 0} (1 + n)^{1/n}, \quad e^r = \lim_{n\to 0} (1 + n/r)^{r/n}. \tag{1.103}$$

- *Power series*:

$$e = \sum_{m=0}^{\infty} \frac{1}{m!}. \tag{1.104}$$

- *Hyperbolic functions*:

$$e = \sinh(1) + \cosh(1). \tag{1.105}$$

Convergence of the power series sum in (1.104) with upper limit n instead of infinity is shown in Figure 1.26. As n is varied over the 11 integers $\{0, \dots, 10\}$, we find that the sum quickly approaches e; the first six values are 1, 2, 2.5, 2.6667, 2.7083, and 2.7167.

The exponential function based on Napier's constant is defined next, which is discussed further in Chapter 5.

Definition: Exponential Function The *exponential function* is

$$\exp(x) \triangleq e^x. \tag{1.106}$$

It has domain $\mathcal{R}$ and range $\mathcal{R}^+$.

The exponential function has the following properties.

- *Product*:

$$\exp(x)\exp(y) = \exp(xy). \tag{1.107}$$

- *Ratio*:

$$\frac{\exp(x)}{\exp(y)} = \exp(x/y). \tag{1.108}$$

- *Derivative*:

$$\frac{d}{dx}\exp(x) = \exp(x). \tag{1.109}$$

- *Integrals*:

$$\int \exp(x)dx = \exp(x) + c, \quad \int_{-\infty}^{x} \exp(v)dv = \exp(x), \quad \int_{0}^{x} \exp(v)dv = \exp(x) - 1. \tag{1.110}$$

- *Power series*:

$$\exp(x) = \sum_{n=0}^{\infty} \frac{x^n}{n!}. \tag{1.111}$$

- *Hyperbolic functions*:

$$\exp(x) = \cosh(x) + \sinh(x), \quad \exp(-x) = \cosh(x) - \sinh(x). \tag{1.112}$$

The last property gives $\cosh(x) = (1/2)[\exp(x) + \exp(-x)]$ and $\sinh(x) = (1/2)[\exp(x) - \exp(-x)]$, which is similar to Euler's formula for complex numbers discussed in Chapter 4. The exponential functions in (1.112) and their hyperbolic components are plotted in Figure 1.27.

The exponential function arises *naturally* in many engineering problems because of its unique derivative and integral properties. This is demonstrated by the following example in probability.

Example 1.20 The exponential probability density function (pdf) is

$$f_X(x) = \begin{cases} \alpha \exp(-\alpha x), & x \geq 0 \\ 0, & x < 0, \end{cases} \tag{1.113}$$

where the uppercase notation X denotes a *random variable* with outcomes x, and the parameter $\alpha > 0$ determines the mean and variance of X. This pdf has domain $\mathcal{R}$,

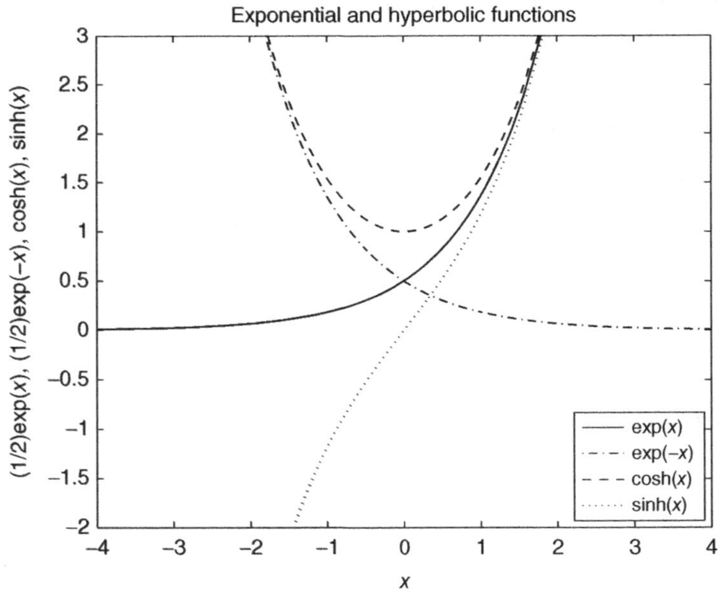

Figure 1.27 Exponential and hyperbolic functions.

support $\mathcal{R}^+$, and range $\mathcal{R}^+$. A valid pdf satisfies the following two conditions:

$$f_X(x) \geq 0, \qquad \int_{-\infty}^{\infty} f_X(x)dx = 1. \tag{1.114}$$

These are obviously true for the exponential pdf:

$$\alpha \exp(-\alpha x) \geq 0, \qquad \int_0^{\infty} \alpha \exp(-\alpha x)dx = -\exp(-\alpha x)|_0^{\infty} = 1. \tag{1.115}$$

Suppose instead that we are interested in another decaying function such as $f_X(x) = ba^{-x} \geq 0$ for $a, b \geq 0$ and $x \in \mathcal{R}^+$. The integral of this function is

$$b\int_0^{\infty} a^{-x}dx = -\frac{ba^{-x}}{\ln(a)}\bigg|_0^{\infty} = \frac{b}{\ln(a)}, \tag{1.116}$$

where $\ln(\cdot)$ is the natural logarithm defined next. In order for the integral to be 1, it is necessary that $b = \ln(a)$, and so, we must have $a > 1$, yielding the following valid pdf:

$$f_X(x) = \ln(a)a^{-x}, \quad x \in \mathcal{R}^+, \tag{1.117}$$

which has a maximum value of $\ln(a)$ at $x = 0$. Thus, other exponential-like decaying functions are possible, but they require a leading coefficient, and so, they are not the "natural" choice as is $a = e$ with $\ln(a) = 1$. The derivative and integral properties of $\exp(x)$ eliminate such multiplicative scaling of the function. The same reasoning can be used to justify e in the *Gaussian* pdf:

$$f_X(x) = \frac{1}{\sqrt{2\pi}\sigma}\exp(-(x-\mu)^2/2\sigma^2), \tag{1.118}$$

where μ and σ are its mean and standard deviation, respectively. Likewise, the pdf of the *Laplace* random variable is

$$f_X(x) = \frac{1}{2\alpha}\exp(-|x|/\alpha), \tag{1.119}$$

with parameter $\alpha > 0$, which determines the variance $2\alpha^2$. The support for these last two pdfs is the entire real line $\mathcal{R}$.

Finally, we consider logarithms and their connection to e.

Definition: Logarithm The *logarithm* of x is the exponent y with base b such that $b^y = x$. It is written as $\log_b(x) = y$ with domain $\mathcal{R}^+$ and range $\mathcal{R}$.

Perhaps the most familiar base is $b = 10$, which yields *common logarithms*. *Binary logarithms* with $b = 2$ are used in the analysis of digital systems. Note that

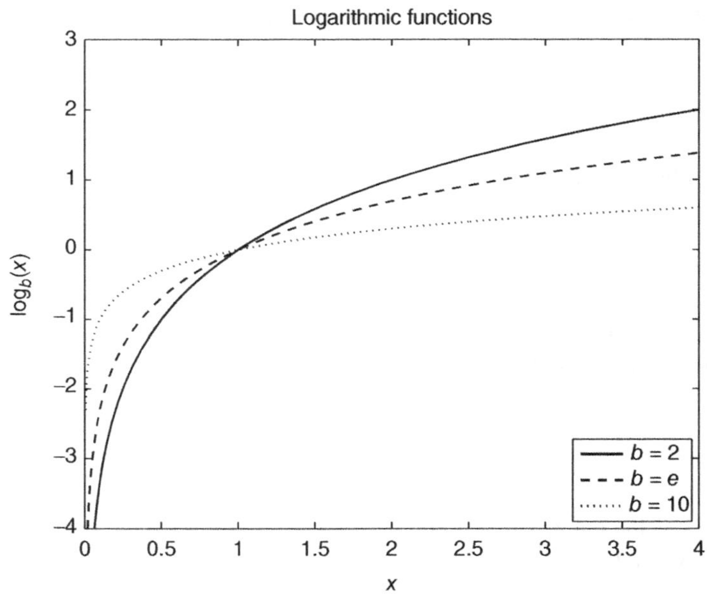

Figure 1.28 Logarithmic functions with different base b.

$\log_b(1) = 0$ for any b as depicted in Figure 1.28 where the base is varied from 2 to 10. The conversion formula of a logarithm from base b_1 to base b_2 is

$$\log_{b_2}(x) = \log_{b_1}(x)/\log_{b_1}(b_2). \tag{1.120}$$

Example 1.21 For $b = 10$, the subscript is often omitted: $\log(x)$ (though in MATLAB log has base e and log10 has base 10). Examples include $\log(1000) = 3$ and $\log(0.1) = -1$. Integer powers of 2 are important numbers in digital systems because their logic is based on the binary number system, usually represented by $\{0, 1\}$. Thus, $b = 2$ such that $\log_2(8) = 3$, $\log_2(64) = 6$, $\log_2(1/2) = -1$, and so on.

The following logarithm appears frequently in engineering applications.

Definition: Natural Logarithm The *natural logarithm* is

$$\ln(x) \triangleq \log_e(x), \tag{1.121}$$

which has domain $\mathcal{R}^+$ and range $\mathcal{R}$. It is also defined by the definite integral:

$$\ln(x) \triangleq \int_1^x (1/v)dv. \tag{1.122}$$

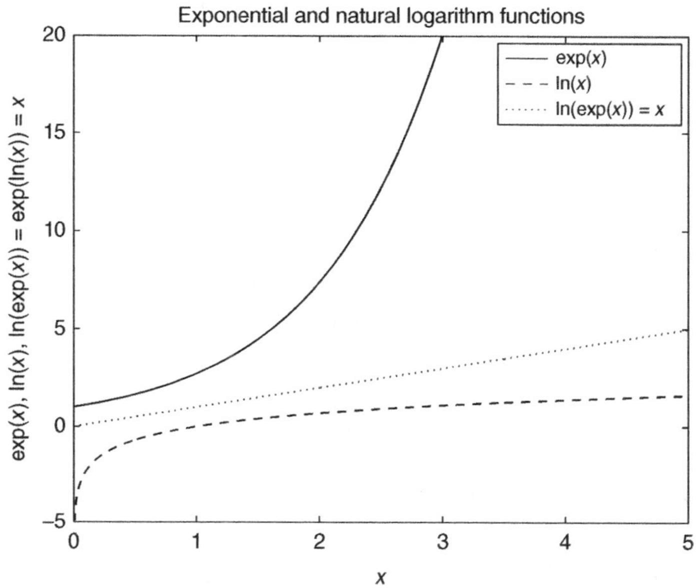

Figure 1.29 Exponential and natural logarithm functions.

This is not an improper integral of the pseudofunction $1/v$ because the limits of integration do not include the origin. From (1.121), we have

$$\ln(\exp(x)) = x, \quad \exp(\ln(x)) = x, \tag{1.123}$$

where it is assumed that $x > 0$ in the second equation. The exponential and natural logarithm functions are plotted in Figure 1.29, where the vertical axis has been limited to 20 because the exponential function increases rapidly (e.g., $\exp(5) \approx 148.41$). Observe the following properties: (i) $\ln(x)$ increases much more slowly than $\exp(x)$ and (ii) $\ln(x) \to -\infty$ as $x \to 0$. We have also included the straight-line plot for $\ln(\exp(x)) = \exp(\ln(x)) = x$, demonstrating that they are in fact inverse functions of each other.

Logarithms have the following properties.

- *Integrals*:

$$\int \log_b(x)dx = x[\log_b(x) - 1/\ln(b)] + c, \quad \int \ln(x)dx = x\ln(x) - x + c. \tag{1.124}$$

- *Sum*:

$$\log_b(x) + \log_b(y) = \log_b(xy). \tag{1.125}$$

- *Difference*:

$$\log_b(x) - \log_b(y) = \log_b(x/y). \tag{1.126}$$

- *Exponent*:

$$\log_b(x^n) = n\log_b(x). \tag{1.127}$$

- *Derivatives*:

$$\frac{d}{dx}\log_b(x) = \frac{1}{x\ln(b)}, \quad \frac{d}{dx}\ln(x) = 1/x. \tag{1.128}$$

- *Limit*:

$$\ln(x) = \lim_{n\to\infty} n(x^{1/n} - 1). \tag{1.129}$$

- *Power series*:

$$\ln(x) = \sum_{n=1}^{\infty} \frac{(-1)^{n+1}}{n}(x-1)^n, \quad \ln(x+1) = \sum_{n=1}^{\infty} \frac{(-1)^{n+1}}{n}x^n. \tag{1.130}$$

Example 1.22 From the identity $\alpha = \exp(\ln(\alpha))$, we can write

$$\alpha^v = \exp(v\ln(\alpha)). \tag{1.131}$$

Suppose v is a function of x such that

$$\alpha^{v(x)} = \exp(v(x)\ln(\alpha)). \tag{1.132}$$

The right-hand side and the chain rule can be used to find the derivative of functions of this form with x in the exponent:

$$\begin{aligned}\frac{d}{dx}\alpha^{v(x)} &= \frac{d}{dx}\exp(v(x)\ln(\alpha)) \\ &= \exp(v(x)\ln(\alpha))\ln(\alpha)\frac{d}{dx}v(x) \\ &= \ln(\alpha)\alpha^{v(x)}\frac{d}{dx}v(x),\end{aligned} \tag{1.133}$$

where (1.131) has been substituted in the final expression. This result is not the same as the more commonly used derivative

$$\frac{d}{dx}v^n(x) = nv^{n-1}(x)\frac{d}{dx}v(x), \tag{1.134}$$

where n in the exponent is a constant.

We conclude this section with proofs of the derivatives in (1.109) and (1.128) using the limit definition of the derivative in (1.44). For the natural logarithm:

$$\frac{d}{dx}\ln(x) = \lim_{\Delta x\to 0}\frac{\ln(x+\Delta x)-\ln(x)}{\Delta x} = \lim_{\Delta x\to 0}\frac{\ln((x+\Delta x)/x)}{\Delta x}. \tag{1.135}$$

Multiplying and dividing by x and then using the exponent property yield

$$\frac{d}{dx}\ln(x) = \lim_{\Delta x\to 0}\frac{(x/\Delta x)\ln((x+\Delta x)/x)}{x} = (1/x)\lim_{\Delta x\to 0}\ln((1+\Delta x/x)^{x/\Delta x}), \tag{1.136}$$

where $1/x$ has been brought outside the limit. The second form of the limit for e in (1.103) (with x in place of n) gives the final result:

$$\frac{d}{dx}\ln(x) = (1/x)\ln(e) = 1/x. \tag{1.137}$$

The derivative of $\exp(x)$ is obtained from the derivative of the natural logarithm and the chain rule:

$$\frac{d}{dx}\ln(\exp(x)) = \frac{1}{\exp(x)}\frac{d}{dx}\exp(x) \implies \frac{d}{dx}\exp(x) = \exp(x), \tag{1.138}$$

where we have used the fact that the left-hand side equals 1.

PROBLEMS

MATHEMATICAL MODELS

1.1 Sketch the following transfer characteristic:

$$y = \begin{cases} 0, & x < 0 \\ x^2, & 0 \le x < 3 \\ 2x+3, & 3 \le x < 5 \\ 0, & x \ge 5, \end{cases} \tag{1.139}$$

and sketch its output $y(t)$ when the input is the exponential function $x(t) = \exp(t)u(t)$.

1.2 Repeat the previous problem for

$$y = \begin{cases} 0, & x < 0 \\ 2x, & 0 \le x < 2 \\ 4, & 2 \le x < 4 \\ 4\exp(-2(x/2-4)), & x \ge 4, \end{cases} \tag{1.140}$$

and $x(t) = 2tu(t)$.

1.3 (a) For the linear system of equations in (1.5) and (1.6) with

$$\mathbf{A} = \begin{bmatrix} 2 & 1 \\ 1 & 2 \end{bmatrix}, \tag{1.141}$$

solve for $\{y_1, y_2\}$ given $x_1 = x_2 = 1$. (b) Repeat part (a) for

$$\mathbf{A} = \begin{bmatrix} 2 & 2 \\ 2 & 2 \end{bmatrix}. \tag{1.142}$$

1.4 (a) For the nonlinear system in Example 1.4, iteratively solve for y_1 and y_2 by using the same parameter values and inputs, except let the exponential parameter be $\alpha = 5$. (b) Describe the behavior of the iterations for $\alpha = 2$.

1.5 A diode circuit using an exponential model with a series resistor is represented by the following system of equations:

$$a_{11} \exp(\alpha y) + a_{12} = x, \quad a_{21} + a_{22} y = x, \tag{1.143}$$

where the input x is a current (A) and the output y is a voltage (V). The coefficients $\{a_{mn}\}$ depend on the series resistor, the voltage source, and the diode parameters. Iteratively solve for $\{x, y\}$ using the following parameter values: $a_{11} = 10^{-15}$, $a_{12} = -a_{11}$, $a_{21} = 10^{-3}$, $a_{22} = -a_{21}$, and $\alpha = 40$. Let the initial value be $y = 0.6$ V.

1.6 A transistor circuit with a series resistor is represented by the following system of equations:

$$a_{11} y^2 = x, \quad a_{21} y + a_{22} = x, \tag{1.144}$$

where the input x is a current (A) and y is an output voltage (V). The coefficients $\{a_{mn}\}$ depend on the series resistor, the voltage source, and the transistor parameters. Let the parameter values be $a_{11} = 0.5 \times 10^{-3}$, $a_{21} = -10^{-3}$, and $a_{22} = 5 \times 10^{-3}$. Iteratively solve for $\{x, y\}$ assuming the initial value $y = 1$ V.

FUNCTIONS AND PROPERTIES

1.7 Specify the domain, range, and support for the following functions, assuming that x and y are real-valued. (a) $y = \sqrt{x^2 - 1}$. (b) $y = u(x - 2)$ (shifted unit step function). (c) $y = 1/|x - 1|$.

1.8 For real-valued $f(x) = 1/\sqrt{x + 2}$ and $g(x) = |x|$, give the domain, range, and support for the following functions. (a) $y_1 = f(x)/g(x)$. (b) $y_2 = g(x)/f(x)$. (c) $y_3 = f(x)g(x)$.

1.9 Specify the inverse image for each function. (a) $y_1 = x^2 - 5$. (b) $y_2 = |x - 1|/\sqrt{x}$. (c) $y_3 = \text{sgn}(x)u(x + 2)$.

1.10 Find the range of values for x. (a) $|(x-1)/x| < 2$. (b) $|x+2| > 3x$. (c) $x^2 + |x| - 1 > 0$.

1.11 Determine the values of x for which the following functions are continuous. (a) $y_1 = 2x^3 - x^2 + x$. (b) $y_2 = x/(x^2-1)$. (c) $y_3 = \text{sgn}(x-2)\text{sgn}(x+2)$.

1.12 Let $\{x_1, x_2\}$ be the roots of the quadratic equation $ax^2 + bx + c = 0$. Prove that $x_1 + x_2 = -b/a$ and $x_1x_2 = c/a$.

1.13 It can be shown that if $f(x)$ is a polynomial with real coefficients such that $f(x_1) < 0$ and $f(x_2) > 0$ for real $\{x_1, x_2\}$, then $f(x) = 0$ for some x between x_1 and x_2. Determine if this is the case for the following functions. (a) $f_1(x) = x^3 - 5x^2 + 2x + 8$ with $x_1 = 1$ and $x_2 = 3$. (b) $f_2(x) = x^3 + 2x^2 - 5x - 6$ with $x_1 = -2$ and $x_2 = 0$.

1.14 If function $f(x)$ has a derivative at x_o, then show using the following expression that it is also continuous at x_o:

$$\lim_{x \to x_o} |f(x) - f(x_o)| = \lim_{x \to x_o} \left| (x - x_o) \frac{f(x) - f(x_o)}{x - x_o} \right|. \tag{1.145}$$

DERIVATIVES AND INTEGRALS

1.15 Find the derivative of $y = x^3 + 2x$ using the limit definition.

1.16 Repeat the previous problem for $y = \sqrt{x+2}$.

1.17 Repeat Problem 1.15 for $y = x^2 + 1/x$ assuming $x \neq 0$.

1.18 (a) Use the product and chain rules to write an expression for

$$y = \frac{d}{dx} g^m(x) h^n(f(x)), \tag{1.146}$$

where $\{m, n\}$ are constants. (b) Find the derivative of $x^{2m} \exp(\alpha n \sin(x))$.

1.19 (a) Extend the chain rule to find an expression for

$$y = \frac{d^2}{dx^2} f(g(x)). \tag{1.147}$$

(b) Find the second derivative of $\exp(\alpha \sin(x))$.

1.20 Determine which of the following improper integrals converge.

$$\text{(a)} \int_0^\infty \exp(-\alpha x) \sin(x) dx. \quad \text{(b)} \int_2^4 \frac{dx}{(x-2)^3}. \quad \text{(c)} \int_0^\infty \frac{dx}{x^2+4}. \tag{1.148}$$

1.21 Consider the integral transform

$$X(\sigma) \triangleq \int_0^\infty x(t) \exp(-\sigma t) dt, \tag{1.149}$$

where σ is real-valued. Find $X(\sigma)$ for $x(t) = \exp(-t)u(t) + u(t)$ and specify the range of values for σ such that the integral is convergent.

1.22 The current $i(t)$ in a series circuit with resistance R and inductance L is modeled by the following first-order ODE:

$$L\frac{d}{dt}i(t) + Ri(t) = 0. \tag{1.150}$$

(a) Verify that the solution of this equation has the form $i(t) = i(0)\exp(-\alpha t)u(t)$ where $i(0)$ is the initial current, and specify the constant α. (b) Find the value of t such that the current is $1/2e$ its initial value $i(0)$.

1.23 Repeat the previous problem for the integral equation

$$(1/L)\int v(t)dt + v(t)/R = 0, \tag{1.151}$$

where $v(t)$ is a voltage with initial value $v(0)$.

SINE, COSINE, AND π

1.24 Prove the identity $\cos(\theta_2 - \theta_1) = \cos(\theta_1)\cos(\theta_2) + \sin(\theta_1)\sin(\theta_2)$ using an illustration on the unit circle.

1.25 (a) Repeat the previous problem for the double angle formula $\cos(2\theta) = \cos^2(\theta) - \sin^2(\theta)$, and (b) verify this result algebraically using the identity in that problem.

1.26 Find the minimum and maximum of $y = 3\cos(x) + 2\sin(x/2)$ on the interval $x \in [0, \pi]$.

1.27 Solve $\sin^2(\theta) + 2\sin(\theta) - 1 = 0$ for $\theta \in [-\pi/2, \pi/2]$.

1.28 For a general triangle whose sides have lengths x, y, and r, prove the law of cosines:

$$x^2 + y^2 - 2xy\cos(\theta) = r^2, \tag{1.152}$$

where θ is the angle formed by the x and y sides.

1.29 Show that

$$\frac{d}{dx}\sin^{-1}(x) = \frac{1}{\sqrt{1-x^2}}. \tag{1.153}$$

NAPIER'S CONSTANT e AND LOGARITHMS

1.30 (a) Use the natural logarithm to find an expression for the derivative of $y = f^{g(x)}(x)$ with respect to x. (b) Find the derivative of $y = x^{\ln(x)}$.

1.31 From the logarithm sum property $\ln(y_1 y_2) = \ln(y_1) + \ln(y_2)$, prove the product property $\exp(x_1 + x_2) = \exp(x_1)\exp(x_2)$ where $y = \exp(x)$.

1.32 Use the fact that $d\ln(y)/dx = (1/y)dy/dx$ to find the derivative of the following functions. (a) $y_1 = x^2\sqrt{x-1}$. (b) $y_2 = x^{2\cos(x)}$.

1.33 Prove that the minimum of $y = x^x$ is located at $x = e^{-1}$.

1.34 Solve $\ln(x-1) - 2\ln(x) = \ln(2)$ for x.

1.35 The time constant of the exponential function $y = 2\exp(-t/\tau)u(t)$ is $\tau > 0$. It is the value of t such that y is $1/e$ times its initial value of 2. (a) Give the number of time constants such that $y = 1/5$. (b) Repeat part (a) by approximating the exponential function using the first two terms of the Maclaurin series expansion in Appendix E.

COMPUTER PROBLEMS

1.36 For the model in Problem 1.1, use MATLAB to plot the input and output for $x(t) = 6\sin(2\pi t)$ on the interval $t \in [0, 1]$.

1.37 A transistor has the following input/output voltage transfer characteristic:

$$y = \begin{cases} A, & x < \alpha \\ A - \beta(x-\alpha)^2, & \alpha \le x < y + \alpha \\ \text{complicated}, & x > y + \alpha. \end{cases} \tag{1.154}$$

Find the upper bound for x in the second region of the transfer characteristic, and approximate the third region using the exponential function $y = y_b\exp(-\beta(x - x_b))u(x - x_b)$, where y_b is the output when the input is $x = x_b$. Repeat the previous problem using this model with input $x(t) = 2\sin(2\pi t) + 2$ on the interval $t \in [0, 1]$. Let the parameters be $A = 5$, $\alpha = 1$, and $\beta = 2$.

1.38 For the model in Problem 1.5, use MATLAB to plot the two functions and show the first few results of the iterative approach for finding the solution for $\{x, y\}$.

PART I

CIRCUITS, MATRICES, AND COMPLEX NUMBERS

2

CIRCUITS AND MECHANICAL SYSTEMS

2.1 INTRODUCTION

In this chapter, we describe mathematical models for some circuit devices and describe basic laws for voltages and currents in a circuit. The properties of resistance, inductance, and capacitance are assumed to be due only to devices at specific locations in a circuit; the connecting wires are ideal conductors. Such *lumped parameter* circuit models yield linear ordinary differential equations (ODEs) with constant coefficients (as opposed to partial differential equations (PDEs), which are more difficult to analyze). We also cover some mechanical systems that are described by similar ODEs, which should provide physical intuition for analogous circuits and their components. These mathematical models represent the input/output characteristics of the circuit devices without requiring information about their underlying physics. They can be derived from measurements of actual devices, and they usually apply only over a limited operating range for the current and voltage. Although factors such as humidity and temperature can influence the behavior of these devices, we assume ideal models.

A thorough analysis of the many types of circuits is beyond the scope of this book. Instead, we focus on simple circuits that are modeled by first- and second-order ODEs. The goal is to illustrate how ODEs arise in circuits and mechanical systems, and in subsequent chapters, we describe techniques for solving for the unknown variables. In Chapters 7 and 8 on the Laplace and Fourier transforms, we consider

Mathematical Foundations for Linear Circuits and Systems in Engineering, First Edition. John J. Shynk.

Companion Website: http://www.wiley.com/go/linearcircuitsandsystems

higher order system models without specifying the underlying physical systems. The material on first- and second-order circuits and systems covered in this chapter should provide some insights into the behavior of higher order system models.

2.2 VOLTAGE, CURRENT, AND POWER

We begin with some basic definitions.

Definition: Electric Circuit An *electric circuit* is a network of electrical devices whose terminals are connected together by ideal conducting wires.

The linear circuit elements considered in this book are *resistors*, *capacitors*, and *inductors*. We also briefly discuss *diodes*, which are nonlinear semiconductor elements. Each of these devices can be represented by the system model given previously in Figure1.1, where the input and output correspond to the current through or the voltage across the device.

Definition: Elementary Charge The *elementary charge* $q_e \approx 1.6021 \times 10^{-19}$ coulombs (C) is the charge of a proton.

(C for coulomb should not be confused with italic C used later for the capacitor.) The total charge q stored in an electric device such as a capacitor is the sum of all elementary charges, and so the total positive charge is an integer multiple of q_e. Most books on electric circuits assume by convention that current is the flow of positive charge (proton charge), even though, in fact, electrons move through the device; electron charge is the negative of proton charge. An example of a simple circuit is shown in Figure 2.1, consisting of a battery (voltage source) and one of the devices to be described later.

Definition: Current The *current* through a circuit device is the time rate of change of charge:

$$i \triangleq \frac{dq}{dt}, \tag{2.1}$$

which has units of amperes (A) defined as coulombs/second (C/s).

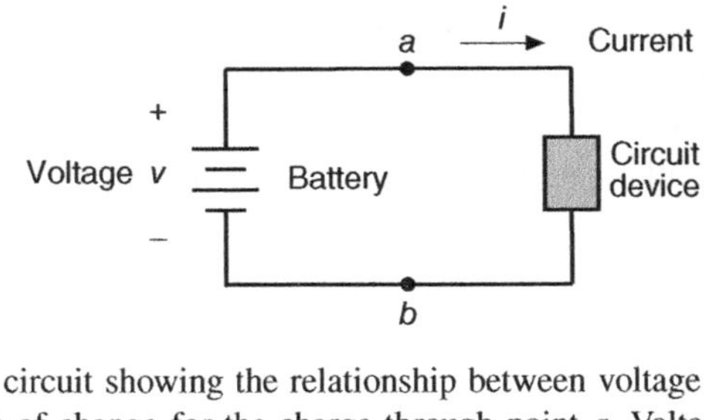

Figure 2.1 Simple circuit showing the relationship between voltage and current. Current i (C/s) is the time rate of change for the charge through point a. Voltage v (J/C) is the work needed to move charge q from point b to point a.

The current in Figure 2.1 is provided by the charge stored in the battery, and the amount of i depends on the voltage v and the type of circuit device. The model in (2.1) for current is more intuitive than the model given next for voltage.

Definition: Voltage The *voltage* across a circuit device is the work (energy) w in joules (J) required to move charge q through the device:

$$v \triangleq \frac{dw}{dq}, \tag{2.2}$$

which has units of volts (V) defined as joules/coulomb (J/C).

Since work and energy have the same units, the voltage is the *potential energy*, and so, it is also called the *electric potential*. Voltage is always defined across two points in a circuit, whereas current is the flow of charge through a single point.

Energy in a circuit implies that power is associated with each of the circuit elements.

Definition: Power The *instantaneous power* of a circuit device is the rate of energy delivered or absorbed:

$$p \triangleq \frac{dw}{dt}, \tag{2.3}$$

which has units of watts (W) defined as joules/second (J/s). The *average power* for duration T is

$$P \triangleq \frac{1}{T}\int_0^T p(t)dt, \tag{2.4}$$

which also has units of watts.

From the definitions of voltage and current, (2.3) can be rewritten so that the power associated with a circuit element is the product of v and i:

$$p = \frac{dw}{dq}\frac{dq}{dt} = vi. \tag{2.5}$$

Power is absorbed by a device when $p > 0$; otherwise, it is delivered by a device. In Figure 2.1, the battery is an *active* device that provides power to the circuit. The shaded device in the figure may be an active or *passive* element and may deliver or absorb power.

A resistor always absorbs power, dissipating the energy as heat. Ideal capacitors and inductors are capable of absorbing and delivering power because they are *energy-storage* devices. Voltage and current sources can absorb or deliver power depending on their placement in a circuit. Figure 2.2 shows a block diagram of a circuit element with two terminals. By convention, when the current i (flow of positive

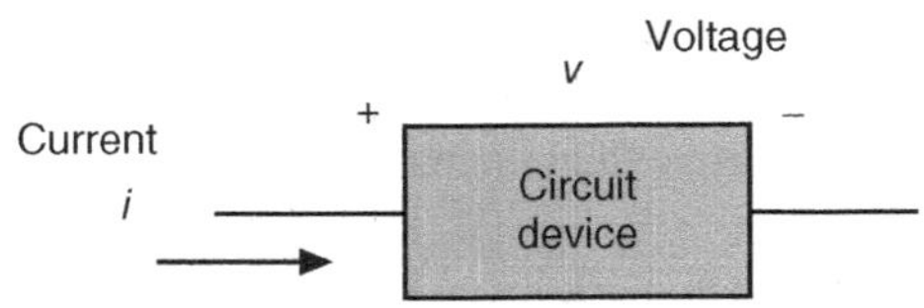

Figure 2.2 Conventional labels of a circuit element. When i enters the positive terminal, power is absorbed by the device; otherwise, it is delivering power.

charge) enters the positive terminal defined by the voltage v, the device is absorbing power. However, if after analyzing the circuit it turns out that the value of i is negative, then the device is actually delivering power. We use the standard voltage polarity and current direction shown in the figure, and the analysis will yield negative values if v has the opposite polarity or i is flowing in the opposite direction. The current entering and leaving a two-terminal device must be the same.

For convenience, we have summarized the units of the various electrical quantities in Table 2.1. Different notations for the voltage, current, and power are used depending on whether or not they are time-varying (lowercase) or constant (uppercase). These are summarized in Table 2.2. The reader should note that the same letter may be used for different quantities, though usually with different fonts. For example, italic W is used to denote constant work, whereas roman W is the abbreviation for watts. Similarly, italic R denotes resistance while calligraphic $\mathcal{R}$ is the symbol representing all real numbers (see Table 1.1). The lowercase energy symbol e should not be confused with Napier's constant.

Example 2.1 Suppose the current through a device is sinusoidal:

$$i(t) = A_m \sin(\omega_o t), \qquad t \in \mathcal{R}, \tag{2.6}$$

where A_m is the maximum amplitude and ω_o is angular frequency. For the resistor mentioned in Chapter 1 and discussed in the next section, the voltage and current are

TABLE 2.1 Electrical Symbols and Units

Property	Symbol	Units	Related Units
Charge	q, Q	Coulomb (C)	A s, F V
Current	i, I	Ampere (A)	C/s
Voltage	v, V	Volt (V)	J/C
Work	w, W	Joule (J)	W s, C V
Energy	e, E	Joule (J)	W s, C V
Power	p, P	Watt (W)	J/s
Resistance	R	Ohm (Ω)	J s/C^2
Capacitance	C	Farad (F)	C^2/J
Inductance	L	Henry (H)	J s^2/C^2

TABLE 2.2 Circuit Notation

Type	Notation
Time-varying quantities	$e(t)$, $i(t)$, $p(t)$, $q(t)$, $v(t)$, $w(t)$
Constant quantities	E, I, P, Q, V, W
Fixed device parameters	C, L, R

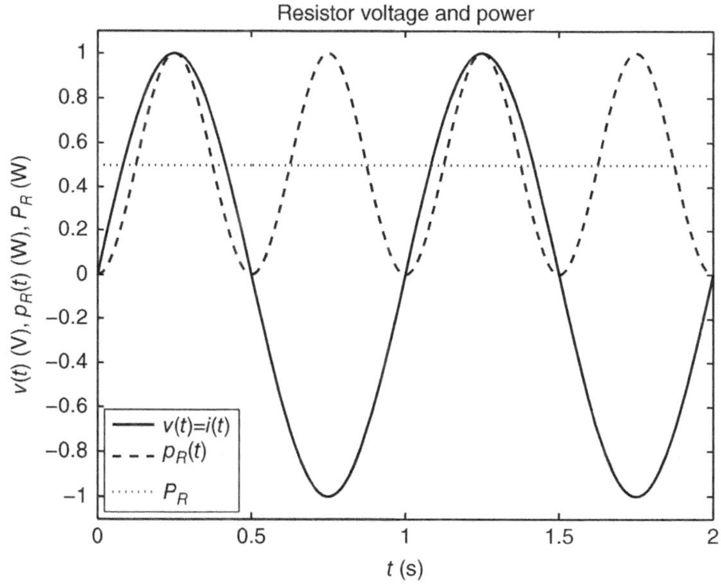

Figure 2.3 Voltage and power results in Example 2.1 for a sinusoidal current through a resistor with $T_o = 1$ s, $\omega_o = 2\pi$ rad/s ($f_o = 1$ Hz), $A_m = 1$ A, and $R = 1\ \Omega$.

related as $v(t) = Ri(t)$ (Ohm's law), and the instantaneous power is

$$p_R(t) = A_m^2 R \sin^2(\omega_o t) = (1/2)\, A_m^2 R[1 - \cos(2\omega_o t)], \tag{2.7}$$

where the subscript is sometimes used to denote the particular device. Examples of this voltage and power are illustrated in Figure 2.3 for $\omega_o = 2\pi$ rad/s and $f_o = 1$ Hz. Sinusoidal voltage and current are always in phase for a resistor, and, in this case, they have the same value because $R = 1\ \Omega$. The power is nonnegative, which means the resistor absorbs power; it always dissipates energy in the form of heat as mentioned earlier. The frequency of $p_R(t)$ is twice that of the voltage because the current and voltage are perfectly aligned (in phase), and the product $v(t)i(t)$ causes the negative portions of the waveform to become positive (the dashed line). The average power is

$$P_R = \frac{A_m^2 R}{2T_o} \int_0^{T_o} [1 - \cos(2\omega_o t)]dt, \tag{2.8}$$

where $T_o \triangleq 1/f_o$ is the period. Since the integral is performed over one period, the term containing cosine is 0, yielding $P_R = A_m^2 R/2$. As expected, the power increases with increasing waveform amplitude A_m or a larger resistance R. For the waveforms in Figure 2.3, $P_R = 1/2$ W (the dotted line) because all parameter values are 1.

Since power in many applications can have a wide range of values, it is often convenient to represent it using logarithms.

Definition: Decibel (dB) The *decibel (dB)* is the logarithm of the ratio of two powers:

$$P_{\text{dB}} \triangleq 10\log_{10}(P_1/P_0), \tag{2.9}$$

where P_1 and P_0 have the same units.

Although average power is used in the definition, it also applies to instantaneous power. If the units of P_0 and P_1 are both milliwatts, for example, then it is not necessary to convert into watts because the ratio handles common multiplier prefixes (10^{-3} in this case). The prefix "dec" of decibel means that it is one-tenth of a *bel*, which is a unit rarely used in practice. In the next section, we show that the power dissipated by a resistor R is $P_R = RV^2$, and so in decibels, we have

$$P_{\text{dB}} = 10 \log_{10}(RV_1^2/RV_0^2) = 20 \log_{10}(V_1/V_0), \tag{2.10}$$

where the exponent property of logarithms has been used to give the multiplier 20. This demonstrates that it is possible to write the ratio of *amplitudes* in decibels, but we must use the multiplicative factor 20 instead of 10. The square of an amplitude is proportional to power.

A ratio is used in (2.9) so that the argument of the logarithm is dimensionless. If P_0 is not explicitly given in (2.9) then $P_0 = 1$ W is assumed. We often write $P_{\text{dB}} = 10 \log_{10}(P)$ with the understanding that the denominator of the argument is 1. Sometimes it is convenient that the units of the denominator be milliwatts, in which case dBm is used and we would write $P_{\text{dBm}} = 10 \log_{10}(P)$ where it is implied that P is relative to 10^{-3} W.

The dB plot in Figure 2.4 illustrates how the dB formula compresses the horizontal axis; for example, 20 W is mapped to 13.0103 dB. This compression becomes more dramatic for large numbers: for example, 1 megawatt (MW) maps to 60 dB. Of course, the quantity in dB is simply the exponent of the prefix (mega in this case) scaled by 10. Observe that each doubling of power corresponds to a 3 dB increase on the vertical

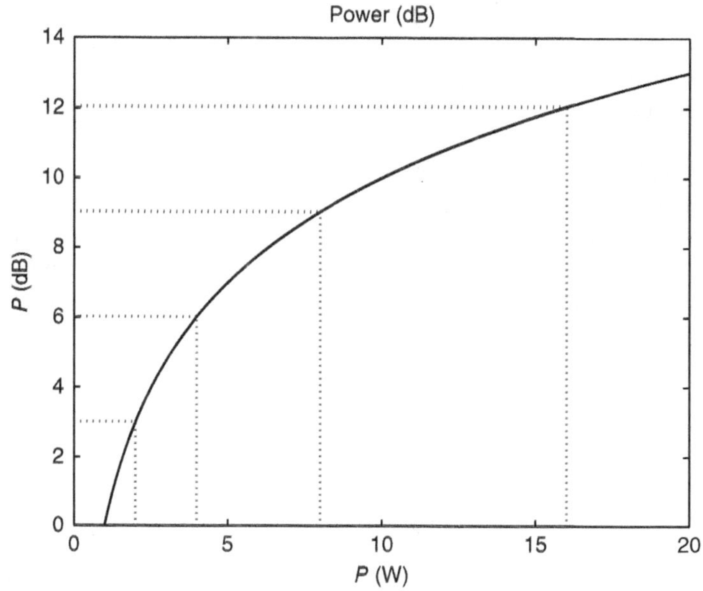

Figure 2.4 Power in decibels. The dotted lines illustrate that doubling P from 1 to 2 W corresponds to a 3 dB increase, quadrupling to 4 W yields a 6 dB increase, an eightfold increase is 9 dB, and so on.

TABLE 2.3 Decimal Prefixes and Multipliers

Prefix	Multiplier	Prefix	Multiplier
Atto (a)	10^{-18}	Exa (E)	10^{18}
Femto (f)	10^{-15}	Peta (P)	10^{15}
Pico (p)	10^{-12}	Tera (T)	10^{12}
Nano (n)	10^{-9}	Giga (G)	10^{9}
Micro (μ)	10^{-6}	Mega (M)	10^{6}
Milli (m)	10^{-3}	Kilo (k)	10^{3}
Centi (c)	10^{-2}	Hecto (h)	10^{2}
Deci (d)	10^{-1}	Deca (da)	10

axis. This plot illustrates that multiplication is transformed to addition when using the logarithm, which is a property discussed in Chapter 1.

Table 2.3 provides a summary of several decimal prefixes and their multipliers. For example, 1 kV equals 1×10^3 V and 1 mA equals 1×10^{-3} A. The prefixes hecto and deca are seldom used; instead, we would simply write 10 and 100 V, for example. The very small and very large prefixes are useful when describing the wavelengths and frequency bands of the high-energy end of the electromagnetic spectrum (see Chapter 8).

2.3 CIRCUIT ELEMENTS

For the system model in Figure 1.1, we may choose the input to be the current $x(t) = i$ through one device and the output to be the voltage $y(t) = v$ across another device in a particular circuit. This is frequently done in circuit analysis for which it is possible to derive a mathematical expression for $y(t)$ in terms of $x(t)$. Similarly, we can choose $x(t) = v$ and $y(t) = i$ in order to examine how a current varies due to changes in some voltage. Such a system model provides the *current-voltage characteristic* (I-V) of the circuit, which is perhaps the most widely used description for circuit devices. For the resistor, capacitor, and inductor, the voltage and current are related to each other by the following linear mathematical models:

$$\text{resistor:} \quad v = Ri, \tag{2.11}$$

$$\text{capacitor:} \quad i = C\frac{dv}{dt}, \tag{2.12}$$

$$\text{inductor:} \quad v = L\frac{di}{dt}, \tag{2.13}$$

where R is resistance in ohms (Ω), C is capacitance in farads (F), and L is inductance in henries (H). The I-V equation for a resistor is known as *Ohm's law*. The device symbols are summarized in Figure 2.5. Equations (2.11)–(2.13) are accurate models based on their physical properties and experiments using actual devices, and they are used to represent the elements in various circuits. The voltage across a resistor is proportional to the current, whereas for an inductor, the voltage is proportional to the *rate of change* of the current. Similarly, the current through a capacitor is proportional to the rate of change of the voltage. As mentioned earlier, these equations apply only over some limited range of values for v and i. For example, if the current through a resistor exceeds some threshold, the device will be damaged and the relation $v = Ri$ no longer applies. For notational convenience, the time argument of $i(t)$ and $v(t)$ is often suppressed as is the case in (2.11)–(2.13). (Note that D for the diode in Figure 2.5 is symbolic only; its I-V model is nonlinear and depends on the saturation current and thermal voltage described later in this section.)

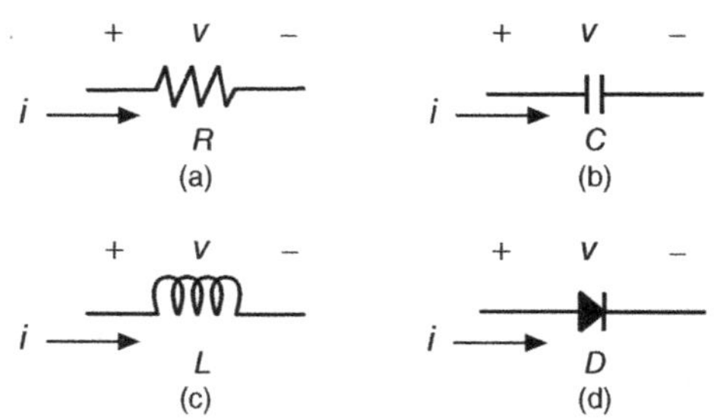

Figure 2.5 Circuit elements. (a) Resistor R. (b) Capacitor C. (c) Inductor L. (d) Diode D.

A resistor impedes electron flow because it is constructed of materials such as carbon, which are not as conductive as copper or silver. This impedance causes electron collisions and so heat is generated and a voltage appears across the device, which means work is required to move electrons from one end of the resistor to the other end. A capacitor stores electron charge when a voltage is applied across its terminals. The simplest model of a capacitor consists of two parallel plates that are closely spaced next each other but are not connected. When a voltage source is attached to the terminals, electric charge accumulates on the plates until its voltage matches that of the source. Current does *not* flow between the two plates. When the voltage source varies, there is current flow in a capacitor circuit only due to charge flowing to and from the plates through the external connecting circuit. When the voltage is fixed, a capacitor acts like an open circuit and there is no current flow.

The energy stored in a capacitor is (Problem 2.11)

$$E_C = \frac{1}{2}CV^2 = \frac{Q^2}{2C}, \tag{2.14}$$

where V is the fixed voltage across the capacitor. The last expression is due to the fact that the total charge on the plates is related to the voltage as follows:

$$Q = CV. \tag{2.15}$$

The voltage across an inductor is due to fluctuations in its magnetic field as the current varies. If the current is constant through an inductor, then there is no voltage across the device, and it is a short circuit that functions simply as an ideal wire. However, with a constant current I, the following energy is stored in the inductor (Problem 2.12):

$$E_L = \frac{1}{2}LI^2. \tag{2.16}$$

For the capacitor and inductor, the stored energies E_C and E_L can be used to deliver power to other parts of the circuit. When the voltage in (2.14) or the current in (2.16) varies with time, the instantaneous power equations are

$$p_C(t) = \frac{d}{dt}e_C(t) = Cv(t)\frac{d}{dt}v(t), \tag{2.17}$$

$$p_L(t) = \frac{d}{dt}e_L(t) = Li(t)\frac{d}{dt}i(t), \tag{2.18}$$

where $e_C(t)$ and $e_L(t)$ are time-varying versions of E_C and E_L. The expressions in (2.17) and (2.18) are also derived from the product of $v(t)$ and $i(t)$, using (2.12) for the capacitor current and (2.13) for the inductor voltage. As mentioned earlier, if these quantities are positive, the devices are absorbing power; otherwise, they are delivering power to the circuit.

Example 2.2 For the current in (2.6), the instantaneous power of inductor L is

$$p_L(t) = A_m^2 L\omega_o \sin(\omega_o t)\cos(\omega_o t) = (1/2)\, A_m^2 L\omega_o \sin(2\omega_o t), \tag{2.19}$$

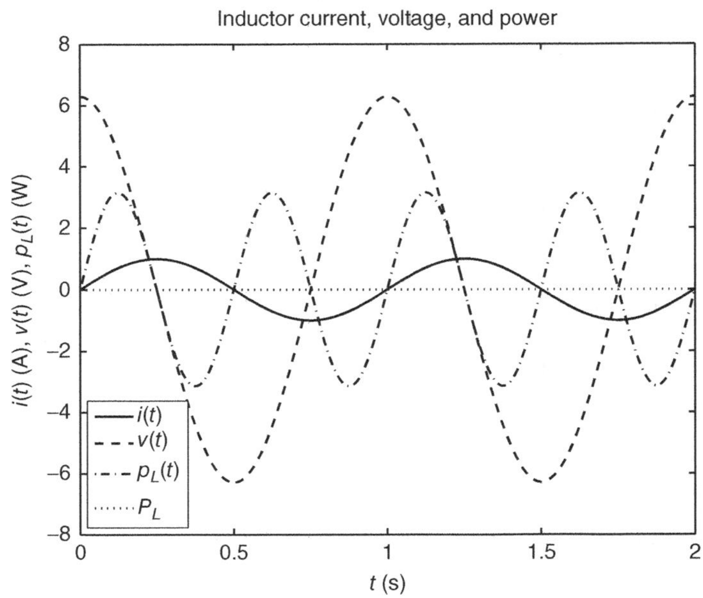

Figure 2.6 Voltage and power results in Example 2.1 for a sinusoidal current through an inductor with $\omega_o = 2\pi$ rad/s ($f_o = 1$ Hz), $A_m = 1$ A, and $L = 1$ H.

where a trigonometric identity from Appendix C has been used to write the last result. The current, voltage, and power are plotted in Figure 2.6 for $A_m = 1$ A, $L = 1$ H, and $\omega_o = 2\pi$ rad/s. Observe that the voltage and current are 90° out of phase relative to each other, which of course is due to the derivative in the inductor voltage model of (2.13). The frequency of the instantaneous power is twice that of the current and the voltage, and we see that the inductor absorbs power ($p_L(t) > 0$) for two intervals during one period of the current. Similarly, it delivers power for two intervals over the same period. This occurs because the voltage $v(t)$ and current $i(t)$ are out of phase, and it is their product that determines the sign of $p_L(t)$. The average power over one period is obviously 0 (the dotted line) because an inductor absorbs and delivers equal amounts of instantaneous power. Similar results can be shown for a capacitor.

Since the underlying physics of each device are not important for the material covered in this book, it is only necessary that the reader understand expressions of the form in (2.11)–(2.13). They will be used to develop ODEs that model the behavior of linear circuits and systems. It is possible to rewrite the device model for the capacitor in (2.12) as an integral by integrating both sides with respect to t:

$$C \int_{t_o}^{t} \frac{dv(t)}{dt} dt = \int_{t_o}^{t} i(t) dt. \tag{2.20}$$

The left-hand side becomes

$$C\int_{t_o}^{t}\frac{dv(t)}{dt}dt = C\int_{t_o}^{t}dv(t) = C[v(t) - v(t_o)], \tag{2.21}$$

and the voltage is

$$v(t) = \frac{1}{C}\int_{t_o}^{t} i(t)dt + v(t_o), \tag{2.22}$$

where $v(t_o)$ is the initial voltage across the capacitor at time t_o (usually $t_o = 0$). This model shows that the voltage across the capacitor increases or decreases depending on the *area* of the current waveform on the interval $[t_o, t]$. If the area is negative, then overall the current exits the capacitor and the voltage drops (thus, it delivers power to the circuit). The opposite result occurs for positive area. The corresponding equation for the inductor is

$$i(t) = \frac{1}{L}\int_{t_o}^{t} v(t)dt + i(t_o), \tag{2.23}$$

where $i(t_o)$ is the initial current through the device. It is interesting that the capacitor and the inductor have a *dual relationship*: interchanging $v(t)$ and $i(t)$ and replacing C with L in the capacitor model yield the inductor model. This property is exploited in the design of circuits to have a dynamic behavior that would not be readily achieved with the capacitor or the inductor alone.

Example 2.3 Suppose the current through a device has the triangular waveform

$$i(t) = \begin{cases} 2t, & 0 \le t < 1/2 \\ -2t + 2, & 1/2 \le t < 3/2 \\ 2t - 4, & 3/2 \le t \le 2, \end{cases} \tag{2.24}$$

where the units of t are seconds (s) and those of $i(t)$ are milliamperes (mA). If the device is a resistor, then its voltage waveform is identical to that of $i(t)$, but scaled as $v_R(t) = Ri(t)$. The voltage across an inductor is the derivative of (2.24), scaled by L:

$$v_L(t) = \begin{cases} 2L, & 0 \le t < 1/2 \\ -2L, & 1/2 \le t < 3/2 \\ 2L, & 3/2 \le t \le 2, \end{cases} \tag{2.25}$$

which is a rectangular waveform. The voltage across a capacitor is the integral of (2.24), scaled by $1/C$:

$$v_C(t) = \begin{cases} t^2/C, & 0 \le t < 1/2 \\ (-t^2 + 2t - 1/2)/C, & 1/2 \le t < 3/2 \\ (t^2 - 4t + 4)/C, & 3/2 \le t \le 2, \end{cases} \tag{2.26}$$

which is a quadratic waveform (we have assumed that $v_C(0) = 0$). Note that the value of $v_C(t)$ at the end of the first time interval is used as the initial condition for the equation that describes $v_C(t)$ for the second time interval, and similarly for the third time interval. These results are depicted in Figure 2.7(a) for specific values of $\{R, L, C\}$. The plot for the inductor shows an abrupt change in voltage, which is due to the derivative in (2.13). The voltage for the capacitor is much smoother because it is derived as the integral of the current in (2.22). The corresponding energy waveforms for the inductor and the capacitor are shown in Figure 2.7(b), along with $i_L(t)$ and $v_C(t)$ for comparison. Observe that the energy waveforms are just scaled versions of the voltage squared (for the capacitor) and the current squared (for the inductor), as given by (2.14) and (2.16), respectively. Of course, the energy is always nonnegative, and it is 0 only for zero current through the inductor and zero voltage across the capacitor.

The previous example motivates additional properties of C and L.

- The voltage across capacitor C cannot change instantaneously.
- The current through inductor L cannot change instantaneously.

These properties follow from the integral equations in (2.22) and (2.23). It takes time for the charge to accumulate in a capacitor, and so its voltage does not have any discontinuities. (We assume there are no impulsive voltage or current sources modeled by the Dirac delta function, which is introduced in Chapter 5.) Similarly, it takes time for the inductor magnetic field to build up, and so its current does not have any discontinuities. On the other hand, because of the derivative models for C and L in (2.12) and (2.13), respectively, the voltage across an inductor and the current through a capacitor *can* change instantaneously.

If a voltage has the waveform in (2.24), then the same curves in Figure 2.7(a) are obtained for the current along the vertical axis (especially since all device values are 1 in this example). However, the waveforms for the capacitor and the inductor would be interchanged. This duality property of C and L is also evident from the units of the device models in (2.11)–(2.13), given by ohms (Ω), farads (F), and henries (H), respectively. These are related to the voltage in volts (V) and the current in amperes (A) as follows:

$$R = v/i \implies \Omega = \text{V}/\text{A}, \tag{2.27}$$

$$C = i(dt/dv) \implies \text{F} = \text{A(s/V)} = \text{s}/\Omega, \tag{2.28}$$

$$L = v(dt/di) \implies \text{H} = \text{V(s/A)} = \Omega\ \text{s}, \tag{2.29}$$

where s is seconds. Thus, H is proportional to Ω, whereas F is proportional to its inverse $1/\Omega$, again showing the duality of the two devices.

The diode is a semiconductor device that has a *nonlinear* I-V characteristic. Several models with increasing complexity have been developed for the diode. One of

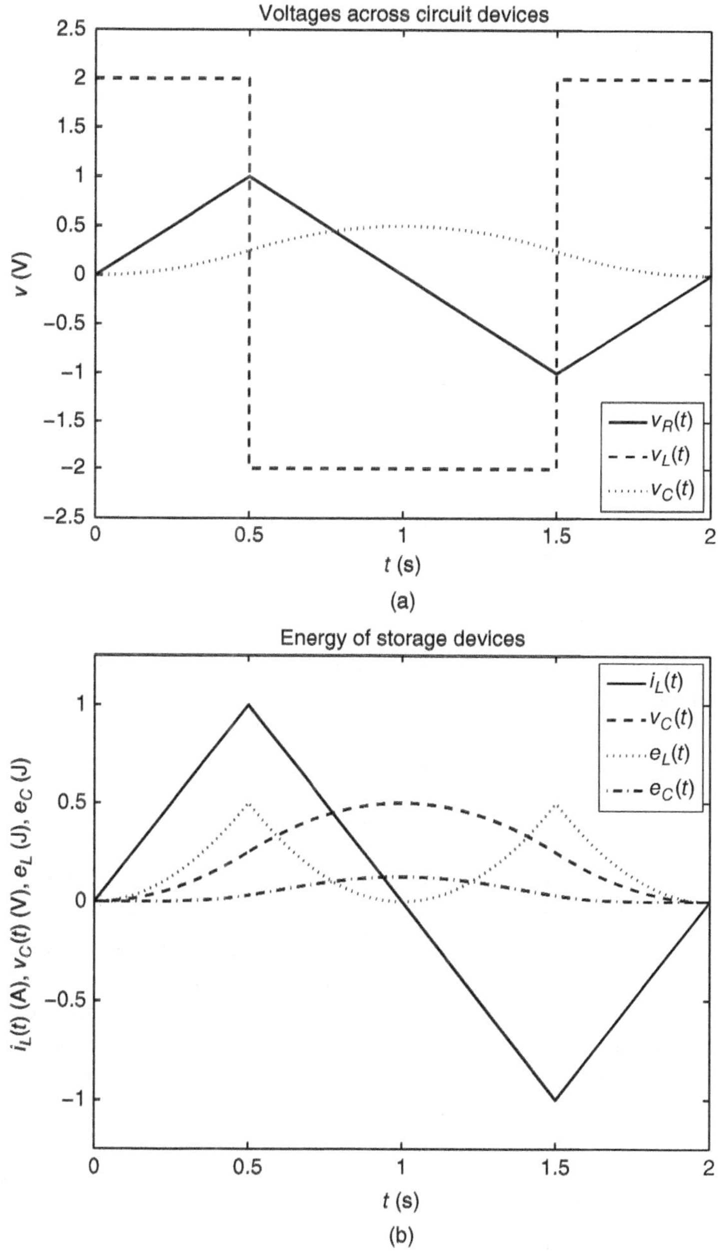

Figure 2.7 Device results for the time-varying current in Example 2.3 with $R = 1\ \Omega$, $L = 1$ H, and $C = 1$ F. (a) Voltage waveforms. (b) Inductor and capacitor energy.

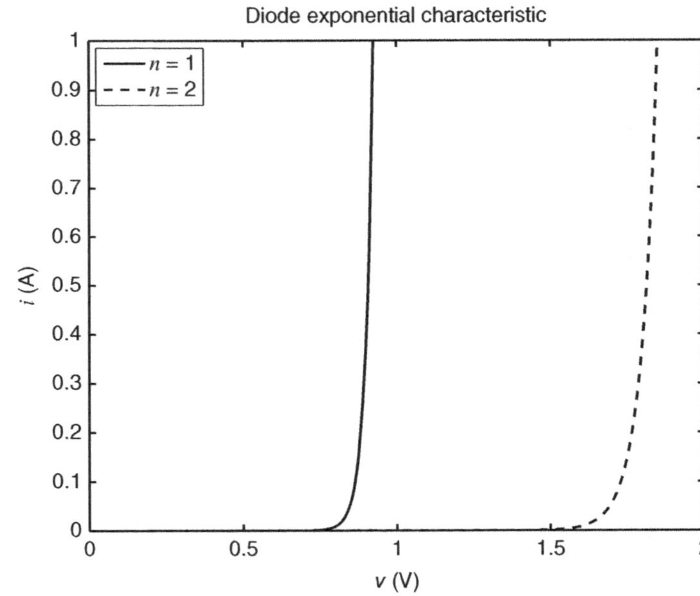

Figure 2.8 Diode exponential characteristic in (2.30).

the simplest models has the piecewise linear characteristic shown in Figure 1.12(b) and described by (1.28). Another model is based on the exponential function:

$$i = \begin{cases} I_s[\exp\,(v/nV_T) - 1], & v \geq 0 \\ -I_s, & v < 0, \end{cases} \tag{2.30}$$

where $I_s \approx 10^{-15}$ A is the reverse-biased *saturation current*, $V_T \approx 0.026$ V (26 mV) is the *thermal voltage*, and $n \in [1, 2]$ depends on the device fabrication. This model is illustrated in Figure 2.8 for two values of n. The diode is "on" (forward-biased) for a positive voltage, and it is "off" (reverse-biased) for a negative voltage. The large arrow of the diode symbol in Figure 2.5 indicates the forward-biased direction. The curve in (2.30) can be approximated reasonably well by the piecewise linear diode model shown in Figure 1.12(b). Typically, $n = 1$ is assumed in circuit courses such that the voltage drop across the diode is approximately a constant 0.7 V, as seen in Figure 2.8 (the solid line).

The circuit elements in Figure 2.5 do not provide any *net* energy to a circuit, and so they are *passive* devices. Capacitors and inductors store energy derived from a power source, but they do not provide any additional power. Though certain types of diodes provide amplification and are considered to be "active" on that basis (as are transistors), we assume that the diode is passive. The two active devices considered in this book are voltage and current sources whose symbols are shown in Figure 2.9. These power sources are ideal: V_s and I_s remain constant when attached to any circuit.

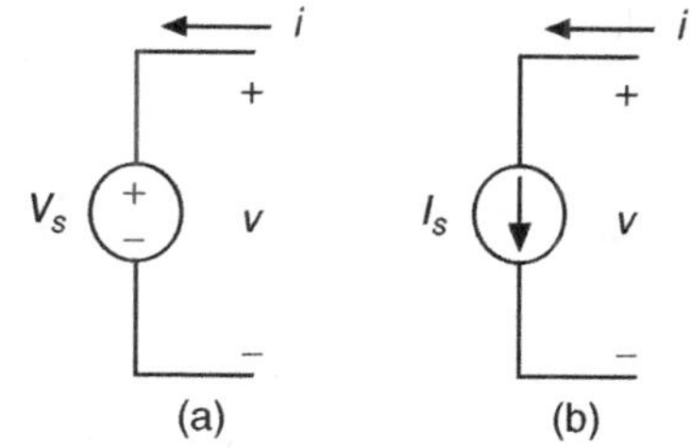

Figure 2.9 Power sources. (a) Voltage source V_s. (b) Current source I_s.

This means that the I-V characteristic where i is plotted versus v (as in Figure 2.2) is a vertical line at $v = V_s$ for a voltage source, and a horizontal line at $i = I_s$ for a current source. An example of a voltage source is the household battery, though it is not ideal: its voltage actually drops with increasing current. A current source can be implemented using transistors and operational amplifiers, which are active devices covered in basic electronics courses. The terminals of a voltage source should not be connected without some series resistance (or capacitance) in order to avoid a short circuit and a large current (an infinite current in the ideal model). On the other hand, since I_s is fixed, the terminals of a current source must be connected to some circuit so that its current flow is not interrupted.

2.4 BASIC CIRCUIT LAWS

Consider the simple resistive circuit shown in Figure 2.10, which can be viewed as a system with input $x(t) = V_s$ and output $y(t) = v$ across resistor R_3. Of course, other output variations are possible; for example, we might be interested in the voltage across R_2 or the current through R_3. Two basic circuit laws, known as *Kirchoff's circuit laws*, are used to derive an equation for $y(t)$ in terms of $x(t)$,

- *Kirchoff's voltage law* (*KVL*): The sum of all voltages across elements around any closed loop is 0.
- *Kirchoff's current law* (*KCL*): The sum of all currents entering any node of connecting wires is 0.

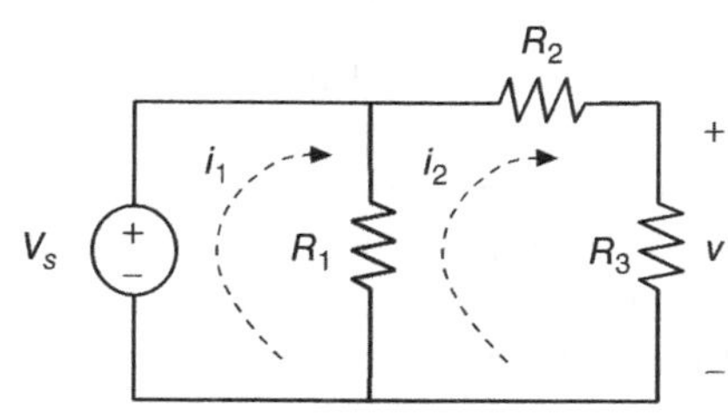

Figure 2.10 Resistive circuit and voltage source V_s.

Mathematically, the two laws are

$$\text{KVL:} \quad \sum_{n=1}^{N} v_n = 0, \qquad \text{KCL:} \quad \sum_{n=1}^{N} i_n = 0, \tag{2.31}$$

where $\{v_n\}$ are voltages across N devices in a loop and $\{i_n\}$ are N currents entering a node. In order to use these laws, we label the voltage polarity and give the current direction for each device. If any of these polarities/directions are incorrect, those quantities will turn out to have a negative sign at the end of the analysis.

Example 2.4 For the circuit in Figure 2.10, there are two loops with currents labeled $\{i_1, i_2\}$. KVL yields two equations:

$$-V_s + (i_1 - i_2)R_1 = 0, \qquad (i_2 - i_1)R_1 + i_2 R_2 + i_2 R_3 = 0, \tag{2.32}$$

where by convention we have assumed that a current enters the + terminal for each resistor (see Figure 2.5). In the first loop, the current entering R_1 is the difference of the two labeled currents: $i_1 - i_2$, and the + terminal is located at the top of R_1. For the second loop, the situation is reversed: the + terminal is located at the bottom of R_1 and the current entering there is $i_2 - i_1$. The reverse situation occurs because we have chosen both loop currents to flow in a clockwise direction. This example shows that the actual current through a device is often a combination of the defined loop currents. Solving the first equation for i_1 and substituting it into the second equation yields

$$i_1 = V_s/R_1 + i_2, \qquad (i_2 - V_s/R_1 - R_1 i_2)R_1 + i_2(R_2 + R_3) = 0, \tag{2.33}$$

and the currents are

$$i_2 = \frac{V_s}{R_2 + R_3}, \qquad i_1 = \frac{R_1 + R_2 + R_3}{R_1(R_2 + R_3)} V_s. \tag{2.34}$$

From Ohm's law, the output voltage is $v = R_3 i_2$, and the input/output (transfer) characteristic of the circuit is

$$v = \frac{R_3}{R_2 + R_3} V_s. \tag{2.35}$$

This result is an example of *voltage division*, which is discussed in the next section. The voltage across R_1 is V_s, and this divides across R_2 and R_3 depending on their relative values as given by the ratio in (2.35). The current expressions in (2.34) can also be derived using KCL. For the node at the top of the circuit just before R_2, the currents are summed:

$$i_1 + i_3 - i_2 = 0, \tag{2.36}$$

where we have defined i_3 to be the current entering the node from R_1. Since i_2 exits the node, it has a minus sign in this expression. Using Ohm's law, these currents are rewritten in terms of the voltages V_s and v. Since the voltage across R_1 is V_s, we have

$$i_3 = -\frac{V_s}{R_1}, \qquad i_2 = \frac{V_s}{R_2 + R_3}, \tag{2.37}$$

which shows that i_3 is actually exiting the node because of the minus sign. The current in the first loop is derived from (2.36):

$$i_1 = i_2 - i_3 = \frac{V_s}{R_2 + R_3} + \frac{V_s}{R_1} = \frac{R_1 + R_2 + R_3}{R_1(R_2 + R_3)} V_s, \tag{2.38}$$

and so the same current results as in (2.34) are derived.

The analysis of all-resistive circuits yields a system of linear equations with the number of equations equal to the number of loops or nodes. The matrix equation for the currents in (2.32) is

$$\begin{bmatrix} R_1 & -R_1 \\ -R_1 & R_1 + R_2 + R_3 \end{bmatrix} \begin{bmatrix} i_1 \\ i_2 \end{bmatrix} = \begin{bmatrix} V_s \\ 0 \end{bmatrix}. \tag{2.39}$$

The unknown variables $\{i_1, i_2\}$ are solved by applying Cramer's rule or Gaussian elimination, both of which are described in Chapter 3.

Example 2.5 For the resistive circuit in Figure 2.10, assume that $V_s = 10$ V and the resistors are all equal: $R_1 = R_2 = R_3 = 100\ \Omega$. From the previous example, we immediately find that the loop currents are $i_1 = 0.15$ A and $i_2 = 0.05$ A, and the voltage across R_3 is $v = 5$ V. This result is expected because the voltage across R_1 is 10 V, and from voltage division, V_s splits equally across the other two resistors because $R_2 = R_3$. Observe also that the current through R_1 is $i_1 - i_2 = 0.1$ A, which verifies that the voltage across R_1 is $0.1 \times 100 = 10$ V.

2.4.1 Mesh-Current and Node-Voltage Analysis

The basic circuit laws KVL and KCL can be extended to more complicated circuits by using techniques called *mesh-current analysis* and *node-voltage analysis*.

Definition: Mesh and Node A *mesh* is a closed loop in a circuit that does not enclose any other loop. A *node* is a point in a circuit where two or more circuit elements are connected.

These two techniques are illustrated by finding the voltage v in the circuit in Figure 2.11.

Example 2.6 Observe in the figure that there are three meshes with currents labeled $\{i_1, i_2, i_3\}$. A mesh-current analysis uses KVL around each mesh to write voltage

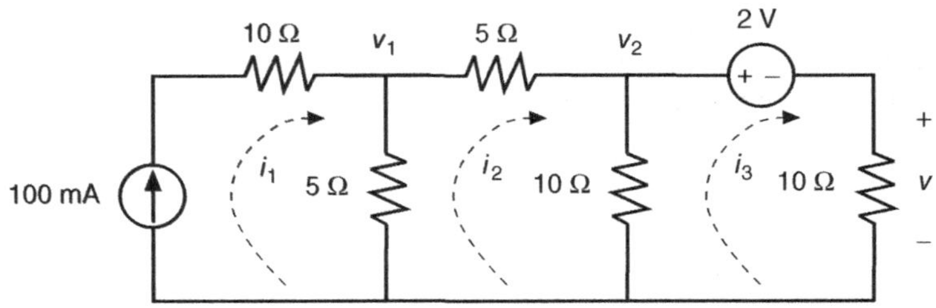

Figure 2.11 Resistive circuit for the mesh-current and node-voltage analysis in Example 2.6.

equations in terms of these currents via Ohm's law. However, since $i_1 = 100$ mA, only the two meshes on the right need to be examined:

$$5(i_2 - i_1) + 5i_2 + 10(i_2 - i_3) = 0, \tag{2.40}$$

$$10(i_3 - i_2) + 2 + 10i_3 = 0, \tag{2.41}$$

where we have used the conventional voltage polarity for each of the resistors. For example, in the middle mesh, i_2 enters the positive terminal of the vertical 10 Ω resistor, and so i_3 enters the negative terminal, resulting in the voltage $10(i_2 - i_3)$. For the third mesh, the polarity of that resistor is reversed, and the voltage is $10(i_3 - i_2)$ as given in (2.41). Substituting $i_1 = 0.1$ A in (2.40) yields two equations in two unknowns:

$$20i_2 - 10i_3 - 1/2 = 0, \qquad 20i_3 - 10i_2 + 2 = 0, \tag{2.42}$$

where the coefficients of $\{i_2, i_3\}$ have been combined. Solving these equations yields $i_2 = -1/30$ A and $i_3 = -7/60$ A, demonstrating that these currents actually flow counterclockwise in the circuit. The output voltage is $v = 10i_3 = -7/6$ V. For a node-voltage analysis, technically there are five nodes, but only three of them are *essential nodes* where three or more elements are connected. Two of these are labeled with voltages $\{v_1, v_2\}$, both of which are defined relative to the common node at the bottom called the *reference node*. Using an alternative convention that all currents exit a node, KCL and Ohm's law yield

$$-i_1 + v_1/5 + (v_1 - v_2)/5 = 0, \qquad (v_2 - v_1)/5 + v_2/10 + i_3 = 0. \tag{2.43}$$

Substituting $i_1 = 0.1$ A, $i_3 = v/10$, and $v_2 = v + 2$, we have two equations and two unknowns:

$$2v_1/5 - v/5 - 1/2 = 0, \qquad 2v/5 - v_1/5 + 3/5 = 0. \tag{2.44}$$

Solving these yields $v = -7/6$ V, $v_1 = 2/3$ V, and $v_2 = 5/6$ V. This example demonstrates that one of the analysis techniques is usually easier to implement. Because of the 2 V source, we are not able to directly write an expression for i_3 exiting the

v_2 node; it is necessary that the third voltage v be brought into the equations. The mesh analysis is slightly easier because i_1 is known, and as a result, only two mesh equations are needed.

2.4.2 Equivalent Resistive Circuits

Two special cases of KVL and KCL arise in a circuit (or part of a circuit) involving two resistors.

- *Voltage division*: For two resistors $\{R_1, R_2\}$ in *series*, the overall voltage V_s across them divides as

$$v_{R_1} = \frac{R_1}{R_1 + R_2} V_s, \qquad v_{R_2} = \frac{R_2}{R_1 + R_2} V_s. \tag{2.45}$$

- *Current division*: For two resistors $\{R_1, R_2\}$ in *parallel*, the overall current I_s entering a common node divides as

$$i_{R_1} = \frac{R_2}{R_1 + R_2} I_s, \qquad i_{R_2} = \frac{R_1}{R_1 + R_2} I_s. \tag{2.46}$$

The corresponding circuits are shown in Figure 2.12. Voltage division follows directly from KVL and Ohm's law:

$$i = \frac{1}{R_1 + R_2} V_s \Rightarrow v_{R_1} = R_1 i = \frac{R_1}{R_1 + R_2} V_s, \tag{2.47}$$

and similarly for v_{R_2}. Observe that the resistor numerators are interchanged for current division in (2.46) compared with voltage division in (2.45). This result is due to KCL and Ohm's law:

$$v_{R_1} = v_{R_2} \Rightarrow i_{R_1} R_1 = i_{R_2} R_2 \Rightarrow i_{R_2} = (R_1/R_2) i_{R_1}. \tag{2.48}$$

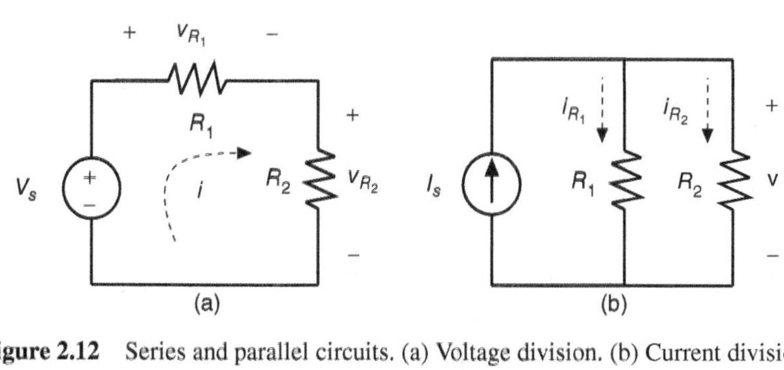

Figure 2.12 Series and parallel circuits. (a) Voltage division. (b) Current division.

Substituting this expression into $i_{R_1} + i_{R_2} = I_s$ yields

$$i_{R_1} + (R_1/R_2)\, i_{R_1} = I_s \Rightarrow i_{R_1} = \frac{R_2}{R_1 + R_2} I_s, \tag{2.49}$$

and similarly for i_{R_2}.

Example 2.7 From the previous results, we can determine how to combine two resistors that are in series or in parallel with each other, resulting in an *equivalent resistance*. For the series circuit in Figure 2.12(a), KVL shows that the voltage across both resistors together must be V_s. Since they have the same current i, we can write

$$V_s/i = R_1 + R_2 = R_{\text{series}}, \tag{2.50}$$

showing that resistors in series add together. It is important to note that they must have the same current in order to be considered in series. For the parallel circuit in Figure 2.12(b), we have from KCL that

$$I_s = i_{R_1} + i_{R_2} = v/R_{\text{parallel}}, \tag{2.51}$$

where v is the same voltage across each resistor. Applying Ohm's law to the middle expression yields:

$$v/R_1 + v/R_2 = v/R_{\text{parallel}}. \tag{2.52}$$

Cancelling v and solving for R_{parallel}, the equivalent resistance is

$$R_{\text{parallel}} = \frac{1}{1/R_1 + 1/R_2} = \frac{R_1 R_2}{R_1 + R_2}. \tag{2.53}$$

In order to combine parallel resistors as in (2.53), they must have the same voltage across them, which is v in this example. The equations in (2.50) and (2.53) are easily extended to three or more resistors (see Problem 2.18).

Finally, we introduce two equivalent circuits that are used to represent an all-resistive circuit by a single power source (voltage or current) and a single resistor. They are known as Thévenin and Norton equivalent circuits, which are depicted in Figure 2.13.

- *Thévenin open circuit voltage*: V_{oc} is computed at the two terminals of interest.
- *Norton short circuit current*: I_{sc} is computed at the two terminals of interest.
- *Thévenin resistance*: $R_{\text{th}} = V_{\text{oc}}/I_{\text{sc}}$ is the same for both equivalent circuits.

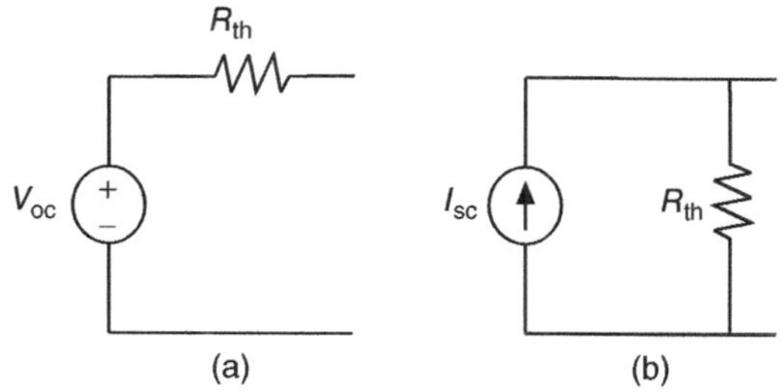

Figure 2.13 Equivalent resistive circuits. (a) Thévenin. (b) Norton.

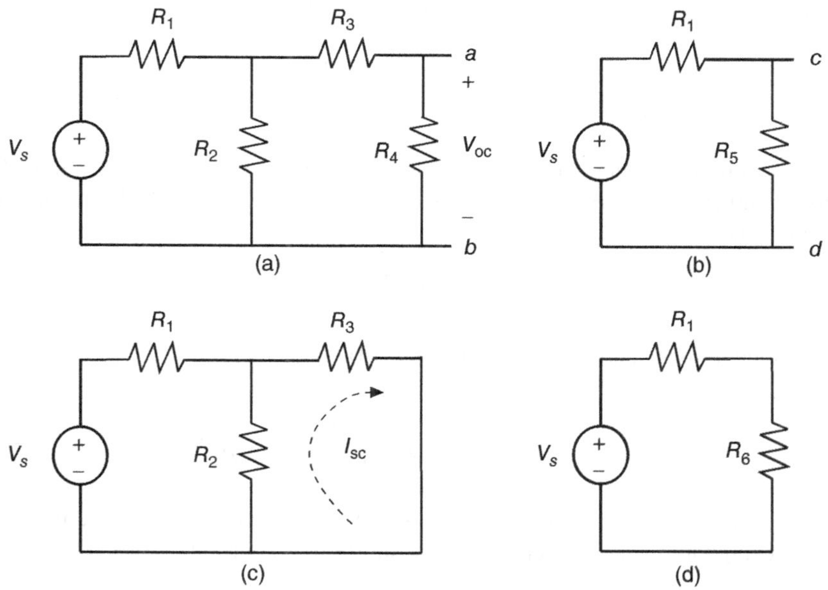

Figure 2.14 Resistive circuit used in Example 2.8. (a) Original open circuit showing V_{oc}. (b) Equivalent open circuit. (c) Original short circuit showing I_{sc}. (d) Equivalent short circuit.

The resistance R_{th} is also derived by replacing all voltage sources with short circuits and all current sources with open circuits. The equivalent resistance at the terminals of interest is then computed. Each type of equivalent circuit is derived from the other using Ohm's law: $V_{th} = I_{sc}R_{th}$ and $I_{sc} = V_{th}/R_{th}$.

Example 2.8 Consider the circuit in Figure 2.14(a) consisting of four resistors and a voltage source. The goal in this example is to replace the circuit with a Thévenin equivalent as seen from the *a*–*b* terminals. The open-circuit voltage across R_4 is derived by first combining the resistors as follows: R_3 and R_4 are in series, and together they are in parallel with R_2. Those three resistors have the following equivalent resistance:

$$R_5 = \frac{R_2(R_3 + R_4)}{R_2 + R_3 + R_4}, \tag{2.54}$$

which is shown in Figure 2.14(b). Note that the voltage at terminals *c–d* is *not* the same as that at terminals *a–b*. Voltage division across R_5 yields

$$V_{R_5} = \frac{R_5}{R_1 + R_5} V_s = \frac{R_2(R_3 + R_4)}{R_1(R_2 + R_3 + R_4) + R_2(R_3 + R_4)} V_s, \tag{2.55}$$

which also happens to be the voltage across R_2 in the original open circuit. Thus, voltage division across R_4 gives the open-circuit voltage:

$$V_{\text{oc}} = \frac{R_4}{R_3 + R_4} V_{R_5} = \frac{R_2 R_4}{R_1(R_2 + R_3 + R_4) + R_2(R_3 + R_4)} V_s. \tag{2.56}$$

The short-circuit current is derived by connecting the *a–b* terminals such that R_4 is ignored and R_3 is in parallel with R_2, as shown in Figure 2.14(c). Their equivalent resistance is

$$R_6 = \frac{R_2 R_3}{R_2 + R_3}, \tag{2.57}$$

and voltage division across R_6 in Figure 2.14(d) yields

$$V_{R_6} = \frac{R_6}{R_1 + R_6} V_s = \frac{R_2 R_3}{R_1(R_2 + R_3) + R_2 R_3} V_s. \tag{2.58}$$

Since V_{R_6} is also the voltage across R_3 of the original short circuit, we obtain its current using Ohm's law, which is also the short-circuit current:

$$I_{\text{sc}} = V_{R_3}/R_3 = \frac{R_2}{R_1(R_2 + R_3) + R_2 R_3} V_s. \tag{2.59}$$

Finally, the Thévenin resistance is the ratio of these two results:

$$R_{\text{th}} = V_{\text{oc}}/I_{\text{sc}} = \frac{R_1 R_4(R_2 + R_3) + R_2 R_3 R_4}{R_1(R_2 + R_3 + R_4) + R_2(R_3 + R_4)}, \tag{2.60}$$

where V_s has cancelled. This last expression can also be derived by shorting the voltage source and finding the overall equivalent resistance. In this case, R_1 and R_2 are in parallel, which together are in series with R_3, resulting in the equivalent resistance:

$$R_7 = \frac{R_1 R_2}{R_1 + R_2} + R_3 = \frac{R_1 R_2 + R_3(R_1 + R_2)}{R_1 + R_2}. \tag{2.61}$$

TABLE 2.4 Properties of Resistive Circuits

Property	Formula
Series resistance	$R_{eq} = R_1 + R_2$
	$R_{eq} = R_1 + R_2 + R_3$
Parallel resistance	$R_{eq} = R_1R_2/(R_1 + R_2)$
	$R_{eq} = R_1R_2R_3/(R_1R_2 + R_1R_3 + R_2R_3)$
Voltage division (series resistors)	$V_{R_1} = R_1V_s/(R_1 + R_2)$
	$V_{R_1} = R_1V_s/(R_1 + R_2 + R_3)$
Current division (parallel resistors)	$I_{R_1} = R_2I_s/(R_1 + R_2)$
	$I_{R_1} = R_2R_3I_s/(R_1R_2 + R_1R_3 + R_2R_3)$
Thévenin and Norton equivalents	$V_{oc} = R_{th}I_{sc}$

Combining this expression with the parallel resistor R_4 yields

$$R_{th} = \frac{R_4R_7}{R_4 + R_7} = \frac{R_1R_2R_4 + R_3R_4(R_1 + R_2)}{R_4(R_1 + R_2) + R_1R_2 + R_3(R_1 + R_2)}, \tag{2.62}$$

which is the same as (2.60). For a numerical example, let $V_s = 10$ V and assume that all four resistors are 100 Ω. These yield $R_{th} = 60$ Ω, $V_{oc} = 2$ V, and $I_{sc} = 1/30 \approx 0.0333$ A.

Properties for two and three resistors are summarized in Table 2.4. In the previous examples, there is no time variation in an all-resistive circuit if the voltage or current source remains fixed. Even if the voltage source were to change suddenly, the currents through and the voltages across all devices in the circuit would theoretically adjust instantaneously, without any rise time or fall time. When a circuit contains a capacitor or an inductor, we find in the next two sections that the currents and voltages require time to reach steady-state values in response to changes in the power sources or changes to the circuit configuration due, for example, to a switch opening or closing. They also depend on any nonzero initial voltage across or initial current through L and C.

2.4.3 RC and RL Circuits

An example *first-order* circuit is shown in Figure 2.15 where R_3 in Figure 2.10 has been replaced by capacitor C. The order of such circuits is generally determined by the number of capacitors and inductors, which is also the order of the ODE model. From KVL, we can write

$$V_s = R_2i_2 + v. \tag{2.63}$$

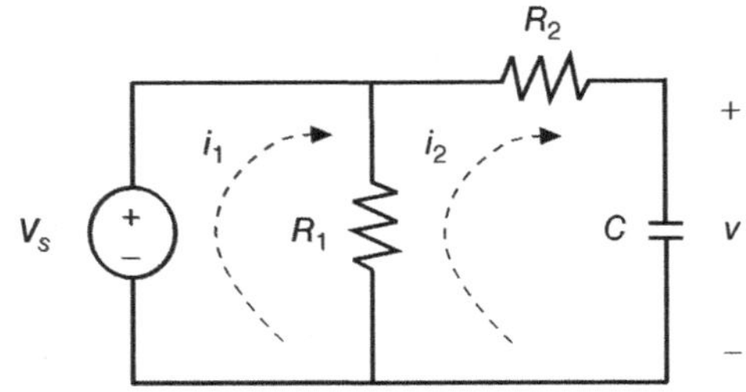

Figure 2.15 First-order circuit with capacitor C.

Substituting the current equation for the capacitor in (2.12) given by $i_2 = Cdv/dt$ yields a first-order linear ODE with constant coefficients:

$$R_2 C\frac{dv}{dt} + v = V_s. \qquad (2.64)$$

If $V_s = 0$ and the initial voltage across the capacitor is $v(0)$, then using the techniques in Chapter 6 we find that the solution is

$$v(t) = v(0)\exp(-t/R_2 C)u(t), \qquad (2.65)$$

where $u(t)$ is the unit step function mentioned in Chapter 1, which equals 1 for $t \in \mathcal{R}^+$ and is 0 otherwise. An exponentially decaying function is the characteristic behavior of the voltages and currents of a first-order circuit with nonzero initial conditions. The corresponding capacitor current is derived using (2.12):

$$i_2(t) = -[v(0)/R_2]\exp(-t/R_2 C)\,u(t) = -[v(t)/R_2]u(t), \qquad (2.66)$$

which we find is in the opposite direction of that shown in the figure. The initial charge in the capacitor dissipates as heat through resistor R_2. The last result in (2.66) is due to Ohm's law because the voltage across the capacitor is the same as that across R_2. There are no oscillations as there can be for the second-order RLC circuit described in the next section. Since $V_s = 0$, which means the voltage source is replaced by a short circuit, none of the current flows through R_1 because it must also have zero volts by KVL. Thus, the two currents shown in the figure are actually equal: $i_1 = i_2$. When V_s is nonzero, and especially if it is time-varying, the capacitor voltage and current expressions are more complicated, as shown later in Chapter 6.

Example 2.9 For the RC circuit in Figure 2.15, assume that $v(0) = 1$ V and $C = 100$ μF. Figure 2.16 shows the exponential result in (2.65) for two different values of R_2. As discussed in Chapter 5, the time constant for the exponentially decreasing waveform $\exp(-t/\tau)u(t)$ is τ in the exponent. Observe that $t = \tau$ gives $\exp(-1) = 1/e$, and so, one time constant is the time required for the function to decrease by a factor of $1/e \approx 0.3679$ times its initial value $v(0)$. Since the time constant for this

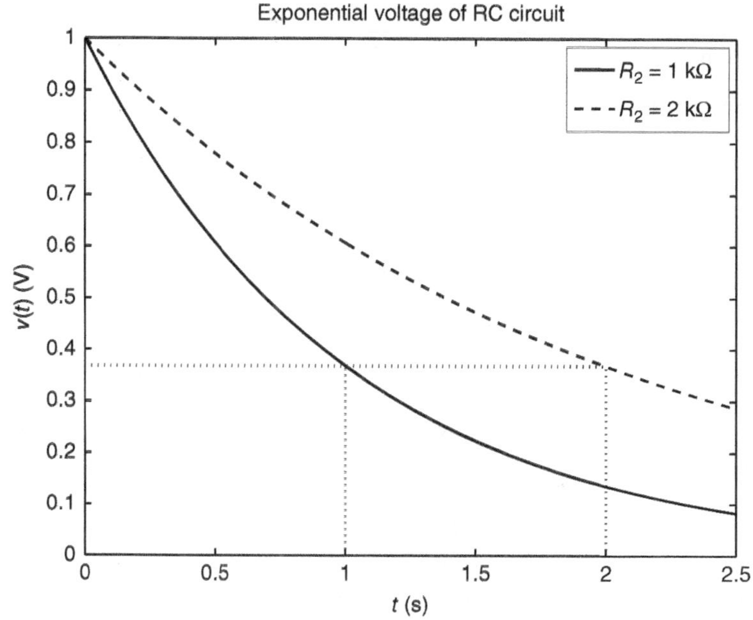

Figure 2.16 Exponentially decreasing voltage for the RC circuit in Example 2.9 with $C = 100\ \mu$F. The dotted lines denote one time constant for each curve: $\tau = 1$ s and $\tau = 2$ s.

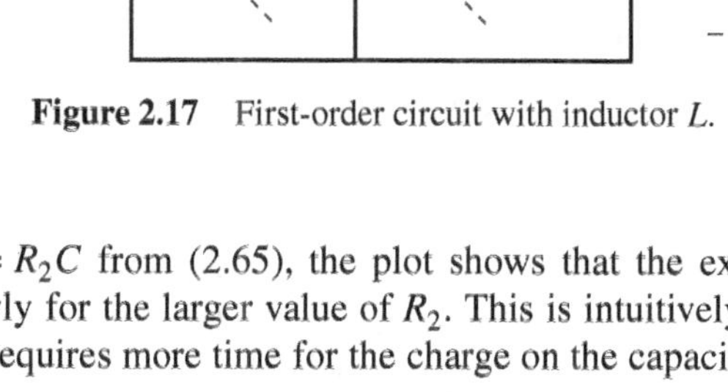

Figure 2.17 First-order circuit with inductor L.

RC circuit is $\tau = R_2C$ from (2.65), the plot shows that the exponential function decays more slowly for the larger value of R_2. This is intuitively correct because a larger resistance requires more time for the charge on the capacitor to be dissipated as heat.

If the capacitor in Figure 2.15 is replaced by inductor L, similar equations are derived for the current and voltage in Figure 2.17 using KVL and (2.63):

$$V_s = R_2 i_2 + L\frac{d}{dt}i_2. \tag{2.67}$$

TABLE 2.5 Equivalent Inductance and Capacitance

Property	Formula
Series inductance	$L_{\rm eq} = L_1 + L_2$
	$L_{\rm eq} = L_1 + L_2 + L_3$
Parallel inductance	$L_{\rm eq} = L_1L_2/(L_1 + L_2)$
	$L_{\rm eq} = L_1L_2L_3/(L_1L_2 + L_1L_3 + L_2L_3)$
Series capacitance	$C_{\rm eq} = C_1C_2/(C_1 + C_2)$
	$C_{\rm eq} = C_1C_2C_3/(C_1C_2 + C_1C_3 + C_2C_3)$
Parallel capacitance	$C_{\rm eq} = C_1 + C_2$
	$C_{\rm eq} = C_1 + C_2 + C_3$

Assuming that $V_s = 0$ and the initial current through the inductor is $i_2(0)$ in the direction shown in the figure, the solution of this first-order ODE is also exponential:

$$i_2(t) = i_2(0)\exp(-R_2t/L)u(t). \tag{2.68}$$

The time constant is $\tau = L/R_2$, which obviously increases with increasing inductance or decreasing resistance. As in the case of the RC circuit, the initial energy stored in the inductor is dissipated as heat through R_2. The corresponding inductor voltage is derived from (2.13):

$$v(t) = -R_2i_2(0)\exp(-R_2t/L)u(t) = -R_2i_2(t)u(t), \tag{2.69}$$

and so v in the circuit is actually negative as it approaches 0.

Table 2.5 summarizes the series and parallel combinations for inductors and capacitors that are derived in Problems 2.23 and 2.24. The equations for equivalent inductance are similar to those for equivalent resistance.

2.4.4 Series RLC Circuit

An example *second-order* circuit is shown in Figure 2.18, which has two energy storage elements: capacitor C and inductor L. This is a series circuit because the same

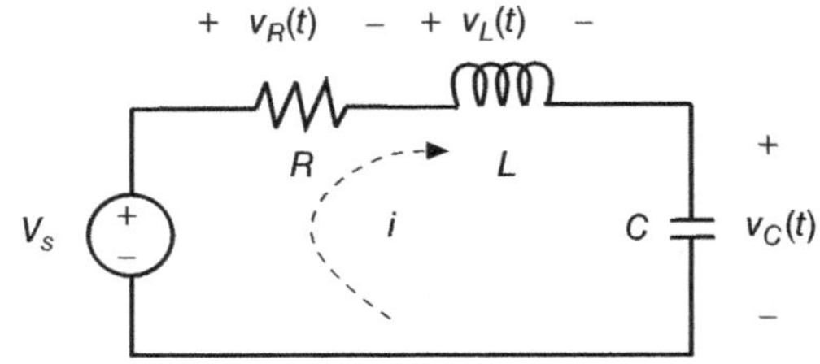

Figure 2.18 Second-order series circuit with resistor R, inductor L, and capacitor C.

current $i(t)$ flows through each device; an example of a parallel circuit where each device has the same voltage is discussed in Chapter 6. KVL gives an equation that models this circuit:

$$v_R(t) + v_L(t) + v_C(t) = V_s, \tag{2.70}$$

where the subscripts on v denote the three passive circuit elements. Substituting the models in (2.11) and (2.12) for the resistor and inductor voltages yields

$$Ri(t) + L\frac{d}{dt}i(t) + \frac{q(t)}{C} = V_s. \tag{2.71}$$

The voltage of the capacitor has been written in terms of the total charge $q(t)$, which follows from the model in (2.22):

$$\begin{aligned} v_C(t) &= \frac{1}{C}\int_{t_o}^{t} \frac{dq(t)}{dt}dt + v(t_o) \\ &= \frac{q(t)}{C} - \frac{q(t_o)}{C} + v(t_o) = \frac{q(t)}{C}, \end{aligned} \tag{2.72}$$

where the last two terms have cancelled because $q(t_o)/C = v(t_o)$. Substituting $i(t) = dq(t)/dt$ for the current in (2.71) gives a second-order ODE for the charge:

$$L\frac{d^2}{dt^2}q(t) + R\frac{d}{dt}q(t) + \frac{1}{C}q(t) = V_s. \tag{2.73}$$

The corresponding equation for the current is derived from (2.71) by substituting (2.22) in place of $q(t)/C$:

$$Ri(t) + L\frac{d}{dt}i(t) + \frac{1}{C}\int_{t_o}^{t} i(t)dt + v(t_o) = V_s. \tag{2.74}$$

Differentiating and rearranging this expression yield another second-order ODE:

$$L\frac{d^2}{dt^2}i(t) + R\frac{d}{dt}i(t) + \frac{1}{C}i(t) = \frac{d}{dt}V_s. \tag{2.75}$$

Observe that (2.73) and (2.75) have the same form and coefficients, except that the ODE for the charge $q(t)$ depends on V_s, whereas that for the current $i(t)$ depends on the derivative of V_s.

It turns out that the solutions for equations such as (2.73) and (2.75) can take on one of three possible forms depending on the relative values of the three parameters $\{R, L, C\}$. Assuming $V_s = 0$ so that the ODE is *homogeneous*, all forms contain exponentials as follows:

Overdamped: $$i(t) = [c_1 \exp(-\alpha_1 t) + c_2 \exp(-\alpha_2 t)]u(t), \tag{2.76}$$

Underdamped: $$i(t) = \exp(-\alpha t)\,[c_1 \cos(\omega_d t) + c_2 \sin(\omega_d t)]\,u(t), \tag{2.77}$$

Critically damped: $$i(t) = [c_1 + c_2 t] \exp(-\alpha t)u(t). \tag{2.78}$$

The constant coefficients $\{c_1, c_2\}$ are determined by the initial conditions $\{i(0), i'(0)\}$, and the parameters $\{\alpha, s_1, s_2, \omega_d\}$ depend on the specific values of $\{R, L, C\}$ and the type of *damping*. Derivations of these results and a description of the three types of damping are provided in Chapter 6. They are mentioned here in order to qualitatively describe the behavior of an RLC circuit, so we can see the similarity of the results compared with those for the mechanical systems described in the next section. Damping refers to the exponential function weighting the sinusoidal waveforms in (2.77). Observe that for the overdamped case, the current decays exponentially to 0, and with two different time constants $1/\alpha_1$ and $1/\alpha_2$. The underdamped case also decays exponentially to 0, but with one time constant $1/\alpha$, and it does so sinusoidally with *damped angular frequency* ω_d. The exponential function in this case determines the *envelope* of sine and cosine because they are multiplied together. The critically damped case has an exponentially decaying term with time constant $1/\alpha$, but also a term that is the product of t and an exponential function. Overall, $i(t)$ tends to 0 for this case because $\exp(-\alpha t) \to 0$ faster than t increases.

The resistor voltages $v_R(t)$ for the three cases are easily obtained by multiplying (2.76)–(2.78) by R. The inductor voltages $v_L(t)$ are derived by taking derivatives of $i(t)$ and multiplying the terms by L. The capacitor voltages $v_C(t)$ are derived by integrating the current, dividing by C, and adding any initial voltage $v_C(0)$.

Example 2.10 For the series RLC circuit with $V_s = 0$, suppose that $R = 2.5$ kΩ, $C = 1$ μF, and $L = 1$ H. From the results in Chapter 6, it can be shown that this is an overdamped system with $\alpha_1 = 500$ and $\alpha_2 = 2000$. The initial conditions for this case yield the following system of equations:

$$i(0) = c_1 + c_2, \qquad i'(0) = -c_1\alpha_1 - c_2\alpha_2, \tag{2.79}$$

which must be solved simultaneously for $\{c_1, c_2\}$ given $\{i(0), i'(0)\}$ and $\{\alpha_1, \alpha_2\}$. Assuming that $i(0) = 1$ mA and $i'(0) = 1$ mA/s, we find for the given values of $\{\alpha_1, \alpha_2\}$ that the coefficients are $c_1 = 1.3340$ and $c_2 = -0.3340$ (with units mA), and the overall solution is

$$i(t) = [1.3340\exp(-500t) - 0.3340\exp(-2000t)]u(t) \text{ mA}. \tag{2.80}$$

This response and the two individual components are plotted in Figure 2.19(a). If the resistor value is decreased to 250 Ω, the underdamped case occurs and it takes longer for the energy in the inductor and capacitor to dissipate. The parameters are $\alpha = 125$ and $\omega_d = 992.16$ rad/s. The initial conditions for the underdamped case yield the following equations:

$$i(0) = c_1, \qquad i'(0) = \omega_d c_2 - \alpha c_1. \tag{2.81}$$

For $\{i(0), i'(0)\}$ used earlier, $c_1 = 1$ and $c_2 = (1 + 125)/992.16 \approx 0.1270$ (with units mA), and the underdamped solution is

$$i(t) = \exp(-125t)[\cos(992.16t) + 0.1270\sin(992.16t)]u(t) \text{ mA}. \tag{2.82}$$

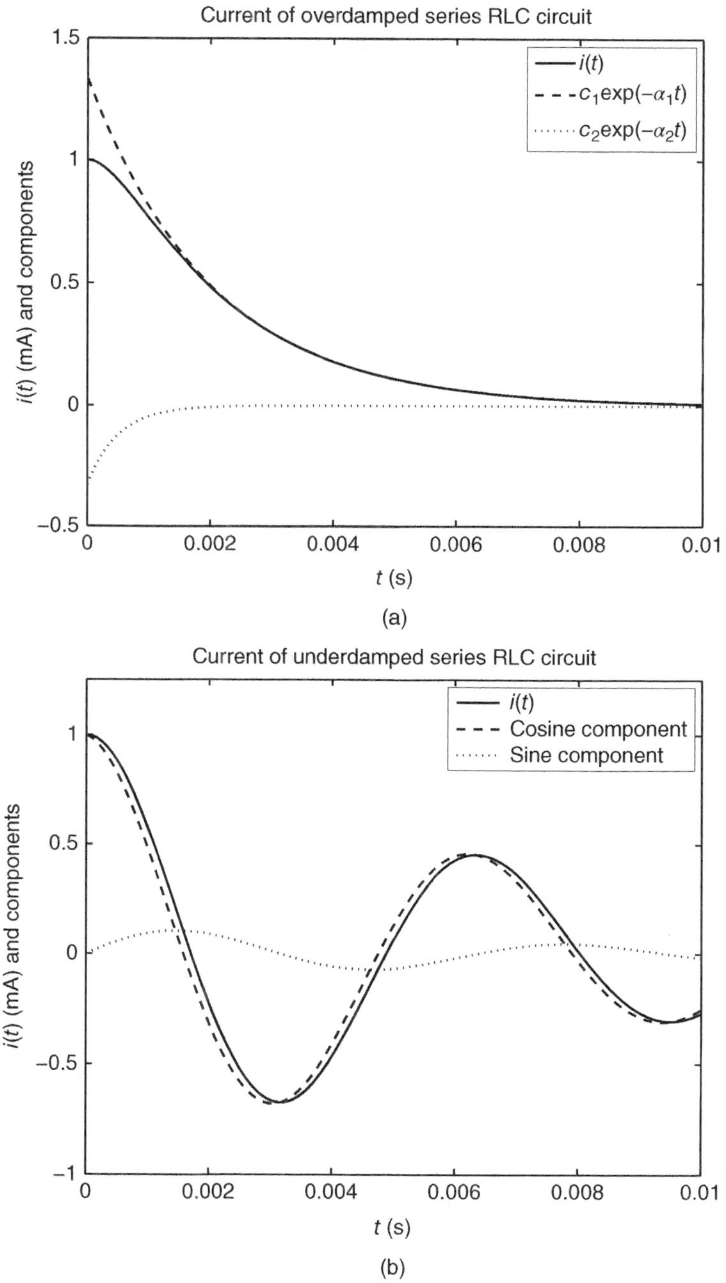

Figure 2.19 Current for the series RLC circuit in Example 2.10. (a) Overdamped circuit. (b) Underdamped circuit.

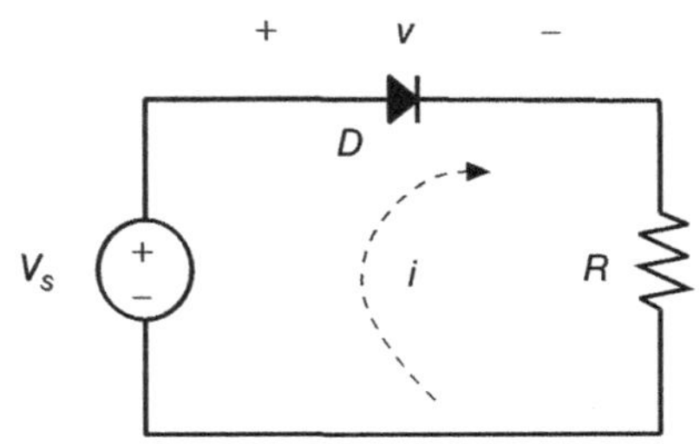

Figure 2.20 Series diode circuit with resistor.

This current along with its components are plotted in Figure 2.19(b). Note that the exponential envelope of $i(t)$ and those of its individual components decay more slowly than in the overdamped case because the exponent is -125 versus -500 and -2000. Similar results can be derived for the critically damped case using the techniques in Chapter 6.

2.4.5 Diode Circuits

Next, we consider the diode circuit in Figure 2.20 to illustrate once again the difficulty encountered when solving systems that have nonlinear components. KVL yields the following expression for the current:

$$-V_s + v + Ri = 0 \implies i = (V_s - v)/R. \tag{2.83}$$

In order to continue, we need to incorporate one of the I-V models for the diode D. The exponential model in (2.30) with $n = 1$ and $i \gg I_s$ is given approximately by

$$i \approx I_s \exp(v/V_T), \tag{2.84}$$

which can be rearranged as

$$v = V_T \ln(i/I_s). \tag{2.85}$$

Although we have two equations and two unknowns for this circuit, it is not possible to explicitly solve for v and i in terms of ordinary functions because of the natural logarithm. An iterative procedure (Sedra and Smith, 2004) can be applied as discussed in Chapter 1 where an estimate of v is used in (2.83), and the resulting i is substituted into (2.85) to refine the estimate of v. This procedure is repeated until v and i converge, as illustrated next in Example 2.11.

For the piecewise linear model in (1.28), the current is

$$i = \begin{cases} (v - v_c)/R_D, & v \geq v_c \\ 0, & v < v_c, \end{cases} \tag{2.86}$$

TABLE 2.6 Iterative Solution for Diode Circuit

Iteration	Current i (A)	Voltage v (V)
1	0.0050	0.7603
2	0.0044	0.7569
3	0.0044	0.7571
4	0.0044	0.7571

where v_c is the cutoff voltage and R_D is the diode resistance, which is usually much smaller than R of the circuit. Assuming $v > v_c$, the circuit is modeled by two linear equations with two unknowns. Thus, equating the first expression in (2.86) with (2.83) yields

$$(v - v_c)/R_D = (V_s - v)/R \Rightarrow v = \frac{Rv_c + R_D V_s}{R + R_D} \tag{2.87}$$

and

$$i = \frac{V_s - v_c}{R + R_D}. \tag{2.88}$$

If $v \geq v_c$ in (2.87) as assumed, then this solution is valid; otherwise, the diode is off (reverse-biased) such that $i \approx 0$ and $v = V_s$.

Example 2.11 For the series diode circuit in Figure 2.20, let $V_s = 1.2$ V and $R = 100\ \Omega$. The diode parameters for the exponential model are $I_s = 10^{-15}$ A and $V_T = 0.026$ V. From (2.83) and (2.85) with initial estimate $v = 0.7$ V, MATLAB provides the results in Table 2.6. Since there is no change in the last iteration, those values are the current through and the voltage across the diode. For the piecewise linear model with $R_D = 10\ \Omega$ and $v_c = 0.6$ V, (2.87) and (2.88) give $i = 0.0055$ A and $v = 0.6545$ V, which is a valid solution because $v > v_c$. The curves for the exponential and piecewise linear models are illustrated in Figure 2.21. The circuit *load line* is of the form $i = av + b$ (affine) with slope $a = -1/R$ and ordinate $b = V_s/R$:

$$i = -v/R + V_s/R \Rightarrow i = -0.01v + 0.012. \tag{2.89}$$

Observe that the coordinates where the load line intersects the two diode model curves match those derived earlier: for the exponential model $i = 0.0044$ A and $v = 0.7571$ V, and for the piecewise linear model $i = 0.0055$ A and $v = 0.6545$ V. Appendix F shows how to find an explicit expression for the current of the diode exponential model using the Lambert W-function.

Another iterative technique that can be used to solve a nonlinear equation is Newton's method (NM) (Kreyszig, 1979). For the diode circuit, we equate the

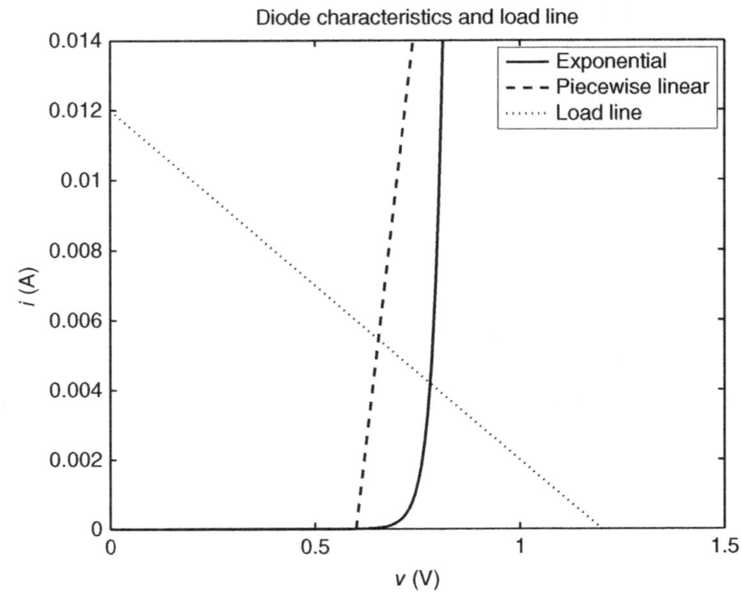

Figure 2.21 Diode models and circuit load line used in Example 2.11.

current equations in (2.83) and (2.84), and then define the function $f(v)$ for the unknown voltage:

$$(V_s - v)/R = I_s \exp(v/V_T) \Rightarrow f(v) \triangleq I_s \exp(v/V_T) + (v - V_s)/R. \tag{2.90}$$

Starting with an initial voltage estimate denoted by v_0, NM computes the next estimate v_1 as follows:

$$v_1 = v_0 - \frac{f(v_0)}{f'(v_0)}, \tag{2.91}$$

where for the series diode circuit

$$f'(v) = (I_s/V_T)\exp(v/V_T) + 1/R. \tag{2.92}$$

The algorithm in (2.91) is repeated until v no longer changes, and the corresponding current i is derived using either (2.83) or (2.84).

Example 2.12 For the circuit parameters used in Example 2.11, Figure 2.22 shows a plot of function $f(v)$ (the solid line) and the ratio $f(v)/f'(v)$. These two curves intersect at $v \approx 0.7571$ V (the dotted line) because both functions are 0 when the voltage is the solution for this nonlinear circuit. With initial estimate $v_0 = 0.7$ V, the following voltage values are obtained: $\{0.8557, 0.8302, 0.8056, 0.7834, 0.7666, 0.7571\}$ V with final current value 0.0044 A. NM has the advantage that only one equation for

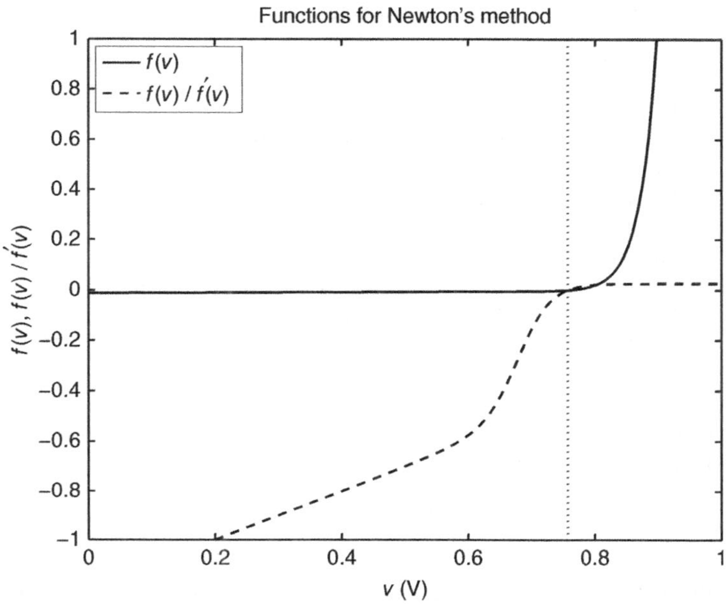

Figure 2.22 Functions for Newton's method in Example 2.12. The vertical dotted line is the voltage solution.

the voltage v is evaluated for each iteration, and then i is computed at the end. The disadvantage is that NM is a gradient technique, and so, convergence can be slow because the derivative of $f(v)$ is quite flat for voltages below the solution as shown in Figure 2.22.

In the rest of this chapter, we describe some *mechanical systems* that have the same behavior as first- and second-order circuits, with parameters that are the mechanical equivalents of $\{R, L, C\}$. These mechanical systems should be familiar to most readers, and they should provide intuition about linear circuits. These are examples of many systems (natural and human-made) that exhibit similar waveforms, such as an exponential decay or sinusoidal oscillatory motion.

2.5 MECHANICAL SYSTEMS

First, we define momentum and review a basic law in mechanics.

Definition: Momentum The *momentum* of a body with mass M and velocity v is Mv.

(Velocity v should not be confused with voltage.) Newton's second law states that the change in momentum of an object is proportional to the net force F applied to it:

$$\frac{d}{dt}Mv = \alpha F, \tag{2.93}$$

TABLE 2.7 Mechanical Symbols and Units

Property	Symbol	Units	Related Units
Length	L	Meter (m)	
Velocity	v	Meters/s (m/s)	
Acceleration	a	Meters/s^2 (m/s^2)	
Mass	M	Kilogram (kg)	
Force	F	Newton (N)	kg m/s^2
Weight	W	Newton (N)	kg m/s^2
Energy	E	Joule (J)	N m
Power	P	Watt (W)	J/s

where α is the proportionality constant. Assuming that M is constant, this expression can be written in terms of acceleration a as follows:

$$a \triangleq \frac{d}{dt}v \implies F = Ma, \tag{2.94}$$

which is the well-known form of Newton's second law. The units of the various quantities have been chosen such that $\alpha = 1$; they are t in seconds (s), M in grams (g), distance in centimeters (cm), a in cm/s^2, and F in dynes (g cm/s^2). Since F represents the *net* force acting on the mass, it is necessary that all forces be added together with the appropriate signs and angles, indicating the directions from which they are applied to the mass. The units for various mechanical quantities are summarized in Table 2.7, where force F has units of newtons (N) (kg m/s^2).

2.5.1 Simple Pendulum

A simple pendulum consists of a point object with mass M attached to a rigid horizontal surface by a light string or rod as depicted in Figure 2.23. The pendulum oscillates about a *pivot point* where the string is connected to the surface. (A compound pendulum is a rigid body in place of the mass and string that rotates about some fixed pivot

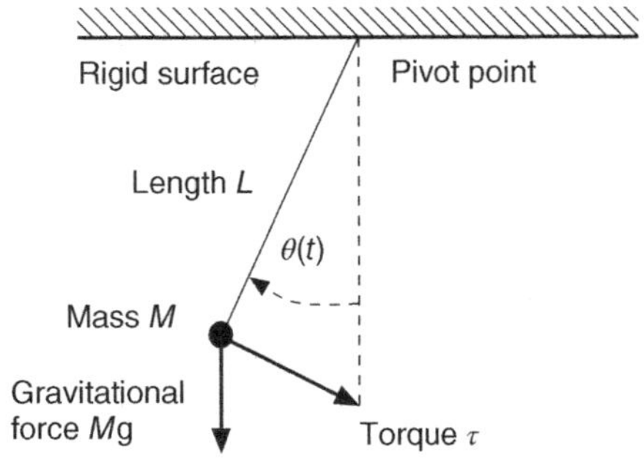

Figure 2.23 Simple pendulum.

point.) The weight of the pendulum is Mg where $g = 9.80665$ m/s^2 is the acceleration due to gravity. In the following model, we assume that the resistance caused by air is negligible so that in theory the pendulum would continue to oscillate indefinitely. The *torque* (rotational force) of the pendulum about the pivot point is

$$\tau = -MgL\sin(\theta(t)), \tag{2.95}$$

where the angle $\theta(t)$ is defined by the string and the vertical line perpendicular to the surface, and L is the length of the string (not to be confused with inductance). Since this expression is the *weight* Mg of the object multiplied by $L\sin(\theta(t))$, it is the force in the direction of motion as shown in the figure. The minus sign is included because this torque, which is called a *restoring force*, is in the opposite direction of the defined angle $\theta(t)$. The force $Mg\cos(\theta(t))$ is exactly balanced by the tension in the string, and so, those two forces can be ignored because they do not affect the motion of the pendulum. As a result, (2.95) is F on the left-hand side of Newton's equation in (2.94). The right-hand side is given by

$$Ma = I\frac{d^2}{dt^2}\theta(t), \tag{2.96}$$

where $I \triangleq ML^2$ is the moment of inertia and a is proportional to the angular acceleration $d^2\theta(t)/dt^2$. Combining (2.95) and (2.96) yields the following ODE:

$$I\frac{d^2}{dt^2}\theta(t) + MgL\sin(\theta(t)) = Lx(t), \tag{2.97}$$

where an external force $Lx(t)$ has been included on the right-hand side, which is in the opposite direction of $MgL\sin(\theta(t))$. This is a *nonlinear* ODE because of the sinusoidal term, and so, it does not have a simple functional solution. However, it is possible to linearize this equation by using the approximation $\sin(\theta(t)) \approx \theta(t)$ for small $\theta(t)$, such that

$$\frac{d^2}{dt^2}\theta(t) + (g/L)\theta(t) = (1/ML)\,x(t), \tag{2.98}$$

where I has been substituted and then divided on both sides of the equation. This result is a second-order linear ODE as given previously in (1.17) with output $y(t) = \theta(t)$, input $(1/ML)x(t)$, and constant coefficients $\{a_0 = g/L, a_1 = 0\}$. The form of this ODE is similar to that in (2.75) derived for the series RLC circuit.

For the homogeneous ODE where the right-hand side of (2.98) is 0, the solution is similar to that in (2.77) for the voltages and currents of an underdamped series RLC circuit. Since there is no damping (the air resistance has been ignored), the solution is sinusoidal with a constant envelope ($\alpha = 0$ in (2.98)):

$$\theta(t) = c_1\cos(\omega_o t) + c_2\sin(\omega_o t), \tag{2.99}$$

where $\omega_o \triangleq \sqrt{g/L}$ is used in place of ω_d, and the coefficients $\{c_1, c_2\}$ depend on the initial conditions $\{\theta(0), \theta'(0)\}$. By ignoring frictional effects, this solution is technically *undamped* and the pendulum will oscillate indefinitely. It is also called *free, undamped, harmonic oscillation*. Because (2.98) does not include the term $d\theta(t)/dt$ ($a_1 = 0$ in the standard ODE notation), the result in (2.99) is the only type of solution; the overdamped and critically damped solutions cannot happen. The oscillation frequency $f_o = \omega_o/2\pi$ increases with smaller L and larger g, which are physically intuitive results. The period of oscillation is $T_o = 1/f_o = 2\pi\sqrt{L/g}$. Observe that

$$\theta(0) = A\cos(0) = c_1, \tag{2.100}$$

$$\theta'(0) = -c_1\omega_o \sin(0) + c_2\omega_o \cos(0) = c_2\omega_o, \tag{2.101}$$

and so the linearized solution is

$$\theta(t) = [\theta(0)\cos(\omega_o t) + [\theta'(0)/\omega_o]\sin(\omega_o t)]\, u(t). \tag{2.102}$$

This expression can be written in terms of a single sinusoid as follows:

$$\theta(t) = \left[\sqrt{\theta^2(0) + [\theta'(0)/\omega_o]^2}\cos(\omega_o t + \phi)\right] u(t), \tag{2.103}$$

where $\phi = -\tan^{-1}(\theta'(0)/\theta(0)\omega_o)$. Suppose at $t = 0$ the angle of the pendulum is at its maximum and the initial velocity is $\theta'(0) = 0$. Then the solution simplifies to

$$\theta(t) = \theta(0)\cos(\omega_o t)u(t), \tag{2.104}$$

which is the expected result: for small $\theta(t)$, the pendulum angle is approximately sinusoidal with maximum value given by the initial angle $\theta(0)$. If $\theta(0) = 0$, which means the pendulum is perfectly vertical, then the initial angular velocity $\theta'(0)$ must be nonzero in order to have oscillations. In that case, $\phi = -90°$ and

$$\theta(t) = [\theta'(0)/\omega_o]\cos(\omega_o t - 90°)u(t) = [\theta'(0)/\omega_o]\sin(\omega_o t)u(t), \tag{2.105}$$

which also follows from (2.102). Of course, if $\theta(0) = \theta'(0) = 0$, then the pendulum is at rest.

This mechanical example demonstrates that similar dynamic behavior occurs in different types of physical systems. For the underdamped series RLC circuit, the current is oscillatory because of the exchange of energy between the capacitor and the inductor. The charge on the capacitor and the magnetic field of the inductor continually increase and decrease. For the pendulum, there is continuous transformation

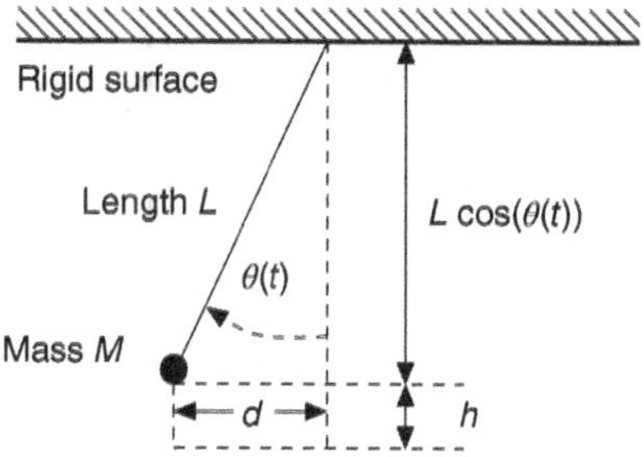

Figure 2.24 Height and horizontal distance of the pendulum relative to the lowest point of its trajectory.

between potential energy and kinetic energy. The potential energy is

$$E_p = Mgh, \tag{2.106}$$

where h is the height of the pendulum above the lowest point of its trajectory. The potential energy is maximum when the pendulum is at its maximum height, and it is 0 at the lowest point of its trajectory. From Figure 2.24, the height h is derived using trigonometry as follows:

$$L = L\cos(\theta(t)) + h \Rightarrow h = L[1 - \cos(\theta(t))]. \tag{2.107}$$

The kinetic energy is

$$E_k = (1/2)\,Mv^2, \tag{2.108}$$

where v is the velocity of the pendulum. The kinetic energy is maximum when the angle is 0, and it is 0 when the pendulum is at its maximum height. The total energy at any angle is a constant:

$$E_t = (1/2)\,Mv^2 + MgL[1 - \cos(\theta(t))], \tag{2.109}$$

and for θ_{max} where $v = 0$:

$$E_t = MgL\,[1 - \cos(\theta_{\text{max}})]. \tag{2.110}$$

This angle would typically be the starting point for the pendulum where it is held and then released. Equating these two equations for E_t and solving for $(1/2)Mv^2$, we can write an expression for the kinetic energy in (2.108) and the velocity for any angle:

$$\begin{aligned} E_k &= MgL[1 - \cos(\theta_{\text{max}})] - MgL[1 - \cos(\theta(t))] \\ &= MgL[\cos(\theta(t)) - \cos(\theta_{\text{max}})], \end{aligned} \tag{2.111}$$

and

$$v(t) = \sqrt{2gL[\cos(\theta(t)) - \cos(\theta_{\max})]}. \tag{2.112}$$

This equation does not include any directional information in $v(t)$ because it was derived from the kinetic energy, and so, it is actually the speed of the pendulum. The maximum velocity occurs at $\theta(t) = 0$:

$$v_{\max} = \sqrt{2gL[1 - \cos(\theta_{\max})]}. \tag{2.113}$$

It should be noted that the *tangential velocity* is related to the angular velocity $d\theta(t)/dt$ as follows:

$$v(t) = L\frac{d}{dt}\theta(t) = -L\omega_o\theta(0)\sin(\omega_o t), \tag{2.114}$$

where (2.104) has been substituted and $\theta(0)$ is in radians. This expression contains the correct sign for $v(t)$ as demonstrated in the next example.

Example 2.13 Let $M = 1$ kg such that $Mg = 9.80665$ newtons (N). Assume that $L = 1$ meter (m) and $\theta_{\max} = 0.0873$ radians, which corresponds to 5°. This angle is sufficiently small for the pendulum approximation to be valid because $\sin(0.0873) \approx 0.0872$. The total energy of the system is derived from the maximum potential energy:

$$E_{p,\max} = MgL[1 - \cos(\theta_{\max})] = 9.80665 \times 0.0038 \approx 0.0373 \text{ J}, \tag{2.115}$$

where J = N m denotes joules. Figure 2.25(a) shows trajectories for the potential and kinetic energies in (2.106) and (2.111). The corresponding pendulum velocity is shown in Figure 2.25(b). Observe that the magnitude of $v(t)$ is maximum when E_k is maximum, as expected; it is given by $v_{\max} = 0.2732$ m/s. The frequency and period of oscillation for the pendulum are $\omega_o \approx 3.1316$ rad/s and $T_o = 2\pi\sqrt{1/9.80665} \approx 2.0064$ s. This period is verified by the energy curves in the plots, which have a period twice that of T_o because e_k and e_p have two maximums (and minimums) per period of the pendulum. Finally, it is possible to calculate the maximum vertical distance of the pendulum trajectory from (2.107):

$$h_{\max} = L[1 - \cos(5°)] \approx 0.0038 \text{ m}, \tag{2.116}$$

and the maximum horizontal distance from the vertical dashed line in Figure 2.24 is

$$d_{\max} = L\sin(5°) \approx 0.0872 \text{ m}. \tag{2.117}$$

The trajectories for the height h and horizontal distance d are provided in Figure 2.25(c). Note that d can be negative because the vertical dashed line in Figure 2.24 is located at 0 on the horizontal axis. Thus, the total horizontal distance traveled is $2d_{\max} \approx 0.1743$ m. Of course, $d_{\max}$ is much larger than $h_{\max}$ for the small initial angle of 5°.

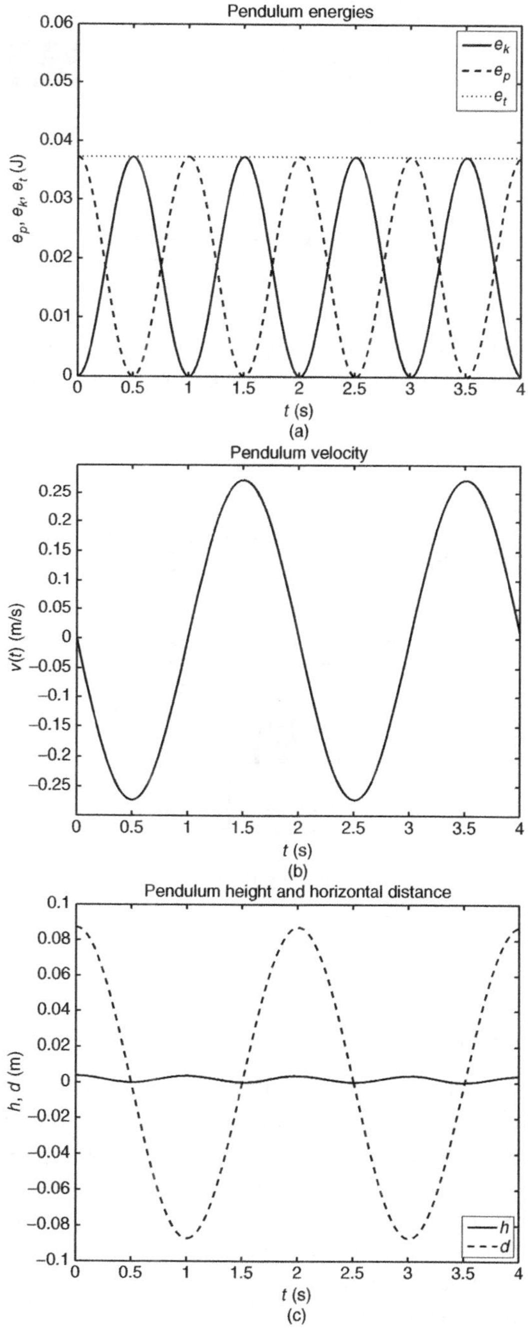

Figure 2.25 Pendulum energy, velocity, and distances in Example 2.13. (a) Potential energy, kinetic energy, and total energy. (b) Velocity v. (c) Height h and horizontal distance d.

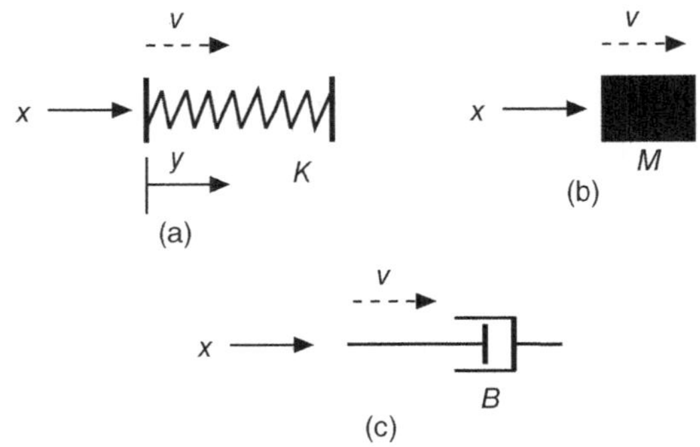

Figure 2.26 Elements of a mechanical spring system, where v is velocity, y is displacement of the spring, and x is an external force. (a) Spring constant K. (b) Mass M. (c) Damping constant B.

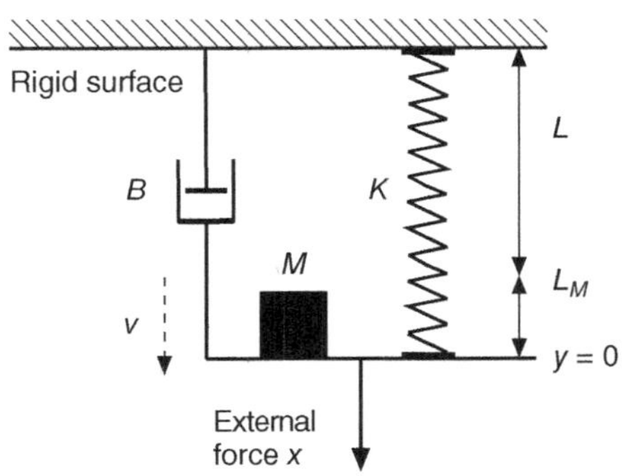

Figure 2.27 Mass on a spring with damping device, which is the mechanical analog of the series RLC circuit in Figure 2.18.

2.5.2 Mass on a Spring

Next, we derive the ODE for a system with a fixed mass on a spring and the components summarized in Figure 2.26. Each element is characterized by a single parameter: spring constant K (which depends on the type of spring), mass M, and damping factor B (a type of resistance). Observe that each element has velocity $v(t)$ (in a single direction) and the spring has displacement $y(t)$. An external force $x(t)$ is also shown for each element, though when the components are connected to each other, this force would typically be applied only to the mass.

Figure 2.27 shows a system consisting of the mass attached to the spring and a damping device, which in turn are attached to a rigid horizontal surface. The weight of the mass is Mg, where g is the acceleration due to gravity given in the previous section. The natural length of the spring without the mass attached is L. The spring has a restoring force described by *Hooke's law*:

$$F = KL_M, \tag{2.118}$$

where L_M is the additional length of the spring with the mass attached, and K is a proportionality constant with units newtons/meter (N/m). When the mass is at rest, the force due to gravity and the spring restoring force must be equal:

$$KL_M = Mg. \tag{2.119}$$

The ODE derived next represents the displacement $y(t)$ of the mass from its resting position. When the mass is above the horizontal solid line labeled $y = 0$ in Figure 2.27, the displacement $y(t)$ is negative, and $y(t) > 0$ when it is below.

Assume initially that there is no damping B. When the spring is stretched to length $L_M + y(t)$ with $y(t) > 0$, the restoring force is

$$F_1 = K[L_M + y(t)] = Mg + K\,y(t), \tag{2.120}$$

where (2.119) has been substituted. Let there be another force $F_2 = x(t)$ operating in the same direction as the gravitational force. From Newton's second law $F = Ma$, all forces are added together with the appropriate signs as follows:

$$Mg - F_1 + F_2 = Mg - [Mg + K\,y(t)] + x(t) = m\frac{d^2}{dt^2}y(t), \tag{2.121}$$

where $a = d^2y(t)/dt^2$ is the acceleration of the mass. This yields the second-order ODE:

$$M\frac{d^2}{dt^2}y(t) + K\,y(t) = x(t) \;\Rightarrow\; \frac{d^2}{dt^2}\,y(t) + (K/M)y(t) = (1/M)\,x(t). \tag{2.122}$$

This equation is identical to the linearized ODE for the simple pendulum in (2.98), except for the different parameters. It is an undamped system because there is no term containing the first derivative $dy(t)/dt$. As a result, we can use the solution given earlier but with the appropriate change of parameters. When $x(t) = 0$, the solution is

$$y(t) = \sqrt{y^2(0) + [y'(0)/\omega_o]^2}\,\cos(\omega_o t + \phi)u(t), \tag{2.123}$$

where $\phi = -\tan^{-1}(y'(0)/y(0)\omega_o)$ and $\omega_o = \sqrt{K/M}$. The frequency of oscillation increases with a larger spring constant and a smaller mass.

Suppose now that the damping element is included in the system as illustrated in Figure 2.27. The damper is called a *dashpot* whose simplified model consists of a piston inside a cylinder. It is similar to a shock absorber used in automobiles to mitigate up-and-down oscillations when the vehicle moves along a bumpy road. The damping force is proportional to the velocity of the mass:

$$F_3 = B\frac{d}{dt}y(t), \tag{2.124}$$

with damping constant B, which has units N s/m. The direction of this force is always opposite that of the movement of the mass, so it is subtracted on the left-hand side of (2.121). Thus, (2.122) becomes

$$M\frac{d^2}{dt^2}y\,(t) + B\frac{d}{dt}y(t) + K\,y(t) = x(t), \tag{2.125}$$

which we rewrite as

$$\frac{d^2}{dt^2}y(t) + (B/M)\frac{d}{dt}y(t) + (K/M)y(t) = (1/M)x(t). \tag{2.126}$$

This equation has a term with $dy(t)/dt$ because of the damping element. The ODE has the same form as in (2.71) for the series RLC circuit in Figure 2.27, and so there is a connection between the parameters of that circuit to those of the damped spring system as summarized in Table 2.8. This electrical/mechanical analogy is known as the *force–voltage model* because it assumes that a force acting on the mechanical system is analogous to a voltage across a circuit device (Harman and Lytle, 1962). The voltages across the three circuit elements are related to the forces associated with the components of the mechanical system as follows.

TABLE 2.8 Electrical and Mechanical Analogs, Force-Voltage Model: Series RLC Circuit and a Mass/Spring System with Damping

Electrical (units)	Mechanical (units)
Charge $q(t)$ (C)	Displacement $y(t)$ (m)
Current $i(t)$ (A)	Velocity $v(t) = dy(t)/dt$ (m/s)
Voltage $v(t)$ (V)	External force $x(t)$ (N)
Resistance R (Ω)	Damping constant B (N s/m)
Inductance L (H)	Mass M (kg)
Capacitance C (F)	Inverse of spring constant K (N/m)
Ohm's law $v(t) = Ri(t)$	Frictional force $F = Bv(t)$
Inductor voltage $v(t) = Ldi/dt$	Inertia $F = Mdv/dt$
Capacitor voltage $v(t) = q(t)/C$	Hooke's law $F = Ky(t)$
Resistor power $Ri^2(t)$	Frictional power $Bv^2(t)$
Inductor energy $(1/2)Li^2(t)$ (J)	Kinetic energy $(1/2)Mv^2(t)$ (J)
Capacitor energy $(1/2)Cv^2(t)$ (J) or $(1/2)q^2(t)/C$ (J)	Potential energy $(1/2K)x^2(t)$ (J) or $(1/2)Ky^2(t)$ (J)

- resistor $v = Ri \Leftrightarrow$ damping element $F = Bv$.
- capacitor $v = q/C = (1/C) \int i dt \Leftrightarrow$ spring $F = Ky = K \int v dt$.
- inductor $v = L\, di/dt \Leftrightarrow$ mass $F = Ma = M\, dv/dt$.

(The initial conditions associated with the integrals have been ignored in this comparison.)

It is clear that the circuit resistance R and the damping constant B serve the same purpose in the two systems. A resistor impedes the flow of charge $q(t)$, and the damping element tends to reduce the variations of $y(t)$. The current is the rate at which the charge varies with time, and the velocity of the mass is the rate at which its position changes. The equivalence of the capacitance C and the spring constant K follows from the voltage and force equations. Recall that a capacitor is an energy-storage device with energy $(1/2)Cv^2$. Similarly, the potential energy stored in a spring when it is stretched by an amount y is $(1/2)Ky^2$, which can be rewritten as $(1/2)(Ky)^2/K = (1/2)F^2/K$ where $F = Ky$ is the force associated with the spring (Hooke's law). The equivalence of the inductance L and the mass M also follows from the voltage and force equations because di/dt is the electrical analog of acceleration $a = dv/dt$. The energy stored in an inductor is $(1/2)\, Li^2$, and the kinetic energy of the mass is $(1/2)\, Mv^2$; current is analogous to velocity in the force-voltage model.

2.5.3 Electrical and Mechanical Analogs

The previous results for a series RLC circuit and a damped mass/spring system describe properties of a second-order system, with the relationships summarized in Table 2.8. It is possible to derive similar results for a first-order RC circuit described by the homogeneous ODE:

$$Ri + (1/C) \int i dt = 0 \Rightarrow \frac{d}{dt} i + (1/RC)\, i = 0, \tag{2.127}$$

where $i = dq/dt$ is the current. From Table 2.8, the ODE for an analogous mechanical system is

$$Bv + K \int v dt = 0 \Rightarrow \frac{d}{dt} v + (K/B) v = 0, \tag{2.128}$$

where $v = dy/dt$ is the velocity of a point at the end of the spring. Figure 2.28 illustrates the two systems, where the mechanical system consists of a spring and a damper; there is no mass as in the previous second-order case. Since there are no driving forces, the solutions of these first-order ODEs are exponentially decaying functions with time constants $\tau = 1/RC$ and $\tau = K/B$, respectively. The energy stored in the capacitor dissipates as heat through the resistor; there are no oscillations as is possible in a second-order system. Similarly, the energy stored in the spring is dissipated as heat in the damping device until there is no more movement. Since the system is horizontal, the gravitational force has no impact on the velocity.

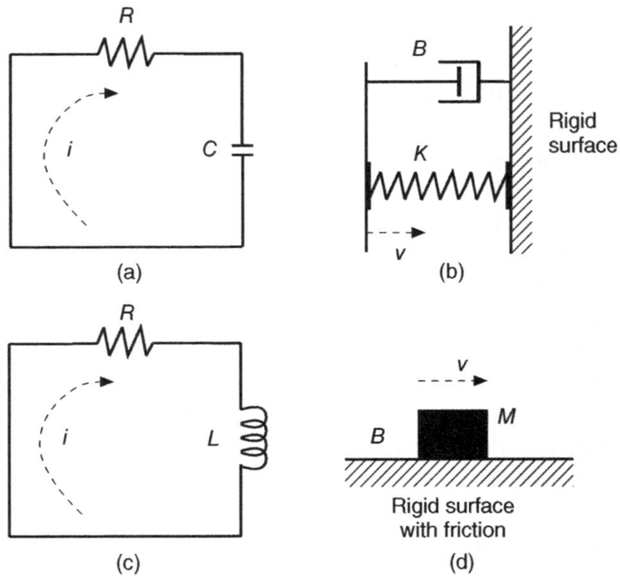

Figure 2.28 First-order system analogs. (a) RC circuit. (b) Horizontal spring with damping. (c) RL circuit. (d) Mass and frictional surface.

From Table 2.8, a mechanical analog for the series RL circuit can also be derived. The ODE for the first-order RL circuit in Figure 2.28(c) is

$$\frac{d}{dt}i + (R/L)i = 0, \tag{2.129}$$

and the mechanical analog is

$$\frac{d}{dt}v + (B/M)v = 0. \tag{2.130}$$

In this example, the mass M is moving along a frictional surface (instead of being connected to a damping device) as depicted in Figure 2.28(d). The force due to the coefficient of friction B is proportional to the velocity:

$$F = Bv, \tag{2.131}$$

whose direction is opposite that of the velocity shown in the figure. From Newton's second law, we have

$$F = Ma \Rightarrow -Bv = M\frac{d}{dt}v, \tag{2.132}$$

which is the ODE in (2.130).

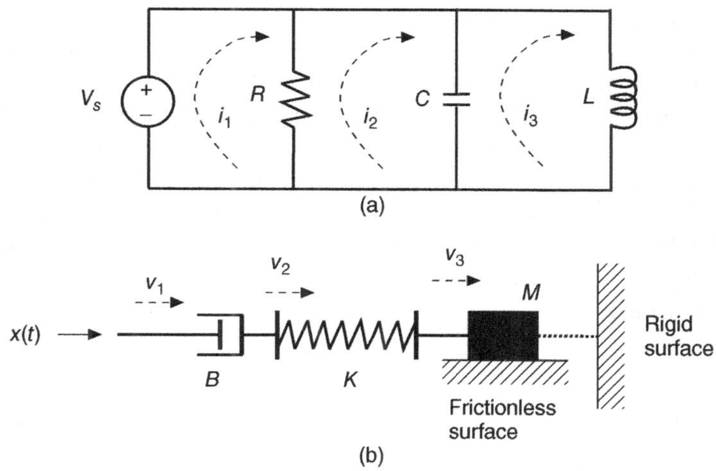

Figure 2.29 (a) Parallel RLC circuit and (b) its mechanical analog. The dotted line is not a connection in (b): it indicates that the vertical surface is a frame of reference.

We conclude this chapter by showing how to convert the *parallel* RLC circuit in Figure 2.29(a) to its mechanical analog by using Table 2.8. KVL yields three equations in terms of the labeled currents:

$$(i_1 - i_2)\,R = V_s, \tag{2.133}$$

$$(i_2 - i_1)\,R + \frac{1}{C}\int_0^t (i_2 - i_3)\,dt = 0, \tag{2.134}$$

$$\frac{1}{C}\int_0^t (i_3 - i_2)\,dt + L\frac{d}{dt}i_3 = 0, \tag{2.135}$$

where we assume a zero initial voltage on the capacitor associated with the integrals in the second and third equations. Since current translates to velocity, the first mechanical equation is

$$(v_1 - v_2)\,B = x(t). \tag{2.136}$$

For the second and third equations, the integral yields charge, and so, the mechanical analog is displacement y:

$$(v_2 - v_1)\,B + K(y_2 - y_3) = 0, \tag{2.137}$$

$$K(y_3 - y_2) + M\frac{d}{dt}v_3 = 0. \tag{2.138}$$

The mechanical analog of the parallel RLC circuit is shown in Figure 2.29(b). The equations in (2.136)–(2.138) could also have been derived starting with the mechanical system in the figure. The last term in (2.138) is Newton's second law $F = Ma$ where $a = dv_3/dt$ is acceleration.

Finally, we mention that there is another model for electrical/mechanical systems known as the *force–current model*. It can be viewed as the dual of the force-voltage model with the following equivalences: mechanical force $x(t) \leftrightarrow$ current $i(t)$, velocity $v(t) \leftrightarrow$ voltage $v(t)$, mass $M \leftrightarrow$ capacitance C, and inverse of spring constant $1/K \leftrightarrow$ inductance L. The damping constant B and the resistance R are analogous in both force models. Because of duality, the series RLC circuit in Figure 2.18 is the electrical analog of the force-current model of the mechanical system in Figure 2.29(b). Similarly, the parallel RLC circuit in Figure 2.29(a) is the electrical analog of the mechanical system in Figure 2.27. The mechanical system problems given later consider only the force-voltage model.

PROBLEMS

Voltage, Current, and Power

2.1 The current in a circuit is $i(t) = 10\exp(-t)u(t)$ mA. Calculate the amount of charge delivered after (a) 50 ms and (b) 200 ms.

2.2 The charge in a circuit varies over time as follows:

$$q(t) = \begin{cases} 0, & t < 0 \\ 2t^2 + 1, & 0 \leq t < 1 \\ 3, & 1 \leq t < 2 \\ -2t + 7 & 2 \leq t < 3.5 \\ 0, & t \geq 3.5, \end{cases} \tag{2.139}$$

with units C for $q(t)$ and seconds (s) for t. (a) Find an expression for the current $i(t)$. (b) Suppose the voltage across a device in the circuit is

$$v(t) = \begin{cases} 0, & t < 0 \\ 1, & 0 \leq t < 2 \\ t/2, & 2 \leq t < 3.5 \\ 0, & t \geq 3.5, \end{cases} \tag{2.140}$$

which has units V. Give an expression for the instantaneous power $p(t)$.

2.3 The voltage across a 60 W incandescent light bulb is sinusoidal with amplitude peaks ± 120 V. (a) Determine the amount of energy dissipated as heat every 15 minutes. (b) The frequency of the sinusoidal voltage is 60 Hz. Assuming the current is in phase with the voltage, find expressions for the current $i(t)$ and the instantaneous power $p(t)$.

2.4 The current through resistor $R = 10\ \Omega$ is

$$i(t) = \cos(2\pi t) + 2\sin(2\pi t). \tag{2.141}$$

(a) Give expressions for the instantaneous power $p(t)$ and the average power P. (b) Repeat part (a) if the argument of sine is changed to $4\pi t$.

2.5 Find the average power P with units W for each of the following cases. (a) 31 dB. (b) −14 dB. (c) 23 dBm.

Circuit Elements

2.6 The initial voltage on capacitor $C = 0.01$ μF is $v(0) = 1$ V. Find an expression for the voltage on the interval $t \in [0, 4]$ s if the current is

$$i(t) = \begin{cases} 0, & t < 0 \\ 0.1t, & 0 \leq t < 1 \\ 0.1t^2, & 1 \leq t < 2 \\ -0.2t + 0.8, & 2 \leq t < 4 \\ 0, & t \geq 4, \end{cases} \tag{2.142}$$

with units μA. Give the voltage at $t = 3$ s and 4 s.

2.7 Suppose two capacitors $C_1 = 1$ and $C_2 = 2$ μF are connected in parallel with a 10 V source. (a) Specify the total charge on each capacitor. (b) Repeat part (a) for C_1 and C_2 placed in series with the 10 V voltage source.

2.8 The initial current through inductor $L = 2$ mH is $i(0) = 1$ mA. Find an expression for the current on the interval $t \in [0, 10]$ s if the voltage is

$$v(t) = \begin{cases} 0, & t < 0 \\ 0.2t, & 0 \leq t < 3 \\ 0.6, & 3 \leq t < 8 \\ -0.3t + 3, & 8 \leq t < 10 \\ 0, & t \geq 10, \end{cases} \tag{2.143}$$

with units mV. Give the current at $t = 5$ s and 9 s.

2.9 The voltage across inductor $L = 1$ H in a second-order RLC circuit is exponential as follows:

$$v(t) = [2\exp(-10t) - \exp(-5t)]u(t)\ \text{V}. \tag{2.144}$$

(a) Find an expression for the instantaneous power. (b) Determine when the inductor is absorbing power and delivering power.

2.10 Repeat the previous problem for

$$v(t) = 5\exp(-2t)\cos(2\pi t)u(t)\ \text{V}. \tag{2.145}$$

2.11 The energy stored in a capacitor can be derived from the equation for power $p = vi$ because p is the rate of change of energy:

$$p = \frac{d}{dt}e_C = vC\frac{d}{dt}v, \tag{2.146}$$

where the capacitor current model has been substituted. From this expression, derive the time-varying energy $e_C(t)$, giving a result similar to (2.14).

2.12 Use the approach in the previous problem to derive the energy $e_L(t)$ stored in an inductor that is similar to (2.16).

2.13 Let the current through a capacitor be $i(t) = 2\exp(-t)u(t)$ mA. (a) Find the power absorbed by the capacitor. (b) Give an expression for the energy $e_c(t)$ as a function of time.

2.14 Find the voltage drop across a diode using the exponential model with $n = 1$ and the parameters in (2.30) for (a) $i = 2$ mA and (b) $i = 50$ mA.

2.15 A diode and resistor R are in series with a 10 V voltage supply. Find R such that the voltage drop across the diode is 0.7 V using the exponential model in (2.30) with $n = 1$.

2.16 Suppose two identical diodes are placed in series. (a) Using the piecewise linear model with $v_c = 0.6$ V and $R_D = 10\ \Omega$, find the current through these diodes if the voltage across the combination is 5 V. (b) Repeat part (a) for 1 V across the combination.

Basic Circuit Laws

2.17 (a) Prove that for two resistors in parallel, R_{eq} is always less than R_1 and R_2 individually. Give an intuitive reason why this is the case. (b) Show that R_{eq} for n equal-valued resistors R in parallel is R/n.

2.18 (a) Use KVL to find R_{series} for three resistors $\{R_1, R_2, R_3\}$ in series. (b) Use KCL to find R_{parallel} for the three resistors in parallel.

2.19 For the resistive circuit in Figure 2.14, let $R_1 = 100\ \Omega$, $R_2 = 50\ \Omega$, $R_3 = 200\ \Omega$, and $R_4 = 75\ \Omega$. (a) Find the equivalent resistance R_{eq} at terminals *a–b*. (b) Use this result to find the Thévenin equivalent circuit.

2.20 (a) Find the Norton equivalent circuit at terminals *a–b* for the resistive circuit in Figure 2.30. (b) Give the corresponding Thévenin equivalent circuit.

2.21 Repeat the previous problem if the circuit is modified to have a 100 Ω resistor connecting terminals *a–b*.

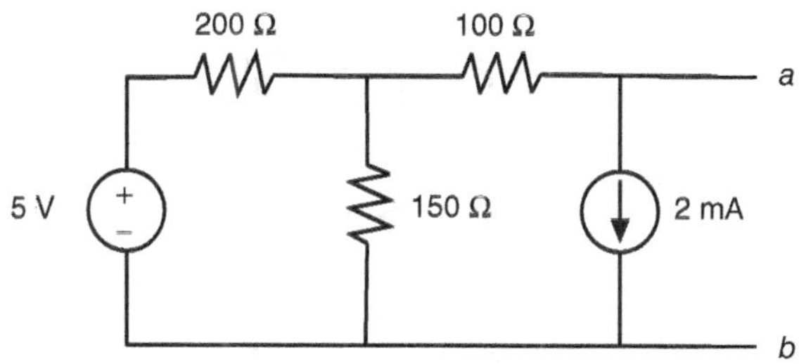

Figure 2.30 Resistive circuit used in Problems 2.20 and 2.21.

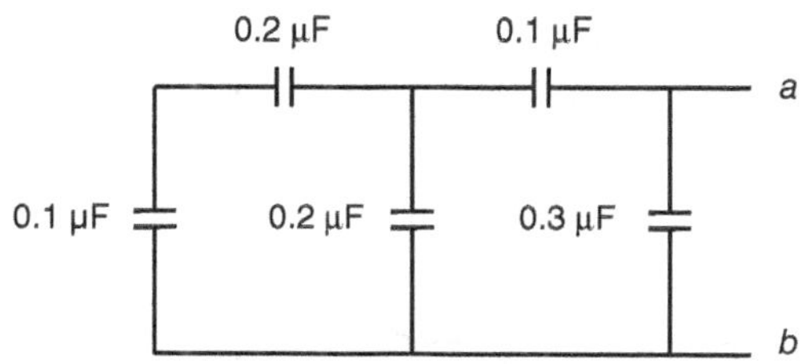

Figure 2.31 Capacitor circuit used in Problem 2.22.

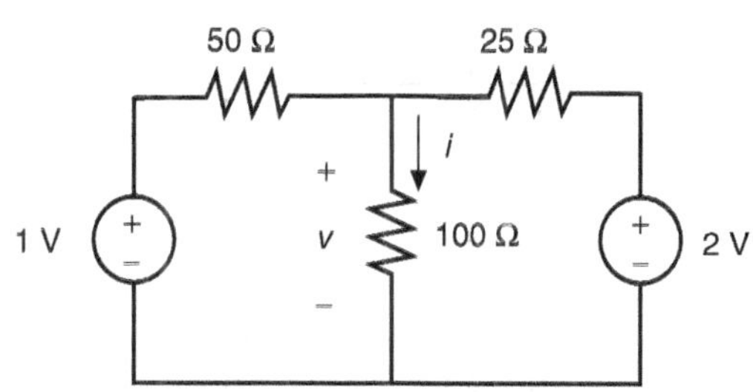

Figure 2.32 Resistive circuit used in Problem 2.25.

2.22 Find C_{eq} at terminals *a–b* for the capacitor circuit in Figure 2.31.

2.23 Using the mathematical model for the inductor in (2.13) and the circuit laws KVL and KCL, derive formulas for the equivalent inductance in Table 2.5.

2.24 Repeat the previous problem for the equivalent capacitance using the model in (2.12).

2.25 Find v and i for the circuit in Figure 2.32 using (a) mesh-current analysis and (b) node-voltage analysis.

2.26 Repeat the previous problem for the circuit in Figure 2.33.

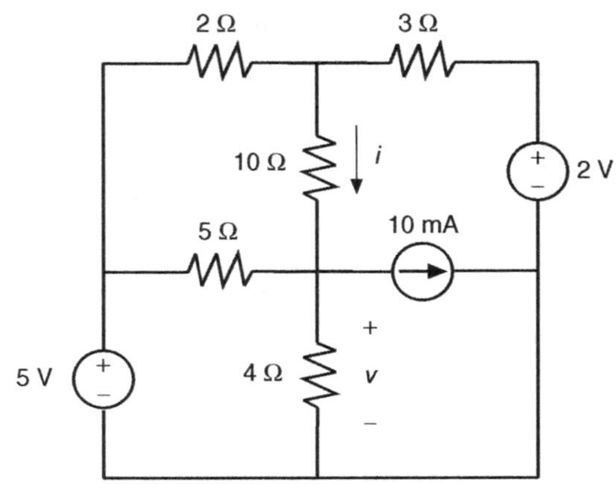

Figure 2.33 Resistive circuit used in Problem 2.26.

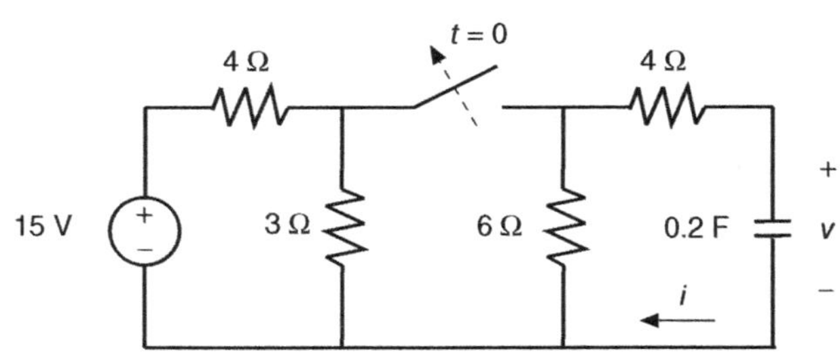

Figure 2.34 RC circuit used in Problem 2.27.

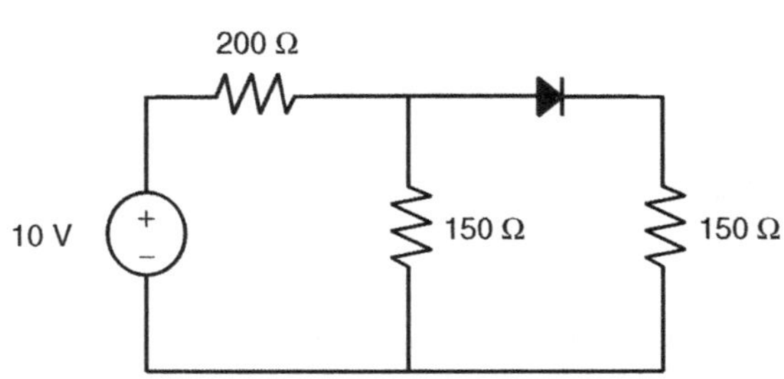

Figure 2.35 Diode circuit used in Problem 2.28.

2.27 The circuit in Figure 2.34 is in steady state before the switch is opened at $t = 0^-$, which means the capacitor is an open circuit and $i(0^-) = 0$ A. (a) Find $v(0^-)$. Since the capacitor voltage cannot change instantaneously, $v(0^+) = v(0^-)$ just after the switch is opened. However, the current i can change instantaneously. (b) Find $i(0^+)$.

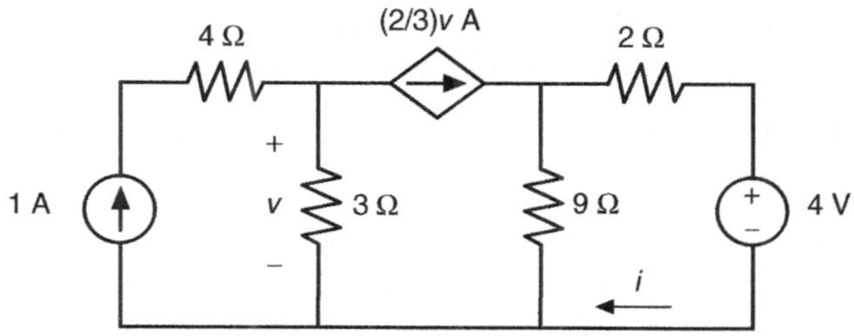

Figure 2.36 Circuit with dependent current source used in Problem 2.29.

2.28 Let the voltage drop across the ideal diode in Figure 2.35 be fixed at 0.7 V. (a) Find the current through the diode using KVL. (b) Repeat part (a) using Norton's equivalent circuit for the voltage source in order to combine the first two resistors and then convert back to a Thévenin equivalent circuit, resulting in a single circuit loop.

2.29 The circuit in Figure 2.36 has a *dependent* current source whose value is a function of the voltage v across the 3 Ω resistor. Use any analysis technique to find the voltage v and current i.

Mechanical Systems

2.30 For the linearized angle model in (2.102) for the simple pendulum, let $\theta(0) = 3°$ and $\theta'(0) = 0.1$ rad/s. Find $\theta_{\max}$ and $v_{\max}$ for $M = 1$ kg and $L = 1$ m.

2.31 Suppose the angular motion in degrees for an unforced simple pendulum is

$$\theta(t) = [5\cos(\pi t) + 10\sin(\pi t)]u(t). \tag{2.147}$$

(a) Give the initial angle and initial angular velocity. (b) Specify the length L of the string and the maximum velocity $v_{\max}$.

2.32 For a spring constant of $K = 2$ N/m, (a) determine the mass M such that the spring is extended by 0.1 m and (b) find the period of oscillation T_o for an undamped system.

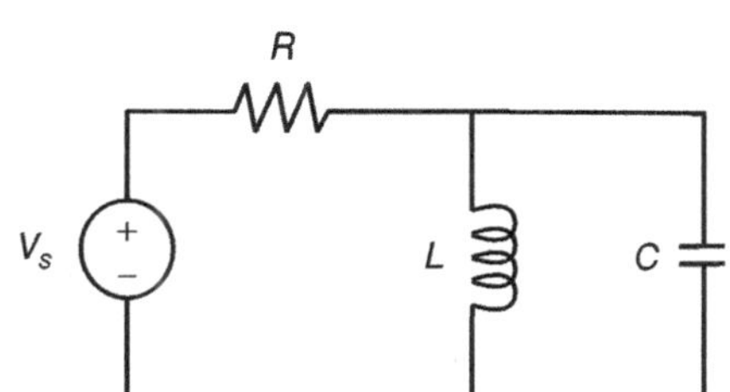

Figure 2.37 RLC circuit used in Problems 2.33 and 2.34.

2.33 Using the force-voltage model summarized in Table 2.8, sketch the mechanical system analog for the RLC circuit shown in Figure 2.37.

2.34 Repeat Problem 2.33 for the same circuit but with R and C interchanged.

2.35 Consider the ODE in (2.126) for the mass/spring system with damping. (a) Assuming $x(t) = 0$ such that the ODE is homogeneous, show by direct substitution that $y(t) = A\exp(-\alpha t)u(t)$ is a solution, resulting in a second-order algebraic equation in α. (b) Setting this equation equal to 0, find expressions in terms of the parameters $\{B, K, M\}$ such that the solutions for α are (i) real and (ii) complex. This is one approach taken in Chapter 6 to solve linear ODEs with constant parameters.

Computer Problems

2.36 Use MATLAB to generate plots for the three types of damped current responses given in (2.76)–(2.78) for $c_1 = c_2 = 1$ mA and $\omega_d = 2\pi$ rad/s. Choose values for $\{\alpha, \alpha_1, \alpha_2\}$ so that the exponential components have time constants in the range $\tau \in (0, 0.5)$ s.

2.37 Repeat the computer simulation for the series diode circuit in Example 2.11 using MATLAB with $V_s = 1.5$ V and $R = 125\ \Omega$. (a) Use the same diode parameters and let the initial estimate of the voltage be $v = 0.75$ V for the exponential model. (b) Generate the corresponding results for the piecewise linear model.

2.38 Repeat Example 2.13 by using the same parameters except let $\theta_{\max} = 6°$. Specify the total energy E_t, the period T_o, $v_{\max}$, $h_{\max}$, and $d_{\max}$. Use MATLAB to generate three plots similar to those in Figure 2.25.

3

LINEAR EQUATIONS AND MATRICES

3.1 INTRODUCTION

In this chapter, we investigate the properties of *systems of linear equations* that are represented by matrix equations. Such systems contain unknown variables and known variables, and the goal is to solve for the unknown variables, given the coefficients (parameters) that define the system. Gaussian elimination (GE) is an efficient method that recursively solves for the unknown variables without having to find the inverse of a matrix. We also discuss the *determinant* of a matrix, which is often used in circuit courses instead of GE to solve for unknown voltages and currents.

The reader may want to study only the early parts of this chapter, which also covers *Cramer's rule*, and then return later to some of the more advanced topics such as the four *subspaces* of a matrix and its *eigendecomposition*. The results in this chapter are readily applied to the all-resistive circuits discussed in Chapter 2, which do not have an exponential or sinusoidal time response like RC, RL, and RLC circuits that are modeled by linear ODEs. However, these matrix techniques can also be applied to the s-domain models discussed in Chapter 7 where ODEs are transformed to algebraic equations. This procedure based on the *Laplace transform* is depicted in Figure 3.1.

Algebraic equations in the s-domain are similar to the equations derived for an all-resistive circuit except the variables are *complex-valued*. Since these equations

Mathematical Foundations for Linear Circuits and Systems in Engineering, First Edition. John J. Shynk.

Companion Website: http://www.wiley.com/go/linearcircuitsandsystems

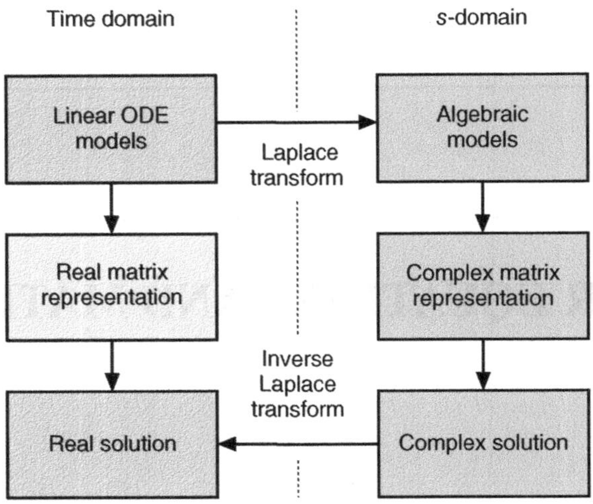

Figure 3.1 Time- and s-domain approaches to solving linear ODEs using matrix representations.

are algebraic, it is straightforward to rearrange them as a matrix equation, which can be solved using the techniques discussed in this chapter. The complex solutions are then transformed to the time domain, yielding the usual real-valued functions of time. Since the transformations are invertible, these solutions are identical to those that would have been obtained using only time-domain techniques. This transformation approach is advantageous because the time-domain methods for solving linear ODEs are generally more difficult, especially for high-order systems as discussed in Chapters 6 and 7.

Later in this book, we focus on the solution path to the right of the vertical dotted line in Figure 3.1. Although it is possible to express ODEs as systems of equations using a matrix representation, as depicted by the path containing the lightly shaded box, this topic is beyond the scope of the book and is only briefly discussed at the end of Chapter 6.

3.2 VECTOR SPACES

We begin with a discussion of fields and vector spaces.

Definition: Field A *field* $\mathcal{F}$ is a set of numbers that is closed under addition and multiplication:

$$x, y \in \mathcal{F} \implies x + y \in \mathcal{F}, \qquad x, y \in \mathcal{F} \implies xy \in \mathcal{F}. \tag{3.1}$$

These conditions imply several other properties of a field such as the identities $0 + x = x$ and $1 \times x = x$. In this book, either $\mathcal{F} = \mathcal{R}$ (all real numbers) or $\mathcal{F} = \mathcal{C}$ (all complex numbers).

TABLE 3.1 Vector Space Properties for $\mathbf{v}, \mathbf{v}_1, \mathbf{v}_2, \mathbf{v}_3 \in \mathcal{V}$ and $\alpha, \alpha_1, \alpha_2 \in \mathcal{F}$

Property	Operation
Addition	$\mathbf{v}_1 + \mathbf{v}_2 \in \mathcal{V}$
Scalar multiplication	$\alpha\mathbf{v} \in \mathcal{V}$
Commutative	$\mathbf{v}_1 + \mathbf{v}_2 = \mathbf{v}_2 + \mathbf{v}_1$
Associative (vectors)	$\mathbf{v}_1 + (\mathbf{v}_2 + \mathbf{v}_3) = (\mathbf{v}_1 + \mathbf{v}_2) + \mathbf{v}_3$
Associative (scalars)	$\alpha_2(\alpha_1\mathbf{v}) = \alpha_1(\alpha_2\mathbf{v})$
Distributive (one scalar)	$\alpha(\mathbf{v}_1 + \mathbf{v}_2) = \alpha\mathbf{v}_1 + \alpha\mathbf{v}_2$
Distributive (two scalars)	$(\alpha_1 + \alpha_2)\mathbf{v} = \alpha_1\mathbf{v} + \alpha_2\mathbf{v}$
Identity (0)	$\mathbf{0} \in \mathcal{V} \implies \mathbf{v} + \mathbf{0} = \mathbf{v}$
Identity (1)	$1 \in \mathcal{F} \implies 1 \times \mathbf{v} = \mathbf{v}$
Inverse	$-\mathbf{v} \in \mathcal{V} \implies \mathbf{v} + (-\mathbf{v}) = \mathbf{0}$

Definition: Vector Space A *vector space* $\mathcal{V}$ over a field $\mathcal{F}$ is a collection of elements that is closed under addition and multiplication by a scalar.

The elements are called *vectors* even though they need not be the usual vectors in Euclidean space (as discussed in the following example). Let $\mathbf{v}_1$ and $\mathbf{v}_2$ represent two vectors in vector space $\mathcal{V}$. Then the definition states

$$\mathbf{v}_1 + \mathbf{v}_2 \in \mathcal{V}, \qquad \alpha\mathbf{v}_1 \in \mathcal{V}, \tag{3.2}$$

where $\alpha \in \mathcal{F}$ is a scalar. Since α can be negative, the definition of a vector space also implies subtraction of vectors. Additional properties of a vector space are given in Table 3.1.

Example 3.1 The column vectors in N-dimensional $\mathcal{R}^N$ or $\mathcal{C}^N$ form a vector space: $\mathbf{v} \triangleq [v_1, \ldots, v_N]^T$, where each $v_n \in \mathcal{R}$ (or $\in \mathcal{C}$) and the superscript T denotes vector transpose:

$$\mathbf{v} \triangleq \begin{bmatrix} v_1 \\ \vdots \\ v_N \end{bmatrix} \implies \mathbf{v}^T = [v_1, \ldots, v_N]. \tag{3.3}$$

Matrices in $\mathcal{R}^{M\times N}$ or $\mathcal{C}^{M\times N}$ with the form given later in (3.6) also form a vector space, which are the focus of this chapter. A vector space may also consist of functions $\{f_n(x)\}$, provided they satisfy the properties in Table 3.1. For example, they may be the set of all polynomials of the form $f(x) = a_2x^2 + a_1x + a_0$ with $\{a_0, a_1, a_2\} \in \mathcal{R}$. The quaternions discussed in Chapter 4 comprise a four-dimensional vector space over $\mathcal{R}$.

Generally, vectors in this chapter are defined to be column vectors, and row vectors explicitly include the superscript T.

3.3 SYSTEM OF LINEAR EQUATIONS

Consider the following system of linear equations:

$$\begin{aligned} a_{11}y_1 + a_{12}y_2 + \ldots + a_{1N}y_N &= x_1 \\ a_{21}y_1 + a_{22}y_2 + \ldots + a_{2N}y_N &= x_2 \\ \vdots \qquad\qquad &\qquad \vdots \\ a_{M1}y_1 + a_{M2}y_2 + \ldots + a_{MN}y_N &= x_M, \end{aligned} \tag{3.4}$$

where $\{y_n\}$ are unknown variables, $\{x_m\}$ are known variables, and $\{a_{mn}\}$ are known fixed coefficients (derived from the parameters of an underlying physical system, such as a resistive circuit). This system of equations is the natural extension of one equation with N independent variables $\{y_n\}$ to M equations with the same set of independent variables. Although many books on linear algebra use the reverse notation with $\{x_n\}$ and $\{y_m\}$ interchanged, we are interested in the arrangement in (3.4) because from a system viewpoint, $\{x_n\}$ are multiple inputs and $\{y_m\}$ are multiple outputs for which we want to solve. As described in Chapter 1, the equations in (3.4) represent a multiple-input, multiple-output (MIMO) system with parameters $\{a_{mn}\}$. Generally, the equations (rows) have different coefficients on the left-hand side and different known variables on the right-hand side. The equations are linear because all terms are only added or subtracted, and the exponent of each y_n is 1; there are no higher powers of y_n, square roots, transcendental functions, and so on.

It is notationally and operationally convenient to write this set of equations in matrix/vector form by defining the following column vectors:

$$\mathbf{y} \triangleq \begin{bmatrix} y_1 \\ \vdots \\ y_N \end{bmatrix}, \qquad \mathbf{x} \triangleq \begin{bmatrix} x_1 \\ \vdots \\ x_M \end{bmatrix}. \tag{3.5}$$

If the elements of $\mathbf{y}$ are real-valued, then we can write $\mathbf{y} \in \mathcal{R}^{N\times 1}$, and so its transpose is $\mathbf{y}^T \in \mathcal{R}^{1\times N}$. When the dimensions are clear in a discussion, we will often write $\mathbf{y} \in \mathcal{R}^N$ for simplicity, in the same way we write the real-valued scalar as $y \in \mathcal{R}$ instead of $y \in \mathcal{R}^1$.

Definition: Matrix *Matrix* $\mathbf{A}$ is a rectangular array of numbers or coefficients representing numbers:

$$\mathbf{A} \triangleq \begin{bmatrix} a_{11} & a_{12} & \cdots & a_{1N} \\ a_{21} & a_{22} & \cdots & a_{2N} \\ \vdots & \vdots & \vdots & \vdots \\ a_{M1} & a_{M2} & \cdots & a_{MN} \end{bmatrix}, \tag{3.6}$$

with M rows and N columns. For real elements, $\mathbf{A} \in \mathcal{R}^{M\times N}$, and for complex elements $\mathbf{A} \in \mathcal{C}^{M\times N}$.

A square matrix with $M = N$ is an important case discussed throughout this chapter. Using $\mathbf{A}$ and the vector definitions in (3.5), it is possible to write (3.4) more compactly as follows:

$$\mathbf{A}\mathbf{y} = \mathbf{x}. \tag{3.7}$$

The mth element of $\mathbf{x}$ is generated by multiplying the mth row of $\mathbf{A}$ by $\mathbf{y}$, element by element, and the product terms are summed in the same way as in (3.4). To be more specific, if we define the mth row vector of $\mathbf{A}$ as $\overline{\mathbf{a}}_m^T \triangleq [a_{m1}, \ldots, a_{mN}]$ (with the overbar notation), then the mth element of $\mathbf{x}$ is

$$\overline{\mathbf{a}}_m^T\mathbf{y} = a_{m1}y_1 + \ldots + a_{mN}y_N = \sum_{n=1}^{N} a_{mn}y_n = x_m. \tag{3.8}$$

The sum is performed over the second subscript of a_{mn} and, hence, over the columns of $\mathbf{A}$.

When a row vector multiplies a column vector, which is necessarily of the same size, the result is a scalar as in (3.8); this is called an inner product of the two vectors.

Definition: Inner and Outer Products For vectors $\mathbf{x} \in \mathcal{R}^N$ and $\mathbf{y} \in \mathcal{R}^N$, the *inner product* $\mathbf{x}^T\mathbf{y} = \mathbf{y}^T\mathbf{x} = c$ is a scalar. For vectors $\mathbf{x} \in \mathcal{R}^M$ and $\mathbf{y} \in \mathcal{R}^N$, the *outer product* $\mathbf{x}\mathbf{y}^T = \mathbf{M}$ is an $M \times N$ matrix and $\mathbf{y}\mathbf{x}^T = \mathbf{N}$ is an $N \times M$ matrix.

The inner product of two vectors is also known as the *dot product* and is sometimes written as $\mathbf{x} \cdot \mathbf{y}$. The vectors in an inner product commute because the result is a scalar: $c^T = (\mathbf{x}^T\mathbf{y})^T = \mathbf{y}^T\mathbf{x} = c$. This is not the case in general for an outer product whose vectors need not have the same dimensions: $(\mathbf{x}\mathbf{y}^T)^T = \mathbf{y}\mathbf{x}^T \neq \mathbf{x}\mathbf{y}^T$. The inner product is used to describe the length (size) of a vector.

Definition: Vector Norm The *vector norm* of $\mathbf{x}$ is

$$\|\mathbf{x}\| \triangleq \sqrt{\mathbf{x}^T\mathbf{x}} = \sqrt{x_1^2 + \cdots + x_N^2}. \tag{3.9}$$

The *squared norm* is written as $\|\mathbf{x}\|^2 = \mathbf{x}^T\mathbf{x}$.

An example of a vector with $N = 2$ elements is shown graphically in Figure 3.2, where the two axes of $\mathbf{x} \in \mathcal{R}^2$ in the Cartesian coordinate system correspond to the two elements $\{x_1, x_2\}$. Of course, $\|\mathbf{x}\|$ provides no information about the location of the vector on the plane. It is easy to verify that $\|\mathbf{x}\|$ is, in fact, the length of the vector using trigonometry. The length of the horizontal dashed line is $x_1 = r\cos(\phi)$ where r is the length of the hypotenuse of the right triangle, and the length of the vertical dashed line is $x_2 = r\sin(\phi)$. The squared norm is

$$\|\mathbf{x}\|^2 = x_1^2 + x_2^2 = r^2[\cos^2(\phi) + \sin^2(\phi)] = r^2, \tag{3.10}$$

and so $r = \|\mathbf{x}\|$.

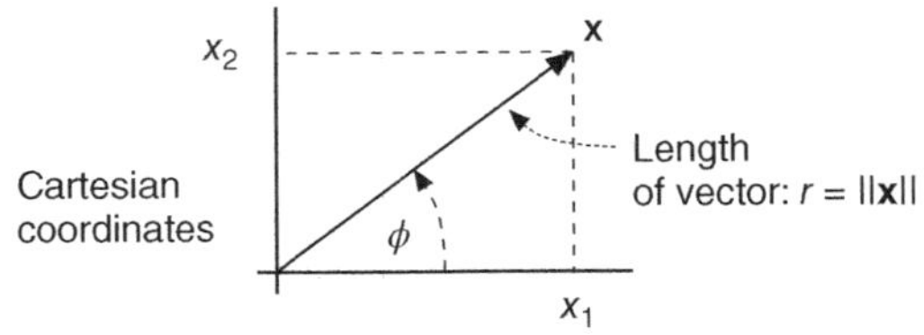

Figure 3.2 Example vector $\mathbf{x} \in \mathcal{R}^2$ with components $\{x_1, x_2\}$. Its length is $\|\mathbf{x}\| = \sqrt{x_1^2 + x_2^2}$. The angle ϕ is given by $x_1 = r\cos(\phi)$ or $x_2 = r\sin(\phi)$.

TABLE 3.2 Basic Matrix Properties for $\mathbf{A} \in \mathcal{R}^{M\times N}$, $\mathbf{B} \in \mathcal{R}^{P\times Q}$, and $\mathbf{C} \in \mathcal{R}^{R\times S}$

Property	Operation	Restrictions
Addition	$\mathbf{F} = \mathbf{A} + \mathbf{B} \implies f_{mn} = a_{mn} + b_{mn}$	$M = P$ and $N = Q$
Subtraction	$\mathbf{F} = \mathbf{A} - \mathbf{B} \implies f_{mn} = a_{mn} - b_{mn}$	$M = P$ and $N = Q$
Product	$\mathbf{F} = \mathbf{AB} \implies f_{mn} = \overline{\mathbf{a}}_m^T \mathbf{b}_n$	$N = P$
Reverse product	$\mathbf{F} = \mathbf{BA} \implies f_{mn} = \overline{\mathbf{b}}_m^T \mathbf{a}_n$	$M = Q$
Associative	$(\mathbf{AB})\mathbf{C} = \mathbf{A}(\mathbf{BC})$	$N = P$ and $Q = R$
Distributive	$\mathbf{A}(\mathbf{B} + \mathbf{C}) = \mathbf{AB} + \mathbf{AC}$	$N = P = R$ and $Q = S$
Commutative	$\mathbf{AB} \neq \mathbf{BA}$	Generally not commutative
Transpose	$(\mathbf{AB})^T = \mathbf{B}^T\mathbf{A}^T$	$N = P$
Conjugate transpose	$(\mathbf{AB})^H = \mathbf{B}^H\mathbf{A}^H$	$N = P$
Inverse	$(\mathbf{AB})^{-1} = \mathbf{B}^{-1}\mathbf{A}^{-1}$	$M = N = P = Q$
Determinant	$\det(\mathbf{AB}) = \det(\mathbf{A})\det(\mathbf{B})$	$M = N = P = Q$
Trace	$\mathrm{tr}(\mathbf{AB}) = \mathrm{tr}(\mathbf{BA})$	$N = P$ and $M = Q$

Matrices are multiplied together using inner products of their rows and columns. Let $\overline{\mathbf{a}}_m^T$ be the mth row of $\mathbf{A}$ and $\mathbf{b}_n$ the nth column of $\mathbf{B}$. Then, the mnth element of matrix $\mathbf{C} = \mathbf{AB}$ is the inner product $c_{mn} = \overline{\mathbf{a}}_m^T \mathbf{b}_n$. Matrices $\mathbf{A}$ and $\mathbf{B}$ need not be square; however, the number of columns of $\mathbf{A}$ must match the number of rows of $\mathbf{B}$. For example, if $\mathbf{A} \in \mathcal{R}^{M\times N}$ and $\mathbf{B} \in \mathcal{R}^{P\times Q}$, then we must have $N = P$ for $\mathbf{C} = \mathbf{AB} \in \mathcal{R}^{M\times Q}$ to be well defined. This result is unlike that for matrix addition where both matrices must have the same dimensions because the mnth element of $\mathbf{F} = \mathbf{A} + \mathbf{B}$ is $f_{mn} = a_{mn} + b_{mn}$. Basic properties of matrices are summarized in Table 3.2.

Usually when given a set of linear equations represented by (3.7), the goal is to solve for the unknown vector $\mathbf{y}$. In many problems, the number of equations equals the number of variables ($M = N$) such that $\mathbf{A}$ is a *square matrix*. However, two other types of systems also arise in practice:

- *Overdetermined system $M > N$*: more equations than unknown variables.
- *Underdetermined system $M < N$*: more unknown variables than equations.

Both of these systems have a rectangular matrix $\mathbf{A} \in \mathcal{R}^{M \times N}$. Depending on the specific coefficients $\{a_{mn}\}$ and the values for M and N, a system of equations may have (i) no solutions, (ii) a unique solution, or (iii) an infinity of solutions.

Definition: Consistent System A system of equations is *consistent* if at least one solution exists; otherwise, it is *inconsistent.*

Definition: Trivial Solution A solution is *trivial* if all unknown variables are 0. If at least one variable is nonzero, then the solution is *nontrivial.*

Definition: Nonsingular System A system of equations is *nonsingular* if it has only one solution; otherwise, it is *singular* with no solution or an infinity of solutions.

Example 3.2 Consider the simplest nontrivial system of equations with $M = 1$ equation (one row) and $N = 2$ variables (two columns):

$$a_{11}y_1 + a_{12}y_2 = x_1 \implies y_2 = -(a_{12}/a_{11})y_1 + x_1/a_{11}, \tag{3.11}$$

which has been rewritten as the equation of a line with slope $-a_{12}/a_{11}$ and ordinate x_1/a_{11} (assuming $a_{11} \neq 0$). The ordinate is the value where the line intersects the vertical axis on the y_1–y_2 plane. This underdetermined "system" of equations has an infinity of solutions: for every $y_1 \in \mathcal{R}$, there is a unique value for y_2. An example of this line is plotted in Figure 3.3 for $a_{11} = a_{12} = 1$ and $x_1 = 2$ (the solid line). Suppose a second equation is included in the system:

$$a_{21}y_1 + a_{22}y_2 = x_2 \implies y_2 = -(a_{21}/a_{22})y_1 + x_2/a_{22}, \tag{3.12}$$

such that along with (3.11), matrix $\mathbf{A}$ is square with dimensions $M = N = 2$. This line with $a_{21} = 1$, $a_{22} = -1$, and $x_2 = 1/2$ is also plotted in Figure 3.3 (the dashed line). Observe that a unique solution exists where the two lines (solid and dashed) intersect. Consider including a third equation:

$$a_{31}y_1 + a_{32}y_2 = x_3 \implies y_2 = -(a_{31}/a_{32})y_1 + x_3/a_{32}, \tag{3.13}$$

with $a_{31} = a_{32} = 1$ and $x_3 = 1$. This is the dotted line in the figure, which does not intersect the solid line, and so, the system of two equations given by (3.11) and (3.13) has no solution. This last case obviously occurs because the two lines are parallel to each other. They have the same slope but different ordinates: $y_2 = -y_1 + 2$ versus $y_2 = -y_1 + 1$. On the other hand, the system of equations represented by the dashed and dotted lines for (3.12) and (3.13) also have a unique solution because they intersect at a single point. Another case is obtained when (3.12) is proportional to (3.11), which means there exists $c \neq 0$ such that they are *collinear*:

$$c(a_{21}y_1 + a_{22}y_2) = a_{11}y_1 + a_{12}y_2 \text{ and } cx_1 = x_2. \tag{3.14}$$

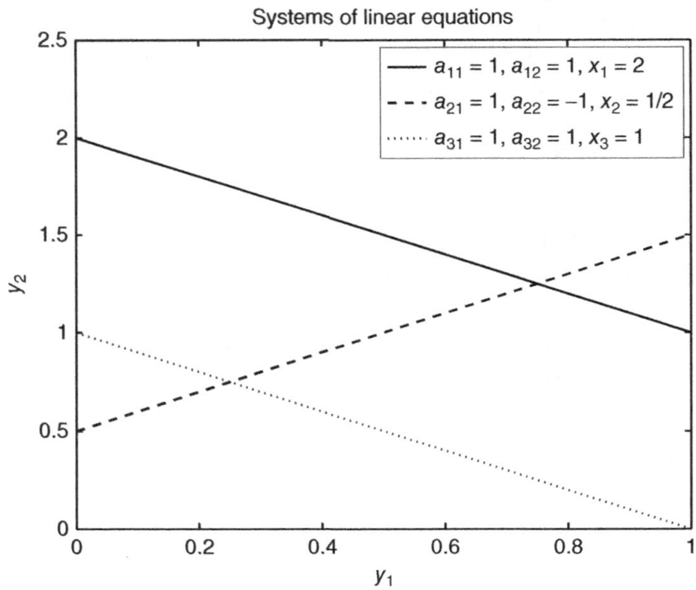

Figure 3.3 Systems of linear equations with a unique solution (solid and dashed lines, dashed and dotted lines) and no solution (solid and dotted lines).

These equations are *linearly dependent*, and as a result, there is an infinity of solutions as was the case for the single equation in (3.11). Collinear lines when plotted as in Figure 3.3 are actually the same line. The two parallel lines shown in the figure do *not* represent collinear equations. Parallel lines describe different properties of a system, whereas collinear lines provide essentially the same information about a system. Finally, if all three equations represented by the lines in Figure 3.3 are included in the system of equations, then there is no solution: the three lines do not intersect at any single point on the $\mathcal{R}^2$ plane. The system of three equations is inconsistent.

Example 3.3 Consider the resistive circuit in Figure 3.4. Kirchoff's voltage law (KVL) from Chapter 2 yields the following two equations for the loop currents:

$$100i_1 + 100(i_1 - i_2) = 10, \qquad 100(i_2 - i_1) + 200i_2 = 0, \tag{3.15}$$

where we have combined the two series 100 Ω resistors on the right-hand side of the circuit. Later in the chapter, we solve such systems of equations using determinants and GE. Since there are only two variables, it is straightforward to find those quantities by first solving for i_2 in the second equation and then substituting it into the first equation:

$$i_2 = i_1/5 \implies 100i_1 + 100(2/3)i_1 = 10. \tag{3.16}$$

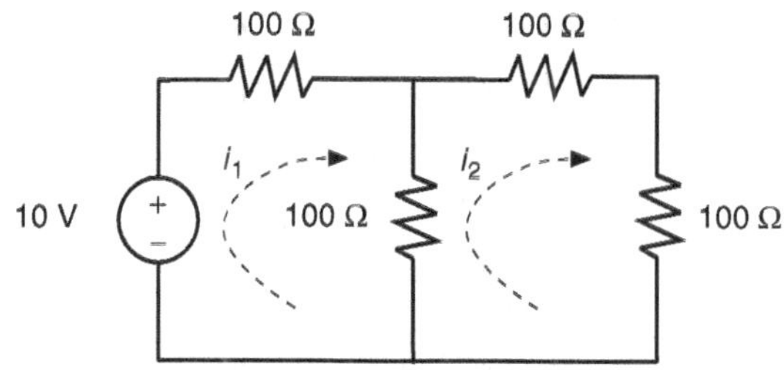

Figure 3.4 Resistive circuit for Example 3.3.

Thus, $i_1 = 60$ mA and $i_2 = 20$ mA, which can then be used to determine the voltages across each of the resistors via Ohm's law $v = Ri$. The current through the vertical resistor on the left-hand side is $i_1 - i_2 = 40$ mA downward, and so, the voltage is $(i_1 - i_2)100 = 4$ V with positive polarity on top. The voltage across the horizontal resistor on the left-hand side is 6 V with positive polarity on the left. From voltage division, we find that the voltages across the other two resistors are both 2 V.

3.4 MATRIX PROPERTIES AND SPECIAL MATRICES

Next, we summarize some general properties of matrices, including those that apply only to *square* matrices. Assume that **A**, **B**, and **C** have appropriate dimensions for matrix multiplication and addition as given in Table 3.2. Although matrices generally do *not* commute, $\mathbf{AB} \neq \mathbf{BA}$, one exception occurs when both matrices are diagonal.

Definition: Diagonal Matrix Matrix **D** is *diagonal* if the only nonzero elements are located along the main descending diagonal from upper left to lower right.

Example 3.4 An example of the product of two diagonal matrices with $M = N = 3$ is

$$\begin{bmatrix} 1 & 0 & 0 \\ 0 & 2 & 0 \\ 0 & 0 & 3 \end{bmatrix}\begin{bmatrix} 4 & 0 & 0 \\ 0 & 5 & 0 \\ 0 & 0 & 6 \end{bmatrix} = \begin{bmatrix} 4 & 0 & 0 \\ 0 & 5 & 0 \\ 0 & 0 & 6 \end{bmatrix}\begin{bmatrix} 1 & 0 & 0 \\ 0 & 2 & 0 \\ 0 & 0 & 3 \end{bmatrix} = \begin{bmatrix} 4 & 0 & 0 \\ 0 & 10 & 0 \\ 0 & 0 & 18 \end{bmatrix}. \qquad (3.17)$$

These matrices obviously commute.

The elements along the main diagonal of a diagonal matrix need not all be of the same value. If these elements are identical, then $\mathbf{A} = c\mathbf{I}$ for $c \in \mathcal{R}$ $(c \neq 0)$ where **I** is the identity matrix.

Definition: Identity Matrix The *identity matrix* **I** is a square matrix whose elements are 0 except for 1s along the main descending diagonal.

For $N = 3$, the identity matrix is

$$\mathbf{I} = \begin{bmatrix} 1 & 0 & 0 \\ 0 & 1 & 0 \\ 0 & 0 & 1 \end{bmatrix}. \tag{3.18}$$

Sometimes a subscript on $\mathbf{I}_N$ is used to indicate the size of the identity matrix, which is necessarily square of size $N \times N$. Likewise, we may use a subscript for the square zero matrix $\mathbf{0}_N$. If the zero matrix is rectangular, then the notation $\mathbf{0}_{MN}$ is used. Any matrix multiplied by $\mathbf{I}$ is unchanged:

$$\mathbf{AI} = \mathbf{A}, \qquad \mathbf{IA} = \mathbf{A}, \tag{3.19}$$

assuming proper matrix dimensions. For $\mathbf{A} \in \mathcal{R}^{M \times N}$, the identity matrix is $\mathbf{I}_N$ in the first expression, whereas it is $\mathbf{I}_M$ in the second expression. The identity matrix is obtained by concatenating N *unit vectors* as follows:

$$\mathbf{I} = [\mathbf{e}_1, \ldots, \mathbf{e}_N], \tag{3.20}$$

where the elements of column vector $\mathbf{e}_n \in \mathcal{R}^N$ are all 0 except for 1 in the nth row.

Definition: Matrix Trace The *trace* of square matrix $\mathbf{A}$ is the sum of its diagonal elements:

$$\text{tr}(\mathbf{A}) = \sum_{n=1}^{N} a_{nn}. \tag{3.21}$$

Matrix $\mathbf{A}$ need not be diagonal.

Properties of trace are summarized in Table 3.3.

TABLE 3.3 Matrix Trace Properties for $\mathbf{A} \in \mathcal{R}^{N \times N}$ and $\mathbf{B} \in \mathcal{R}^{N \times N}$

Property	Operation
Definition	$\text{tr}(\mathbf{A}) = \sum_{n=1}^{N} a_{nn}$
Identity matrix	$\text{tr}(\mathbf{I}) = N$
Transpose	$\text{tr}(\mathbf{A}^T) = \text{tr}(\mathbf{A})$
Scalar product	$\text{tr}(c\mathbf{A}) = c\text{tr}(\mathbf{A})$ $(c \in \mathcal{R})$
Matrix product	$\text{tr}(\mathbf{AB}) = \text{tr}(\mathbf{BA})$
Addition	$\text{tr}(\mathbf{A} + \mathbf{B}) = \text{tr}(\mathbf{A}) + \text{tr}(\mathbf{B})$
Eigenvalues	$\text{tr}(\mathbf{A}) = \sum_{n=1}^{N} \lambda_n$

A triangular matrix is an important type of square matrix that appears in some matrix *decompositions.*

Definition: Triangular Matrix The elements of a *lower triangular matrix* $\mathbf{L}$ are 0 above the main descending diagonal, and the elements of an *upper triangular matrix* $\mathbf{U}$ are 0 below the main descending diagonal.

A diagonal matrix is a special case of a triangular matrix with nonzero elements only on the main descending diagonal: it is both lower and upper triangular.

Definition: Matrix Inverse If it exists, the *matrix inverse* of square matrix $\mathbf{A}$ denoted by $\mathbf{A}^{-1}$ satisfies $\mathbf{A}\mathbf{A}^{-1} = \mathbf{I} = \mathbf{A}^{-1}\mathbf{A}$.

In this case, $\mathbf{A}$ and $\mathbf{A}^{-1}$ commute to give $\mathbf{I}$, whereas as mentioned before, matrices do not commute in general. If the rows and columns are linearly independent, then square matrix $\mathbf{A} \in \mathcal{R}^{N\times N}$ has an inverse.

Definition: Linearly Independent Columns The columns $\{\mathbf{a}_n\}$ of matrix $\mathbf{A}$ are *linearly dependent* if nonzero $\{c_n\}$ exist such that

$$\sum_{n=1}^{N} c_n \mathbf{a}_n = \mathbf{0}. \tag{3.22}$$

Otherwise, they are *linearly independent.*

Dependent columns can be written as a linear combination of each other. This means there are "extra" columns in the system of equations that provide the same information about the system as some combination of the other columns. The same definition applies to the rows of $\mathbf{A}$.

When $\mathbf{A}$ is square and its inverse exists, the solution to (3.7) is

$$\mathbf{y} = \mathbf{A}^{-1}\mathbf{x}. \tag{3.23}$$

The solution $\mathbf{y}$ can also be derived using GE without finding the inverse of $\mathbf{A}$, as described later in this chapter.

Definition: Permutation Matrix A *permutation matrix* is a square matrix whose elements are all 0 except for a single 1 in each row and column.

For $N = 3$, there are $N! = 6$ permutation matrices:

$$\mathbf{P}_{213} = \begin{bmatrix} 0 & 1 & 0 \\ 1 & 0 & 0 \\ 0 & 0 & 1 \end{bmatrix}, \quad \mathbf{P}_{321} = \begin{bmatrix} 0 & 0 & 1 \\ 0 & 1 & 0 \\ 1 & 0 & 0 \end{bmatrix}, \quad \mathbf{P}_{132} = \begin{bmatrix} 1 & 0 & 0 \\ 0 & 0 & 1 \\ 0 & 1 & 0 \end{bmatrix}, \tag{3.24}$$

$$\mathbf{P}_{231} = \begin{bmatrix} 0 & 0 & 1 \\ 1 & 0 & 0 \\ 0 & 1 & 0 \end{bmatrix}, \quad \mathbf{P}_{312} = \begin{bmatrix} 0 & 1 & 0 \\ 0 & 0 & 1 \\ 1 & 0 & 0 \end{bmatrix}, \quad \mathbf{P}_{123} = \begin{bmatrix} 1 & 0 & 0 \\ 0 & 1 & 0 \\ 0 & 0 & 1 \end{bmatrix}, \tag{3.25}$$

where the subscript indicates the row numbers for the 1 in each column. All permutation matrices for arbitrary N can be obtained by interchanging one or more columns of the identity matrix **I**, and so, the subscript also indicates the order of the columns taken from **I** to generate **P**. Permutation matrices can also be derived systematically from the unit vectors: there are N choices of $\mathbf{e}_n$ for the first column, $N - 1$ for the second column, and so on until the last column for which there is only one unused unit vector. Thus, there are

$$N \times (N-1) \times \ldots \times 1 = N! \tag{3.26}$$

permutation matrices including the identity matrix $\mathbf{I} = \mathbf{P}_{12\ldots N}$. Multiplying **A** on the left by **P** causes the rows of **A** to be rearranged according to the subscript; similarly, the columns of **A** are rearranged when it is multiplied by **P** on the right.

Example 3.5 Multiplying **A** on the left by $\mathbf{P}_{213}$ interchanges rows 1 and 2:

$$\begin{bmatrix} 0 & 1 & 0 \\ 1 & 0 & 0 \\ 0 & 0 & 1 \end{bmatrix} \begin{bmatrix} a_{11} & a_{12} & a_{13} \\ a_{21} & a_{22} & a_{23} \\ a_{31} & a_{32} & a_{33} \end{bmatrix} = \begin{bmatrix} a_{21} & a_{22} & a_{23} \\ a_{11} & a_{12} & a_{13} \\ a_{31} & a_{32} & a_{33} \end{bmatrix}, \tag{3.27}$$

and multiplying **A** on the right interchanges columns 1 and 2:

$$\begin{bmatrix} a_{11} & a_{12} & a_{13} \\ a_{21} & a_{22} & a_{23} \\ a_{31} & a_{32} & a_{33} \end{bmatrix} \begin{bmatrix} 0 & 1 & 0 \\ 1 & 0 & 0 \\ 0 & 0 & 1 \end{bmatrix} = \begin{bmatrix} a_{12} & a_{11} & a_{13} \\ a_{22} & a_{21} & a_{23} \\ a_{32} & a_{31} & a_{33} \end{bmatrix}. \tag{3.28}$$

Definition: Exchange Matrix *Exchange matrix* **J** is similar to the identity matrix, except that the 1s are on the main counter-diagonal from upper right to lower left. It is the permutation matrix with the subscripts written in reverse order: $\mathbf{J} \triangleq \mathbf{P}_{N\ldots 21}$. Sometimes the notation $\tilde{\mathbf{I}}$ is used.

Multiplication of **A** on the left by **J** causes all rows of **A** to be rearranged in reverse order, and likewise for the columns of **A** when it is multiplied on the right by **J**.

Definition: Elementary Matrix *Elementary matrix* **E** is a square matrix that performs one of the following three types of row operations when multiplying **A** on the left: (i) scales row m by c, (ii) interchanges rows m and n, or (iii) adds scaled row m to row n. Multiplying **A** on the right by an elementary matrix modifies the columns in the same way.

An elementary matrix is derived by an *elementary operation* on the identity matrix. Examples of the three types of elementary matrices and their notation for $M = N = 3$ are (i) scaling a row:

$$\mathbf{E}_2(c) = \begin{bmatrix} 1 & 0 & 0 \\ 0 & c & 0 \\ 0 & 0 & 1 \end{bmatrix} \implies \mathbf{E}_2(c)\mathbf{A} = \begin{bmatrix} a_{11} & a_{12} & a_{13} \\ ca_{21} & ca_{22} & ca_{23} \\ a_{31} & a_{32} & a_{33} \end{bmatrix}. \tag{3.29}$$

The subscript specifies which row of $\mathbf{I}$ is scaled by $c \in \mathcal{R}$ to give the elementary matrix. (ii) Interchanging two rows:

$$\mathbf{E}_{23} = \begin{bmatrix} 1 & 0 & 0 \\ 0 & 0 & 1 \\ 0 & 1 & 0 \end{bmatrix} \implies \mathbf{E}_{23}\mathbf{A} = \begin{bmatrix} a_{11} & a_{12} & a_{13} \\ a_{31} & a_{32} & a_{33} \\ a_{21} & a_{22} & a_{23} \end{bmatrix}. \tag{3.30}$$

The subscripts denote which rows of $\mathbf{I}$ are interchanged to give the elementary matrix. From this result, we see that a permutation matrix derived by interchanging exactly *two* rows of $\mathbf{I}$ is also an elementary matrix. For $N = 3$, these correspond to $\mathbf{P}_{213}$, $\mathbf{P}_{321}$, and $\mathbf{P}_{132}$. For general N, the following number of the $N!$ permutation matrices are elementary matrices:

$$(N-1) + (N-2) + \ldots + 2 + 1, \tag{3.31}$$

which from a closed-form expression in Appendix C is

$$\sum_{n=1}^{N-1} n = (1/2)(N-1)N. \tag{3.32}$$

(iii) Adding scaled row m to row n:

$$\mathbf{E}_{31}(c) = \begin{bmatrix} 1 & 0 & 0 \\ 0 & 1 & 0 \\ c & 0 & 1 \end{bmatrix} \implies \mathbf{E}_{31}(c)\mathbf{A} = \begin{bmatrix} a_{11} & a_{12} & a_{13} \\ a_{21} & a_{22} & a_{23} \\ ca_{11} + a_{31} & ca_{12} + a_{32} & ca_{13} + a_{33} \end{bmatrix}. \tag{3.33}$$

The subscripts indicate the row and column of $\mathbf{I}$ where the scalar $c \in \mathcal{R}$ is placed to define the elementary matrix. When multiplied on the left by $\mathbf{E}_{31}(c)$, the first row of $\mathbf{A}$ scaled by c is added to the third row.

In summary, we find from the notation used in (3.29)–(3.33) that it is possible to uniquely determine the type of elementary matrix from the subscripts and whether or not there is an argument c. It is straightforward to verify that the elementary matrices have the following inverses (see Problem 3.9):

$$\mathbf{E}_m^{-1}(c) = \mathbf{E}_m(1/c), \quad \mathbf{E}_{mn}^{-1} = \mathbf{E}_{mn}, \quad \mathbf{E}_{mn}^{-1}(c) = \mathbf{E}_{mn}(-c). \tag{3.34}$$

Since only the mth row of a matrix is scaled by c when multiplied on the left by $\mathbf{E}_m(c)$, it follows that if $\mathbf{E}_m(c)\mathbf{E}_m(d) = \mathbf{I}$, then we must have $d = 1/c$. The inverse for $\mathbf{E}_{mn}$ is itself because interchanging rows m and n twice yields the original identity matrix: $\mathbf{E}_{mn}\mathbf{E}_{mn} = \mathbf{I} \implies \mathbf{E}_{mn}^{-1} = \mathbf{E}_{mn}$.

Elementary matrices can be used to find the inverse of a nonsingular matrix as follows. Suppose we premultiply matrix $\mathbf{A}$ successively by elementary matrices until the resulting row operations yield the identity matrix $\mathbf{I}$. Let the product of those elementary matrices be denoted by $\mathbf{E}$. Observe that

$$\mathbf{EA} = \mathbf{I} \implies \mathbf{EAA}^{-1} = \mathbf{IA}^{-1} \implies \mathbf{EI} = \mathbf{A}^{-1}, \tag{3.35}$$

which demonstrates that the inverse of matrix $\mathbf{A}$ is generated by premultiplying the identity matrix by the same product $\mathbf{E}$ of elementary matrices.

Example 3.6 In this example, we show for the following matrix how to obtain its inverse by using elementary matrices:

$$\mathbf{A} = \begin{bmatrix} 1 & 2 \\ 3 & 1 \end{bmatrix}. \tag{3.36}$$

The approach is facilitated by using the following *augmented* matrix:

$$[\mathbf{A} \mid \mathbf{I}] = \left[\begin{array}{cc|cc} 1 & 2 & 1 & 0 \\ 3 & 1 & 0 & 1 \end{array}\right], \tag{3.37}$$

and applying the elementary matrices to both components. First, $\mathbf{E}_{12}(-3)$ adds row 1 scaled by -3 to the second row:

$$[\mathbf{E}_{12}(-3)\mathbf{A} \mid \mathbf{E}_{12}(-3)\mathbf{I}] = \left[\begin{array}{cc|cc} 1 & 2 & 1 & 0 \\ 0 & -5 & -3 & 1 \end{array}\right]. \tag{3.38}$$

Next, $\mathbf{E}_2(-5)$ scales the second row in the last expression:

$$[\mathbf{E}_2(-5)\mathbf{E}_{12}(-3)\mathbf{A} \mid \mathbf{E}_2(-5)\mathbf{E}_{12}(-3)\mathbf{I}] = \left[\begin{array}{cc|cc} 1 & 2 & 1 & 0 \\ 0 & 1 & 3/5 & -1/5 \end{array}\right]. \tag{3.39}$$

Finally, $\mathbf{E}_{21}(-2)$ eliminates 2 in the upper right corner of $\mathbf{A}$, yielding the desired result:

$$[\mathbf{E}_{21}(-2)\mathbf{E}_2(-5)\mathbf{E}_{12}(-3)\mathbf{A} \mid \mathbf{E}_{21}(-2)\mathbf{E}_2(-5)\mathbf{E}_{12}(-3)\mathbf{I}] = \left[\begin{array}{cc|cc} 1 & 0 & -1/5 & 2/5 \\ 0 & 1 & 3/5 & -1/5 \end{array}\right], \tag{3.40}$$

and so, the inverse is

$$\mathbf{A}^{-1} = \begin{bmatrix} -1/5 & 2/5 \\ 3/5 & -1/5 \end{bmatrix}. \tag{3.41}$$

A similar approach is used in GE, where instead of computing the inverse of a matrix, *back-substitution* is used to find the unknown variables $\mathbf{y}$ in $\mathbf{Ay} = \mathbf{x}$.

The product $\mathbf{EAy} = \mathbf{Ex}$ yields an upper triangular matrix $\mathbf{U} \triangleq \mathbf{EA}$, for which it is straightforward to solve for y_N because the last row of $\mathbf{U}$ has only one nonzero element. Once y_N has been found, back-substitution successively solves for y_{N-1}, y_{N-2}, and so on until reaching y_1. It is not necessary to convert $\mathbf{A}$ to the identity matrix in order to find $\mathbf{y}$ as was done to find $\mathbf{A}^{-1}$ in the previous example.

Definition: Matrix Rank The *rank* R of matrix $\mathbf{A}$ is the number of independent columns, which is also the number of independent rows.

This definition does not require that the matrix be square.

Example 3.7 For the three equations/lines described in Example 3.2, the 3×2 matrix is

$$\mathbf{A} = \begin{bmatrix} 1 & 1 \\ 1 & -1 \\ 1 & 1 \end{bmatrix}, \tag{3.42}$$

whose rank obviously cannot exceed $\min(M, N) = \min(3, 2) = 2$. The first and third rows are linearly dependent, whereas the first and second rows are not, and so, the rank of this matrix is $R = 2$. It is also clear that the two columns are linearly independent. We emphasize that the rank is not necessarily equal to $\min(M, N)$. For example, if -1 in (3.42) is changed to 1, then all three rows are linearly dependent as are both columns, and the rank is $R = 1$.

Definition: Symmetric Matrix Square matrix $\mathbf{A}$ is *symmetric* if $\mathbf{A}^T = \mathbf{A}$ and *skew-symmetric* if $\mathbf{A}^T = -\mathbf{A}$.

A diagonal matrix is always symmetric.

Definition: Hermitian Matrix Square matrix $\mathbf{A}$ is *Hermitian* if $\mathbf{A}^H = \mathbf{A}$ and *skew-Hermitian* if $\mathbf{A}^H = -\mathbf{A}$ where $\mathbf{A}^H \triangleq (\mathbf{A}^T)^* = (\mathbf{A}^*)^T$.

The superscript $*$ denotes complex conjugation of every element in the matrix. A complex number has the two-dimensional form $c = a + jb$ with $j \triangleq \sqrt{-1}$; its complex conjugate is $c^* = a - jb$, which is discussed further in Chapter 4. The real part of c is a and its imaginary part is b. The real part of c^* is also a, but its imaginary part is $-b$. A Hermitian (skew-Hermitian) matrix is equivalent to a symmetric (skew-symmetric) matrix if all its elements are real.

Example 3.8 Example symmetric and skew-symmetric matrices are

$$\mathbf{A} = \begin{bmatrix} 1 & 2 & 3 \\ 2 & 4 & 5 \\ 3 & 5 & 6 \end{bmatrix}, \quad \mathbf{B} = \begin{bmatrix} 0 & 2 & 3 \\ -2 & 0 & 5 \\ -3 & -5 & 0 \end{bmatrix}, \tag{3.43}$$

from which we find that the diagonal elements of a skew-symmetric matrix must all be 0. This result follows from the defining property $b_{mn} = -b_{nm}$ and the fact that this equation holds only when the elements are 0 for $m = n$. There is no similar restriction for a symmetric matrix whose elements satisfy $a_{mn} = a_{nm}$. Example Hermitian and skew-Hermitian matrices are

$$\mathbf{A} = \begin{bmatrix} 1 & 2+j & 3-j \\ 2-j & 4 & 5+2j \\ 3+j & 5-2j & 6 \end{bmatrix}, \quad \mathbf{B} = \begin{bmatrix} 0 & 2+j & 3-j \\ -2+j & 0 & 5+2j \\ -3-j & -5+2j & 0 \end{bmatrix}, \tag{3.44}$$

from which we find that the diagonal elements of a Hermitian matrix must all be real. This occurs due to the defining property $a^*_{mn} = a_{nm}$, which is valid for $m = n$ only when the imaginary part is 0. The corresponding equation for a skew-Hermitian matrix is $b^*_{mn} = -b_{nm}$, which shows that the real parts of transposed elements have the opposite sign and their imaginary parts have the same sign.

Definition: Idempotent Matrix Matrix **A** is *idempotent* if $\mathbf{A}^2 = \mathbf{A}$ which implies $\mathbf{A}^k = \mathbf{A}$ for $k \in \mathcal{N}$.

Definition: Nilpotent Matrix Matrix **A** is *nilpotent* if $\mathbf{A}^k = \mathbf{0}$ for $k > N$ and some integer $N \in \mathcal{N}$.

Example 3.9 An example of an idempotent matrix is the *projection matrix* that arises in *least-squares problems*:

$$\mathbf{B} = \mathbf{A}(\mathbf{A}^T\mathbf{A})^{-1}\mathbf{A}^T, \tag{3.45}$$

where $(\mathbf{A}^T\mathbf{A})^{-1}$ is the inverse of $\mathbf{A}^T\mathbf{A}$, and $\mathbf{A} \in \mathcal{R}^{M\times N}$ need not be a square matrix. It is straightforward to verify that **B** is idempotent from the multiplication rules for matrices:

$$\mathbf{B}^2 = \mathbf{A}(\mathbf{A}^T\mathbf{A})^{-1}\mathbf{A}^T\mathbf{A}(\mathbf{A}^T\mathbf{A})^{-1}\mathbf{A}^T = \mathbf{A}(\mathbf{A}^T\mathbf{A})^{-1}\mathbf{A}^T = \mathbf{B}. \tag{3.46}$$

A numerical example is

$$\mathbf{B} = \begin{bmatrix} 1 & 0 & 0 \\ 0 & 1 & 0 \\ 2 & 0 & 0 \end{bmatrix} \Longrightarrow \mathbf{B}^2 = \begin{bmatrix} 1 & 0 & 0 \\ 0 & 1 & 0 \\ 2 & 0 & 0 \end{bmatrix}. \tag{3.47}$$

The identity matrix is the only idempotent matrix that is nonsingular. An example nilpotent matrix is

$$\mathbf{A} = \begin{bmatrix} 0 & 0 & 0 \\ 3 & 0 & 0 \\ 2 & 1 & 0 \end{bmatrix} \Longrightarrow \mathbf{A}^2 = \begin{bmatrix} 0 & 0 & 0 \\ 0 & 0 & 0 \\ 3 & 0 & 0 \end{bmatrix}, \tag{3.48}$$

with $\mathbf{A}^k = \mathbf{0}$ for $k \geq 3$. It turns out that the *eigenvalues* of a nilpotent matrix, which are defined later in this chapter, are all 0.

Definition: Orthogonal and Unitary Matrices Square matrix $\mathbf{A}$ is *orthogonal* if $\mathbf{A}^T\mathbf{A} = \mathbf{A}\mathbf{A}^T = \mathbf{I}$, which implies $\mathbf{A}^T = \mathbf{A}^{-1}$. Square matrix $\mathbf{A}$ is *unitary* if $\mathbf{A}^H\mathbf{A} = \mathbf{A}\mathbf{A}^H = \mathbf{I}$, which implies $\mathbf{A}^H = \mathbf{A}^{-1}$.

Of course, a unitary matrix is equivalent to an orthogonal matrix when all its elements are real.

Example 3.10 An important example of an orthogonal matrix is the following *rotation matrix*:

$$\mathbf{A} = \begin{bmatrix} \cos(\phi) & -\sin(\phi) \\ \sin(\phi) & \cos(\phi) \end{bmatrix}, \tag{3.49}$$

from which we find

$$\begin{aligned} \mathbf{A}^T\mathbf{A} &= \begin{bmatrix} \cos(\phi) & \sin(\phi) \\ -\sin(\phi) & \cos(\phi) \end{bmatrix} \begin{bmatrix} \cos(\phi) & -\sin(\phi) \\ \sin(\phi) & \cos(\phi) \end{bmatrix} \\ &= \begin{bmatrix} \cos^2(\phi) + \sin^2(\phi) & -\cos(\phi)\sin(\phi) + \sin(\phi)\cos(\phi) \\ -\sin(\phi)\cos(\phi) + \cos(\phi)\sin(\phi) & \sin^2(\phi) + \cos^2(\phi) \end{bmatrix} = \mathbf{I}. \end{aligned} \tag{3.50}$$

This result holds for any angle $\phi \in [0, 2\pi]$. Equation (3.49) is a rotation matrix because $\mathbf{A}$ in $\mathbf{Ay} = \mathbf{x}$ rotates $\mathbf{y}$ by angle ϕ to give $\mathbf{x}$, and without any change to its length: $\|\mathbf{x}\| = \|\mathbf{y}\|$. This is seen as follows:

$$\mathbf{x} = \mathbf{Ay} = \begin{bmatrix} \cos(\phi)x_1 - \sin(\phi)x_2 \\ \sin(\phi)x_1 + \cos(\phi)x_2 \end{bmatrix}, \tag{3.51}$$

with squared norm

$$\begin{aligned} \mathbf{x}^T\mathbf{x} &= \cos^2(\phi)x_1^2 + \sin^2(\phi)x_2^2 + 2\sin(\phi)\cos(\phi)x_1x_2 + \sin^2(\phi)x_1^2 \\ &\quad + \cos^2(\phi)x_2^2 - 2\sin(\phi)\cos(\phi)x_1x_2 \\ &= x_1^2 + x_2^2 = \mathbf{y}^T\mathbf{y}. \end{aligned} \tag{3.52}$$

For positive ϕ, $\mathbf{A}$ rotates $\mathbf{y}$ counterclockwise on the plane to $\mathbf{x}$ as depicted in Figure 3.5. The coordinates of $\mathbf{x}$ are $x_1 = \cos(\phi)y_1 - \sin(\phi)y_2 < y_1$ and $x_2 = \sin(\phi)y_1 + \cos(\phi)y_2 > y_2$. An example unitary matrix is

$$\mathbf{B} = \frac{1}{\sqrt{2}} \begin{bmatrix} 1 & \exp(-j\phi) \\ -\exp(j\phi) & 1 \end{bmatrix}, \tag{3.53}$$

where $\exp(j\phi) = \cos(\phi) + j\sin(\phi)$ is the complex exponential function discussed in Chapter 4.

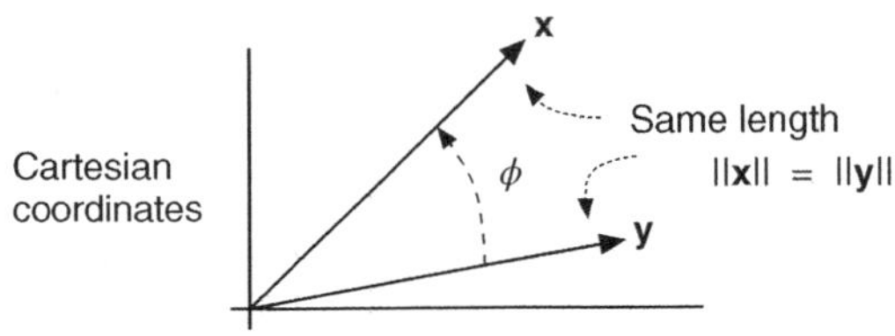

Figure 3.5 Vector $\mathbf{y}$ rotated counterclockwise by ϕ radians to $\mathbf{x} = \mathbf{A}\mathbf{y}$ using the rotation matrix in (3.49).

Definition: Toeplitz and Circulant Matrices Matrix $\mathbf{A}$ is *Toeplitz* if the elements for every descending diagonal from upper left to lower right are identical within the same diagonal. Toeplitz matrix $\mathbf{A}$ is *circulant* if each row can be derived by circularly shifting the previous row by one element to the right.

For square circulant matrix $\mathbf{A} \in \mathcal{R}^{N\times N}$, there are at most N unique elements, as opposed to N^2 for general square matrix $\mathbf{A}$.

Example 3.11 A Toeplitz matrix for $N = 3$ has the following form:

$$\mathbf{A} = \begin{bmatrix} a & b & c \\ d & a & b \\ e & d & a \end{bmatrix}. \tag{3.54}$$

This matrix is circulant if $d = c$ and $e = b$:

$$\mathbf{A} = \begin{bmatrix} a & b & c \\ c & a & b \\ b & c & a \end{bmatrix}, \tag{3.55}$$

which is uniquely specified by any row or column. A Toeplitz matrix is also specified by any row or column provided it is symmetric (or Hermitian in the complex case), as shown by making (3.54) symmetric with $d = b$ and $e = c$:

$$\mathbf{A} = \begin{bmatrix} a & b & c \\ b & a & b \\ c & b & a \end{bmatrix}. \tag{3.56}$$

Various special *square* matrices and their properties are summarized in Table 3.4. (Although technically we could apply the Toeplitz property to rectangular matrices, they are assumed to be square in the table.)

3.5 DETERMINANT

Next, we discuss the *determinant* of a square matrix, which is useful for writing an explicit solution to the system of equations $\mathbf{A}\mathbf{y} = \mathbf{x}$ provided that $\mathbf{A}$ is nonsingular. It is also used to find the eigenvalues of $\mathbf{A}$ as described at the end of this chapter.

TABLE 3.4 Special Square Matrices and Properties

Matrix	Property
Diagonal **D**	All elements are 0 except on the main descending diagonal
Lower triangular **L**	All elements above the main descending diagonal are 0
Upper triangular **U**	All elements below the main descending diagonal are 0
Identity **I**	All elements are 0 except for 1s on the main diagonal
Permutation **P**	Permutation of rows or columns of **I**
Exchange **J**	All elements are 0 except for 1s on the main counter-diagonal
Elementary **E**	Three types derived from **I** or **P**
Minor **M**	Minor matrix of **A** with a row and column deleted
Inverse $\mathbf{A}^{-1}$	Rows/columns are linearly independent
Symmetric	$\mathbf{A}^T = \mathbf{A}$
Skew-symmetric	$\mathbf{A}^T = -\mathbf{A}$
Hermitian	$\mathbf{A}^H = \mathbf{A}$
Skew-Hermitian	$\mathbf{A}^H = -\mathbf{A}$
Idempotent	$\mathbf{A}^k = \mathbf{A}$ for $k \in \mathcal{N}$
Nilpotent	$\mathbf{A}^k = \mathbf{0}$ for $k > N$ and some integer $N \in \mathcal{N}$
Toeplitz	Identical elements along the same descending diagonals
Circulant	Each row is generated as a circular shift of the previous row

Definition: Determinant The *determinant* of square matrix $\mathbf{A} \in \mathcal{R}^{N\times N}$ is

$$\det(\mathbf{A}) = \sum_{n=1}^{N} a_{mn} A_{mn}, \tag{3.57}$$

where a_{mn} is the element in row m and column n. The scalar

$$A_{mn} = (-1)^{m+n} \det(\mathbf{M}_{mn}) \tag{3.58}$$

is called a *cofactor*, and the *minor* matrix $\mathbf{M}_{mn}$ is derived from **A** by deleting its mth row and nth column.

The determinant of **A** is computed by finding the determinants of successively smaller minor matrices using cofactors defined along any row until 2×2 matrices are reached.

Example 3.12 The basic 2×2 matrix

$$\mathbf{A} = \begin{bmatrix} a_{11} & a_{12} \\ a_{21} & a_{22} \end{bmatrix}, \tag{3.59}$$

has the following minor "matrices":

$$\mathbf{M}_{11} = a_{22}, \quad \mathbf{M}_{12} = a_{21}, \quad \mathbf{M}_{21} = a_{12}, \quad \mathbf{M}_{22} = a_{11}, \tag{3.60}$$

with cofactors

$$A_{11} = a_{22}, \quad A_{12} = -a_{21}, \quad A_{21} = -a_{12}, \quad A_{22} = a_{11}, \tag{3.61}$$

and so, its determinant is

$$\det(\mathbf{A}) = a_{11}A_{11} + a_{12}A_{12} = a_{11}a_{22} - a_{12}a_{21}. \tag{3.62}$$

If instead cofactors for the second row are used, we obtain the same result:

$$\det(\mathbf{A}) = a_{21}A_{21} + a_{22}A_{22} = -a_{21}a_{12} + a_{22}a_{11}. \tag{3.63}$$

Example 3.13 For the 3×3 matrix

$$\mathbf{A} = \begin{bmatrix} a_{11} & a_{12} & a_{13} \\ a_{21} & a_{22} & a_{23} \\ a_{31} & a_{32} & a_{33} \end{bmatrix}, \tag{3.64}$$

the minor matrices are

$$\mathbf{M}_{11} = \begin{bmatrix} a_{22} & a_{23} \\ a_{32} & a_{33} \end{bmatrix}, \quad \mathbf{M}_{12} = \begin{bmatrix} a_{21} & a_{23} \\ a_{31} & a_{33} \end{bmatrix}, \quad \mathbf{M}_{13} = \begin{bmatrix} a_{21} & a_{22} \\ a_{31} & a_{32} \end{bmatrix}, \tag{3.65}$$

$$\mathbf{M}_{21} = \begin{bmatrix} a_{12} & a_{13} \\ a_{32} & a_{33} \end{bmatrix}, \quad \mathbf{M}_{22} = \begin{bmatrix} a_{11} & a_{13} \\ a_{31} & a_{33} \end{bmatrix}, \quad \mathbf{M}_{23} = \begin{bmatrix} a_{11} & a_{12} \\ a_{31} & a_{32} \end{bmatrix}, \tag{3.66}$$

$$\mathbf{M}_{31} = \begin{bmatrix} a_{12} & a_{13} \\ a_{22} & a_{23} \end{bmatrix}, \quad \mathbf{M}_{32} = \begin{bmatrix} a_{11} & a_{13} \\ a_{21} & a_{23} \end{bmatrix}, \quad \mathbf{M}_{33} = \begin{bmatrix} a_{11} & a_{12} \\ a_{21} & a_{22} \end{bmatrix}. \tag{3.67}$$

Using cofactors for the first row, the determinant is

$$\begin{aligned} \det(\mathbf{A}) &= a_{11} \det \mathbf{M}_{11} - a_{12} \det \mathbf{M}_{12} + a_{13} \det \mathbf{M}_{13} \\ &= a_{11}(a_{22}a_{33} - a_{32}a_{23}) - a_{12}(a_{21}a_{33} - a_{31}a_{23}) \\ &\quad + a_{13}(a_{21}a_{32} - a_{31}a_{22}). \end{aligned} \tag{3.68}$$

It is easy to verify that the same determinant is obtained by using cofactors for either the second or third row.

The determinant of **A** is unique, and it has the properties in Table 3.5. The property for a triangular matrix also holds for a diagonal matrix because it is a special type of triangular matrix.

TABLE 3.5 Matrix Determinant Properties for $\mathbf{A} \in \mathcal{R}^{N\times N}$ and $\mathbf{B} \in \mathcal{R}^{N\times N}$

Property	Operation
Identity matrix	$\det(\mathbf{I}) = 1$
Transpose	$\det(\mathbf{A}^T) = \det(\mathbf{A})$
Inverse	$\det(\mathbf{A}^{-1}) = 1/\det(\mathbf{A})$
Scalar product	$\det(c\mathbf{A}) = c^N \det(\mathbf{A})$ $(c \in \mathcal{R})$
Matrix product	$\det(\mathbf{AB}) = \det(\mathbf{A})\det(\mathbf{B})$
Addition	$\det(\mathbf{A}+\mathbf{B}) \neq \det(\mathbf{A}) + \det(\mathbf{B})$
Eigenvalues	$\det(\mathbf{A}) = \prod_{n=1}^{N} \lambda_n$
Triangular matrix	$\det(\mathbf{A}) = \prod_{n=1}^{N} a_{nn}$
Dependent rows/columns	$\det(\mathbf{A}) = 0$

Definition: Adjugate Matrix The *adjugate matrix* of $\mathbf{A}$ contains its cofactors as follows:

$$\mathrm{adj}(\mathbf{A}) \triangleq \begin{bmatrix} A_{11} & \dots & A_{N1} \\ \vdots & \dots & \vdots \\ A_{1N} & \dots & A_{NN} \end{bmatrix}, \tag{3.69}$$

where the subscripts are transposed relative to $\{a_{mn}\}$ of the original matrix $\mathbf{A}$.

If we replace a_{mn} in the determinant expression of (3.57) with a_{pn} (for $p \neq m$) and keep A_{mn} in the sum as follows:

$$\sum_{n=1}^{N} a_{pn} A_{mn} = 0, \quad p \neq m, \tag{3.70}$$

then this change is equivalent to replacing the mth row of $\mathbf{A}$ with its pth row. As a result, (3.70) is the determinant of a matrix with two identical rows, which is singular, and as a result, the sum across columns is 0. Thus, the elements of the matrix product $\mathbf{A}\mathrm{adj}(\mathbf{A})$ are all 0 except along the main descending diagonal where $p = m$:

$$\mathbf{A}\mathrm{adj}(\mathbf{A}) = \det(\mathbf{A})\mathbf{I}. \tag{3.71}$$

The diagonal elements on the right-hand side are all of the form in (3.57) as m of a_{mn} and A_{mn} is varied from 1 to N, giving $\det(\mathbf{A})$. For the off-diagonal elements, the first subscript of a_{mn} and A_{mn} always differ, and (3.70) yields 0 for those terms. This is easily verified for the 2×2 matrix in (3.59) where

$$\mathrm{adj}(\mathbf{A}) = \begin{bmatrix} a_{22} & -a_{12} \\ -a_{21} & a_{11} \end{bmatrix} \tag{3.72}$$

and

$$\mathbf{A}\text{adj}(\mathbf{A}) = \begin{bmatrix} a_{22} & -a_{12} \\ -a_{21} & a_{11} \end{bmatrix} \begin{bmatrix} a_{11} & a_{12} \\ a_{21} & a_{22} \end{bmatrix}$$

$$= \begin{bmatrix} a_{22}a_{11} - a_{12}a_{21} & a_{22}a_{12} - a_{12}a_{22} \\ -a_{21}a_{11} + a_{11}a_{21} & -a_{21}a_{12} + a_{11}a_{22} \end{bmatrix}$$

$$= \begin{bmatrix} \det(\mathbf{A}) & 0 \\ 0 & \det(\mathbf{A}) \end{bmatrix}. \tag{3.73}$$

For nonsingular $\mathbf{A}$, the product of (3.71) and $\mathbf{A}^{-1}$ yields

$$\mathbf{A}^{-1} = \frac{1}{\det(\mathbf{A})}\text{adj}(\mathbf{A}), \tag{3.74}$$

and an explicit solution for $\mathbf{Ay} = \mathbf{x}$ is

$$\mathbf{y} = \frac{1}{\det(\mathbf{A})}\text{adj}(\mathbf{A})\mathbf{x}. \tag{3.75}$$

As mentioned earlier, it is possible to solve $\mathbf{Ay} = \mathbf{x}$ using determinants and a technique known as *Cramer's rule* (although GE is more efficient). Cramer's rule is often used in circuit courses because it yields an explicit expression for each element of $\mathbf{y}$. Since $\mathbf{A}$ usually is not large in such circuits problems, complexity and numerical stability issues associated with computing $\mathbf{A}^{-1}$ are not of concern when using this approach. The mth element of $\mathbf{y}$ is

$$y_m = \frac{\det(\mathbf{A}_m)}{\det(\mathbf{A})}, \tag{3.76}$$

where the square matrix $\mathbf{A}_m$ is generated from $\mathbf{A}$ by replacing its mth column with $\mathbf{x}$:

$$\mathbf{A}_m \triangleq \begin{bmatrix} a_{11} & \cdots & a_{1,m-1} & x_1 & a_{1,m+1} & \cdots & a_{1N} \\ \vdots & \vdots & \vdots & \vdots & \vdots & \vdots & \vdots \\ a_{N1} & \cdots & a_{N,m-1} & x_N & a_{N,m+1} & \cdots & a_{NN} \end{bmatrix} \tag{3.77}$$

The result in (3.76) follows from (3.75) by examining the column vector

$$\text{adj}(\mathbf{A})\mathbf{x} = \begin{bmatrix} A_{11} & \cdots & A_{N1} \\ \vdots & \cdots & \vdots \\ A_{1N} & \cdots & A_{NN} \end{bmatrix} \begin{bmatrix} x_1 \\ \vdots \\ x_N \end{bmatrix}, \tag{3.78}$$

whose mth element is

$$\sum_{n=1}^{N} A_{nm}x_n = \det \begin{bmatrix} a_{11} & \cdots & a_{1,m-1} & x_1 & a_{1,m+1} & \cdots & a_{1N} \\ \vdots & \cdots & \vdots & \vdots & \vdots & \cdots & \vdots \\ a_{N1} & \cdots & a_{N,m-1} & x_N & a_{N,m+1} & \cdots & a_{NN} \end{bmatrix} = \det(\mathbf{A}_m). \tag{3.79}$$

Comparing the left-hand side with (3.57), we find that x_n has replaced a_{mn} and the subscripts of A_{mn} have been interchanged to A_{nm}. The middle expression of (3.79) follows because the sum is over the first subscript of A_{nm}, instead of the second subscript as in (3.57). The mth column of $\mathbf{A}_m$ is vector $\mathbf{x}$, and so, this expression is $\det(\mathbf{A}_m)$. Since this result holds for every row of $\text{adj}(\mathbf{A})\mathbf{x}$, we have the result in (3.76). Thus, the solution $\mathbf{y}$ for the nonsingular system of equations $\mathbf{Ay} = \mathbf{x}$ is derived by computing $N + 1$ determinants $\{\det(\mathbf{A}), \det(\mathbf{A}_1), \ldots, \det(\mathbf{A}_N)\}$ and taking their ratios according to (3.76).

Example 3.14 Consider the following system of linear equations:

$$\begin{bmatrix} 1 & 2 & 4 \\ 3 & 4 & 1 \\ 2 & 3 & 1 \end{bmatrix} \begin{bmatrix} y_1 \\ y_2 \\ y_3 \end{bmatrix} = \begin{bmatrix} 3 \\ 1 \\ 2 \end{bmatrix}, \tag{3.80}$$

for which the determinant is

$$\det(\mathbf{A}) = \det \begin{bmatrix} 1 & 2 & 4 \\ 3 & 4 & 1 \\ 2 & 3 & 1 \end{bmatrix} = 1(4 \times 1 - 3 \times 1) - 2(3 \times 1 - 2 \times 1) + 4(3 \times 3 - 2 \times 4) = 3. \tag{3.81}$$

The column-modified matrices containing $\mathbf{x}$ are

$$\mathbf{A}_1 = \begin{bmatrix} 3 & 2 & 4 \\ 1 & 4 & 1 \\ 2 & 3 & 1 \end{bmatrix}, \quad \mathbf{A}_2 = \begin{bmatrix} 1 & 3 & 4 \\ 3 & 1 & 1 \\ 2 & 2 & 1 \end{bmatrix}, \quad \mathbf{A}_3 = \begin{bmatrix} 1 & 2 & 3 \\ 3 & 4 & 1 \\ 2 & 3 & 2 \end{bmatrix}, \tag{3.82}$$

which have determinants $\{-15, 12, 0\}$, respectively, such that the solution of $\mathbf{Ay} = \mathbf{x}$ is $\mathbf{y} = [-5, 4, 0]^T$.

It is important to note that Cramer's rule can be used only when there is a unique solution, which is often the case if a physical system such as an all-resistive circuit is modeled correctly by $\mathbf{Ay} = \mathbf{x}$. The next example illustrates what happens when there is an infinity of solutions because $\mathbf{A}$ is singular.

Example 3.15 Suppose instead that the system in Example 3.14 is

$$\begin{bmatrix} 1 & 2 & 4 \\ 3 & 4 & 1 \\ 2 & 4 & 8 \end{bmatrix} \begin{bmatrix} y_1 \\ y_2 \\ y_3 \end{bmatrix} = \begin{bmatrix} 3 \\ 1 \\ 6 \end{bmatrix}, \tag{3.83}$$

where the last row of $\mathbf{A}$ is a scaled version of the first row, and likewise for the last element of $\mathbf{x}$ on the right-hand side. Since this matrix is singular, we immediately find that

$$\det(\mathbf{A}) = \det\begin{bmatrix} 1 & 2 & 4 \\ 3 & 4 & 1 \\ 2 & 4 & 8 \end{bmatrix} = 1(4\times 8 - 4\times 1) - 2(3\times 8 - 2\times 1) + 4(3\times 4 - 2\times 4) = 0, \tag{3.84}$$

which shows that Cramer's rule cannot be used. When the system of equations is inconsistent such that there are no solutions, matrix **A** is singular and again Cramer's rule cannot be used.

Example 3.16 Continuing with the resistive circuit in Example 3.3, the matrix equation is

$$\begin{bmatrix} 200 & -100 \\ -100 & 300 \end{bmatrix}\begin{bmatrix} i_1 \\ i_2 \end{bmatrix} = \begin{bmatrix} 10 \\ 0 \end{bmatrix}, \tag{3.85}$$

whose right-hand side is a vector of voltages. The determinant is $\det(\mathbf{A}) = 50,000$, and using Cramer's rule, the two currents are

$$i_1 = \det\begin{bmatrix} 10 & -100 \\ 0 & 300 \end{bmatrix}/50,000 = 3000/50,000 = 60\ \text{mA}, \tag{3.86}$$

$$i_2 = \det\begin{bmatrix} 200 & 10 \\ -100 & 0 \end{bmatrix}/50,000 = 1000/50,000 = 20\ \text{mA}, \tag{3.87}$$

which are the same results obtained earlier by a direct substitution of variables.

3.6 MATRIX SUBSPACES

There are four subspaces associated with matrix $\mathbf{A} \in \mathcal{R}^{M\times N}$.

Definition: Subspace A *subspace* $\mathcal{S}$ of the vector space $\mathcal{V} = \mathcal{R}^N$ is a subset of those vectors such that

$$\mathbf{v} \in \mathcal{S}, c \in \mathcal{R} \implies c\mathbf{v} \in \mathcal{S}, \qquad \mathbf{v}_1, \mathbf{v}_2 \in \mathcal{S} \implies \mathbf{v}_1 + \mathbf{v}_2 \in \mathcal{S}. \tag{3.88}$$

The first condition implies that the zero vector is always included in subspace $\mathcal{S}$ (which, of course, is also in the vector space $\mathcal{V}$).

Example 3.17 The following vectors describe various subspaces of $\mathcal{V} = \mathcal{R}^2$:

$$\left\{\begin{bmatrix} v \\ 0 \end{bmatrix} : v \in \mathcal{R}\right\}, \quad \left\{\begin{bmatrix} v_1 \\ -v_1 \end{bmatrix} : v_1 \in \mathcal{R}\right\}, \quad \left\{\begin{bmatrix} v_1 \\ -v_2 \end{bmatrix} : v_1, v_2 \in \mathcal{R}\right\}. \tag{3.89}$$

The first two cases describe lines on a plane, and the third case is equivalent to $\mathcal{R}^2$. The following do *not* describe subspaces of $\mathcal{R}^2$ for $c \neq 0$:

$$\left\{\begin{bmatrix} v \\ c \end{bmatrix} : v \in \mathcal{R}\right\}, \quad \left\{\begin{bmatrix} v_1 \\ -v_1 + c \end{bmatrix} : v_1 \in \mathcal{R}\right\}, \quad \left\{\begin{bmatrix} v_1 \\ v_1^2 \end{bmatrix} : v_1 \in \mathcal{R}\right\}. \tag{3.90}$$

Although the first two cases describe lines, they do not pass through the origin: $[0, 0]^T$ is not possible for nonzero c. The third case is nonlinear and does not satisfy (3.88) as demonstrated by the following example:

$$\mathbf{v}_1 = \begin{bmatrix} 1 \\ 1 \end{bmatrix}, \mathbf{v}_2 = \begin{bmatrix} 2 \\ 4 \end{bmatrix} \implies \mathbf{v}_1 + \mathbf{v}_2 = \begin{bmatrix} 3 \\ 5 \end{bmatrix} \neq \begin{bmatrix} 3 \\ 9 \end{bmatrix}, \tag{3.91}$$

where the last vector is due to the form of the last set of vectors in (3.90).

For the system of linear equations $\mathbf{Ay} = \mathbf{x}$ with $\mathbf{A} \in \mathcal{R}^{M \times N}$, the right-hand side $\mathbf{x}$ is a linear combination of the *columns* of $\mathbf{A}$:

$$\mathbf{x} = \sum_{n=1}^{N} y_n \mathbf{a}_n, \tag{3.92}$$

where $\mathbf{a}_n$ denotes the nth column vector of $\mathbf{A}$ and y_n is the nth element of $\mathbf{y}$.

Definition: Column Space The *column space* $C(\mathbf{A})$ of matrix $\mathbf{A}$ consists of all vectors $\mathbf{x} \in \mathcal{R}^M$ derived as linear combinations of the columns of $\mathbf{A}$. The column space is also called the *range* of $\mathbf{A}$.

Equation (3.92) is an important property showing that $\mathbf{x}$ must be in $C(\mathbf{A})$ for a solution of $\mathbf{Ay} = \mathbf{x}$ to exist. If $\mathbf{x}$ cannot be expressed as some linear combination of the columns of $\mathbf{A}$, then there is no $\mathbf{y}$ satisfying $\mathbf{Ay} = \mathbf{x}$.

Example 3.18 For $\mathbf{A} \in \mathcal{R}^3$ (a vector), the system of linear equations is

$$\begin{bmatrix} a_{11} \\ a_{21} \\ a_{21} \end{bmatrix} y_1 = \begin{bmatrix} x_1 \\ x_2 \\ x_3 \end{bmatrix}, \tag{3.93}$$

which means the column space is a *line* in $\mathcal{R}^3$ given by the left-hand side with $y_1 \in \mathcal{R}$. If the matrix is expanded to two columns as $\mathbf{A} \in \mathcal{R}^{3 \times 2}$:

$$\begin{bmatrix} a_{11} & a_{12} \\ a_{21} & a_{22} \\ a_{21} & a_{23} \end{bmatrix} \begin{bmatrix} y_1 \\ y_2 \end{bmatrix} = \begin{bmatrix} x_1 \\ x_2 \\ x_3 \end{bmatrix}, \tag{3.94}$$

then $C(\mathbf{A})$ is a *plane* in $\mathcal{R}^3$ provided the two columns of $\mathbf{A}$ are linearly independent; otherwise, $C(\mathbf{A})$ is again a line in $\mathcal{R}^3$. Finally, for $\mathbf{A} \in \mathcal{R}^{3 \times 3}$:

$$\begin{bmatrix} a_{11} & a_{12} & a_{13} \\ a_{21} & a_{22} & a_{23} \\ a_{21} & a_{23} & a_{33} \end{bmatrix} \begin{bmatrix} y_1 \\ y_2 \\ y_3 \end{bmatrix} = \begin{bmatrix} x_1 \\ x_2 \\ x_3 \end{bmatrix}, \tag{3.95}$$

and $C(\mathbf{A}) \equiv \mathcal{R}^3$ provided all three columns of $\mathbf{A}$ are linearly independent. In this case, $\mathbf{x}$ can be any vector in $\mathcal{R}^3$ and a unique solution $\mathbf{y} \in \mathcal{R}^3$ exists because $\mathbf{A}$ has an inverse: $\mathbf{y} = \mathbf{A}^{-1}\mathbf{x}$. For the two *overdetermined* cases in (3.93) and (3.94), a solution exists provided $\mathbf{x} \in C(\mathbf{A})$, but the solution is not unique as demonstrated in the next example.

Example 3.19 For an overdetermined system with matrix ($M = 3, N = 2$)

$$\mathbf{A} = \begin{bmatrix} 1 & 0 \\ 1 & 1 \\ 0 & 1 \end{bmatrix}, \tag{3.96}$$

a vector in the column space has the form

$$\mathbf{x} = \begin{bmatrix} x_1 \\ x_2 \\ x_3 \end{bmatrix} = \mathbf{Ay} = \begin{bmatrix} y_1 \\ y_1 + y_2 \\ y_2 \end{bmatrix}. \tag{3.97}$$

Since $\mathbf{x}$ has three components, it lies in the three-dimensional subspace $\mathcal{R}^3$. However, as the components $\{x_1, x_2\}$ are varied over $\mathcal{R}^2$, there is a restriction on the other component: $x_3 = x_2 - x_1$. This follows from the right-hand side: since $x_1 = y_1$ and $x_2 = y_1 + y_2$, we must have

$$x_3 = y_2 \implies x_3 = x_2 - y_1 \implies x_3 = x_2 - x_1. \tag{3.98}$$

This equation for x_3 defines the column space of $\mathbf{A}$ in $\mathcal{R}^3$, which is the tilted *plane* shown in Figure 3.6 that passes through the origin $\mathbf{x} = \mathbf{0}$. The system $\mathbf{Ay} = \mathbf{x}$ has a solution $\mathbf{y}$ for any vector $\mathbf{x}$ located in this plane. Examples of vectors in $C(\mathbf{A})$ generated from the right-hand side of (3.97) are

$$\begin{bmatrix} 0 \\ 0 \\ 0 \end{bmatrix}, \quad \begin{bmatrix} 0 \\ y_2 \\ y_2 \end{bmatrix}, \quad \begin{bmatrix} y_1 \\ y_1 \\ 0 \end{bmatrix}, \quad \begin{bmatrix} 2 \\ 4 \\ 2 \end{bmatrix}, \tag{3.99}$$

where $\{y_1, y_2\}$ are any real numbers for the middle two examples. Of course, the zero vector $\mathbf{0}$ is always in the column space for any matrix $\mathbf{A}$. The solid line in Figure 3.6 is defined by $x_3 = x_2 - x_1 = 0$ for $x_1 = x_2$, and so, it lies in the column space. Examples of vectors that are *not* in the column space of $\mathbf{A}$ are

$$\begin{bmatrix} 1 \\ 1 \\ 1 \end{bmatrix}, \quad \begin{bmatrix} 0 \\ y_2 \\ -y_2 \end{bmatrix}, \quad \begin{bmatrix} y_1 \\ y_1 + 1 \\ 0 \end{bmatrix}, \quad \begin{bmatrix} 2 \\ 4 \\ 4 \end{bmatrix}, \tag{3.100}$$

assuming that $y_2 \neq 0$ in the second case. The dashed line in Figure 3.6 is the first vector in (3.100) scaled by 3, which we see does not lie in the column space represented by the plane.

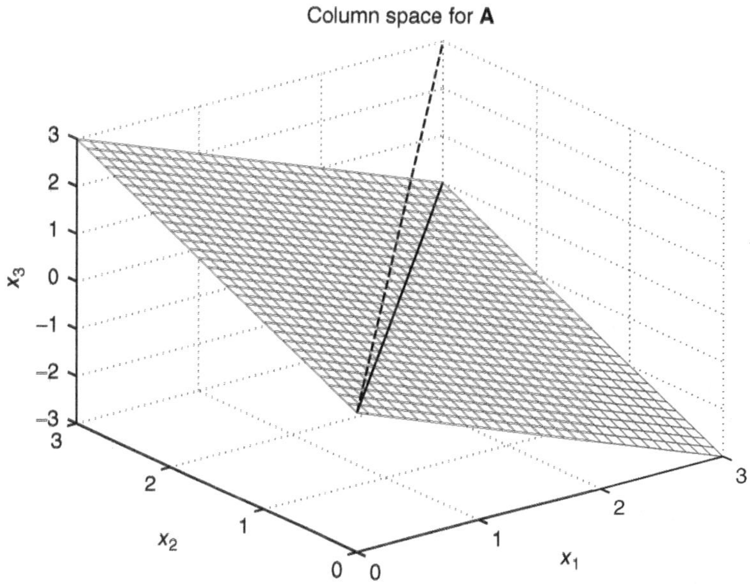

Figure 3.6 Column space of **A** in Example 3.19 is a plane in $\mathcal{R}^3$. The solid line on the plane for **x** lies in the column space, whereas the dashed line does not.

Example 3.20 Continuing with Example 3.18, suppose $\mathbf{A} \in \mathcal{R}^{3\times4}$ such that the system of linear equations is *underdetermined*:

$$\begin{bmatrix} a_{11} & a_{12} & a_{13} & a_{14} \\ a_{21} & a_{22} & a_{23} & a_{24} \\ a_{21} & a_{23} & a_{33} & a_{34} \end{bmatrix} \begin{bmatrix} y_1 \\ y_2 \\ y_3 \\ y_4 \end{bmatrix} = \begin{bmatrix} x_1 \\ x_2 \\ x_3 \end{bmatrix}, \tag{3.101}$$

which means that at most only three columns of **A** are linearly independent. The column space is a subspace in $\mathcal{R}^3$: either a line, a plane, or $\mathcal{R}^3$ depending on the *rank* of **A**: $R = 1, 2,$ or 3, respectively. Consider

$$\mathbf{A} = \begin{bmatrix} 1 & 0 & 0 & 1 \\ 0 & 1 & 0 & 0 \\ 0 & 0 & 1 & 0 \end{bmatrix} \begin{bmatrix} y_1 \\ y_2 \\ y_3 \\ y_4 \end{bmatrix} = \begin{bmatrix} y_1 + y_4 \\ y_2 \\ y_3 \end{bmatrix} = \begin{bmatrix} x_1 \\ x_2 \\ x_3 \end{bmatrix}, \tag{3.102}$$

where the fourth column of **A** is identical to the first column and the matrix has rank $R = 3$. This matrix is in the so-called *row-echelon form* described later in connection with GE. The right-hand side defines the column space, which is $C(\mathbf{A}) = \mathcal{R}^3$ because $y_1, y_2, y_3, y_4 \in \mathcal{R}$. Thus, every **x** is in the column space, and the solution for

$\mathbf{y}$ is not unique: $y_3 = x_3$, $y_2 = x_2$, and $y_1 + y_4 = x_1$, which we write as

$$\begin{bmatrix} y_1 \\ y_2 \\ y_3 \\ y_4 \end{bmatrix} = \begin{bmatrix} x_1 - y_4 \\ x_2 \\ x_3 \\ y_4 \end{bmatrix}. \tag{3.103}$$

The elements $\{y_2, y_3\}$ are fixed by $\{x_2, x_3\}$, whereas once the *free variable* y_4 (defined later) is chosen to be any number in $\mathcal{R}$, the first variable is fixed at $y_1 = x_1 - y_4$. As a result, there is an infinity of solutions for this particular $\mathbf{A}$. Underdetermined systems have an infinity of solutions provided $\mathbf{x}$ is in the column space $C(\mathbf{A})$.

Definition: Null Space The *null space* $N(\mathbf{A})$ of matrix $\mathbf{A}$ consists of all column vectors $\mathbf{y} \in \mathcal{R}^N$ that satisfy

$$\mathbf{Ay} = \mathbf{0}. \tag{3.104}$$

The null space is also called the *kernel*.

If $\mathbf{A}$ is square and nonsingular, then the null space is trivial: it contains only $\mathbf{0}$, and $\mathbf{y} = \mathbf{0}$ is the unique solution satisfying (3.104). This result follows because $\mathbf{A}^{-1}$ exists and we can write

$$\mathbf{Ay} = \mathbf{0} \implies \mathbf{y} = \mathbf{A}^{-1}\mathbf{0} = \mathbf{0}. \tag{3.105}$$

Example 3.21 Consider the following matrix for an underdetermined system of equations:

$$\mathbf{A} = \begin{bmatrix} 1 & 1 & 0 \\ 0 & 1 & 1 \end{bmatrix}, \tag{3.106}$$

and observe that

$$\mathbf{Ay} = \begin{bmatrix} y_1 + y_2 \\ y_2 + y_3 \end{bmatrix}. \tag{3.107}$$

Setting this expression equal to $\mathbf{0}$, we find there are three variables $\{y_1, y_2, y_3\}$ but only two equations: $y_1 + y_2 = 0$ and $y_2 + y_3 = 0$, which means that $\mathbf{Ay} = \mathbf{0}$ has an infinity of solutions. Since $\mathbf{y}$ has three components, the null space of $\mathbf{A}$ is a subspace of $\mathcal{R}^3$ defined by

$$y_2 = -y_1, \quad y_3 = -y_2 \implies y_3 = y_1. \tag{3.108}$$

As a result, the null space is given by all vectors of the form

$$\mathbf{y} = \begin{bmatrix} 1 \\ -1 \\ 1 \end{bmatrix} y_1, \tag{3.109}$$

for any $y_1 \in \mathcal{R}$. This line in $\mathcal{R}^3$ is depicted in Figure 3.7.

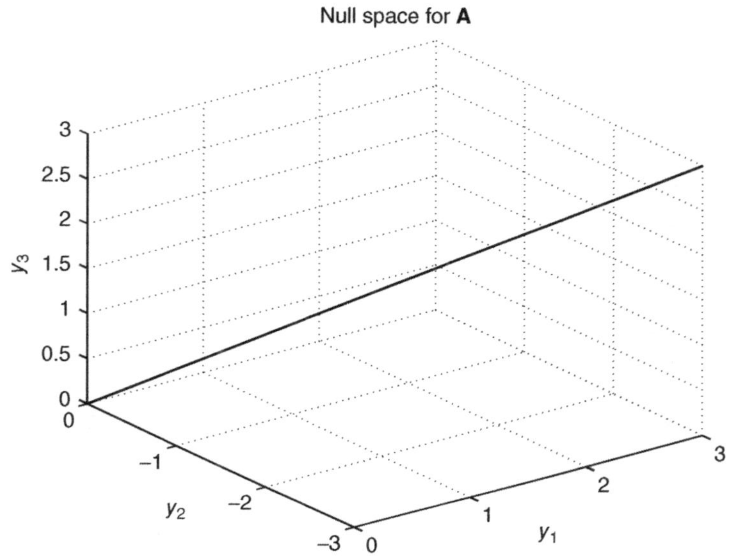

Figure 3.7 Null space for **A** in Example 3.21 which is a line in $\mathcal{R}^3$.

The zero vector $\mathbf{y} = \mathbf{0}$ is in both the null space and the column space of **A**, and those two subspaces intersect at **0** as illustrated in the next example.

Example 3.22 The columns of the following matrix are not linearly independent:

$$\begin{bmatrix} 1 & 3 & 1 \\ 0 & 2 & 1 \\ 2 & 6 & 2 \end{bmatrix} \mathbf{y} = \begin{bmatrix} x_1 \\ x_2 \\ x_3 \end{bmatrix}, \tag{3.110}$$

because the third row is twice the first row:

$$\begin{bmatrix} y_1 + 3y_2 + y_3 \\ 2y_2 + y_3 \\ 2y_1 + 6y_2 + 2y_3 \end{bmatrix} = \begin{bmatrix} x_1 \\ x_2 \\ x_3 \end{bmatrix}. \tag{3.111}$$

In order for a solution to exist, we must have $x_3 = 2x_1$, in which case the column space for this matrix is $\mathbf{x} = [x_1, x_2, 2x_1]^T$ with $x_1, x_2 \in \mathcal{R}$. The null space is derived by setting (3.111) equal to **0** and finding expressions for $\{y_1, y_2, y_3\}$. Since the third row is a scaled version of the first row, it can be ignored. Thus, we have

$$y_1 + 3y_2 + y_3 = 0, \qquad 2y_2 + y_3 = 0, \tag{3.112}$$

which means that the null space is defined by $\mathbf{y} = [y_3/2, -y_3/2, y_3]^T$ for $y_3 \in \mathcal{R}$. The column space for this matrix is a plane and the null space is a vector, both in

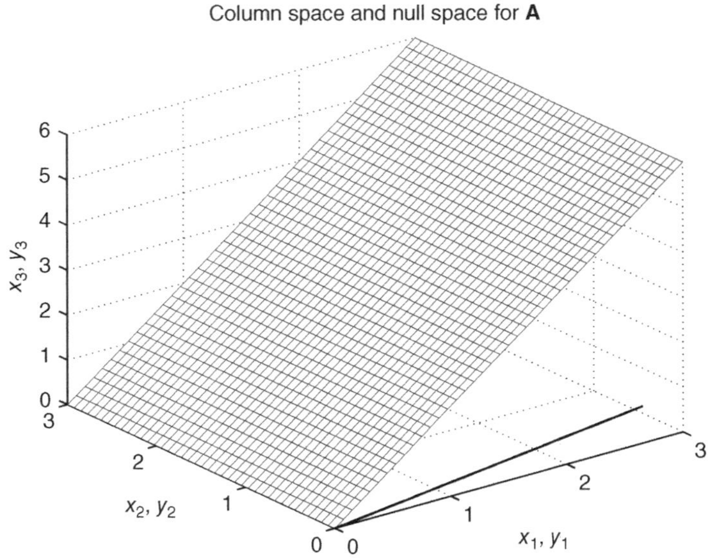

Figure 3.8 Column space and null space for **A** in Example 3.22, which are a plane and a line, respectively, in $\mathcal{R}^3$.

$\mathcal{R}^3$ because **A** has three columns. These subspaces are depicted in Figure 3.8. This property is true in general: the sum of the dimensions of the column space and the null space equals the number of *columns* of $\mathbf{A} \in \mathcal{R}^{M \times N}$.

Matrix **A** has two additional subspaces.

Definition: Row Space The *row space* $R(\mathbf{A})$ of matrix **A** consists of all row vectors $\mathbf{x}^T \in \mathcal{R}^N$ derived as linear combinations of the rows of **A**. This is equivalent to $R(\mathbf{A}) = C(\mathbf{A}^T)$.

Definition: Left Null Space The *left null space* $L(\mathbf{A})$ of matrix **A** consists of all row vectors $\mathbf{y}^T \in \mathcal{R}^M$ that satisfy $\mathbf{y}^T\mathbf{A} = \mathbf{0}^T$. This is equivalent to $L(\mathbf{A}) = N(\mathbf{A}^T)$.

The row space and left null space of a matrix involve linear combinations of the *rows* of **A**. These four subspaces are discussed again after GE is covered in the next section. It turns out that the column space and the left null space are orthogonal complements in $\mathcal{R}^M$, and the row space and null space are orthogonal complements in $\mathcal{R}^N$.

Definition: Orthogonal Complement The *orthogonal complement* of subspace $\mathcal{S} \in \mathcal{R}^N$, denoted by $\mathcal{S}^\perp$, consists of all vectors in $\mathcal{R}^N$ that are orthogonal to the vectors in $\mathcal{S}$.

It is clear from the previous definition that subspace $\mathcal{S}$ and $\mathcal{S}^{\perp}$ together comprise every vector in $\mathcal{R}^N$: $\mathcal{S} \cup \mathcal{S}^{\perp} = \mathcal{R}^N$. (A brief review of set theory and set operations is provided in Appendix C.)

Definition: Basis and Span A *basis* for a vector space in $\mathcal{R}^N$ is a set of vectors that are linearly independent and *span* the vector space.

To span means that it is possible to write any other vector in the vector space as a linear combination of the basis vectors. Methods of finding a basis for each of the four subspaces are described below.

The four matrix subspaces and their dimensions are summarized in Table 3.6. Note that the union $R(\mathbf{A}) \cup N(\mathbf{A}) = \mathcal{R}^N$ comprises all N-dimensional vectors, and similarly, $C(\mathbf{A}) \cup L(\mathbf{A}) = \mathcal{R}^M$ comprises all M-dimensional vectors.

3.7 GAUSSIAN ELIMINATION

GE is a technique for solving a system of linear equations that does not explicitly produce the inverse of matrix $\mathbf{A}$, only $\mathbf{y}$ of the system $\mathbf{Ay} = \mathbf{x}$. This approach is advantageous because finding $\mathbf{y} = \mathbf{A}^{-1}\mathbf{x}$ requires more computations and can be numerically unstable. Let $M = N = 3$ such that the system of linear equations $\mathbf{Ay} = \mathbf{x}$ has the form

$$\begin{bmatrix} a_{11} & a_{12} & a_{13} \\ a_{21} & a_{22} & a_{23} \\ a_{31} & a_{32} & a_{33} \end{bmatrix} \begin{bmatrix} y_1 \\ y_2 \\ y_3 \end{bmatrix} = \begin{bmatrix} x_1 \\ x_2 \\ x_3 \end{bmatrix}. \tag{3.113}$$

GE performs row operations simultaneously on $\mathbf{A}$ and the right-hand side $\mathbf{x}$ to obtain an *upper triangular matrix*. This is done by subtracting scaled versions of the first row from the second and third rows where $\overline{\mathbf{a}}_m^T$ defines the mth row:

$$\overline{\mathbf{a}}_2^T - (a_{21}/a_{11})\overline{\mathbf{a}}_1^T \longrightarrow \text{new second row}, \tag{3.114}$$

$$\overline{\mathbf{a}}_3^T - (a_{31}/a_{11})\overline{\mathbf{a}}_1^T \longrightarrow \text{new third row}. \tag{3.115}$$

TABLE 3.6 Summary of Four Subspaces for Matrix $\mathbf{A} \in \mathcal{R}^{M \times N}$ with Rank R

Subspace	Number of Basis Vectors	Size of Basis Vectors
Column space $C(\mathbf{A}) = L^{\perp}(\mathbf{A})$	R	M
Row space $R(\mathbf{A}) = C(\mathbf{A}^T)$	R	N
Null space $N(\mathbf{A}) = R^{\perp}(\mathbf{A})$	$N - R$	N
Left null space $L(\mathbf{A}) = N(\mathbf{A}^T)$	$M - R$	M

The system of equations with the modified second row is

$$\begin{bmatrix} a_{11} & a_{12} & a_{13} \\ 0 & a_{22}-(a_{21}/a_{11})a_{12} & a_{23}-(a_{21}/a_{11})a_{13} \\ a_{31} & a_{32} & a_{33} \end{bmatrix} \begin{bmatrix} y_1 \\ y_2 \\ y_3 \end{bmatrix} = \begin{bmatrix} x_1 \\ x_2-(a_{21}/a_{11})x_1 \\ x_3 \end{bmatrix}. \tag{3.116}$$

Observe the following results:

- This operation produces a 0 for the first element of the second row. If a_{21} happens to be 0, then it is not necessary to perform this row operation. If a_{11} happens to be 0, then the original first two rows should be interchanged. The order of rows in a system of linear equations is not important, as long as the corresponding elements of **x** are also interchanged. Of course, the elements of **y** are *not* affected by row exchanges because they are weighted by the columns of **A**.
- Element a_{11} in this row operation is called a *pivot*.

For the third row using again pivot a_{11}, the modified system of equations is

$$\begin{bmatrix} a_{11} & a_{12} & a_{13} \\ 0 & a_{22}-(a_{21}/a_{11})a_{12} & a_{23}-(a_{21}/a_{11})a_{13} \\ 0 & a_{32}-(a_{31}/a_{11})a_{12} & a_{33}-(a_{31}/a_{11})a_{13} \end{bmatrix} \begin{bmatrix} y_1 \\ y_2 \\ y_3 \end{bmatrix} = \begin{bmatrix} x_1 \\ x_2-(a_{21}/a_{11})x_1 \\ x_3-(a_{31}/a_{11})x_1 \end{bmatrix}. \tag{3.117}$$

It is straightforward to see how this technique is extended to a matrix with any number of rows. For notational convenience, we use tildes on the elements of the modified system of equations:

$$\begin{bmatrix} a_{11} & a_{12} & a_{13} \\ 0 & \tilde{a}_{22} & \tilde{a}_{23} \\ 0 & \tilde{a}_{32} & \tilde{a}_{33} \end{bmatrix} \begin{bmatrix} y_1 \\ y_2 \\ y_3 \end{bmatrix} = \begin{bmatrix} x_1 \\ \tilde{x}_2 \\ \tilde{x}_3 \end{bmatrix}. \tag{3.118}$$

This matrix equation has the same solution as the original system in (3.113). Next, we perform a row operation on the third row, by subtracting a scaled version of the modified second row. The pivot in this case is $\tilde{a}_{22} \triangleq a_{22}-(a_{21}/a_{11})a_{12}$ (if $\tilde{a}_{22}$ happens to be 0, then the second and third rows should be interchanged). This yields an upper triangular matrix on the left-hand side:

$$\begin{bmatrix} a_{11} & a_{12} & a_{13} \\ 0 & \tilde{a}_{22} & \tilde{a}_{23} \\ 0 & 0 & \tilde{a}_{33}-(\tilde{a}_{32}/\tilde{a}_{22})\tilde{a}_{23} \end{bmatrix} \begin{bmatrix} y_1 \\ y_2 \\ y_3 \end{bmatrix} = \begin{bmatrix} x_1 \\ \tilde{x}_2 \\ \tilde{x}_3-(\tilde{a}_{32}/\tilde{a}_{22})\tilde{x}_2 \end{bmatrix}, \tag{3.119}$$

whose modified elements in the third row can be rewritten using additional tildes:

$$\begin{bmatrix} a_{11} & a_{12} & a_{13} \\ 0 & \tilde{a}_{22} & \tilde{a}_{23} \\ 0 & 0 & \tilde{\tilde{a}}_{33} \end{bmatrix} \begin{bmatrix} y_1 \\ y_2 \\ y_3 \end{bmatrix} = \begin{bmatrix} x_1 \\ \tilde{x}_2 \\ \tilde{\tilde{x}}_3 \end{bmatrix}. \tag{3.120}$$

From this expression where the matrix is in *row-echelon form*, we can easily solve for the variables using a sequence of operations called *back-substitution* starting with the last row:

$$y_3 = \tilde{\tilde{x}}_3/\tilde{\tilde{a}}_{33}, \tag{3.121}$$

which is used to find y_2:

$$y_2 = (\tilde{x}_2 - \tilde{a}_{23}y_3)/\tilde{a}_{22}, \tag{3.122}$$

and then y_1:

$$y_1 = (x_1 - a_{13}y_3 - a_{12}y_2)/a_{11}. \tag{3.123}$$

These are the elements of $\mathbf{y}$ that satisfy the system of equations $\mathbf{Ay} = \mathbf{x}$. Equations (3.113) and (3.120) are examples of *equivalent* systems of linear equations because they have the same solution. Due to the mechanism used to derive the row-echelon form, the rows of one system can be derived as linear combinations of the rows of the other system, and the solution is unchanged.

Definition: Row-Echelon Form A matrix is in *row-echelon form* if the first nonzero element of every row is located to the right of the first nonzero element of the previous row.

This definition implies that the elements of a matrix in row-echelon form below the leading nonzero element of a row are all zero. As a result, any all-zero rows are located below rows with nonzero elements. The nonzero elements of this matrix have an *upper triangular form* (sometimes called an upper *trapezoidal form*).

Since the vector of unknown variables $\mathbf{y}$ is not affected by GE operations, for notational convenience, we can group $\mathbf{A}$ and $\mathbf{x}$ together in an *augmented* matrix:

$$\left[\begin{array}{ccc|c} a_{11} & a_{12} & a_{13} & x_1 \\ a_{21} & a_{22} & a_{23} & x_2 \\ a_{31} & a_{32} & a_{33} & x_3 \end{array}\right]. \tag{3.124}$$

Suppose that GE yields the following row-echelon form for the augmented matrix:

$$\left[\begin{array}{ccc|c} a_{11} & a_{12} & a_{13} & x_1 \\ 0 & \tilde{a}_{22} & \tilde{a}_{23} & \tilde{x}_2 \\ 0 & 0 & 0 & 0 \end{array}\right]. \tag{3.125}$$

This means that the three rows of $\mathbf{A}$ are not linearly independent: the matrix is singular and the solution is not unique. Any real number can be chosen for y_3, called a *free variable*, which is used in back-substitution to solve for $\{y_1, y_2\}$. Such a variable whose row has a zero pivot is "free" because it can take on any value in $\mathcal{R}$. Those variables with nonzero pivots are called *basic variables*. If another value is chosen for y_3, then $\{y_1, y_2\}$ necessarily change, but the solution is still valid for this system.

Definition: Basic and Free Variables For a matrix in row-echelon form, those variables of $\mathbf{y}$ with nonzero pivots are *basic variables*. All other variables of $\mathbf{y}$ are *free variables*.

The number of nonzero pivots equals the rank R of matrix $\mathbf{A}$. If GE yields:

$$\left[\begin{array}{ccc|c} a_{11} & a_{12} & a_{13} & x_1 \\ 0 & 0 & 0 & 0 \\ 0 & 0 & 0 & 0 \end{array}\right], \tag{3.126}$$

then any values can be chosen in $\mathcal{R}^2$ for the free variables $\{y_2, y_3\}$, and these are used to solve for y_1. Thus, for square matrix $\mathbf{A}$ with $M = N$, there will be a unique solution if the final upper triangular matrix has all nonzero pivots, or there can be an infinity of solutions because of at least one zero pivot. If GE yields the following form:

$$\left[\begin{array}{ccc|c} a_{11} & a_{12} & a_{13} & x_1 \\ 0 & \tilde{a}_{22} & \tilde{a}_{23} & \tilde{x}_2 \\ 0 & 0 & 0 & \tilde{x}_3 \end{array}\right], \tag{3.127}$$

then again the matrix is singular, but there are no solutions for *nonzero* $\tilde{x}_3$. The system of equations is *inconsistent*, which means the planes in $\mathcal{R}^3$ defined by the three original equations do not intersect at any common points.

Example 3.23 An augmented matrix for the underdetermined system $\mathbf{Ay} = \mathbf{x}$ transformed to row-echelon form has the following symbolic representation:

$$\left[\begin{array}{cccccc|c} p_1 & \cdot & \cdot & \cdot & \cdot & \cdot & \cdot \\ 0 & p_2 & \cdot & \cdot & \cdot & \cdot & \cdot \\ 0 & 0 & 0 & p_4 & \cdot & \cdot & \cdot \\ 0 & 0 & 0 & 0 & 0 & p_6 & \cdot \\ 0 & 0 & 0 & 0 & 0 & 0 & \cdot \end{array}\right], \tag{3.128}$$

where the leading $\{p_m\}$ are nonzero pivots. Clearly, there can be at most one nonzero pivot in each row and column. The other entries denoted by $\cdot$ may or may not be 0. The column located to the right of the vertical line is the transformed right-hand side $\mathbf{x}$ of the original system. For this example, we find that $\{y_1, y_2, y_4, y_6\}$ are basic variables and $\{y_3, y_5\}$ are free variables. Assuming the system is consistent, which means that $\cdot$ in the last row is necessarily 0, the solution for $\mathbf{y}$ is not unique because of the two free variables. *Row-reduced echelon form* is derived from row-echelon form by dividing each row by its nonzero pivot such that the leading nonzero element is 1. For the augmented matrix in (3.128), the row-reduced echelon form is

$$\left[\begin{array}{cccccc|c} 1 & \cdot & \cdot & \cdot & \cdot & \cdot & \cdot \\ 0 & 1 & \cdot & \cdot & \cdot & \cdot & \cdot \\ 0 & 0 & 0 & 1 & \cdot & \cdot & \cdot \\ 0 & 0 & 0 & 0 & 0 & 1 & \cdot \\ 0 & 0 & 0 & 0 & 0 & 0 & 0 \end{array}\right], \tag{3.129}$$

where each row with a nonzero pivot has been divided by p_m in that row. Of course, the $\cdot$ entries in (3.128) are modified by these divisions (except for $\cdot$ in the last row, which we have replaced with zero to have a consistent system). In order to solve for the unknown vector $\mathbf{y}$, it is not usually necessary to convert to row-reduced echelon form because back-substitution is straightforward even for a nonzero pivot $p_m \neq 1$, as demonstrated in the next example.

Example 3.24 Consider again the system in Example 3.14, which is written as follows using an augmented matrix:

$$\left[\begin{array}{ccc|c} 1 & 2 & 4 & 3 \\ 3 & 4 & 1 & 1 \\ 2 & 3 & 1 & 2 \end{array}\right]. \tag{3.130}$$

GE results in the following sequence of augmented matrices:

$$\left[\begin{array}{ccc|c} 1 & 2 & 4 & 3 \\ 0 & -2 & -11 & -8 \\ 0 & -1 & -7 & -4 \end{array}\right] \rightarrow \left[\begin{array}{ccc|c} 1 & 2 & 4 & 3 \\ 0 & -2 & -11 & -8 \\ 0 & 0 & -3/2 & 0 \end{array}\right], \tag{3.131}$$

and back-substitution yields

$$y_3 = 0, \qquad y_2 = 4, \qquad y_1 = -5. \tag{3.132}$$

The 0 in the transformed $\mathbf{x}$ column in the last augmented matrix does not lead to an infinity of solutions, as is the case when an entire row of the augmented matrix is $\mathbf{0}$. Suppose that the last row of $\mathbf{A}$ in (3.130) is replaced as follows:

$$\left[\begin{array}{ccc|c} 1 & 2 & 4 & 3 \\ 3 & 4 & 1 & 1 \\ 6 & 8 & 2 & 2 \end{array}\right]. \tag{3.133}$$

Then (3.131) becomes

$$\left[\begin{array}{ccc|c} 1 & 2 & 4 & 3 \\ 0 & -2 & -11 & -8 \\ 0 & -4 & -22 & -16 \end{array}\right] \rightarrow \left[\begin{array}{ccc|c} 1 & 2 & 4 & 3 \\ 0 & -2 & -11 & -8 \\ 0 & 0 & 0 & 0 \end{array}\right], \tag{3.134}$$

and back-substitution yields

$$y_2 = 4 - (11/2)y_3, \qquad y_1 = -5 + (3/2)y_3, \tag{3.135}$$

where the free variable y_3 is any value in $\mathcal{R}$. This infinity of solutions occurs because the third row of (3.133) is an integer multiple of the second row, and so, they are dependent and the original matrix is singular.

Next, we examine the row-echelon form for rectangular matrices and the possible solutions for underdetermined and overdetermined systems of equations. For an overdetermined system with $M = 3$ and $N = 2$, the possible row-echelon forms are

$$\left[\begin{array}{cc|c} a_{11} & a_{12} & x_1 \\ 0 & \tilde{a}_{22} & \tilde{x}_2 \\ 0 & 0 & 0 \end{array}\right], \quad \left[\begin{array}{cc|c} a_{11} & a_{12} & x_1 \\ 0 & \tilde{a}_{22} & \tilde{x}_2 \\ 0 & 0 & \tilde{\tilde{x}}_3 \end{array}\right]. \tag{3.136}$$

The system on the left-hand side has a solution because $N = 2$ and there can be at most two linearly independent rows, which is the reason the third row is 0 after the row operations. Since there are no free variables, the unique solution is $y_2 = \tilde{x}_2/\tilde{a}_{21}$ and $y_1 = (x_1 - a_{12}y_2)/a_{11}$. The system on the right-hand side has no solution if $\tilde{\tilde{x}}_3$ is nonzero.

Example 3.25 For the system in (3.80), suppose the third column is dropped such that the row-echelon form for the remaining system is

$$\begin{bmatrix} 1 & 2 \\ 3 & 4 \\ 2 & 3 \end{bmatrix} \begin{bmatrix} y_1 \\ y_2 \end{bmatrix} = \begin{bmatrix} 3 \\ 1 \\ 2 \end{bmatrix} \rightarrow \left[\begin{array}{cc|c} 1 & 2 & 3 \\ 0 & -2 & -8 \\ 0 & 0 & 0 \end{array}\right]. \tag{3.137}$$

This overdetermined system has unique solution $y_2 = 4$ and $y_1 = -5$ because the columns are linearly independent, as illustrated by the three intersecting lines in Figure 3.9. The system of equations represented by the three lines in Figure 3.3 is

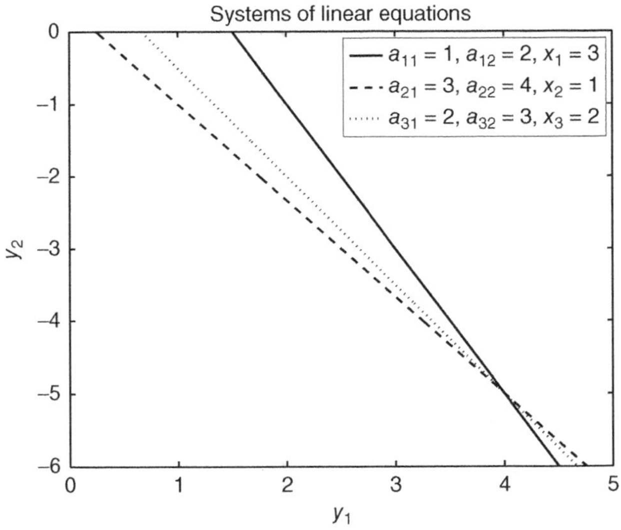

Figure 3.9 Overdetermined system of linear equations in Example 3.25 with a unique solution.

an example of an overdetermined system with no solution. The row-echelon form for that case is

$$\begin{bmatrix} 1 & 1 \\ 1 & -1 \\ 1 & 1 \end{bmatrix} \begin{bmatrix} y_1 \\ y_2 \end{bmatrix} = \begin{bmatrix} 2 \\ 1/2 \\ 1 \end{bmatrix} \rightarrow \left[\begin{array}{cc|c} 1 & 1 & 2 \\ 0 & -2 & -3/2 \\ 0 & 0 & -1 \end{array}\right], \tag{3.138}$$

which has the form of the right-hand side of (3.136).

Example 3.26 The following overdetermined system has been written in row-echelon form:

$$\left[\begin{array}{cc|c} 1 & 0 & x_1 \\ 1 & 1 & x_2 \\ 0 & 1 & x_3 \end{array}\right] \rightarrow \left[\begin{array}{cc|c} 1 & 0 & x_1 \\ 0 & 1 & x_2 - x_1 \\ 0 & 1 & 3 \end{array}\right] \rightarrow \left[\begin{array}{cc|c} 1 & 0 & x_1 \\ 0 & 1 & x_2 - x_1 \\ 0 & 0 & x_3 - x_2 + x_1 \end{array}\right]. \tag{3.139}$$

In order for a solution to exist, the last element on the right-hand side must satisfy $x_3 = x_2 - x_1$, which is the same condition on x_3 given earlier in (3.97) and was derived by examining the column space. There are no restrictions on $\{x_1, x_2\}$, and the solution is $y_1 = x_1$ and $y_2 = x_2 - x_1$.

Example 3.27 Assume that $\mathbf{x} = \mathbf{0}$ for the following system in row-echelon form:

$$\left[\begin{array}{ccc|c} 1 & 3 & 1 & x_1 \\ 0 & 2 & 1 & x_2 \\ 0 & 0 & 0 & x_3 - 2x_1 \end{array}\right] = \left[\begin{array}{ccc|c} 1 & 3 & 1 & 0 \\ 0 & 2 & 1 & 0 \\ 0 & 0 & 0 & 0 \end{array}\right], \tag{3.140}$$

such that all solutions for $\mathbf{y}$ are vectors in the null space. These are easily found by back-substitution: the free variable $y_3 \in \mathcal{R}$ can be any value, $y_2 = -y_3/2$, $y_1 = -3y_2 - y_3 = y_3/2$, and so, the null space consists of all vectors of the form: $\mathbf{y} = y_3[1/2, -1/2, 1]^T$. This corresponds to a *line* in $\mathcal{R}^3$ that passes through the origin.

The possible row-echelon forms for an underdetermined system with $M = 2$ and $N = 3$ are

$$\left[\begin{array}{ccc|c} a_{11} & a_{12} & a_{13} & x_1 \\ 0 & \tilde{a}_{22} & \tilde{a}_{23} & \tilde{x}_2 \end{array}\right], \quad \left[\begin{array}{ccc|c} a_{11} & a_{12} & a_{13} & x_1 \\ 0 & 0 & \tilde{a}_{23} & \tilde{x}_2 \end{array}\right], \tag{3.141}$$

$$\left[\begin{array}{ccc|c} a_{11} & a_{12} & a_{13} & x_1 \\ 0 & 0 & 0 & 0 \end{array}\right], \quad \left[\begin{array}{ccc|c} a_{11} & a_{12} & a_{13} & x_1 \\ 0 & 0 & 0 & \tilde{x}_2 \end{array}\right]. \tag{3.142}$$

There is an infinity of solutions for the first case in (3.141): any $y_3 \in \mathcal{R}$ is a solution, and the other variables are $y_2 = (\tilde{x}_2 - \tilde{a}_{23}y_3)/\tilde{a}_{22}$ and $y_1 = (x_1 - a_{13}y_3 - a_{12}y_2)/a_{11}$. For the second case in (3.141), $y_3 = \tilde{x}_2/\tilde{a}_{23}$ is fixed, any value for the free variable $y_2 \in \mathcal{R}$ is a solution, and then $y_1 = (x_1 - a_{13}y_3 - a_{12}y_2)/a_{11}$. The two equations for the first case in (3.142) are dependent (collinear), and so, there is an infinity of solutions: y_2 and y_3 are free variables, which yield $y_1 = (x_1 - a_{13}y_3 - a_{12}y_2)/a_{11}$. The second case in (3.142) has no solutions for nonzero $\tilde{x}_2$.

Example 3.28 The following system of equations and its row-echelon form are underdetermined:

$$\begin{bmatrix} 1 & 2 & -2 \\ 2 & -1 & 1 \end{bmatrix} \begin{bmatrix} y_1 \\ y_2 \\ y_3 \end{bmatrix} = \begin{bmatrix} 4 \\ 3 \end{bmatrix} \rightarrow \left[\begin{array}{ccc|c} 1 & 2 & -2 & 4 \\ 0 & -5 & 5 & -5 \end{array}\right], \quad (3.143)$$

which is similar to the first case in (3.141), such that it has an infinity of solutions because y_3 is a free variable. This result can be visualized by rewriting the original two equations as follows:

$$y_1 = 4 - 2y_2 + 2y_3, \qquad y_1 = (3 + 2y_2 - y_3)/2, \quad (3.144)$$

and then varying $\{y_2, y_3\}$ to generate two *planes* in three dimensions as illustrated in Figure 3.10(a). The solutions correspond to the *line* where the two planes intersect. From the row-echelon form in (3.143):

$$-5y_2 + 5y_3 = -5 \implies y_2 = 1 + y_3, \quad (3.145)$$

$$y_1 + 2y_2 - 2y_3 = 4 \implies y_1 = 4 + 2y_3 - 2y_2 = 2, \quad (3.146)$$

where y_2 of (3.145) has been substituted into (3.146) so that y_3 cancels to give $y_1 = 2$. Thus, the solution is

$$\mathbf{y} = \begin{bmatrix} 2 \\ 1 + y_3 \\ y_3 \end{bmatrix}, \quad (3.147)$$

for any $y_3 \in \mathcal{R}$. This result is consistent with the line where the two planes intersect in Figure 3.10(a): y_1 is a constant 2 and $y_2 = y_3 + 1$ increases with increasing y_3. If the second equation in (3.143) is modified so that

$$\begin{bmatrix} 1 & 2 & -2 \\ 1 & 2 & -2 \end{bmatrix} \begin{bmatrix} y_1 \\ y_2 \\ y_3 \end{bmatrix} = \begin{bmatrix} 4 \\ -4 \end{bmatrix} \rightarrow \left[\begin{array}{ccc|c} 1 & 2 & -2 & 4 \\ 0 & 0 & 0 & -8 \end{array}\right], \quad (3.148)$$

then there is no solution. The plot in Figure 3.10(b) shows that the two planes derived from (3.148) and the following equations do not intersect:

$$y_1 = 4 - 2y_2 + 2y_3, \qquad y_1 = -4 - 2y_2 + 2y_3. \quad (3.149)$$

From the previous discussions, we find that there is a connection between over- and underdetermined systems and the column space and null space of $\mathbf{A}$. If square matrix $\mathbf{A}$ is nonsingular, then there is a unique solution: $\mathbf{x}$ lies in the column space of $\mathbf{A}$ and the null space is trivial, containing only $\mathbf{0}$. For underdetermined systems, the matrix has fewer rows than columns, and it is not possible to have a unique solution. Either there is an infinity of solutions or there are no solutions. These two

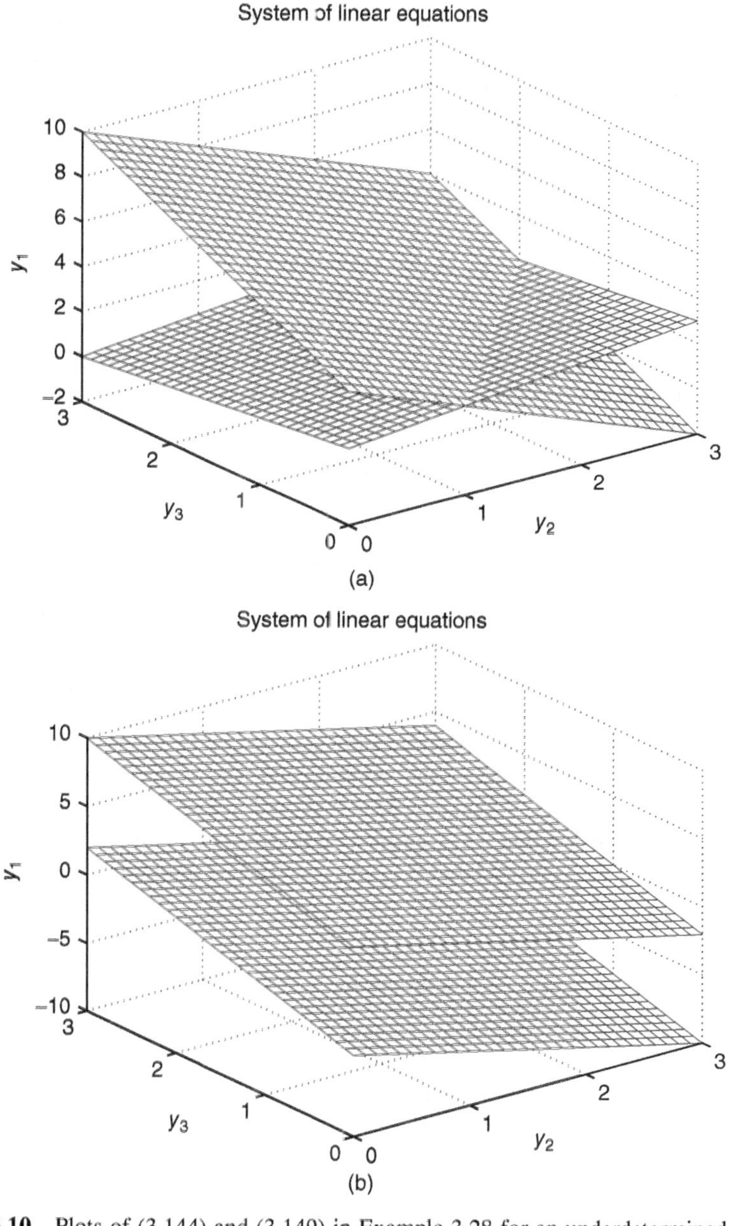

Figure 3.10 Plots of (3.144) and (3.149) in Example 3.28 for an underdetermined system. (a) An infinity of solutions. (b) No solutions.

cases occur because there are always free variables in the row-echelon form and the system of equations may be inconsistent, respectively. In the first case, the null space is nontrivial, and any vector in the null space that is added to a solution is also a valid solution. This is seen as follows, assuming that $\mathbf{y}_1$ is a solution and $\mathbf{y}_2$ is a vector in the null space of $\mathbf{A}$:

$$\mathbf{A}(\mathbf{y}_1 + \mathbf{y}_2) = \mathbf{x} \implies \mathbf{A}\mathbf{y}_1 + \mathbf{A}\mathbf{y}_2 = \mathbf{x} \implies \mathbf{A}\mathbf{y}_1 + \mathbf{0} = \mathbf{x}, \tag{3.150}$$

which demonstrates that $\mathbf{y}_1 + \mathbf{y}_2$ is also a solution.

For an overdetermined system, the matrix has fewer columns than rows, and as a result, three types of solutions are possible: (i) unique, (ii) infinite, or (iii) none. In the first case, there are no free variables, so the solution is unique provided $M - R$ rows of the modified $\mathbf{x}$ in row-echelon form are 0. Otherwise, we have the third case: no solutions. For the second case, there are $N - R$ free variables, and there is an infinity of solutions provided that $M - R$ rows of the modified $\mathbf{x}$ are 0. Otherwise, we again have no solutions. When there is an infinity of solutions, the null space is nontrivial and the property in (3.150) holds. For the unique overdetermined solution, the null space is trivial, containing only $\mathbf{0}$.

Table 3.7 summarizes the types of solutions for $\mathbf{Ay} = \mathbf{x}$ when $\mathbf{A}$ is square and rectangular. The size of the column space $C(\mathbf{A})$ is determined by the number of basic variables and the rank R, and the size of the null space is the number of free variables, which is $N - R$. The constraints on $\mathbf{x}$ mentioned in the table require that the elements of the transformed $\mathbf{x}$ in row-echelon form be 0 for every zero row of the upper triangular matrix. If the constraints are satisfied, at least one solution exists.

These results are also summarized symbolically next using augmented matrices with dimensions $M = N = 3$ (square), $M = 2, N = 3$ (underdetermined), and $M = 3$, $N = 2$ (overdetermined). The $\{p_m\}$ entries are nonzero pivots, and the $\cdot$ entries may or may not be 0. Those entries labeled c_n must be 0, corresponding to a *constraint* on

TABLE 3.7 Summary of Ay = x Solutions for $\mathbf{A} \in \mathcal{R}^{M \times N}$ with Rank R

Matrix Dimensions	Rank	Number of Basic Variables	Number of Free Variables	Number of Solutions
Square:				
$M = N$	$R = M$	R	0	One: unique
	$R < M$	R	$N - R$	Infinite with $M - R$ constraints on $\mathbf{x}$
Underdetermined:				
$M < N$	$R = M$	R	$N - R$	Infinite
	$R < M$	R	$N - R$	Infinite with $M - R$ constraints on $\mathbf{x}$
Overdetermined:				
$M > N$	$R = N$	R	0	One: unique with $M - R$ constraints on $\mathbf{x}$
	$R < N$	R	$N - R$	Infinite with $M - R$ constraints on $\mathbf{x}$

x needed for the system of equations to be consistent. If any c_n is nonzero, then that system of equations has no solutions.

- *Square*:

$$\text{(i)} \begin{bmatrix} p_1 & \cdot & \cdot & \cdot \\ 0 & p_2 & \cdot & \cdot \\ 0 & 0 & p_3 & \cdot \end{bmatrix}, \quad \text{(ii)} \begin{bmatrix} p_1 & \cdot & \cdot & \cdot \\ 0 & p_2 & \cdot & \cdot \\ 0 & 0 & 0 & c_1 \end{bmatrix}. \tag{3.151}$$

$$\text{(iii)} \begin{bmatrix} p_1 & \cdot & \cdot & \cdot \\ 0 & 0 & p_3 & \cdot \\ 0 & 0 & 0 & c_1 \end{bmatrix}, \quad \text{(iv)} \begin{bmatrix} p_1 & \cdot & \cdot & \cdot \\ 0 & 0 & 0 & c_1 \\ 0 & 0 & 0 & c_2 \end{bmatrix}. \tag{3.152}$$

Number of solutions: (i) one: unique, (ii) infinite if $c_1 = 0$, (iii) infinite if $c_1 = 0$, and (iv) infinite if $c_1 = c_2 = 0$.

- *Underdetermined*:

$$\text{(i)} \begin{bmatrix} p_1 & \cdot & \cdot & \cdot \\ 0 & p_2 & \cdot & \cdot \end{bmatrix}, \quad \text{(ii)} \begin{bmatrix} p_1 & \cdot & \cdot & \cdot \\ 0 & 0 & p_3 & \cdot \end{bmatrix}, \quad \text{(iii)} \begin{bmatrix} p & \cdot & \cdot & \cdot \\ 0 & 0 & 0 & c_1 \end{bmatrix}. \tag{3.153}$$

Number of solutions: (i) infinite, (ii) infinite, and (iii) infinite if $c_1 = 0$.

- *Overdetermined*:

$$\text{(i)} \begin{bmatrix} p_1 & \cdot & \cdot \\ 0 & p_2 & \cdot \\ 0 & 0 & c_1 \end{bmatrix}, \quad \text{(ii)} \begin{bmatrix} p & \cdot & \cdot \\ 0 & 0 & c_1 \\ 0 & 0 & c_2 \end{bmatrix}. \tag{3.154}$$

Number of solutions: (i) one: unique if $c_1 = 0$ and (ii) infinite if $c_1 = c_2 = 0$.

Thus, we can say in general that underdetermined systems *tend* to have infinitely many solutions because there are always free variables, whereas overdetermined systems *tend* to have no solutions because there are always constraints on **x**.

Example 3.29 Continuing with the resistive circuit in Examples 3.3 and 3.16 where it was solved by direct substitution and Cramer's rule, respectively, the augmented matrix and its row-echelon form are

$$\left[\begin{array}{cc|c} 100 & 200 & 10 \\ 200 & 300 & 0 \end{array}\right] \Longrightarrow \left[\begin{array}{cc|c} 100 & 200 & 10 \\ 0 & -100 & -20 \end{array}\right]. \tag{3.155}$$

Back-substitution yields

$$i_2 = 20/100 = 20 \text{ mA}, \quad 100i_1 + 200i_2 = 10 \Longrightarrow i_1 = (10-4)/100 = 60 \text{ mA}, \tag{3.156}$$

which are the same results as before.

3.7.1 LU and LDU Decompositions

From the previous set of examples, we find that GE yields an upper triangular matrix and then back-substitution is applied to solve for the unknown dependent variable vector **y**. This procedure is equivalent to factoring matrix **A** into the product of a lower (L) triangular matrix **L** and an upper (U) triangular matrix **U**, called an LU decomposition. The steps used to derive **U** actually correspond to successive multiplication of **A** by *elementary matrices* of the form $\mathbf{E}_{mn}(c)$ in (3.33) (shown for $N = 3$), which adds a scaled row to another row. If a modified row has a zero pivot, then that row is interchanged with the next row, which means **A** is multiplied by the appropriate permutation matrix **P**.

Example 3.30 GE applied to the system in Example 3.24 is repeated here, except that **A** is premultiplied by elementary matrices to achieve its row-echelon form. Since we are interested in an LU decomposition for **A**, the right-hand side **x** is ignored in this example. The three row operations are achieved successively as follows:

$$\mathbf{E}_{21}(-3)\begin{bmatrix} 1 & 2 & 4 \\ 3 & 4 & 1 \\ 2 & 3 & 1 \end{bmatrix} = \begin{bmatrix} 1 & 2 & 4 \\ 0 & -2 & -11 \\ 2 & 3 & 1 \end{bmatrix}, \tag{3.157}$$

$$\mathbf{E}_{31}(-2)\begin{bmatrix} 1 & 2 & 4 \\ 0 & -2 & -11 \\ 2 & 3 & 1 \end{bmatrix} = \begin{bmatrix} 1 & 2 & 4 \\ 0 & -2 & -11 \\ 0 & -1 & -7 \end{bmatrix}, \tag{3.158}$$

and

$$\mathbf{E}_{32}(-1/2)\begin{bmatrix} 1 & 2 & 4 \\ 0 & -2 & -11 \\ 0 & -1 & -7 \end{bmatrix} = \begin{bmatrix} 1 & 2 & 4 \\ 0 & -2 & -11 \\ 0 & 0 & -3/2 \end{bmatrix} \triangleq \mathbf{U}, \tag{3.159}$$

which is the desired upper triangular matrix. It is important to note that in general these matrices must be multiplied in this specific order so as to match the sequence of GE operations. The overall product of these elementary is

$$\mathbf{E}_{32}(-1/2)\mathbf{E}_{31}(-2)\mathbf{E}_{21}(-3) = \begin{bmatrix} 1 & 0 & 0 \\ -3 & 1 & 0 \\ -1/2 & -1/2 & 1 \end{bmatrix}, \tag{3.160}$$

which is a lower triangular matrix with 1s along the main descending diagonal. Since the inverse of $\mathbf{E}_{mn}(c)$ is also an elementary matrix $\mathbf{E}_{mn}^{-1}(c) = \mathbf{E}_{mn}(-c)$, multiplying the matrix inverses in reverse order yields

$$\mathbf{E}_{21}^{-1}(-3)\mathbf{E}_{31}^{-1}(-2)\mathbf{E}_{32}^{-1}(-1/2) = \mathbf{E}_{21}(3)\mathbf{E}_{31}(2)\mathbf{E}_{32}(1/2) = \begin{bmatrix} 1 & 0 & 0 \\ 3 & 1 & 0 \\ 2 & 1/2 & 1 \end{bmatrix} \triangleq \mathbf{L}, \tag{3.161}$$

which is the desired lower triangular matrix. Observe that the subscripts of the elementary matrices give the locations of their arguments in this product matrix. The LU decomposition is $\mathbf{A} = \mathbf{LU}$ where $\mathbf{U}$ is the upper triangular matrix in (3.159) and $\mathbf{L}$ is the lower triangular matrix in (3.161), derived as the product of the inverses of the elementary matrices used to generate $\mathbf{U}$, multiplied in *reverse order*. This is verified as follows:

$$\begin{aligned}\mathbf{A} = \mathbf{IA} &= \left[\mathbf{E}_{21}^{-1}(-3)\,\mathbf{E}_{31}^{-1}(-2)\,\mathbf{E}_{32}^{-1}(-1/2)\right]\left[\mathbf{E}_{32}(-1/2)\,\mathbf{E}_{31}(-2)\,\mathbf{E}_{21}(-3)\right]\mathbf{A}\\ &= \left[\mathbf{E}_{21}(3)\,\mathbf{E}_{31}(2)\,\mathbf{E}_{32}(1/2)\right]\left[\mathbf{E}_{32}(-1/2)\,\mathbf{E}_{31}(-2)\,\mathbf{E}_{21}(-3)\right]\mathbf{A}\\ &= \mathbf{L}\left[\mathbf{E}_{32}(-1/2)\,\mathbf{E}_{31}(-2)\,\mathbf{E}_{21}(-3)\right]\mathbf{A} = \mathbf{LU}.\end{aligned} \tag{3.162}$$

If a zero pivot occurs after multiplication by an elementary matrix, then it is necessary that $\mathbf{A}$ be premultiplied by a permutation matrix $\mathbf{P}$ to obtain its LU decomposition. For such cases, the decomposition actually is of the form $\mathbf{PA} = \mathbf{LU}$. Finally, the corresponding LDU decomposition is obtained by writing $\mathbf{U}$ as the product of a diagonal matrix $\mathbf{D}$ and another upper triangular matrix with 1s along the main descending diagonal. The diagonal elements of $\mathbf{D}$ are the diagonal elements of the original $\mathbf{U}$, and the new upper triangular matrix is derived from $\mathbf{U}$ by dividing each row by its diagonal element. For this example, the LDU decomposition is

$$\mathbf{A} = \begin{bmatrix}1 & 0 & 0\\ 3 & 1 & 0\\ 2 & 1/2 & 1\end{bmatrix}\begin{bmatrix}1 & 0 & 0\\ 0 & -2 & 0\\ 0 & 0 & -3/2\end{bmatrix}\begin{bmatrix}1 & 2 & 4\\ 0 & 1 & 11/2\\ 0 & 0 & 1\end{bmatrix} \triangleq \mathbf{LDU}. \tag{3.163}$$

LU and LDU decompositions can also be generated for rectangular matrices (see Problem 3.28). For $\mathbf{A} \in \mathcal{R}^{M\times N}$, the dimensions of the matrices are $\mathbf{U} \in \mathcal{R}^{M\times N}$, $\mathbf{D} \in \mathcal{R}^{M\times M}$, and $\mathbf{L} \in \mathcal{R}^{M\times M}$. The upper triangular matrix $\mathbf{U}$ is always the same size as the original matrix (which, of course, follows from GE). The diagonal matrix must be square and its size is determined by the number of rows of $\mathbf{A}$ such that the diagonal elements of the product $\mathbf{DU}$ match those of the LU decomposition. As a result, $\mathbf{L}$ must also be square and of the same size as $\mathbf{D}$.

Example 3.31 For the undetermined system of equations in Example 3.28, row-echelon form is derived as follows:

$$\mathbf{E}_{21}(-2)\begin{bmatrix}1 & 2 & -2\\ 2 & -1 & 1\end{bmatrix} = \begin{bmatrix}1 & 2 & -2\\ 0 & -5 & 5\end{bmatrix}. \tag{3.164}$$

The diagonal and lower triangular matrices are

$$\mathbf{D} = \begin{bmatrix}1 & 0\\ 0 & -5\end{bmatrix}, \qquad \mathbf{L} = \mathbf{E}_{21}^{-1}(-2) = \mathbf{E}_{21}(2) = \begin{bmatrix}1 & 0\\ 2 & 1\end{bmatrix}, \tag{3.165}$$

and the LDU decomposition is

$$\mathbf{A} = \begin{bmatrix}1 & 0\\ 2 & 1\end{bmatrix}\begin{bmatrix}1 & 0\\ 0 & -5\end{bmatrix}\begin{bmatrix}1 & 2 & -2\\ 0 & 1 & -1\end{bmatrix}, \tag{3.166}$$

whose product is given in (3.164).

3.7.2 Basis Vectors

A basis for a subspace S is a set of vectors that are linearly independent and span the subspace, so that every vector in S can be written as a linear combination of the basis vectors. A basis is not unique because an equivalent basis can be derived as a linear combination of another basis. Although the basis vectors can be normalized so that each has unit length and the basis is unique, this is not necessary in practice. In this section, a basis is found for each of the four subspaces of matrix $\mathbf{A} \in \mathcal{R}^{M\times N}$, which has rank R. The procedure involves first transforming $\mathbf{A}$ to row-echelon form $\mathbf{U}$ (Strang, 1980).

- *Column space*: The basis vectors for the column space of $\mathbf{A}$ are the columns of $\mathbf{A}$ corresponding to those of $\mathbf{U}$ with nonzero pivots, which are associated with the basic variables. There are R such columns, and the resulting vectors in $C(\mathbf{A})$ form a subset of $\mathcal{R}^M$.
- *Row space*: Since the row operations used to derive $\mathbf{U}$ do not change the row space, the R nonzero rows of $\mathbf{U}$ form a basis for $R(\mathbf{A})$, which is a subset of $\mathcal{R}^N$.
- *Null space*: The dimension of the null space is equal to the number of free variables of $\mathbf{U}$, which is $N - R$ (assuming $N > R$ for nontrivial $N(\mathbf{A})$). The basis vectors, which form a subset of $\mathcal{R}^N$, are derived by sequentially setting each free variable to 1 and the other free variables to 0, and then solving for the basic variables of this modified $\mathbf{y}$ in the homogeneous system $\mathbf{Uy} = \mathbf{0}$.
- *Left null space*: From the LU decomposition $\mathbf{A} = \mathbf{LU}$, we can write $\mathbf{L}^{-1}\mathbf{A} = \mathbf{U}$ because the lower triangular matrix is square and invertible. Since the last $M - R$ rows of $\mathbf{U}$ are zero (assuming $M > R$ for nontrivial $L(\mathbf{A})$), the corresponding rows of $\mathbf{L}^{-1}$ form a basis for the left-null space. These rows are independent because the elementary matrices that generate $\mathbf{L}$ are nonsingular.

As mentioned earlier, the row space and null space are orthogonal complements, and these two vector subspaces together contain every vector in $\mathcal{R}^N$. Since any vector in these subspaces can be derived as a linear combination of their basis vectors, we need to only show that the two sets of basis vectors are orthogonal in order to verify this property. Let $\mathbf{v}_1$ be a vector in $R(\mathbf{A})$, which means $\mathbf{v}_1$ is a vector in the column space $C(\mathbf{A}^T)$, and so it can be written as a linear combination of the columns of $\mathbf{A}^T$:

$$\mathbf{v}_1 = \mathbf{A}^T\mathbf{c}, \tag{3.167}$$

for some coefficient vector $\mathbf{c}$. Let $\mathbf{v}_2$ be a vector in $N(\mathbf{A})$ and write the inner product

$$\mathbf{v}_2^T\mathbf{v}_1 = \mathbf{v}_2^T\mathbf{A}^T\mathbf{c}, \tag{3.168}$$

where (3.167) has been substituted. By combining the first two terms according to the rules of matrix transposition, we have $\mathbf{v}_2^T\mathbf{A}^T = (\mathbf{Av}_2)^T = \mathbf{0}^T$ because $\mathbf{v}_2$ is in the null space of $\mathbf{A}$. As a result, (3.168) is 0, proving that $\mathbf{v}_1 \in R(\mathbf{A})$ and $\mathbf{v}_2 \in N(\mathbf{A})$ are orthogonal.

Likewise, the column space and left null space are orthogonal complements, and these two vector subspaces together contain every vector in $\mathcal{R}^M$. The proof is similar to the one used earlier. Let $\mathbf{v}_1$ be a vector in $C(\mathbf{A})$ such that

$$\mathbf{v}_1 = \mathbf{Ac}, \tag{3.169}$$

for some coefficient vector $\mathbf{c}$. If $\mathbf{v}_2^T$ is in the left null space of $\mathbf{A}$, the inner product is 0:

$$\mathbf{v}_2^T\mathbf{v}_1 = \mathbf{v}_2^T\mathbf{Ac} = 0, \tag{3.170}$$

where $\mathbf{v}_2^T\mathbf{A} = \mathbf{0}^T$ has been substituted. The following example illustrates how to find a basis for each of the four subspaces and demonstrates their orthogonality properties.

Example 3.32 The system of linear equations

$$\begin{bmatrix} 1 & 2 & 2 \\ 2 & 4 & 4 \end{bmatrix} \begin{bmatrix} y_1 \\ y_2 \\ y_3 \end{bmatrix} = \begin{bmatrix} x_1 \\ x_2 \end{bmatrix} \tag{3.171}$$

has the following augmented matrix and row-echelon form:

$$\left[\begin{array}{ccc|c} 1 & 2 & 2 & x_1 \\ 2 & 4 & 4 & x_2 \end{array}\right] \rightarrow \left[\begin{array}{ccc|c} 1 & 2 & 2 & x_1 \\ 0 & 0 & 0 & x_2 - 2x_1 \end{array}\right]. \tag{3.172}$$

There is a row of zeros because the two rows of the matrix are linearly dependent, yielding a nontrivial left null space (the null space is also nontrivial because $N > M$). The elementary matrix used to derive $\mathbf{U}$ is $\mathbf{E}_{21}(-2) \equiv \mathbf{L}^{-1}$. In order to have a consistent set of equations, we must have $x_2 - 2x_1 = 0$, in which case there is an infinity of solutions. This is an example of the underdetermined case in Table 3.7 with $M = 2$, $N = 3$, and $R = 1$, and so there are two free variables and only one basic variable. A basis for the column space consists of the column of $\mathbf{A}$ corresponding to the nonzero pivot column in $\mathbf{U}$:

$$\text{column space: } \begin{bmatrix} 1 \\ 2 \end{bmatrix} \in \mathcal{R}^2. \tag{3.173}$$

The left null space is derived from the last row of $\mathbf{E}_{21}(-2)$:

$$\mathbf{E}_{21}(-2) = \begin{bmatrix} 1 & 0 \\ -2 & 1 \end{bmatrix} \implies \text{left null space: } \begin{bmatrix} -2 \\ 1 \end{bmatrix} \in \mathcal{R}^2. \tag{3.174}$$

These basis vectors for the column space and the left null space are orthogonal. The row space is the nonzero row in $\mathbf{U}$:

$$\text{row space: } \begin{bmatrix} 1 \\ 2 \\ 2 \end{bmatrix} \in \mathcal{R}^3. \tag{3.175}$$

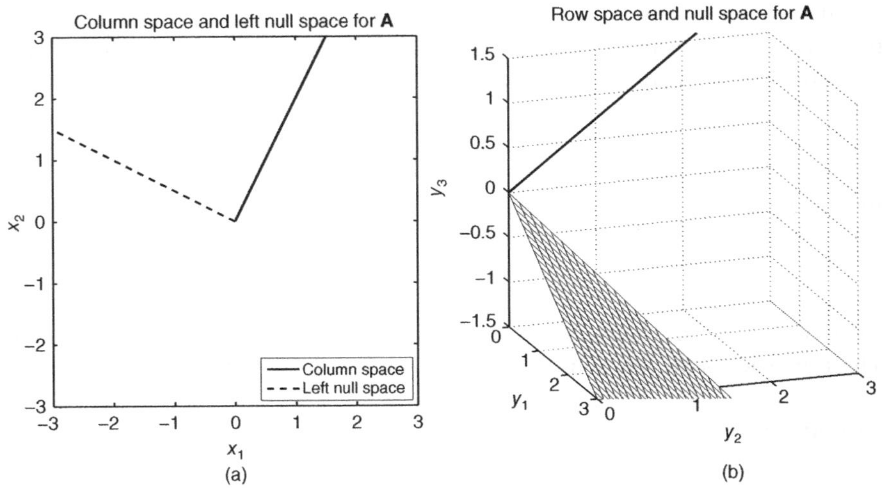

Figure 3.11 Orthogonal subspaces of matrix **A** in Example 3.32. (a) Column space and left null space. (b) Row space (plane) and null space (line).

The null space is obtained by solving for **y** in $\mathbf{Ay} = \mathbf{0}$ twice, first with $\{y_2 = 1, y_3 = 0\}$ and then with $\{y_2 = 0, y_3 = 1\}$:

$$\begin{bmatrix} 1 & 2 & 2 \\ 0 & 0 & 0 \end{bmatrix} \begin{bmatrix} y_1 \\ 1 \\ 0 \end{bmatrix} = \begin{bmatrix} 0 \\ 0 \\ 0 \end{bmatrix} \implies y_1 = -2, \tag{3.176}$$

$$\begin{bmatrix} 1 & 2 & 2 \\ 0 & 0 & 0 \end{bmatrix} \begin{bmatrix} y_1 \\ 0 \\ 1 \end{bmatrix} = \begin{bmatrix} 0 \\ 0 \\ 0 \end{bmatrix} \implies y_1 = -2. \tag{3.177}$$

The null space has two vectors, which are derived by substituting these values for y_1 into the two column vectors on the left-hand sides of (3.176) and (3.177):

$$\text{null space:} \begin{bmatrix} -2 \\ 1 \\ 0 \end{bmatrix}, \begin{bmatrix} -2 \\ 0 \\ 1 \end{bmatrix} \in \mathcal{R}^3, \tag{3.178}$$

which we see are orthogonal to the basis vector for the row space. The orthogonal basis vectors for the column space and the left null space are plotted on the plane $\mathcal{R}^2$ in Figure 3.11(a). The vectors comprising the column space are of the form $c[1, 2]^T$ for $c \in \mathcal{R}$, and so they are all collinear. Likewise, the vectors comprising the left null space are of the form $c[-2, 1]^T$. These subspaces together give every vector in $\mathcal{R}^2$ as the linear combination $c_1[1, 2]^T + c_2[-2, 1]^T$ for $c_1, c_2 \in \mathcal{R}$. The two basis vectors for the null space form a plane in $\mathcal{R}^3$, and the basis vector for the row space is a line as illustrated in Figure 3.11(b). These are orthogonal, and it is clear that any vector in $\mathcal{R}^3$

can be derived as the linear combination $c_1[-2,1,0]^T + c_2[-2,0,1]^T + c_3[1,2,2]^T$ for $c_1, c_2, c_3 \in \mathcal{R}$.

3.7.3 General Solution of $\mathbf{Ay} = \mathbf{x}$

The general solution of the system of equations $\mathbf{Ay} = \mathbf{x}$ has two components:

$$\mathbf{y} = \mathbf{y}_p + \mathbf{y}_h, \tag{3.179}$$

where $\mathbf{y}_h$ is a solution to the *homogeneous* system $\mathbf{Ay} = \mathbf{0}$, and $\mathbf{y}_p$ is the *particular solution* of the system with nonzero $\mathbf{x}$ on the right-hand side. Thus, if the null space of $\mathbf{A}$ is nontrivial, $\mathbf{y}_h$ is nonzero and there is an infinity of solutions (as mentioned earlier) given by $\mathbf{y}_p + c\mathbf{y}_h$:

$$\mathbf{A}(\mathbf{y}_p + c\mathbf{y}_h) = \mathbf{Ay}_p + c\mathbf{Ay}_h = \mathbf{Ay}_p + \mathbf{0} = \mathbf{x}, \tag{3.180}$$

where $c \in \mathcal{R}$ is any real scalar. The particular solution is derived from the row-echelon form:

$$\mathbf{Uy}_p = \tilde{\mathbf{x}}, \tag{3.181}$$

where $\tilde{\mathbf{x}} = \mathbf{L}^{-1}\mathbf{x}$ because $\mathbf{U} = \mathbf{L}^{-1}\mathbf{A}$. Back-substitution yields the elements of $\mathbf{y}_p$, where typically the free variables (which are associated with zero pivots) are set to 0. This is illustrated in the next example.

Example 3.33 The system in (3.133) has the row-echelon form

$$\begin{bmatrix} 1 & 2 & 4 \\ 0 & -2 & -11 \\ 0 & 0 & 0 \end{bmatrix} \begin{bmatrix} y_1 \\ y_2 \\ y_3 \end{bmatrix} = \begin{bmatrix} 3 \\ -8 \\ 0 \end{bmatrix}, \tag{3.182}$$

with free variable y_3, which means the dimension of the null space is 1. Setting $y_3 = 0$, back-substitution yields

$$-2y_2 - 11y_3 = -8 \implies y_2 = 4, \quad y_1 + 2y_2 + 4y_3 = 3 \implies y_1 = -5, \tag{3.183}$$

and the particular solution is $\mathbf{y}_p = [-5, 4, 0]^T$. The homogeneous solution is derived by setting $y_3 = 1$ in (3.182), replacing the right-hand side with $\mathbf{0}$ and solving for $\mathbf{y}$:

$$-2y_2 - 11 = 0 \implies y_2 = -11/2, \quad y_1 + 2y_2 + 4 = 0 \implies y_1 = 7, \tag{3.184}$$

which yields $\mathbf{y}_h = [7, -11/2, 1]^T$. Thus, the general solution is

$$\mathbf{y} = \begin{bmatrix} -5 \\ 4 \\ 0 \end{bmatrix} + c \begin{bmatrix} 7 \\ -11/2 \\ 1 \end{bmatrix}, \tag{3.185}$$

for $c \in \mathcal{R}$. Although it is convenient to set the free variables to zero when deriving $\mathbf{y}_p$, this is not necessary because a free variable can take on any value. Of course, this follows from the form in (3.182), where if y_3 is left as a variable, the solution is

$$-2y_2 - 11y_3 = -8 \implies y_2 = -(11/2)y_3 + 4,$$
$$y_1 + 2y_2 + 4y_3 = 3 \implies y_1 = 7y_3 - 5, \qquad (3.186)$$

which gives $\mathbf{y} = [7y_3 - 5, -(11/2)y_3 + 4, y_3]^T$. The reader can verify that this is a solution for any $y_3 \in \mathcal{R}$.

The expression in (3.179) can be viewed as a decomposition of the general solution into two components that reveals basis vectors for the column space and the null space of $\mathbf{A}$. We find a similar decomposition in Chapter 6 for linear ODEs, in terms of a homogeneous solution and a particular solution for the time-varying waveform $y(t)$.

3.8 EIGENDECOMPOSITION

In this section, we describe another matrix decomposition that illustrates additional properties of a system of linear equations.

Definition: Eigenvalues The *eigenvalues* of square matrix $\mathbf{A}$ are those λ satisfying the following equation:

$$\det(\mathbf{A} - \lambda\mathbf{I}) = 0. \qquad (3.187)$$

The resulting expression is an Nth-order polynomial in λ called the *characteristic equation* of matrix $\mathbf{A}$:

$$\lambda^N + \alpha_{N-1}\lambda^{N-1} + \ldots + \alpha_1\lambda + \alpha_0 = 0, \qquad (3.188)$$

where the coefficients $\{\alpha_n\}$ depend on the specific elements of $\mathbf{A}$.

Definition: Eigenvectors The *eigenvectors* of square matrix $\mathbf{A}$ are the column vectors $\mathbf{v}$ that satisfy the following equation for each unique eigenvalue λ:

$$\mathbf{A}\mathbf{v} = \lambda\mathbf{v}. \qquad (3.189)$$

This expression is equivalent to

$$(\mathbf{A} - \lambda\mathbf{I})\mathbf{v} = \mathbf{0}, \qquad (3.190)$$

such that $\mathbf{A} - \lambda\mathbf{I}$ is a singular matrix for any eigenvalue.

From (3.187), we find that the eigenvalues are those values such that matrix $\mathbf{A} - \lambda\mathbf{I}$ is *singular*, meaning that one or more of its rows (and columns) are linearly dependent. The expression in (3.190) shows that an eigenvector for a particular λ is a nonzero vector such that a linear combination of the columns of $\mathbf{A} - \lambda\mathbf{I}$ is zero. This means that $\mathbf{A} - \lambda\mathbf{I}$ has a *nontrivial null space*. Since a scaled version $c\mathbf{v}$ of the eigenvector $\mathbf{v}$ for $c \neq 0$ is also an eigenvector (as is easily verified in (3.190)), unique *normalized* eigenvectors are often used in practice. These are obtained by scaling $\mathbf{v}$ as follows:

$$\mathbf{q} \triangleq \frac{\mathbf{v}}{\sqrt{\mathbf{v}^H\mathbf{v}}} = \frac{\mathbf{v}}{\|\mathbf{v}\|}, \tag{3.191}$$

which has unit squared norm:

$$\mathbf{q}^H\mathbf{q} = \frac{\mathbf{v}^H\mathbf{v}}{\sqrt{\mathbf{v}^H\mathbf{v}}\sqrt{\mathbf{v}^H\mathbf{v}}} = 1, \tag{3.192}$$

where the superscript H for the *complex conjugate transpose* has been used because in general the eigenvectors and eigenvalues could be complex. The complex conjugate transpose of a vector is defined as follows: $\mathbf{v}^H \triangleq (\mathbf{v}^T)^* = (\mathbf{v}^*)^T$, where the superscript $*$ denotes complex conjugation of every element in the vector (see Chapter 4).

An eigenvector is a special vector in the sense that when it postmultiplies $\mathbf{A}$, the same vector is obtained though it is scaled by a constant: $\lambda\mathbf{v}$. Of course, this property is not true for other vectors: $\mathbf{A}\mathbf{y} = \mathbf{x}$ yields vector $\mathbf{x}$ in the column space of $\mathbf{A}$, which generally is not proportional to $\mathbf{y}$. We will see a similar property for linear systems modeled by ODEs with constant coefficients for which the sinusoidal functions are *eigenfunctions*. This means that if the input of a linear system is $\cos(\omega_o t)$ with angular frequency ω_o, then its output is also cosine with the same frequency, but possibly with a different magnitude and phase shift: $A\cos(\omega_o t + \phi)$.

The eigenvalues and eigenvectors form another decomposition of matrix $\mathbf{A}$.

Definition: Eigendecomposition The *eigendecomposition* of $\mathbf{A}$ is

$$\mathbf{A} = \mathbf{Q}\boldsymbol{\Lambda}\mathbf{Q}^{-1}, \tag{3.193}$$

where $\boldsymbol{\Lambda}$ is a diagonal matrix containing its eigenvalues and $\mathbf{Q}$ is a matrix containing the associated eigenvectors:

$$\boldsymbol{\Lambda} \triangleq \begin{bmatrix} \lambda_1 & 0 & \dots & 0 \\ 0 & \ddots & & \vdots \\ \vdots & & \ddots & 0 \\ 0 & \dots & 0 & \lambda_N \end{bmatrix}, \qquad \mathbf{Q} \triangleq [\mathbf{q}_1, \dots, \mathbf{q}_N], \tag{3.194}$$

where each eigenvector has unit squared norm $\mathbf{q}_n^H\mathbf{q}_n = 1$, resulting in a unique eigendecomposition.

Example 3.34 The eigenvalues of

$$\mathbf{A} = \begin{bmatrix} 2 & 1 \\ 1 & 2 \end{bmatrix} \tag{3.195}$$

are derived by solving

$$\det\left(\begin{bmatrix} 2-\lambda & 1 \\ 1 & 2-\lambda \end{bmatrix}\right) = 0 \implies (\lambda-2)^2 - 1 = \lambda^2 - 4\lambda + 3 = 0, \tag{3.196}$$

yielding $\lambda_1 = 1$ and $\lambda_2 = 3$. The unit-norm eigenvectors are generated from (3.190):

$$\begin{bmatrix} (2-1) & 1 \\ 1 & (2-1) \end{bmatrix} \mathbf{q}_1 = \mathbf{0} \implies \mathbf{q}_1 = (1/\sqrt{2}) \begin{bmatrix} 1 \\ -1 \end{bmatrix}, \tag{3.197}$$

and

$$\begin{bmatrix} (2-3) & 1 \\ 1 & (2-3) \end{bmatrix} \mathbf{q}_2 = \mathbf{0} \implies \mathbf{q}_2 = (1/\sqrt{2}) \begin{bmatrix} 1 \\ 1 \end{bmatrix}, \tag{3.198}$$

resulting in the following unique matrix of eigenvectors:

$$\mathbf{Q} = \left(1/\sqrt{2}\right) \begin{bmatrix} 1 & 1 \\ -1 & 1 \end{bmatrix}. \tag{3.199}$$

For this example, the eigenvalues and eigenvectors happen to be real, which is not the case in general. Finally, we verify this eigendecomposition by multiplying together all three matrices:

$$\begin{aligned} \mathbf{Q\Lambda Q}^{-1} &= (1/2) \begin{bmatrix} 1 & 1 \\ -1 & 1 \end{bmatrix} \begin{bmatrix} 1 & 0 \\ 0 & 3 \end{bmatrix} \begin{bmatrix} 1 & -1 \\ 1 & 1 \end{bmatrix} = (1/2) \begin{bmatrix} 1 & 3 \\ -1 & 3 \end{bmatrix} \begin{bmatrix} 1 & -1 \\ 1 & 1 \end{bmatrix} \\ &= (1/2) \begin{bmatrix} 4 & 2 \\ 2 & 4 \end{bmatrix} = \begin{bmatrix} 2 & 1 \\ 1 & 2 \end{bmatrix} = \mathbf{A}. \end{aligned} \tag{3.200}$$

Example 3.35 The eigenvalues of the rotation matrix **A** in (3.49) are derived by solving the following equation:

$$\det\left(\begin{bmatrix} \cos(\phi)-\lambda & -\sin(\phi) \\ \sin(\phi) & \cos(\phi)-\lambda \end{bmatrix}\right) = [\cos(\phi)-\lambda]^2 + \sin^2(\phi) = 0. \tag{3.201}$$

Rearranging this expression gives

$$\cos^2(\phi) - 2\lambda\cos(\phi) + \lambda^2 + \sin^2(\phi) = \lambda^2 - 2\lambda\cos(\phi) + 1 = 0, \tag{3.202}$$

which is quadratic with solutions

$$\begin{aligned} \lambda_1, \lambda_2 &= \cos(\phi) \pm \sqrt{\cos^2(\phi) - 1} = \cos(\phi) \pm j\sin(\phi) \\ &= \exp(\pm j\phi), \end{aligned} \tag{3.203}$$

where $\exp(\pm j\phi)$ is the complex exponential function given earlier in (1.100) (the identity in (3.203) is *Euler's formula* and is discussed in Chapter 4). The two

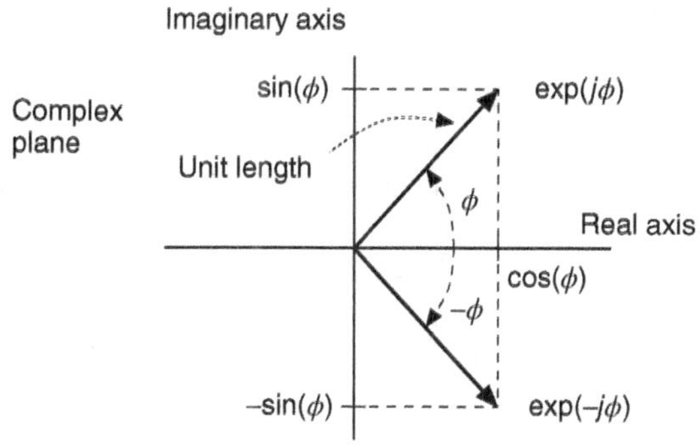

Figure 3.12 Complex eigenvalues on the complex plane for the rotation matrix in (3.203) of Example 3.35.

eigenvalues are represented on the *complex plane* in Figure 3.12, which is defined by perpendicular axes for the real and imaginary components. Observe that the eigenvalues have symmetry on the complex plane where $\exp(-j\phi)$ is a mirror image of $\exp(j\phi)$ about the real axis. They form a *complex conjugate pair*, which is necessary because the elements of $\mathbf{A}$ are real. This conjugate symmetry is discussed further in Chapter 4. For two values of ϕ, the eigenvalues are real: (i) $\phi = 0 \implies \lambda = \{1, 0\}$ and (ii) $\phi = \pi \implies \lambda = \{-1, 0\}$. When $\phi = \pi/2$ or $3\pi/2$, the eigenvalues are strictly imaginary, and for all other values of $\phi \in [0, 2\pi]$, they are complex. The eigenvectors $\{\mathbf{q}_1, \mathbf{q}_2\}$ are derived from

$$\begin{bmatrix} \cos(\phi) - \exp(j\phi) & -\sin(\phi) \\ \sin(\phi) & \cos(\phi) - \exp(j\phi) \end{bmatrix} \mathbf{q}_1 = \mathbf{0}, \tag{3.204}$$

$$\begin{bmatrix} \cos(\phi) - \exp(-j\phi) & -\sin(\phi) \\ \sin(\phi) & \cos(\phi) - \exp(-j\phi) \end{bmatrix} \mathbf{q}_2 = \mathbf{0}. \tag{3.205}$$

Substituting $\exp(\pm j\phi)$ from (3.203) causes cosine to cancel in both equations such that sine can be factored:

$$\sin(\phi) \begin{bmatrix} -j & -1 \\ 1 & -j \end{bmatrix} \mathbf{q}_1 = \mathbf{0}, \quad \sin(\phi) \begin{bmatrix} j & -1 \\ 1 & j \end{bmatrix} \mathbf{q}_2 = \mathbf{0}. \tag{3.206}$$

For nonzero $\sin(\phi)$, the normalized eigenvectors are

$$\mathbf{q}_1 = \frac{1}{\sqrt{2}} \begin{bmatrix} j \\ 1 \end{bmatrix}, \quad \mathbf{q}_2 = \frac{1}{\sqrt{2}} \begin{bmatrix} 1 \\ j \end{bmatrix}, \tag{3.207}$$

and the eigendecomposition is

$$\mathbf{A} = \mathbf{Q}\boldsymbol{\Lambda}\mathbf{Q}^{-1} = \frac{1}{\sqrt{2}} \begin{bmatrix} j & 1 \\ 1 & j \end{bmatrix} \begin{bmatrix} \exp(j\phi) & 0 \\ 0 & \exp(-j\phi) \end{bmatrix} \begin{bmatrix} -j & 1 \\ 1 & -j \end{bmatrix} \frac{1}{\sqrt{2}}. \tag{3.208}$$

Multiplying these complex matrices together and using (3.203) give the original real matrix **A**.

Example 3.36 It can be shown that the eigenvalues of an idempotent matrix are either 0 or 1 (see Problem 3.16). This property can be used to construct idempotent matrix **B** as follows:

$$\mathbf{B} = \mathbf{A}\mathbf{D}\mathbf{A}^{-1}, \tag{3.209}$$

where **A** is any invertible matrix and **D** is a diagonal matrix with only 0s and 1s along the main descending diagonal. It is straightforward to show that **B** constructed in this manner is idempotent:

$$\mathbf{B}^2 = (\mathbf{A}\mathbf{D}\mathbf{A}^{-1})\,(\mathbf{A}\mathbf{D}\mathbf{A}^{-1}) = \mathbf{A}\mathbf{D}^2\mathbf{A}^{-1}, \tag{3.210}$$

and it follows that $\mathbf{D}^2 = \mathbf{D}$ because of the 1s and 0s on the main diagonal. Matrix **D** is a trivial form of an idempotent matrix (recall that **I** is the only nonsingular idempotent matrix). For example, let

$$\mathbf{A} = \begin{bmatrix} 1 & 2 \\ 2 & 1 \end{bmatrix} \implies \mathbf{A}^{-1} = \begin{bmatrix} -1/3 & 2/3 \\ 2/3 & -1/3 \end{bmatrix}. \tag{3.211}$$

These yield the following two idempotent matrices:

$$\mathbf{D} = \begin{bmatrix} 1 & 0 \\ 0 & 0 \end{bmatrix} \implies \mathbf{B} = \begin{bmatrix} -1/3 & 2/3 \\ -2/3 & 1/3 \end{bmatrix}, \tag{3.212}$$

$$\mathbf{D} = \begin{bmatrix} 0 & 0 \\ 0 & 1 \end{bmatrix} \implies \mathbf{B} = \begin{bmatrix} 1/3 & -2/3 \\ 2/3 & -1/3 \end{bmatrix}. \tag{3.213}$$

3.9 MATLAB FUNCTIONS

We conclude this chapter with a brief summary of some of the MATLAB functions that are useful for the preceding material. Detailed information can be found in the online MATLAB documentation. (Unlike MATLAB, we use italic and bold symbols in the following functions.)

- *Identity matrix, zero matrix*: eye(m) (size $m \times m$), zeros(m, n) (size $m \times n$).
- *Transpose*: $\mathbf{A}.'$ (nonconjugated), $\mathbf{A}'$ (conjugate transpose).
- *Norm*: norm($\mathbf{v}$).
- *Trace*: trace($\mathbf{A}$).
- *Determinant*: det($\mathbf{A}$).
- *Matrix inverse*: inverse($\mathbf{A}$).
- *Gaussian elimination*: $\mathbf{x}$ = linsolve($\mathbf{A}, \mathbf{y}$).

- *LU decomposition*: $[\mathbf{L}, \mathbf{U}] = \text{lu}(\mathbf{A})$.
- *Eigendecomposition*: $[\mathbf{V}, \mathbf{D}] = \text{eig}(\mathbf{A})$. The columns of $\mathbf{V}$ are the eigenvectors of $\mathbf{A}$, and $\mathbf{D}$ is a diagonal matrix containing the eigenvalues.

PROBLEMS

System of Linear Equations

3.1 Determine which of the following subsets of $\mathcal{R}^3$ are subspaces for $a, b \in \mathcal{R}$. (a) $\mathbf{v}_1 = [a, b, 1]^T$. (b) $\mathbf{v}_2 = [a, b, a-b]^T$. (c) $\mathbf{v}_3 = [a, b, c]^T$ for $c \geq 0$. (d) $\mathbf{v}_4 = [a, 0, a+b]^T$.

3.2 (a) Write the following system of linear equations in matrix form $\mathbf{Ay} = \mathbf{x}$:

$$\begin{aligned} y_1 + 2y_2 - y_3 &= 4, \\ 2y_1 + 3y_2 + y_3 &= 2, \\ y_1 - 4y_2 + 2y_3 &= -1, \end{aligned} \tag{3.214}$$

and find the vector norm $\|\mathbf{x}\|$. (b) Determine if any two rows of $\mathbf{A}$ are orthogonal where $\overline{\mathbf{a}}_m^T \overline{\mathbf{a}}_n = 0$ for some $m \neq n$.

3.3 Repeat the previous problem for

$$\begin{aligned} 2y_1 - y_2 + 3y_3 &= 1, \\ y_1 + 2y_2 + 4y_3 &= -1, \\ 2y_1 - 3y_2 + y_3 &= 2. \end{aligned} \tag{3.215}$$

3.4 Find a solution for the following upper triangular system of linear equations by a back-substitution of variables starting with y_3:

$$\begin{aligned} y_1 - 2y_2 + 2y_3 &= 1, \\ y_2 - y_3 &= 2, \\ 3y_3 &= 6. \end{aligned} \tag{3.216}$$

Verify your answer by writing the system in matrix form and show that $\mathbf{Ay} = \mathbf{x}$.

3.5 Repeat the previous problem for

$$\begin{aligned} 2y_1 + y_2 - 4y_3 &= 2, \\ y_2 + 2y_3 &= 1, \\ 2y_2 - y_3 &= -2. \end{aligned} \tag{3.217}$$

3.6 (a) For column vectors $\mathbf{v}_1 = [1, 2, 4]^T$ and $\mathbf{v}_2 = [3, 5, 4]^T$, give the outer products $\mathbf{v}_1\mathbf{v}_2^T \triangleq \mathbf{A}_1$ and $\mathbf{v}_2\mathbf{v}_2^T \triangleq \mathbf{A}_2$. (b) Demonstrate using $\mathbf{w}_1 = [c_1, c_2, c_3]^T$ and $\mathbf{w}_2 = [1, 2, 3]^T$ that the columns of $\mathbf{B}_1 \triangleq \mathbf{w}_1\mathbf{w}_2^T$ and $\mathbf{B}_2 \triangleq \mathbf{w}_2\mathbf{w}_1^T$ are linearly dependent.

Matrix Properties

3.7 For the following matrices, show that (a) $\mathbf{AB} \neq \mathbf{BA}$ and (b) $(\mathbf{AB})^T = \mathbf{B}^T\mathbf{A}^T$:

$$\mathbf{A} = \begin{bmatrix} 1 & 3 & -2 \\ 2 & -4 & 1 \\ -1 & 1 & -3 \end{bmatrix}, \quad \mathbf{B} = \begin{bmatrix} 2 & -1 & 3 \\ 4 & 1 & 2 \\ 1 & -2 & 5 \end{bmatrix}. \tag{3.218}$$

3.8 Prove the following properties of the trace of a matrix: (a) $\text{tr}(c\mathbf{A}) = c\text{tr}(\mathbf{A})$, (b) $\text{tr}(\mathbf{AB}) = \text{tr}(\mathbf{BA})$, and (c) $\text{tr}(\mathbf{A} + \mathbf{B}) = \text{tr}(\mathbf{A}) + \text{tr}(\mathbf{B})$.

3.9 Verify the matrix inverses in (3.34) for each type of elementary matrix.

3.10 Show that any square matrix can be written as the sum $\mathbf{A} = \mathbf{B} + \mathbf{C}$ where $\mathbf{B}$ is a symmetric matrix and $\mathbf{C}$ is a skew-symmetric matrix.

3.11 Determine if the columns of the following matrix are linearly independent:

$$\mathbf{A} = \begin{bmatrix} 1 & 2 & 1 \\ 3 & 1 & 2 \\ 1 & -3 & 0 \end{bmatrix}, \tag{3.219}$$

by solving for $c_1\mathbf{a}_1 + c_2\mathbf{a}_2 + c_3\mathbf{a}_3 = \mathbf{0}$ where $\mathbf{a}_n$ is the nth column of $\mathbf{A}$.

3.12 Find the inverse of $\mathbf{A}$ using elementary matrices and the augmented matrix form as was done in Example 3.6:

$$\mathbf{A} = \begin{bmatrix} 2 & 3 & -1 \\ -1 & -2 & 4 \\ 2 & 1 & -3 \end{bmatrix}. \tag{3.220}$$

3.13 Verify the following property for vectors $\{\mathbf{x}, \mathbf{y}\}$ and nonsingular matrix $\mathbf{A}$:

$$(\mathbf{A} + \mathbf{x}\mathbf{y}^T)^{-1} = \mathbf{A}^{-1} - \frac{\mathbf{A}^{-1}\mathbf{x}\mathbf{y}^T\mathbf{A}^{-1}}{1 + \mathbf{y}^T\mathbf{A}^{-1}\mathbf{x}}. \tag{3.221}$$

3.14 Determine if either of the following matrices is idempotent or nilpotent:

$$\text{(a) } \mathbf{A} = \begin{bmatrix} 1 & -1 & 0 \\ 1 & 0 & -1 \\ -1 & 1 & 0 \end{bmatrix}, \quad \text{(b) } \mathbf{B} = \begin{bmatrix} -4 & 2 & 1 \\ -10 & 5 & 2 \\ 0 & 0 & 1 \end{bmatrix}. \tag{3.222}$$

3.15 Let $\mathbf{A}$ be an idempotent matrix. Specify if any of the following matrices are idempotent: (a) $\mathbf{I} - \mathbf{A}$, (b) $\mathbf{A} - \mathbf{I}$, (c) $\mathbf{A}^T$, and (d) $\mathbf{A}\mathbf{A}^T$.

3.16 Prove that the eigenvalues of an idempotent matrix are either 0 or 1.

Determinant and Matrix Subspaces

3.17 Find the determinant for each matrix in (3.218).

3.18 Find the adjugate matrix for each of the following matrices:

$$\text{(a) } \mathbf{A} = \begin{bmatrix} 3 & -1 & 2 \\ 2 & 4 & -2 \\ 1 & -1 & 2 \end{bmatrix}, \qquad \text{(b) } \mathbf{B} = \begin{bmatrix} 2 & 1 & 2 & 2 \\ -1 & 3 & 4 & 1 \\ -2 & 1 & -3 & 5 \\ 1 & 2 & 1 & -2 \end{bmatrix}. \tag{3.223}$$

3.19 Use Cramer's rule to find the three currents labeled in Figure 3.13.

3.20 Repeat the previous problem for the four currents labeled in Figure 3.14.

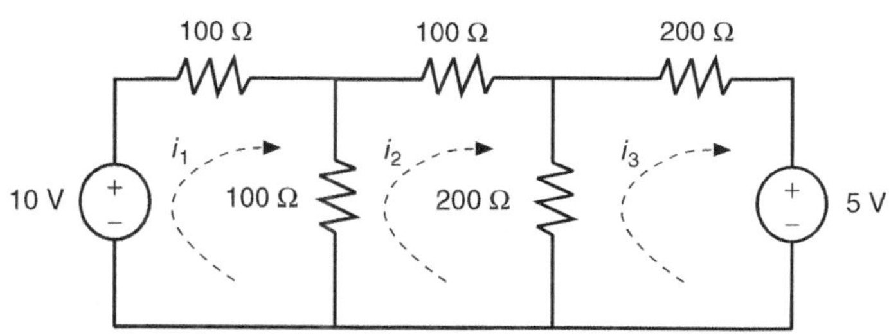

Figure 3.13 Resistive circuit for Problem 3.19.

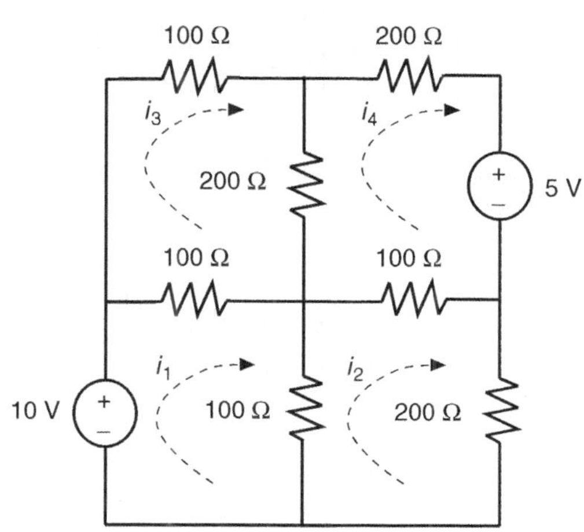

Figure 3.14 Resistive circuit for Problem 3.20.

3.21 Find a basis and the dimension for each of the four matrix subspaces for **A** in (3.223).

3.22 Repeat the previous problem for **A** in (3.225).

3.23 (a) Show that the null space of **B** is a subset of the null space of **AB**. (b) Show that the column space of **AB** is a subset of the column space of **A**.

Gaussian Elimination

3.24 Repeat Problem 3.19 using GE.

3.25 Repeat Problem 3.20 using GE.

3.26 For the following system of equations in augmented form, find all values of α such that there are (a) infinite solutions, (b) no solutions, and (c) a unique solution.

$$[\mathbf{A} \mid \mathbf{x}] = \left[\begin{array}{ccc|c} 1 & 1 & -1 & 2 \\ 2 & -1 & 2 & 1 \\ 1 & 1 & \alpha^2 - 2 & \alpha + 1 \end{array}\right]. \tag{3.224}$$

3.27 Find LU and LDU decompositions for the square matrices in Problem 3.18.

3.28 Repeat the previous problem for the rectangular matrices

$$\text{(a) } \mathbf{A} = \begin{bmatrix} 2 & 1 & 3 & 1 & 1 \\ 4 & 1 & -1 & -2 & 3 \\ 1 & 4 & -1 & 2 & 5 \end{bmatrix}, \qquad \text{(b) } \mathbf{B} = \begin{bmatrix} 3 & 1 & -2 \\ 2 & 6 & 1 \\ 4 & -1 & 1 \\ -1 & 2 & 3 \\ 2 & 1 & -4 \end{bmatrix}. \tag{3.225}$$

3.29 Solve for $\{y_1, y_2, y_3\}$ without multiplying together the two matrices:

$$\begin{bmatrix} 2 & 0 & 0 \\ 1 & 2 & 0 \\ 1 & 1 & 2 \end{bmatrix} \begin{bmatrix} 1 & 3 & 0 \\ 0 & 2 & 1 \\ 0 & 0 & 1 \end{bmatrix} \begin{bmatrix} y_1 \\ y_2 \\ y_3 \end{bmatrix} = \begin{bmatrix} 1 \\ 2 \\ 3 \end{bmatrix}. \tag{3.226}$$

3.30 Derive the particular solution $\mathbf{y}_p$ and the homogeneous solution $\mathbf{y}_h$ for the following system of equations:

$$\begin{bmatrix} 1 & 0 & -1 & 2 \\ -2 & 1 & 3 & -4 \\ 0 & 2 & 5 & 1 \end{bmatrix} \begin{bmatrix} y_1 \\ y_2 \\ y_3 \\ y_4 \end{bmatrix} = \begin{bmatrix} 2 \\ -1 \\ 1 \end{bmatrix}. \tag{3.227}$$

3.31 Repeat the previous problem for **A** in (3.225) and $\mathbf{y} = [\, y_1, y_2, y_3, y_4, y_5]^T$.

3.32 Let **A** be a nonsingular matrix with rank R. Show that the determinant of its adjugate matrix is

$$\det(\text{adj}(\mathbf{A})) = [\det(\mathbf{A})]^{R-1}. \tag{3.228}$$

Eigendecomposition

3.33 Use the product property of the determinant and the eigendecomposition of **A** to show that $\det(\mathbf{A}) = \prod_{n=1}^{N} \lambda_n$.

3.34 Write the eigenvalues of (a) $\mathbf{A}^n$ for $n \in \mathcal{N}$ and (b) $\mathbf{A}^{-1}$, both in terms of the eigenvalues $\{\lambda_1, \ldots, \lambda_N\}$ of nonsingular **A**.

3.35 The characteristic equation for matrix **A** in (3.188) is an Nth-order polynomial in λ. Show that $\alpha_0 = \det(\mathbf{A})$.

3.36 Find the eigendecomposition of **A** in (3.220).

3.37 Repeat the previous problem for **A** in (3.223).

3.38 When $\mathbf{A} = \mathbf{A}^T$ is a symmetric matrix, show that the eigendecomposition is $\mathbf{A} = \mathbf{Q}\mathbf{D}\mathbf{Q}^H$ and the normalized eigenvectors are orthogonal: $\mathbf{Q}^H\mathbf{Q} = \mathbf{I}$.

Computer Problems

3.39 Use eig in MATLAB to derive the eigendecomposition for **B** in (3.223).

3.40 Use mesh in MATLAB to generate the two planes in Figure 3.10(a) representing the linear equations in (3.144).

3.41 Use det in MATLAB and Cramer's rule to find **y** in (3.227).

4

COMPLEX NUMBERS AND FUNCTIONS

4.1 INTRODUCTION

We first provide a brief review of different number systems. The *natural numbers* consist of the set $\{1, 2, \ldots\}$ and are denoted by $\mathcal{N}$. They are "natural" because they are used to represent various numbers of objects in nature, such as the number of apples on a tree. The natural numbers are closed under addition, which means

$$x \in \mathcal{N}, y \in \mathcal{N} \implies x + y \in \mathcal{N}. \tag{4.1}$$

With subtraction, however, 0 and negative numbers must be included in the set in order for it to be closed under addition, thus yielding the *integers* $\mathcal{Z} \triangleq \{\ldots, -2, -1, 0, 1, 2, \ldots\}$. By including negative numbers in $\mathcal{Z}$, a sense of *direction* is implied. For example, we can take two steps forward or two steps backward, which are represented by +2 and −2, respectively. We denote the nonnegative integers by $\mathcal{Z}^+$, consisting of the union of $\mathcal{N}$ and 0: $\mathcal{Z}^+ \triangleq \mathcal{N} \cup \{0\}$.

Division of the integers leads to the set of *rational numbers* $\mathcal{Q}$, which have the form a/b with $a, b \in \mathcal{Z}$ and $b \neq 0$. The fractional part of a rational number has a finite number of digits, as is the case for $1/2 = 0.5$, or it repeats such as $1/3 = 0.3333 \ldots$ The *irrational numbers* comprise all other real numbers: they cannot be expressed as the ratio a/b. This set, written as the difference $\mathcal{R} - \mathcal{Q}$, includes Napier's constant $e = 2.7182818284\ldots$, $\pi = 3.1415926535\ldots$, $\sqrt{2} = 1.4142135623\ldots$, and so

Mathematical Foundations for Linear Circuits and Systems in Engineering, First Edition. John J. Shynk.

Companion Website: http://www.wiley.com/go/linearcircuitsandsystems

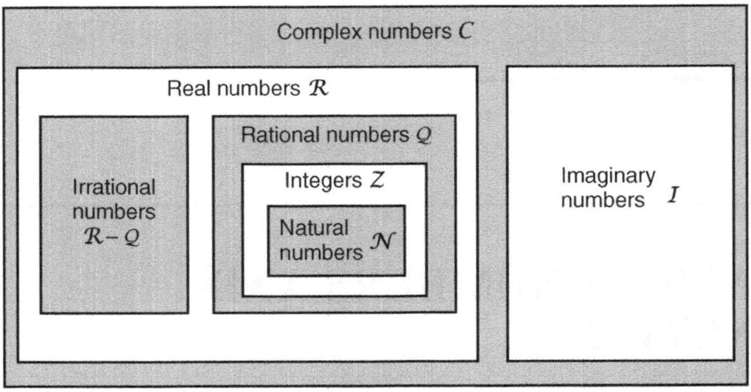

Figure 4.1 Subsets of the complex numbers $\mathcal{C}$. (The imaginary numbers include the same subsets as the real numbers, multiplied by j, but these are not shown for visual clarity. The rectangles do not indicate the relative sizes of the subsets of $\mathcal{C}$.)

TABLE 4.1 Comparison of Infinite Sets

Type	Infinite Sets with Equivalent Cardinality
Countable	$\mathcal{Z}, \mathcal{Z}^+, \mathcal{N}, \mathcal{Q}$
Uncountable	$\mathcal{R}, \mathcal{R}^+, \mathcal{R}-\mathcal{Q}, [0, 1], \mathcal{I}, \mathcal{C}$

on. By definition, the fractional part of an irrational number does not have a finite number of digits, nor does it repeat. The set of nonnegative real numbers (which includes 0) is denoted by $\mathcal{R}^+$. The notation for various sets of numbers was summarized earlier in Table 1.1. Figure 4.1 shows several subsets of the complex numbers $\mathcal{C}$. The rational numbers $\mathcal{Q}$, the real numbers $\mathcal{R}$, and the complex numbers $\mathcal{C}$ are *fields* (see Chapter 3), whereas the natural numbers $\mathcal{N}$ and the integers $\mathcal{Z}$ are not because they do not have multiplicative inverses.

Table 4.1 indicates the relative sizes of several subsets of $\mathcal{C}$; it also includes the set of imaginary numbers $\mathcal{I}$ discussed in the next section. A set of numbers is countable if it is possible to place its elements in one-to-one correspondence with the integers. Although $\mathcal{N}$ appears to have fewer elements than $\mathcal{Z}$, these two sets actually have the same size.

Definition: Cardinality The *cardinality* of a set is its number of elements. The cardinality of set A is denoted by $|A|$.

The fact that $\mathcal{N}$ and $\mathcal{Z}$ have the same cardinality is due to the nature of sets with an infinity of elements. The infinite set consisting of the real numbers is quite different, however, because for any two numbers n_1 and n_2, there exists an infinity of numbers in between, as in the unit interval $[0, 1]$. This is not the case for the integers: for example, there are no integers between 0 and 1.

4.2 IMAGINARY NUMBERS

Next, we show how number systems can be extended beyond the set $\mathcal{R}$. Consider the following quadratic equation:

$$x^2 - 1 = 0 \implies x^2 = 1, \tag{4.2}$$

which has two real solutions: $x = \pm 1$. They are also called the *roots* of the equation. Suppose now that (4.2) is modified slightly to

$$x^2 + 1 = 0 \implies x^2 = -1. \tag{4.3}$$

Obviously, there is no real number such that when it is squared, -1 is obtained. In order to solve such an equation, we must extend the set of numbers beyond $\mathcal{R}$, analogous to the extension of natural numbers $\mathcal{N}$ to the integers $\mathcal{Z}$ for subtraction, and similarly from $\mathcal{Z}$ to the rational numbers $\mathcal{Q}$ for division.

Definition: Imaginary Number An *imaginary number* is of the form jb where

$$j \triangleq \sqrt{-1}, \tag{4.4}$$

and $b \in \mathcal{R}$.

Using this definition, (4.3) has two imaginary solutions: $x = \pm j$ because $j^2 = -1$ and $(-j)^2 = (-1)^2 j^2 = -1$. (Although i is used to represent $\sqrt{-1}$ in mathematics courses, j is usually used in engineering because a circuit current is denoted by i.) The designation "imaginary" is actually a misnomer; the imaginary numbers $\mathcal{I} = j\mathcal{R}$ have as much significance and validity as the real numbers. As discussed in the next section, an imaginary number can be viewed as one *coordinate* of a *complex number*, which is an extension of the real line to the *complex plane*.

If (4.2) and (4.3) are plotted as functions $y = x^2 - 1$ and $y = x^2 + 1$, as shown in Figure 4.2(a), their solutions are found graphically by determining where the curves intersect the x-axis defined by $y = 0$ (the dotted line). For $y = x^2 - 1$, the *zero-crossings* occur at $x = \pm 1$ as expected. Since $y = x^2 + 1$ does not cross the dotted horizontal line, $x^2 + 1 = 0$ has no *real* solutions. It is clear from this graphical interpretation that a real function without any zero-crossings has no real solutions. Consider the cubic equations plotted in Figure 4.2(b), each of which has a single zero-crossing. This means that those functions have only one real solution. They factor as follows:

$$y = x^3 - 1 = (x - 1)(x^2 + x + 1), \tag{4.5}$$

$$y = x^3 + 1 = (x + 1)(x^2 - x + 1), \tag{4.6}$$

where the leading factors give real roots: $x - 1 \implies x = 1$ and $x + 1 \implies x = -1$. The other factors, which are quadratic, are plotted in Figure 4.2(c), demonstrating that they do not have any real solutions. It turns out that they have *complex roots* as described in the next section.

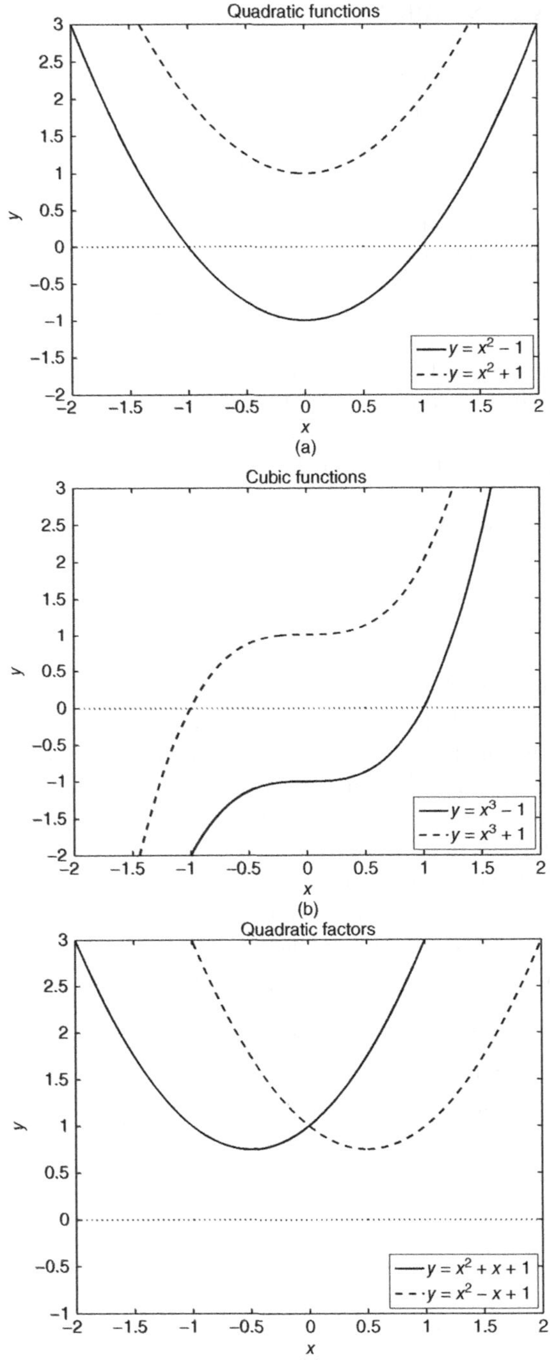

Figure 4.2 (a) Quadratic functions in (4.2) and (4.3). (b) Cubic functions in (4.5) and (4.6). (c) Quadratic factors of the cubic functions.

Example 4.1 Consider another quadratic equation:

$$x^2 + 4 = 0 \implies x^2 = -4, \tag{4.7}$$

which also has no real solutions. By allowing the set of imaginary numbers $\mathcal{I}$ to be included in the number system, it is clear that the two solutions are $x = \pm j2$, which are easily verified: $(j2)^2 = j^2 4 = -4$ and $(-j2)^2 = j^2(-2)^2 = -4$.

We summarize the basic properties of j.

- *Multiplication and division*:

$$j^2 = -1, \quad j^3 = -j, \quad 1/j = -j. \tag{4.8}$$

- *Addition and subtraction*:

$$j + j = 2j, \quad j - j = 0, \quad jb \pm j = j(b \pm 1), \tag{4.9}$$

for $b \in \mathcal{R}$.

Under algebraic operations, j behaves like any real number with the important exception that $j^2 = -1$ by definition.

4.3 COMPLEX NUMBERS

Consider one of the quadratic factors mentioned in the previous section:

$$y = x^2 + x + 1, \tag{4.10}$$

which, using the quadratic formula for $y = 0$, we find does not have any real roots:

$$x_1, x_2 = \frac{-1 \pm \sqrt{1^2 - 4 \times 1 \times 1}}{2} = -\frac{1}{2} \pm \frac{\sqrt{-3}}{2}. \tag{4.11}$$

By using j to "handle" $\sqrt{-3} = \sqrt{-1}\sqrt{3} = j\sqrt{3}$, we can write these solutions as the following two *complex numbers*:

$$x_1, x_2 = -\frac{1}{2} \pm \frac{j\sqrt{3}}{2}, \tag{4.12}$$

which have real part $-1/2$ and imaginary parts $\pm\sqrt{3}/2$. The other quadratic factor in Figure 4.2(c), given by

$$x^2 - x + 1 = 0, \tag{4.13}$$

has solutions

$$x_1, x_2 = \frac{1 \pm \sqrt{(-1)^2 - 4 \times 1 \times 1}}{2} = \frac{1}{2} \pm \frac{j\sqrt{3}}{2}. \tag{4.14}$$

From these two examples, we have the following definition of a complex number.

Definition: Complex Number A *complex number* has the form:

$$c = a + jb, \tag{4.15}$$

where $\{a, b\} \in \mathcal{R}$ and $j \triangleq \sqrt{-1}$.

A complex number has *two coordinates*: a real part and an imaginary part that are perpendicular to each other as illustrated in Figure 4.3. We use $C = \mathcal{R} + j\mathcal{R}$ to represent all complex numbers on the *complex plane*, which is also known as an *Argand diagram*. The following notation is used for the real and imaginary parts of complex number c:

$$\text{Re}(c) = a, \quad \text{Im}(c) = b. \tag{4.16}$$

Multiplication of two complex numbers yields

$$(a_1 + jb_1)(a_2 + jb_2) = a_1a_2 - b_1b_2 + j(a_1b_2 + a_2b_1), \tag{4.17}$$

where $j^2 = -1$ has been used to produce the term $-b_1b_2$. Consider the product

$$(a + jb)(a - jb) = a^2 + b^2, \tag{4.18}$$

which is strictly real. The multiplier $a - jb$, known as the *complex conjugate* of $c = a + jb$, is denoted by $c^* = a - jb$ and is depicted in Figure 4.3 (sometimes the notation $\overline{c}$ is used). It is important to note that $-c = -a - jb$ is not the same as c^*. The complex conjugate is useful for rewriting the ratio of complex numbers in the *standard form* of (4.15):

$$\frac{1}{a + jb} = \frac{a - jb}{(a + jb)(a - jb)} = \frac{a - jb}{a^2 + b^2} = \frac{a}{a^2 + b^2} - j\frac{b}{a^2 + b^2}, \tag{4.19}$$

which has real and imaginary parts $a/(a^2 + b^2)$ and $-b/(a^2 + b^2)$, respectively. Observe that j is not included as part of the imaginary component; it simply denotes

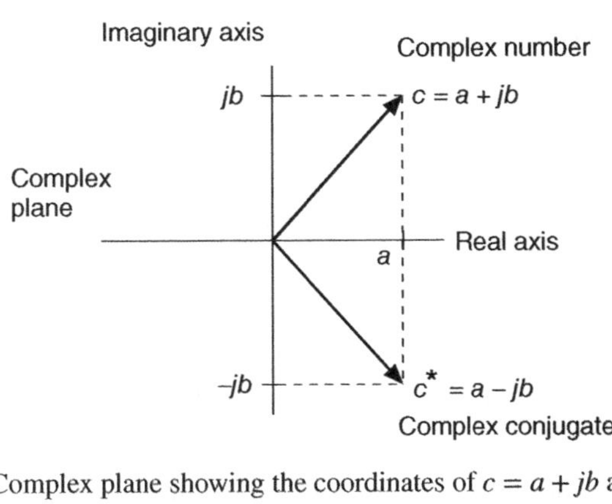

Figure 4.3 Complex plane showing the coordinates of $c = a + jb$ and $c^* = a - jb$.

that the quantity multiplying it is the imaginary part of the complex number. As a result, we can view j in such an equation as a "marker" that indicates which part of the expression is imaginary, corresponding to the vertical axis on the complex plane. (The quantity j by itself, depending on the context, can also refer to the complex number with real part 0 and imaginary part 1.) From the previous result, the ratio of two complex numbers is written in standard form as follows:

$$\frac{a+jb}{c+jd} = \frac{(a+jb)(c-jd)}{(c+jd)(c-jd)} = \frac{ac+bd}{c^2+d^2} + j\frac{bc-ad}{c^2+d^2}. \tag{4.20}$$

For real function $y = f(x)$, like those considered at the beginning of this chapter, complex roots must occur as *complex conjugate pairs*. Thus, a polynomial with odd degree must have at least one real root, as we found for the cubic equations in (4.5) and (4.6). The other two roots may also be real or they could be complex, but they must be a conjugate pair of the form shown in Figure 4.3. The roots of a polynomial with even degree could be some combination of real roots and complex conjugate pairs.

4.4 TWO COORDINATES

Returning to the interpretation that a complex number has two coordinates and j operates as a marker for the imaginary part, the solution to equations such as (4.10) can be viewed as a generalization where x now has *two components*. This additional *degree of freedom*, represented by the imaginary part of a complex number, allows for a solution to exist in equations that have no real solutions.

Example 4.2 For example, substitute $x = a + jb$ into (4.10) and rewrite it in terms of its real and imaginary parts:

$$y = (a+jb)^2 + (a+jb) + 1 \implies y = (a^2 - b^2 + a + 1) + j(2ab + b). \tag{4.21}$$

There are now two parameters $\{a, b\}$, whereas before we simply had real $x = a$. The real part of y in (4.21) is plotted in Figure 4.4(a) versus a and b, where we find that it crosses the plane defined by $\text{Re}(y) = 0$. Since the original equation in (4.10) is real, y must also be real, and the imaginary part in (4.21) must be constrained to 0:

$$2ab + b = 0 \implies b = 0 \text{ or } a = -1/2. \tag{4.22}$$

We are not interested in the case $b = 0$ because that reduces the problem to real-valued $x = a$. Thus, in order for y to be real, we must have $x = -1/2 + jb$ for the particular quadratic equation in (4.10). Substituting $a = -1/2$ into (4.21) gives

$$y = 3/4 - b^2 = 0 \implies b = \pm\sqrt{3}/2 \implies a + jb = -1/2 \pm \sqrt{3}/2, \tag{4.23}$$

which is exactly the solution in (4.12). The function in (4.21) for $a = -1/2$ is shown in Figure 4.4(b), which is a "slice" of the three-dimensional plot defined by the

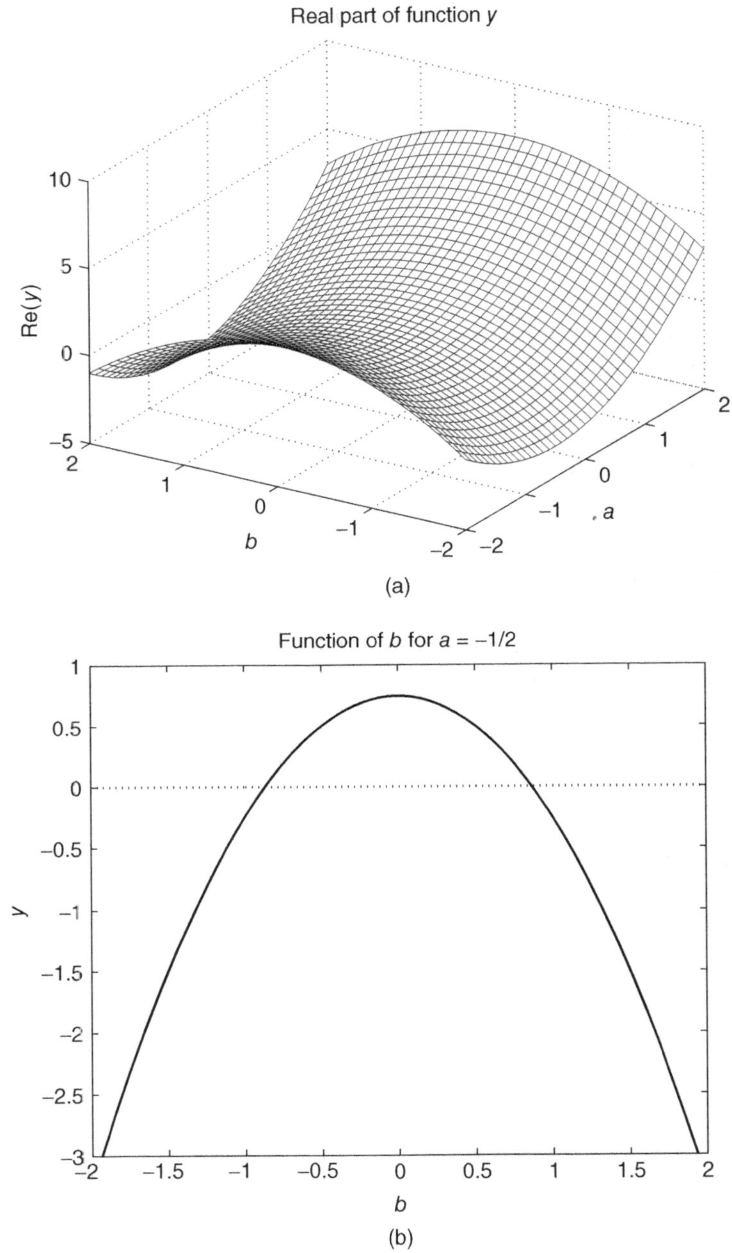

Figure 4.4 Plots of the quadratic function in (4.21) versus the $\{a, b\}$ components of $x = a + jb$. (a) Real part of y. (b) y for $a = -1/2$.

vertical plane at $a = -1/2$. This quadratic function has two solutions where it crosses the dotted line at $y = 0$.

From the previous discussions, we can view complex solutions to equations in two ways: (i) when $\sqrt{-1}$ is encountered, j is used as a marker and we proceed to find solutions using the algebra of complex numbers. (ii) Alternatively, the variable x has two coordinates (real and imaginary parts) with an additional degree of freedom so that solutions are possible. For the previous example, we found that y is real if the solutions are of the form $x = -1/2 + jb$ with $b \in \mathcal{R}$. The plot of y versus b in Figure 4.4(b) is located on the vertical dashed line in Figure 4.5 where $a = -1/2$, and the two complex solutions where y crosses the dotted line are represented by the solid circles on the complex plane.

Example 4.3 Consider the quartic equation:

$$\begin{aligned} x^4 - 2x^3 + x^2 + 2x - 2 &= (x^2 - 1)[(x-1)^2 + 1] \\ &= (x-1)(x+1)(x-1+j)(x-1-j), \end{aligned} \tag{4.24}$$

which has two complex roots and two real roots. Figure 4.6 shows that this function (the solid line) has two zero-crossings. The quadratic factor $x^2 - 1$ has two zero-crossings (two real roots), and the other quadratic factor $(x-1)^2 + 1 = x^2 - 2x + 2$ does not have any zero-crossings, and so its two roots must be a complex conjugate pair. Of course, the two zero-crossings of the quartic function coincide with those of the quadratic factor $x^2 - 1$ in (4.24).

4.5 POLAR COORDINATES

The previous examples shown on the complex plane are represented in the Cartesian coordinate system defined by the perpendicular real and imaginary axes. In this

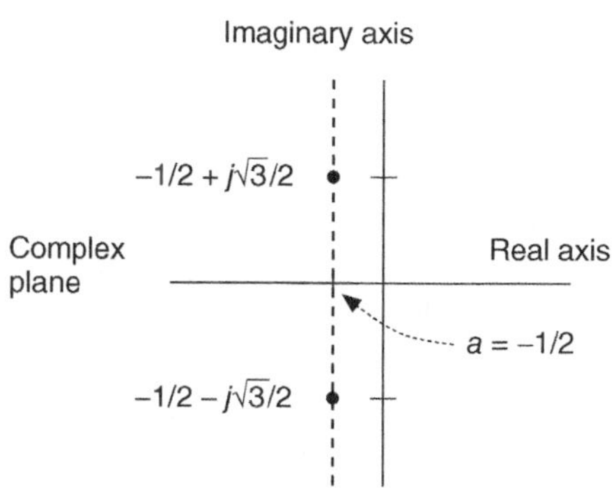

Figure 4.5 Complex plane showing the vertical axis (the dashed line) where $y = x^2 + x + 1$ is real for complex x and the two complex solutions (the solid circles) of $x^2 + x + 1 = 0$.

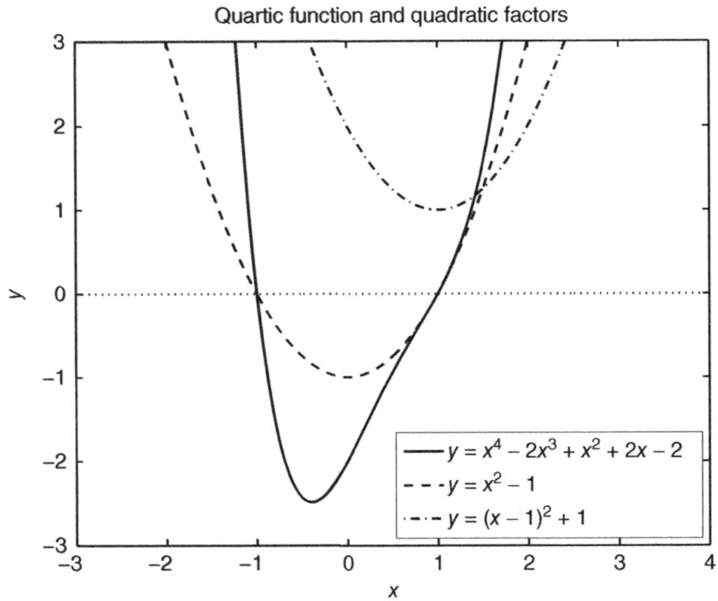

Figure 4.6 Quartic function and factors for Example 4.3.

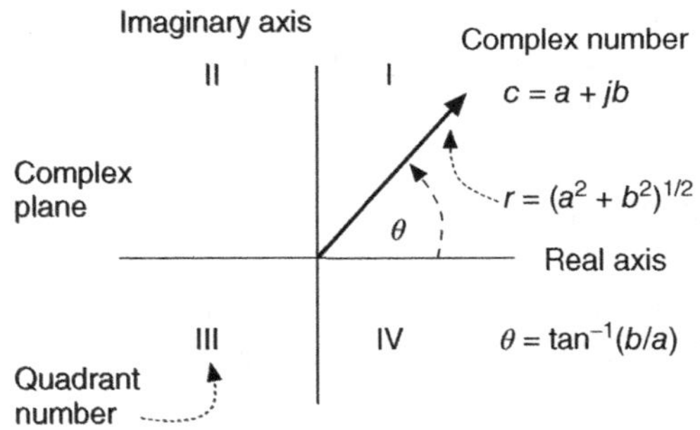

Figure 4.7 Complex plane showing the polar coordinates for complex number c.

section, we describe how to represent a complex number using *polar coordinates*. Figure 4.7 shows a diagonal line connecting the origin and the point representing the complex number $c = a + jb$. This same number is described using polar coordinates, in terms of the *radius* r of the line and its *angle* θ relative to the positive real axis. From trigonometry, we know

$$r^2 = a^2 + b^2 \implies r = \sqrt{a^2 + b^2}, \tag{4.25}$$

where the positive square root is retained because r cannot be negative. Note that b^2 is used in this equation, not $(jb)^2$, because as mentioned previously, j is a marker used to identify the imaginary part of a complex number. The angle θ is derived from the usual trigonometric formulas:

$$r\cos(\theta) = a \implies \theta = \cos^{-1}(a/r), \quad r\sin(\theta) = b \implies \theta = \sin^{-1}(b/r), \tag{4.26}$$

and the ratio of these two expressions yields

$$\frac{r\sin(\theta)}{r\cos(\theta)} = \frac{b}{a} = \tan(\theta) \implies \theta = \tan^{-1}(b/a). \tag{4.27}$$

Thus, a complex number can be written using a second formulation.

Definition: Polar Form The *polar form* of a complex number is

$$c = a + jb = r\angle\theta = |c|\angle\arg(c), \tag{4.28}$$

where the notation $r\angle\theta$ specifies the radius $r = |c|$ and angle $\theta = \arg(c)$ (*argument* of c) on the complex plane relative to the positive real axis. The quantities $|c|$ and $\arg(c)$ are the *magnitude* and *phase* of c.

We demonstrate later how to perform algebraic operations with complex numbers written in polar form. The *rectangular form* $a + jb$ in Cartesian coordinates is generated from the polar form using (4.26).

Example 4.4 Consider again the quadratic factor $x^2 + x + 1 = 0$, which has roots $x = -1/2 \pm j\sqrt{3}/2$. The polar coordinates of the roots are

$$r^2 = (-1/2)^2 + (\pm\sqrt{3}/2)^2 = 1/4 + 3/4 \implies r = 1, \tag{4.29}$$

and

$$\theta = \tan^{-1}(\pm(\sqrt{3}/2)/(-1/2)) = \tan^{-1}(\mp\sqrt{3}) = \mp\tan^{-1}(\sqrt{3}) = \pm 120°. \tag{4.30}$$

As a result, the following expressions are equivalent representations for the complex solutions:

$$x_1, x_2 = -1/2 \pm j\sqrt{3}/2 \quad \text{(rectangular form)}, \tag{4.31}$$

$$x_1, x_2 = 1\angle \pm 120° \quad \text{(polar form)}. \tag{4.32}$$

This example illustrates that we must determine in which *quadrant* of the complex plane the complex number is located in order to obtain the correct angle. These quadrants are labeled I–IV counterclockwise in Figure 4.7, and the corresponding signs of the components $\{a, b\}$ for a complex number are summarized in Table 4.2. For quadrants II–IV, either 180° or 360° is added to $\tan^{-1}(b/a)$.

TABLE 4.2 Angle θ Based on the Quadrant in the Complex Plane

Quadrant	$\{a, b\}$	Angle θ
I	$a > 0, b > 0$	$\tan^{-1}(b/a)$
II	$a < 0, b > 0$	$\tan^{-1}(b/a) + 180°$
III	$a < 0, b < 0$	$\tan^{-1}(b/a) + 180°$
IV	$a > 0, b < 0$	$\tan^{-1}(b/a) + 360°$

Example 4.5 The complex number $x = -1/2 + j\sqrt{3}/2$ is located in quadrant II. The inverse tangent function gives $-60°$, and so, the actual angle is $(-60 + 180)° = 120°$. Likewise, $x = -1/2 - j\sqrt{3}/2$ is located in quadrant III such that $\tan^{-1}(b/a) = 60°$ and $\theta = 60° + 180° = 240°$, which is the same as $-120°$.

Next, we describe the difference between angles specified in degrees versus *radians*. Figure 4.8 shows the *unit circle* on the complex plane consisting of all complex numbers with radius $r = 1$. Recall from geometry that the circumference of a circle is πd radians, where $d = 2r$ is its diameter. The unit circle has circumference 2π, which is why angles on the complex plane are specified in radians. This is also the reason why the trigonometric functions repeat with period 2π; sine and cosine are defined in terms of the horizontal and vertical axes for a unit circle (though not necessarily on the complex plane). Angles in degrees are simply the corresponding values of the unit circle divided into 360 equal intervals ("pie slices"). As a result, $2\pi \Leftrightarrow 360°$, $\pi \Leftrightarrow 180°$, $\pi/2 \Leftrightarrow 90°$, and so on. For convenience, we have provided in Table 4.3 the conversions for several common angles, as well as the corresponding tangent values.

We conclude this section with a summary of the basic algebraic properties of complex numbers in polar form, which are readily verified.

- *Multiplication*:

$$c_1 c_2 = r_1 r_2 \angle(\theta_1 + \theta_2). \tag{4.33}$$

- *Division*:

$$c_1 / c_2 = (r_1 / r_2) \angle(\theta_1 - \theta_2). \tag{4.34}$$

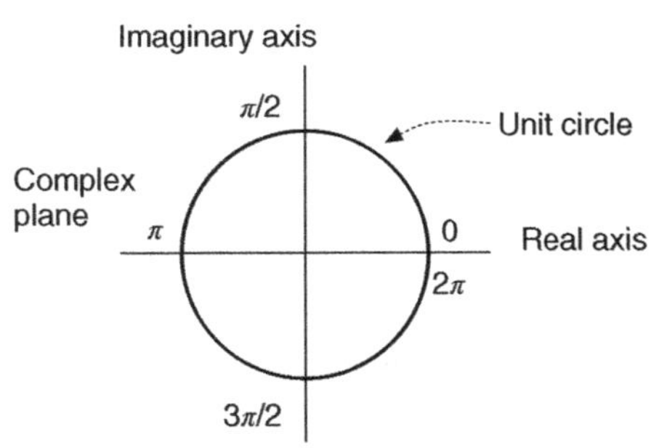

Figure 4.8 Angles in radians along the unit circle.

TABLE 4.3 Angle θ in Radians and Degrees

Radians	Degrees	tan(θ)	Radians	Degrees	tan(θ)
$0, 2\pi$	$0°, 360°$	0	π	180°	0
$\pi/6$	30°	$1/\sqrt{3}$	$7\pi/6$	210°	$1/\sqrt{3}$
$\pi/4$	45°	1	$5\pi/4$	225°	1
$\pi/3$	60°	$\sqrt{3}$	$4\pi/3$	240°	$\sqrt{3}$
$\pi/2$	90°	∞	$3\pi/2$	270°	∞
$2\pi/3$	120°	$-\sqrt{3}$	$5\pi/3$	300°	$-\sqrt{3}$
$3\pi/4$	135°	-1	$7\pi/4$	315°	-1
$5\pi/6$	150°	$-1/\sqrt{3}$	$11\pi/6$	330°	$-1/\sqrt{3}$

These are much easier to calculate than when c is expressed in rectangular form. In order to add and subtract two complex numbers, it is necessary that they be converted to rectangular form. In the next section, we show that it is more convenient to represent complex numbers in polar form using the exponential function.

4.6 EULER'S FORMULA

Complex numbers expressed in polar form can also be written as

$$c = r\exp(j\theta), \tag{4.35}$$

where $\exp(\theta)$ is the ordinary exponential function, $\exp(j\theta)$ is the *complex exponential* function, and the units of θ are radians. The exponential function with exponent j has a special identity known as *Euler's formula*:

$$\exp(j\theta) = \cos(\theta) + j\sin(\theta), \tag{4.36}$$

which is a complex number with real part $\cos(\theta)$ and imaginary part $\sin(\theta)$. (This equation is similar to the expression in (1.112) for the exponential function written in terms of hyperbolic functions, except here the exponential function is complex.) Observe that (4.36) gives the complex number on the unit circle of the complex plane at angle θ. It has squared magnitude

$$|\exp(j\theta)|^2 = \cos^2(\theta) + \sin^2(\theta) = 1, \tag{4.37}$$

and the angle follows from the ratio of the imaginary and real parts:

$$\theta = \tan^{-1}\left(\frac{\sin(\theta)}{\cos(\theta)}\right) = \tan^{-1}(\tan(\theta)). \tag{4.38}$$

Thus, any complex number with angle θ and magnitude r can be written by using the exponential function in (4.35), and it is located on a circle of radius r on the complex

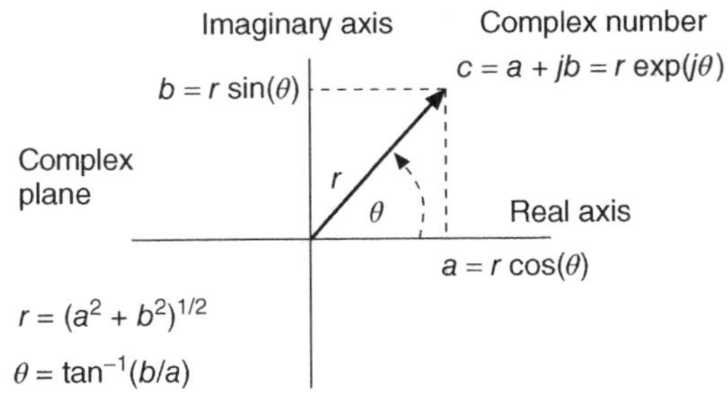

Figure 4.9 Complex plane showing rectangular coordinates and polar coordinates using the complex exponential function for complex number c.

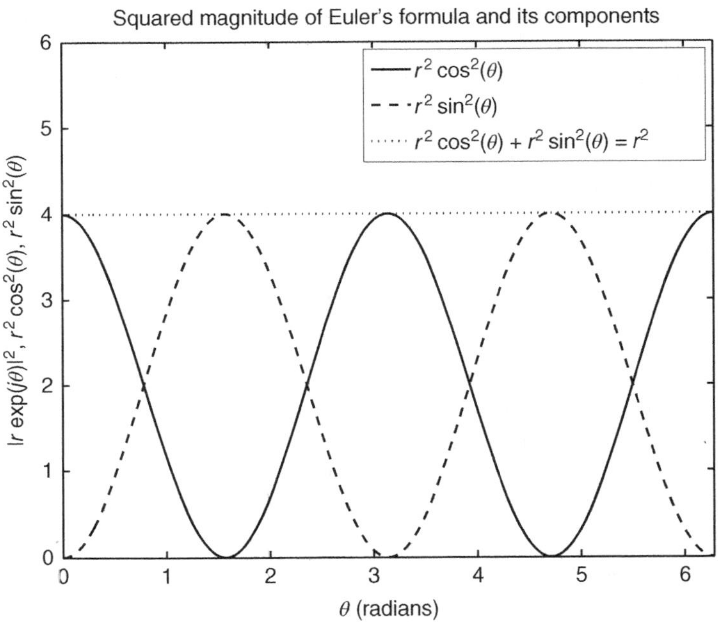

Figure 4.10 Euler's formula showing $r^2\cos^2(\theta)$, $r^2\sin^2(\theta)$, and $r^2\cos^2(\theta) + r^2\sin^2(\theta) = r^2$ for $r = 2$.

plane at angle θ with respect to the positive real axis as illustrated in Figure 4.9. A plot of the squared magnitude $|c|^2 = r^2\cos^2(\theta) + r^2\sin^2(\theta) = r^2$ and its components is shown in Figure 4.10, verifying that they in fact sum to a constant.

Euler's formula *simultaneously* describes the sine and cosine of angle θ by using *two coordinates*. Chapter 1 provided a review of the trigonometric definitions $\sin(\theta) \triangleq y/r$ and $\cos(\theta) \triangleq x/r$, where x is the projection of the hypotenuse of a right triangle onto the horizontal axis, and y is its projection onto the vertical axis. Euler's formula represents both axes together as $c = a + jb$ with real and imaginary components $a = r\cos(\theta)$ and $b = r\sin(\theta)$. In the next chapter, Euler's formula is used to write the complex exponential as a function of time (a waveform) as $\exp(j\omega_o t)$ where ω_o is a constant angular frequency with units radians/second (rad/s).

In order to prove Euler's formula, the derivative properties of the exponential function can be used. First, write the following product by substituting (4.36):

$$\exp(-j\theta)\exp(j\theta) = 1 = \exp(-j\theta)[\cos(\theta) + j\sin(\theta)] \triangleq f(\theta). \tag{4.39}$$

Differentiating $f(\theta)$ with respect to θ, the product rule yields

$$\frac{d}{d\theta}f(\theta) = -j\exp(-j\theta)[\cos(\theta) + j\sin(\theta)] + \exp(-j\theta)[-\sin(\theta) + j\cos(\theta)]. \tag{4.40}$$

By factoring $\exp(-j\theta)$, this expression is rearranged using the basic algebraic properties of j:

$$\exp(-j\theta)([\sin(\theta) - \sin(\theta)] + j[\cos(\theta) - \cos(\theta)]) = 0. \tag{4.41}$$

Since the derivative of (4.39) is 0, $f(\theta)$ must be a constant for every θ. If we can find a value for $f(\theta)$ for some θ, called a *boundary condition*, then we know $f(\theta)$ for every θ. Since $f(0) = \exp(-j0)[\cos(0) + j\sin(0)] = 1$, the function is $f(\theta) = 1$, which verifies the left-hand side of (4.39) and proves (4.36). It is also clear that

$$\exp(-j\theta) = \cos(\theta) - j\sin(\theta), \tag{4.42}$$

because $\cos(-\theta) = \cos(\theta)$ (an even function) and $\sin(-\theta) = -\sin(\theta)$ (an odd function). From this expression, we have further verification of Euler's formula:

$$\begin{aligned}\exp(j\theta)\exp(-j\theta) &= \cos^2(\theta) + \sin^2(\theta) + j\sin(\theta)\cos(\theta) - j\cos(\theta)\sin(\theta)\\ &= \cos^2(\theta) + \sin^2(\theta) = 1.\end{aligned} \tag{4.43}$$

The sine and cosine functions can be written in terms of complex exponentials as follows:

$$\exp(j\theta) + \exp(-j\theta) = 2\cos(\theta) \implies \cos(\theta) = (1/2)[\exp(j\theta) + \exp(-j\theta)], \tag{4.44}$$

$$\exp(j\theta) - \exp(-j\theta) = 2j\sin(\theta) \implies \sin(\theta) = (1/2j)[\exp(j\theta) - \exp(-j\theta)], \tag{4.45}$$

which are called Euler's inverse formulas. An interesting result known as *Euler's identity* is obtained when $\theta = \pi$:

$$\exp(j\pi) = \cos(\pi) + j\sin(\pi) \implies \exp(j\pi) + 1 = 0. \tag{4.46}$$

Since $\exp(j\pi) = e^{j\pi}$, this simple equation ties together the five fundamental numbers in mathematics: 0, 1, $j = \sqrt{-1}$, π, and e. Of course, this result is readily visible on the unit circle in Figure 4.8 at $\theta = \pi$ where the complex number $c = a + jb$ has components $a = -1$ and $b = 0$. Multiplication and division are easily performed using the complex exponential because the exponents add and subtract, respectively.

- *Multiplication*:

$$c_1c_2 = r_1 \exp(j\theta_1)r_2 \exp(j\theta_2) = r_1r_2 \exp(j(\theta_1 + \theta_2)). \tag{4.47}$$

- *Division*:

$$c_1/c_2 = \frac{r_1 \exp(j\theta_1)}{r_2 \exp(j\theta_2)} = (r_1/r_2) \exp(j(\theta_1 - \theta_2)). \tag{4.48}$$

These operations use actual functions, whereas (4.33) and (4.34) show multiplication/division in terms of the *notation* $\angle$ for the angle.

Example 4.6 Consider the following equality:

$$c^n = r^n[\cos(n\theta) + j\sin(n\theta)], \tag{4.49}$$

which is easily verified from its polar form:

$$c^n = [r\exp(j\theta)]^n = r^n \exp(jn\theta). \tag{4.50}$$

Applying Euler's formula to the complex exponential with angle $n\theta$ yields $[\cos(\theta) + j\sin(\theta)]^n = \cos(n\theta) + j\sin(n\theta)$, known as *de Moivre's formula*.

Several properties of complex numbers are summarized in Table 4.4. An expression for the nth root of a complex number is also included, which is derived by letting the complex quantity $d \triangleq r_d \exp(\theta_d)$ be represented in the form $c = r\exp(\theta)$ by defining d to be the nth root of complex c:

$$d \triangleq \sqrt[n]{c} \implies c = d^n. \tag{4.51}$$

As a result:

$$c = r\exp(\theta) = r_d^n \exp(n\theta_d), \tag{4.52}$$

such that $r_d = \sqrt[n]{r}$. From Euler's formula, we know that equality is achieved when $n\theta_d = \theta + 2m\pi$ for $m = 0, \ldots, n-1$. This occurs because the sine and cosine functions are periodic with period 2π, and adding an integer multiple of 2π to the argument gives the same value for the complex exponential. As a result, $\theta_d = (\theta + 2m\pi)/n$ and the nth root is

$$\sqrt[n]{c} = \sqrt[n]{r}[\cos((\theta + 2m\pi)/n) + j\sin((\theta + 2m\pi)/n)], \quad m = 0, \ldots, n-1. \tag{4.53}$$

A special case of (4.53) with $r = 1$, $\theta = 0$, and $c = 1$ is known as the *nth root of unity*:

$$\sqrt[n]{1} = \cos(2m\pi/n) + j\sin(m\pi/n), \quad m = 0, \ldots, n-1. \tag{4.54}$$

The right-hand side defines n equally spaced points on the unit circle of the complex plane. For $n = 2$, the two points have angles $\theta = \{0, \pi\}$, and for $n = 3$, they have

TABLE 4.4 Properties of Complex Numbers

Property	Equation
Conjugation	$c^* = a - jb$, $(c_1 \pm c_2)^* = c_1^* \pm c_2^*$, $(c_1c_2)^* = c_1^*c_2^*$, $(c_1/c_2)^* = c_1^*/c_2^*$
Squared magnitude	$\|c\|^2 = cc^* = a^2 + b^2$
Negative	$-c = -a - jb$
Inverse	$c^{-1} = c^*/\|c\|^2$
Identities	$c \times 1 = c$, $c + 0 = c$
Polar form	$c = r\exp(j\theta)$ where $r = \sqrt{a^2+b^2} = \|c\|$ and $\theta = \tan^{-1}(b/a)$
Euler's formula	$\exp(j\theta) = \cos(\theta) + j\sin(\theta)$
Euler's inverse formulas	$\cos(\theta) = [\exp(j\theta) + \exp(-j\theta)]/2$ $\sin(\theta) = [\exp(j\theta) - \exp(-j\theta)]/2j$
Euler's identity	$\exp(j\pi) + 1 = 0$
de Moivre's formula	$[\cos(\theta) + j\sin(\theta)]^n = \cos(n\theta) + j\sin(n\theta)$ such that $c^n = r^n[\cos(n\theta) + j\sin(n\theta)]$
nth Root	$\sqrt[n]{c} = \sqrt[n]{r}[\cos(\theta_m/n) + j\sin(\theta_m/n)]$, where $\theta_m = \theta + 2m\pi$ for $m = 0,\ldots,n-1$
nth Root of unity	$\sqrt[n]{1} = [\cos(\theta_m/n) + j\sin(\theta_m/n)]$, where $\theta_m = 2m\pi$ for $m = 0,\ldots,n-1$
Complex logarithm	$z = \ln(c) = \ln(r) + j\theta$
Addition	$c_1 + c_2 = (a_1 + a_2) + j(b_1 + b_2)$
Subtraction	$c_1 - c_2 = (a_1 - a_2) + j(b_1 - b_2)$
Multiplication	$c_1c_2 = r_1r_2\exp(j(\theta_1 + \theta_2))$ $= (a_1a_2 - b_1b_2) + j(a_1b_2 + a_2b_1)$
Division	$c_1/c_2 = (r_1/r_2)\exp(j(\theta_1 - \theta_2))$ $= [(a_1a_2 + b_1b_2) + j(a_2b_1 - a_1b_2)]/(a_2^2 + b_2^2)$
Commutative	$c_1c_2 = c_2c_1$, $c_1 + c_2 = c_2 + c_1$
Associative	$c_1c_2c_3 = (c_1c_2)c_3 = c_1(c_2c_3)$ $c_1 + c_2 + c_3 = (c_1 + c_2) + c_3 = c_1 + (c_2 + c_3)$
Distributive	$c_1(c_2 + c_3) = c_1c_2 + c_1c_3$

angles $\theta = \{0, 2\pi/3, 4\pi/3\}$. The roots are easily remembered because they form the vertices of a *regular polygon* on the unit circle with one vertex located at $c = 1$ where $\theta = 0$. This is illustrated in Figure 4.11 for $n = 4$ where the vertices form a square (the dashed lines). The result in (4.53) also gives the vertices of a regular polygon, except they are located on a circle with radius $\sqrt[n]{r}$, and the polygon is rotated counterclockwise by angle θ about the origin. When $\theta = 0$ such that the polygon is not rotated, c is obviously a real number. For example, when $c = 2$ and $n = 4$, we have the same square as in Figure 4.11, except the roots (vertices) lie on a circle with radius $\sqrt[4]{2} \approx 1.1892$.

Example 4.7 Euler's formula can be used to perform rotations of vectors on the complex plane. For any complex number $c = r\exp(j\theta)$, the following multiplication

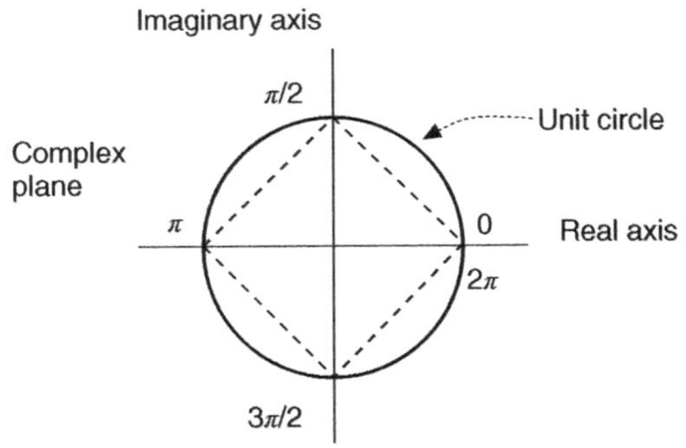

Figure 4.11 Roots of unity for $n = 4$, which form the vertices of a square.

causes c to be rotated counterclockwise by ϕ radians:

$$\exp(j\phi)c = \exp(j\phi)r\exp(j\theta) = r\exp(j(\phi + \theta)). \tag{4.55}$$

The radius is also changed by multiplying c by $\alpha\exp(j\phi)$ instead of $\exp(j\phi)$. Of course, these results follow from the multiplication property in (4.47).

The trigonometric identities in Appendix C can be proved using Euler's formula.

Example 4.8 For example, consider

$$\begin{aligned}\exp(jx)\exp(jy) &= [\cos(x) + j\sin(x)][\cos(y) + j\sin(y)] \\ &= \cos(x)\cos(y) - \sin(x)\sin(y) \\ &\quad + \ j[\sin(x)\cos(y) + \cos(x)\sin(y)].\end{aligned} \tag{4.56}$$

The left-hand side is

$$\begin{aligned}\exp(jx)\exp(jy) &= \exp(j(x + y)) \\ &= \cos(x + y) + j\sin(x + y).\end{aligned} \tag{4.57}$$

Equating the real and imaginary components of (4.56) with those of (4.57) yields identities for the cosine/sine sum of angles:

$$\cos(x + y) = \cos(x)\cos(y) - \sin(x)\sin(y), \tag{4.58}$$

$$\sin(x + y) = \sin(x)\cos(y) + \cos(x)\sin(y). \tag{4.59}$$

Similar results can be derived for the other trigonometric identities (see Problems 4.13 and 4.14).

Next, we provide some insights into the connection between e, sine, and cosine of Euler's formula. In Chapter 1, we mentioned that the exponential function $\exp(x)$ is motivated by a compound interest problem, corresponding to exponential growth or decay depending on the sign of x. The complex exponential $\exp(jx)$ does *not* exhibit real exponential growth or decay: it has a constant magnitude of 1. Earlier, the derivative property of the exponential function was used to prove Euler's formula. Here, we show that the complex exponential is the *only* function that can be used to represent the two-dimensional complex function $\cos(x) + j\sin(x)$. Define

$$f(jx) \triangleq \cos(x) + j\sin(x), \tag{4.60}$$

whose derivative exists because sine and cosine are smooth differentiable functions. Thus:

$$\frac{d}{dx}f(jx) = -\sin(x) + j\cos(x), \tag{4.61}$$

which can be rewritten as

$$\frac{d}{dx}f(jx) = j[\cos(x) + j\sin(x)] = \frac{d}{d(jx)}f(jx)\frac{d}{dx}jx = jf'(jx), \tag{4.62}$$

where the chain rule has been used on the right-hand side, $f'(\cdot)$ is the ordinary derivative of $f(\cdot)$, and $-1 = j^2$ has been substituted into the second expression. Cancelling j yields

$$f'(jx) = \cos(x) + j\sin(x) = f(jx). \tag{4.63}$$

Since the exponential function is the only function whose ordinary derivative is itself, we must have $f(jx) = \exp(jx)$. As a result, sine, cosine, and e are connected because of the *derivative properties* of these three functions.

We conclude this section with a definition of the logarithm for the complex exponential function.

Definition: Complex Natural Logarithm The *complex natural logarithm* of c is the complex number z such that $\exp(z) = c$. Substituting $c = r\exp(j\theta)$ yields $z = \ln(r\exp(j\theta)) = \ln(r) + j\theta$.

Observe that z is not unique: adding integer multiples of $j2\pi$ yields the same value for c:

$$\exp(z) = \exp(\ln(r) + j\theta + j2\pi n) = r\exp(j\theta)\exp(j2\pi n) = r\exp(j\theta) = c, \tag{4.64}$$

because $\exp(j2\pi n) = 1$ for every $n \in \mathcal{Z}$. This, of course, occurs because of the cyclical nature of the unit circle as the angle defined relative to the real axis exceeds 2π. In order to avoid this ambiguity, we often take the *principal value* of $z = \ln(r) + j(\theta + 2\pi n)$ such that the imaginary part $\theta + 2\pi n \in [-\pi, \pi]$.

4.7 MATRIX REPRESENTATION

From the matrix material in Chapter 3, we find that a complex number $c = a + jb$ can also be represented in matrix form as follows (Eves, 1980):

$$\mathbf{C} = \begin{bmatrix} a & -b \\ b & a \end{bmatrix}, \tag{4.65}$$

where the marker j is implied for the *off-diagonal* terms and we have used a bold uppercase letter to be consistent with the notation in the previous chapter. It is clear that the addition and subtraction of two complex numbers using this representation yields the correct complex form:

$$\begin{aligned} \mathbf{C}_1 \pm \mathbf{C}_2 &= \begin{bmatrix} a_1 & -b_1 \\ b_1 & a_1 \end{bmatrix} \pm \begin{bmatrix} a_2 & -b_2 \\ b_2 & a_2 \end{bmatrix} \\ &= \begin{bmatrix} a_1 \pm a_2 & -(b_1 \pm b_2) \\ b_1 \pm b_2 & a_1 \pm a_2 \end{bmatrix}, \end{aligned} \tag{4.66}$$

and so does multiplication:

$$\begin{aligned} \mathbf{C}_1\mathbf{C}_2 &= \begin{bmatrix} a_1 & -b_1 \\ b_1 & a_1 \end{bmatrix} \begin{bmatrix} a_2 & -b_2 \\ b_2 & a_2 \end{bmatrix} \\ &= \begin{bmatrix} a_1a_2 - b_1b_2 & -(a_1b_2 + b_1a_2) \\ a_1b_2 + b_1a_2 & a_1a_2 - b_1b_2 \end{bmatrix}. \end{aligned} \tag{4.67}$$

These matrices commute, which is not true in general:

$$\begin{aligned} \mathbf{C}_2\mathbf{C}_1 &= \begin{bmatrix} a_2 & -b_2 \\ b_2 & a_2 \end{bmatrix} \begin{bmatrix} a_1 & -b_1 \\ b_1 & a_1 \end{bmatrix} \\ &= \begin{bmatrix} a_1a_2 - b_1b_2 & -(a_1b_2 + b_1a_2) \\ a_1b_2 + b_1a_2 & a_1a_2 - b_1b_2 \end{bmatrix}. \end{aligned} \tag{4.68}$$

This property is evident from the form on the right-hand side of (4.67) where interchanging the subscripts yields the same matrix in (4.68).

From $c = \text{Re}(c) + j\text{Im}(c)$ and Euler's formula with $c = \exp(j\theta)$, we can write

$$\mathbf{C} = \begin{bmatrix} \cos(\theta) & -\sin(\theta) \\ \sin(\theta) & \cos(\theta) \end{bmatrix}, \tag{4.69}$$

which is the *rotation* matrix discussed in Chapter 3. This matrix has determinant $\cos^2(\theta) + \sin^2(\theta) = 1$, and it causes a two-dimensional vector to be rotated counterclockwise by angle θ on the plane defined by the two coordinates:

$$\begin{bmatrix} y_1 \\ y_2 \end{bmatrix} = \begin{bmatrix} \cos(\theta) & -\sin(\theta) \\ \sin(\theta) & \cos(\theta) \end{bmatrix} \begin{bmatrix} x_1 \\ x_2 \end{bmatrix} = \begin{bmatrix} x_1\cos(\theta) - x_2\sin(\theta) \\ x_1\sin(\theta) + x_2\cos(\theta) \end{bmatrix}. \tag{4.70}$$

This result using a matrix is consistent with that derived when multiplying a complex number by the complex exponential:

$$\begin{aligned}\exp(j\theta)(a+jb) &= [\cos(\theta)+j\sin(\theta)](a+jb) \\ &= [a\cos(\theta)-b\sin(\theta)]+j[a\sin(\theta)+b\cos(\theta)]. \qquad (4.71)\end{aligned}$$

Finally, note that because of the form of the matrix representation, the squared magnitude of c is generated from

$$\begin{aligned}\mathbf{C}\mathbf{C}^T &= \begin{bmatrix} a & -b \\ b & a \end{bmatrix}\begin{bmatrix} a & b \\ -b & a \end{bmatrix} = \begin{bmatrix} a^2+b^2 & 0 \\ 0 & a^2+b^2 \end{bmatrix} \\ &= (a^2+b^2)\mathbf{I} = |c|^2\mathbf{I}, \qquad (4.72)\end{aligned}$$

where we find that the matrix representing c^* is the *transpose* of $\mathbf{C}$. This result follows because the imaginary elements of the matrix in (4.72) are 0. The squared magnitude is also derived from the determinant (see Chapter 3) of the original matrix $\mathbf{C}$:

$$|c|^2 = \det(\mathbf{C}) = \det\begin{bmatrix} a & -b \\ b & a \end{bmatrix} = a^2+b^2, \qquad (4.73)$$

and we also have $|c|^2 = \sqrt{\det(\mathbf{C}\mathbf{C}^T)}$.

4.8 COMPLEX EXPONENTIAL ROTATION

In this section, we explore further the rotation properties of $\exp(j)$ on the complex plane (Needham, 1999). If we start with the vector defined by $1+j0$ on the horizontal real axis and multiply it by $\exp(j)$ (with $\theta = 1$ rad), then it is rotated counterclockwise on the unit circle to $\exp(j) = \cos(1)+j\sin(1) \approx 0.5403 + j0.8415$. Similar to the real exponential function discussed in Chapter 1, we examine the following limit for finite n:

$$\lim_{n\to\infty}\left(1+\frac{j}{n}\right)^n = e^j, \qquad (4.74)$$

where from (1.102), $x_o = 1$ has been substituted and real r has been replaced with imaginary j. For integer values of n, the left-hand side is

$$n=0:\ 1, \qquad (4.75)$$

$$n=1:\ 1+j, \qquad (4.76)$$

$$n=2:\ (1+j/2)(1+j/2) = 3/4+j, \qquad (4.77)$$

$$\begin{aligned}n=3:\ &(1+j/3)(1+j/3)(1+j/3) = (8/9+j2/3)(1+j/3) \\ &= 2/3 - j26/27, \qquad (4.78)\end{aligned}$$

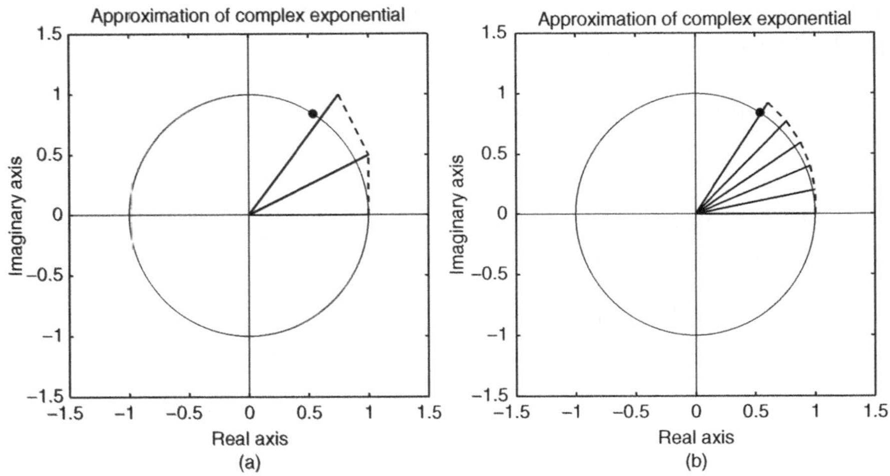

Figure 4.12 Components of the product $(1 + j/n)^n$ on the complex plane. The point $\exp(j) = 0.5403 + j0.8415$ on the unit circle is denoted by •. (a) $n = 2$. (b) $n = 5$.

and so on for $n \in \mathcal{Z}^+$. The approximation $(1 + j/n)^n$ for each n is a complex number that can be plotted on the complex plane as a vector starting at the origin. This is depicted in Figure 4.12(a) for $n = 2$ and the partial products leading up to the vector in (4.77):

$$1 + j0, \;\; 1 + j/2, \;\; (1 + j/2)^2 = 3/4 + j. \tag{4.79}$$

The three solid lines are connected together by two dashed lines, which turn out to be *perpendicular* to the lower two solid lines.

This orthogonality property of the dashed lines is verified as follows. The lower solid line extends along the horizontal axis to the point $1 + j0$. Since the middle solid line extends to $1 + j/2$, the lower dashed line is obviously perpendicular to $1 + j0$ because the two vectors have the same real part = 1. For the upper solid line, consider the triangle described by the three points: $0 + j0$ (the origin), $1 + j/2$, and $3/4 + j$. We demonstrate that it is a right triangle by showing that the squared magnitude of the hypotenuse (the solid line defined by $3/4 + j$) equals the sum of the squared magnitudes of the other two sides. Since the middle solid line is $1 + j/2$, the magnitude of the lower dashed line is derived from the difference $(3/4 + j) - (1 + j/2) = -1/4 + j/2$. Thus, the vector lengths are

$$\text{hypotenuse:} \;\; (3/4)^2 + 1^2 = 25/16, \tag{4.80}$$

$$\text{sum of other two sides:} \;\; 1^2 + (1/2)^2 + (-1/4)^2 + (1/2)^2 = 25/16, \tag{4.81}$$

such that the angle between the upper dashed line and the middle solid line forming the triangle is 90°. This result can be shown for every such triangle with increasing n, as is apparent in Figure 4.12(b) where $n = 5$, with the solid lines given by the six partial products

$$1 + j0, \quad 1 + j/5, \quad (1 + j/5)^2 = 24/25 + j2/5 = 0.96 + j0.4,$$
$$(1 + j/5)^3 = 22/25 + j74/125 = 0.88 + j0.592,$$
$$(1 + j/5)^4 = 476/625 + j96/125 \approx 0.7616 + j0.7680,$$
$$(1 + j/5)^5 = 380/625 + j2876/3125 \approx 0.6080 + j0.9203. \tag{4.82}$$

When the vector represented by $1 + j0$ on the horizontal axis is multiplied by $\exp(j)$, it is rotated counterclockwise exactly along the unit circle on the complex plane. The approximation in (4.74) with finite n yields a series of vectors from the partial products that form adjacent right triangles. The fact that each dashed line is perpendicular to the immediate lower solid line forming the right triangle causes a *rotation in two dimensions* rather an exponential growth in one dimension. As n is increased, the triangles become smaller and they more closely follow the unit circle. This is confirmed in Figure 4.13 where the magnitude of (4.74) starts to approach 1 for relatively small n. In the limit as $n \to \infty$, the rotation takes $1 + j0$ to $0.5403 + j0.8415$, corresponding to angle $\theta = \tan^{-1}(0.8415/0.5403) = 1$ rad, which is 57.2968° and is denoted by • in Figure 4.12. Of course, this angle is also evident from $\exp(j) = \cos(1) + j\sin(1)$.

The rotation of $1 + j0$ can be generalized to any angle $\theta \in [0, 2\pi]$; for example, $\exp(j\pi/2)$ rotates $1 + j0$ to be aligned with the vertical axis at $0 + j$ on the complex

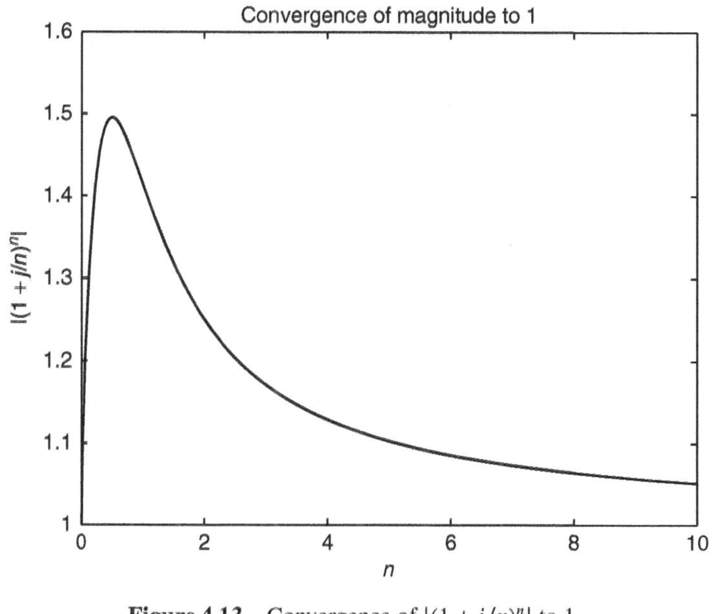

Figure 4.13 Convergence of $|(1 + j/n)^n|$ to 1.

plane. As θ is increased, the vector moves counterclockwise along the unit circle until $1 + j0$ is reached when $\theta = 2\pi$, at which point it starts to traverse the unit circle again. Rather than increase or decrease like the real exponential function, the complex exponential is restricted by the successive right triangles to rotate counterclockwise in two dimensions. For continuous rotation as $n \to \infty$, the complex exponential lies exactly on the unit circle and repeats itself with period 2π. Similar behavior occurs for negative θ, except that the rotation is clockwise. Such rotations can also be performed along any circle of radius r by using $r\exp(j\theta)$.

Figure 4.14 illustrates the two types of exponential functions: complex $\exp(\pm j\omega_o t)$ and real $\exp(\pm\sigma t)$, where $\sigma > 0$ and $\omega_o > 0$ are real parameters, and we have included time t in the exponents. Scaled complex exponential growth/rotation is derived by multiplying these two functions:

$$\exp(\sigma t)\exp(j\omega_o t) = \exp((\sigma + j\omega_o)t) = \exp(st), \tag{4.83}$$

where $s \triangleq \sigma + j\omega_o$ is a complex variable (which is notation used extensively in subsequent chapters). From the previous results, we find that if the complex exponential is plotted in three dimensions by including the time axis t, it has a spiral trajectory as it follows a circle with time-varying radius. For $\sigma > 0$, the radius increases, and for $\sigma < 0$, it decreases, as depicted in Figure 4.15.

Based on the previous observations, it is straightforward to once again connect $\exp(j\theta)$ to $\sin(\theta)$ and $\cos(\theta)$. From Figure 4.12, we find using trigonometry that

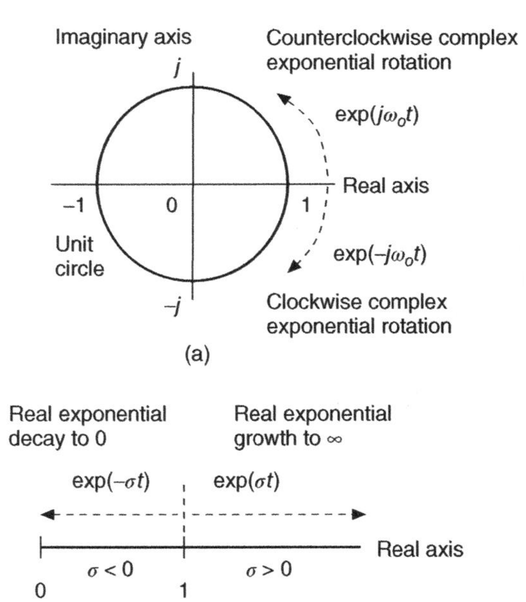

Figure 4.14 (a) Complex exponential rotation on the unit circle. (b) Real exponential growth on the real axis.

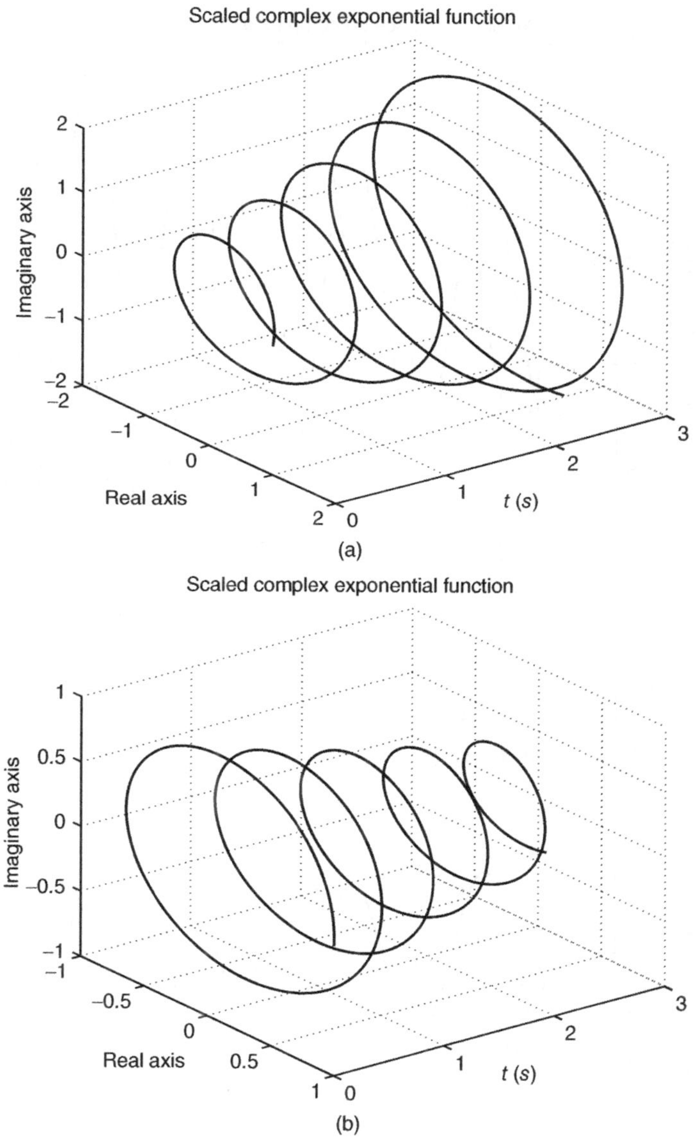

Figure 4.15 Scaled complex exponential function $\exp(\sigma t)\exp(j\omega_o t)$ with $\omega_o = 10$ rad/s. (a) $\sigma = 0.3$. (b) $\sigma = -0.3$.

$\sin(\theta)$ is the projection of $\exp(j\theta)$ onto the imaginary axis, and $\cos(\theta)$ is its projection onto the real axis. Using j for the imaginary axis and the notation for a complex number, these results lead directly to Euler's formula. Projecting the increasing spiral in Figure 4.15(a) onto the real axis yields the exponentially increasing cosine function $\exp(\sigma t)\cos(\omega_o t)$ shown in Figure 4.16(a), and likewise, the projection onto the

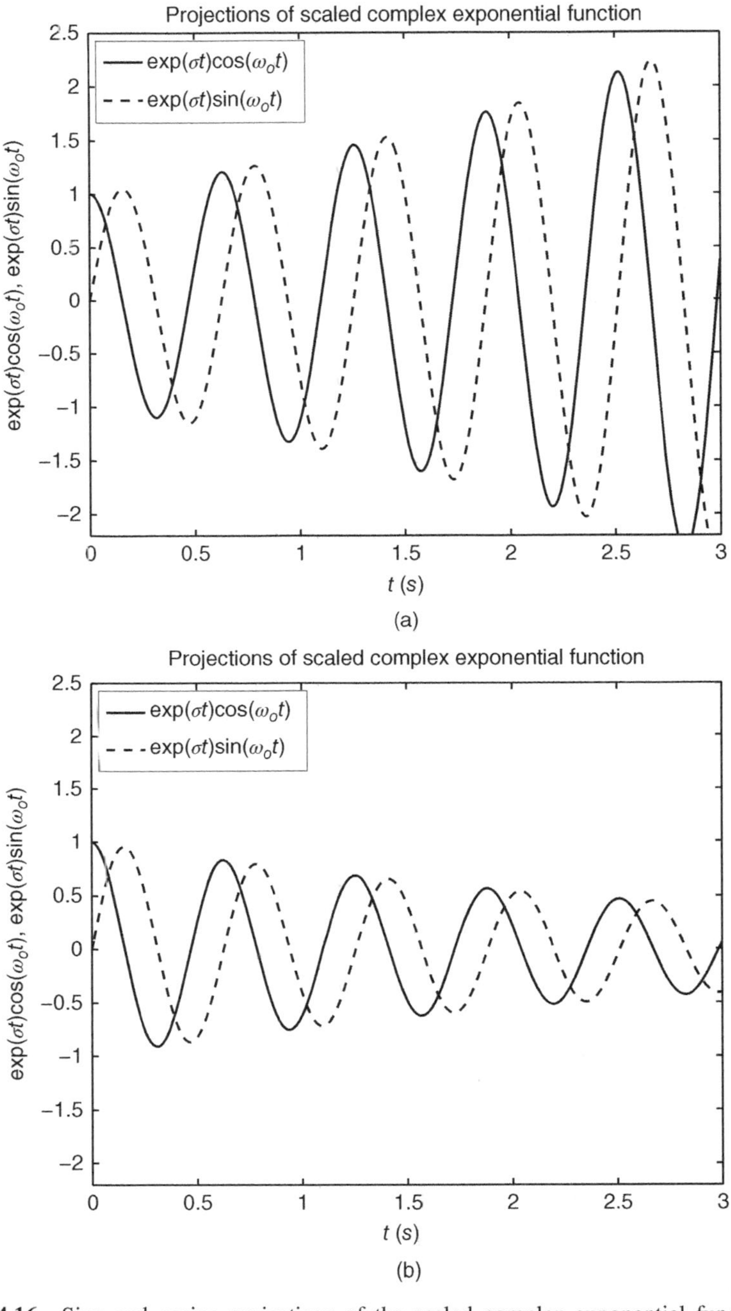

Figure 4.16 Sine and cosine projections of the scaled complex exponential functions in Figure 4.15 with $\omega_o = 10$ rad/s. (a) $\sigma = 0.3$. (b) $\sigma = -0.3$.

imaginary axis is the exponentially increasing sine function $\exp(\sigma t)\sin(\omega_o t)$ shown in Figure 4.16(b). Of course, these weighted sine and cosine waveforms are necessarily 90° out of phase with respect to each other. The decreasing functions in Figure 4.16(b) are important in circuit and system analysis, as this decaying response arises often in practical systems, such as the second-order RLC circuits discussed in Chapter 2.

4.9 CONSTANT ANGULAR VELOCITY

As $r\exp(j\omega_o t)$ traverses a circle with fixed radius r and *constant angular velocity* ω_o, its projections onto the real and imaginary axes are $r\cos(\omega_o t)$ and $r\sin(\omega_o t)$, respectively. Suppose we want to trace different geometric objects on the complex plane with constant angular velocity, such as the square shown in Figure 4.17(a). This can be done using $r(t)\exp(j\omega_o t)$, which now has a *time-varying radius*. As this vector moves from angle 0 to $\pi/4$, the length of the radius varies from 1 to $\sqrt{2}$. Since the real part is fixed at 1 for this range of angles, we can use trigonometry on the right triangle formed by the rotating vector to find an expression for $r(t)$:

$$r(t)\cos(\omega_o t) = 1 \implies r(t) = 1/\cos(\omega_o t). \tag{4.84}$$

The projection onto the imaginary axis is still sine, but scaled by $r(t)$:

$$r(t)\sin(\omega_o t) = \sin(\omega_o t)/\cos(\omega_o t) = \tan(\omega_o t). \tag{4.85}$$

When the angle is in the interval $(\pi/4, 3\pi/4]$, the projection of the vector onto the imaginary axis is a constant 1; likewise, it is a constant -1 for the interval $(5\pi/4, 7\pi/4]$. In the second quadrant, the radius decreases from $\sqrt{2}$ to 1 over $(3\pi/4, \pi]$, yielding

$$r(t)\cos(\pi - \omega_o t) = 1 \implies r(t) = 1/\cos(\pi - \omega_o t) = -1/\cos(\omega_o t), \tag{4.86}$$

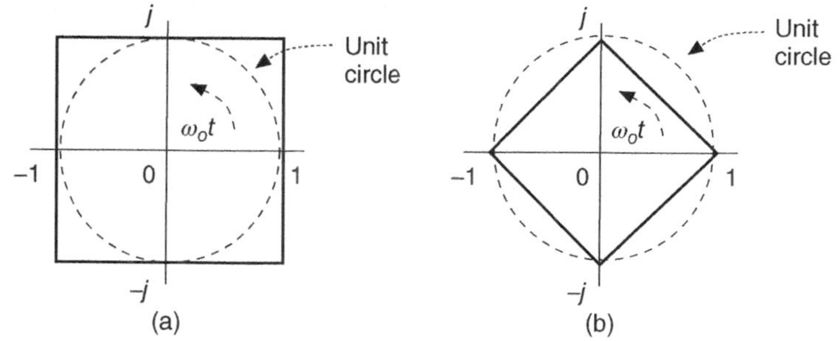

Figure 4.17 Traversing geometric objects on the complex plane with constant angular velocity ω_o. (a) Square. (b) Diamond.

and

$$r(t)\sin(\omega_o t) = -\tan(\omega_o t). \tag{4.87}$$

By traversing the square, the overall projection onto the *imaginary* axis is

$$f_1(\omega_o t) = \begin{cases} \tan(\omega_o t), & 0 \leq \omega_o t < \pi/4 \\ 1, & \pi/4 \leq \omega_o t < 3\pi/4 \\ -\tan(\omega_o t), & 3\pi/4 \leq \omega_o t < 5\pi/4 \\ -1, & 5\pi/4 \leq \omega_o t < 7\pi/4 \\ \tan(\omega_o t), & 7\pi/4 \leq \omega_o t < 0. \end{cases} \tag{4.88}$$

The resulting periodic function is shown in Figure 4.18(a) for $\omega_o t \in [0, 2\pi]$ (one period). We have also included the sine wave for comparison which, of course, is generated by tracing the unit circle and projecting it onto the imaginary axis. A waveform similar to (4.88) is obtained via a projection of the square onto the real axis; as in the case of the unit circle, this waveform has a phase shift of $\pi/2$ relative to that in (4.88).

Obviously, the projection of $r(t)\exp(j\omega_o t)$ for time-varying $r(t)$ is more complicated than the sine waveform where $r(t)$ is a constant. A *circle* is the only object on the complex plane that produces sine on the imaginary axis and cosine on the real axis. For geometric objects other than the circle, the projection does not have such a simple harmonic behavior. In fact, it can be shown from the *Fourier series* representation discussed in Chapter 5 that such projections can be expressed as the sum of weighted sines and cosines with frequencies that are integer multiples of the fundamental frequency ω_o. The waveform in Figure 4.18(a) has a zero DC component and, since it is an odd function, only sine terms appear in its Fourier series. The resulting *harmonics* given by $n\omega_o$ for $n \in \mathcal{Z}^+$ are caused by the product $r(t)\sin(\omega_o t)$, which can be viewed as a *time-varying* system with input $\sin(\omega_o t)$. This is in contrast to a linear time-invariant (LTI) system with a sinusoidal input, whose output is also sinusoidal with the same single frequency, but possibly with a different amplitude and phase. Harmonics do not appear in the output of an LTI system.

Similar results are obtained for other geometric objects on the complex plane, such as the diamond in Figure 4.17(b), which has the projection onto the imaginary axis shown in Figure 4.18(b). It is somewhat more difficult to derive this projection because there are no regions where $r(t)$ is constant as the diamond is traversed. Using trigonometry, it can be shown that the projection onto the imaginary axis for constant angular velocity ω_o is (see Problem 4.25)

$$f_2(\omega_o t) = \begin{cases} \sin(\omega_o t)/[\sin(\omega_o t) + \cos(\omega_o t)], & 0 \leq \omega_o t < \pi/2 \\ \sin(\omega_o t)/[\sin(\omega_o t) - \cos(\omega_o t)], & \pi/2 \leq \omega_o t < \pi \\ -\sin(\omega_o t)/[\sin(\omega_o t) + \cos(\omega_o t)], & \pi \leq \omega_o t < 3\pi/2 \\ -\sin(\omega_o t)/[\sin(\omega_o t) - \cos(\omega_o t)], & 3\pi/2 \leq \omega_o t < 2\pi. \end{cases} \tag{4.89}$$

A similar waveform is derived for the projection onto the real axis, but it is shifted by $\pi/2$.

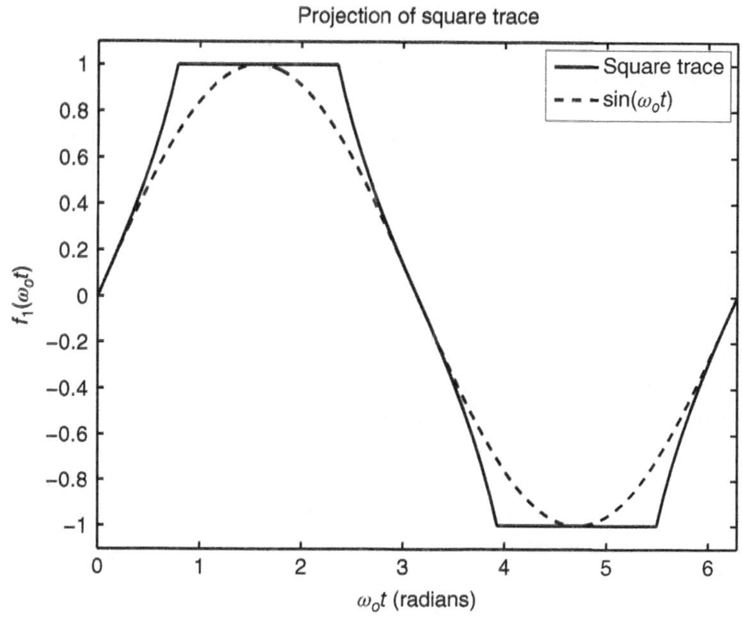

(a)

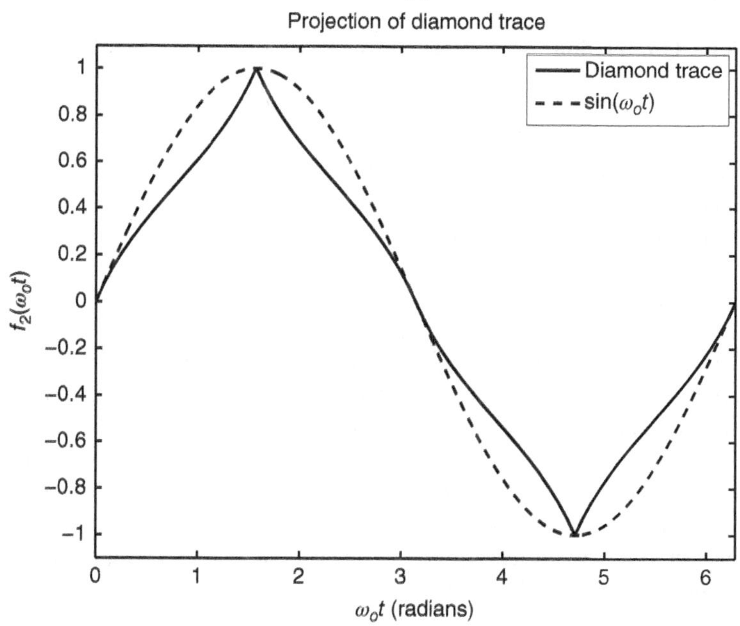

(b)

Figure 4.18 Projection of $r(t)\exp(j\omega_o t)$ onto the imaginary axis when traversing geometric objects on the complex plane. (a) $f_1(\omega_o t)$ when tracing a square. (b) $f_2(\omega_o t)$ when tracing a diamond.

Observe in Figure 4.18(a) that the magnitude of the sine waveform does not exceed that of the square trace, as expected because the unit circle lies inside the square on the complex plane in Figure 4.17(a). For the second case, the magnitude of the diamond trace in Figure 4.18(b) does not exceed that of the sine waveform because the diamond lies inside the unit circle in Figure 4.17(b). This diamond projection also does not have a DC component, and its Fourier series has only sine terms because it is an odd function.

4.10 QUATERNIONS

In the final section of this chapter, we briefly describe an extension of complex numbers $\mathcal{C}$, which are defined on a plane (two coordinates), to *quaternions* $\mathcal{H}$ defined on a four-dimensional subspace (Hanson, 2006; Goldman, 2010). Although quaternions will not be used later in this book, we describe their properties to emphasize that complex numbers are simply a two-dimensional extension of the real numbers: $\mathcal{R} \Longrightarrow \mathcal{C} = \mathcal{R} + j\mathcal{R}$, and so they form a subset of quaternions.

Unlike complex numbers, which can arise when solving algebraic equations, quaternions were devised as a means to extend the rotation property of complex numbers to three-dimensional space. This was achieved by including two additional coordinates.

Definition: Quaternion A *quaternion* is a four-dimensional number of the form

$$h = a + ib_1 + jb_2 + kb_3, \tag{4.90}$$

where $\{i, j, k\}$ are markers for the three components beyond a, and $\{a, b_1, b_2, b_3\}$ are real numbers. These markers are all defined to be $i = j = k \triangleq \sqrt{-1}$, and the set of quaternions can be expressed as $\mathcal{H} = \mathcal{R} + i\mathcal{R} + j\mathcal{R} + k\mathcal{R}$.

We have used i for one of the markers because $\{i, j, k\}$ is the standard notation for quaternions (there will be no confusion with the symbol i for current because we do not return to quaternions in subsequent chapters). The three components using $\{i, j, k\}$ are called the *extended imaginary part* of the quaternion, and a is the usual real part. It turns out that it is not sufficient to include only one additional coordinate of the form $a + ib_1 + jb_2$ because there are inconsistencies with multiplication and division. It is necessary that the fourth coordinate kb_3 be included. (Quaternions can likewise be extended to have additional coordinates. In order for multiplication to be consistent, this extension to *octonions* has eight coordinates: one real and seven imaginary.)

The basic properties of the markers are

$$i^2 = j^2 = k^2 = -1, \quad ij = k, \quad jk = i, \quad ki = j. \tag{4.91}$$

Unlike the other sets of numbers that we have considered, multiplication is *not* commutative:

$$ji = -k, \quad kj = -i, \quad ik = -j. \tag{4.92}$$

From the products in (4.91) and (4.92), it is straightforward to show that

$$ijk = jki = kij = -1, \quad ikj = jik = kji = 1. \tag{4.93}$$

The product of two quaternions is

$$\begin{aligned} h_1 h_2 &= (a_1 + ib_{11} + jb_{12} + kb_{13})(a_2 + ib_{21} + jb_{22} + kb_{23}) \\ &= a_1 a_2 - b_{11}b_{21} - b_{12}b_{22} - b_{13}b_{23} + i(a_1 b_{21} + a_2 b_{11} + b_{12}b_{23} - b_{13}b_{22}) \\ &\quad + j(a_1 b_{22} + a_2 b_{12} + b_{13}b_{21} - b_{11}b_{23}) \\ &\quad + k(a_1 b_{23} + a_2 b_{13} + b_{11}b_{22} - b_{12}b_{21}), \end{aligned} \tag{4.94}$$

whereas the reverse product is

$$\begin{aligned} h_2 h_1 &= (a_2 + ib_{21} + jb_{22} + kb_{23})(a_1 + ib_{11} + jb_{12} + kb_{13}) \\ &= a_1 a_2 - b_{11}b_{21} - b_{12}b_{22} - b_{13}b_{23} + i(a_1 b_{21} + a_2 b_{11} - b_{12}b_{23} + b_{13}b_{22}) \\ &\quad + j(a_1 b_{22} + a_2 b_{12} - b_{13}b_{21} + b_{11}b_{23}) \\ &\quad + k(a_1 b_{23} + a_2 b_{13} - b_{11}b_{22} + b_{12}b_{21}). \end{aligned} \tag{4.95}$$

The signs of the last two terms of the resulting $\{i, j, k\}$ multipliers are reversed for $h_2 h_1$ compared with those of $h_1 h_2$. The quaternion conjugate is

$$h^* = a - ib_1 - jb_2 - kb_3, \tag{4.96}$$

and similar to complex numbers, the following product is real:

$$hh^* \triangleq |h|^2 = a^2 + b_1^2 + b_2^2 + b_3^2. \tag{4.97}$$

All cross-terms in the squared magnitude have cancelled because of the properties in (4.91) and (4.92).

The matrix representation for a quaternion is

$$\mathbf{H} \triangleq \begin{bmatrix} a & b_1 & b_2 & b_3 \\ -b_1 & a & -b_3 & b_2 \\ -b_2 & b_3 & a & -b_1 \\ -b_3 & -b_2 & b_1 & a \end{bmatrix}, \tag{4.98}$$

and like the matrix representation for a complex number, the quaternion conjugate h^* is the transpose $\mathbf{H}^T$ of this matrix. The squared magnitude $|h|^2$ is derived from

$$\begin{aligned} \mathbf{H}\mathbf{H}^T &= \begin{bmatrix} a^2 + b_1^2 + b_2^2 + b_3^2 & 0 & 0 & 0 \\ 0 & a^2 + b_1^2 + b_2^2 + b_3^2 & 0 & 0 \\ 0 & 0 & a^2 + b_1^2 + b_2^2 + b_3^2 & 0 \\ 0 & 0 & 0 & a^2 + b_1^2 + b_2^2 + b_3^2 \end{bmatrix} \\ &= (a^2 + b_1^2 + b_2^2 + b_3^2)\mathbf{I} = |h|^2 \mathbf{I}, \end{aligned} \tag{4.99}$$

which yields $|h|^2 = \sqrt[4]{\det(\mathbf{H}\mathbf{H}^T)}$. It is also generated from the determinant of (4.98):

$$|h|^2 = \det(\mathbf{H}). \tag{4.100}$$

The fact that h_1 and h_2 do not commute can also be verified from their matrix representations. However, $hh^* = h^*h$, resulting in the same diagonal matrix $\mathbf{H}\mathbf{H}^T = \mathbf{H}^T\mathbf{H}$ in (4.99). A quaternion can also be expressed as a 2×2 matrix using *complex* numbers as follows:

$$\mathbf{H}_c = \begin{bmatrix} a + jb_1 & b_2 + jb_3 \\ -b_2 + jb_3 & a - jb_1 \end{bmatrix}, \tag{4.101}$$

where the subscript c emphasizes that it is a complex matrix of lower dimension than $\mathbf{H}$. We find that h^* is represented by $\mathbf{H}_c^H$, where the superscript denotes complex conjugation and transpose of its elements: $\mathbf{H}_c^H = (\mathbf{H}_c^T)^* = (\mathbf{H}_c^*)^T$ (as discussed in Chapter 3). Thus,

$$\mathbf{H}_c\mathbf{H}_c^H = \mathbf{H}_c^H\mathbf{H}_c = \begin{bmatrix} a^2 + b_1^2 + b_2^2 + b_3^2 & 0 \\ 0 & a^2 + b_1^2 + b_2^2 + b_3^2 \end{bmatrix} = |h|^2\mathbf{I}, \tag{4.102}$$

from which we conclude

$$|h|^2 = \sqrt{\det(\mathbf{H}_c\mathbf{H}_c^H)}. \tag{4.103}$$

In order to understand rotations in three dimensions, we examine the three *spherical coordinates* defined by

$$\begin{bmatrix} x_1 \\ x_2 \\ x_3 \end{bmatrix} = \begin{bmatrix} r\cos(\theta)\sin(\phi) \\ r\sin(\theta)\sin(\phi) \\ r\cos(\phi) \end{bmatrix}, \tag{4.104}$$

where θ is the *azimuth* angle in the x_1–x_2 plane, and ϕ is the *inclination* angle that extends along the x_3 axis as illustrated in Figure 4.19. Observe that the projection of the vector onto the x_1–x_2 plane is $r\sin(\phi)$, which is the length of the horizontal solid line defined by θ. The component of that line along the x_1 axis is $r\sin(\phi)\cos(\theta)$, and

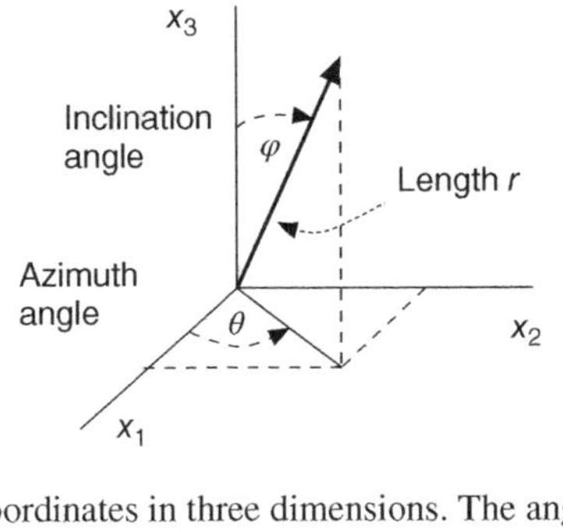

Figure 4.19 Spherical coordinates in three dimensions. The angle of elevation is $90° - \phi$.

along the x_2 axis it is $r\sin(\phi)\sin(\theta)$. The component of the vector along the x_3 axis is determined only from the inclination angle of $r\cos(\phi)$.

Rotations in three dimensions are performed as follows:

$$\begin{bmatrix} y_1 \\ y_2 \\ y_3 \end{bmatrix} = \begin{bmatrix} 1 & 0 & 0 \\ 0 & \cos(\theta) & -\sin(\theta) \\ 0 & \sin(\theta) & \cos(\theta) \end{bmatrix} \begin{bmatrix} x_1 \\ x_2 \\ x_3 \end{bmatrix} = \begin{bmatrix} x_1 \\ x_2\cos(\theta) - x_3\sin(\theta) \\ x_2\sin(\theta) + x_3\cos(\theta) \end{bmatrix}, \tag{4.105}$$

$$\begin{bmatrix} y_1 \\ y_2 \\ y_3 \end{bmatrix} = \begin{bmatrix} \cos(\theta) & 0 & \sin(\theta) \\ 0 & 1 & 0 \\ -\sin(\theta) & 0 & \cos(\theta) \end{bmatrix} \begin{bmatrix} x_1 \\ x_2 \\ x_3 \end{bmatrix} = \begin{bmatrix} x_3\sin(\theta) + x_1\cos(\theta) \\ x_2 \\ x_3\cos(\theta) - x_1\sin(\theta) \end{bmatrix}, \tag{4.106}$$

$$\begin{bmatrix} y_1 \\ y_2 \\ y_3 \end{bmatrix} = \begin{bmatrix} \cos(\theta) & -\sin(\theta) & 0 \\ \sin(\theta) & \cos(\theta) & 0 \\ 0 & 0 & 1 \end{bmatrix} \begin{bmatrix} x_1 \\ x_2 \\ x_3 \end{bmatrix} = \begin{bmatrix} x_1\cos(\theta) - x_2\sin(\theta) \\ x_1\cos(\theta) + x_2\sin(\theta) \\ x_3 \end{bmatrix}, \tag{4.107}$$

where the first set of matrices with a 1 on the main diagonal are *rotation matrices* denoted by $\mathbf{R}_1$, $\mathbf{R}_2$, and $\mathbf{R}_3$, respectively. In each case, one coordinate remains unchanged so that a vector is rotated only in the plane defined by the other two coordinates. Furthermore, the length of the vector remains fixed: for the first rotation with vectors $\mathbf{y}$ and $\mathbf{x}$, we have

$$\begin{aligned} \|\mathbf{y}\|^2 &= x_1^2 + [x_2\cos(\theta) - x_3\sin(\theta)]^2 + [x_2\sin(\theta) + x_3\cos(\theta)]^2 \\ &= x_1^2 + x_2^2\cos^2(\theta) + x_3\sin^2(\theta) - 2x_2x_3\cos(\theta)\sin(\theta) \\ &\quad + x_2^2\sin^2(\theta) + x_3^2\cos^2(\theta) + 2x_2x_3\sin(\theta)\cos(\theta) \\ &= x_1^2 + x_2^2 + x_3^2 = \|\mathbf{x}\|^2. \end{aligned} \tag{4.108}$$

Of course, this result follows from the fact that for the rotation matrix in (4.105), $\mathbf{R}_1^T\mathbf{R}_1 = \mathbf{I}$ and $\|\mathbf{y}\|^2 = \mathbf{x}^T\mathbf{R}_1^T\mathbf{R}_1\mathbf{x}^T = \|\mathbf{x}\|^2$. The same results are obtained for $\mathbf{R}_2$ and $\mathbf{R}_3$. It is possible to rotate a column vector anywhere in three dimensions with arbitrary angles by successively premultiplying it by these matrices. The final overall rotation depends on the order that the matrices are multiplied because these matrices do not commute.

Example 4.9 Examples of these rotations are illustrated in Figure 4.20. The original vector is $\mathbf{x} = [1, 1, 1]^T$ (the solid line) and the angle of rotation is 30°. The three rotation matrices for this angle are

$$\mathbf{R}_1 = \begin{bmatrix} 1 & 0 & 0 \\ 0 & 0.8660 & -0.5 \\ 0 & 0.5 & 0.8660 \end{bmatrix}, \quad \mathbf{R}_2 = \begin{bmatrix} 0.8660 & 0 & 0.5 \\ 0 & 1 & 0 \\ -0.5 & 0 & 0.8660 \end{bmatrix},$$

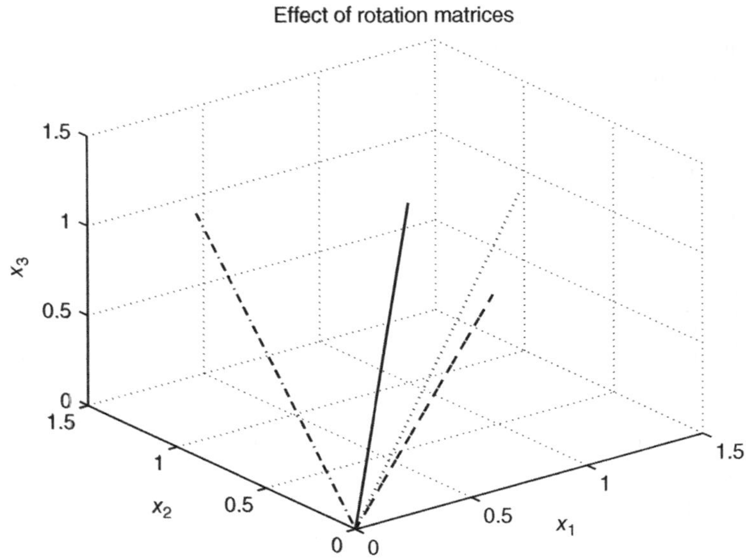

Figure 4.20 Rotation of vector $\mathbf{x} = [1, 1, 1]^T$ (the solid line) in Example 4.9.

$$\mathbf{R}_3 = \begin{bmatrix} 0.8660 & -0.5 & 0 \\ 0.5 & 0.8660 & 0 \\ 0 & 0 & 1 \end{bmatrix}, \qquad (4.109)$$

and the rotated vectors generated by individually applying the matrices are

$$\mathbf{y}_1 = \begin{bmatrix} 1 \\ 0.3660 \\ 1.3660 \end{bmatrix}, \quad \mathbf{y}_2 = \begin{bmatrix} 1.3660 \\ 1 \\ 0.3660 \end{bmatrix}, \quad \mathbf{y}_3 = \begin{bmatrix} 0.3660 \\ 1.3660 \\ 1 \end{bmatrix}. \qquad (4.110)$$

These are shown in Figure 4.20 as the dotted, dashed, and dash-dotted lines, respectively. The squared norm of each rotated vector is 3, which is the squared norm of the original vector: $\|\mathbf{x}\|^2 = 3$.

Returning to the notation for quaternions, we can write

$$h = a + ib_1 + jb_2 + kb_3 \triangleq a + \mathbf{b}, \qquad (4.111)$$

where $\mathbf{b} \triangleq ib_1 + jb_2 + kb_3$. Since $\mathbf{b}$ has three components, it is similar to a vector, but it does not have the same properties. The notation $(a, \mathbf{b})$ is often used to represent quaternions. An extension of Euler's formula for quaternions is derived by letting $b_1 = b_2 = b_3 = 1$ in (4.111) and defining $\mathbf{j} \triangleq i + j + k$, yielding

$$\exp(\mathbf{j}\theta) \triangleq \cos(\theta) + \mathbf{j}\sin(\theta) = \cos(\theta) + [i\sin(\theta) + j\sin(\theta) + k\sin(\theta)], \qquad (4.112)$$

TABLE 4.5 Properties of Quaternions

Property	Equation
Conjugate	$h^* = (a, -\mathbf{b})$
Squared magnitude	$\lvert h\rvert^2 = hh^* = a^2 + \mathbf{b}\mathbf{b}^* = a^2 + b_1^2 + b_2^2 + b_3^2$
Product	$h_1h_2 = (a_1a_2 - \mathbf{b}_1\mathbf{b}_2, a_1\mathbf{b}_2 + a_2\mathbf{b}_1 + \mathbf{b}_1 \times \mathbf{b}_2)$
Inverse	$h^{-1} = h^*/\lvert h\rvert^2$

where **j** likewise is not a conventional vector in this formulation. Quaternions have the properties summarized in Table 4.5, which match those given earlier in (4.94)–(4.97). The *cross product* in the table is

$$\mathbf{b}_1 \times \mathbf{b}_2 = i(b_{12}b_{23} - b_{13}b_{22}) + j(b_{13}b_{21} - b_{11}b_{23}) + k(b_{11}b_{22} - b_{12}b_{21}), \quad (4.113)$$

which is not commutative: $\mathbf{b}_2 \times \mathbf{b}_1 \neq \mathbf{b}_1 \times \mathbf{b}_2$.

Suppose that we would like to rotate the three-dimensional column vector $\mathbf{x}$ using quaternions. This procedure is summarized in the following steps.

- Write $\mathbf{x}$ in the form $\mathbf{x} = ix_1 + jx_2 + kx_3$, and using the quaternion notation in (4.111), define $x = (0, \mathbf{x})$ with real part 0.
- Let θ be the desired angle of rotation and define the quaternion

$$h = (\cos(\theta/2), \mathbf{q}\sin(\theta/2)), \quad (4.114)$$

  with $\mathbf{q} \triangleq (iq_1 + jq_2 + kq_3)$.
- The rotation is achieved by the product $y = hxh^*$.

Various rotations can be performed by choosing different values for the $\{q_m\}$.

Example 4.10 Let $\mathbf{x} = ix_1 + jx_2 + kx_3$ represent a vector in $\mathcal{R}^3$ that is to be rotated by $\theta = 90°$ with respect to the i axis. For this case, $\mathbf{q} = i$ and

$$\begin{aligned} h &= (\cos(45°), \mathbf{q}\sin(45°)) = (1/\sqrt{2}, \mathbf{q}/\sqrt{2}) \\ &= (1/\sqrt{2})(1 + i), \end{aligned} \quad (4.115)$$

which yields

$$\begin{aligned} y = hxh^* &= (1/2)(1 + i)(ix_1 + jx_2 + kx_3)(1 - i) \\ &= (1/2)[-x_1 + ix_1 + j(x_2 - x_3) + k(x_2 + x_3)](1 - i) \\ &= (1/2)(i2x_1 - j2x_3 + k2x_2) = (0, ix_1 - jx_3 + kx_2), \end{aligned} \quad (4.116)$$

where the marker multiplication rules in (4.91) and (4.92) have been used. We mention again that the order of marker multiplications must be taken into account to achieve the proper signs. Thus, the general form for the rotated vector is

$\mathbf{y} = ix_1 - jx_3 + kx_2$. In order to illustrate the behavior of this rotation, consider four cases:

$$\mathbf{x} = i \implies \mathbf{y} = i, \tag{4.117}$$

$$\mathbf{x} = j \implies \mathbf{y} = k, \tag{4.118}$$

$$\mathbf{x} = k \implies \mathbf{y} = -j, \tag{4.119}$$

$$\mathbf{x} = i + j \implies \mathbf{y} = i + k, \tag{4.120}$$

which are depicted in Figure 4.21. In the last case, $\mathbf{x}$ lies in the i–j plane and is rotated to $\mathbf{y}$ in the i–k plane. For $h = (1/\sqrt{2})(1 + i)$, the rotation is about the i axis and the corresponding result is given in (4.105), which we repeat here for $\theta = 90°$:

$$\begin{bmatrix} x_1 \\ x_2\cos(\theta) - x_3\sin(\theta) \\ x_2\sin(\theta) + x_3\cos(\theta) \end{bmatrix} = \begin{bmatrix} x_1 \\ -x_3 \\ x_2 \end{bmatrix}. \tag{4.121}$$

This expression gives the same rotations from $\mathbf{x}$ to $\mathbf{y}$ as in Figure 4.21. Similar results can be shown for different angles and rotations about the other axes (see Problems 4.29 and 4.30).

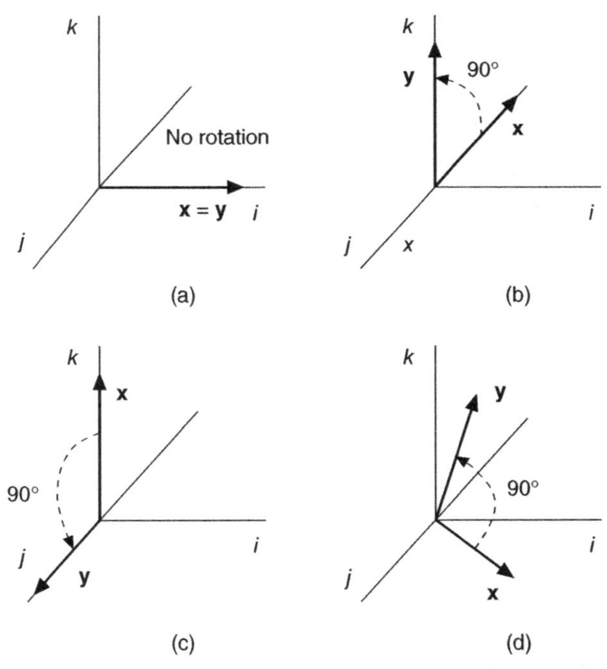

Figure 4.21 Quaternion rotations for $h = (1/\sqrt{2})(1 + i)$. (a) $\mathbf{x} = i$. (b) $\mathbf{x} = j$. (c) $\mathbf{x} = k$. (d) $\mathbf{x} = i + j$.

PROBLEMS

Complex Numbers

4.1 Rewrite the following ratios in the standard complex form $a + jb$:

$$\text{(a) } x = \frac{2+j3}{4-j2}, \quad \text{(b) } y = \frac{5-j2}{1+2j}, \quad \text{(c) } z = \frac{4+j}{1-j}. \tag{4.122}$$

4.2 For x and y in the previous problem, find expressions for (a) xy, (b) $x - y$, and (c) xy^*, writing them all in the standard complex form $a + jb$.

4.3 Give the range of values for α such that the following real-valued functions have complex roots:

$$\text{(a) } f(x) = x^2 + \alpha x + 1, \quad \text{(b) } g(x) = x^2 + x + \alpha, \quad \text{(c) } h(x) = \alpha x^2 + x + 1. \tag{4.123}$$

4.4 Prove the triangle inequality for complex $\{x_1, x_2\}$:

$$|x_1 + x_2| \leq |x_1| + |x_2|. \tag{4.124}$$

4.5 Complex $c = a + jb$ in rectangular form has squared magnitude $|c|^2 = a^2 + b^2$. Use this property to show $|a| + |b| \leq \alpha|c|$ for $\alpha = \sqrt{2}$.

4.6 Describe the regions on the complex plane defined by the following real functions of complex x:

$$\text{(a) } f(x) = ||x| - 1| \leq 1, \quad \text{(b) } g(x) = ||x| + 1| \geq 1. \tag{4.125}$$

4.7 The discriminant of the cubic equation $x^3 + ax + b = 0$ is $D = b^2/4 + a^3/27$. A function has complex conjugate roots when $D > 0$ and real roots when $D < 0$. Verify this property for

$$\text{(a) } f(x) = (x-2)(x^2 + 2x + 2), \quad \text{(b) } g(x) = (x+1)(x^2 - x - 2). \tag{4.126}$$

Polar Coordinates

4.8 Rewrite the complex numbers $\{x, y, z\}$ in polar form:

$$\text{(a) } x = 2 - j3, \quad \text{(b) } y = \frac{1+j}{1-j}, \quad \text{(c) } z = \frac{2+j}{4+j}. \tag{4.127}$$

4.9 Convert the following complex numbers $\{x, y\}$ into polar form, compute (a) $z_1 = xy$, (b) $z_2 = x/y$, and (c) $z_3 = xy^*$, and then convert the results back to rectangular form:

$$x = 3 - 2j, \quad y = 1 + j. \tag{4.128}$$

Verify your results by performing these operations using the rectangular form.

4.10 Repeat the previous problem for

$$x = \frac{1+j}{2-j}, \quad y = \frac{3+j}{1+j2}. \tag{4.129}$$

4.11 Find the distance between two complex numbers written in polar form: (a) $\exp(j2)$ and $3\exp(j)$. (b) $2\exp(-3j)$ and $\exp(2j)$.

4.12 The equation for a shifted circle centered at d on the complex plane is $f(\theta) = |\exp(j\theta) - d|$. Find θ such that $f(\theta) = 0$ for (a) $d = 1$ and (b) $d = 1 + j$.

Euler's Formula

4.13 Use Euler's formula to verify the trigonometric identities:

$$\text{(a) } \sin(x-y) = \sin(x)\cos(y) - \cos(x)\sin(y), \quad \text{(b) } \cos(2x) = \cos^2(x) - \sin^2(x). \tag{4.130}$$

4.14 Repeat the previous problem for

$$\cos(x) - \cos(y) = -2\sin((x+y)/2)\sin((x-y)/2). \tag{4.131}$$

4.15 Use Euler's inverse formula for $\cos(x)$ to find the indefinite integrals of (a) $\cos^2(\alpha x)$ and (b) $\sin(x)\cos(x)$.

4.16 The exponential function is written in (1.111) as the power series

$$\exp(x) = \sum_{n=1}^{\infty} \frac{x^n}{n!}. \tag{4.132}$$

Use this expression and the power series expansions for sine and cosine in Appendix C to verify Euler's formula.

4.17 Write the following expressions in standard complex form using de Moivre's formula:

$$\text{(a) } x = (2+j)^6, \quad \text{(b) } y = \frac{1}{(1-3j)^4}. \tag{4.133}$$

4.18 Find the roots for the following equations using the nth root formula in Table 4.4:

$$\text{(a) } x^3 = 64, \quad \text{(b) } y^3 = 8j. \tag{4.134}$$

4.19 Find the square root for each of the following functions:

$$\text{(a) } x = -16j, \quad \text{(b) } y = 2 - j, \quad \text{(c) } z = 4 + 3j. \tag{4.135}$$

Matrix Representation

4.20 For the complex numbers $\{x, y\}$ in (4.128), write them as matrices $\mathbf{C}_1$ and $\mathbf{C}_2$. (a) Demonstrate that these matrices commute in a product as shown in (4.67) and (4.68). (b) Verify that $\mathbf{C}_1\mathbf{C}_1^T = |x|^2\mathbf{I}$ and $\mathbf{C}_2\mathbf{C}_2^T = |y|^2\mathbf{I}$.

4.21 The matrix representation for complex numbers can be expressed as

$$c = a + jb \Leftrightarrow a\mathbf{I} + b\mathbf{R}(\pi/2) = \mathbf{C}, \tag{4.136}$$

where the notation $\mathbf{R}(\pi/2)$ refers to the rotation matrix in (4.69) with $\theta = \pi/2$:

$$\mathbf{R}(\pi/2) = \begin{bmatrix} 0 & -1 \\ 1 & 0 \end{bmatrix}. \tag{4.137}$$

(a) Verify that $\mathbf{C}^T\mathbf{C} = (a^2 + b^2)\mathbf{I}$ by writing it in terms of the matrices in (4.136).
(b) Find an expression for $\mathbf{C}^2$ using (4.136).

4.22 In order to examine additional properties of complex numbers written in matrix form, expand the notation as follows for $c_n = a_n + jb_n$:

$$\mathbf{C}_{a_n,b_n} = \begin{bmatrix} a_n & -b_n \\ b_n & a_n \end{bmatrix}. \tag{4.138}$$

(a) Let the matrix inverse $\mathbf{C}^{-1}_{a_n,b_n}$ represent

$$\frac{1}{c_n} = \frac{1}{a_n + jb_n}. \tag{4.139}$$

Specify the subscripts $\{\alpha, \beta\}$ such that $\mathbf{C}^{-1}_{a_n,b_n} = \mathbf{C}_{\alpha,\beta}$. (b) Find $\{\alpha, \beta\}$ for the expression $\mathbf{C}_{a_1,b_2}\mathbf{C}^{-1}_{a_2,b_2} = \mathbf{C}_{\alpha,\beta}$ representing the ratio $c_1/c_2 = (a_1 + jb_1)/(a_2 + jb_2)$.

Complex Exponential Rotation and Constant Angular Velocity

4.23 The complex function $\exp((\sigma + j\omega)t) = \exp(\sigma t)[\cos(\omega t) + j\sin(\omega t)]$ has increasing sinusoidal components for $\sigma > 0$. Describe the behavior of the ratio $\exp((\sigma_1 + j\omega_1)t)/\exp((\sigma_2 + j\omega_2)t)$ relative to that of $\exp((\sigma_1 + j\omega_1)t)$ alone.

4.24 Consider the rectangle defined by $x_1 \leq x \leq x_2$ and $y_1 \leq y \leq y_2$ in Cartesian coordinates. (a) Describe how the rectangle maps to the complex plane via the transformation $\exp(z)$ for $z = x + jy$. (b) Suppose $\{x_2, y_2\}$ increase with time t. Describe how the mapping to the complex plane changes.

4.25 Derive the function $f_2(\omega_o t)$ in (4.89), generated when tracing a diamond on the complex plane.

4.26 Derive the projection $f(\omega_o t)$ of $r(t)\exp(j\omega_o t)$ onto the imaginary axis when tracing the rectangle in Figure 4.22, assuming constant angular velocity ω_o.

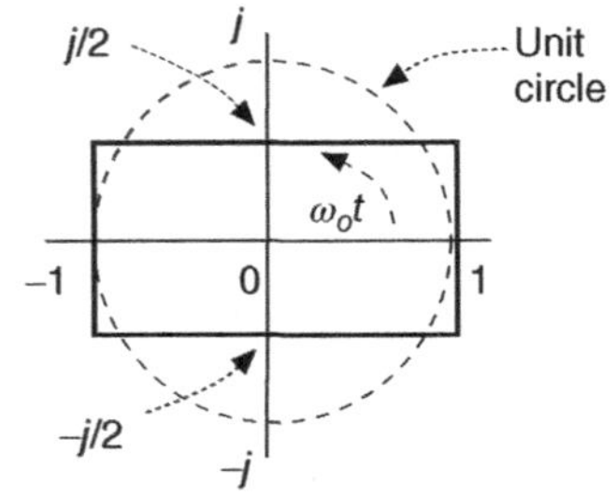

Figure 4.22 Rectangle on the complex plane for Problem 4.26.

Quaternions

4.27 Consider the quaternions $h_1 = 1 + i - j2 + k$ and $h_2 = 2 + 3i + j - 2k$. (a) Write the quaternion matrices $\{\mathbf{H}_1, \mathbf{H}_2\}$ and determine if their product $\mathbf{H}_1\mathbf{H}_2$ gives the same result as h_1h_2. (b) Repeat part (a) using the complex quaternion matrices $\{\mathbf{H}_{c_1}\mathbf{H}_{c_2}\}$.

4.28 For the quaternions $\{h_1, h_2\}$ below, find (a) h_1h_2, (b) h_2h_1, and (c) $h_1^{-1}h_2$:

$$h_1 = 1 + i - j - k, \quad h_2 = 2 - i + 2j + k. \tag{4.140}$$

4.29 Examine the rotations for the four cases of $\mathbf{x}$ in Example 4.10 for $\theta = 45°$ and $\mathbf{q} = j$. Verify your results using the appropriate rotation matrix.

4.30 Consider a unit cube with one end point at the origin in $\mathcal{R}^3$ and extending into positive $\{x_1, x_2, x_3\}$ with the furthest end point at (1, 1, 1). Determine how it is rotated by a quaternion with $\mathbf{q} = i + j$ and $\theta = 90°$.

Computer Problems

4.31 Plot $(1 + 2j/n)^n$ on the complex plane using MATLAB and verify that it approaches $\exp(2j) \approx -0.4161 + j0.9093$ with increasing n.

4.32 Using MATLAB, plot $f(\omega_o t)$ derived in Problem 4.26, along with $\sin(\omega_o t)$, and explain how this projection differs from the square trace in Figure 4.18(a).

4.33 Perform the rotations in Example 4.10 using quatrotate in MATLAB for different combinations of the angles $\{45°, 90°, 120°\}$ for (a) $\mathbf{q} = i$ and (b) $\mathbf{q} = i + j$.

PART II

SIGNALS, SYSTEMS, AND TRANSFORMS

5

SIGNALS, GENERALIZED FUNCTIONS, AND FOURIER SERIES

5.1 INTRODUCTION

In this chapter, we describe several functions of a continuous variable that are used to represent *signal waveforms* in many engineering applications. For the rest of the book, we are interested in functions of the independent variable time t, which we refer to as *signals*, such as the input $x(t)$ and output $y(t)$ of a linear system. The following special function is useful for defining the *support* of another function when they are multiplied together.

Definition: Indicator Function The *indicator function* is

$$I_{[a,b]}(t) \triangleq \begin{cases} 1, & t \in [a,b] \\ 0, & \text{else}, \end{cases} \tag{5.1}$$

where $[a,b]$ is a closed interval: $t \in [a,b]$ means $a \leq t \leq b$. Other intervals are possible such as semi-open $[a,b) \implies a \leq t < b$ and $(a,b] \implies a < t \leq b$, open $(a,b) \implies a < t < b$, and even a set of discrete values $\{a, \ldots, b\} \implies t \in \{a, \ldots, b\}$.

Symbols for sets of numbers such as $\mathcal{R}^+$ and $\mathcal{N}$ can also be used for the subscript of I, which should not be confused with the identity matrix $\mathbf{I}$ (which has bold font in this book). The support, range, and domain of a function are defined in Chapter 1.

Mathematical Foundations for Linear Circuits and Systems in Engineering, First Edition. John J. Shynk.

Companion Website: http://www.wiley.com/go/linearcircuitsandsystems

Example 5.1 The sinusoidal waveform $\sin(\omega_o t)I_{[0,2\pi]}(t)$ is nonzero only for $0 \leq t \leq 2\pi$, and the exponential waveform $\exp(-t)I_{[0,\infty)}(t)$ is nonzero only for $t \in \mathcal{R}^+$. If an indicator function is not used, such as $\cos(\omega_o t)$, then the support is assumed to be the entire real line $t \in \mathcal{R}$ unless otherwise specified.

5.2 ENERGY AND POWER SIGNALS

Let $x(t)$ be a real signal with domain $t \in \mathcal{R}$.

Definition: Energy and Power The *energy* of a signal is the area under the squared function:

$$E \triangleq \int_{-\infty}^{\infty} x^2(t)dt. \tag{5.2}$$

The *average power* of a signal is

$$P = \lim_{T\to\infty} \frac{1}{2T} \int_{-T}^{T} x^2(t). \tag{5.3}$$

(For circuits, the average power was defined in Chapter 2 in terms of the instantaneous power $p(t)$, and the energy was defined in terms voltage and charge.) For a particular signal, only one of these quantities is finite and nonzero: $0 < P < \infty \implies E \to \infty$ or $0 < E < \infty \implies P = 0$. Some signals have infinite power (and thus infinite energy). Thus, a signal can be classified into one of three types: (i) an energy signal, (ii) a power signal, or (iii) an infinite power signal.

Definition: Energy Signal A waveform is an *energy signal* if $0 < E < \infty$.

The average power P of an energy signal is necessarily zero.

Example 5.2 The rectangular function $x(t) = I_{[0,1]}(t)$ has finite energy:

$$E = \int_{0}^{1} x^2(t)dt = 1, \tag{5.4}$$

and it has zero average power:

$$P = \lim_{T\to\infty} \frac{1}{2T} \int_{-T}^{T} I_{[0,1]}(t)dt = \lim_{T\to\infty} \frac{1}{2T} \int_{0}^{1} dt = 0. \tag{5.5}$$

The one-sided exponential function $x(t) = \exp(-t)I_{[0,\infty)}(t)$ has finite energy:

$$E = \int_{0}^{\infty} \exp(-2t)dt = -(1/2)\exp(-2t)|_0^{\infty} = 1/2, \tag{5.6}$$

and zero power:

$$P = \lim_{T\to\infty} \frac{1}{2T} \int_0^T \exp(-2t)dt$$
$$= \lim_{T\to\infty} \frac{1}{4T}[1 - \exp(-2T)] = 0. \tag{5.7}$$

It is clear from the previous example that ordinary finite-duration waveforms are energy signals. Some infinite duration signals are energy signals, but often they are power signals.

Definition: Power Signal A waveform is a *power signal* if $0 < P < \infty$.

The energy E of a power signal is necessarily infinite.

Example 5.3 The cosine waveform with support $t \in \mathcal{R}$ is a power signal:

$$P = \lim_{T\to\infty} \frac{1}{2T} \int_{-T}^T \cos^2(\omega_o t)dt$$
$$= \lim_{T\to\infty} \frac{1}{4T} \int_{-T}^T [1 + \cos(2\omega_o t)]dt. \tag{5.8}$$

The cosine term divided by $4T$ is 0 in the limit, which gives $P = 1/2$. Since the area under $\cos^2(\omega_o t)$ is infinite, the energy of $\cos(\omega_o t)$ is $E \to \infty$.

Example 5.4 The unit step function is a power signal, but the ramp function is neither an energy signal nor a power signal: it has infinite power. For the unit step function:

$$\int_{-\infty}^{\infty} u^2(t)dt = \int_0^{\infty} dt \implies E \to \infty, \tag{5.9}$$

$$\lim_{T\to\infty} \frac{1}{2T} \int_{-T}^T u^2(t)dt = \lim_{T\to\infty} \frac{1}{2T} \int_0^T dt \implies P = 1/2, \tag{5.10}$$

and for the ramp function:

$$\lim_{T\to\infty} \frac{1}{2T} \int_{-T}^T r^2(t)dt = \lim_{T\to\infty} \frac{1}{2T} \int_0^T t^2 dt$$
$$= \lim_{T\to\infty} \left.\frac{t^3}{6T}\right|_0^T \implies P \to \infty. \tag{5.11}$$

Generally, energy signals have finite duration or they decay to 0 “sufficiently fast.” Power signals are typically periodic, and signals with infinite power tend to be infinitely increasing (positively or negatively). This classification of signals will be useful in Chapter 8 when we cover the Fourier transform. Energy signals always have a Fourier transform, whereas power signals and signals with infinite power may have a Fourier transform provided *singular generalized functions* are used in the frequency-domain representation. Several important functions in engineering are summarized in Appendix A, which includes their classification as energy, power, or infinite power signals.

5.3 STEP AND RAMP FUNCTIONS

Similar to the indicator function, the unit step function is often used in engineering to define the support of a function.

Definition: Unit Step Function The *unit step function* is

$$u(t) \triangleq I_{[0,\infty)}(t). \tag{5.12}$$

It is also called the *Heaviside step function*, and sometimes the symbol $H(t)$ is used.

Although $u(0) = 1/2$ in some applications, $u(0) = 1$ is used in this definition, and so the unit step function is continuous from the right as discussed in Chapter 1. More general step functions are obtained by scaling and shifting $u(t)$:

$$\alpha u(t-\tau) = \alpha I_{[\tau,\infty)}(t), \tag{5.13}$$

where α is the amplitude and τ is the delay. Examples are shown in Figure 5.1. The location of the discontinuity is found by examining the argument of the function:

$$u(t-\tau) = 1 \text{ when } t-\tau \geq 0 \implies t \geq \tau. \tag{5.14}$$

When τ is positive, the step function is shifted to the right, and when it is negative, the function is shifted to the left. The reverse situation occurs for argument $t+\tau$. Of course, this shifting applies to any function of the form $f(t-\tau)$ or $f(t+\tau)$. Step functions are used to model the effect of turning on a device, such as a voltage source in a circuit.

Example 5.5 The sinusoidal waveform $\sin(\omega_o t)u(t)$ is nonzero only for $t \in \mathcal{R}^+$, and the exponential waveform $\exp(-t)[u(t) - u(t-1)]$ is nonzero only for the finite interval $t \in [0, 1)$. Although the upper limit for t is a strict inequality (the semi-open interval $0 \leq t < 1$), in practice, we can generally include equality: $t \in [0, 1]$. This is done for most functions such as the rectangle function described in the next section.

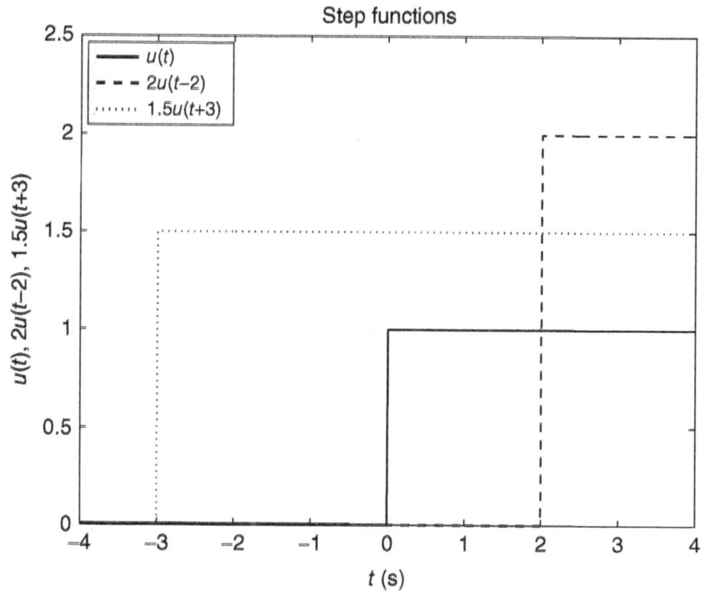

Figure 5.1 Example step functions.

The following two-sided function is related to the unit step function.

Definition: Signum Function The *signum function* is

$$\mathrm{sgn}(t) \triangleq \begin{cases} 1, & t > 0 \\ 0, & t = 0 \\ -1, & t < 0, \end{cases} \tag{5.15}$$

which is also known as the *sign* function. It can be written as the difference of two unit step functions:

$$\mathrm{sgn}(t) = u(t) - u(-t). \tag{5.16}$$

The signum function is related to the absolute value function as follows:

$$\mathrm{sgn}(t) = \frac{t}{|t|}, \tag{5.17}$$

and it is the derivative of $|t|$ except at $t = 0$ where the derivative is not defined. These functions are shown in Figure 5.2.

Definition: Ramp Function The *ramp function* is

$$r(t) \triangleq tu(t), \tag{5.18}$$

which can also be written in terms of the absolute value function: $r(t) = |t|u(t)$.

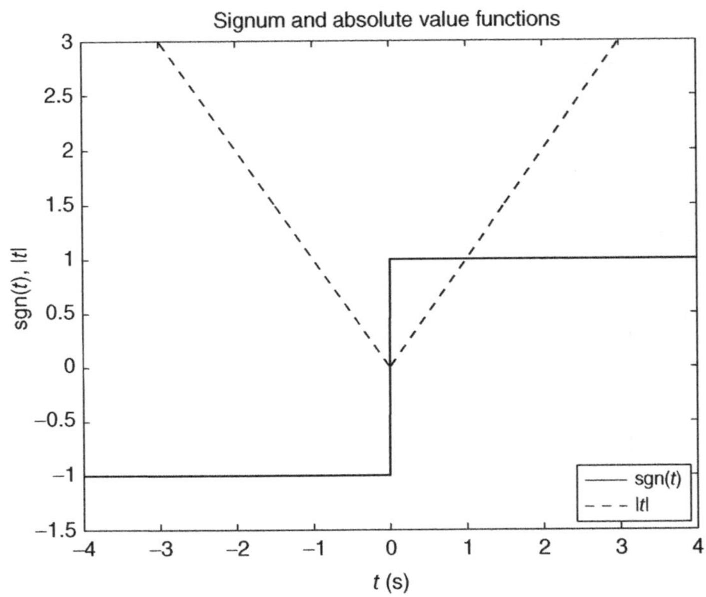

Figure 5.2 Signum and absolute value functions.

It is related to the unit step function as follows:

$$r(t) = \int_0^t u(\tau)d\tau, \qquad u(t) = \frac{d}{dt}r(t). \tag{5.19}$$

In order to find the derivative of $r(t)$ in (5.19), the product rule of differentiation should be used

$$\frac{d}{dt}r(t) = \left(\frac{d}{dt}t\right)u(t) + t\left(\frac{d}{dt}u(t)\right) = u(t) + t\delta(t) = u(t), \tag{5.20}$$

where $\delta(t) \triangleq du(t)/dt$ is the Dirac delta function defined later. The so-called sampling property of $\delta(t)$ when it is multiplied by continuous function $x(t)$ is $x(t)\delta(t) = x(0)\delta(t)$, such that the second term in the derivative is $t\delta(t) = 0$. In order to properly discuss the derivative of the unit step function, we need to expand ordinary functions to include *generalized functions*.

Under integrals, the unit step function defines the range of integration, and it also serves to define the support of the resulting integral. Thus, for the first expression in (5.19):

$$r(t) = \int_{-\infty}^t u(\tau)d\tau = u(t)\int_0^t d\tau = tu(t). \tag{5.21}$$

Similar techniques are used for the indicator function when it defines the support of a function. Example ramp functions obtained by integrating the step functions in Figure 5.1 are shown in Figure 5.3.

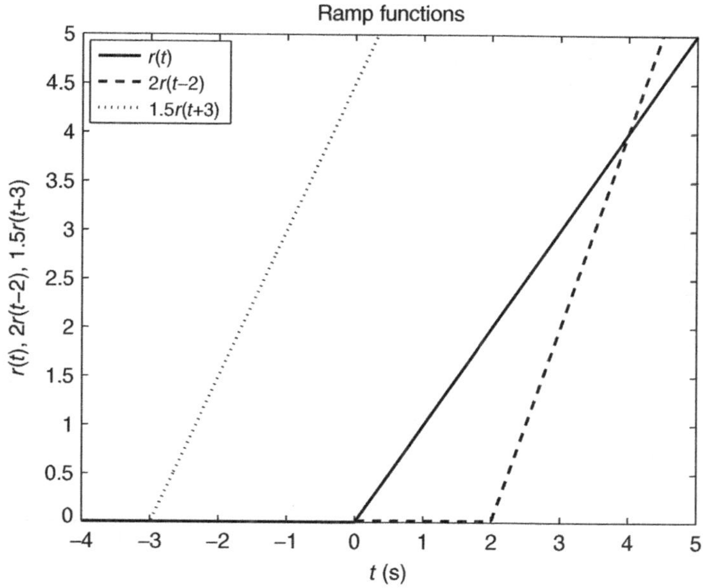

Figure 5.3 Example ramp functions.

5.4 RECTANGLE AND TRIANGLE FUNCTIONS

The two functions described in this section have finite support.

Definition: Rectangle Function The *rectangle function* is

$$\text{rect}(t) \triangleq I_{[-1/2,1/2]}(t), \tag{5.22}$$

which has unit width and unit height.

It is the solid waveform in Figure 5.4. The rectangle function can also be written as the difference of two unit step functions:

$$\text{rect}(t) \triangleq u(t + 1/2) - u(t - 1/2), \tag{5.23}$$

where it is assumed that the right-hand side equals 1 at $t = \pm 1/2$. The rectangle function is used to represent switching operations where a device is turned on and off, such as a voltage source in a circuit. Like the unit step function, rect(t) is often scaled and shifted; for example, the support of rect($t - 1/2$) is $[0, 1]$. Sometimes the rectangle function is defined as follows:

$$\text{rect}(t) \triangleq \begin{cases} 1, & |t| < 1/2 \\ 1/2, & |t| = 1/2 \\ 0, & |t| < 1/2, \end{cases} \tag{5.24}$$

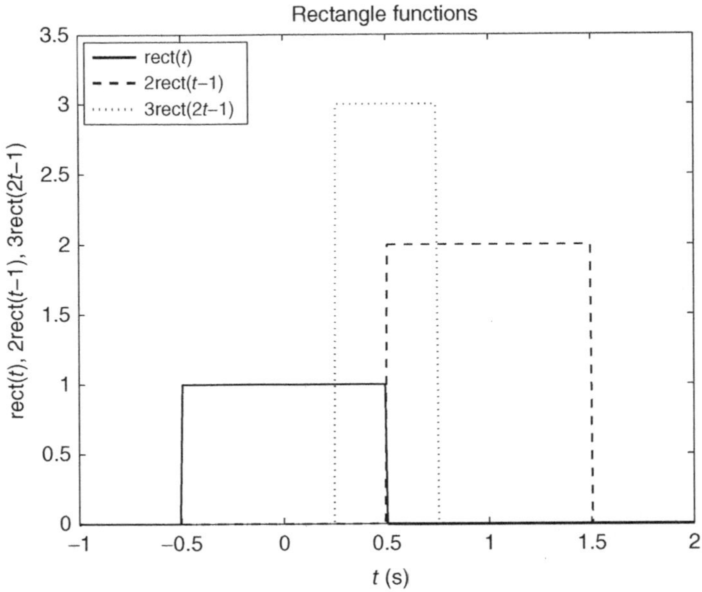

Figure 5.4 Example rectangle functions.

which has the value $1/2$ at the discontinuities (similar to the alternative definition of the unit step function). Generally, we use the definition in (5.22).

Example 5.6 The rectangle function $2\text{rect}(t-1)$ has height 2, width 1, and is centered at $t = 1$. The width of a rectangle function is always 1, except when the variable t is scaled. For example, $3\text{rect}(2t-1)$ has height 3, is centered at $2t - 1 = 0 \implies t = 1/2$, and its width is found by determining the values of t such that the argument of the function is $\pm 1/2$:

$$2t - 1 = 1/2 \implies t = 3/4, \qquad 2t - 1 = -1/2 \implies t = 1/4. \tag{5.25}$$

Subtracting these two quantities gives a width of $1/2$. These two examples are also shown in Figure 5.4.

Definition: Triangle Function The *triangle function* is

$$\text{tri}(t) \triangleq (1 - |t|)I_{[-1,1]}(t), \tag{5.26}$$

which has unit area.

Observe that by combining a rectangle function and the signum function, the triangle function is generated from the following integral:

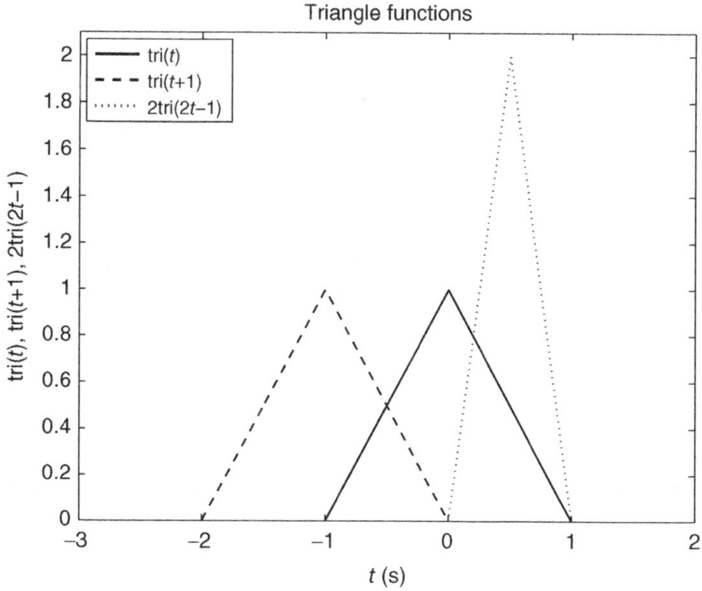

Figure 5.5 Example triangle functions.

$$\int_{-\infty}^{t} \text{rect}(\tau/2)\text{sgn}(-\tau)d\tau = I_{[-1,1]}(t) \int_{-1}^{t} \text{sgn}(-\tau)d\tau$$
$$= \begin{cases} \int_{-1}^{t} d\tau, & -1 \le t < 0 \\ \int_{-1}^{0} d\tau + \int_{0}^{t}(-1)d\tau, & 0 \le t \le 1, \end{cases}$$
$$= \begin{cases} t+1, & -1 \le t < 0 \\ 1-t, & 0 \le t \le 1, \end{cases} \tag{5.27}$$

which is the same as (5.26). The reversed signum function sgn($-t$) serves to change the sign of the rectangle function for $t \ge 0$. It was also necessary to scale τ in the rectangle function so that the width of the resulting triangle function is 2. Scaling t by $1/2$ causes rect($t/2$) to have support $[-1, 1]$, which is verified as follows:

$$t/2 = \pm 1/2 \implies t = \pm 1. \tag{5.28}$$

Example triangle functions are shown in Figure 5.5.

It turns out that the triangle function is also obtained as the *convolution* of two unit rectangle functions:

$$\text{tri}(t) = \text{rect}(t) * \text{rect}(t) = \int_{-\infty}^{\infty} \text{rect}(t-\tau)\text{rect}(\tau)d\tau$$
$$= \int_{\max(-1/2,t-1/2)}^{\min(t+1/2,1/2)} d\tau = \min(t+1/2, 1/2) - \max(-1/2, t-1/2), \tag{5.29}$$

which has support $t \in [-1, 1]$. Evaluating this expression over two finite intervals for t, given by $[-1, 0)$ and $[0, 1]$, we have

$$\text{rect}(t) * \text{rect}(t) = \begin{cases} t + 1/2 - (-1/2), & -1 \le t < 0 \\ 1/2 - (t - 1/2), & 0 \le t \le 1, \end{cases} \tag{5.30}$$

which is the same as (5.27). The convolution operator $*$ should not be confused with the superscript $*$ for the conjugate of a complex number. In the first line of (5.29), $\text{rect}(t - \tau)$ is a reversed rectangle function because τ is the variable of integration, and it is shifted by t. This is not the same function as $\text{rect}(\tau - t)$, which is not reversed. Convolution is discussed in greater detail in Chapters 6 and 7.

The previous results illustrate the importance of choosing the appropriate argument of a function in order to properly scale and shift it in time when representing a signal of interest.

Example 5.7 Consider two more cases for the rectangle function: $\text{rect}((t - 1)/2)$ and $\text{rect}(t/2 - 1)$. For the first case:

$$(t - 1)/2 = \pm 1/2 \implies t - 1 = \pm 1 \implies t \in [0, 2]. \tag{5.31}$$

The right-hand side is first scaled by 2 and then it is shifted by 1. For the second case:

$$t/2 - 1 = \pm 1/2 \implies t/2 = 1/2, 3/2 \implies t \in [1, 3], \tag{5.32}$$

and so the right-hand side is first shifted by 1 and then scaled by 2. Both of these rectangular functions have width 2, but their end points are quite different.

5.5 EXPONENTIAL FUNCTION

The exponential function is used to model the behavior of many systems, both natural and human-made. The standard exponential function was defined in Chapter 1, which we repeat here but with independent variable t for continuous time:

$$\exp(t) \triangleq e^t, \tag{5.33}$$

where Napier's constant $e = 2.71828182845\ldots$ is the base of the natural logarithm. Technically any function with the following form is called exponential:

$$x(t) = a^t, \tag{5.34}$$

where $a > 0$ and $a \neq 1$. We are generally interested only in the form of (5.33), which has the following unique properties:

$$\frac{d}{dt}\exp(t) = \exp(t), \qquad \int_{-\infty}^{t} \exp(t)dt = \exp(t). \tag{5.35}$$

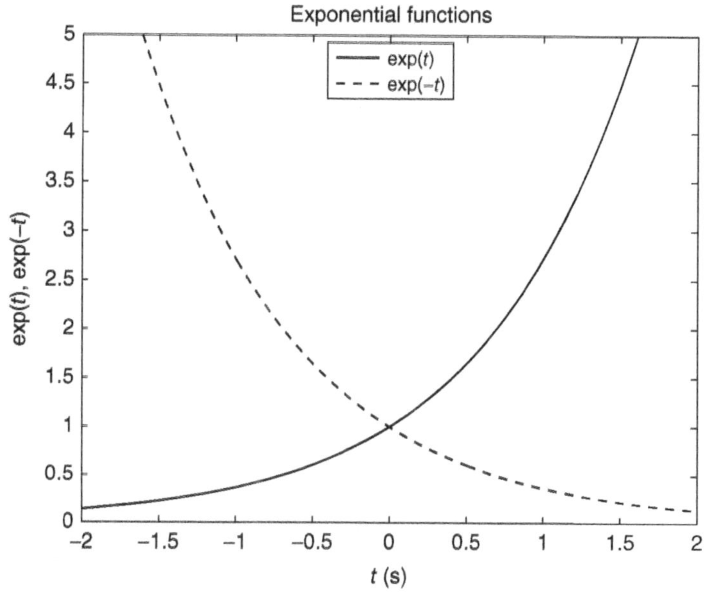

Figure 5.6 Increasing and decreasing exponential functions.

The standard exponential function is illustrated in Figure 5.6, along with a decaying exponential whose exponent is negative: $\exp(-t)$.

Example 5.8 The derivative property in (5.35) is proved using a power series representation of the exponential function (see Appendix E):

$$\exp(t) = 1 + t + t^2/2! + t^3/3! + \cdots \tag{5.36}$$

Differentiating each term on the right-hand side with respect to t yields

$$\frac{d}{dt}\exp(t) = 0 + 1 + 2t/2! + 3t^2/3! + \cdots = \exp(t). \tag{5.37}$$

The derivative of the general exponential form in (5.34) is not the same function for $a \neq e$:

$$\frac{d}{dt}a^t = a^t \ln(a), \tag{5.38}$$

where $\ln(\cdot)$ is the natural logarithm. The power series in (5.36) can also be used to prove the integral property of the exponential function (see Problem 5.9).

A decaying exponential starting at the origin can be written using the unit step function as follows:

$$x(t) = \exp(-t)u(t), \tag{5.39}$$

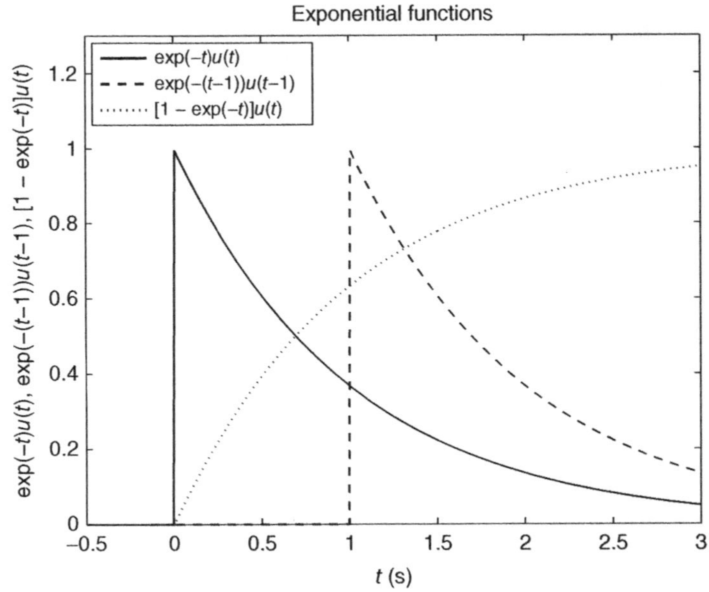

Figure 5.7 Example right-sided exponential functions.

and a delayed version is given by

$$x(t - t_o) = \exp(-(t - t_o))u(t - t_o). \tag{5.40}$$

The function $\exp(-t)u(t - t_o)$ is not the same as the delayed version in (5.40); this exponential function has *not* been shifted, and only its support has been changed to $t \geq t_o$. An exponential function that increases to a constant 1 is written as follows:

$$x(t) = [1 - \exp(-t)]u(t). \tag{5.41}$$

Such an expression is a model for signals in first-order RL and RC circuits that have a voltage or current source. These right-sided exponential functions are illustrated in Figure 5.7.

Definition: Time Constant The *time constant* of $x(t) = \exp(-\alpha t)u(t)$ with $\alpha > 0$ is the time $t = \tau$ such that the amplitude of the function has decreased to $1/e \approx 0.3679$ of its original value:

$$\exp(-\alpha\tau) = 1/e \implies \tau = 1/\alpha. \tag{5.42}$$

An exponential function has decreased to $< 5\%$ (≈ 0.0498) of its original value by 3τ. Several time constants (the vertical dotted lines) are illustrated in Figure 5.8 for $\exp(-t)u(t)$. A decaying exponential function is often written in terms of its time constant τ as follows:

$$x(t) = \exp(-t/\tau)u(t). \tag{5.43}$$

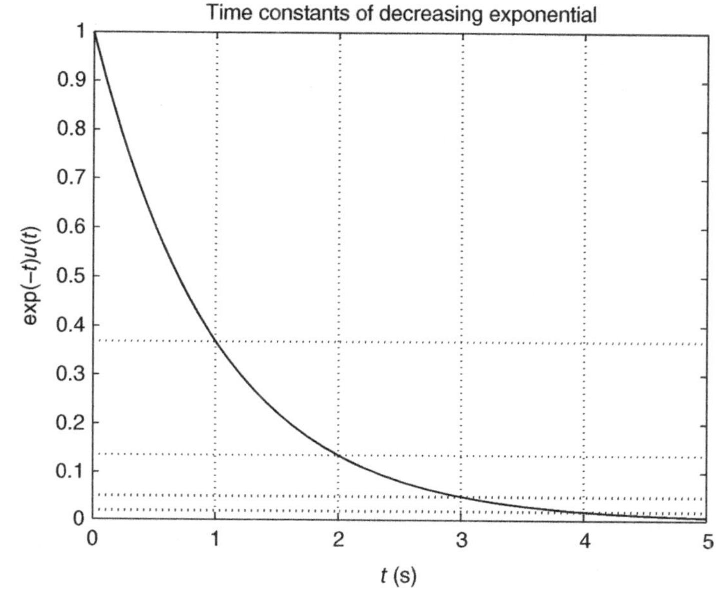

Figure 5.8 Multiple time constants for $\exp(-t)u(t)$ are denoted by the vertical dotted lines for $t = \tau, 2\tau, 3\tau, 4\tau$, and 5τ. The horizontal dotted lines are the corresponding values of the function: 0.3679, 0.1353, 0.0498, 0.0183, and 0.0067.

With this form, integer values of t yield the function at integer multiples of the time constant (as shown in Figure 5.8 for $\tau = 1$).

5.6 SINUSOIDAL FUNCTIONS

Sinusoidal functions appear in many applications, and any *periodic* signal can be represented by an infinite sum of weighted sines and cosines (the Fourier series expansion discussed later in this chapter). Generalizing the sinusoids considered in Chapter 4 to be functions of time, we have

$$x_1(t) = A\sin(\omega_o t + \phi), \qquad x_2(t) = A\cos(\omega_o t + \phi), \tag{5.44}$$

where A is the amplitude, ω_o is *angular frequency*, and ϕ is a phase shift. These expressions can be written in terms of the ordinary frequency f_o by substituting

$$\omega_o \triangleq 2\pi f_o. \tag{5.45}$$

As discussed previously for other functions, $\phi > 0$ causes sine and cosine to be shifted to the left, and they are shifted to the right for $\phi < 0$. The units of ω_o are rad/s, and those of f_o are hertz (Hz) = second^{-1} (sometimes called cycles/s). Thus, the arguments of the functions in (5.44) are in radians as was the case for $\sin(\theta)$ and $\cos(\theta)$

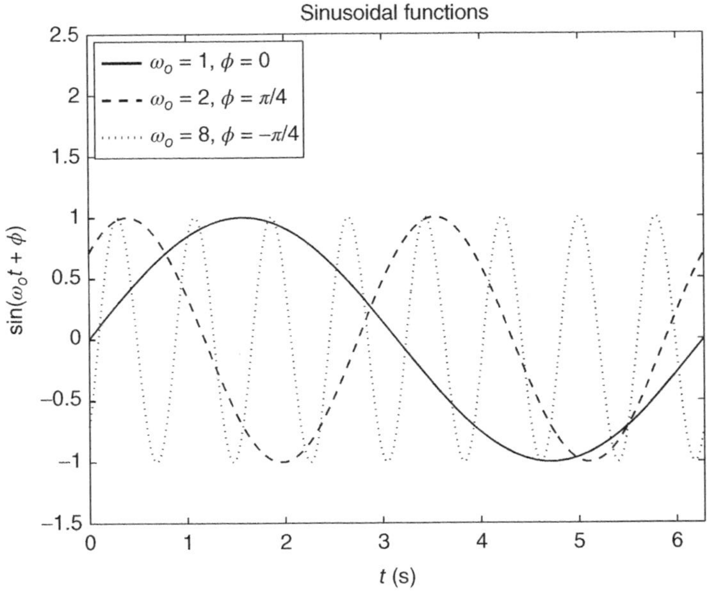

Figure 5.9 Sinusoidal functions with amplitude $A = 1$.

in Chapter 4. The period (one cycle) of a sinusoid is $T_o = 1/f_o$ with units of seconds. Examples of sinusoidal waveforms with $A = 1$ are shown in Figure 5.9. Observe that for $\omega_o = 1$ rad/s, the period is $T_o = 1/f_o = 2\pi/\omega_o = 2\pi$ s (the solid line in the figure).

Next, we describe a time-varying version of the *complex* exponential function introduced in Chapter 4. The general form is

$$x(t) = r\exp(j(\omega_o t + \phi)) = r\exp(j\phi)\exp(j\omega_o t), \tag{5.46}$$

where ω_o is angular frequency as used earlier for the sinusoidal waveforms, $r > 0$ is a constant magnitude, and ϕ is a constant phase. Using Euler's formula and assuming $\phi = 0$, we have

$$x(t) = r\cos(\omega_o t) + jr\sin(\omega_o t). \tag{5.47}$$

The squared magnitude of this function is a constant for any ω_o and all t:

$$|x(t)|^2 = r^2[\cos^2(\omega_o t) + \sin^2(\omega_o t)] = r^2, \tag{5.48}$$

which means that $x(t)$ is located on a circle with radius r on the complex plane. The sine and cosine functions are 90° ($\pi/2$ radians) out of phase with respect to each other, such that when sine is maximum or minimum, cosine is 0. This was depicted

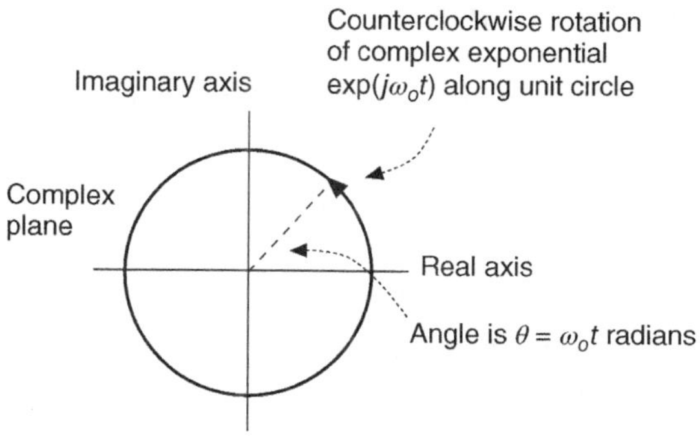

Figure 5.10 Time-varying complex exponential function on the unit circle.

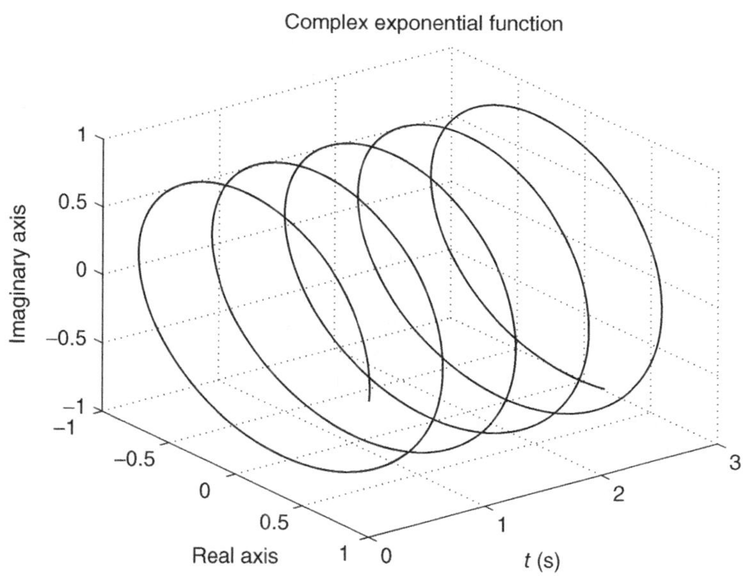

Figure 5.11 Trajectory of complex exponential with $r = 1$ and $\omega_o = 10$ rad/s.

previously in Figure 4.10 where the argument is a fixed angle θ. For $r = 1$, the function rotates counterclockwise on the *unit circle* as shown in Figure 5.10. It makes a complete rotation when $\omega_o t$ is an integer multiple of 2π, which corresponds to $t = 2\pi n/\omega_o = n/f = nT_o$ for $n \in \mathcal{Z}$. This result follows because complete rotations are achieved for integer multiples of the period T_o. The three-dimensional plot in Figure 5.11 shows the spiral trajectory of (5.47) for $r = 1$ and $\omega_o = 10$ rad/s as the function rotates counterclockwise along the unit circle. Similar plots were shown previously in Figure 4.15, but with exponential weighting $\exp(\sigma t)$ that caused the spiral to increase ($\sigma > 0$) or decrease ($\sigma < 0$) with increasing t.

5.7 DIRAC DELTA FUNCTION

The derivative of the unit step function is the *Dirac delta function*:

$$\delta(t) \triangleq \frac{d}{dt}u(t). \tag{5.49}$$

It can be defined as follows.

Definition: Dirac Delta Function The *Dirac delta function* is an impulse located at $t = 0$ that has zero width and unit area:

$$\delta(t) = \begin{cases} 0, & t \neq 0 \\ \text{undefined}, & t = 0, \end{cases} \qquad \int_{-\infty}^{\infty} \delta(t)dt = 1. \tag{5.50}$$

Although technically $\delta(t)$ is not defined at $t = 0$, some books assume infinity. The Dirac delta function is not an ordinary function because the area of an ordinary function is 0 if it is nonzero only at a countable number of points. We can view $\delta(t)$ as a *symbol* for a particular *generalized function*, and its most important feature is how it behaves *under an integral* as discussed in the next section.

We can also view the Dirac delta function as the limit of rectangle functions, which is an approach frequently used to describe its properties:

$$\delta(t) = \lim_{a\to 0}(1/a)\text{rect}(t/a). \tag{5.51}$$

Since the support of the standard rectangle function is $t \in [-1/2, 1/2]$, we find that the support of the right-hand side of (5.51) is $-1/2 \leq t/a \leq 1/2 \implies -a/2 \leq t \leq a/2$. In the limit as $a \to 0$, the width of the right-hand side approaches 0 and its height approaches infinity, but its area is fixed at $(1/a)[a/2 - (-a/2)] = 1$. Examples of the right-hand side of (5.51) for finite values of a are shown in Figure 5.12, from which we can visualize the rectangles approaching an impulse as $a \to 0$.

The Dirac delta function is scaled and shifted according to $\alpha\delta(t - \tau)$, which has area α and is located at $t = \tau$. Multiplication by a constant and shifting in time are handled in the same way as ordinary functions, except that for the Dirac delta function, the *area* is scaled by α. This is readily seen when scaling (5.51) by α. An arrow is used to represent the Dirac delta function as depicted in Figure 5.13, and its height corresponds to the *area*. The delta functions in the figure, which are necessarily nonoverlapping, can be written as a composite signal consisting of all three impulses simply by adding them together:

$$x(t) = \delta(t) - 2\delta(t - 1) + 2.5\delta(t + 2). \tag{5.52}$$

When a delta function is preceded by a minus sign, it is denoted by a downward pointing arrow when plotted. It still has zero width, but its area is defined to be negative; this interpretation also follows by using a rectangle function with height $-1/a$ in (5.51) and letting $a \to 0$.

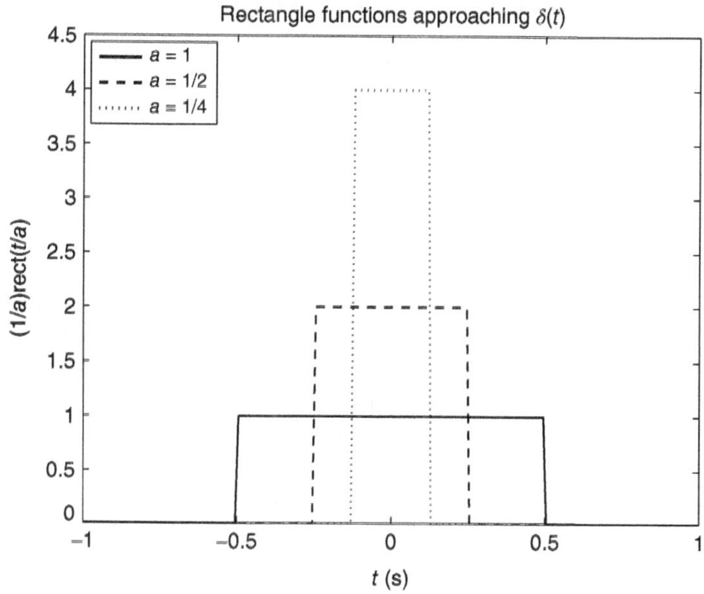

Figure 5.12 Rectangle functions in (5.51) approaching $\delta(t)$ in the limit.

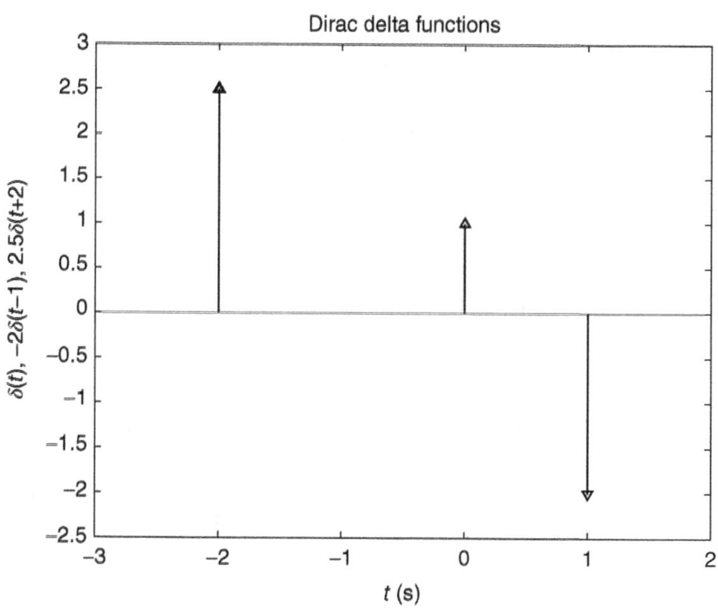

Figure 5.13 Dirac delta functions.

The Dirac delta function has two useful properties involving continuous function $f(t)$.

- *Sampling property*:
$$\delta(t-\tau)f(t) = \delta(t-\tau)f(\tau). \tag{5.53}$$

- *Sifting property*:
$$\int_{-\infty}^{\infty} \delta(t-\tau)f(t)dt = f(\tau). \tag{5.54}$$

When $\delta(t)$ multiplies continuous function $f(t)$, the sampling property yields another Dirac delta function at the same location, but with area given by the value of the function at $t = \tau$. The sifting property describes the behavior of the Dirac delta function under an integral where the value of the function $f(t)$ at $t = \tau$ is "sifted out." The result in (5.53) is still a Dirac delta function, whereas the result in (5.54) is a real number.

Example 5.9 In this example, we prove the sifting property for $\delta(t)$ in (5.54) with $\tau = 0$, starting with a rectangle function:

$$\begin{aligned}(1/a)\int_{-\infty}^{\infty} \text{rect}(t/a)f(t)dt &= (1/a)\int_{-a/2}^{a/2} f(t)dt \\ &= (1/a)[F(a/2) - F(-a/2)],\end{aligned} \tag{5.55}$$

where $F(t)$ is the *antiderivative* of $f(t)$. Since the last expression in (5.55) is a finite approximation of the derivative of $F(t)$ at $t = 0$, we have

$$\lim_{a\to 0}(1/a)[F(a/2) - F(-a/2)] = \left.\frac{d}{dt}F(t)\right|_{t=0} = f(0), \tag{5.56}$$

which shows

$$\int_{-\infty}^{\infty} \delta(t)f(t) = f(0). \tag{5.57}$$

The sampling property can also be proved by starting with the rectangle function:

$$\lim_{a\to 0}(1/a)\text{rect}(t/a)f(t) = f(0)\delta(t). \tag{5.58}$$

As the rectangle function becomes increasingly narrow about $t = 0$, the fixed area is scaled by $f(0)$ so that in the limit we have an impulse with area $f(0)$. For a delta function at another point in time $t = \tau$, the appropriate shifted version of the rectangle function is used to prove both properties (see Problem 5.13).

The sampling property of the Dirac delta function does not hold if $f(t) = \delta(t)$; the isolated product $\delta(t)\delta(t)$ is *not* defined. On the other hand, from the sifting property:

$$\int_{-\infty}^{\infty} \delta(t-\tau)\delta(t)dt = \delta(\tau), \tag{5.59}$$

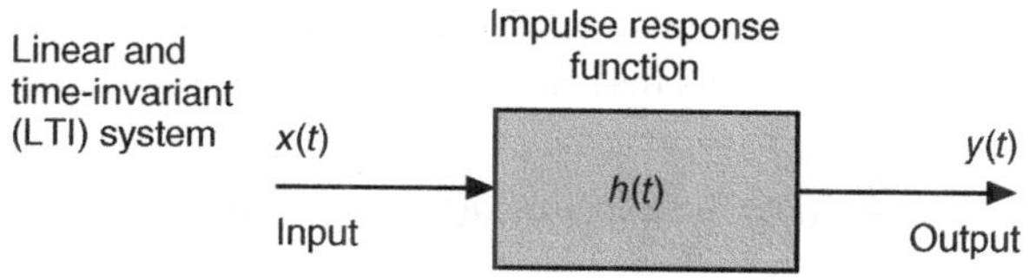

Figure 5.14 Linear time-invariant system with input $x(t)$, output $y(t)$, and impulse response function $h(t)$.

which is valid because this product of two delta functions is evaluated under an integral (which is actually a convolution). This will be evident from the definition of generalized functions.

The Dirac delta function is a useful model in engineering for impulsive-type signals, and it is used to generate the *impulse response function* of a linear and time-invariant (LTI) system, which in turn describes the response of the system for other types of input signals. Figure 5.14 shows a block diagram of a system with input $x(t)$ and output $y(t)$. The impulse response $h(t)$ for an LTI system is the output generated when $x(t) = \delta(t)$. It turns out that the output for any input $x(t)$ is generated by the convolution integral

$$y(t) = \int_{-\infty}^{\infty} h(\tau)x(t-\tau)d\tau = h(t) * x(t), \tag{5.60}$$

which was mentioned earlier. This important integral is widely used in courses on linear systems, and it is described further in Chapter 7, which also gives a precise definition of an LTI system.

5.8 GENERALIZED FUNCTIONS

In this section, we provide a brief overview of generalized functions (Kanwal, 2004; Strichartz, 1994), which is an extension of ordinary functions to include nonfunctions such as the Dirac delta function and its derivative the *unit doublet*. As previously mentioned, the defining characteristic of $\delta(t)$ is its behavior under an integral:

$$\int_{-\infty}^{\infty} \delta(t-\tau)dt = 1, \qquad \int_{-\infty}^{\infty} \delta(t-\tau)f(\tau)d\tau = f(t), \tag{5.61}$$

where it is assumed that function $f(t)$ is continuous at $t = \tau$. The first integral shows that $\delta(t)$ has unit area, and the second integral is the sifting property where $\delta(t)$ extracts the value of $f(t)$ at $t = \tau$. (The first integral is a special case of the sifting property where $f(t) = 1$ for an interval that includes $t = \tau$.) It is important to note that these integrals are only *symbolic*; they are not obtained in the limit from a Riemann sum. Instead, we *define* $\delta(t)$ to have these properties represented by the two integrals.

Example 5.10 Consider the right-sided function $f(t) = \exp(-\alpha t)u(t)$ whose support is $\mathcal{R}^+$. Using the product rule, its derivative is

$$\frac{d}{dt}f(t) = -\alpha \exp(-\alpha t)u(t) + \exp(-\alpha t)\delta(t),$$
$$= \delta(t) - \alpha \exp(-\alpha t)u(t), \tag{5.62}$$

where the sampling property of the Dirac delta function has been used to give $\exp(-\alpha t)\delta(t) = \exp(0)\delta(t) = \delta(t)$. This result is not unexpected because of the discontinuity at $t = 0$. The derivative of this function at the origin actually does not exist in the usual sense; it is handled by including the Dirac delta function. The product rule indirectly uses the theory of generalized functions by substituting $\delta(t) = du(t)/dt$.

Before describing generalized functions, we need some background definitions. Recall that the ordinary function $f(t)$ is a mapping of the real number t (the input) to a unique real number denoted notationally by $f(t)$ (the output). Thus, a function can be written as the *ordered pair* $\{t, f(t)\}$. This representation is extended to *functionals* where instead of the number t, the function $\phi(t)$ is used.

Definition: Functional *Functional* $F(\phi)$ is a mapping of function ϕ to a real number denoted by $F(\phi)$. It can be expressed as the ordered pair $\{\phi, F(\phi)\}$.

Although t is suppressed in this definition, ϕ is a function of t that we could write explicitly as $\phi(t)$, though do not for notational convenience. In this book, we are interested in *linear functionals* that satisfy the following two properties:

$$F(\phi_1 + \phi_2) = F(\phi_1) + F(\phi_2), \qquad F(c\phi) = cF(\phi), \tag{5.63}$$

where $\{\phi, \phi_1, \phi_2\}$ are functions of t and c is a constant. In particular, we focus on integrals of the form

$$F(\phi) = \int_{-\infty}^{\infty} f(t)\phi(t)dt, \tag{5.64}$$

where uppercase F is the functional of ϕ associated with lowercase function f under the integral. Since the integration is performed over the independent variable t, the functional $F(\phi)$ depends on the particular ϕ and not t, which is why ϕ appears explicitly as an argument of $F(\phi)$.

Definition: Locally Integrable Function $\phi(t)$ is *locally integrable* on $\mathcal{R}$ if the following integral exists:

$$\int_T |\phi(t)|dt < \infty, \tag{5.65}$$

where T is any closed interval on the real line $\mathcal{R}$.

Existence means that the integral is finite as mentioned earlier. Since T is a closed interval $[a, b]$ with $a < b$, this definition eliminates open and semi-open intervals of the form $(-\infty, \infty)$, $(-\infty, a]$, and $[b, \infty)$.

Example 5.11 It is clear that any continuous function is locally integrable. However, not all such functions are *globally* integrable. For example, the integral of the constant function $\phi(t) = 1$ is finite for any closed interval, but clearly it is not finite over $\mathcal{R}^+ = [0, \infty)$. Likewise, $\phi(t) = \exp(t)u(t)$ and $\phi(t) = tu(t)$ are locally integrable but not globally integrable.

The basic definition of a generalized function requires that function $\phi(t)$ in (5.64) be locally integrable and it must have compact support.

Definition: Compact Support Function $\phi(t)$ has *compact support* T if it is 0 for $|t| > K$ for some finite $K < \infty$, and so T is a bounded set on $t \in \mathcal{R}$.

Example 5.12 The support of the exponential function $\exp(-t)u(t)$ is $T = \mathcal{R}^+$, and that of the sinusoidal function $\cos(\omega_o t)$ is the entire real line $T = \mathcal{R}$. Neither of these functions has a compact support. The support of the rectangle function is the bounded interval $T = [-1/2, 1/2]$, and so it is compact.

Definition: Smooth Function Function $f(t)$ is *smooth* if it is infinitely differentiable on its support: $d^n f(t)/dt^n$ exists for all $n \in \mathcal{N}$.

Of course, this definition includes functions whose derivatives are 0 after some value for n.

Example 5.13 The sinusoidal waveforms $\cos(\omega_o t)$ and $\sin(\omega_o t)$ are smooth, whereas the unit step and the rectangle functions are not. The quadratic function $x(t) = t^2$ for $t \in \mathcal{R}$ is an example of a smooth function whose derivatives are 0 for $n > 2$.

Definition: Test Function A *test function* $\phi(t)$ has the following two properties: (i) compact support T and (ii) smooth on $t \in T$.

Example 5.14 The following truncated exponential is a test function:

$$\phi(t) = \exp(-\alpha^2/(\alpha^2 - t^2))I_{[-\alpha,\alpha]}(t). \tag{5.66}$$

It has compact support $T = [-\alpha, \alpha]$, and its first derivative is

$$\frac{d}{dt}\phi(t) = -\frac{2t}{(\alpha^2 - t^2)^2}\exp(-\alpha^2/(\alpha^2 - t^2))I_{[-\alpha,\alpha]}(t). \tag{5.67}$$

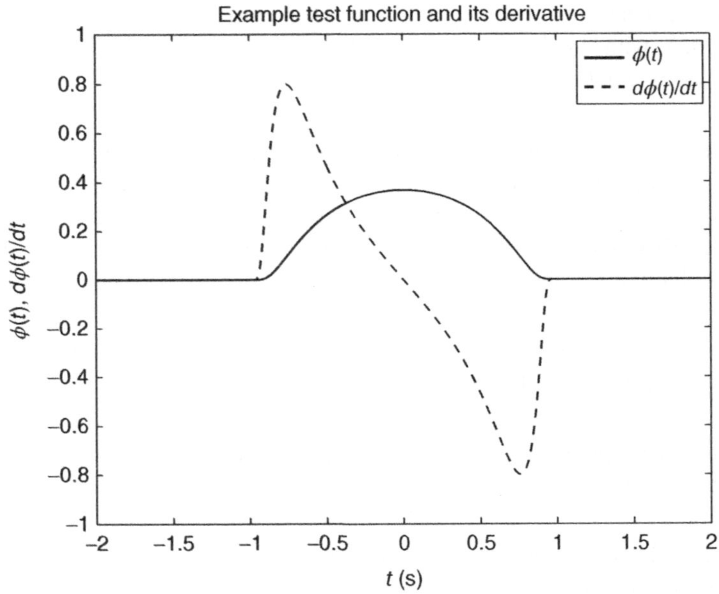

Figure 5.15 Test function in (5.66) and its derivative in (5.67) of Example 5.14 with $\alpha = 1$.

These are plotted in Figure 5.15 for $\alpha = 1$. It is clear that this function is infinitely differentiable on $\mathcal{T}$. Other possible test functions are considered in Problem 5.18. The following rectangular function is not infinitely differentiable:

$$f(t) = \begin{cases} \alpha, & |t| \leq 1/2\alpha \\ 0, & |t| > 1/2\alpha, \end{cases} \tag{5.68}$$

and so it is not a test function even though $f(t)$ has compact support. We mention, however, that it is possible to describe the derivatives of the rectangle function in terms of generalized functions. For example, its first derivative is a pair of Dirac delta functions:

$$f'(t) = \alpha\delta(t + 1/2) - \alpha\delta(t - 1/2), \tag{5.69}$$

which follows intuitively from the derivative of the unit step function $u'(t) = \delta(t)$.

Let the set $\mathcal{D}$ consist of all test functions (smooth with compact support) that have the following properties:

- *Linearity*: $\phi_1(t), \phi_2(t) \in \mathcal{D} \implies c_1\phi_1(t) + c_2\phi_2(t) \in \mathcal{D}$ for every $\{c_1, c_2\} \in \mathcal{R}$.
- *Derivatives*: $\phi(t) \in \mathcal{D} \implies d^n\phi(t)/dt^n \in \mathcal{D}$ for every $n \in \mathcal{N}$.
- *Product*: $\phi(t) \in \mathcal{D} \implies f(t)\phi(t) \in \mathcal{D}$ for smooth function $f(t)$.

It is not necessary that the test functions all have the same support T, but they should all be compact. The set of test functions $\mathcal{D}$ is defined for some domain Ω, in which case we could write $\mathcal{D}(\Omega)$ to be more precise. For example, the domain could be $\Omega = \mathcal{R}^n$, $\Omega = \mathcal{R}$, $\Omega = [0, 1]$, and so on. We generally use the real line $\Omega = \mathcal{R}$, and thus, we simply write $\mathcal{D}$ for the set of test functions.

For $\Omega = \mathcal{R}$, the following operations on $\phi(t)$ yield another function in $\mathcal{D}$:

- *Translation*: $\phi(t) \in \mathcal{D} \implies \phi(t - t_o) \in \mathcal{D}$ for finite t_o.
- *Time scale*: $\phi(t) \in \mathcal{D} \implies \phi(\alpha t) \in \mathcal{D}$ for finite $\alpha \neq 0$.
- *Product*: $\phi(t) \in \mathcal{D} \implies g(t)\phi(t) \in \mathcal{D}$ for smooth function $g(t)$.

The product rule of differentiation for the last property yields a function in $\mathcal{D}$:

$$\frac{d}{dt}g(t)\phi(t) = g(t)\frac{d}{dt}\phi(t) + \phi(t)\frac{d}{dt}g(t). \tag{5.70}$$

The first term on the right hand-side is a function in $\mathcal{D}$ because $d\phi(t)/dt$ has compact support, and so that term also has compact support and is infinitely differentiable. Similarly, the second term on the right-hand side is in $\mathcal{D}$ because $g(t)$ is infinitely differentiable by assumption and, of course, its product with $\phi(t)$ has compact support.

With the previous definitions and properties, we now define a generalized function.

Definition: Generalized Function The linear functional $F(\phi)$ on the set $\mathcal{D}$ of test functions is a *generalized function* provided it is continuous, satisfying

$$\lim_{m\to\infty} F(\phi_m) = F\left(\lim_{m\to\infty} \phi_m\right) = F(\phi), \tag{5.71}$$

where $\{\phi_m\}$ is any sequence of test functions such that $\lim_{m\to\infty}\phi_m = \phi$. A generalized function is also called a *distribution*, and the commonly used notation is

$$\langle f, \phi\rangle \triangleq \int_{-\infty}^{\infty} f(t)\phi(t)dt, \qquad \phi(t) \in \mathcal{D}, \tag{5.72}$$

where on the left-hand side, the variable of integration t is usually suppressed.

Equation (5.71) states that if a sequence of test functions $\{\phi_m\} \in \mathcal{D}$ converges to test function $\phi \in \mathcal{D}$, then the functional is continuous if it converges to the real number $F(\phi)$. It can be shown that this property holds for the integral in (5.64).

From these definitions, we find that generalized functions are defined relative to a set of test functions and how their product behaves under an integral. Whereas the support for an ordinary function consists of *points* on the real line, the "support" for a generalized function consists of the test functions. As a result, for such "functions" like $\delta(t)$, which are not well defined for points on $\mathcal{R}$, they can be defined in terms of how they operate under an integral when multiplying smooth functions. In summary:

$$\text{ordinary function: point } t \implies \text{function } \{t, f(t)\}, \tag{5.73}$$

$$\text{generalized function: test function } \phi(t) \implies \text{functional } \{\phi, F(\phi)\} \triangleq \langle f, \phi \rangle, \tag{5.74}$$

where $\{t, f(t)\}$ and $\{\phi, F(\phi)\}$ are ordered pairs for a function and functional, respectively, and $\langle f, \phi \rangle$ means the integral in (5.72) with integrand $f(t)\phi(t)$.

We use the notation in (5.72) instead of uppercase letter F because it is more convenient to manipulate as shown next. The uppercase function $F(\cdot)$ is used in the subsequent chapters on Fourier and Laplace transforms, which also have the integral form in (5.64). In those chapters, generalized functions are defined on *different classes* of test functions, which do not have compact support but decrease to 0 sufficiently fast as $t \to \pm\infty$.

The left-hand side of (5.72) is a *number* for $f(t)$ and a specific test function $\phi(t) \in \mathcal{D}$. For a different $\phi(t)$, a different number $\langle f, \phi \rangle$ is usually produced, and it is the set of these numbers for all $\phi(t)$ that describe the distribution of $f(t)$.

Definition: Dual Space The *dual space* of $\mathcal{D}$, denoted by $\mathcal{D}'$, is the set of all distributions defined on $\mathcal{D}$. It is a generalization of ordinary functions that includes both regular and singular distributions.

(Of course, the reader should not confuse $\mathcal{D}'$ with the ordinary derivative. This is the standard notation for the dual space.)

Next, we provide some useful properties of generalized functions and then explain the difference between regular distributions and singular distributions.

- *Product*: For smooth functions $f(t)$ and $g(t)$:

$$\langle fg, \phi \rangle = \langle f, g\phi \rangle = \langle g, f\phi \rangle. \tag{5.75}$$

Proof: These expressions follow because $g(t)\phi(t) \in \mathcal{D}$ and $f(t)\phi(t) \in \mathcal{D}$:

$$\begin{aligned} \langle fg, \phi \rangle &= \int_{-\infty}^{\infty} [f(t)g(t)]\phi(t)dt = \int_{-\infty}^{\infty} f(t)[g(t)\phi(t)]dt = \langle f, g\phi \rangle \\ &= \int_{-\infty}^{\infty} g(t)[f(t)\phi(t)]dt = \langle g, f\phi \rangle. \end{aligned} \tag{5.76}$$

- *Derivative*:

$$\langle f', \phi \rangle = -\langle f, \phi' \rangle. \tag{5.77}$$

Proof: This result is verified using integration by parts (see Appendix C):

$$\begin{aligned} \int_{-\infty}^{\infty} \frac{df(t)}{dt}\phi(t)dt &= f(t)\phi(t)|_{-\infty}^{\infty} - \int_{-\infty}^{\infty} \frac{d\phi(t)}{dt}f(t)dt \\ &= 0 - \langle f, \phi' \rangle. \end{aligned} \tag{5.78}$$

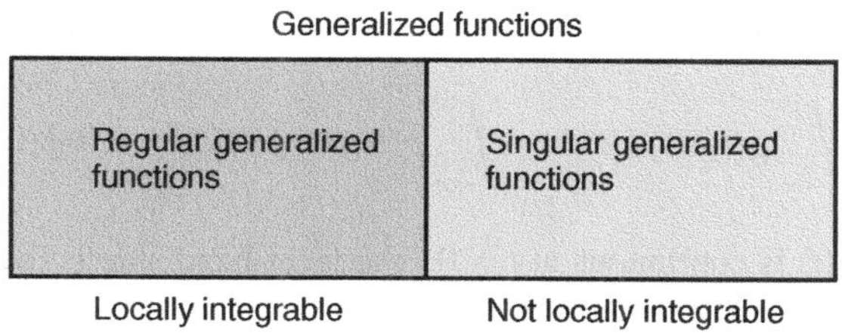

Figure 5.16 Types of generalized functions. (The rectangles do not indicate the relative sizes of the subsets.)

Since test functions have compact support, the first term on the right-hand side is 0 and the second term is the desired result. The derivative property is especially useful because we can describe the derivative of nonfunctions like the Dirac delta "function" by using the right-hand side and the fact that the test functions are infinitely differentiable.

- *High-order derivatives*: The previous result is readily extended as follows:

$$\langle f^{(n)}, \phi \rangle = (-1)^n \langle f, \phi^{(n)} \rangle, \tag{5.79}$$

where the superscript $^{(n)}$ denotes the nth ordinary derivative.

When $f(t)$ is a locally integrable function, $\langle f, \phi \rangle$ is called a *regular generalized function*. As mentioned earlier, the definition of a generalized function expands the concept of a function to include nonfunctions like $\delta(t)$. This expanded space of functions is depicted in Figure 5.16 where the additional elements, which are not locally integrable functions, are called *singular generalized functions*.

Example 5.15 The unit step function $u(t)$ and the ramp function $r(t)$ are locally integrable, and so they are regular distributions. The Dirac delta function $\delta(t)$ and its derivatives $\delta^{(n)}(t)$ are not locally integrable, and so they are singular distributions. Regular generalized functions include ordinary functions like $\exp(-t)$, as well as functions such as $\exp(-t)u(t)$ whose derivative has a Dirac delta function at the origin. The rectangle function is a regular distribution, but its derivative $\delta(t+2) - \delta(t-2)$ is a singular distribution.

Example 5.16 For $f(t) = \delta(t)$, we have from its sifting property:

$$\langle \delta, \phi \rangle = \int_{-\infty}^{\infty} \delta(t)\phi(t)dt \triangleq \phi(0), \tag{5.80}$$

where the right-hand side gives the distribution consisting of all test functions evaluated at $t = 0$. Again, this integral and the notation $\delta(t)$ are only symbolic because obviously we cannot partition the t axis into subintervals and define a Riemann sum that converges to $\phi(0)$.

From (5.75), we have

$$\int_{-\infty}^{\infty} g(t)\delta(t)\phi(t)dt = \int_{-\infty}^{\infty} \delta(t)[g(t)\phi(t)]dt = g(0)\phi(0), \tag{5.81}$$

assuming that $g(t)$ is continuous at $t = 0$. The right-hand side is the distribution for the product $g(t)\delta(t)$. Observe also that

$$g(0)\phi(0) = g(0)\int_{-\infty}^{\infty} \delta(t)\phi(t)dt = \int_{-\infty}^{\infty} g(0)\delta(t)\phi(t)dt. \tag{5.82}$$

Comparing the first integral in (5.81) and the second integral in (5.82), we find that

$$g(t)\delta(t) = g(0)\delta(t), \tag{5.83}$$

which is the sampling property of the Dirac delta function. This last result shows how the notation $\langle f, \phi \rangle$ can be used to find expressions for such quantities as $g(t)\delta(t)$. A similar expression is easily derived for $g(t)\delta(t - t_o)$ using the same approach (see Problem 5.19).

Example 5.17 Consider again the unit step function $u(t)$, which is a regular distribution:

$$\langle u, \phi \rangle = \int_{-\infty}^{\infty} u(t)\phi(t)dt = \int_{0}^{\infty} \phi(t)dt, \tag{5.84}$$

where $u(t)$ determines the lower limit of integration. The right-hand side gives the distribution consisting of the area of every test function defined on $\mathcal{R}^+$ (since the test functions have compact support, there is actually a finite upper limit of integration). From the *generalized derivative* property:

$$\begin{aligned}\int_{-\infty}^{\infty} \frac{du(t)}{dt}\phi(t)dt &= -\int_{-\infty}^{\infty} u(t)\frac{d\phi(t)}{dt}dt = -\int_{0}^{\infty} \frac{d\phi(t)}{dt}dt \\ &= -\int_{0}^{\infty} d\phi(t) = -\phi(t)|_0^{\infty} = \phi(0). \end{aligned}\tag{5.85}$$

The last result follows because $\phi(t)$ has a compact support: $\lim_{t\to\infty} \phi(t) = 0$. Since $\phi(0)$ equals the expression in (5.80), we find from the left-hand side of (5.85) that the derivative of the unit step function is the Dirac delta function:

$$\int_{-\infty}^{\infty} \frac{du(t)}{dt}\phi(t)dt = \phi(0) \implies \frac{d}{dt}u(t) = \delta(t). \tag{5.86}$$

This example shows that the derivative of a regular generalized function can be a singular generalized function. It also illustrates how such operations as the derivative of a function can be extended to nonfunctions by using test functions under integrals.

TABLE 5.1 Basic Distributions

Generalized Function $f(t)$	$\langle f, \phi \rangle$	Type
Dirac $\delta(t)$	$\phi(0)$	Singular
Unit doublet $\delta'(t)$	$-\phi'(0)$	Singular
Unit triplet $\delta''(t)$	$\phi''(0)$	Singular
Unit step $u(t)$	$\int_0^\infty \phi(t)dt$	Regular
Ramp $r(t)$	$\int_0^\infty t\phi(t)dt$	Regular
Absolute value $\lvert t\rvert$	$\int_0^\infty t\phi(t)dt - \int_{-\infty}^0 t\phi(t)dt$	Regular
Signum sgn(t)	$\int_0^\infty \phi(t)dt - \int_{-\infty}^0 \phi(t)dt$	Regular

TABLE 5.2 Properties of Generalized Functions at $\Omega = \mathcal{R}$

Property	Distributions
Equality	$\langle f, \phi \rangle = \langle g, \phi \rangle \implies f = g$
Linearity	$\langle f + g, \phi \rangle = \langle f, \phi \rangle + \langle g, \phi \rangle$
Product	$\langle gf, \phi \rangle = \langle f, g\phi \rangle = \langle g, f\phi \rangle$
Time shift	$\langle f(t - \tau), \phi \rangle = \langle f, \phi(t + \tau) \rangle$
Time scale	$\langle f(\alpha t), \phi \rangle = (1/\lvert\alpha\rvert)\langle f, \phi(t/\alpha) \rangle$
Derivatives	$\langle f^{(n)}, \phi \rangle = (-1)^n \langle f, \phi^{(n)} \rangle$
Even	$\langle f(-t), \phi(t) \rangle = \langle f(t), \phi(-t) \rangle = \langle f(t), \phi(t) \rangle$
Odd	$\langle f(-t), \phi(t) \rangle = \langle f(t), \phi(-t) \rangle = -\langle f(t), \phi(t) \rangle$

Some basic distributions are summarized in Table 5.1. The derivative of the ramp function is the unit step function, and the derivative of the absolute value function is the signum function, all of which are regular distributions. The derivative of the signum function is a singular distribution. These are covered in some of the problems at the end of this chapter. Various properties of generalized functions are summarized in Table 5.2 where $g(t)$ is a smooth function and α is nonzero.

The utility of the theory for generalized functions is evident from the table of properties, where we find that an operation on function $f(t)$ is "transferred" to the test function, which is a smooth function with compact support. For example, the general derivative property is $\langle f^{(n)}, \phi \rangle = (-1)^n \langle f, \phi^{(n)} \rangle$, which has the derivative $\phi^{(n)}(t)$ on the right-hand side. It may be that $f^{(n)}(t)$ is only symbolic, as is the case for $\delta'(t)$, but the right-hand side is well defined because $\phi(t)$ is infinitely differentiable.

It is straightforward to extend the definition of a generalized function to complex numbers $\Omega = C$. This domain for the test functions is needed when generalized functions are encountered in subsequent chapters. For the Fourier transform in Chapter 8, generalized functions are called *tempered distributions* based on a different class of test functions. Similarly, another class of test functions is assumed for the Laplace transform in Chapter 7.

Example 5.18 Consider the quadratic function $f(t) = t^2u(t)$ whose derivative we can write using the product rule:

$$\begin{aligned}\frac{d}{dt}f(t) &= u(t)\frac{d}{dt}t^2 + t^2\frac{d}{dt}u(t)\\ &= 2tu(t) + t^2\delta(t) = 2tu(t),\end{aligned} \tag{5.87}$$

where the sampling property of the Dirac delta function has been used to drop $t^2\delta(t)$. The same result is derived using the theory of generalized functions and the derivative property in Table 5.2:

$$\langle dt^2u(t)/dt, \phi\rangle = -\langle t^2u(t), d\phi(t)/dt\rangle = -\int_0^\infty t^2\frac{d\phi(t)}{dt}dt, \tag{5.88}$$

where the unit step function gives the lower limit of integration. Integration by parts yields

$$-\int_0^\infty t^2\frac{d\phi(t)}{dt}dt = -t^2\phi(t)|_{t=0}^\infty + \int_0^\infty 2t\phi(t)dt. \tag{5.89}$$

The first term on the right-hand side is 0 because $\phi(t)$ has compact support, and the second term is the distribution $\langle 2tu(t), \phi\rangle$ of the ramp function. Thus, from the equality property of distributions in Table 5.2:

$$\langle dt^2u(t)/dt, \phi\rangle = \langle 2tu(t), \phi\rangle \implies \frac{d}{dt}t^2u(t) = 2tu(t). \tag{5.90}$$

Example 5.19 Returning to the exponential function in Example 5.10, we find its derivative using the notation for generalized functions. From the generalized derivative property:

$$\begin{aligned}\langle d\exp(-\alpha t)u(t)/dt, \phi\rangle &= -\int_0^\infty \exp(-\alpha t)\frac{d\phi(t)}{dt}dt\\ &= -\exp(-\alpha t)\phi(t)|_{t=0}^\infty - \alpha\int_0^\infty \exp(-t)\phi(t)dt\\ &= \phi(0) - \alpha\langle\exp(-t)u(t), \phi\rangle.\end{aligned} \tag{5.91}$$

Thus,

$$\frac{d}{dt}\exp(-\alpha t)u(t) = \delta(t) - \alpha\exp(-\alpha t)u(t), \tag{5.92}$$

which is the same result as in (5.62).

TABLE 5.3 Distribution and Test Function Notation

Test Functions $\{\phi(t)\}$	Property	Dual Space	Application
$\mathcal{D}$ compact support	$\phi(t) = 0$ beyond $\lvert t\rvert > K$	$\mathcal{D}'$	Conventional
$\mathcal{E}$ exponential decay	$\lvert \exp(\alpha t) d^n\phi(t)/dt^n\rvert \leq c$	$\mathcal{E}' \subset \mathcal{D}'$	Laplace transform
$\mathcal{S}$ rapid decay	$\lvert t^p d^n\phi(t)/dt^n\rvert \leq c_{n,p}$	$\mathcal{S}' \subset \mathcal{D}'$	Fourier transform

The product of two singular generalized functions is *not* defined. For example, from the multiplication property, we might be tempted to write $\langle f_1 f_2, \phi\rangle = \langle f_1, f_2\phi\rangle$ where $f_1(t)$ and $f_2(t)$ are singular functions. However, the product $f_2(t)\phi(t)$ is no longer a test function because it may not be smooth even though it still has compact support. This problem does not occur with $g(t)\phi(t)$ for any smooth function $g(t)$ as given in Table 5.2.

Finally, we mention again that distributions can be defined for different types of test functions as summarized in Table 5.3. In this chapter, we focused on test functions with compact support, but it turns out that this is too restrictive for the Laplace transform and the Fourier transform covered later. In all three cases, $\{\phi(t)\}$ must be smooth; only the support changes as indicated in the table. The test functions in $\mathcal{E}$ are defined on the entire real line $\mathcal{R}$, but these functions and their derivatives must decay to 0 faster than exponential functions for every $\alpha \in \mathcal{R}$, $c > 0$, and $n \in \mathcal{Z}^+$. Similarly, the test functions in $\mathcal{S}$ are defined on $\mathcal{R}$, but these functions and their derivatives must decay to 0 faster than the reciprocal of polynomials; there must be some finite $c_{n,p}$ for every $n, p \in \mathcal{Z}^+$. Since the space of test functions for these two cases has been expanded from the conventional set $\mathcal{D}$ with compact support, the dual space of each is a subset of $\mathcal{D}'$ as indicated in the table and discussed in Chapters 7 and 8.

5.9 UNIT DOUBLET

In order to discuss the derivative of the Dirac delta function, we first use the limit of a sequence of triangle functions to represent $\delta(t)$. This approach is similar to the sequence of rectangle functions used previously, except that the triangle function is smoother. The standard triangle function has unit area:

$$\text{tri}(t) = (1 - |t|)I_{[-1,1]}(t). \tag{5.93}$$

Scaling the argument, the Dirac delta function is obtained as the following limit:

$$\delta(t) = \lim_{a\to 0}(1/a)\text{tri}(t/a). \tag{5.94}$$

When a is increased, the width of the triangle decreases and its height increases while maintaining unit area. Examples are shown in Figure 5.17.

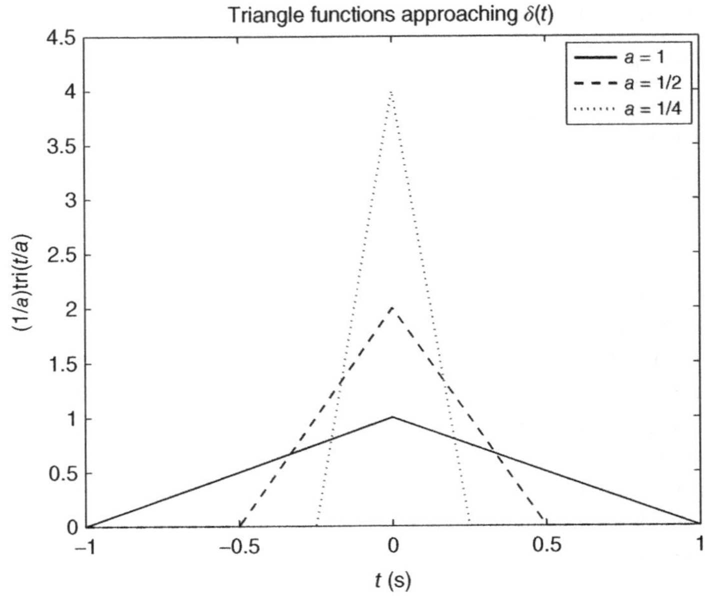

Figure 5.17 Triangle functions approaching $\delta(t)$ in the limit.

The derivative of (5.94) is a pair of rectangle functions:

$$\delta'(t) = \lim_{a\to 0}[(1/a^2)\text{rect}(t/a + 1/2) - (1/a^2)\text{rect}(t/a - 1/2)], \tag{5.95}$$

which are shown in Figure 5.18 for the three values of a used in Figure 5.17. These rectangle functions do *not* have unit area. The area of each rectangle approaches infinity because their width is a, but the scale factor is $1/a^2$ (whereas $1/a$ is used in (5.94) for the limit that yields $\delta(t)$). However, although the area of each rectangle increases as $a \to 0$, the overall area of the function is 0 because the rectangles have opposite signs about the origin. The symbol used for the unit doublet $\delta'(t)$ has two arrows with opposite directions as depicted in Figure 5.19 for

$$x(t) = \delta'(t) - 2\delta'(t-1) + 2.5\delta'(t+2), \tag{5.96}$$

which is the derivative of (5.52). The upward arrow is located "just to the left" of the time instant defined by the argument, and the downward arrow is located "just to the right." Recall that scaled delta functions are depicted with height α given by their areas. We likewise vary the height of the arrows representing the doublet to indicate the scale factor α, but the height does *not* represent the area (which is infinite as mentioned earlier). Even though the overall area is 0, we must keep track of any factor that scales $\delta'(t)$. The two arrows of the doublet are coupled; they cannot be separated into two delta-like functions. Also note that when a doublet is preceded by a minus sign, the two arrows are interchanged as shown in Figure 5.19 for $-2\delta'(t-1)$.

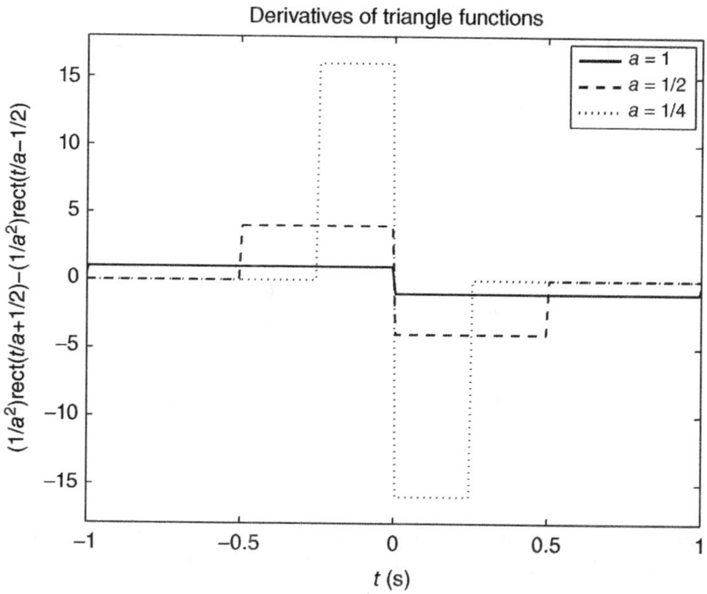

Figure 5.18 Derivatives of the triangle functions in Figure 5.17.

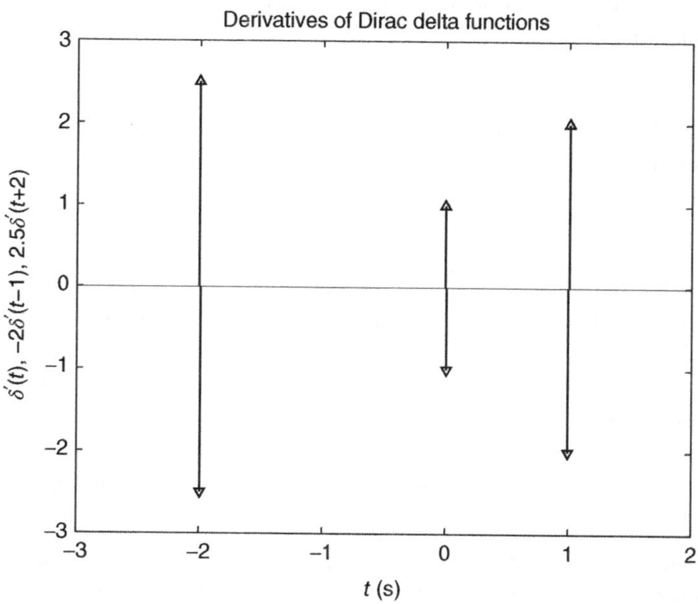

Figure 5.19 Derivatives of the Dirac delta functions in Figure 5.13.

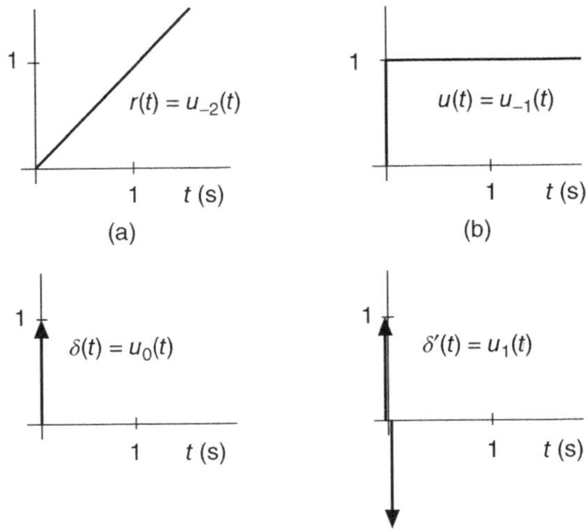

Figure 5.20 The ramp function and its derivatives. (a) Ramp $r(t)$. (b) Unit step $u(t) = dr(t)/dt$. (c) Dirac delta $\delta(t) = d^2r(t)/dt^2$. (d) Unit doublet $\delta'(t) = dr^3(t)/dt^3$.

Like the Dirac delta function, the unit doublet $\delta'(t)$ is a singular generalized function that is properly defined by its behavior under an integral.

There is a compact notation for various derivatives of the Dirac delta function. For its first derivative:

$$u_1(t) \triangleq \delta'(t) \tag{5.97}$$

is often used for the unit doublet. By varying the subscript, we have the following related notation:

$$u_0(t) \triangleq \delta(t), \qquad u_{-1}(t) \triangleq u(t), \qquad u_2(t) = \delta''(t), \tag{5.98}$$

and so on for the nth derivative. The second derivative of $\delta(t)$ is called the *unit triplet*. The ramp function and its derivatives using this notation are summarized in Figure 5.20.

Next, we consider the derivative of $\delta(t)$ using the properties of generalized functions. Observe that

$$\langle \delta', \phi \rangle = \int_{-\infty}^{\infty} \delta'(t)\phi(t)dt = -\int_{-\infty}^{\infty} \delta(t)\phi'(t)dt = -\phi'(0), \tag{5.99}$$

where in the second integral, the sifting property of $\delta(t)$ has been used at $t = 0$. Thus, the first integral is the sifting property of $\delta'(t)$ at $t = 0$. Suppose we multiply the unit doublet $\delta'(t)$ by the smooth function $f(t)$. Integration by parts yields

$$\langle \delta' f, \phi \rangle = \int_{-\infty}^{\infty} \delta'(t)f(t)\phi(t)dt = \delta(t)f(t)\phi(t)|_{-\infty}^{\infty} - \int_{-\infty}^{\infty} \delta(t)\frac{d[f(t)\phi(t)]}{dt}dt. \tag{5.100}$$

The first term on the right-hand side is 0 because $\phi(t)$ has compact support, and the product rule applied to the second term gives

$$\begin{aligned}\langle \delta' f, \phi \rangle &= -\int_{-\infty}^{\infty} \delta(t) \left[f(t)\frac{d\phi(t)}{dt} + \phi(t)\frac{df(t)}{dt} \right] dt \\ &= -f(0)\phi'(0) - \phi(0)f'(0), \end{aligned} \tag{5.101}$$

where the last expression is due to the sifting property of $\delta(t)$. Substituting the integrals in (5.80) and (5.99) yields

$$\begin{aligned}\langle \delta' f, \phi \rangle &= f(0) \int_{-\infty}^{\infty} \delta'(t)\phi(t)dt - f'(0) \int_{-\infty}^{\infty} \delta(t)\phi(t)dt \\ &= \int_{-\infty}^{\infty} [f(0)\delta'(t) - f'(0)\delta(t)]\phi(t)dt, \end{aligned} \tag{5.102}$$

from which we have the *sampling property* of $\delta'(t)$:

$$\delta'(t)f(t) = f(0)\delta'(t) - f'(0)\delta(t). \tag{5.103}$$

Additional properties of the unit doublet are summarized next.

- *Area*:

$$\int_{-\infty}^{\infty} \delta'(t)dt = 0. \tag{5.104}$$

  Proof: This result can be inferred from the discussion following (5.95). A derivation based on other properties of the doublet is considered in Problem 5.28.

- *Sifting property*:

$$\int_{-\infty}^{\infty} \delta'(t-\tau)f(\tau)d\tau = f'(t), \tag{5.105}$$

  provided that $f(t)$ is continuous at t. Symbolically we can write this convolution expression as $f'(t) = u_1(t) * f(t)$ where $u_1(t)$ is the alternative symbol mentioned earlier for the unit doublet. For the nth derivative of the Dirac delta function, it can be shown that this sifting property extends as

$$f^{(n)}(t) = u_n(t) * f(t) = \underbrace{u_1(t) * \cdots * u_1(t)}_{n \text{ times}} * f(t). \tag{5.106}$$

  Proof: Using integration by parts, (5.105) is verified as follows:

$$\int_{-\infty}^{\infty} \delta'(t-\tau)f(\tau)d\tau = -\delta(\tau)f(t-\tau)|_{-\infty}^{\infty} + \int_{-\infty}^{\infty} \delta(t-\tau)f'(\tau)d\tau = f'(t). \tag{5.107}$$

The first term in the middle equation is 0 because $\delta(t)$ is 0 for $\tau \neq 0$, and the last result follows from the sifting property of the Dirac delta function. Thus, $\delta'(t) * f(t) = \delta(t) * f'(t) = f'(t)$. Note also that for $t = 0$:

$$\int_{-\infty}^{\infty} \delta'(-\tau) f(\tau) d\tau = f'(0), \tag{5.108}$$

and since $\delta'(t)$ is an odd function, we have

$$\int_{-\infty}^{\infty} \delta'(\tau) f(\tau) d\tau = -f'(0). \tag{5.109}$$

The previous results can also be derived using the generalized function approach (see Problem 5.27).

- *Product with t*:

$$t\delta'(t) = -\delta(t). \tag{5.110}$$

Proof: This property follows immediately from (5.103) with $f(0) = 0$ and $f'(0) = 1$. It is also verified by using the rectangle functions in (5.95) and the fact that t is an odd function. Figure 5.21 shows that the product of the rectangle functions and t are truncated ramp functions with negative amplitudes. Since the rectangle functions are 0 beyond the interval $[-a, a]$, we find that the ramps

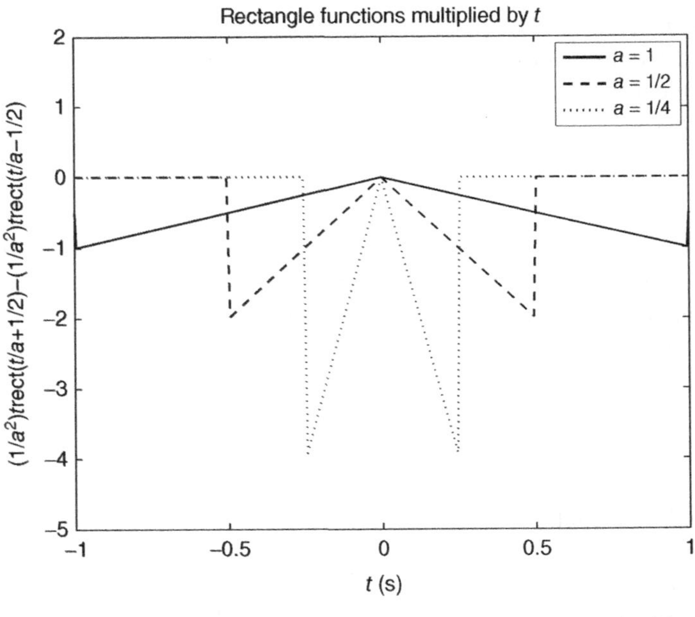

Figure 5.21 Multiplication of rectangle functions in Figure 5.18 with t.

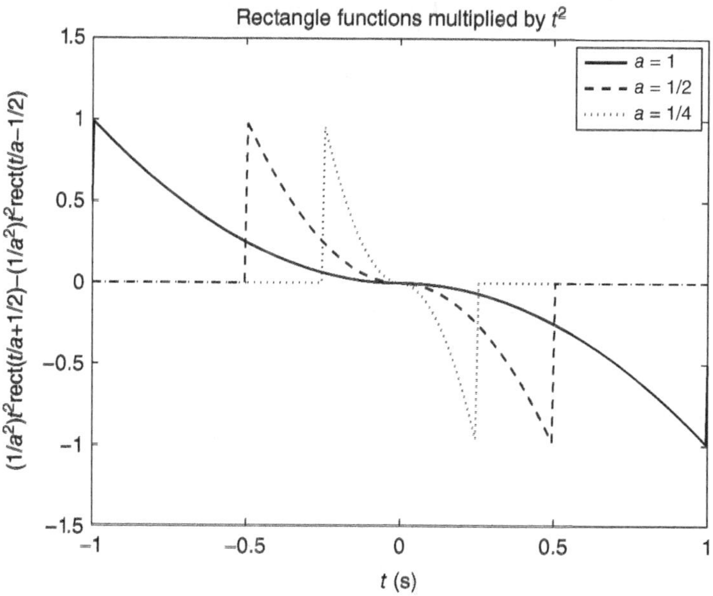

Figure 5.22 Multiplication of rectangle functions in Figure 5.18 with t^2.

are truncated to have minimum value $-1/a$. Thus, the product is actually a truncated absolute value function:

$$(t/a^2)\text{rect}(t/a + 1/2) - (t/a^2)\text{rect}(t/a - 1/2) = -|t/a^2|I_{[-a,a]}(t), \quad (5.111)$$

which has area -1 for every a. As $a \to 0$, the height approaches minus infinity, and the width specified by the indicator function approaches 0, yielding $-\delta(t)$.

- *Product with t^2*:

$$t^2\delta'(t) = 0. \quad (5.112)$$

Proof: This property also follows from (5.103) with $f(0) = f'(0) = 0$. It is verified by multiplying the rectangle functions in (5.95) by t^2 and taking the limit as $a \to 0$. This is depicted in Figure 5.22 where we see that as a is decreased toward 0, the area of each component of the product decreases. This is due to the fact that the magnitude of the product is always fixed at 1 because the exponent of t matches that of a:

$$(t^2/a^2)\text{rect}(t/a + 1/2) - (t^2/a^2)\text{rect}(t/a - 1/2) = (t/a)^2 I_{[-a,0]}(t) - (t/a)^2 I_{[0,a]}. \quad (5.113)$$

Since the width of the function decreases according to the indicator functions, the product approaches 0 as $a \to 0$.

A summary of several properties of the Dirac delta function and its first and second derivatives is provided in Table 5.4.

TABLE 5.4 Properties of the Dirac Delta Function and Its Derivatives

Property	Expression
$\delta(t)$ symmetry	$\delta(t) = \delta(-t)$ (even)
$\delta(t)$ sifting	$\int_{-\infty}^{\infty} \delta(t-t_o)f(t)dt = f(t_o)$
$\delta(t)$ convolution	$\int_{-\infty}^{\infty} \delta(t_o-t)f(t)dt = f(t_o)$
$\delta(t)$ product	$\delta(t-t_o)f(t) = \delta(t-t_o)f(t_o)$ (sampling)
$\delta(t)$ area	$\int_{-\infty}^{\infty} \delta(t-t_o)dt = 1$ (sifting with $f(t) = 1$)
$\delta(t)$ moment	$\int_{-\infty}^{\infty} t\delta(t)dt = 0$ (sifting with $f(t) = t$ and $t_o = 0$)
$\delta'(t)$ symmetry	$\delta'(t) = -\delta'(-t)$ (odd)
$\delta'(t)$ sifting	$\int_{-\infty}^{\infty} \delta'(t-t_o)f(t)dt = -f'(t_o)$
$\delta'(t)$ convolution	$\int_{-\infty}^{\infty} \delta'(t_o-t)f(t)dt = f'(t_o)$
$\delta'(t)$ product	$\delta'(t-t_o)f(t) = \delta'(t-t_o)f(t_o) - f'(t_o)\delta(t-t_o)$ (sampling)
$\delta'(t)$ area	$\int_{-\infty}^{\infty} \delta'(t-t_o)dt = 0$ (sifting with $f(t) = 1$)
$\delta'(t)$ moment	$\int_{-\infty}^{\infty} t\delta'(t)dt = -1$ (sifting with $f(t) = t$ and $t_o = 0$)
$\delta'(t)$ product 1	$t\delta'(t) = -\delta(t)$ (sampling with $f(t) = t$ and $t_o = 0$)
$\delta'(t)$ product 2	$t^2\delta'(t) = 0$ (sampling with $f(t) = t^2$ and $t_o = 0$)
$\delta^{(2)}(t)$ symmetry	$\delta^{(2)}(t) = \delta^{(2)}(-t)$ (even)
$\delta^{(2)}(t)$ sifting	$\int_{-\infty}^{\infty} \delta^{(2)}(t-t_o)f(t)dt = f^{(2)}(t_o)$
$\delta^{(2)}(t)$ convolution	$\int_{-\infty}^{\infty} \delta^{(2)}(t_o-t)f(t)dt = f^{(2)}(t_o)$
$\delta^{(2)}(t)$ product	$\delta^{(2)}(t)f(t) = f(0)\delta^{(2)}(t) - 2f'(0)\delta'(t) + f^{(2)}(0)\delta(t)$ (sampling)
$\delta^{(2)}(t)$ area	$\int_{-\infty}^{\infty} \delta^{(2)}(t-t_o)dt = 0$ (sifting with $f(t) = 1$)
$\delta^{(2)}(t)$ moment	$\int_{-\infty}^{\infty} t\delta^{(2)}(t)dt = 0$ (sifting with $f(t) = t$ and $t_o = 0$)
$\delta^{(2)}(t)$ product 1	$t\delta^{(2)}(t) = -2\delta'(t)$ (sampling with $f(t) = t$ and $t_o = 0$)
$\delta^{(2)}(t)$ product 2	$t^2\delta^{(2)}(t) = 2\delta(t)$ (sampling with $f(t) = t^2$ and $t_o = 0$)

5.10 COMPLEX FUNCTIONS AND SINGULARITIES

In this section, we consider functions of the complex variable $z = x + jy$ and *singularities* of a function (Brown and Churchill, 2009), which will be useful when the Laplace transform is covered in Chapter 7.

Definition: Analytic Function A function $f(z)$ of complex variable z is *analytic* at z_o if it is finite and infinitely differentiable at z_o. This means $f(z_o)$ can be represented by a Laurent series expansion with terms of the form $(z - z_o)^n$ for $n \in \mathcal{Z}$.

This definition is consistent with our notion of a continuous function that has no discontinuities or points where $f(z)$ or its derivatives are not defined. Analytic functions that have no singularities are also called *well-behaved* and *smooth*.

Definition: Entire Function Function $f(z)$ that is analytic everywhere on the finite complex x–y plane is called an *entire function.*

The finite complex plane C consists of all $z = x + jy$ such that $|x| < \infty$ and $|y| < \infty$.

Example 5.20 The following are entire analytic functions:

$$f(z) = \exp(z), \qquad f(z) = z + z^2, \qquad f(z) = \sin(z). \tag{5.114}$$

It is clear from the definition that all polynomials of z are analytic functions. The Laurent series for exp(z) is

$$\exp(z) = \sum_{n=0}^{\infty} z^n/n!, \tag{5.115}$$

which is identical to the power series expansion of real-valued exp(x) with x replaced by z. The Laurent series is discussed further in Appendix E. Likewise for sin(z):

$$\sin(z) = \sum_{n=0}^{\infty} (-1)^n z^{2n+1}/(2n+1)!, \tag{5.116}$$

which can also be written as Euler's inverse formula with complex z:

$$\sin(z) = (1/2j)[\exp(jz) - \exp(-jz)]. \tag{5.117}$$

Note that unlike (5.115), j appears in the argument of the exponential functions in this expression.

The function $f(t) = 1/z$ is analytic for all z except at $z = 0$, which is a singularity.

Definition: Singularity A *singularity* of function $f(z)$ is a value of z where the function or its derivatives are not defined. This value is also called a *singular point*. A singular point z_o is *isolated* if there exists a neighborhood $0 < |z - z_o| < \epsilon$ for some $\epsilon > 0$ where the function is analytic.

For a singularity at $z = z_o$, the Laurent series for the function about that point is

$$f(z) = \sum_{n=-\infty}^{\infty} c_n(z - z_o)^n = \sum_{n=0}^{\infty} c_n(z - z_o)^n + \sum_{m=1}^{\infty} \frac{c_{-m}}{(z - z_o)^m}, \tag{5.118}$$

where we have split the first summation into two sums and then changed variables to $n \to -m$ in the last sum to explicitly show the terms in the denominator. If the last sum over m has only a finite number of nonzero $\{c_{-m}\}$, then the singularity associated with the kth term $c_{-k}/(z-z_o)^k$ is called a *pole of order k* at $z = z_o$. It is called a *simple pole* if $k = 1$. Examples of functions with poles include

$$f(z) = 1/z, \qquad f(z) = 1/(z+1), \qquad f(z) = 1/z(z-1), \tag{5.119}$$

which have singularities at $z_o = 0$, $z_o = -1$, and $z_o = \{0, 1\}$, respectively. Poles are discussed further in Chapter 7 where ODEs are solved using the Laplace transform.

If the last sum in (5.118) has an infinite number of terms, then the singularity at $z = z_o$ is called an *essential singular point*. Examples of functions with essential singular points include

$$f(z) = \sin(1/z), \qquad f(z) = \exp(1/z). \tag{5.120}$$

If the last sum in (5.118) has no terms, then the singularity is *removable*, which means the function is actually analytic at $z = z_o$. Thus, we can write

$$\lim_{z\to z_o} (z - z_o)f(z) = 0, \tag{5.121}$$

which obviously follows from (5.118) when the last sum is 0. Examples include

$$f(z) = \sin(z)/z, \qquad f(z) = [1 - \cos(z)]/z. \tag{5.122}$$

L'Hôpital's rule applied to the first function evaluated at 0 yields

$$\left.\frac{d\sin(z)/dz}{dz/dz}\right|_{z=0} = \cos(z)|_{z=0} = 1, \tag{5.123}$$

and for the second function:

$$\left.\frac{d[1-\cos(z)]/dz}{dz/dz}\right|_{z=0} = \sin(z)|_{z=0} = 0. \tag{5.124}$$

Plots of the three basic types of singularities are shown in Figure 5.23. In Chapter 7, we will again see functions of a complex variable that have poles on the complex plane.

5.11 CAUCHY PRINCIPAL VALUE

When performing integrations like those of the Fourier and Laplace transforms covered later, it is important that the integrals be properly defined for functions with singularities. For example, $f(t) = 1/t$ is not defined at $t = 0$, in which case the domain $(0, \infty)$ is often assumed. However, a problem arises if we attempt to integrate this function as is done in (5.72) for a distribution:

$$\langle 1/t, \phi\rangle = \int_{-\infty}^{\infty} (1/t)\phi(t)dt. \tag{5.125}$$

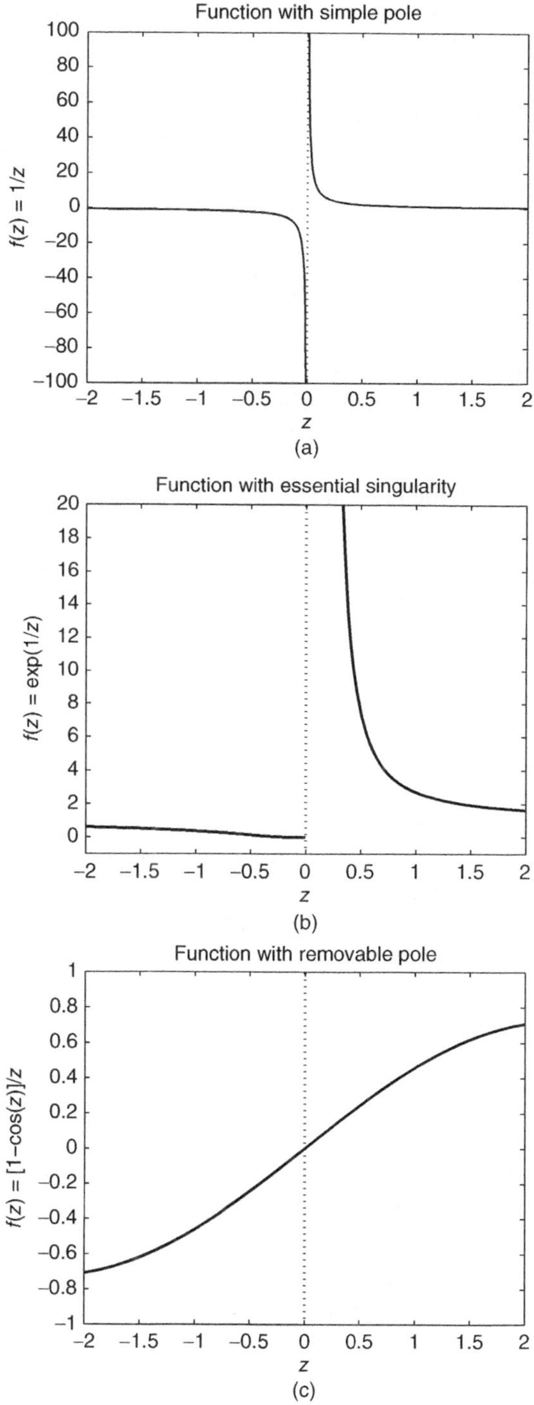

Figure 5.23 Functions with singularities on the real axis at $z = 0$. (a) Simple pole: $f(z) = 1/z$. (b) Essential singularity: $f(z) = \exp(1/z)$. (c) Removable pole at $z = 0$ with $f(0) = 0$: $f(z) = [1 - \cos(z)]/z$.

This integral is not well defined, and so we need to place some restriction on how it is performed. Suppose the integral is evaluated as follows:

$$\langle 1/t, \phi \rangle = \lim_{\epsilon \to 0} \left[\int_{-\infty}^{-\epsilon} (1/t)\phi(t)dt + \int_{\epsilon}^{\infty} (1/t)\phi(t)dt \right]. \tag{5.126}$$

It turns out that a different result is obtained using

$$\langle 1/t, \phi \rangle = \lim_{\epsilon \to 0} \left[\int_{-\infty}^{-\epsilon} (1/t)\phi(t)dt + \int_{2\epsilon}^{\infty} (1/t)\phi(t)dt \right], \tag{5.127}$$

where 2ϵ appears in the second integral. There is actually an infinity of results depending on how the integral is calculated near the singularity at $t = 0$. In order to handle this problem, the Cauchy principal value (CPV) is used.

Definition: Cauchy Principal Value The CPV for the integral of function $f(t)$ with singularity at t_o is

$$\lim_{\epsilon \to 0} \left[\int_{-\infty}^{t_o - \epsilon} f(\tau)d\tau + \int_{t_o + \epsilon}^{\infty} f(\tau)d\tau \right], \tag{5.128}$$

where both limits proceed at the same rate toward t_o as ϵ is varied.

This definition is symmetric about t_o, unlike the form in (5.127) where the limits include ϵ and 2ϵ. In order to be reminded that caution should be exercised when integrating functions with singularities, the notation $\mathcal{P}(f(t))$ is used, indicating that the CPV is computed for an integral. Thus, we would write $\langle \mathcal{P}(1/t), \phi \rangle$ on the left-hand side of (5.126). Other examples include $\mathcal{P}(1/(t - t_o))$, $\mathcal{P}(\text{sgn}(t)/t^2)$, and so on. Some examples are plotted in Figure 5.24. The CPVs for $\int_{-\infty}^{\infty}(1/t)dt$ and $\int_{-\infty}^{\infty}(1/t^3)dt$ are both 0, whereas $1/t^2$ is not integrable at $t = 0$.

Example 5.21 Consider integrating $\mathcal{P}(1/t^3)$ on the interval $[-1, 1]$:

$$\begin{aligned} \int_{-1}^{-\epsilon} (1/t^3)dt + \int_{\epsilon}^{1} (1/t^3)dt &= (-1/2t^2)\Big|_{-1}^{-\epsilon} + (-1/2t^2)\Big|_{\epsilon}^{1} \\ &= -1/2\epsilon^2 + 1/2 - 1/2 + 1/2\epsilon^2 = 0. \end{aligned} \tag{5.129}$$

Although the definition in (5.128) is symmetric about t_o, it may turn out that the function is one-sided such that only one term is included. For example, $\mathcal{P}(u(t)/\sqrt{t})$ is 0 for $t < 0$ and has a singularity at $t = 0$. Applying (5.128) for $t \in [0, 1/4]$ yields

$$\lim_{\epsilon \to 0} \int_{\epsilon}^{1/4} (1/\sqrt{t})dt = \lim_{\epsilon \to 0} 2\sqrt{t}\Big|_{\epsilon}^{1/4} = 1. \tag{5.130}$$

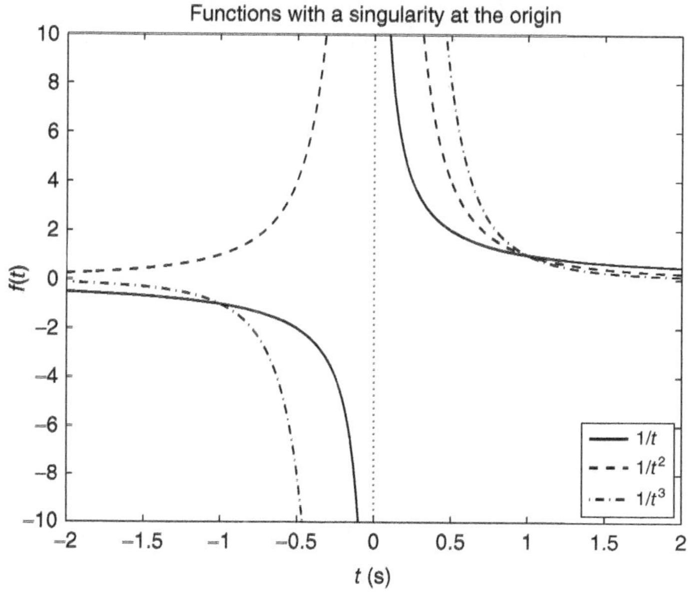

Figure 5.24 Functions with a singularity at $t = 0$.

5.12 EVEN AND ODD FUNCTIONS

The following properties of functions are useful in many applications, and in particular for the Fourier series representation of a signal that is presented later in this chapter.

Definition: Even and Odd Functions *Even function* $f_E(t)$ and *odd function* $f_O(t)$ have the following identities about $t = 0$:

$$f_E(t) = f_E(-t), \qquad f_O(t) = -f_O(-t). \tag{5.131}$$

An even function is symmetric about $t = 0$, and an odd function is antisymmetric.

Any ordinary function can be decomposed into the sum of an even function and an odd function:

$$f(t) = f_E(t) + f_O(t), \tag{5.132}$$

where

$$f_E(t) \triangleq [f(t) + f(-t)]/2, \qquad f_O(t) \triangleq [f(t) - f(-t)]/2. \tag{5.133}$$

The odd component is necessarily 0 at $t = 0$, and by definition, it has zero area:

$$\int_{-\infty}^{\infty} f_O(t)dt = 0. \tag{5.134}$$

TABLE 5.5 Properties of Even and Odd Functions

Function	Property
$f_E(t)f_O(t)$	Odd
$f_{1E}(t)f_{2E}(t)$	Even
$f_{1O}(t)f_{2O}(t)$	Even
$f_E(t)+f_O(t)$	Neither
$f_{1E}(t)+f_{2E}(t)$	Even
$f_{1O}(t)+f_{2O}(t)$	Odd
$df_E(t)/dt$	Odd
$df_O(t)/dt$	Even

This is not the case for the even component:

$$\int_{-\infty}^{\infty} f_E(t)dt \triangleq A_E \neq 0. \tag{5.135}$$

As a result, the even component can be decomposed further as the sum of an even component $\tilde{f}_E(t) \triangleq f_E(t) - A_E$ that is shifted on the vertical axis so that it has zero area and the constant A_E, yielding

$$f(t) = A_E + \tilde{f}_E(t) + f_O(t), \tag{5.136}$$

where A_E is the *DC component* of $f(t)$. The Fourier series decomposition shown later for a periodic signals has a similar form. Several properties of even and odd functions are summarized in Table 5.5.

Example 5.22 An example of an even/odd decomposition is shown in Figure 5.25 for the following piecewise linear function:

$$f(t) = \begin{cases} 0, & t \leq -3 \\ -t, & -3 < t \leq 0 \\ 0, & 0 < t \leq 2 \\ -3t+6, & 2 < t \leq 3 \\ 0, & t > 3, \end{cases} \tag{5.137}$$

whose mirror image about the vertical axis is

$$f(-t) = \begin{cases} 0, & t \leq -3 \\ 3t+6, & -3 < t \leq -2 \\ 0, & -2 < t \leq 0 \\ t, & 0 < t \leq 3 \\ 0, & t > 3. \end{cases} \tag{5.138}$$

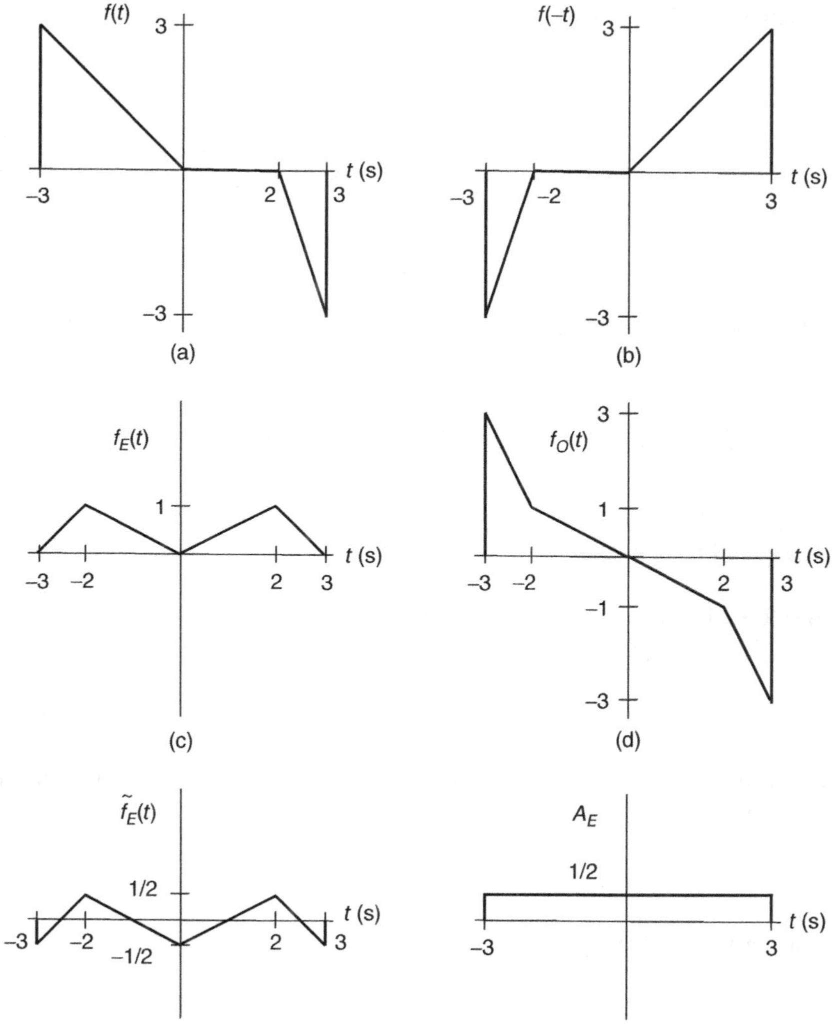

Figure 5.25 Example even and odd parts of a piecewise linear function. (a) $f(t)$. (b) $f(-t)$. (c) $f_E(t)$. (d) $f_O(t)$. (e) $\tilde{f}_E(t)$. (f) A_E.

The even part of the function is derived by considering four nonzero intervals for t:

$$f_E(t) = [f(t) + f(-t)]/2 = \begin{cases} 0, & t \leq -3 \\ t+3, & -3 < t \leq -2 \\ -t/2, & -2 < t \leq 0 \\ t/2, & 0 < t \leq 2 \\ -t+3, & 2 < t \leq 3 \\ 0, & t > 3, \end{cases} \tag{5.139}$$

and likewise for the odd part:

$$f_O(t) = [f(t) - f(-t)]/2 = \begin{cases} 0, & t \leq -3 \\ -2t-3, & -3 < t \leq -2 \\ -t/2, & -2 < t \leq 0 \\ -t/2, & 0 < t \leq 2 \\ -2t+3, & 2 < t \leq 3 \\ 0, & t > 3. \end{cases} \tag{5.140}$$

The line for $f_O(t)$ is continuous on $t \in [-2, 2]$ with value $-t/2$ on the two inner regions for t. The area of the even component is

$$A_E = 2\int_0^2 (t/2)dt + 2\int_2^3 (-t+3)dt = 2t|_0^2 + [(-t^2/2) + 3t]|_2^3 = 1/2. \tag{5.141}$$

Thus, $f_E(t)$ is decomposed as $\tilde{f}_E(t) + A_E$ where $\tilde{f}_E(t)$ is shifted downward by $1/2$, as illustrated in Figure 5.25(e).

Example 5.23 In this example, we illustrate techniques for proving some properties in Table 5.5. For the first property, define the product function

$$g(t) \triangleq f_E(t)f_O(t) = -f_E(-t)f_O(-t) = -g(-t), \tag{5.142}$$

where the definitions of an even and odd function have been used to generate the third expression, which follows because

$$g(-t) = f_E(-t)f_O(-t). \tag{5.143}$$

For the derivative property, define

$$h(t) \triangleq \frac{d}{dt}f_E(t) = \frac{d}{dt}f_E(-t) = \frac{d}{d\tau}f_E(\tau)\frac{d\tau}{dt}, \tag{5.144}$$

where we have used the fact that $f_E(t)$ is even, defined $\tau \triangleq -t$, and applied the product rule for differentiation. The last derivative is $d\tau/dt = -1$, and substituting $f_E(\tau) = f_E(-\tau)$ yields

$$h(t) = -\frac{d}{d\tau}f_E(-\tau) = -h(\tau) = -h(-t), \tag{5.145}$$

demonstrating that the derivative of an even function is odd.

5.13 CORRELATION FUNCTIONS

The cross-correlation function of two signals is a useful measure of their *similarity* and is widely used in signal processing and communications.

Definition: Cross-Correlation Function The *cross-correlation* function of $f(t)$ and $g(t)$ is

$$c_{fg}(\tau) = f(t) \star g(t) \triangleq \int_{-\infty}^{\infty} f(\tau + t)g(\tau)d\tau = \int_{-\infty}^{\infty} g(\tau - t)f(\tau)d\tau, \qquad (5.146)$$

where in general $\tau \in \mathcal{R}$.

This is not a symmetric operation: $f(t) \star g(t) \neq g(t) \star f(t)$; instead

$$f(t) \star g(t) = g(-t) \star f(t). \qquad (5.147)$$

The *autocorrelation function* is obtained when $g(t) = f(t)$, which is a symmetric (even) function: $f(t) \star f(t) = c_{ff}(\tau)$.

Example 5.24 The cross-correlation function of two one-sided exponential functions $f(t) = \exp(-t/\tau_1)u(t)$ and $g(t) = \exp(-t/\tau_2)u(t)$ with different time constants is

$$\begin{aligned} c_{fg}(\tau) &= \int_{-\infty}^{\infty} \exp(-(\tau + t)/\tau_1)u(\tau + t)\exp(-t/\tau_2)u(t)dt \\ &= \exp(-\tau/\tau_1)\int_{\max(0,-\tau)}^{\infty} \exp(-t(\tau_1 + \tau_2)/\tau_1\tau_2)dt, \end{aligned} \qquad (5.148)$$

where the lower limit of integration has been determined from $u(t)$ and $u(\tau + t) = 1$ for $\tau + t > 0 \implies t \geq -\tau$. Thus,

$$c_{fg}(\tau) = \frac{\tau_1\tau_2}{\tau_1 + \tau_2}\exp(-\tau/\tau_1)\exp(-\max(0, -\tau)(\tau_1 + \tau_2)/\tau_1\tau_2). \qquad (5.149)$$

The cross-correlation function $c_{gf}(\tau)$ is derived from this last expression by interchanging τ_1 and τ_2, which requires only that τ_1 be changed to τ_2 in the first exponential. The autocorrelation function is derived by setting $\tau_1 = \tau_2 \triangleq \tau_o$ so that $f(t) = g(t)$:

$$\begin{aligned} c_{ff}(\tau) &= (\tau_o/2)\exp(-\tau/\tau_o)\exp(-2\max(0, -\tau)/\tau_o) \\ &= (\tau_o/2)\exp(-|\tau|/\tau_o), \end{aligned} \qquad (5.150)$$

which has simplified by considering the two cases $\tau \geq 0$ and $\tau \leq 0$. These correlation functions are shown in Figure 5.26 for $\tau_1 = 1$, $\tau_2 = 2$, and $\tau_o = 2$. Observe that with increasing τ, all three correlation functions decrease, which is intuitive because the exponential functions become less similar with greater relative shifts in time. The autocorrelation function is always symmetric (the dotted line), whereas the cross-correlation functions are not.

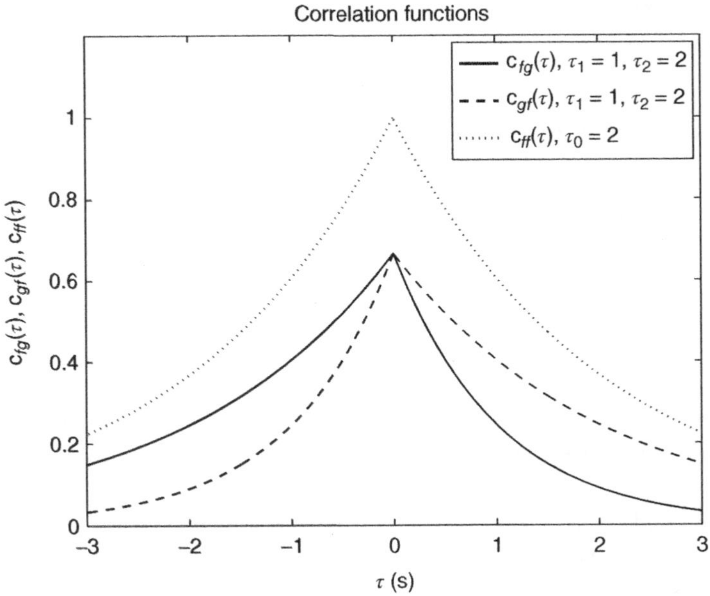

Figure 5.26 Correlation functions in Example 5.24

Definition: Orthogonal Functions Two functions $f(t)$ and $g(t)$ are *orthogonal* if

$$\int_{-\infty}^{\infty} f(t)g(t)dt = 0, \tag{5.151}$$

which is equivalent to $c_{fg}(0) = c_{gf}(0) = 0$.

Example 5.25 Sine and cosine functions are orthogonal when integrated over an integer multiple of π:

$$\begin{aligned}\int_0^{\pi} \sin(t)\cos(t)dt &= (1/2)\int_0^{\pi} [\sin(2t) + \sin(0)]dt \\ &= -(1/4)\cos(2t)|_0^{\pi} = 0.\end{aligned} \tag{5.152}$$

From Table 5.5, we know that the product of an even function and an odd function is odd. The area of an odd function is zero provided we integrate over the whole function, or in the case of a periodic function, over one period. If sine and cosine are shifted relative to each other, the integral is nonzero:

$$\begin{aligned}\int_0^{2\pi} \sin(t+\tau)\cos(t)dt &= (1/2)\int_0^{2\pi} [\sin(2t+\tau) + \sin(\tau)]dt \\ &= -(1/4)\cos(2t+\tau)|_0^{2\pi} + \pi\sin(\tau) \\ &= -(1/4)[\cos(2t)\cos(\tau)|_0^{2\pi} - \sin(2t)\sin(\tau)|_0^{2\pi}] + \pi\sin(\tau).\end{aligned} \tag{5.153}$$

The first two terms in the last expression are 0, and the cross-correlation function is

$$\int_0^{2\pi} \sin(t+\tau)\cos(t)dt = \pi \sin(\tau), \tag{5.154}$$

which is 0 when $\tau = n\pi$ for $n \in \mathcal{Z}$. The waveforms have a maximum positive correlation of π for $\tau = \pi/2 \pm 2\pi n$ with $n \in \mathcal{Z}$, as expected because the periodic waveforms are exactly aligned for such shifts, and a maximum negative correlation of $-\pi$ for $\tau = -\pi/2 \pm 2\pi n$ with $n \in \mathcal{Z}$.

5.14 FOURIER SERIES

Let $f(t)$ be a periodic function with period T_o such that the shape of the waveform repeats every T_o seconds. A basic example is the sinusoid

$$f(t) = \sin(2\pi f_o t), \qquad t \in \mathcal{R}, \tag{5.155}$$

where $f_o = 1/T_o$ is ordinary frequency (Hz) and $\omega_o = 2\pi f_o = 2\pi/T_o$ is angular frequency (rad/s). It is straightforward to show using a trigonometric identity that this function repeats for integer multiples of T_o:

$$\begin{aligned}\sin(2\pi f_o(t+nT_o)) &= \sin(2\pi f_o t)\cos(2\pi f_o nT_o) + \cos(2\pi f_o t)\sin(2\pi f_o nT_o)\\ &= \sin(2\pi f_o t),\end{aligned} \tag{5.156}$$

because $f_o T_o = 1$, and so $\cos(2\pi n) = 1$ and $\sin(2\pi n) = 0$ for $n \in \mathcal{Z}$.

Definition: Periodic Function A *periodic function* $f(t)$ with period T_o has the following property for $t \in \mathcal{R}$:

$$f(t) = f(t + nT_o), \tag{5.157}$$

and for every $n \in \mathcal{Z}$.

This definition assumes that $f(t)$ extends *infinitely* in *both* directions of time. A Fourier series representation of $f(t)$ requires this condition; it does not apply to functions that are "periodic" over some subset of $\mathcal{R}$, nor to aperiodic functions like the exponential function. These types of waveforms (as well as periodic waveforms) are represented in the frequency domain using the *Fourier transform* discussed in Chapter 8.

We introduce the Fourier series as a definition and later demonstrate that the representation holds for any periodic function.

Definition: Fourier Series The *Fourier series* of periodic function $f(t)$ with period T_o is the following infinite sum of terms:

$$f(t) = a_0 + \sum_{n=1}^{\infty} a_n \cos(n\omega_o t) + \sum_{n=1}^{\infty} b_n \sin(n\omega_o t), \tag{5.158}$$

with *Fourier coefficients* derived as

$$a_0 \triangleq \frac{1}{T_o} \int_{t_o}^{t_o+T_o} f(t)dt, \tag{5.159}$$

$$a_n \triangleq \frac{2}{T_o} \int_{t_o}^{t_o+T_o} f(t)\cos(n\omega_o t)dt, \tag{5.160}$$

$$b_n \triangleq \frac{2}{T_o} \int_{t_o}^{t_o+T_o} f(t)\sin(n\omega_o t)dt, \tag{5.161}$$

where $\omega_o = 2\pi f_o$ and $f_o = 1/T_o$.

This is the *trigonometric form* of the Fourier series; later the exponential form is given, which is often easier to compute. The lower limit t_o is chosen to facilitate evaluating the integrals; for many waveforms, $t_o = 0$ or $-T_o/2$. Since the first coefficient a_0 is the average of the function over one period, it is the DC component of $f(t)$. The other integrals can be viewed as cross-correlations between $f(t)$ and sines and cosines having integer multiples of the *fundamental frequency* ω_o. The coefficients $\{a_n\}$ and $\{b_n\}$ indicate the degree to which $f(t)$ is similar to cosine and sine, respectively, as $n\omega_o$ is varied. For integer $n > 1$, the frequencies $\{n\omega_o\}$ are called *harmonics* of ω_o, and the period of the nth harmonic is T_o/n.

As discussed previously, any function (periodic or aperiodic) can be decomposed into the sum of a constant, an even function, and an odd function. The Fourier series in (5.158) is a specific example of a DC/even/odd decomposition where a_0 is the DC component, $\sum_{n=1}^{\infty} a_n \cos(n\omega_o t)$ is the even component, and $\sum_{n=1}^{\infty} b_n \sin(n\omega_o t)$ is the odd component. If $f(t)$ happens to be an even function, then all $\{b_n\}$ are 0, and likewise, all $\{a_n\}$ are 0 if $f(t)$ is an odd function. The DC component $a_0 = 0$ if the area of $f(t)$ over one period is 0, which means the function has symmetry about the *horizontal* axis. In general, all components with coefficients a_0, $\{a_n\}$, and $\{b_n\}$ are needed to represent a periodic function using a Fourier series.

Next, we verify the expressions in (5.159)–(5.161). For convenience and without loss of generality, assume that $t_o = -T_o/2$. Substituting (5.158) for a_0 yields

$$\frac{1}{T_o} \int_{-T_o/2}^{T_o/2} f(t)dt = \frac{1}{T_o} \int_{-T_o/2}^{T_o/2} a_0 dt + \sum_{n=1}^{\infty} a_n \int_{-T_o/2}^{T_o/2} \cos(n\omega_o t)dt$$
$$+ \sum_{n=1}^{\infty} b_n \int_{-T_o/2}^{T_o/2} \sin(n\omega_o t)dt = a_0, \tag{5.162}$$

because the area of sine and cosine over integer multiples of the period is 0. For the $\{a_n\}$ coefficients, we must use a different integer m when substituting (5.158):

$$\frac{2}{T_o}\int_{-T_o/2}^{T_o/2} f(t)\cos(n\omega_o t)dt = \frac{2}{T_o}a_0\int_{-T_o/2}^{T_o/2}\cos(n\omega_o t)dt$$
$$+\frac{2}{T_o}\sum_{m=1}^{\infty} a_m\int_{-T_o/2}^{T_o/2}\cos(n\omega_o t)\cos(m\omega_o t)dt$$
$$+\frac{2}{T_o}\sum_{m=1}^{\infty} b_m\int_{-T_o/2}^{T_o/2}\cos(n\omega_o t)\sin(m\omega_o t)dt. \quad (5.163)$$

The first term on the right-hand side is 0. The trigonometric identities needed for the other terms are the product formulas:

$$\cos(n\omega_o t)\cos(m\omega_o t) = (1/2)[\cos((n-m)\omega_o t) + \cos((n+m)\omega_o t)] \quad (5.164)$$
$$\cos(n\omega_o t)\sin(m\omega_o t) = (1/2)[\sin((n+m)\omega_o t) - \sin((n-m)\omega_o t)]. \quad (5.165)$$

Since $n-m$ and $n+m$ are also integers, all terms in the two summations of (5.163) integrate to 0 except when $m = n$. Thus,

$$\frac{2}{T_o}\int_{-T_o/2}^{T_o/2} f(t)\cos(n\omega_o t)dt = \frac{a_n}{T_o}\int_{-T_o/2}^{T_o/2}\cos(0)dt + \frac{b_n}{T_o}\int_{-T_o/2}^{T_o/2}\sin(0)dt = a_n. \quad (5.166)$$

A similar derivation is used to verify the expression for b_n (see Problem 5.37).

Example 5.26 Suppose we modify the rectangle function as follows:

$$g(t) \triangleq \text{rect}(t/T) = \begin{cases} 1, & -1/2 \le t/T \le 1/2 \\ 0, & \text{else.} \end{cases} \quad (5.167)$$

Dividing the argument t by T causes the standard rectangle function to be scaled so that it has support $t \in [-T/2, T/2]$. A periodic function is generated from this modified rectangle function by defining the following sum:

$$f(t) \triangleq \sum_{n=-\infty}^{\infty} g(t - nT_o) = \sum_{m=-\infty}^{\infty} \text{rect}((t - mT_o)/T), \quad (5.168)$$

where it is assumed that $T < T_o$, so the shifted rectangles do not overlap. An example is shown in Figure 5.27(a) where the duration of $1/2$ s between rectangles is the same as their width. The rectangle for a value of m is centered at $(t - mT_o)/T = 0 \implies t = mT_o$, and so the rectangles are centered at integer multiples of T_o. The Fourier series

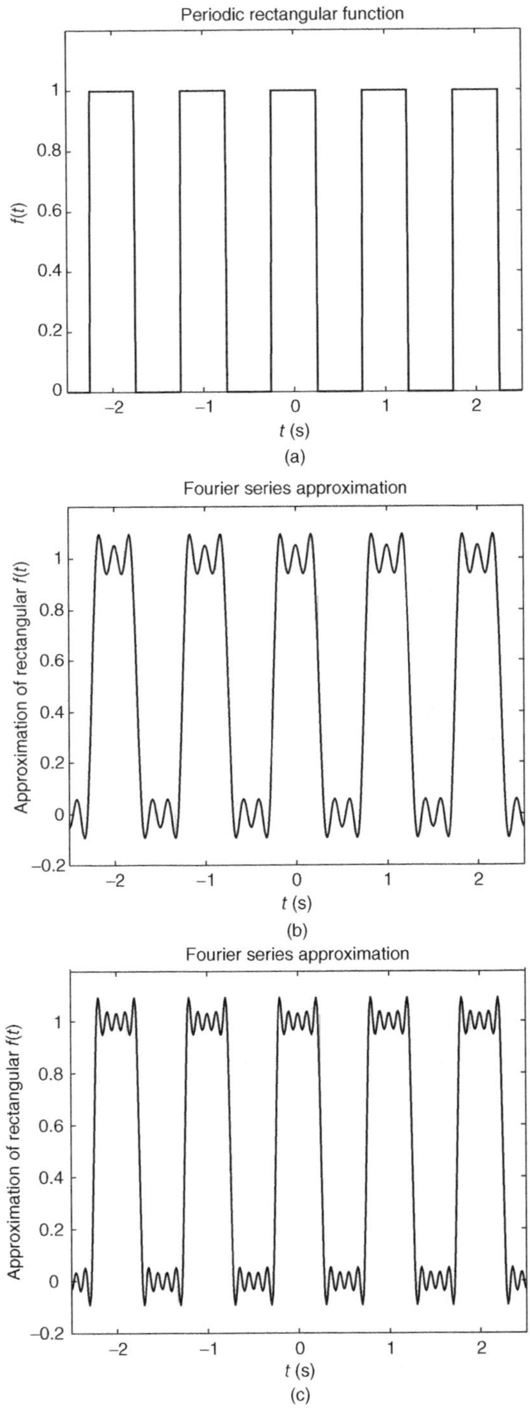

Figure 5.27 (a) Periodic function of rectangles with $T_o = 1$ and $T = 1/2$. Fourier series approximation including the DC term, (b) 5 cosine terms, and (c) 10 cosine terms.

coefficients for this periodic function are readily computed by integrating over the rectangle at the origin:

$$a_0 = \frac{1}{T_o}\int_{-T/2}^{T/2}(1) = T/T_o, \tag{5.169}$$

and

$$\begin{aligned} a_n &= \frac{2}{T_o}\int_{-T/2}^{T/2}\cos(2\pi nt/T_o)dt = \frac{2}{T_o}\frac{T_o}{2\pi n}\sin(2\pi nt/T_o)\Big|_{t=-T/2}^{t=T/2} \\ &= \frac{1}{\pi n}[\sin(\pi nT/T_o) - \sin(-\pi nT/T_o)]. \end{aligned} \tag{5.170}$$

The limits of integration are usually $\pm T_o/2$, but the rectangle at the origin is 0 beyond $\pm T/2$. Since sine is an odd function, we can combine the two terms in (5.170):

$$\begin{aligned} a_n &= \frac{2}{\pi n}\sin(\pi nT/T_o) = \frac{2T}{T_o}\text{sinc}(nT/T_o) \\ &= \text{sinc}(n/2), \end{aligned} \tag{5.171}$$

where

$$\text{sinc}(t) \triangleq \frac{\sin(\pi t)}{\pi t} \tag{5.172}$$

is the *sinc function*, which is shown in Figure 5.28 as a function of continuous time t (note that π is implicit in this definition). The sinc function in (5.171) for a_n is

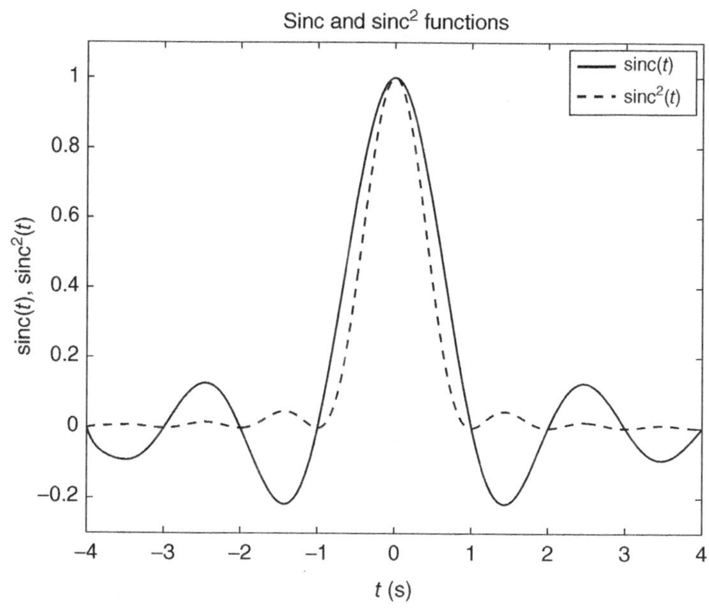

Figure 5.28 Functions sinc(t) and sinc^2(t).

evaluated only at *integer multiples* of $T/T_o = 1/2$. Finally, since rect(t) is an even function and the integral for b_n is performed over $[-T/2, T/2]$, it is clear that every $b_n = 0$ because the product of rect(t) and sine is an odd function. Collecting together all terms, the Fourier series representation for the periodic rectangular waveform in (5.168) is

$$f(t) = (1/2)\left[1 + 2\sum_{n=1}^{\infty} \text{sinc}(n/2)\cos(2\pi nt/T_o)\right]. \tag{5.173}$$

Figure 5.27(b) and (c) shows approximations of $f(t)$ in Figure 5.27(a) when the upper limit in (5.173) is $n = 5$ and 10, respectively. The weighting sinc($n/2$) is not a continuous function of time t; it is defined only for $n \in \mathcal{N}$. From the form of the sinc function, we see that less weight is placed on higher harmonics, which is why a small number of terms provide a good approximation of the original waveform. However, in order to entirely remove the *ripples* (known as *Gibbs phenomenon*) in Figure 5.27(b) and (c), all Fourier series terms must be included.

The Fourier series for a periodic signal is closely related to the *Fourier transform* of an aperiodic waveform covered in Chapter 8. The Fourier transform of the rectangle function, given by the waveform centered at the origin in Figure 5.27(a) but with $T = 1$, is

$$\begin{aligned} F(\omega) &= \int_{-\infty}^{\infty} \text{rect}(t)\exp(-j\omega t)dt \\ &= \frac{\sin(\omega/2)}{\omega/2} = \text{sinc}(\omega/2\pi), \end{aligned} \tag{5.174}$$

which is a continuous function of radian frequency ω, and whose magnitude decreases according to $1/\omega$. The Fourier coefficients of the periodic rectangular waveform decrease in a similar manner according to $1/n$ because $\text{sinc}(n/2) = 2\sin(\pi n/2)/\pi n$.

Example 5.27 A periodic version of the triangle function is given by

$$f(t) = \sum_{m=-\infty}^{\infty} \text{tri}((t - mT_o)/T), \tag{5.175}$$

with $T_o = 1$ and $T = 1/2$ such that the repeated triangles are adjacent to each other as shown in Figure 5.29(a). Its Fourier series coefficients are (see Problem 5.39):

$$a_0 = T, \quad a_n = \frac{2T_o}{n^2\pi^2 T} \ (n \text{ odd}), \quad a_n = 0 \ (n \text{ even}), \quad b_n = 0. \tag{5.176}$$

These coefficients decrease according to $1/n^2$, which is faster than those of the rectangle function. Using the trigonometric identity $\sin^2(\theta/2) = (1/2)[1 - \cos(\theta)]$,

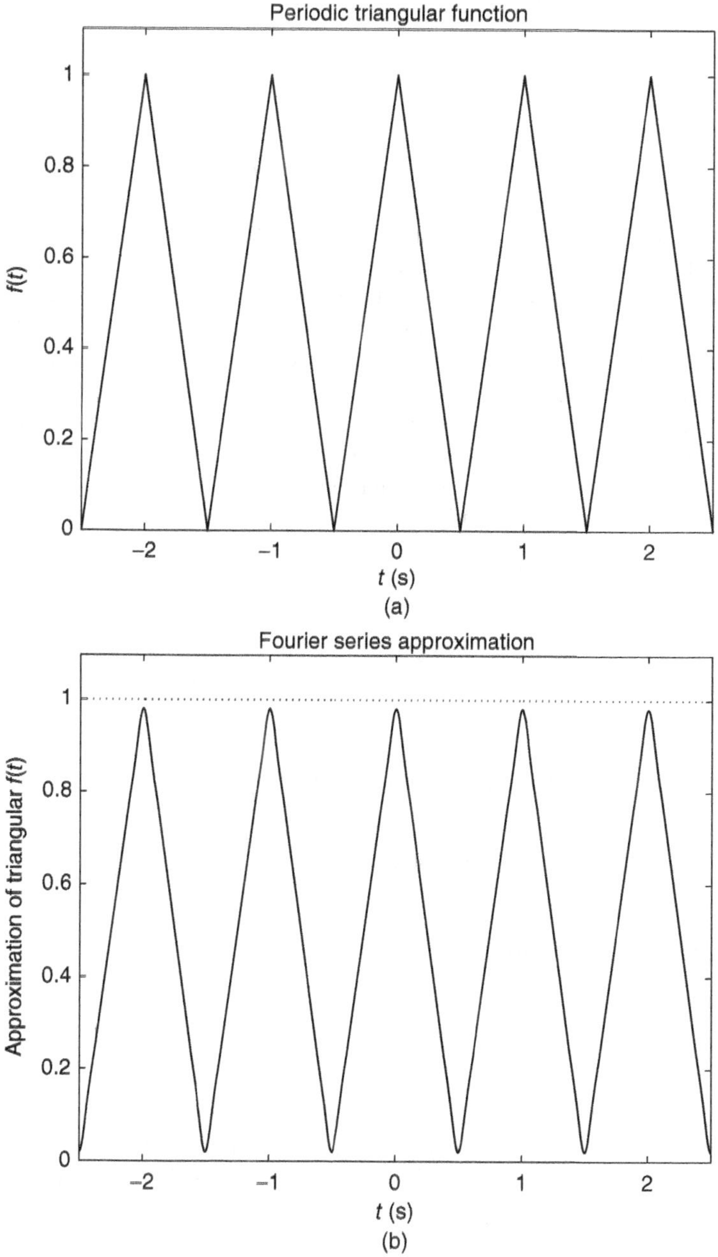

Figure 5.29 (a) Periodic function of triangles with $T_o = 1$ and $T = 1/2$. (b) Fourier series approximation including only the DC term and four cosine terms.

coefficient a_n can be rewritten as

$$a_n = \frac{2T_o}{\pi^2 n^2 T}\sin^2(\pi n/2) = (T_o/2T)\text{sinc}^2(n/2)$$
$$= \text{sinc}^2(n/2), \tag{5.177}$$

which is 0 for even n because the argument is $n/2$. The squared sinc function is shown in Figure 5.28. Observe in Figure 5.29(b) that the Fourier series approximation is quite accurate for only a few terms, unlike the rectangular waveform, which has a ripple effect. This occurs because the triangle function more closely resembles the cosine waveform, and it does not have any discontinuities as does the rectangle function. The factor $1/n^2$ means that relatively few terms are needed for a good approximation because the $\{a_n\}$ become small rather quickly with increasing n. The waveform approximation in Figure 5.29(b) does not quite reach 0 or 1 (denoted by the dotted line); additional Fourier series terms are needed to reach the minimum and maximum values. The Fourier transform of the triangle function is

$$F(\omega) = \frac{\sin^2(\omega/2)}{\omega^2/4} = \text{sinc}^2(\omega/2\pi), \tag{5.178}$$

which decreases according to $1/\omega^2$, and so again we see a connection between the Fourier series of a periodic function and the Fourier transform of the waveform for one period.

The complex exponential form of the Fourier series is

$$f(t) = \sum_{n=-\infty}^{\infty} c_n \exp(jn\omega_o t), \tag{5.179}$$

where c_n is a *complex* Fourier coefficient:

$$c_n = (1/T_o)\int_{t_o}^{t_o+T_o} f(t)\exp(-j\omega_o nt)dt. \tag{5.180}$$

The lower limit of the summation for this representation is $-\infty$ (whereas it is 1 for the trigonometric form of the Fourier series). It is straightforward to show that the exponential form is equivalent to the definition in (5.158) by first rewriting (5.179) as follows:

$$f(t) = c_0 + \sum_{n=1}^{\infty} c_n \exp(jn\omega_o t) + \sum_{n=-\infty}^{-1} c_n \exp(jn\omega_o t)$$
$$= c_0 + \sum_{n=1}^{\infty} c_n \exp(jn\omega_o t) + \sum_{n=1}^{\infty} c_n^* \exp(-jn\omega_o t), \tag{5.181}$$

where we have changed variables to $n \to -n$ in the second sum and used the fact that c_n for negative n is the complex conjugate of c_n for positive n because $f(t)$ is real. Thus, $c_0 = a_0$ must be real, and substituting $c_n \triangleq (a_n - jb_n)/2$ we have

$$f(t) = a_0 + (1/2)\sum_{n=1}^{\infty} a_n[\exp(jn\omega_o t) + \exp(-jn\omega_o t)]$$
$$+(1/2j)\sum_{n=1}^{\infty} b_n[\exp(jn\omega_o t) - \exp(-jn\omega_o t)], \tag{5.182}$$

where we have used $j/2 = -1/2j$. Applying Euler's inverse formulas to each term in both sums for $n \geq 1$ yields the Fourier series expansion in (5.158).

Example 5.28 For the periodic rectangle function in Example 5.26, the complex Fourier series coefficients are

$$\begin{aligned} c_n &= (1/T_o)\int_{-T/2}^{T/2} \exp(-jn\omega_o t)dt \\ &= (-1/jn\omega_o T)[\exp(-jn\omega_o T/2) - \exp(jn\omega_o T/2)] \\ &= (2/n\omega_o T_o)\sin(n\omega_o T/2). \end{aligned} \tag{5.183}$$

This can be rewritten in terms of the sinc function by substituting $\omega_o = 2\pi/T_o$:

$$c_n = (T/T_o)\text{sinc}(nT/T_o). \tag{5.184}$$

Since $b_n = 0$ for this example, we can also use $c_n = (a_n - jb_n)/2 = a_n/2$ and substitute a_n from (5.171) to produce the same result. These coefficients, which are all real for this example, are plotted in Figure 5.30 for $T_o = 1$ and $T = 1/2$. The exponential Fourier series representation is

$$\begin{aligned} f(t) &= (T/T_o)\sum_{n=-\infty}^{\infty} \text{sinc}(nT/T_o)\exp(n\omega_o t) \\ &= T/T_o + (T/T_o)\sum_{n=1}^{\infty} \text{sinc}(nT/T_o)\cos(n\omega_o t), \end{aligned} \tag{5.185}$$

where from (5.182), we have $a_0 = T/T_o$, $a_n = 2c_n$, and $b_n = 0$. From (5.184), we conclude that $a_n = 0$ for even n. The coefficients in Figure 5.30 for $n \in \{-5, \ldots, 0, \ldots, 5\}$ were used to generate the approximation in Figure 5.27(b).

Example 5.29 The Fourier series for $\cos^2(t)$ has only two terms. Since it is an even function, the sine coefficients $\{b_n\}$ are 0 and the DC component is

$$a_0 = \frac{1}{T_o}\int_{-T_o/2}^{T_o/2} \cos^2(t)dt = \frac{1}{2T_o}\int_{-T_o/2}^{T_o/2} [1 + \cos(2t)]dt. \tag{5.186}$$

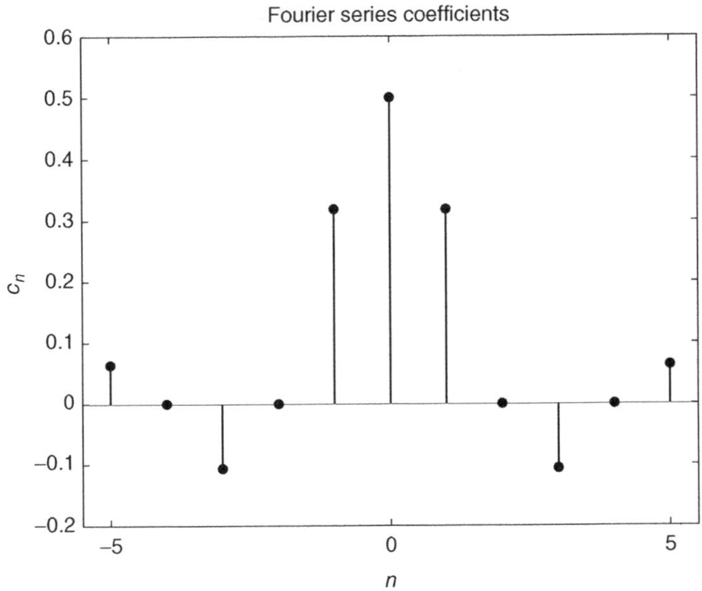

Figure 5.30 Fourier series coefficients for Example 5.28.

Since the period of $\cos(2t)$ is $T_o/2$, it integrates to 0 and $a_0 = 1/2$. For the cosine coefficients, we have

$$a_n = \frac{2}{T_o}\int_{-T_o/2}^{T_o/2} \cos^2(t)\cos(2\pi nt/T_o)dt = \frac{1}{T_o}\int_{-T_o/2}^{T_o/2} [\cos(2nt) + \cos(2t)\cos(2nt)]dt, \tag{5.187}$$

where $\omega_o = 2\pi/T_o = 2$ has been substituted because $T_o = \pi$ for $\cos^2(t)$. The first term in the last expression is 0 for all $n \in \mathcal{N}$, and the second term can be rewritten as

$$\cos(2t)\cos(2nt) = (1/2)[\cos(2(n-1)t) + \cos(2(n+1)t)]. \tag{5.188}$$

The integral of this expression is 0 except for the first term with $n = 1$. Thus

$$a_1 = \frac{1}{\pi}\int_{-\pi/2}^{\pi/2} (1/2)dt = 1/2, \tag{5.189}$$

and the Fourier series is

$$\cos^2(t) = (1/2)[1 + \cos(2t)]. \tag{5.190}$$

However, this is just the trigonometric identity used in the derivation, and so none of this work was actually necessary. Whenever a function can be written directly as the sum of cosine and sine terms with arguments that are integer multiples of the fundamental frequency, then that result is the Fourier series of the waveform even though the number of terms is finite. The Fourier series for $\sin^2(t)$ is covered in Problem 5.41.

Example 5.30 A half-wave *rectified* sine function for one period has the following form:

$$f(t) = \begin{cases} \sin(t), & 0 \le t \le \pi \\ 0, & \pi < t \le 2\pi, \end{cases} \tag{5.191}$$

with $\omega_o = 1$ rad/s. This is an example of a function that is neither even nor odd, and so all Fourier series coefficients need to be examined, including the DC component. However, because it is derived from a sine wave, it turns out that $b_1 = 1/2$ and all other sine coefficients are $b_n = 0$ for $n \neq 1$. Since the b_1 term of the Fourier series exactly matches $f(t)$ when it is positive, it is clear that the DC component and the cosine terms of the Fourier series are needed to cancel the negative part of the sine cycle and give the rectified waveform. These coefficients are $a_0 = 1/\pi$ and (see Problem 5.42)

$$a_n = \begin{cases} 0, & n \text{ odd} \\ (2/\pi)/(1-n^2), & n \text{ even}, \end{cases} \tag{5.192}$$

which yields the Fourier series

$$f(t) = 1/\pi + (1/2)\sin(t) + (2/\pi) \sum_{n=2,4,\ldots} [1/(1-n^2)]\cos(nt). \tag{5.193}$$

These results are depicted in Figure 5.31. Three of the cosine terms are shown in Figure 5.31(b), which tend to cancel the negative part of the sine wave given by the dashed line in Figure 5.31(a). In Figure 5.31(c), we have included $(1/2)\sin(t)$ (the solid line) and the sum of the three cosine terms and the DC term, which shifts the sum upward by 1/2 (the dashed line). Observe that for one-half of the period, the cosine terms reinforce the positive cycles of the shifted sine wave, bringing them closer to 1. During the negative cycles, the cosine terms add to the shifted sine wave in order to cancel those components, bringing them closer to 0. The dotted line in Figure 5.31(c) is the result when five Fourier series terms are included: $\{b_1, a_0, a_2, a_4, a_6\}$. This last figure can also be viewed as an even/odd decomposition of the function given by the dotted line: the dashed line is the even part plus the DC term, and the solid line is the odd part.

5.15 PHASOR REPRESENTATION

The phasor representation of a sinusoidal signal is essentially a *notation* for representing the waveform using a *complex number*. Consider the cosine waveform in (5.44), which we repeat here for convenience:

$$f(t) = A\cos(\omega_o t + \phi), \tag{5.194}$$

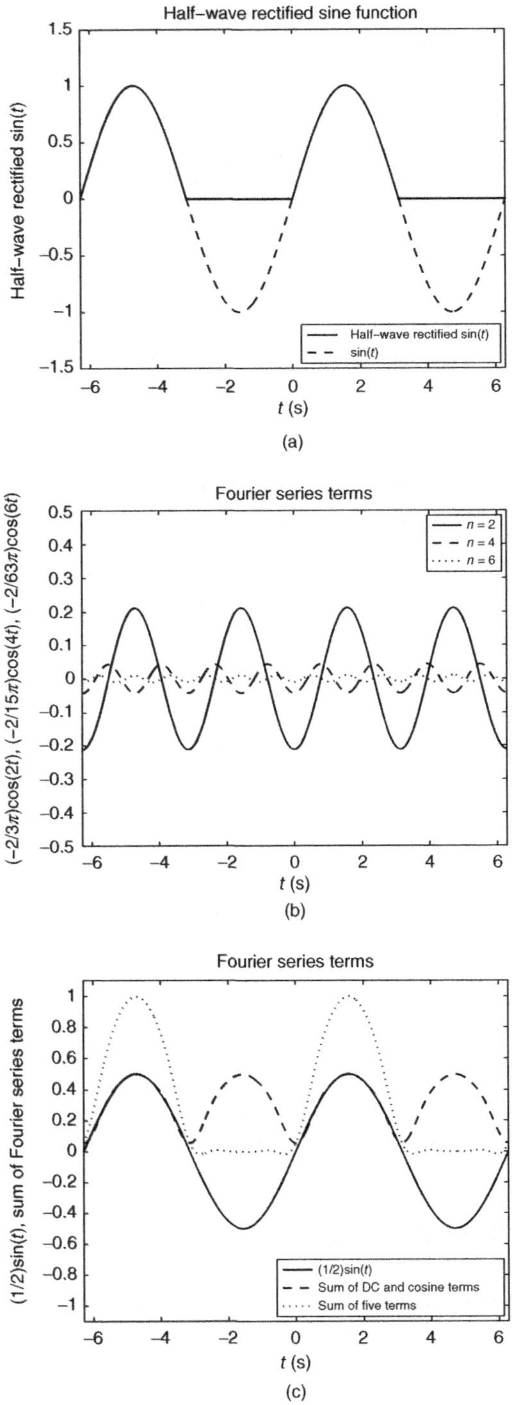

Figure 5.31 Fourier series approximation of the half–wave rectified sine function in Example 5.30. (a) Half-wave rectified sin(t). (b) Three cosine terms. (c) Sum of various terms.

where A, ω_o, and ϕ are the amplitude, angular frequency, and phase, all of which are real-valued. We emphasize that for phasors, the support of $f(t)$ is assumed to be the entire real line $t \in \mathcal{R}$, as was the case for the Fourier series. From Euler's formula:

$$A \exp(j(\omega_o t + \phi)) = A \cos(\omega_o t + \phi) + jA \sin(\omega_o t + \phi), \tag{5.195}$$

such that $f(t)$ can be written as the real part of this expression:

$$f(t) = \text{Re}(A \exp(j\omega_o t + j\phi)). \tag{5.196}$$

For an LTI system with a sinusoidal input (such as the series RLC circuit covered in Chapter 2), the output is also sinusoidal with the *same frequency*, but usually with a different amplitude and phase. For a linear circuit with fixed lumped-parameter elements and a single sinusoidal voltage or current source, all internal voltages and currents are sinusoidal with the same frequency. This property allows us to analyze such systems more easily using phasors. The phasor approach is a preview of the more general Laplace transform method used for LTI systems in Chapter 7, where the voltage and current sources need not be sinusoidal and the support need not be $t \in \mathcal{R}$.

In order to define a phasor, note that (5.196) can be rewritten as

$$f(t) = \text{Re}(A \exp(j\omega_o t) \exp(j\phi)), \tag{5.197}$$

where the product property of exponentials has been used.

Definition: Phasor A *phasor* is a notation that represents the cosine waveform $A \cos(\omega_o t + \phi)$ as the following complex number:

$$\mathbf{F} = A \exp(j\phi). \tag{5.198}$$

The notation $\mathbf{F} = A\angle\phi$ is also used.

Bold uppercase letters are usually used to denote phasors (which should not be confused with the matrices covered in Chapter 3). A phasor retains only the amplitude and phase of the sinusoid. The angular frequency is ignored because, after analyzing a circuit using phasor notation, the corresponding time-domain waveform is generated from (5.197) and (5.198) as follows:

$$f(t) = \text{Re}(\mathbf{F} \exp(j\omega_o t)). \tag{5.199}$$

The phasor $\mathbf{F}$ is multiplied by a complex exponential function with the appropriate angular frequency, and the real part yields the cosine waveform. The real part is computed using Euler's inverse formula:

$$f(t) = (1/2)[\mathbf{F} \exp(j\omega_o t) + \mathbf{F}^* \exp(-j\omega_o t)], \tag{5.200}$$

where the superscript $*$ denotes complex conjugation. However, this calculation is not actually necessary because we know that the real part of (5.199) is a cosine.

Example 5.31 Suppose that $\mathbf{F} = a + jb = \sqrt{a^2 + b^2} \exp(j \tan^{-1}(b/a))$, which are the rectangular and polar representations, respectively, for a complex number. Substituting this expression into (5.200) yields

$$\begin{aligned} f(t) = (1/2) \Big[&\sqrt{a^2 + b^2} \exp(j\omega_o t + j \tan^{-1}(b/a)) \\ &+ \sqrt{a^2 + b^2} \exp(-j\omega_o t - j \tan^{-1}(b/a)) \Big], \end{aligned} \quad (5.201)$$

where the complex conjugate affects only the phase in the second component. Since the square root factors from this expression, Euler's formula yields

$$f(t) = \sqrt{a^2 + b^2} \cos(\omega_o t + \tan^{-1}(b/a)). \quad (5.202)$$

From this result, we find that it is not necessary to write the phasor in polar/exponential form as in (5.200). Once $\mathbf{F}$ is known, its real and imaginary parts $\{a, b\}$ are used in (5.202) to directly write the time-domain waveform.

Phasors are usually defined in terms of a cosine. In order to derive the phasor for the sine waveform in (5.44), we use the fact that sine and cosine are related by a 90° phase shift:

$$f(t) = A \sin(\omega_o t + \phi) = A \cos(\omega_o t + \phi - 90^\circ). \quad (5.203)$$

A cosine waveform is shifted 90° to the *left* of a sine waveform, so that the phasor for (5.203) is

$$\mathbf{F} = A \exp(j(\phi - 90^\circ)). \quad (5.204)$$

If the input of a system consists of several sinusoids with different frequencies, phasors can still be used, by solving for the output for each frequency separately and then adding together the final set of results in the time domain. This is known as *superposition*, which is another characteristic of an LTI system.

Example 5.32 For the phasor of the sinusoidal function

$$f(t) = -A \cos(\omega_o t + \phi), \quad (5.205)$$

one might attempt to take into account the minus sign by using the trigonometric identity $\cos(x + \pi) = -\cos(x)$. This yields

$$f(t) = -A \cos(\omega_o t + \phi) = A \cos(\omega_o t + \pi + \phi) \implies \mathbf{F} = A \exp(j(\pi + \phi)). \quad (5.206)$$

However, this is not necessary (though it is acceptable) because A itself can be negative, and we can immediately just write $\mathbf{F} = -A \exp(j\phi)$. This also follows from

(5.206) because $\exp(j\pi) = -1$. Although the exponential form of a phasor looks similar to the polar form of a complex number, a phasor uses the *amplitude* of the cosine ($-A$ in this example) and not its magnitude $|A|$. Consider another example based on the identity $\cos(x + \pi/2) = -\sin(x)$:

$$f(t) = -A\sin(\omega_o t + \phi) = A\cos(\omega_o t + \pi/2 + \phi) \implies \mathbf{F} = A\exp(j(\pi/2 + \phi)). \tag{5.207}$$

This is the correct form for the phasor. Although $\exp(j\pi/2) = j$ and we could write $f(t) = Aj\exp(j\phi)$, this is not proper phasor form because the leading coefficient should be real; instead, $j\pi/2$ is included in the angle component of the phasor.

5.16 PHASORS AND LINEAR CIRCUITS

Finally, we consider phasors for the voltages and currents of the circuit elements discussed in Chapter 2. For a sinusoidal current $i(t) = A\cos(\omega_o t + \phi)$, the voltage across a resistor from Ohm's law $v = Ri$ is

$$v(t) = RA\cos(\omega_o t + \phi), \tag{5.208}$$

which means the corresponding phasors are $\mathbf{I} = A\exp(j\phi)$ and $\mathbf{V} = RA\exp(j\phi)$. These are complex numbers that specify the amplitude and phase of the real-valued voltage and current waveforms, with the understanding that $v(t)$ and $i(t)$ are cosine functions with the same frequency ω_o.

Definition: Impedance The *impedance* Z of a circuit device is the ratio of its phasor voltage and phasor current: $Z \triangleq \mathbf{V}/\mathbf{I}$. It is a complex number of the form $Z = R + jX$ where R is the *resistance* and Z is the *reactance*.

Impedance is *not* a phasor: Z is not converted to a time-varying waveform as is done in (5.199) for currents and voltages. The impedance is an I-V characterization of a circuit element in the *phasor domain*, when all currents and voltages in a circuit are sinusoidal with the same frequency and have been converted into phasors. The impedance for a resistor is obviously its resistance: $Z_R = R$. For an inductor with sinusoidal current $i(t) = A\cos(\omega_o t + \phi)$:

$$v(t) = L\frac{d}{dt}A\cos(\omega_o t + \phi) = -\omega_o LA\sin(\omega_o t + \phi), \tag{5.209}$$

such that $\mathbf{I} = A\exp(j\phi)$ and $\mathbf{V} = -\omega_o LA\exp(j(\phi - \pi/2))$. Thus, the impedance for an inductor is

$$Z_L = -\omega_o L\exp(-j\pi/2) = j\omega_o L, \tag{5.210}$$

where $\exp(-j\pi/2) = -j$ has been substituted. An ideal inductor has zero resistance and its reactance is always positive. A similar result is obtained for the capacitor with sinusoidal voltage $v(t) = A\cos(\omega_o t + \phi)$:

$$i(t) = C\frac{d}{dt}A\cos(\omega_o t + \phi) = -\omega_o CA\sin(\omega_o t + \phi), \tag{5.211}$$

TABLE 5.6 Phasor Impedance of Linear Circuit Elements

Device	Impedance Z	Resistance R	Reactance X
Resistor	R	R	0
Inductor	$j\omega_o L$	0	$\omega_o L$
Capacitor	$1/j\omega_o C$	0	$-1/\omega_o C$

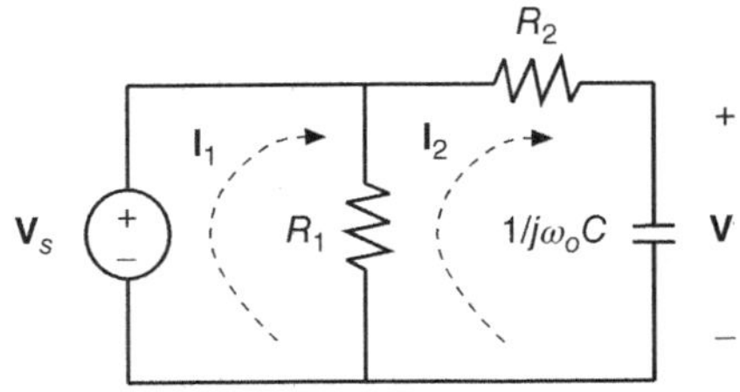

Figure 5.32 First-order circuit with capacitor C and sinusoidal voltage source.

which yields $\mathbf{V} = A\exp(j\phi)$, $\mathbf{I} = -\omega_o CA\exp(j(\phi - \pi/2))$, and

$$Z_C = -1/\omega_o C\exp(-j\pi/2) = 1/j\omega_o C = -j/\omega_o C. \tag{5.212}$$

An ideal capacitor has zero resistance and its reactance is always negative. The impedance results for these three passive circuit elements are summarized in Table 5.6.

Since impedance $Z = \mathbf{V}/\mathbf{I}$ is an extension of Ohm's law to complex quantities, the voltages and currents in an RLC circuit can be determined using algebraic techniques similar to those given earlier for an all-resistive circuit (see Chapter 2). This approach assumes sinusoidal signals (extending to $\pm\infty$), and we must manipulate complex quantities, which makes the analysis somewhat more cumbersome. This is illustrated in the next example for an RC circuit.

Example 5.33 For the first-order circuit in Figure 2.15, assume that the voltage source is sinusoidal $V_s = A\cos(\omega_o t)$ with phasor $\mathbf{V}_s = A$. The modified circuit is shown in Figure 5.32 with the capacitor labeled by its impedance $Z_C = 1/j\omega_o C$. The voltage across the capacitor is given by the result in (2.35) with R_3 replaced by $1/j\omega_o C$:

$$\mathbf{V} = \frac{1/j\omega_o C}{R_2 + 1/j\omega_o C}\mathbf{V}_s = \frac{A}{1 + j\omega_o R_2 C}. \tag{5.213}$$

Rearranging this expression into the standard form for a complex number yields

$$\mathbf{V} = \frac{A}{1 + j\omega_o R_2 C}\left(\frac{1 - j\omega_o R_2 C}{1 - j\omega_o R_2 C}\right) = \frac{A}{1 + (\omega_o R_2 C)^2} - j\frac{\omega_o R_2 CA}{1 + (\omega_o R_2 C)^2}, \tag{5.214}$$

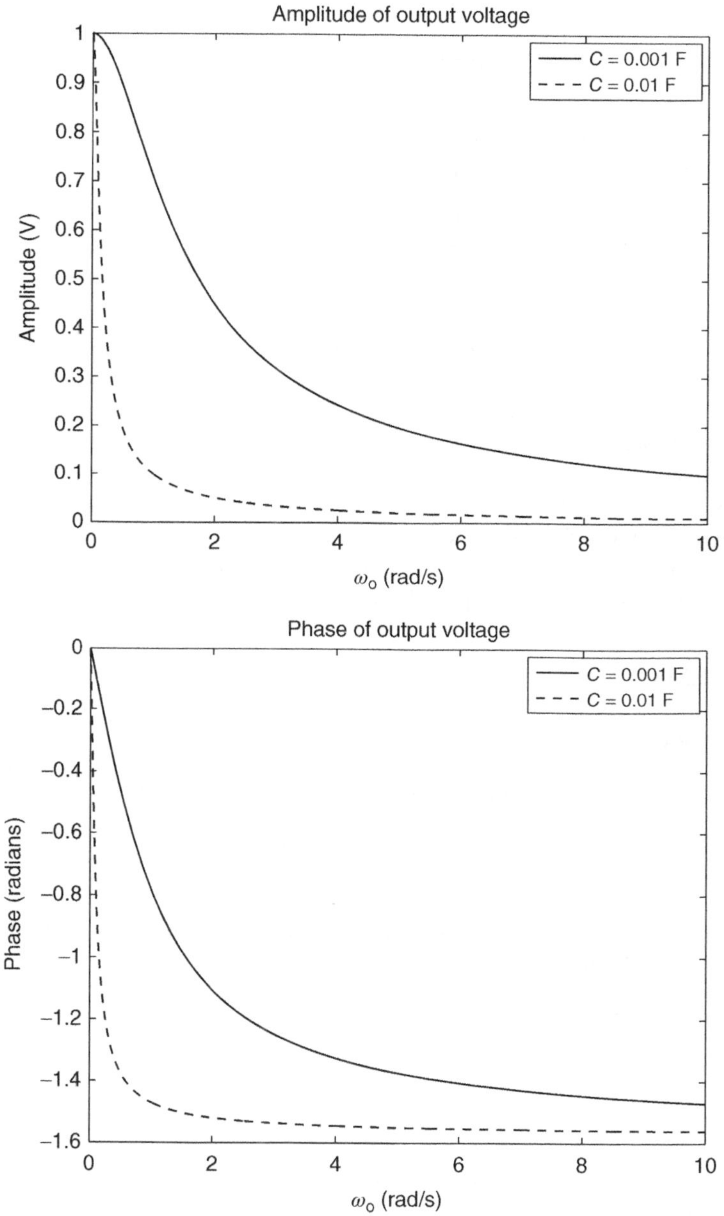

Figure 5.33 Amplitude $A/\sqrt{1+(\omega_o R_2 C)^2}$ and phase $-\tan^{-1}(\omega_o R_2 C)$ of the output voltage for the circuit in Example 5.33 with $A = 1$ V and $R_2 = 1000\ \Omega$. (a) Amplitude. (b) Phase.

which can be expressed as

$$\mathbf{V} = \frac{A}{\sqrt{1+(\omega_o R_2 C)^2}} \exp(\tan^{-1}(-\omega_o R_2 C)). \tag{5.215}$$

The corresponding time-domain waveform is obtained by multiplying this result by $\exp(j\omega_o t)$ and taking the real part:

$$v(t) = \frac{A}{\sqrt{1+(\omega_o R_2 C)^2}} \cos(\omega_o t - \tan^{-1}(\omega_o R_2 C)). \tag{5.216}$$

This result is a scaled and shifted version of V_s, as are the other voltages in this circuit. The amplitude and phase are plotted in Figure 5.33 for $A = 1$ V, $R_2 = 1000\ \Omega$, and two values of C. When $\omega_o = 0$, corresponding to a DC voltage source, the impedance Z_C of the capacitor is infinite (an open circuit), so there is no current through R_2 and all the voltage is across the capacitor with zero phase. At the other extreme as $\omega_o \to \infty$, the impedance $Z_C \to 0$ (a short circuit). The voltage across C approaches 0, and the cosine waveform becomes increasingly shifted to the right because of the negative phase in (5.215), which approaches $-\pi/2$ in the limit. These curves (the dashed lines) are lower for the larger value of C, which is expected because it takes longer to charge a larger capacitor and so its amplitude is smaller for the same frequency ω_o.

PROBLEMS

Step and Ramp Functions

5.1 Specify the support and range for the following functions:

$$\text{(a)}\ f(t) = u(t+4), \quad \text{(b)}\ g(t) = u(t+2) - u(t-3), \quad \text{(c)}\ h(t) = r(t+2) - r(t). \tag{5.217}$$

5.2 The unit step function can be used to create discontinuities in continuous functions. Sketch the following functions:

$$\text{(a)}\ f(t) = \exp(-t)u(t-2), \quad \text{(b)}\ g(t) = \exp(-2|t|)[u(t+1) - u(t-2)], \tag{5.218}$$

$$\text{(c)}\ h(t) = \sin(t)u(t-\pi/2). \tag{5.219}$$

5.3 Verify that the unit step function is derived from the following limits of smooth sigmoidal functions:

$$\text{(a)}\ u(t) = 1/2 + (1/\pi)\lim_{a\to 0} \tan^{-1}(t/a), \tag{5.220}$$

$$\text{(b)}\ u(t) = \lim_{a\to 0} \frac{1}{1+\exp(-t/a)}. \tag{5.221}$$

The last expression with $a = 1$ is the logistic function.

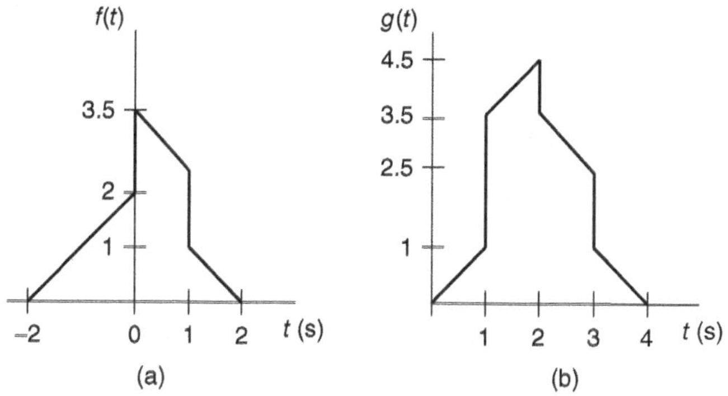

Figure 5.34 Waveforms for Problem 5.7.

5.4 Show that the ramp function can also be written as

$$\text{(a)}\ r(t) = (1/2)(t + |t|), \qquad \text{(b)}\ r(t) = \int_{-\infty}^{\infty} u(\tau)u(t-\tau)d\tau. \tag{5.222}$$

5.5 A sawtooth waveform can be constructed from a sum of weighted and shifted ramp functions. Let the period of the function be $T_o = 1$ s with the first component given by $r(t)[u(t) - u(t-1)]$. (a) Write the sawtooth waveform as an infinite sum of shifted versions of these components. (b) Modify your result such that the period is $T_o = 2$ s and the maximum height of the waveform is still 1.

Rectangle and Triangle Functions

5.6 A series of narrow rectangle functions can be used to sample a continuous waveform. Describe the resulting waveform when the following function multiplies the ramp function $r(t)$:

$$s(t) = \sum_{n=0}^{\infty} \text{rect}(4(t-n) - 1/2). \tag{5.223}$$

5.7 Demonstrate how to write the waveforms in Figure 5.34 in terms of scaled and shifted $\text{rect}(t)$ and $\text{tri}(t)$.

5.8 Express a periodic triangular waveform as an infinite sum of shifted versions of $\text{tri}(t)$, with the first component starting at $t = 0$. The waveform should have a maximum height of 2, a period of $T_o = 1$ s, and the component triangle functions should be adjacent to each other.

Exponential and Sinusoidal Functions

5.9 Prove the integral property of the exponential function using the power series representation for $\exp(t)$.

5.10 Determine the time when the following functions exceed 90% of their maximum values: (a) $f(t) = [2 - \exp(-3t)]u(t)$ and (b) $g(t) = [2 - \exp(-t) - \exp(-2t)]u(t)$.

5.11 Determine if there is any time instant $t \in [0, 2\pi]$ for which $\cos^2(t) + 3\sin(t) = 1$.

5.12 The following function is an example of one type of waveform that can occur for a voltage in an RLC circuit:

$$v(t) = t\exp(-t)\cos(\omega_o t)u(t). \tag{5.224}$$

Find the minimum and maximum values of $v(t)$ as a function of ω_o.

Dirac Delta Function

5.13 Verify the sampling and sifting properties of the shifted Dirac delta function $\delta(t - \tau)$ using the rectangle function as was done in Example 5.9.

5.14 Evaluate the following integrals:

$$\text{(a)} \int_0^1 \delta(t-2)u(t+2)dt, \qquad \text{(b)} \int_0^\infty \delta(\tau - 1)\exp(-(t-\tau))d\tau, \tag{5.225}$$

$$\text{(c)} \int_{-\infty}^\infty \delta(\tau)\cos(t - \tau - 1)d\tau, \qquad \text{(d)} \int_{-\infty}^\infty (\tau + 2)\delta(t - \tau - 2)d\tau. \tag{5.226}$$

5.15 Starting with a rectangle function approximation, find an expression for $\delta(\alpha t)$ in terms of $\delta(t)$.

5.16 Determine how the following functions can be used to represent a shifted Dirac delta function in the limit as the parameter α is varied, and give the location τ of $\delta(t - \tau)$:

$$\text{(a) } f(t) = \alpha\,\text{tri}(\alpha(t+1)), \qquad \text{(b) } g(t) = \frac{1}{\sqrt{2\pi}\alpha}\exp(-(t-2)^2/2\alpha^2). \tag{5.227}$$

5.17 The following "comb" function can be used to generate equally spaced samples of a continuous function:

$$s(t) = \sum_{n=-\infty}^{\infty} \delta(t - n). \tag{5.228}$$

Sketch the samples for (a) $y_1(t) = s(t)\exp(-2|t - 1/2|)$ and (b) $y_2(t) = s(t)\cos(t + 3/4)$.

Generalized Functions

5.18 Determine if the following are valid test functions:

$$\text{(i) } \phi_1(t) = \exp(1/t[t-1])I_{[0,1]}(t), \qquad \text{(ii) } \phi_2(t) = \text{tri}(2t - 1). \tag{5.229}$$

5.19 Using the approach leading to the result in (5.83), demonstrate the sampling property $g(t)\delta(t-t_o) = g(t_o)\delta(t-t_o)$, where $g(t)$ is continuous at t_o.

5.20 (a) Verify that $\langle f, \phi \rangle = \int_0^\infty t\phi(t)dt$ in Table 5.1. (b) Use the properties of generalized functions to find the derivative of the ramp function $r(t)$.

5.21 Based on a derivation similar to that leading to (5.86), derive the following property starting with $u(at+b)$:

$$\delta(at+b) = (1/|a|)\delta(t+b/a). \tag{5.230}$$

5.22 Suppose function $f(t)$ has a step discontinuity of size Δ at $t = t_o$. By writing $f(t)$ as the weighted sum of a unit step function and a smooth function, give an expression for its generalized derivative.

5.23 Use the properties of generalized functions to show that the derivative of the absolute value function $|t|$ is the signum function $\text{sgn}(t)$.

5.24 Repeat the previous problem to show that the second derivative of $|t|$ is $2\delta(t)$, which is the first derivative of the signum function.

5.25 Use the properties of generalized functions to find expressions for the second derivative of (a) $f(t) = \exp(-|t|)$ and (b) $g(t) = \exp(j\omega_o|t|)$.

5.26 Prove the even and odd properties of distributions given in Table 5.2.

Unit Doublet

5.27 Use the generalized function approach to derive the sifting property for the unit doublet.

5.28 Show that $\delta'(t)$ has area zero from the sifting and convolution properties of the unit doublet.

5.29 Prove the following property using integration by parts:

$$f(t)\delta^{(2)}(t) = f^{(2)}(0)\delta(t) - 2f'(0)\delta'(t) + f(0)\delta^{(2)}(t). \tag{5.231}$$

Singularities and Cauchy Principal Value

5.30 Describe the singularities of the following functions. (a) $f_1(z) = (z-1)/(z^2+2)(z+1)$. (b) $f_2(z) = \tanh(z)/z^3$. (c) $f_3(z) = (z^4-1)/(z^2+1)$.

5.31 Find CPVs for

$$\text{(a)} \int_{-\infty}^{\infty} t dt, \qquad \text{(b)} \int_{-\infty}^{\infty} [1/(t-1)]dt, \qquad \text{(c)} \int_{-\infty}^{\infty} [\text{sgn}(t)/t^2]dt. \tag{5.232}$$

5.32 Derive the integral of $(1/x)u(x)$ by splitting it up into two parts on the intervals $[\epsilon, 1]$ and $(1, \infty)$ and then letting $\epsilon \to 0$.

Even and Odd Functions and Correlation

5.33 Decompose the following functions into even and odd components as in (5.136) and sketch the results: (a) $f(t) = \text{tri}(t-1)$ and (b) $g(t) = \text{rect}(t-1) + \text{sgn}(t+1)$.

5.34 Prove the properties in rows two, three, and four of Table 5.5.

5.35 Derive the cross-correlation function $c_{fg}(\tau)$ for $f(t) = \text{rect}(t)$ and $g(t) = \text{tri}(t)$.

5.36 (a) Show that the autocorrelation function of rect(t) is triangular. (b) Find an expression for the autocorrelation function of tri(t).

Fourier Series

5.37 Verify the expression for b_n in (5.161).

5.38 Derive the trigonometric Fourier series coefficients for the periodic rectangular function in Figure 5.27(a) shifted to the right by 1/4 s.

5.39 Find the Fourier series coefficients in (5.176) for the periodic triangular waveform.

5.40 One period of a periodic waveform is defined as follows:

$$x(t) = \begin{cases} 0, & -T_o/2 \le t < -T_o/4 \\ 16t/T_o + 4, & -T_o/4 \le t < -T_o/8 \\ 2, & -T_o/8 \le t < T_o/8 \\ -16t/T_o + 4, & T_o/8 \le t < T_o/4 \\ 0, & T_o/4 \le t < T_o/2. \end{cases} \tag{5.233}$$

Find its complex exponential Fourier series.

5.41 Find the Fourier series for $\sin^2(t)$.

5.42 Derive the Fourier series coefficients in (5.192) for the half-wave rectified sine function.

Phasor Representation and Linear Circuits

5.43 Give phasor representations for the following sinusoidal waveforms, all of which have support $\mathcal{R}$:

$$\text{(a) } f_1(t) = 5\cos(2t - \pi/3), \qquad \text{(b) } f_2(t) = 2\sin(t + \pi/4),$$
$$\text{(c) } f_3(t) = -3\sin(4t - \pi/6). \tag{5.234}$$

5.44 Convert the following phasors to cosine waveforms, all with angular frequency $\omega_o = 5$ rad/s:

$$\text{(a) } \mathbf{F}_1 = 10\exp(j\pi/6), \quad \text{(b) } \mathbf{F}_2 = -2\exp(j\pi/3), \quad \text{(c) } \mathbf{F}_3 = 5\angle\pi/2. \tag{5.235}$$

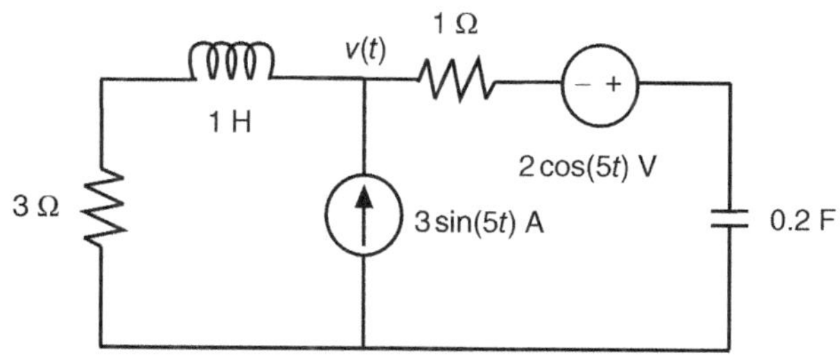

Figure 5.35 Second-order circuit with sinusoidal current and voltage sources for Problem 5.47.

5.45 Replace the capacitor in Figure 5.32 with an inductor L and find an expression for the voltage across L using phasors.

5.46 Repeat the previous problem with a capacitor C in parallel with the inductor L, resulting in a second-order RLC circuit.

5.47 Using phasors, find an expression for $v(t)$ in Figure 5.35.

Computer Problems

5.48 Derive the Fourier series coefficients for a periodic rectangular function similar to that in Figure 5.27(a), but with $T_o = 1$ s and $T = 1/4$ s. Use MATLAB to plot the Fourier series approximation including the DC term, (a) five cosine terms, and (b) ten cosine terms.

5.49 The impedance for an RLC circuit is

$$Z = \frac{1 - LC\omega_o^2 + jRC\omega_o}{jC\omega_o}, \tag{5.236}$$

with $R = 100\ \Omega$, $C = 100\ \mu$F, and $L = 2$ mH. Use MATLAB to plot the magnitude and phase of complex Z as ω_o is varied.

5.50 Plot the functions in (5.139) and (5.140) for $f_E(t)$ and $f_O(t)$, respectively, using Heaviside (the unit step) in MATLAB to truncate the piecewise linear sections. Then add the two functions together and plot the results to verify the original function $f(t)$ in Figure 5.25(a).

6

DIFFERENTIAL EQUATION MODELS FOR LINEAR SYSTEMS

6.1 INTRODUCTION

In this chapter, we describe differential equations (DEs) that are used in engineering to *model* the dynamics of a linear system with input $x(t)$ and output $y(t)$. By solving for the dependent variable of a DE, we obtain an explicit form for $y(t)$ as a function of the independent variable time t. First- and second-order linear ordinary differential equations (ODEs) are considered in this chapter, which model the most widely studied systems in engineering circuits and systems courses. Higher order ODEs are examined in the next chapter when the Laplace transform is covered, where a *transform-domain* approach allows them to be solved more easily than using time-domain methods.

As a preview, we summarize the basic solutions for linear systems, which turn out to be combinations of ordinary functions and singular generalized functions.

- *Decreasing exponential:*

$$y(t) = \exp(-\alpha t)u(t), \tag{6.1}$$

  with $\alpha > 0$.
- *Sine and cosine:*

$$y(t) = \sin(\omega_o t)u(t), \quad y(t) = \cos(\omega_o t)u(t), \tag{6.2}$$

  where ω_o is angular frequency in rad/s.

Mathematical Foundations for Linear Circuits and Systems in Engineering, First Edition. John J. Shynk.

Companion Website: http://www.wiley.com/go/linearcircuitsandsystems

- *Exponentially weighted sine and cosine:*

$$y(t) = \exp(-\alpha t)\sin(\omega_o t)u(t), \quad y(t) = \exp(-\alpha t)\cos(\omega_o t)u(t). \tag{6.3}$$

- *Exponentially weighted and ramped sine and cosine:*

$$y(t) = t\exp(-\alpha t)\sin(\omega_o t)u(t), \quad y(t) = t\exp(-\alpha t)\cos(\omega_o t)u(t). \tag{6.4}$$

- *Unit step function:* $y(t) = u(t)$.
- *Dirac delta function:* $y(t) = \delta(t)$.
- *Unit doublet:* $y(t) = \delta'(t)$.

The majority of solutions in actual circuits tend to be the waveforms in (6.1)–(6.3), as well as the unit step function and the Dirac delta function. Although we can write ODEs for systems with solutions that include the ramp function $r(t)$ and derivatives of the Dirac delta function, they do not occur often in practice. When a function is multiplied by the unit step function, $t < 0$ is excluded from its support. This is done because we are usually interested in *causal* systems with input signals starting at some finite time $t_o \geq 0$. Although any time instant is possible, generally it is convenient to use $t_o = 0$; the time axis can be shifted so that its origin is aligned with the start of the input signal $x(t)$.

6.2 DIFFERENTIAL EQUATIONS

We begin with some basic definitions of different types of DEs and then narrow our discussion to only one kind of ODE examined in this chapter.

Definition: Ordinary Differential Equation A *differential equation* is an equation consisting of two or more variables that includes at least one derivative. It is *ordinary* when dependent variables are functions of only a single independent variable.

For a system with input $x(t)$ and output $y(t)$, the dependent variables are $x(t)$ and $y(t)$, and t is the independent variable. If a dependent variable is a function of two or more independent variables, then we can have a *partial differential equation* (*PDE*) depending on how the derivatives are arranged.

Example 6.1 The following equations are examples of ODEs:

$$\text{first order: } \frac{d}{dt}y(t) + 2y(t) = x(t), \tag{6.5}$$

$$\text{second order: } \frac{d^2}{dt^2}y(t) - y(t) = 2x(t) + \frac{d}{dt}x(t) \triangleq f(t). \tag{6.6}$$

Since the input $x(t)$ is usually a known function in practice, the right-hand side of (6.6) can be replaced by the composite function $f(t)$. The goal is to solve for $y(t)$ as a function of t given $x(t)$ and its derivatives, as well as any nonzero initial conditions for $y(t)$ and its derivatives.

Example 6.2 The following equations are examples of PDEs with independent variables $\{t, v\}$ where slightly different notation is used for the derivatives:

$$\frac{\partial}{\partial t}y(t, v) + 3\frac{\partial}{\partial t}y(t, v) = x(t, v), \tag{6.7}$$

$$\frac{\partial^2}{\partial t \partial v}y(t, v) - 4y(t, v) = x(t, v). \tag{6.8}$$

The goal is to solve for $y(t, v)$ given the known function $x(t, v)$ and any initial conditions. PDEs arise as models for diffusion processes such as heat diffusion through a piece of metal, and they are useful for describing wave phenomena in physics. A simple diffusion equation is

$$\frac{\partial}{\partial t}y(t, v) = \alpha^2 \frac{\partial^2}{\partial v^2}y(t, v), \tag{6.9}$$

where v is position, t is time, and $\alpha > 0$ is a constant. PDEs are generally more difficult to solve than ODEs and are not considered further in this book.

Definition: Linear ODE An ODE is *linear* if the degree of the dependent variable of every term in the sum is 1.

The most general form of the linear ODEs considered in this book and that are used to describe linear time-invariant (LTI) systems is

$$\begin{aligned} a_N \frac{d^N}{dt^N}y(t) + a_{N-1}\frac{d^{N-1}}{dt^{N-1}}y(t) + \cdots + a_1\frac{d}{dt}y(t) + a_0 y(t) \\ = b_M \frac{d^M}{dt^M}x(t) + b_{M-1}\frac{d^{M-1}}{dt^{M-1}}x(t) + \cdots + b_1\frac{d}{dt}x(t) + b_0 x(t), \end{aligned} \tag{6.10}$$

which can be written more compactly using summation notation as

$$\sum_{n=0}^{N} a_n \frac{d^n}{dt^n}y(t) = \sum_{m=0}^{M} b_m \frac{d^m}{dt^m}x(t) \triangleq f(t), \tag{6.11}$$

where $d^0y(t)/dt^0 \triangleq y(t)$ and $d^0x(t)/dt^0 \triangleq x(t)$. For convenience in solving for $y(t)$, we often assume that the coefficient multiplying $d^N y(t)/dt^N$ in (6.10) is $a_N = 1$.

A linear system is represented by a linear ODE, and a time-invariant system has *fixed* coefficients $\{a_n, b_m\}$, which are generally known or can be estimated. The order of the ODE is $\max(M, N)$, and we are interested in finding $y(t)$ for $t \geq t_o$. Observe that the exponents of $\{x(t), y(t)\}$ in (6.11) are all 1, and so the ODE is linear. The maximum degree of the differentials gives the *order* of the ODE; it does *not* specify whether or not the ODE is linear. For most of this chapter, only $x(t)$ will be used on the right-hand side with $M = 0$ and $b_0 = 1$ in order to more easily illustrate how solutions are derived without the added complexity of including the derivatives of $x(t)$.

In order to solve for $y(t)$, we also need $N - 1$ *initial conditions*, which are specific values for the following derivatives evaluated at $t = t_o$:

$$\left.\frac{d^n}{dt^n}y(t)\right|_{t=t_o} \triangleq y^{(n)}(t_o), \quad n = 0, \ldots, N-1. \tag{6.12}$$

Sometimes these initial conditions are all 0. However, if the right-hand side of (6.10) is 0, corresponding to a homogeneous ODE with no input, then at least one initial condition must be nonzero. Otherwise, the solution is trivial: $y(t) = 0$ for $t \geq t_o$.

Definition: Homogeneous ODE A linear ODE is *homogeneous* if we replace the dependent variable $y(t)$ by $cy(t)$ and find that the constant $c \neq 0$ factors and cancels from the equation.

Example 6.3 This definition obviously holds for (6.11) when $f(t) = 0$:

$$\sum_{n=0}^{N} a_n \frac{d^n}{dt^n} cy(t) = c \sum_{n=0}^{N} a_n \frac{d^n}{dt^n} y(t) = 0. \tag{6.13}$$

If $y(t)$ is a solution of the ODE, then $cy(t)$ is also a solution; nonzero initial conditions as described in the next section cause c to have a specific value.

6.3 GENERAL FORMS OF THE SOLUTION

The general solution of the linear ODE

$$\sum_{n=0}^{N} a_n \frac{d^n}{dt^n} y(t) = x(t)u(t - t_o) \tag{6.14}$$

can be partitioned into two parts:

$$y(t) = y_h(t) + y_p(t), \quad t \geq t_o, \tag{6.15}$$

where $y_h(t)$ is the *homogeneous solution* obtained when the right-hand side is $x(t) = 0$, and $y_p(t)$ is the *particular solution* derived for the specific nonzero input $x(t)$. The homogeneous solution is also called the *complementary solution*, and $y(t)$ in (6.15) is called the *complete solution*. As shown later, the homogeneous solution is found first, which is usually straightforward to derive; $y_h(t)$ is the same for any input $x(t)$. The particular solution is generated by starting with $y_h(t)$ and modifying it for the specific input $x(t)$, which is usually more difficult to derive. The homogeneous solution is also called the *natural response* of the system, and the particular solution is known as the *forced response*.

When the input is a step function, the solution of (6.14) can also be arranged to have the following form:

$$y(t) = y_t(t) + y_s, \quad t \geq t_o, \tag{6.16}$$

where $y_t(t)$ is the *transient* solution and y_s is the *steady-state* solution. Of course, (6.15) and (6.16) are the same $y(t)$; the second form is derived from the first form by isolating the steady-state part $y_s = \lim_{t\to\infty} y(t)$. By definition, the transient part decays as $\lim_{t\to\infty} y_t(t) = 0$, because we assume a stable system and so the output is bounded. Even though the first form in (6.15) is derived when solving an ODE, the second form in (6.16) is often more informative for practical systems such as linear circuits. In many problems, we are interested in the steady-state solution for some voltage or current of a circuit when a voltage or current elsewhere in the circuit has changed suddenly and is modeled by a step function.

Definition: Linear System A system is *linear* if the output due to $c_1x_1(t) + c_2x_2(t)$ is $c_1y_1(t) + c_2y_2(t)$ where $y_1(t)$ is the output for input $x_1(t)$, $y_2(t)$ is the output for input $x_2(t)$, and $\{c_1, c_2\}$ are constants.

It is clear that the system modeled by the ODE in (6.14) is linear because

$$\begin{aligned}\sum_{n=0}^{N} a_n \frac{d^n}{dt^n}[c_1y_1(t) + c_2y_2(t)] &= c_1 \sum_{n=0}^{N} a_n \frac{d^n}{dt^n} y_1(t) + c_2 \sum_{n=0}^{N} a_n \frac{d^n}{dt^n} y_2(t) \\ &= c_1x_1(t) + c_2x_2(t).\end{aligned} \tag{6.17}$$

Definition: Linear Time-Invariant System A linear system is *time-invariant* if $y(t - \tau)$ is the output for input $x(t - \tau)$ where $\tau > 0$ is a time delay.

The linear system in (6.14) is also time-invariant because the coefficients $\{a_n\}$ are fixed:

$$\sum_{n=0}^{N} a_n \frac{d^n}{dt^n} y(t - \tau) = x(t - \tau). \tag{6.18}$$

As discussed in Chapter 1, nonlinear systems are generally difficult to solve, which is one reason why many systems are modeled as linear in practice, even though the solution may only be an approximate representation of the actual response to an input. Similarly, time invariance is another property that allows for a relatively straightforward solution. It is evident that (6.14) would be more difficult to solve if $\{a_n\}$ varied with time, even if those variations are precisely known. Later we show that these two properties of a system allow it to be completely specified by its response $h(t)$ when the input is the Dirac delta function: $x(t) = \delta(t)$. Once $h(t)$ is known, it can be used to generate the output $y(t)$ for any input via a *convolution integral*.

6.4 FIRST-ORDER LINEAR ODE

A first-order linear ODE has the following form:

$$\frac{d}{dt}y(t) + ay(t) = x(t)u(t - t_o), \tag{6.19}$$

where a is a fixed known coefficient, time t is the independent variable, $y(t)$ is a dependent variable (the system output), and $x(t)$ is another dependent variable, but is a known function (the system input). The solution of this ODE for $x(t) = 0$ (the homogeneous case) has the exponential form in (6.1), which will be derived in this section.

Examples of circuits that are described by first-order ODEs are shown in Figure 6.1. From Chapter 2, the current through the series capacitor is

$$i(t) = C\frac{dv_C(t)}{dt}, \tag{6.20}$$

where $v_C(t)$ is its voltage. Solving for $v_C(t)$ yields (see (2.22))

$$v_C(t) = (1/C)\int_{t_o}^{t} i(t)dt + v_C(t_o), \tag{6.21}$$

where $v_C(t_o)$ is an initial voltage at time instant $t_o \geq 0$. From Kirchoff's voltage law (KVL), the voltage $v_R(t) = Ri(t)$ across the resistor and $v_C(t)$ together must equal the

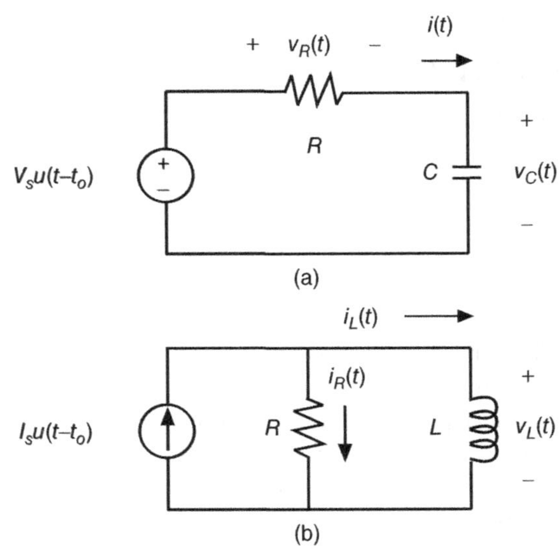

Figure 6.1 First-order circuits. (a) Series RC circuit with voltage source $V_s u(t - t_o)$. (b) Parallel RL circuit with current source $I_s u(t - t_o)$.

source voltage:

$$Ri(t) + (1/C)\int_{t_o}^{t} i(t)dt + v_C(t_o) = V_s u(t - t_o), \tag{6.22}$$

where $u(t - t_o)$ specifies that the voltage source has switched on at $t = t_o$, without explicitly showing a switch in the circuit of Figure 6.1. Differentiating this expression gives a first-order linear ODE for the current:

$$R\frac{d}{dt}i(t) + (1/C)i(t) = V_s\delta(t - t_o), \tag{6.23}$$

where the Dirac delta function is the generalized derivative of $u(t - t_o)$. Rearranging this expression and dividing by R yield

$$\frac{d}{dt}i(t) + (1/RC)i(t) = (V_s/R)\delta(t - t_o). \tag{6.24}$$

When the voltage source switches on, the voltage across the capacitor cannot change instantaneously, which means the voltage across the resistor is $V_s - v_C(t_o)$, and so the initial current is $i(t_o) = [V_s - v_C(t_o)]/R$.

Most books on circuits ignore the delta function (because they usually do not cover generalized functions) and write the homogeneous ODE

$$\frac{d}{dt}i(t) + (1/RC)i(t) = 0, \quad t \geq t_o, \tag{6.25}$$

with the understanding that the initial current $i(t_o)$ is nonzero. This expression has the form in (6.19) with $y(t) = i(t)$, $a = 1/RC$, and $x(t) = 0$. The reason the delta function can be ignored is that the ODE solution is actually defined for $t \geq t_o^+$, which is a time instant chosen so that the solution of (6.22) includes any discontinuities or singular functions at t_o. Thus, when differentiating (6.22), $V_s u(t - t_o)$ is often treated as a constant at t_o^+, and its derivative is 0 leading to (6.25). For simplicity, we will also write such ODEs in homogeneous form with the initial condition specified separately. In the next chapter on the Laplace transform, it will be necessary to distinguish between t_o^- and t_o^+ when solving ODEs, where t_o^- is "just before" any discontinuity at t_o, and t_o^+ is "just after."

The voltage across the capacitor is derived by recognizing that it is the source voltage minus the voltage across the resistor:

$$v_C(t) = V_s u(t - t_o) - Ri(t) = V_s u(t - t_o) - RC\frac{d}{dt}v_C(t), \tag{6.26}$$

where $i(t)$ from (6.20) has been substituted. Rearranging this expression gives a nonhomogeneous ODE with input $x(t) = (V_s/RC)u(t - t_o)$:

$$\frac{d}{dt}v_C(t) + (1/RC)v_C(t) = (V_s/RC)u(t - t_o). \tag{6.27}$$

TABLE 6.1 First-Order RL and RC Circuits

System	Linear ODE Signals and Coefficients
General ODE	$dy(t)/dt + ay(t) = x(t)$
Series RC current	$y(t) = i(t)$, $x(t) = 0$
Series RC resistor voltage	$y(t) = v_R(t)$, $x(t) = 0$
Series RC capacitor voltage	$y(t) = v_C(t)$, $x(t) = (1/RC)V_s u(t - t_o)$
Coefficient	$a = 1/RC$
Parallel RL voltage	$y(t) = v(t)$, $x(t) = 0$
Parallel RL resistor current	$y(t) = i_R(t)$, $x(t) = 0$
Parallel RL inductor current	$y(t) = i_L(t)$, $x(t) = (R/L)I_s u(t - t_o)$
Coefficient	$a = R/L$

These results are summarized in Table 6.1, which also includes the details for the voltage $v_R(t)$ across the resistor (see Problem 6.4). It is important to note that even though there is a voltage source in the circuit, an ODE can be homogeneous or nonhomogeneous depending on the particular dependent variable $y(t)$. From the table, we see that the ODE for $y(t) = v_R(t)$ is homogeneous, whereas it is nonhomogeneous for $y(t) = v_C(t)$. As shown later, this means that the steady-state voltage of the resistor is 0, while it is V_s for the capacitor.

For the parallel RL circuit, the voltage across the inductor is

$$v(t) = L\frac{di_L(t)}{dt}, \tag{6.28}$$

and so its current is (see (2.23))

$$i_L(t) = (1/L)\int_{t_o}^{t} v(t)dt + i_L(t_o), \tag{6.29}$$

where $i_L(t_o)$ is the initial inductor current. The current $v(t)/R$ through the resistor and $i_L(t)$ together must equal the current source:

$$v(t)/R + (1/L)\int_{t_o}^{t} v(t)dt + i_L(t_o) = I_s u(t - t_o). \tag{6.30}$$

Differentiating this expression and multiplying through by R yield a first-order homogeneous ODE for the inductor voltage:

$$\frac{d}{dt}v(t) + (R/L)v(t) = 0, \quad t \geq t_o. \tag{6.31}$$

When the current source is switched on, the current through the inductor cannot change instantaneously, and so all of I_s passes initially through the resistor. This gives an initial voltage of $R[I_s - i_L(t_o)]$ across the parallel inductor. As in the previous RC

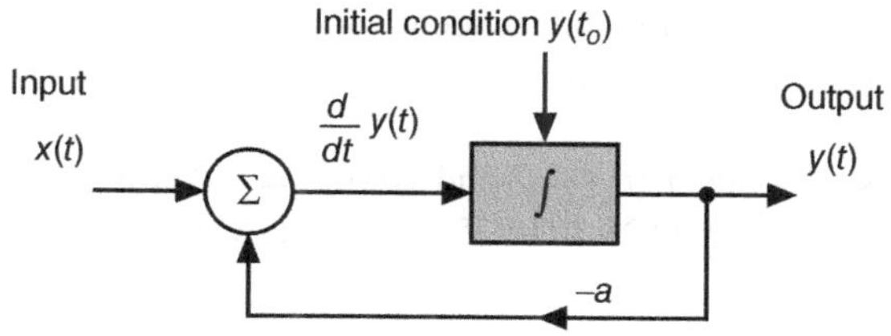

Figure 6.2 Integrator implementation of a first-order ODE.

circuit, we have dropped the Dirac delta function that would have appeared after differentiating (6.30), in favor of specifying the initial condition separately.

The current through the inductor is

$$i_L(t) = I_s u(t - t_o) - v(t)/R = I_s u(t - t_o) - (L/R)\frac{d}{dt}i_L(t), \tag{6.32}$$

where (6.28) has been substituted. Rearranging this expression gives a nonhomogeneous ODE with input $x(t) = (R/L)I_s u(t - t_o)$:

$$\frac{d}{dt}i_L(t) + (R/L)i_L(t) = (R/L)I_s u(t - t_o). \tag{6.33}$$

These results are also summarized in Table 6.1 along with details for the parallel resistor current $i_R(t)$ (see Problem 6.5). Once $i_L(t)$ is found by solving (6.33), the expression in (6.28) can be used to derive the time-varying voltage across the inductor without having to solve an ODE for the voltage. An integrator implementation for these first-order circuits is provided in Figure 6.2, with parameter a, input $x(t)$, and output $y(t)$ given in Table 6.1. Although the initial condition is symbolically shown entering the integrator, it is actually the initial value of the output as shown in the next section.

6.4.1 Homogeneous Solution

For $x(t) = 0$, the first-order ODE in (6.19) can be rearranged as

$$\frac{d}{dt}y(t) = -ay(t), \quad t \geq t_o, \tag{6.34}$$

which is a special case of a separable ODE.

Definition: Separable First-Order ODE A first-order ODE is *separable* if it can be written as the following product:

$$\frac{d}{dt}y(t) = h_1(t)h_2(y(t)), \tag{6.35}$$

where $h_1(t)$ is a function of the independent variable t (it does not depend on $y(t)$), and $h_2(y(t))$ is a function of the dependent variable $y(t)$.

For (6.34), it is clear that $h_1(t) = -a$ and $h_2(y(t)) = y(t)$ (we could also let $h_1(t) = 1$ and $h_2(y(t)) = -ay(t)$). The ODE in (6.35) is solved by dividing it by $h_2(y(t))$, which in our case yields

$$\frac{1}{h_2(t)}\frac{d}{dt}y(t) = h_1(t) \implies \frac{1}{y(t)}\frac{d}{dt}y(t) = -a. \tag{6.36}$$

Integrating both sides gives

$$\int_{y(t_o)}^{y(t)} \frac{1}{y(t)}dy(t) = -a\int_{t_o}^{t} dt \implies \ln(|y(t)|) - \ln(|y(t_o|) = -a(t - t_o), \tag{6.37}$$

where we recognize that the first integral is the natural logarithm. Thus, the solution is an exponential function:

$$\ln(|y(t)/y(t_o)|) = -a(t - t_o) \implies y(t) = y(t_o)\exp(-a(t - t_o))u(t - t_o). \tag{6.38}$$

Since the exponential function is nonnegative, the absolute values can be dropped; $y(t)$ and $y(t_o)$ necessarily have the same sign. If $a > 0$, the exponential function decays to 0; otherwise, it grows to infinity. However, a cannot be negative for the circuits in Figure 6.1 because the parameters $\{R, L, C\}$ are all positive, and so the first-order RL and RC circuits are *stable*. As mentioned earlier, the initial condition $y(t_o)$ is assumed to be nonzero for this homogeneous case.

An alternative approach to solving this ODE is to assume the basic form of the solution and then find the specific parameters. Since the derivative of an exponential function is another exponential function, $y(t) = c\exp(s(t - t_o))$ can be substituted into (6.19) and the equation holds:

$$sc\exp(s(t - t_o)) + ac\exp(s(t - t_o)) = 0, \quad t \geq t_o. \tag{6.39}$$

Canceling common terms, we find that $s + a = 0 \implies s = -a$, which gives

$$y(t) = c\exp(-a(t - t_o)), \quad t \geq t_o. \tag{6.40}$$

The coefficient c is provided by the initial condition $y(t_o)$:

$$y(t_o) = c\exp(-a(t_o - t_o)) \implies c = y(t_o), \tag{6.41}$$

yielding the result in (6.38). The expression $s + a = 0$ is called the *characteristic equation* of the system, which is quite simple for a first-order linear ODE. We find later that the characteristic equation for a second-order linear ODE has more structure and leads to more complicated solutions for the system output $y(t)$.

Example 6.4 A special case of the first-order homogeneous ODE occurs when the coefficient is $a = 0$ such that

$$\frac{d}{dt}y(t) = 0, \quad t \geq t_o, \tag{6.42}$$

whose solution is a constant:

$$y(t) = y(t_o)u(t - t_o). \tag{6.43}$$

For an RC circuit without a voltage source, this means that R is infinite such that there is only a capacitor with a nonzero initial voltage in an open circuit. For an RL circuit without a current source, $R = 0$ and the circuit is shorted such that any initial current in the inductor flows indefinitely around the loop. Obviously, the ODE in (6.34) does not model a practical circuit when $a = 0$.

6.4.2 Nonhomogeneous Solution

The nonhomogeneous first-order ODE is somewhat more difficult to solve. One technique incorporates a function $g(t)$ known as an *integrating factor* that multiplies each term as follows:

$$g(t)\frac{d}{dt}y(t) + ag(t)y(t) = g(t)x(t) = \frac{d}{dt}g(t)y(t), \tag{6.44}$$

where the last expression is a *constraint* on the form of $g(t)$ that allows us to find a solution. Observe that the right-hand side is the derivative of the second term on the left-hand side (excluding the constant a), and so it is possible to cancel terms. Using the product rule of derivatives on the right-hand side, (6.44) becomes

$$g(t)\frac{d}{dt}y(t) + ag(t)y(t) = y(t)\frac{d}{dt}g(t) + g(t)\frac{d}{dt}y(t). \tag{6.45}$$

Canceling the two outer terms of this equation and $y(t)$ of the two inner terms yields a homogeneous equation for $g(t)$:

$$\frac{d}{dt}g(t) - ag(t) = 0, \quad t \geq t_o. \tag{6.46}$$

This expression is solved by rearranging it as follows:

$$\begin{aligned} \frac{dg(t)}{g(t)} = adt \implies & \ln(|g(t)|) - \ln(|g(t_o)|) = \int_{t_o}^{t} adt, \\ \implies & |g(t)/g(t_o)| = \exp\left(\int_{t_o}^{t} adt\right), \end{aligned} \tag{6.47}$$

which yields

$$g(t) = g(t_o)\exp(a(t - t_o)), \quad t \geq t_o. \tag{6.48}$$

The last expression in (6.47) is the reason why $g(t)$ is called an integrating factor. This approach is actually more general because a could be a function of time (see Problem 6.7). However, since we are interested only in LTI systems with constant coefficients, a factors from the integral and we obtain the result in (6.48), which has a form identical to that of the homogeneous solution in (6.38) except that $g(t)$ has replaced $y(t)$. The integrating factor with the constraint in (6.44) has essentially suppressed the input $x(t)$ and caused the nonhomogeneous ODE to become homogeneous, but with variable $g(t)$. Continuing with the derivation, we need to replace $g(t)$ so that the solution is written in terms of the output $y(t)$ and the input $x(t)$. Differentiating (6.48) yields

$$\frac{d}{dt}g(t) = ag(t_o)\exp(a(t - t_o)) = ag(t), \tag{6.49}$$

which can be substituted into the second term on the left-hand side of (6.44):

$$g(t)\frac{d}{dt}y(t) + y(t)\frac{d}{dt}g(t) = g(t)x(t). \tag{6.50}$$

Using the (reverse) product rule on the left-hand side gives

$$\frac{d}{dt}g(t)y(t) = g(t)x(t), \tag{6.51}$$

which has the solution

$$g(t)y(t) = \int_{t_o}^{t} g(t)x(t)dt + g(t_o)y(t_o), \tag{6.52}$$

where $g(t_o)y(t_o)$ is the initial condition of the product $g(t)y(t)$. Finally, substituting $g(t)$ from (6.48) yields

$$g(t_o)\exp(a(t - t_o))y(t) = g(t_o)\int_{t_o}^{t} x(t)\exp(a(t - t_o))dt + g(t_o)y(t_o), \tag{6.53}$$

which can be rearranged to give an explicit expression for the output $y(t)$:

$$y(t) = \exp(-at)\int_{t_o}^{t} x(t)\exp(at)dt + y(t_o)\exp(-a(t - t_o)), \quad t \geq t_o. \tag{6.54}$$

Note that we cannot cancel the first two exponentials in (6.54) because t under the integral is the variable of integration. In such cases, it is preferable to use another variable such as τ to avoid any confusion:

$$\begin{aligned} y(t) &= \exp(-at)\int_{t_o}^{t} x(\tau)\exp(a\tau)d\tau + y(t_o)\exp(-a(t - t_o)) \\ &= \int_{t_o}^{t} x(\tau)\exp(-a(t - \tau))d\tau + y(t_o)\exp(-a(t - t_o)), \quad t \geq t_o, \end{aligned} \tag{6.55}$$

which is the *complete solution* of (6.44). When $x(t) = 0$, the first term on the right-hand side is 0 and this equation reduces to the solution in (6.38) for the homogeneous ODE. The integral in (6.55) is the *particular solution* $y_p(t)$, and the second term is the homogeneous solution $y_h(t)$ derived earlier.

Example 6.5 Continuing with the special case in Example 6.4, let $a = 0$ such that

$$\frac{d}{dt}y(t) = x(t)u(t - t_o). \tag{6.56}$$

The expression in (6.55) shows that the solution is an *integrator*:

$$y(t) = \int_{t_o}^{t} x(\tau)d\tau. \tag{6.57}$$

This result also follows from the integrator implementation in Figure 6.2, which no longer has a feedback path when $a = 0$.

6.4.3 Step Response

Suppose that $x(t) = Ku(t - t_o)$ is a constant due to, for example, a voltage source switching on at time instant $t = t_o$. Substituting this particular input into (6.55) yields

$$\begin{aligned} y(t) &= (K/a)[1 - \exp(-a(t - t_o))] + y(t_o)\exp(-a(t - t_o)) \\ &= [K/a + [y(t_o) - K/a]\exp(-a(t - t_o))]u(t - t_o). \end{aligned} \tag{6.58}$$

The steady-state solution is $y_s = y(\infty) = K/a$, assuming $a > 0$ for a stable system, and the second term is the transient response $y_t(t)$, which decays to 0 as $t \to \infty$. An example is illustrated in Figure 6.3 for $t_o = 0$ with initial condition $y(0) = 1$. In the unlikely event that $y(t_o) = K/a$, the transient response cancels in (6.58) and the solution is a constant K/a. For the example in Figure 6.3, this occurs when $y(0) = 2$: the dashed line would be 0, and both the solid and dotted lines would be horizontal with value 2.

6.4.4 Exponential Input

If $x(t) = K\exp(-b(t - t_o))u(t - t_o)$, then the particular solution from (6.54) with $y(t_o) = 0$ and $a \neq b$ is

$$\begin{aligned} y_p(t) &= K\int_{t_o}^{t} \exp(-b(\tau - t_o))\exp(-a(t - \tau))d\tau \\ &= K\exp(-at + bt_o)\int_{t_o}^{t} \exp((a - b)\tau)d\tau, \end{aligned} \tag{6.59}$$

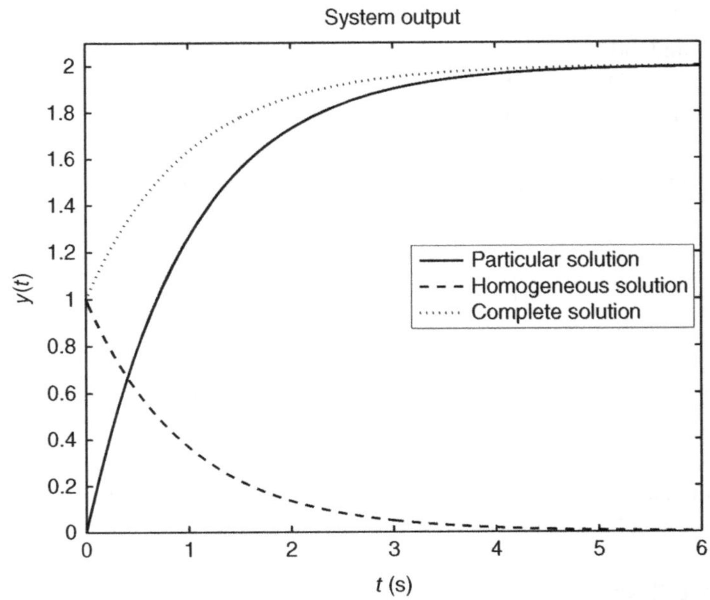

Figure 6.3 First-order system response to a step input with $t_o = 0$, $y(0) = 1$, $a = 1$, and $K = 2$.

which becomes

$$y_p(t) = \frac{K\exp(-at + bt_o)}{a - b}[\exp((a - b)t) - \exp((a - b)t_o)]u(t - t_o). \tag{6.60}$$

Simplifying this expression and including the homogeneous solution from (6.54), which does not depend on the specific $x(t)$, yields the following complete solution:

$$\begin{aligned} y(t) = {} & \frac{K}{a - b}[\exp(-b(t - t_o)) - \exp(-a(t - t_o))]u(t - t_o) \\ & + y(t_o)\exp(-a(t - t_o))u(t - t_o). \end{aligned} \tag{6.61}$$

An example is shown in Figure 6.4 for $t_o = 0$. When $a > 0$ and $b > 0$, the steady-state value is $y_s = 0$, which is intuitive because the input exponential decays to 0. There are two *modes of convergence* to 0 because a and b yield different time constants. When $a = b$, the second line in (6.59) is replaced with

$$y_p(t) = K\exp(-a(t - t_o))\int_{t_o}^{t} d\tau = K(t - t_o)\exp(-a(t - t_o)), \tag{6.62}$$

and the complete solution is

$$y(t) = [K(t - t_o) + y(t_o)]\exp(-a(t - t_o))u(t - t_o). \tag{6.63}$$

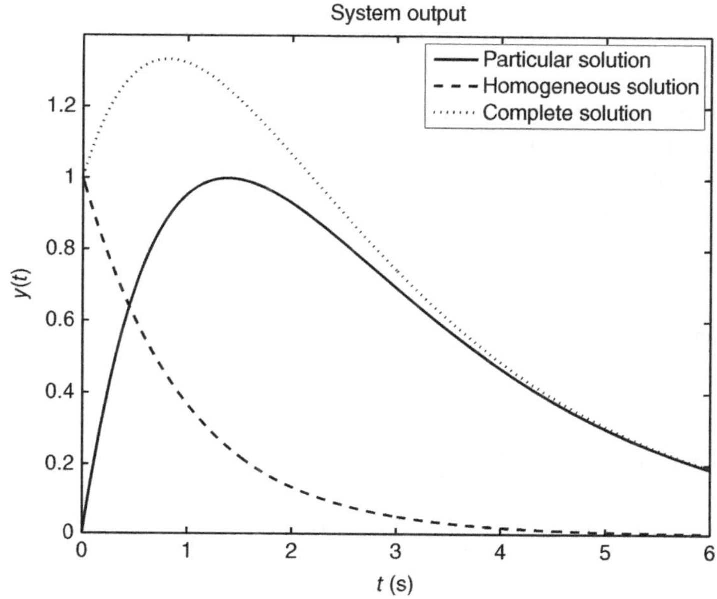

Figure 6.4 First-order system response to an exponential input with $t_o = 0$, $y(0) = 1$, $a = 1$, $b = 0.5$, and $K = 2$.

Since the exponential decays to 0 faster than the ramp $t - t_o$ increases to infinity for $a = b > 0$, the steady-state solution is again $y_s = 0$. For this case, there is only one exponential waveform converging to 0, and $y(t)$ has the appearance of the so-called *critically damped solution* described later for second-order ODEs.

6.4.5 Sinusoidal Input

Suppose now that $x(t) = \cos(\omega_o t)u(t)$ with angular frequency ω_o, and assume zero initial conditions. The output from (6.55) is

$$\begin{aligned} y(t) &= \int_0^t \cos(\omega_o \tau)\exp(-a(t-\tau))d\tau \\ &= \exp(-at)\int_0^t \cos(\omega_o \tau)\exp(a\tau)d\tau, \quad t \geq 0, \end{aligned} \tag{6.64}$$

which is the particular solution; the homogeneous part is 0. This integral is

$$\begin{aligned} y(t) &= \exp(-at)\left.\frac{\exp(a\tau)}{a^2+\omega_o^2}[a\cos(\omega_o\tau) + \omega_o\sin(\omega_o\tau)]\right|_0^t \\ &= \exp(-at)\left(\frac{\exp(at)}{a^2+\omega_o^2}[a\cos(\omega_o t) + \omega_o\sin(\omega_o t)] - \frac{a}{a^2+\omega_o^2}\right) \end{aligned}$$

$$= \frac{1}{a^2 + \omega_o^2}[a\cos(\omega_o t) + \omega_o \sin(\omega_o t)]u(t) - \frac{a\exp(-at)}{a^2 + \omega_o^2}u(t). \tag{6.65}$$

The first term is the steady-state response and the second term is the transient response, which is due to that fact that the cosine waveform starts at $t = 0$. For $a > 0$, the transient part decays to 0, and using a trigonometric identity, we can write the steady-state part as a cosine function with amplitude A and phase ϕ:

$$y_s(t) = A\cos(\omega_o t - \phi)u(t). \tag{6.66}$$

The trigonometric identity is

$$A\cos(v - \phi) = [A\cos(\phi)]\cos(v) + [A\sin(\phi)]\sin(v), \tag{6.67}$$

where in our case from (6.65):

$$v = \omega_o t, \quad A\cos(\phi) = a\alpha, \quad A\sin(\phi) = \omega_o\alpha, \quad \alpha = \frac{1}{a^2 + \omega_o^2}. \tag{6.68}$$

The ratio of the sinusoidal quantities gives the phase on the left-hand side of (6.67):

$$\frac{\sin(\phi)}{\cos(\phi)} = \omega_o/a \implies \phi = \tan^{-1}(\omega_o/a). \tag{6.69}$$

The amplitude is derived as follows:

$$A^2 = A^2\cos(\phi) + A^2\sin(\phi) = a^2\alpha^2 + \omega_o^2\alpha^2, \tag{6.70}$$

which means

$$A = \alpha\sqrt{a^2 + \omega_o^2} = \frac{1}{\sqrt{a^2 + \omega_o^2}}. \tag{6.71}$$

Thus, the following expression is equivalent to the steady-state part of (6.65):

$$y_s(t) = \frac{1}{\sqrt{a^2 + \omega_o^2}}\cos(\omega_o t - \tan^{-1}(\omega_o/a))u(t). \tag{6.72}$$

6.4.6 Impulse Response

For the input $x(t) = \delta(t)$, the first term of (6.55) (with $t_o = 0$) becomes

$$\begin{aligned}\int_0^t \delta(\tau)\exp(-a(t-\tau))d\tau &= \exp(-at)\int_0^t \delta(\tau)\exp(a\tau)d\tau \\ &= \exp(-at)u(t),\end{aligned} \tag{6.73}$$

where the second integral is 1 because of the sifting property of the Dirac delta function. This particular solution of the ODE is known as the *impulse response function* of the system and is usually denoted by $h(t)$. The initial condition, given by the second term in (6.55) with $t_o = 0$, is ignored when computing $h(t)$. Alternatively, when the input is $x(t) = \delta(t)$, as was the case in (6.25) for the first-order RC circuit, we can ignore the first term in (6.55) and immediately derive the impulse response function from the second term in (6.55) with $t_o = 0$ and $y(t_o) = 1$. It turns out that in general for zero initial conditions, the following convolution integral describes how to generate the output of an LTI system from its input $x(t)$:

$$y(t) = \int_0^t x(\tau)h(t-\tau)d\tau, \quad t \geq 0. \tag{6.74}$$

This result actually holds for a linear ODE of any order, although of course $h(t)$ depends on the specific ODE. The following notation is generally used to represent this integral:

$$y(t) = h(t) * x(t). \tag{6.75}$$

(This operation is different from the cross-correlation function covered in Chapter 5, which uses the symbol $\star$.)

For a linear system with zero initial conditions, it is shown later in this chapter that the output can be written as the following integral from the principle of *superposition*

$$y(t) = \int_{t_o}^t h(t_o, t)x(t)dt, \quad t \geq t_o, \tag{6.76}$$

where $h(t_o, t)$ is the notation for the response of the system for delayed input $\delta(t - t_o)$. Superposition is a defining characteristic of a linear system, where the output for the sum of input waveforms is the sum of their individual responses. If the system is also time-invariant, then $h(t_o, t)$ is a function only of the time difference: $h(t_o, t) = h(t - t_o)$, leading to the convolution integral in (6.74). The impulse response function completely specifies an LTI system, and it is used to represent high-order systems as discussed in Chapter 7 where ODEs are solved by using the Laplace transform.

Definition: Causal Linear System A linear time-invariant (LTI) system is *causal* when $h(t - t_o) = 0$ for $t < t_o$.

This property means that the present output of a causal system cannot be a function of a future input. This can be illustrated from the convolution in (6.74) if we let the upper limit extend to infinity and assume $x(\tau) = 0$ for $\tau < 0$:

$$y(t) = \int_0^\infty x(\tau)h(t-\tau)d\tau, \quad t \geq 0. \tag{6.77}$$

Note that because of the upper limit of infinity, the input beyond $\tau = t$ is used to compute $y(t)$. This noncausal (and impractical) situation is handled when $h(t - \tau)$ is 0 for $t < \tau$, which gives upper limit t as in (6.74). In particular for $\tau = 0$, we must have $h(t) = 0$ for $t < 0$, which is the definition for a causal system given in most books on linear systems.

Example 6.6 When the input is $x(t) = u(t)$, the output of the system is its unit step response. In this example, we illustrate graphically how the convolution integral is evaluated for $h(t) = \exp(-t)u(t)$. The integration in (6.74) is performed over the variable τ, and so for $t = 0$, we find that $h(t)$ is reversed about the origin. As t is increased beyond 0, $h(t - \tau)$ is shifted to the right and the integral (area) of the product of the overlapping functions is computed. For $t < 0$, there is no overlap and the integral is 0. Figure 6.5(a) shows the unit step function and the time-reversed and shifted exponential function $\exp(-(t - \tau))u(t - \tau)$ for two values of t. The dashed line shows $\exp(-(1 - \tau))u(-(1 - \tau))$ relative to $u(\tau)$ (the solid line), and the integral is

$$\begin{aligned}\int_0^1 \exp(-(1-\tau))u(\tau)d\tau &= \exp(-1)\int_0^1 \exp(\tau)d\tau \\ &= \exp(-1)[\exp(1) - \exp(0)] \\ &= 1 - \exp(-1) \approx 0.6321. \end{aligned} \tag{6.78}$$

Similarly, for the dotted line representing $\exp(-(2 - \tau))u(-(2 - \tau))$:

$$\begin{aligned}\int_0^2 \exp(-(2-\tau))u(\tau)d\tau &= \exp(-2)\int_0^2 \exp(\tau)d\tau \\ &= \exp(-2)[\exp(2) - \exp(0)] \\ &= 1 - \exp(-2) \approx 0.8647. \end{aligned} \tag{6.79}$$

This example illustrates the *mechanism* for computing a convolution where one of the functions is reversed and shifted relative to the other function. Of course, it is possible to derive $y(t)$ directly from the convolution integral:

$$\begin{aligned}\int_0^t \exp(-(t-\tau))u(\tau)d\tau &= \exp(-t)\int_0^t \exp(\tau)d\tau = \exp(-t)[\exp(t) - 1] \\ &= [1 - \exp(-t)]u(t), \end{aligned} \tag{6.80}$$

which is plotted in Figure 6.5(b). The dotted lines in that plot denote the two values of the integral given in (6.78) and (6.79) for $t = 1$ and 2 s, respectively. Usually, it is helpful to sketch a diagram of the reversed and shifted function when performing a convolution in order to determine the proper limits of integration.

Another convolution example is provided later in this chapter, and we verify using the functions in the previous example that convolution is a symmetric operation.

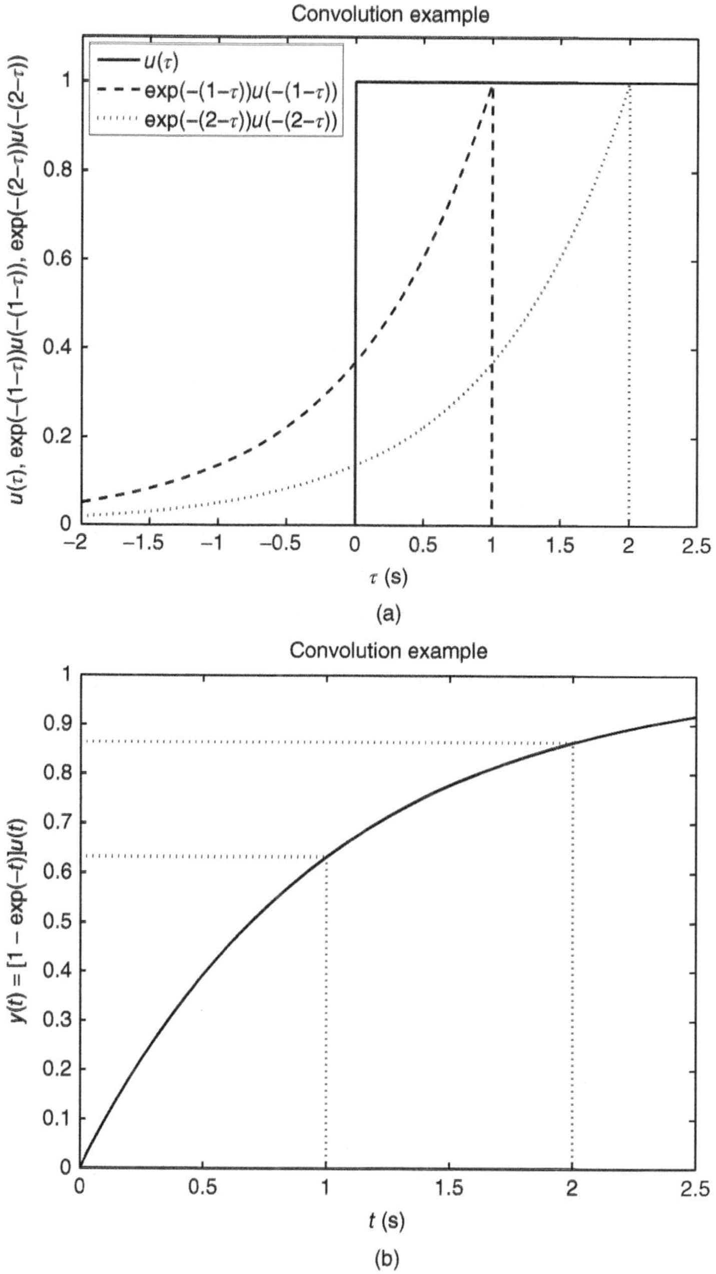

Figure 6.5 Convolution example for a first-order system with impulse response function $h(t) = \exp(-t)u(t)$ and input $x(t) = u(t)$. (a) Reversed exponential function for two values of t (note that the horizontal axis is τ). (b) Overall output $y(t) = x(t) * h(t)$.

6.5 SECOND-ORDER LINEAR ODE

A second-order linear ODE has the following form:

$$\frac{d^2}{dt^2}y(t) + a_1\frac{d}{dt}y(t) + a_0y(t) = x(t)u(t - t_o), \tag{6.81}$$

where the coefficients $\{a_0, a_1\}$ are fixed. The solution to this equation is more complicated to derive than it was for the first-order ODE, and in fact, there are *three types* of solutions depending on the values of the coefficients $\{a_0, a_1\}$. In order to simplify the derivations, we assume $t_o = 0$ throughout this discussion. From the results of the previous section, we know that a simple time shift of the input $x(t)$ yields the same time shift for the output $y(t)$. This is due to the fact that the system represented by the ODE is LTI. Thus, all the solutions derived in this section can be modified to handle a shifted input (and any initial conditions at $t_o > 0$) generally by replacing all instances of t with $t - t_o$ in the final expression for $y(t)$.

Figure 6.6 shows examples of linear circuits that are represented by second-order ODEs. The voltages across the devices in the series circuit sum to 0, which results in an *integro-differential equation* in terms of the current $i(t)$:

$$Ri(t) + L\frac{d}{dt}i(t) + (1/C)\int_0^t i(t)dt + v_C(0) = V_s u(t), \tag{6.82}$$

where $v_C(0)$ is the initial capacitor voltage. Differentiating this expression gives

$$\frac{d^2}{dt^2}i(t) + (R/L)\frac{d}{dt}i(t) + (1/LC)i(t) = (V_s/R)\delta(t), \tag{6.83}$$

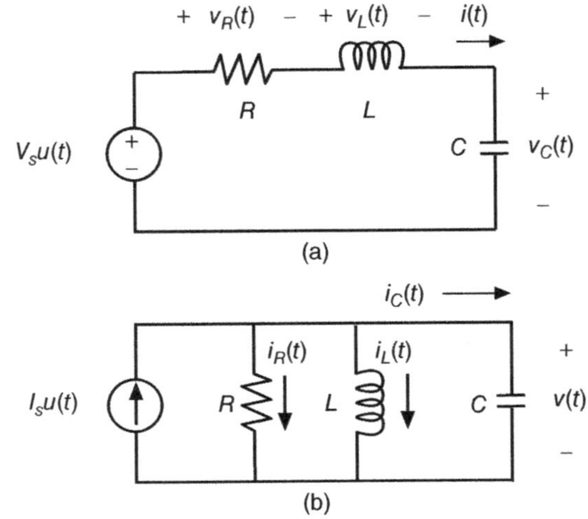

Figure 6.6 Second-order RLC circuits. (a) Series circuit with step voltage source $V_s u(t)$. (b) Parallel circuit with step current source $I_s u(t)$.

where $\delta(t)$ is the generalized derivative of $u(t)$. As in the case of the first-order ODE, we ignore the Dirac delta function on the right-hand side, yielding the homogeneous equation:

$$\frac{d^2}{dt^2}i(t) + (R/L)\frac{d}{dt}i(t) + (1/LC)i(t) = 0, \tag{6.84}$$

and assume a nonzero initial current $i(0)$. Similarly, the currents in the parallel circuit must sum to 0, which leads to an integro-differential equation in terms of the voltage $v(t)$:

$$v(t)/R + (1/L)\int_0^t v(t)dt + i_L(0) + C\frac{d}{dt}v(t) = I_s u(t), \tag{6.85}$$

where $i_L(0)$ is the initial inductor current. Differentiating this equation also yields a homogeneous ODE:

$$\frac{d^2}{dt^2}v(t) + (1/RC)\frac{d}{dt}v(t) + (1/LC)v(t) = 0, \tag{6.86}$$

where again the delta function has been ignored and we assume a nonzero initial voltage $v(0)$. Both of these circuit results are summarized in Table 6.2.

Consider next the voltage across the resistor in the series RLC circuit given by $v_R(t) = Ri(t)$. Replacing $i(t)$ with $v_R(t)/R$ in (6.84) yields an ODE for the resistor voltage:

$$\frac{d^2}{dt^2}v_R(t) + (R/L)\frac{d}{dt}v_R(t) + (1/LC)v_R(t) = 0. \tag{6.87}$$

For the inductor voltage, $i(t) = (1/L)\int_0^t v_L(t)dt + v_L(0)$ is substituted into (6.84):

$$(1/L)\frac{d}{dt}v_L(t) + (R/L^2)v_L(t) + (1/L^2C)\int_0^t v_L(t)dt + (1/LC)v_L(0) = 0. \tag{6.88}$$

TABLE 6.2 Second-Order RLC Circuits

System	Linear ODE Signals and Coefficients
General ODE	$d^2y(t)/dt^2 + a_1 dy(t)/dt + a_0 y(t) = x(t)$
Series current	$y(t) = i(t), x(t) = 0$
Series resistor voltage	$y(t) = v_R(t), x(t) = 0$
Series inductor voltage	$y(t) = v_L(t), x(t) = 0$
Series capacitor voltage	$y(t) = v_C(t), x(t) = (1/LC)V_s u(t)$
Coefficients	$a_0 = 1/LC, a_1 = R/L$
Parallel voltage	$y(t) = v(t), x(t) = 0$
Parallel resistor current	$y(t) = i_R(t), x(t) = 0$
Parallel inductor current	$y(t) = i_L(t), x(t) = (1/LC)I_s u(t)$
Parallel capacitor current	$y(t) = i_C(t), x(t) = 0$
Coefficients	$a_0 = 1/LC, a_1 = 1/RC$

Differentiating this expression gives another homogeneous ODE:

$$\frac{d^2}{dt^2}v_L(t) + (R/L)\frac{d}{dt}v_L(t) + (1/LC)v_L(t) = 0, \tag{6.89}$$

which has the same form as (6.87). For the capacitor voltage, we use the fact that $v_C(t) + v_L(t) + v_R(t) = V_s u(t)$, so that (6.87) and (6.89) are added as follows:

$$\frac{d^2}{dt^2}[v_L(t) + v_R(t)] + (R/L)\frac{d}{dt}[v_L(t) + v_R(t)] + (1/LC)[v_L(t) + v_R(t)] = 0. \tag{6.90}$$

Substituting $v_L(t) + v_R(t) = V_s u(t) - v_C(t)$ yields

$$\frac{d^2}{dt^2}[V_s u(t) - v_C(t)] + (R/L)\frac{d}{dt}[V_s u(t) - v_C(t)] + (1/LC)[V_s u(t) - v_C(t)] = 0. \tag{6.91}$$

The voltage $V_s u(t)$ vanishes in the first two terms after differentiating, again by ignoring the resulting delta function and assuming nonzero initial conditions. Thus, the second-order ODE for the capacitor voltage is nonhomogeneous:

$$\frac{d^2}{dt^2}v_C(t) + (R/L)\frac{d}{dt}v_C(t) + (1/LC)v_C(t) = (1/LC)V_s u(t). \tag{6.92}$$

The ODEs for the remaining quantities of the parallel RLC circuit, one of which is nonhomogeneous, are also summarized in Table 6.2 (see Problem 6.11). An integrator implementation for these second-order circuits is provided in Figure 6.7, which symbolically shows that two initial conditions for $y(t)$ are needed. This system is the same for the series and parallel circuits, only the coefficients and the input and output are different.

6.5.1 Homogeneous Solution

Let $x(t) = 0$ in (6.81) and, as was done for the first-order ODE, assume the solution of the homogeneous second-order ODE has the exponential form $y(t) = c\exp(st)$. Substituting this expression into (6.81) gives

$$s^2 c\exp(st) + sa_1 c\exp(st) + a_0 c\exp(st) = 0, \quad t \geq 0. \tag{6.93}$$

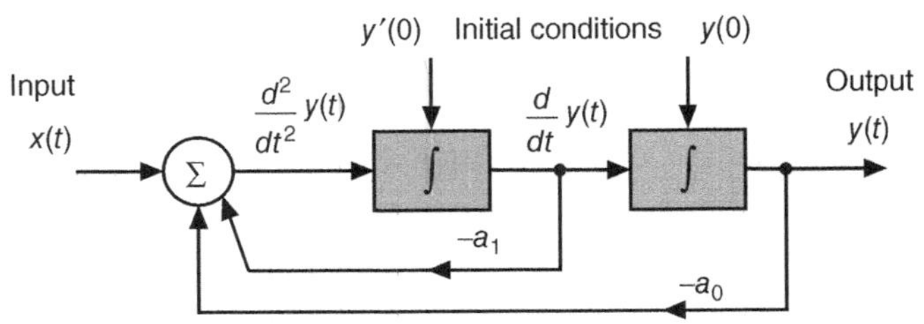

Figure 6.7 Integrator implementation of a second-order ODE.

The result after canceling terms is the second-order *characteristic equation*:

$$s^2 + a_1 s + a_0 = 0, \tag{6.94}$$

which describes the dynamics of the system independently of the specific input and initial conditions of the output. The quadratic formula gives two solutions:

$$s_1, s_2 = -a_1/2 \pm \sqrt{(a_1/2)^2 - a_0}. \tag{6.95}$$

Typically, the following quantities are defined when the second-order ODE is derived for a linear circuit:

$$\alpha \triangleq a_1/2, \quad \omega_o^2 \triangleq a_0, \tag{6.96}$$

where α is *Neper frequency* and ω_o is the *resonant frequency*, both of which have units rad/s (resonance is discussed later in Chapter 8 for a series RLC circuit). There are three possible forms of the solution depending on the *discriminant*, which is the expression under the square root in (6.95). The following names have been given to these system responses:

- *Overdamped*: $a_0 < (a_1/2)^2 \implies \alpha^2 > \omega_o^2 \implies$ real and distinct $\{s_1, s_2\}$:

$$s_1, s_2 = -\alpha \pm \sqrt{\alpha^2 - \omega_o^2}. \tag{6.97}$$

- *Underdamped*: $a_0 > (a_1/2)^2 \implies \alpha^2 < \omega_o^2 \implies$ complex conjugate $\{s_1, s_2\}$:

$$s_1, s_2 = -\alpha \pm j\sqrt{\omega_o^2 - \alpha^2} = -\alpha \pm j\omega_d, \tag{6.98}$$

  where we have defined the *damped angular frequency* $\omega_d \triangleq \sqrt{\omega_o^2 - \alpha^2}$ for nonzero α.

- *Critically damped*: $a_0 = (a_1/2)^2 \implies \alpha^2 = \omega_o^2 \implies$ real and repeated $\{s_1, s_2\}$:

$$s_1 = s_2 = -\alpha. \tag{6.99}$$

Since both roots in the overdamped case satisfy (6.81) when $x(t) = 0$, the most general form of the solution is a linear combination of the two exponentials:

$$y(t) = [c_1 \exp(s_1 t) + c_2 \exp(s_2 t)]u(t). \tag{6.100}$$

The components of this expression are independent of each other, which means that one term cannot be derived from the other by scaling it with a constant. There are two *modes* for the exponentials given by $\{s_1, s_2\}$, and they converge to 0 because the input is 0 and we assume a stable system such that $\text{Re}(s_1) < 0$ and $\text{Re}(s_2) < 0$.

For the underdamped case, (6.98) yields

$$y(t) = \exp(-\alpha t)[d_1 \exp(j\omega_d t) + d_2 \exp(-j\omega_d t)]u(t), \tag{6.101}$$

where the common exponential term has been factored, and $\{d_1, d_2\}$ have been used for the coefficients; these are intermediate quantities needed before defining the final set of coefficients. Substituting Euler's formula for each complex exponential and rearranging the equation give:

$$\begin{aligned} y(t) &= \exp(-\alpha t)[(d_1 + d_2)\cos(\omega_d t) + j(d_1 - d_2)\sin(\omega_d t)]u(t) \\ &= \exp(-\alpha t)[c_1 \cos(\omega_d t) + c_2 \sin(\omega_d t)]u(t), \end{aligned} \tag{6.102}$$

where $c_1 \triangleq d_1 + d_2$ and $c_2 \triangleq j(d_1 - d_2)$. Since d_1 and d_2 must be a complex conjugate pair in order for $y(t)$ to be real, $d_1 - d_2$ is imaginary such that $\{c_1, c_2\}$ are real-valued. The homogeneous solution for complex roots is an exponentially weighted sum of sine and cosine functions; the exponential function forms an *envelope* about the sinusoids as demonstrated later. It is possible to rewrite (6.102) as a single cosine term as follows:

$$y(t) = r\exp(-\alpha t)\cos(\omega_d t - \phi)u(t), \tag{6.103}$$

where $\phi \triangleq \tan^{-1}(c_2/c_1)$ is a phase shift and $r \triangleq \sqrt{c_1^2 + c_2^2}$ is the magnitude. We refer to this solution as the *polar form* due to its similarity to the polar form used for complex numbers in Chapter 4. It is straightforward to verify that (6.103) is the same as (6.102) by using the following trigonometric identities (see Problem 6.12):

$$\cos(x - y) = \cos(x)\cos(y) + \sin(x)\sin(y) \tag{6.104}$$

$$\sin(\tan^{-1}(x)) = \frac{x}{\sqrt{1+x^2}}, \quad \cos(\tan^{-1}(x)) = \frac{1}{\sqrt{1+x^2}}. \tag{6.105}$$

An example is provided in Figure 6.8(a) with $\alpha = 0$ such that $\exp(-\alpha t) = 1$ (the envelopes of the sinusoids are constant). The figure shows the individual components of (6.102) and their sum given by the polar form in (6.103). Figure 6.8(b) has the same $y(t)$ except with $\alpha = 0.06$ rad/s so that it is exponentially weighted. The envelope of $r\exp(-\alpha t)\cos(\omega_d t - \phi)$ is the weighted exponential $r\exp(-\alpha t)$, which we see is an upper bound for the function. The negative function $-r\exp(-\alpha t)$ is also included in envelope plots to show the lower bound.

For the critically damped case, the two roots of the characteristic equation are identical, so it is not possible to use the sum of exponential terms as in the previous cases because they would not be independent. It is easy to verify that $y(t) = c_1 \exp(st)$ with $s = s_1 = s_2 = -\alpha$ is one solution of the homogeneous ODE. The other solution is obtained using a technique where each $y(t)$ in the homogeneous ODE is multiplied by $f(t)$. The goal is to find $f(t)$ such that the product $f(t)y(t) = f(t)c_1 \exp(st)$ is the other solution of the ODE, and by construction, it is independent of $c_1 \exp(st)$.

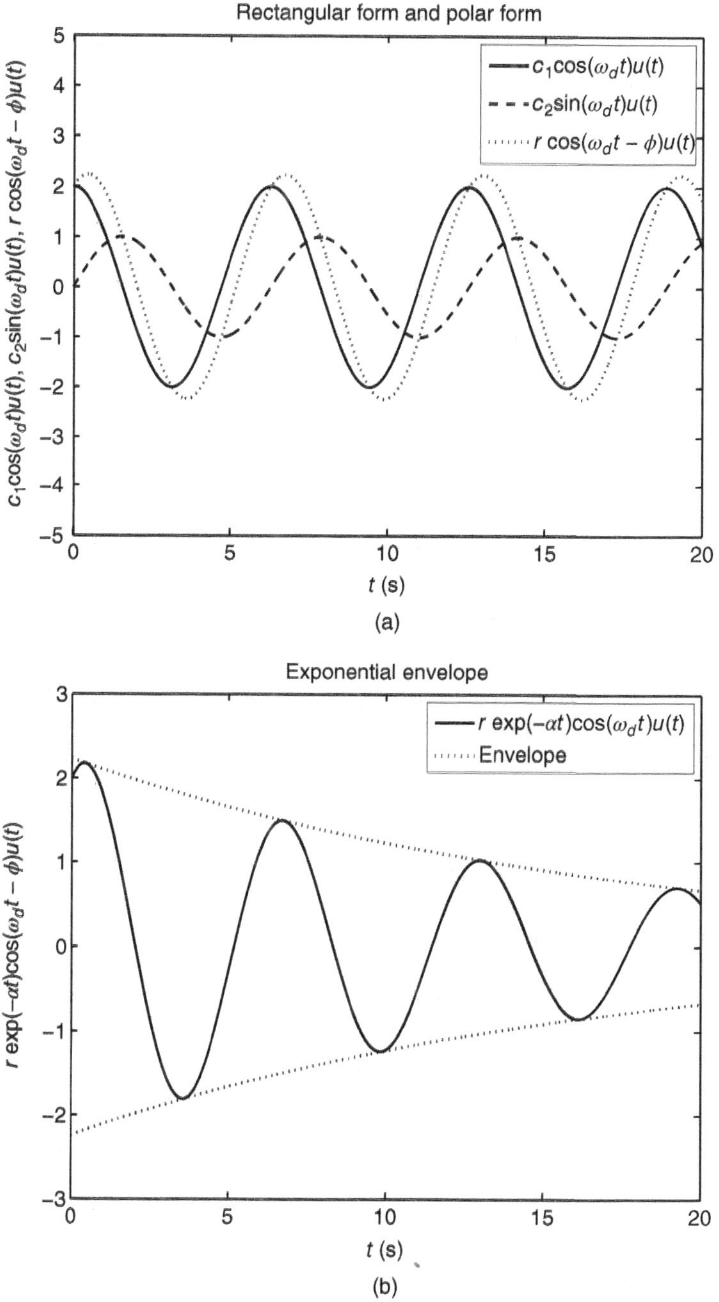

Figure 6.8 The cosine form in (6.103) with $\omega_d = 1$ rad/s, $c_1 = 2$, and $c_2 = 1$ such that $r \approx 2.2361$ and $\phi \approx 0.4636$ rad. (a) $\alpha = 0$ and components of $y(t)$ from (6.102). (b) $\alpha = 0.06$ rad/s and the exponential envelope.

The product rule for the second derivative of $f(t)y(t)$ is

$$\frac{d^2}{dt^2}f(t)y(t) = \frac{d}{dt}\left[f(t)\frac{d}{dt}y(t) + y(t)\frac{d}{dt}f(t)\right]$$
$$= 2\frac{d}{dt}f(t)\frac{d}{dt}y(t) + f(t)\frac{d^2}{dt^2}y(t) + y(t)\frac{d^2}{dt^2}f(t). \tag{6.106}$$

Substituting this expression and the first derivative of $f(t)y(t)$ into (6.81) (with $x(t) = 0$) yields

$$2\frac{d}{dt}f(t)\frac{d}{dt}y(t) + f(t)\frac{d^2}{dt^2}y(t) + y(t)\frac{d^2}{dt^2}f(t)$$
$$+ a_1 f(t)\frac{d}{dt}y(t) + a_1 y(t)\frac{d}{dt}f(t) + a_0 f(t)y(t) = 0. \tag{6.107}$$

Collecting terms together according to the order of the derivative of $f(t)$, we have

$$y(t)\frac{d^2}{dt^2}f(t) + \left[2\frac{d}{dt}y(t) + a_1 y(t)\right]\frac{d}{dt}f(t)$$
$$+ \left[\frac{d^2}{dt^2}y(t) + a_1\frac{d}{dt}y(t) + a_0 y(t)\right]f(t) = 0. \tag{6.108}$$

This ODE for $f(t)$ has the same form as the original ODE for $y(t)$, except that its "coefficients" are functions of time. The expression in the last set of brackets is 0 because it is the original homogeneous ODE and $y(t)$ is a solution. The derivative of the first solution $y(t) = c_1 \exp(-\alpha t)$ is

$$\frac{d}{dt}y(t) = -\alpha c_1 \exp(-\alpha t) = -\alpha y(t) = -(a_1/2)y(t), \tag{6.109}$$

such that the expression in the first set of brackets in (6.108) is also 0. Thus, (6.108) simplifies considerably to

$$\frac{d^2}{dt^2}f(t) = 0, \tag{6.110}$$

where the leading $y(t)$ has canceled. The solution of this equation is $f(t) = t$, so that the second solution of the ODE is the ramped exponential $y(t) = c_2 t \exp(-\alpha t)$. Combining the two results gives the overall solution for the critically damped case:

$$y(t) = (c_1 + c_2 t)\exp(-\alpha t)u(t). \tag{6.111}$$

It is clear that these two components are not linear combinations of each other because t multiplies $c_2 \exp(-\alpha t)$. All three solutions for the homogeneous second-order ODE are summarized in Table 6.3.

Since we are interested in *stable* systems, $\alpha = a_1/2 > 0$ for all three cases so that the exponential functions in each solution decrease to 0. However, it is possible to

TABLE 6.3 Second-Order Homogeneous ODE Solutions

System	Linear ODE Signals and Parameters
Homogeneous ODE	$d^2y(t)/dt^2 + a_1 dy(t)/dt + a_0 y(t) = 0$
Solution	$y(t) = [c_1 y_1(t) + c_2 y_2(t)]u(t)$
Overdamped	$y_1(t) = \exp(s_1 t), y_2(t) = \exp(s_2 t)$
Underdamped	$y_1(t) = \exp(-\alpha t)\cos(\omega_d t), y_2(t) = \exp(-\alpha t)\sin(\omega_d t)$
Critically damped	$y_1(t) = \exp(-\alpha t), y_2(t) = t\exp(-\alpha t)$
Parameters	$s_{1,2} = -\alpha \pm \sqrt{\alpha^2 - \omega_o^2}, \alpha \triangleq a_1/2, \omega_o^2 \triangleq a_0, \omega_d \triangleq \sqrt{\omega_o^2 - \alpha^2}$
Overdamped	$c_1 = [s_2 y(0) - y'(0)]/(s_2 - s_1), c_2 = [y'(0) - s_1 y(0)](s_2 - s_1)$
Underdamped	$c_1 = y(0), c_2 = [y'(0) + \alpha y(0)]/\omega_d$
Critically damped	$c_1 = y(0), c_2 = y'(0) + \alpha y(0)$

have $\alpha = 0$ for the underdamped solution where the sine and cosine terms maintain a constant envelope as in Figure 6.8(a). In this case, $\omega_d = \omega_o$ and the system is called *undamped*. For the overdamped case, we also require $a_0 \geq 0$ for a bounded solution. If $a_0 < 0$, then the square root in (6.95) exceeds $a_1/2$ and it is possible for one or both roots to be positive, resulting in exponentials that increase unbounded. For linear circuits, this restriction is enforced because R, L, and C are all positive, leading to the positive square root of ω_o^2.

A critically damped response occurs for any set of coefficients along the solid curve $a_0 = a_1^2/4$ shown in Figure 6.9. The roots are complex above the curve and real below the curve. All three solutions have the same conditions on $\{a_0, a_1\}$ for boundedness, as indicated by the upper right quadrant formed by the dotted lines in the figure. The shaded region corresponds to bounded solutions for the overdamped case.

Example 6.7 Examples of the three types of solutions $y(t)$ and the components $\{y_1(t), y_2(t)\}$ from Table 6.3 are shown in Figure 6.10. The decay rate of the overdamped solution is dominated by the term with the negative root $s_1 = -0.1$. The underdamped solution is similar to the result in Figure 6.8(a) (the dotted line) except that it has exponential weighting with $\alpha = 0.1$. The shape of the critically damped curve closely follows that of the term with multiplier t (the dashed line), but of course both terms decay to 0 because $\exp(-t)u(t) \to 0$ faster than the ramp $t \to \infty$. In this example, the same coefficients $\{c_1 = 2, c_2 = 1\}$ were used for each solution.

Example 6.8 Consider a special case of (6.81) with $x(t) = 0$ (homogeneous) and $a_1 = 0$:

$$\frac{d^2}{dt^2}y(t) + a_0 y(t) = 0, \tag{6.112}$$

which has the characteristic equation

$$s^2 + a_0 = 0. \tag{6.113}$$

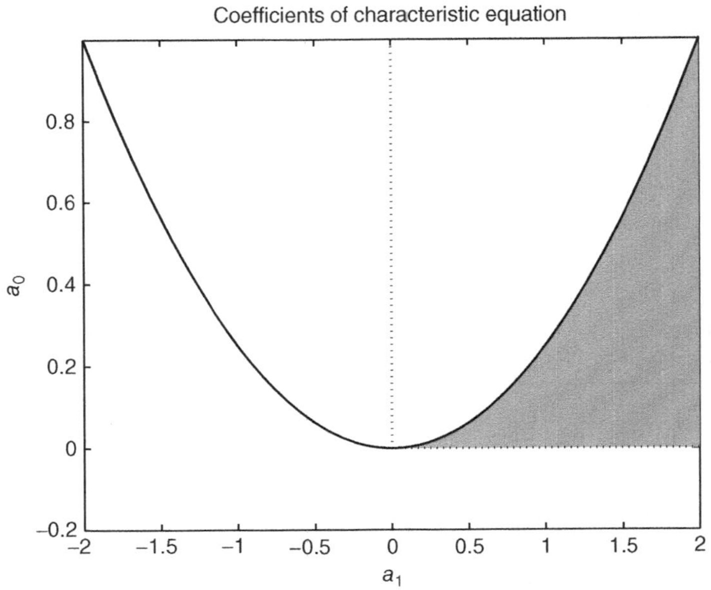

Figure 6.9 Plot of $a_0 = a_1^2/4$ where the discriminant is 0 (shown only for $a_1 \in [-2, 2]$). The roots are real for $\{a_0, a_1\}$ below the curve and complex for $\{a_0, a_1\}$ above the curve. For a bounded solution, all three cases require $a_0 \geq 0$ and $a_1 > 0$ (the upper right quadrant formed by the dotted lines). The bounded overdamped solution is located within the shaded region.

If $a_0 > 0$ (which would be the case for an RLC circuit because $a_0 = 1/LC$), then the roots form a complex conjugate pair $\{s_1, s_2\} = \pm j\sqrt{a_0}$, and the solution $y(t)$ is undamped. This result also follows from Figure 6.9, corresponding to a_0 along the vertical dotted line. Moreover, since the roots are strictly imaginary, the solution is

$$y(t) = [c_1 \cos(\omega_o t) + c_2 \sin(\omega_o t)]u(t), \tag{6.114}$$

which does not decay to 0, similar to the results in Figure 6.8(a). The frequency is $\omega_d = \omega_o = 1/\sqrt{LC}$, and the coefficients $\{c_1, c_2\}$ depend on the initial conditions (as they do for all three types of second-order solutions). From this result, we find that the middle term $a_1 dy(t)/dt$ in the second-order ODE is needed for the solution to decay to 0. This is evident from Table 6.2 for the series and parallel RLC circuits where $a_1 = R/L$ and $a_1 = 1/RC$, respectively. The resistor in each case dissipates the initial circuit energy stored in C or L. Without a resistor, the voltages and currents oscillate sinusoidally without any damping as $t \to \infty$.

Example 6.9 Another special case occurs when $a_0 = 0$ such that

$$\frac{d^2}{dt^2}y(t) + a_1 \frac{d}{dt}y(t) = 0, \tag{6.115}$$

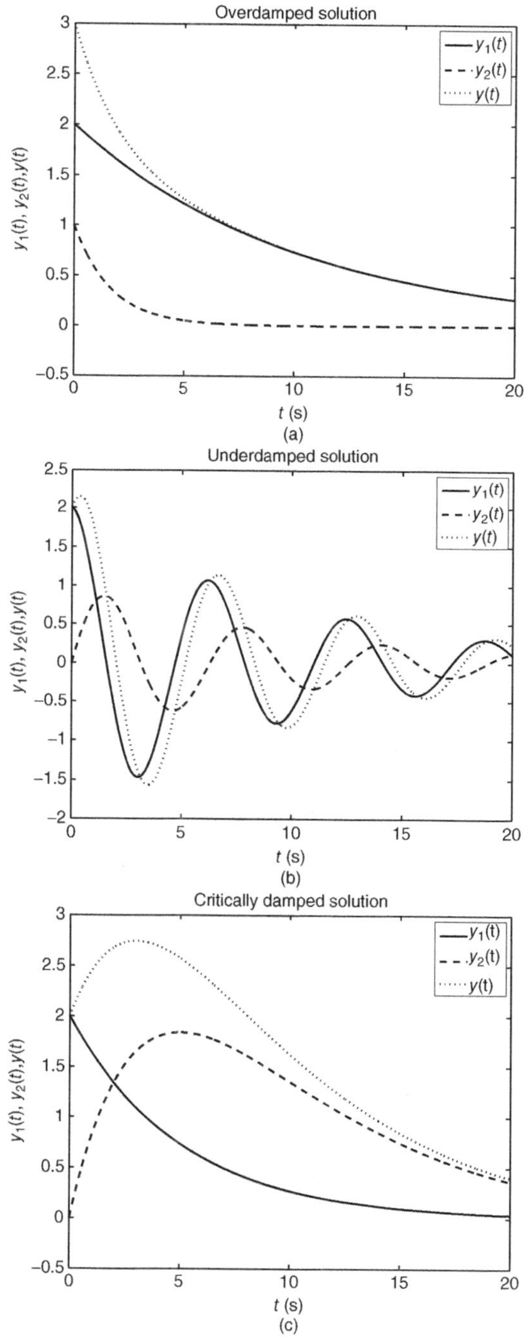

Figure 6.10 Examples of homogeneous solutions with coefficients $c_1 = 2$ and $c_2 = 1$. (a) Overdamped ($s_1 = -0.1$ and $s_2 = -0.6$). (b) Underdamped ($\alpha = 0.1$ rad/s and $\omega_d = 1$ rad/s). (c) Critically damped ($\alpha = 0.2$ rad/s).

which has the characteristic equation

$$s^2 + a_1 s = 0. \tag{6.116}$$

The two roots are $s_1 = 0$ and $s_2 = -a_1 = -2\alpha$, and since they are real and distinct, the overdamped expression in (6.100) is used:

$$y(t) = [c_1 + c_2 \exp(-2\alpha t)]u(t). \tag{6.117}$$

This result has an exponentially decaying component and a fixed component that depends on c_1, which in turn is derived from the initial conditions $y(0)$ and $y'(0)$. Note, however, that this situation would not apply in a practical sense to the second-order RLC circuits in Figure 6.6 because $a_0 = 1/LC = 0$ and nonzero a_1 means $C \to \infty$ and $L \to \infty$ for the series and parallel circuits, respectively (see Table 6.2).

6.5.2 Damping Ratio

In most engineering courses on linear circuits, the second-order ODE in (6.81) is often written as

$$\frac{d^2}{dt^2}y(t) + 2\zeta\omega_o\frac{d}{dt}y(t) + \omega_o^2 y(t) = x(t), \tag{6.118}$$

where ζ is the *damping ratio* and ω_o is the resonant frequency previously given in (6.96). The characteristic equation using this notation is

$$s^2 + 2\zeta\omega_o s + \omega_o^2 = 0, \tag{6.119}$$

which has roots

$$s_1, s_2 = -\zeta\omega_o \pm \omega_o\sqrt{\zeta^2 - 1}. \tag{6.120}$$

The advantage of this notation is that the three types of solutions for the second-order homogeneous ODE are readily determined by the value of ζ.

- *Overdamped* $\zeta > 1$:

$$s_1, s_2 = -\zeta\omega_o \pm \omega_o\sqrt{\zeta^2 - 1}. \tag{6.121}$$

- *Underdamped* $\zeta < 1$:

$$s_1, s_2 = -\zeta\omega_o \pm j\omega_o\sqrt{1 - \zeta^2} = -\zeta\omega_o \pm j\omega_d. \tag{6.122}$$

- *Critically damped* $\zeta = 1$:

$$s_1 = s_2 = -\zeta\omega_o. \tag{6.123}$$

Since $\alpha = \zeta\omega_o$ where $-\alpha$ is the exponent of the exponential in (6.96), ζ is a dimensionless ratio:

$$\zeta = \frac{\alpha}{\omega_o}. \tag{6.124}$$

For fixed ω_o, the damping ratio determines the exponential decay rate for each of the three types of solutions. It is particularly useful for the underdamped case where it indicates the degree to which the sine and cosine terms decrease. For small ζ (close to 0), the solution is highly oscillatory and takes longer to decay than when ζ is close to 1. When $\zeta = 0$, the sinusoids do not decay; this is the undamped solution where the roots are strictly imaginary as discussed in Example 6.8.

Using this notation, we have the following expressions for the three types of homogeneous solutions.

- *Overdamped* $\zeta > 1$:

$$y(t) = [c_1 \exp(\sqrt{\zeta^2 - 1}\omega_o t) + c_2 \exp(-\sqrt{\zeta^2 - 1}\omega_o t)] \exp(-\zeta\omega_o t)u(t). \tag{6.125}$$

- *Underdamped* $\zeta < 1$:

$$y(t) = [c_1 \cos(\omega_d t) + c_2 \sin(\omega_d t)] \exp(-\zeta\omega_o t)u(t). \tag{6.126}$$

- *Critically damped* $\zeta = 1$:

$$y(t) = [c_1 + c_2 t] \exp(-\zeta\omega_o t)u(t). \tag{6.127}$$

These formulations are interesting because they show that all three solutions have a common exponentially decaying term. They differ by the expressions in the brackets: exponential functions for overdamped, sinusoidal functions for underdamped, and step and ramp functions for critically damped.

Example 6.10 Figure 6.11 shows examples of the three types of solutions for a second-order ODE with different values for the damping ratio ζ. For all three cases, $c_1 = c_2 = 1$ and $\omega_o = 0.3$ rad/s. Using the values of ζ in the figure, the two real roots for the overdamped case are $s_1 \approx -0.7854$ and $s_2 \approx -0.1146$. For the critically damped case, $\alpha = 0.3$ rad/s, and for the underdamped case, $\alpha = 0.15$ rad/s and $\omega_d \approx 0.2598$ rad/s. These plots are typical waveforms for the three types of solutions. Overdamped $y(t)$ is the sum of two decaying exponentials, and so it decreases to 0 with two modes (time constants). This is evident by the dashed line where we see relatively rapid decay initially, which is due to the root -0.7854, and then the rest of the curve is dominated by the root -0.1146. The transition between the two modes occurs approximately around $t = 3$ s, and we can see that the curve in that region has a bend, which is not due to a single exponential. Underdamped $y(t)$ (the solid curve) has an oscillatory behavior that is damped down by the exponential weighting. Although $y(t)$ is the sum of sine and cosine, recall that it can be written as a single cosine with

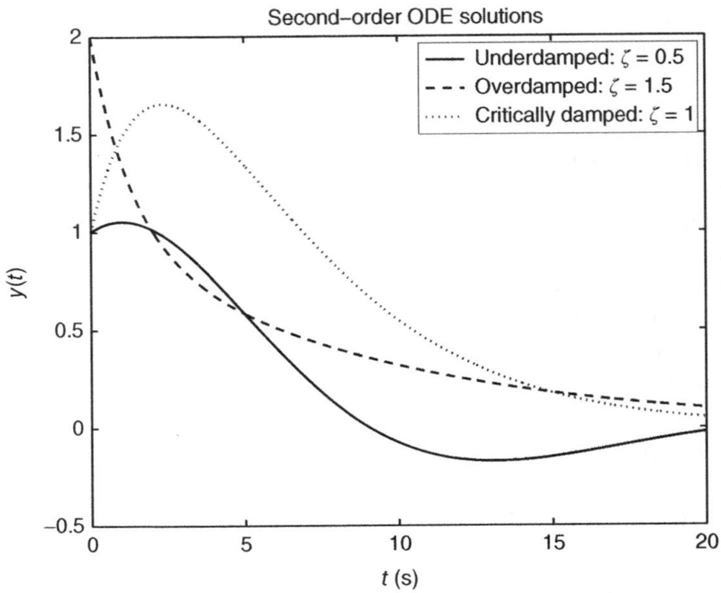

Figure 6.11 Second-order ODE solutions for Example 6.10.

amplitude $\sqrt{c_1^2 + c_2^2} = \sqrt{2}$ and phase shift $\tan^{-1}(c_2/c_1) = 45°$. The curve does not actually reach $\sqrt{2}$ just past $t = 0$ because of the multiplicative exponential function. Critically damped $y(t)$ initially increases because of the ramp t, but eventually the exponential function dominates the solution and brings the output to 0 (the dotted curve).

6.5.3 Initial Conditions

It is straightforward to verify that the two initial conditions for each of the three ODE solutions are as follows with $y'(0) \triangleq dy(t)/dt|_{t=0}$.

- *Overdamped:*

$$y(0) = c_1 + c_2, \quad y'(0) = c_1 s_1 + c_2 s_2. \tag{6.128}$$

- *Underdamped:*

$$y(0) = c_1, \quad y'(0) = \omega_d c_2 - \alpha c_1. \tag{6.129}$$

- *Critically damped:*

$$y(0) = c_1, \quad y'(0) = c_2 - \alpha c_1. \tag{6.130}$$

The quantities on the left-hand side of each pair of equations would be given in a problem statement or they can be determined for a particular circuit or system, from which we solve for $\{c_1, c_2\}$ because there are two equations and two unknowns. It is interesting that the solution for the overdamped case requires solving a second-order system of equations, whereas c_1 is found directly for the other two cases, from which c_2 is also easily found. The coefficients for the overdamped case are derived by inverting a matrix:

$$\begin{bmatrix} 1 & 1 \\ s_1 & s_2 \end{bmatrix} \begin{bmatrix} c_1 \\ c_2 \end{bmatrix} = \begin{bmatrix} y(0) \\ y'(0) \end{bmatrix} \Rightarrow \begin{bmatrix} c_1 \\ c_2 \end{bmatrix} = \frac{1}{s_2 - s_1} \begin{bmatrix} s_2 y(0) - y'(0) \\ y'(0) - s_1 y(0) \end{bmatrix}. \tag{6.131}$$

The equations for the coefficients are also summarized in Table 6.3.

In the next chapter on the Laplace transform, it will be necessary to distinguish between $t = 0^-$ ("just before" $t = 0$) and $t = 0^+$ ("just after" $t = 0$). The function values at these two time instants are usually called initial conditions, though there is actually a difference for some functions. For example, the unit step function has $u(0^-) = 0$ and $u(0^+) = 1$. In order to avoid confusion, we will refer to quantities such as $x(0^-)$ as an *initial state* and $x(0^+)$ as an *initial condition* (or *initial value*). Thus, the equations in (6.128)–(6.130) are technically based on the initial conditions at $t = 0^+$.

6.5.4 Nonhomogeneous Solution

For the nonhomogeneous ODE in (6.81), we start with the general form of the solution for the homogeneous ODE:

$$y(t) = [c_1 y_1(t) + c_2 y_2(t)]u(t), \tag{6.132}$$

where $\{y_1(t), y_2(t)\}$ correspond to one of the three types of solutions in Table 6.3 based on the characteristic equation. The constants $\{c_1, c_2\}$ are replaced with *functions* $\{g_1(t), g_2(t)\}$ so that we can use a technique called *variation of parameters*:

$$y(t) = g_1(t) y_1(t) + g_2(t) y_2(t). \tag{6.133}$$

From the product rule, the derivative yields four terms:

$$\frac{d}{dt} y(t) = g_1(t) \frac{d}{dt} y_1(t) + y_1(t) \frac{d}{dt} g_1(t) + g_2(t) \frac{d}{dt} y_2(t) + y_2(t) \frac{d}{dt} g_2(t). \tag{6.134}$$

In order to solve for $g_1(t)$ and $g_2(t)$, the following condition allows us to cancel terms (similar to that done for the integrating factor of the nonhomogeneous first-order ODE):

$$y_1(t) \frac{d}{dt} g_1(t) + y_2(t) \frac{d}{dt} g_2(t) = 0, \tag{6.135}$$

which simplifies (6.134) to

$$\frac{d}{dt} y(t) = g_1(t) \frac{d}{dt} y_1(t) + g_2(t) \frac{d}{dt} y_2(t). \tag{6.136}$$

Differentiating this result yields

$$\frac{d^2}{dt^2}y(t) = g_1(t)\frac{d^2}{dt^2}y_1(t) + \frac{d}{dt}g_1(t)\frac{d}{dt}y_1(t) + g_2(t)\frac{d^2}{dt^2}y_2(t) + \frac{d}{dt}g_2(t)\frac{d}{dt}y_2(t). \quad (6.137)$$

The expressions in (6.133), (6.136), and (6.137) are substituted into the second-order nonhomogeneous ODE in (6.81), which we rearrange by collecting terms that multiply $\{g_1(t), g_2(t)\}$ and their derivatives:

$$\left[a_0y_1(t) + a_1\frac{d}{dt}y_1(t) + \frac{d^2}{dt^2}y_1(t)\right] g_1(t) + \left[a_0y_2(t) + a_1\frac{d}{dt}y_2(t) + \frac{d^2}{dt^2}y_2(t)\right] g_2(t)$$
$$+ \left[\frac{d}{dt}y_1(t)\right]\frac{d}{dt}g_1(t) + \left[\frac{d}{dt}y_2(t)\right]\frac{d}{dt}g_2(t) = x(t). \quad (6.138)$$

The first two terms are 0 because $\{y_1(t), y_2(t)\}$ are assumed to be solutions for the homogeneous ODE which appears in both brackets, so that (6.138) reduces to

$$\left[\frac{d}{dt}y_1(t)\right]\frac{d}{dt}g_1(t) + \left[\frac{d}{dt}y_2(t)\right]\frac{d}{dt}g_2(t) = x(t). \quad (6.139)$$

This result along with (6.135) are used to find $\{g_1(t), g_2(t)\}$ for a particular $x(t)$, which when substituted into (6.133) give the nonhomogeneous solution. These two equations can be written in matrix form as follows:

$$\begin{bmatrix} y_1(t) & y_2(t) \\ dy_1(t)/dt & dy_2(t)/dt \end{bmatrix} \begin{bmatrix} dg_1(t)/dt \\ dg_2(t)/dt \end{bmatrix} = \begin{bmatrix} 0 \\ x(t) \end{bmatrix}. \quad (6.140)$$

The inverse of the matrix is

$$\begin{bmatrix} y_1(t) & y_2(t) \\ dy_1(t)/dt & dy_2(t)/dt \end{bmatrix}^{-1} = \frac{1}{W(t)} \begin{bmatrix} dy_2(t)/dt & -y_2(t) \\ -dy_1(t)/dt & y_1(t) \end{bmatrix}, \quad (6.141)$$

and the solution of (6.140) is

$$\begin{bmatrix} dg_1(t)/dt \\ dg_2(t)/dt \end{bmatrix} = \frac{1}{W(t)} \begin{bmatrix} -x(t)y_2(t) \\ x(t)y_1(t) \end{bmatrix}, \quad (6.142)$$

where we have defined the determinant

$$W(t) \triangleq y_1(t)\frac{d}{dt}y_2(t) - y_2(t)\frac{d}{dt}y_1(t). \quad (6.143)$$

For this ODE problem, $W(t)$ is called the *Wronskian* of $\{y_1(t), y_2(t)\}$.

Definition: Wronskian The *Wronskian* of N differentiable functions $\{f_n(t)\}$ is the following determinant:

$$W(t) = \det \begin{bmatrix} f_1(t) & \cdots & f_N(t) \\ f_1'(t) & \cdots & f_N'(t) \\ \vdots & \cdots & \vdots \\ f_1^{(N-1)}(t) & \cdots & f_N^{(N-1)}(t) \end{bmatrix}, \tag{6.144}$$

where the matrix contains ordinary derivatives of each function with respect to the independent variable t.

Integrating the elements of (6.142) yields

$$g_1(t) = -\int_0^t x(t)[y_2(t)/W(t)]dt + g_1(0), \tag{6.145}$$

$$g_2(t) = \int_0^t x(t)[y_1(t)/W(t)]dt + g_2(0), \tag{6.146}$$

and substituting these into (6.133) gives the general form of the solution for the second-order nonhomogeneous ODE:

$$y(t) = y_1(t)\left[-\int_0^t x(t)[y_2(t)/W(t)]dt + g_1(0)\right]u(t)$$
$$+ y_2(t)\left[\int_0^t x(t)[y_1(t)/W(t)]dt + g_2(0)\right]u(t). \tag{6.147}$$

It is important to note that $W(t)$ is part of both integrands and must be included when performing the integrations for a specific input $x(t)$; in general, they cannot be factored out. The expression in (6.147) can be rearranged into the sum of the homogeneous solution and the particular solution (given by two integrals with input $x(t)$):

$$y(t) = [c_1y_1(t) + c_2y_2(t)]u(t) - y_1(t)\left[\int_0^t x(t)[y_2(t)/W(t)]dt\right]u(t)$$
$$+ y_2(t)\left[\int_0^t x(t)[y_1(t)/W(t)]dt\right]u(t), \tag{6.148}$$

where the constants $\{g_1(0), g_2(0)\}$ have been replaced with $\{c_1, c_2\}$, which follow from (6.132) and (6.133) for the homogeneous solution.

Next, for each of the three types of solutions for a second-order ODE, we derive expressions for $W(t)$. For the overdamped solution in (6.100):

$$W(t) = s_2 \exp(s_1t)\ \exp(s_2t) - s_1 \exp(s_2t)\exp(s_1t)$$
$$= (s_2 - s_1)\exp((s_1 + s_2)t). \tag{6.149}$$

For the underdamped case in (6.102):

$$\begin{aligned} W(t) &= \exp(-\alpha t)\cos(\omega_d t)[-\alpha\exp(-\alpha t)\sin(\omega_d t) + \omega_d\exp(-\alpha t)\cos(\omega_d t)] \\ &\quad - \exp(-\alpha t)\sin(\omega_d t)[-\alpha\exp(-\alpha t)\cos(\omega_d t) - \omega_d\exp(-\alpha t)\sin(\omega_d t)] \\ &= \exp(-2\alpha t)[-\alpha\cos(\omega_d t)\sin(\omega_d t) + \omega_d\cos^2(\omega_d t) \\ &\quad + \alpha\sin(\omega_d t)\cos(\omega_d t) + \omega_d\sin^2(\omega_d t)]. \end{aligned} \tag{6.150}$$

Since the $\cos(\omega_d t)\sin(\omega_d t)$ terms cancel, this equation simplifies to

$$W(t) = \omega_d\exp(-2\alpha t), \tag{6.151}$$

where $\sin^2(\omega_d t) + \cos^2(\omega_d t) = 1$ has been used. For the critically damped case in (6.111):

$$\begin{aligned} W(t) &= \exp(-\alpha t)[\exp(-\alpha t) - \alpha t\exp(-\alpha t)] + t\exp(-\alpha t)\alpha\exp(-\alpha t) \\ &= \exp(-2\alpha t)(1 - \alpha t + \alpha t) = \exp(-2\alpha t). \end{aligned} \tag{6.152}$$

The Wronskians for the three cases, which are all decaying exponentials, are summarized in Table 6.4 where we have also included expressions for the terms in small brackets multiplying $x(t)$ in the integrands of (6.148).

Substituting the Wronskian results into the general ODE solution in (6.148) for $y(t)$ yields the following complete solutions for the three second-order cases.

- *Overdamped*:

$$\begin{aligned} y(t) &= [c_1\exp(s_1 t) + c_2\exp(s_2 t)]u(t) \\ &\quad + \frac{1}{s_2 - s_1}\left[\int_0^t x(\tau)[\exp(s_2(t-\tau)) - \exp(s_1(t-\tau))]d\tau\right]u(t). \end{aligned} \tag{6.153}$$

TABLE 6.4 Wronskians for Second-Order Linear ODE

System	Wronskian and Integrand Terms
General form	$W(t) = y_1(t)dy_2(t)/dt - y_2(t)dy_1(t)/dt$
Overdamped	$W(t) = (s_2 - s_1)\exp((s_1 + s_2)t)$
	$y_1(t)/W(t) = \exp(-s_2 t)/(s_2 - s_1)$
	$y_2(t)/W(t) = \exp(-s_1 t)/(s_2 - s_1)$
Underdamped	$W(t) = \omega_d\exp(-2\alpha t)$
	$y_1(t)/W(t) = \exp(\alpha t)\cos(\omega_d t)/\omega_d$
	$y_2(t)/W(t) = \exp(\alpha t)\sin(\omega_d t)/\omega_d$
Critically damped	$W(t) = \exp(-2\alpha t)$
	$y_1(t)/W(t) = \exp(\alpha t)$
	$y_2(t)/W(t) = t\exp(\alpha t)$

- *Underdamped*:

$$y(t) = [c_1 \cos(\omega_d t) + c_2 \sin(\omega_d t)] \exp(-\alpha t)u(t)$$
$$+ \frac{1}{\omega_d}\left[\int_0^t x(\tau)\exp(-\alpha(t-\tau))\sin(\omega_d(t-\tau))d\tau\right]u(t). \quad (6.154)$$

- *Critically damped*:

$$y(t) = [c_1 + c_2 t] \exp(-\alpha t)u(t)$$
$$- \left[\int_0^t x(\tau)(t-\tau)\exp(-\alpha(t-\tau))d\tau\right]u(t). \quad (6.155)$$

The initial conditions in (6.128)–(6.130) derived for the three types of homogeneous solutions and summarized in Table 6.3 are also used for the coefficients $\{c_1, c_2\}$ in the previous expressions. The integrals in (6.153)–(6.155) are *convolutions* between the input $x(t)$ and the impulse response functions for the three cases and are discussed later in this chapter.

6.6 SECOND-ORDER ODE RESPONSES

In this section, we examine the responses of the three types of second-order systems to step and Dirac delta functions.

6.6.1 Step Response

When the input $x(t) = Ku(t)$ is a step function, (6.148) becomes

$$y(t) = [c_1 y_1(t) + c_2 y_2(t)]u(t) - Ky_1(t)\left[\int_0^t [y_2(t)/W(t)]dt\right]u(t)$$
$$+ Ky_2(t)\left[\int_0^t [y_1(t)/W(t)]dt\right]u(t), \quad (6.156)$$

where K has been factored from the integrals and the lower limit of integration allows us to drop $u(t)$ from the integrand. Using the results from the previous section, we summarize the complete solutions for each of the three cases as follows.

- *Overdamped:*

$$y(t) = [c_1 \exp(s_1 t) + c_2 \exp(s_2 t)]u(t) + \frac{K}{s_1(s_2 - s_1)}[1 - \exp(s_1 t)]u(t)$$
$$- \frac{K}{s_2(s_2 - s_1)}[1 - \exp(s_2 t)]u(t). \quad (6.157)$$

Combining the terms yields

$$y(t) = [c_1 - K/s_1(s_2 - s_1)] \exp(s_1 t)u(t) + [c_2 + K/s_2(s_2 - s_1)] \exp(s_2 t)u(t) + (K/s_1 s_2)u(t) \quad (6.158)$$

$$= [b_1 \exp(s_1 t) + b_2 \exp(s_2 t) + K/s_1 s_2]u(t), \quad (6.159)$$

where $K/s_1 s_2$ is the steady-state response, the two exponential terms comprise the transient response, and we have defined the constants $b_1 \triangleq c_1 - K/s_1(s_2 - s_1)$ and $b_2 \triangleq c_2 + K/s_2(s_2 - s_1)$. The denominator of the last term simplifies to

$$s_1 s_2 = \left(-\alpha + \sqrt{\alpha^2 - \omega_o^2}\right)\left(-\alpha - \sqrt{\alpha^2 - \omega_o^2}\right)$$

$$= \alpha^2 - \alpha^2 + \omega_o^2 = a_0, \quad (6.160)$$

yielding

$$y(t) = [b_1 \exp(s_1 t) + b_2 \exp(s_2 t) + K/a_0]u(t). \quad (6.161)$$

When substituting (6.158) and $x(t) = Ku(t)$ into the ODE of (6.81), the derivatives remove the constant term K/a_0 so that for a stable system, the third term on the right-hand side of (6.159) is $a_0(K/a_0) = K$ as $t \to \infty$, verifying that the steady-state solution is in fact K/a_0.

- *Underdamped:*

$$y(t) = [c_1 \cos(\omega_d t) + c_2 \sin(\omega_d t)] \exp(-\alpha t)u(t) - \frac{K/\omega_d}{\alpha^2 + \omega_d^2} \times [\alpha \sin(\omega_d t) + \omega_d \cos(\omega_d t)] \exp(-\alpha t)u(t) + \frac{K}{\alpha^2 + \omega_d^2} u(t). \quad (6.162)$$

Combining the terms, we have

$$y(t) = \left(c_1 - \frac{K}{\alpha^2 + \omega_d^2}\right) \exp(-\alpha t) \cos(\omega_d t)u(t) + \left(c_2 - \frac{K\alpha/\omega_d}{\alpha^2 + \omega_d^2}\right) \exp(-\alpha t) \sin(\omega_d t)u(t) + (K/a_0)u(t) \quad (6.163)$$

$$= [b_1 \cos(\omega_d t) + b_2 \sin(\omega_d t)] \exp(-\alpha t)u(t) + (K/a_0)u(t), \quad (6.164)$$

where $b_1 \triangleq c_1 - K/(\alpha^2 + \omega_d^2)$ and $b_2 \triangleq c_2 - K\alpha/\omega_d(\alpha^2 + \omega_d^2)$. The last term on the right-hand side of (6.162) is the steady-state solution, which is identical to the result for the overdamped case in (6.161) when ω_d is substituted:

$$\frac{K}{\alpha^2+\omega_d^2} = \frac{K}{\alpha^2+\omega_o^2-\alpha^2} = K/a_0, \tag{6.165}$$

yielding the final expression in (6.164).

- *Critically damped:*

$$y(t) = [c_1 + c_2 t - (K/\alpha^2)(\alpha t + 1)]\exp(-\alpha t)u(t) + (K/\alpha^2)u(t). \tag{6.166}$$

Combining the terms gives

$$y(t) = [(c_1 - K/\alpha^2) + (c_2 - K/\alpha)t]\exp(-\alpha t)u(t) + (K/a_0)u(t) \tag{6.167}$$

$$= [b_1 + b_2 t]\exp(-\alpha t)u(t) + (K/a_0)u(t), \tag{6.168}$$

where $b_1 \triangleq c_1 - K/\alpha^2$ and $b_2 \triangleq c_2 - K/\alpha$, and the expression has been written in terms of the transient and the steady-state responses. Since $\omega_o^2 = \alpha^2$ for critical damping, $K/\alpha^2 = K/a_0$ in (6.166), which is the same steady-state solution found for the other two cases.

For convenience, we have rearranged the equations based on the initial conditions and summarized the step response results for all three cases in Table 6.5. This table differs from Table 6.3 as follows: (i) nonhomogeneous with step input $x(t) = Ku(t)$, (ii) the solutions include the steady-state output $y_s = K/a_0$, and (iii) the coefficients $\{b_1, b_2\}$ necessarily depend on K (unlike $\{c_1, c_2\}$, which are used in the complete solution of (6.148)). Of course when $K = 0$, all the results in this table reduce to the homogeneous solutions in Table 6.3.

6.6.2 Step Response (Alternative Method)

The coefficients $\{c_1, c_2\}$ for the three types of ODE solutions were derived from the initial conditions $\{y(0), y'(0)\}$ using only the homogeneous part. The particular

TABLE 6.5 Second-Order ODE Solutions for Step Input

System	Linear ODE Signals and Parameters
ODE with step input	$d^2y(t)/dt^2 + a_1 dy(t)/dt + a_0 y(t) = Ku(t)$
Solution	$y(t) = [b_1 y_1(t) + b_2 y_2(t) + K/a_0]u(t)$
Overdamped	$y_1(t) = \exp(s_1 t), y_2(t) = \exp(s_2 t)$
Underdamped	$y_1(t) = \exp(-\alpha t)\cos(\omega_d t), y_2(t) = \exp(-\alpha t)\sin(\omega_d t)$
Critically damped	$y_1(t) = \exp(-\alpha t), y_2(t) = t\exp(-\alpha t)$
Parameters	$s_{1,2} = -\alpha \pm \sqrt{\alpha^2 - \omega_o^2}, \alpha \triangleq a_1/2, \omega_o^2 \triangleq a_0, \omega_d \triangleq \sqrt{\omega_o^2 - \alpha^2}$
Overdamped	$b_1 = [s_2(y(0) - K/a_0) - y'(0)]/(s_2 - s_1)$
	$b_2 = [y'(0) - s_1(y(0) - K/a_0)]/(s_2 - s_1)$
Underdamped	$b_1 = y(0) - K/a_0, b_2 = [y'(0) + \alpha(y(0) - K/a_0)]/\omega_d$
Critically damped	$b_1 = y(0) - K/a_0, b_2 = y'(0) + \alpha(y(0) - K/a_0)$

solution was then added to the homogeneous solution to give the complete solution in each case. In this section, we provide an alternative method for generating the complete solution of the nonhomogeneous ODE when the input $x(t) = Ku(t)$ is a step function. Since the $\{b_1, b_2\}$ multiplying the exponentials in (6.159), (6.164), and (6.168) are just coefficients, they can be derived directly from the initial conditions as follows.

- *Overdamped:*

$$y(0) = b_1 + b_2 + K/a_0, \quad y'(0) = b_1 s_1 + b_2 s_2. \tag{6.169}$$

- *Underdamped:*

$$y(0) = b_1 + K/a_0, \quad y'(0) = \omega_d b_2 - \alpha b_1. \tag{6.170}$$

- *Critically damped:*

$$y(0) = b_1 + K/a_0, \quad y'(0) = b_2 - \alpha b_1. \tag{6.171}$$

Observe that the initial condition $y'(0)$ yields the same equations for $\{b_1, b_2\}$ as in the homogeneous case for $\{c_1, c_2\}$. However, the equations differ for $y(0)$, which include the *steady-state component* K/a_0. For the overdamped case, the coefficients are derived using matrix notation as follows:

$$\begin{bmatrix} 1 & 1 \\ s_1 & s_2 \end{bmatrix} \begin{bmatrix} b_1 \\ b_2 \end{bmatrix} = \begin{bmatrix} y(0) - K/a_0 \\ y'(0) \end{bmatrix}$$
$$\Rightarrow \begin{bmatrix} b_1 \\ b_2 \end{bmatrix} = \frac{1}{s_2 - s_1} \begin{bmatrix} s_2[y(0) - K/a_0] - y'(0) \\ y'(0) - s_1[y(0) - K/a_0] \end{bmatrix}. \tag{6.172}$$

For the underdamped case:

$$b_1 = y(0) - K/a_0, \quad b_2 = [y'(0) + \alpha[y(0) - K/a_0]]/\omega_d, \tag{6.173}$$

and for the critically damped case:

$$b_1 = y(0) - K/a_0, \quad b_2 = y'(0) + \alpha[y(0) - K/a_0]. \tag{6.174}$$

Thus, for the step input $x(t) = Ku(t)$, the coefficients of the complete solution can be derived from $\{y(0), y'(0)\}$ using one of two approaches:

- Compute $\{c_1, c_2\}$ for (6.158), (6.163), or (6.167) using (6.128), (6.129), or (6.130), respectively, which are based on the *homogeneous* solution.
- Compute $\{b_1, b_2\}$ for (6.159), (6.164), or (6.168) using (6.172), (6.173), or (6.174), respectively, which are based on the *complete* solution.

We verify that indeed the two approaches are equivalent by demonstrating that $\{b_1, b_2\}$ can be derived from $\{c_1, c_2\}$ for the coefficient expressions in (6.158), (6.163), and (6.167).

- *Overdamped:*

$$c_1 - K/s_1(s_2 - s_1) = \frac{s_2 y(0) - y'(0)}{s_2 - s_1} - K/s_1(s_2 - s_1)$$

$$= \frac{1}{s_2 - s_1}[s_2 y(0) - y'(0) - s_2 K/a_0] = b_1, \qquad (6.175)$$

$$c_2 + K/s_2(s_2 - s_1) = \frac{y'(0) - s_1 y(0)}{s_2 - s_1} + K/s_2(s_2 - s_1)$$

$$= \frac{1}{s_2 - s_1}[y'(0) - s_1 y(0) + s_1 K/a_0] = b_2. \qquad (6.176)$$

- *Underdamped:*

$$c_1 - \frac{K}{\alpha^2 + \omega_d^2} = y(0) - \frac{K}{\alpha^2 + \omega_d^2} = y(0) - K/a_0 = b_1, \qquad (6.177)$$

$$c_2 - \frac{K\alpha/\omega_d}{\alpha^2 + \omega_d^2} = [y\prime(0) + \alpha y(0)]/\omega_d - \frac{K\alpha/\omega_d}{\alpha^2 + \omega_d^2}$$

$$= [y\prime(0) + \alpha y(0) - K\alpha/a_0]/\omega_d = b_2. \qquad (6.178)$$

- *Critically damped:*

$$c_1 - K/\alpha^2 = y(0) - K/a_0 = b_1, \qquad (6.179)$$

$$c_2 - K/\alpha = y'(0) + \alpha y(0) - \alpha K/a_0 = b_2. \qquad (6.180)$$

Example 6.11 Examples of the step response for the three types of solutions are shown in Figure 6.12. The same set of parameters from Figure 6.10 were used in these computer simulations, with $K/a_0 = 1$ such that $y(t) \to 1$ in all three cases. Since K/a_0 simply adds to $y_1(t) + y_2(t)$, the dotted lines in Figure 6.10 are raised by 1 to produce these results. The same values for $\{b_1, b_2\}$ were used for each of the three cases.

Example 6.12 In this example, we demonstrate that the component terms of a second-order ODE do in fact sum to give the input waveform. Consider the over-damped ODE:

$$\frac{d^2}{dt^2}y(t) + 2\frac{d}{dt}y(t) + 0.5y(t) = 2u(t), \qquad (6.181)$$

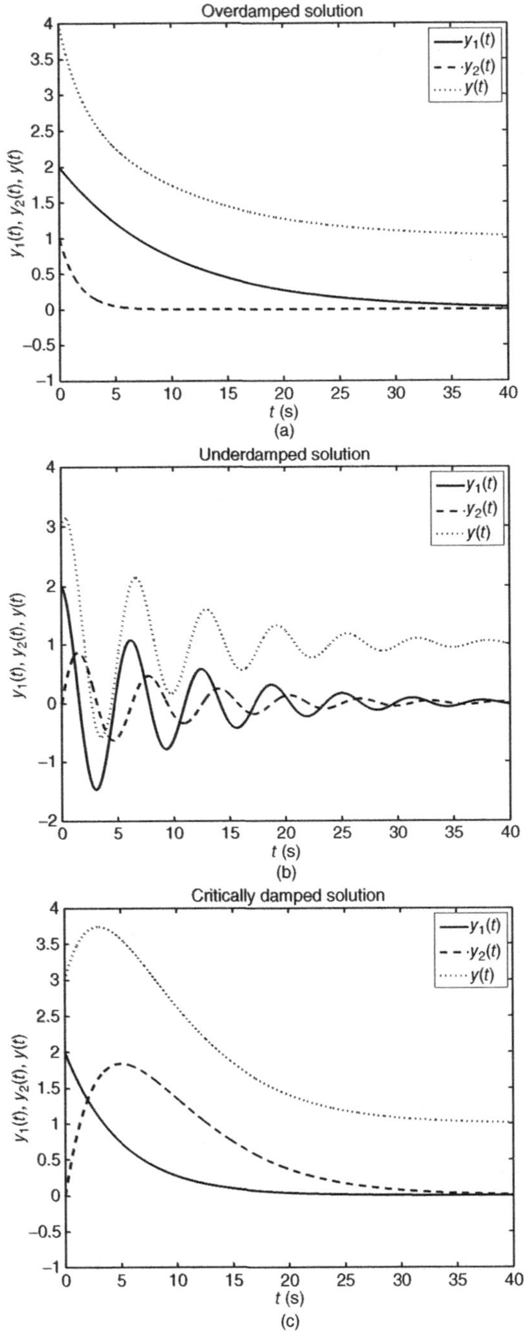

Figure 6.12 Examples of step-response solutions with coefficients $b_1 = 2$, $b_2 = 1$, and $K/a_0 = 1$. (a) Overdamped ($s_1 = -0.1$ and $s_2 = -0.6$). (b) Underdamped ($\alpha = 0.1$ rad/s and $\omega_d = 1$ rad/s). (c) Critically damped ($\alpha = 0.2$ rad/s).

whose characteristic equation has roots

$$s_1, s_2 = -1 \pm \sqrt{1-0.5} = -1 \pm 1/\sqrt{2} \approx -1.7071, -0.2929. \tag{6.182}$$

Since the input is a step function, we have from (6.159):

$$y(t) = [b_1 \exp(s_1 t) + b_2 \exp(s_2 t) + 4]u(t), \tag{6.183}$$

with derivatives

$$\frac{d}{dt}y(t) = [b_1 s_1 \exp(s_1 t) + b_2 s_2 \exp(s_2 t)]u(t), \tag{6.184}$$

$$\frac{d^2}{dt^2}y(t) = [b_1 s_1^2 \exp(s_1 t) + b_2 s_2^2 \exp(s_2 t)]u(t). \tag{6.185}$$

The unit step functions have not been differentiated because we are interested in the solution for $t \geq 0^+$ where they are constant (similar to the reason given earlier when we ignored the Dirac delta functions, resulting in homogeneous ODEs). In this context, $u(t)$ is used to indicate the support of $y(t)$ and its derivatives. Substituting (6.183)–(6.185) into the ODE and collecting the terms on the left-hand side yield

$$\begin{aligned}\frac{d^2}{dt^2}y(t) + 2\frac{d}{dt}y(t) + 0.5y(t) = [&b_1(s_1^2 + 2s_1 + 0.5)\exp(s_1 t)\\ &+ b_2(s_2^2 + 2s_2 + 0.5)\exp(s_2 t) + 0.5(4)]u(t).\end{aligned} \tag{6.186}$$

Both expressions in parentheses are the characteristic equation, and since $\{s_1, s_2\}$ are its roots, these terms are 0, which leaves only the last term $2u(t)$. This is the right-hand side of the ODE, which verifies that the solution is correct. Assume for convenience that $b_1 = b_2 = 1$. Figure 6.13(a) shows the solution $y(t)$ (the solid line) along with its first and second derivatives (the dashed and dotted lines). The weighted sum of these waveforms using the coefficients in the ODE gives exactly $2u(t)$ at every time instant (the dash-dotted line), which is the forcing function $x(t)$ in this example (the right-hand side of the ODE). These results are repeated with coefficients a_0 and a_1 interchanged:

$$\frac{d^2}{dt^2}y(t) + 0.5\frac{d}{dt}y(t) + 2y(t) = 2u(t), \tag{6.187}$$

which corresponds to an underdamped system with parameters $\alpha = 0.25$ rad/s and $\omega_d \approx 1.3919$ rad/s. In this case, the solution is

$$y(t) = [b_1 \exp(-\alpha t)\cos(\omega_d t) + b_2 \exp(-\alpha t)\sin(\omega_d t) + 1]u(t), \tag{6.188}$$

which has derivatives

$$\begin{aligned}\frac{d}{dt}y(t) = &-b_1[\alpha \exp(-\alpha t)\cos(\omega_d t) + \omega_d \exp(-\alpha t)\sin(\omega_d t)]u(t)\\ &+ b_2[-\alpha \exp(-\alpha t)\sin(\omega_d t) + \omega_d \exp(-\alpha t)\cos(\omega_d t)]u(t),\end{aligned} \tag{6.189}$$

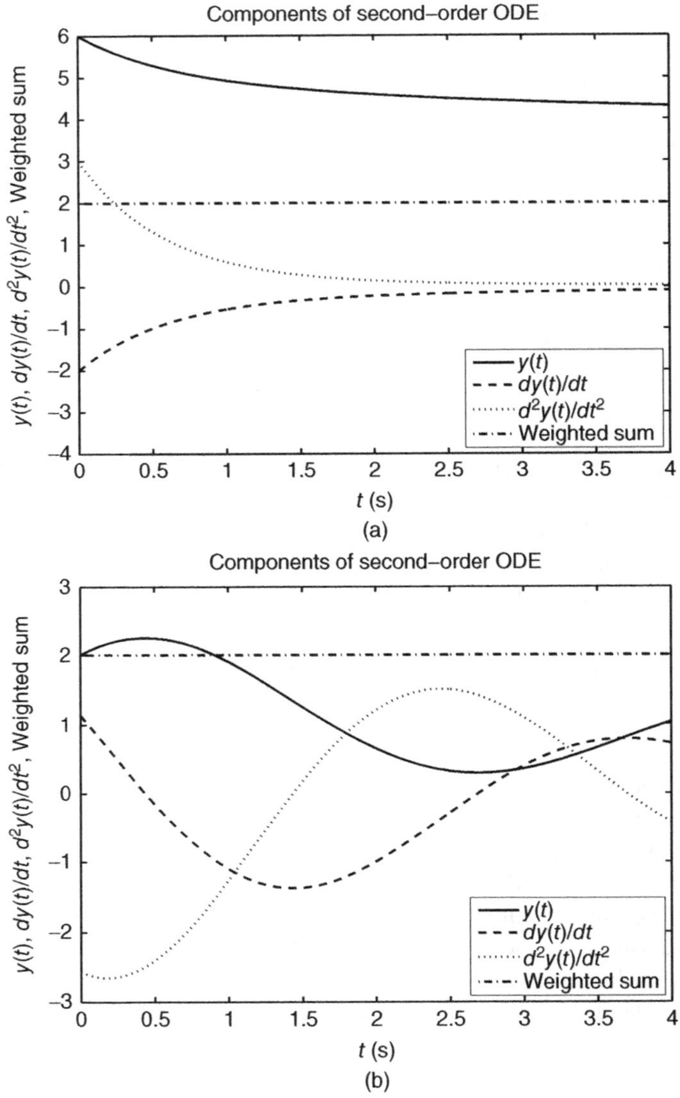

Figure 6.13 Components of the second-order ODE in Example 6.12. (a) Overdamped. (b) Underdamped.

$$\begin{aligned}\frac{d^2}{dt^2}y(t) = {} & b_1[\alpha^2 \exp(-\alpha t)\cos(\omega_d t) + \alpha\omega_d \exp(-\alpha t)\sin(\omega_d t)]u(t) \\ & - b_1[-\alpha\omega_d \exp(-\alpha t)\sin(\omega_d t) + \omega_d^2 \exp(-\alpha t)\cos(\omega_d t)]u(t) \\ & + b_2[\alpha^2 \exp(-\alpha t)\sin(\omega_d t) - \alpha\omega_d \exp(-\alpha t)\cos(\omega_d t)]u(t) \\ & - b_2[\alpha\omega_d \exp(-\alpha t)\cos(\omega_d t) + \omega_d^2 \exp(-\alpha t)\sin(\omega_d t)]u(t). \quad (6.190)\end{aligned}$$

Combining all these terms according to (6.187) yields $2u(t)$ as shown in Figure 6.13(b); all other terms cancel because the characteristic equation is 0 (see Problem 6.19).

6.6.3 Impulse Response

The impulse response function $h(t)$ is derived from (6.148) by ignoring the homogeneous part (the first two terms), which means the initial conditions are 0. Substituting $x(t) = \delta(t)$ into the last two terms of (6.148) gives

$$y(t) = -y_1(t)y_2(0)/W(0) + y_2(t)y_1(0)/W(0), \tag{6.191}$$

where the sifting property of the Dirac delta function has been used. From the expressions for $\{y_1(t), y_2(t)\}$ in Table 6.3 and the Wronskians in Table 6.4, we have the following results generated by substituting $t = 0$ for each of the three second-order cases.

- *Overdamped:* $y_1(0)/W(0) = y_2(0)/W(0) = 1/(s_2 - s_1)$:

$$h(t) = \frac{1}{s_2 - s_1}[\exp(s_2 t) - \exp(s_1 t)]u(t). \tag{6.192}$$

- *Underdamped:* $y_1(0)/W(0) = 1/\omega_d$, $y_2(0)/W(0) = 0$:

$$h(t) = (1/\omega_d)\exp(-\alpha t)\sin(\omega_d t)u(t). \tag{6.193}$$

- *Critically damped:* $y_1(0)/W(0) = 1$, $y_2(0)/W(0) = 0$:

$$h(t) = t\exp(-\alpha t)u(t). \tag{6.194}$$

Note that $\{y_1(0), y_2(0)\}$ are the initial component values from the expressions in Table 6.3; they are not the initial conditions $\{y(0), y'(0)\}$, which are assumed to be 0 when computing the impulse response function $h(t)$. These expressions are summarized in Table 6.6. Using the impulse response function, the complete solution in (6.148) with 0 initial conditions is written more generally as follows:

$$y(t) = \int_0^t x(\tau)h(t - \tau)d\tau, \tag{6.195}$$

where one of the three impulse response functions in (6.192)–(6.194) is used depending on the type of ODE. This is the convolution integral discussed earlier for the three cases in (6.153)–(6.155).

6.7 CONVOLUTION

The integral in (6.195) follows from the fact that superposition holds for an LTI system. The variable of integration is τ, and the resulting output is a function of time t.

TABLE 6.6 Second-Order ODE Impulse Response Function

System	Linear ODE Impulse Response Function and Parameters
ODE with impulse input	$d^2y(t)/dt^2 + a_1dy(t)/dt + a_0y(t) = \delta(t)$
Overdamped	$h(t) = [\exp(s_2t) - \exp(s_1t)]u(t)$
Underdamped	$h(t) = [\exp(-\alpha t)\sin(\omega_dt)/\omega_d]u(t)$
Critically damped	$h(t) = t\exp(-\alpha t)u(t)$
Parameters	$s_{1,2} = -\alpha \pm \sqrt{\alpha^2 - \omega_o^2}, \alpha \triangleq a_1/2, \omega_o^2 \triangleq a_0, \omega_d \triangleq \sqrt{\omega_o^2 - \alpha^2}$

If the output of the system is $y_1(t)$ for input $x_1(t)$ and is $y_2(t)$ for $x_2(t)$, then the output for input $x_1(t) + x_2(t)$ is $y_1(t) + y_2(t)$. As discussed in Chapter 1, nonlinear systems do not have this property. In general, it is convenient to let the convolution integral have infinite limits:

$$y(t) = \int_{-\infty}^{\infty} x(\tau)h(t-\tau)d\tau, \tag{6.196}$$

where t on the left-hand side corresponds to the location of the shifted impulse response function $h(t-\tau)$ on the right-hand side. The support of each function determines the actual limits of integration, as demonstrated later in two examples.

Consider representing the input waveform $x(t)$ approximately by a sum of nonoverlapping rectangles for $t \geq 0$. From Chapter 5, the standard rectangle function is

$$\text{rect}(t) \triangleq \begin{cases} 1, & |t| \leq 1/2 \\ 0, & \text{else.} \end{cases} \tag{6.197}$$

For a shifted rectangle with width Δ and starting at $t = 0$, the argument of the rectangle function is modified to $\text{rect}(t/\Delta - 1/2)$. Similarly, a rectangle of width Δ and starting at $t = \Delta$ is $\text{rect}(t/\Delta - (1 + 1/2))$, which is adjacent to the previous shifted rectangle. The nth shifted rectangle is $\text{rect}(t/\Delta - (n + 1/2))$, and $x(t)$ can be approximated by the following sum of adjacent rectangles:

$$x(t) \approx \sum_{n=0}^{N-1} x(n\Delta)\text{rect}(t/\Delta - (n + 1/2)), \tag{6.198}$$

where N is the number of rectangles under the function up to time instant t. All rectangles have width Δ, and $x(n\Delta)$ is the height of the nth rectangle given by the value of the function at $t = n\Delta$ (the leading edge of the rectangle). This "staircase" approximation is depicted in Figure 6.14 where the first two rectangles are labeled $x(0)\text{rect}(t/\Delta - 1/2)$ and $x(\Delta)\text{rect}(t/\Delta - 3/2)$. Multiplying and dividing by Δ yield

$$x(t) \approx \sum_{n=0}^{N-1} [x(n\Delta)\Delta](1/\Delta)\text{rect}(t/\Delta - (n + 1/2)), \tag{6.199}$$

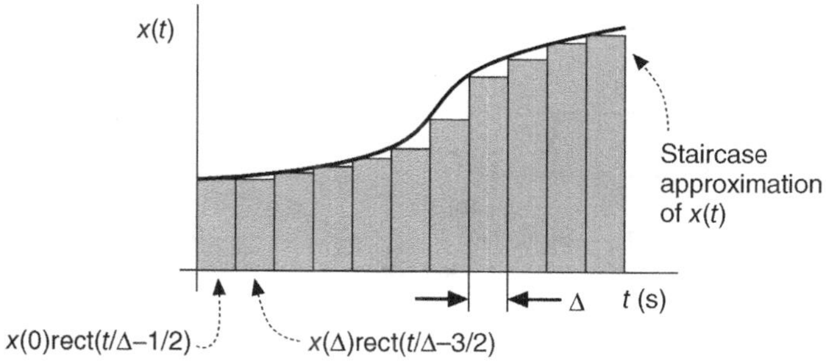

Figure 6.14 Approximation of a continuous-time waveform by a sum of adjacent shifted rectangle functions, each of width Δ.

such that $(1/\Delta)\text{rect}(t/\Delta - (n + 1/2))$ has width Δ, height $1/\Delta$, and unit area. This form explicitly shows that the nth term is a rectangle with area $x(n\Delta)\Delta$. For small Δ, the nth term is approximated by a shifted Dirac delta function with area $x(n\Delta)\Delta$ (see Chapter 5):

$$x(t) \approx \sum_{n=0}^{N-1} [x(n\Delta)\Delta]\delta(t - n\Delta), \tag{6.200}$$

This expression is only an approximation because Δ is not quite 0. Since $h(t)$ is the impulse response function for the LTI system, the output for input $\delta(t - n\Delta)$ is $h(t - n\Delta)$, and so the approximate output for the input model in (6.199) is (Lathi, 1965)

$$y(t) \approx \sum_{n=0}^{N-1} [x(n\Delta)h(t - n\Delta)]\Delta. \tag{6.201}$$

In the limit as $N \to \infty$, $n\Delta \to \tau$, and $\Delta \to d\tau$, this sum becomes the convolution integral in (6.196) (but with the lower limit 0).

Example 6.13 Figure 6.15 illustrates how two rectangular functions are convolved. The functions have different heights but the same support $t \in [0, T]$. From the first integral of (6.196) for the convolution of $x(t)$ and $h(t)$, observe in Figure 6.15(a) that $x(t)$ has been reversed and shifted to give $x(t - \tau)$. As $t > 0$ varies, the function shifts to the right and the convolution is computed as the area of the product of the two functions. This is illustrated by the shaded regions for a value of $t > 0$. Mathematically, it is convenient to write the rectangular functions as indicator functions so that the convolution integral is

$$y(t) = 2\int_{-\infty}^{\infty} I_{[0,T]}(t - \tau)I_{[0,T]}(\tau)d\tau. \tag{6.202}$$

The second indicator function restricts the integration to $\tau \in [0, T]$, and the first indicator function restricts it as follows:

$$t - \tau \geq 0 \implies \tau \leq t, \quad t - \tau \leq T \implies \tau \geq t - T, \tag{6.203}$$

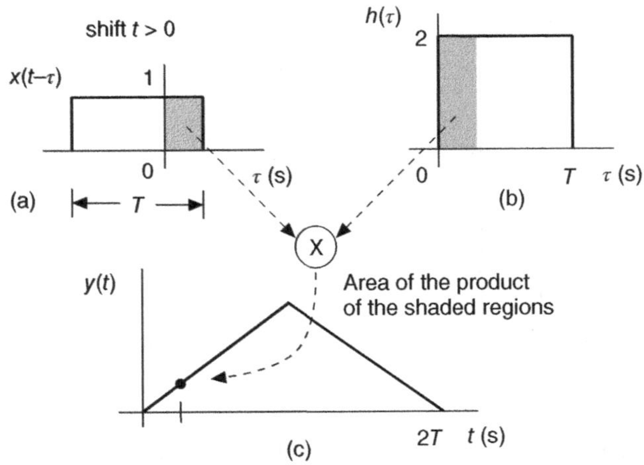

Figure 6.15 Convolution of two functions. (a) Reversed and shifted input $x(t-\tau)$. (b) Impulse response function $h(\tau)$. (c) System output $y(t) = h(t) * x(t)$.

which gives $\tau \in [t-T, t]$. The indicator functions are dropped when the limits of integration are applied:

$$y(t) = \int_{\max(0,t-T)}^{\min(t,T)} d\tau. \tag{6.204}$$

From Figure 6.15, we see four cases for t: (i) $t < 0$, (ii) $0 \le t \le T$, (iii) $T < t \le 2T$, and (iv) $t > 2T$. For cases (i) and (iv), the shifted function $x(t-\tau)$ does not overlap $h(\tau)$, which means $y(t)$ is 0 for those intervals of t. In fact, we find from these two cases that the support for $y(t)$ is $[0, 2T]$. For case (ii), the limits of integration are $\{0, t\}$:

$$y(t) = \int_0^t d\tau, \quad 0 \le t \le T, \tag{6.205}$$

and for case (iii), they are $\{t-T, T\}$:

$$y(t) = \int_{t-T}^{T} d\tau, \quad T \le t \le 2T. \tag{6.206}$$

These integrals are straightforward to evaluate, and so using indicator functions, we have

$$y(t) = tI_{[0,T]}(t) + (2T-t)I_{(T,2T]}(t), \tag{6.207}$$

which is the triangular function shown in Figure 6.15(c).

Example 6.14 In this example, we verify that convolution is a symmetric operation for $x(t) = u(t)$ and $h(t) = \exp(-t)u(t)$, which we examined earlier in Figure 6.5. First,

$h(t)$ is reversed:

$$y(t) = \int_{-\infty}^{\infty} u(\tau)\exp(-(t-\tau))u(t-\tau)d\tau = \int_{0}^{t} \exp(-(t-\tau))d\tau$$
$$= \exp(-t)\exp(\tau)|_0^t = [1-\exp(-t)]u(t), \tag{6.208}$$

where the unit step functions have determined the limits of integration in the second line. The unit step function has been included in the final expression to give the support of $y(t)$. Likewise, by reversing $x(t)$ instead:

$$y(t) = \int_{-\infty}^{\infty} \exp(-\tau)u(\tau)u(t-\tau)d\tau = \int_{0}^{t} \exp(-\tau)d\tau$$
$$= -\exp(-\tau)|_0^t = [1-\exp(-t)]u(t). \tag{6.209}$$

Since $x(t) = u(t)$, this output shown in Figure 6.5(b) is the step response of a system with impulse response function $h(t) = \exp(-t)u(t)$.

6.8 SYSTEM OF ODEs

Finally in this chapter, we show how to write an ODE as a *system of equations* where each equation is a first-order ODE written in terms of *states* of the system. Consider the Nth-order linear ODE with fixed coefficients $\{a_0, \ldots, a_{N-1}\}$:

$$\frac{d^N}{dt^N}y(t) + a_{N-1}\frac{d^{N-1}}{dt^{N-1}}y(t) + \cdots + a_1\frac{d}{dt}y(t) + a_0 y(t) = x(t). \tag{6.210}$$

Define N states as follows:

$$y_0(t) \triangleq y(t), \qquad y_1(t) \triangleq \frac{d}{dt}y(t), \ldots, y_{N-1}(t) \triangleq \frac{d^{N-1}}{dt^{N-1}}y(t), \tag{6.211}$$

such that (6.210) can be rewritten as

$$\frac{d^N}{dt^N}y(t) = x(t) - a_0 y_0(t) - a_1 y_1(t) - \cdots - a_{N-1}y_{N-1}(t). \tag{6.212}$$

(The subscripts on $y(t)$ should not be confused with the different solutions considered earlier for second-order ODEs.) Defining the state vector $\mathbf{y}(t) \triangleq [y_0(t), \ldots, y_{N-1}(t)]^T$ yields the matrix formulation

$$\frac{d}{dt}\begin{bmatrix} y_0(t) \\ \vdots \\ y_{N-2}(t) \\ y_{N-1}(t) \end{bmatrix} = \begin{bmatrix} 0 & 1 & 0 & \cdots & 0 \\ \vdots & 0 & 1 & 0 & \vdots \\ & & & \cdots & 0 \\ 0 & \cdots & & 0 & 1 \\ -a_0 & -a_1 & \cdots & & -a_{N-1} \end{bmatrix} \begin{bmatrix} y_0(t) \\ \vdots \\ y_{N-2}(t) \\ y_{N-1}(t) \end{bmatrix} + \begin{bmatrix} 0 \\ \vdots \\ 0 \\ 0 \\ x(t) \end{bmatrix}, \tag{6.213}$$

which can be written as

$$\dot{\mathbf{y}}(t) = \mathbf{A}\mathbf{y}(t) + \mathbf{b}x(t), \tag{6.214}$$

where $\mathbf{b} \triangleq [0, \ldots, 0, 1]^T$ is the input vector and $\mathbf{A}$ is the *state transition matrix*. This equation is still an Nth-order linear ODE with constant coefficients, but it has been expanded into N equations that together represent the original expression in (6.210). Note that the minus signs in (6.212) are included in the definition of $\mathbf{A}$.

Example 6.15 For $N = 2$, the linear ODE is

$$\frac{d^2}{dt^2}y(t) + a_1\frac{d}{dt}y(t) + a_0 y(t) = x(t), \tag{6.215}$$

for which the state transition matrix and input vector are

$$\mathbf{A} = \begin{bmatrix} 0 & 1 \\ -a_0 & -a_1 \end{bmatrix}, \quad \mathbf{b} = \begin{bmatrix} 0 \\ 1 \end{bmatrix}. \tag{6.216}$$

The eigenvalues of this matrix are derived by solving the following equation:

$$\det\left(\begin{bmatrix} -\lambda & 1 \\ -a_0 & -a_1 - \lambda \end{bmatrix}\right) = \lambda(\lambda + a_1) + a_0 = \lambda^2 + a_1\lambda + a_0 = 0, \tag{6.217}$$

which has the same form as the characteristic equation in (6.94) but with λ in place of s. Thus, the eigenvalues of $\mathbf{A}$ yield the same information about the system as the roots of the characteristic equation. These are also the poles of the system as discussed in Chapter 7 where the Laplace transform is used to solve ODEs.

The matrix equation in (6.214) is homogeneous when $x(t) = 0$, as is the case for the original ODE. For a first-order homogeneous ODE

$$\frac{d}{dt}y(t) + ay(t) = 0, \tag{6.218}$$

we know that the solution is

$$y(t) = y(0)\exp(-at)u(t), \tag{6.219}$$

where $y(0)$ is a nonzero initial condition and we have dropped the subscript on a_0. Likewise, the homogeneous solution for (6.214) can be written as

$$\mathbf{y}(t) = \exp(\mathbf{A}t)\mathbf{y}(0)u(t), \tag{6.220}$$

where

$$\mathbf{y}(0) \triangleq [y(0), y'(0), \ldots, y^{(N-1)}(0)]^T \tag{6.221}$$

contains N initial conditions and $\exp(\mathbf{A}t)$ is the *matrix exponential*. Since $\mathbf{y}(0)$ is a column vector, it must be multiplied on the left by matrix $\exp(\mathbf{A}t)$.

Definition: Matrix Exponential The *matrix exponential* $\exp(\mathbf{A}t)$ for $\mathbf{A} \in \mathcal{R}^{N\times N}$ is the following power series:

$$\exp(\mathbf{A}t) \triangleq \mathbf{I} + \mathbf{A}t + (1/2)\mathbf{A}^2t^2 + \cdots = \sum_{n=0}^{\infty} \mathbf{A}^n t^n/n!, \tag{6.222}$$

where $\mathbf{A}^0 = \mathbf{I}$ is the identity matrix.

The matrix exponential is an extension of the power series expansion for the ordinary exponential function:

$$\exp(at) = \sum_{n=0}^{\infty} (at)^n/n!, \tag{6.223}$$

with $a \in \mathcal{R}$. An integrator implementation of a third-order ODE is shown in Figure 6.16 where $\{y_1(t), y_2(t)\}$ are the internal states. The only state that is directly observable at the output is $y_0(t) = y(t)$, which of course is the overall output of the system. In general, for an Nth-order ODE, $N-1$ states are internal to the system and are not directly observed at the output.

Next, we show that (6.220) is the solution of (6.214) with $x(t) = 0$. Substituting $\mathbf{y}(t)$ and (6.222) yields

$$\frac{d}{dt}\exp(\mathbf{A}t)\mathbf{y}(0) = \frac{d}{dt}\left(\sum_{n=0}^{\infty} \mathbf{A}^n t^n/n!\right)\mathbf{y}(0) = \sum_{n=1}^{\infty}(\mathbf{A}^n n t^{n-1}/n!)\mathbf{y}(0)$$

$$= \mathbf{A}\sum_{n=1}^{\infty}(\mathbf{A}^{n-1}t^{n-1}/(n-1)!)\mathbf{y}(0) = \mathbf{A}\exp(\mathbf{A}t)\mathbf{y}(0), \tag{6.224}$$

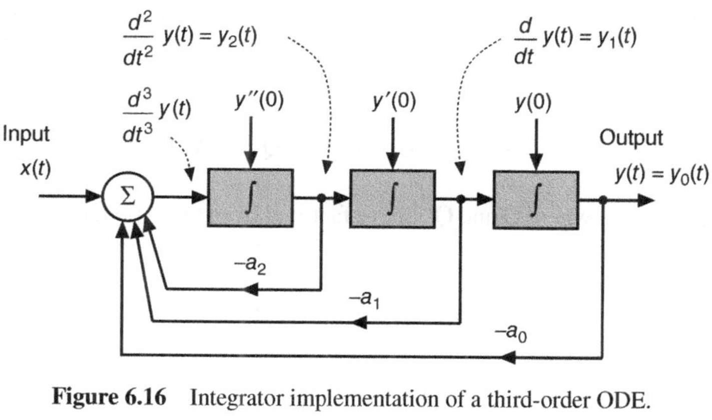

Figure 6.16 Integrator implementation of a third-order ODE.

TABLE 6.7 Properties of the Matrix Exponential

Property	Equation
Derivative	$d\exp(\mathbf{A}t)/dt = \mathbf{A}\exp(\mathbf{A}t)$
Product	$\mathbf{A}\exp(\mathbf{A}t) = \exp(\mathbf{A}t)\mathbf{A}$
Inverse	$[\exp(\mathbf{A}t)]^{-1} = \exp(-\mathbf{A}t)$
Exponent	$\exp(\mathbf{A}(t_1 + t_2)) = \exp(\mathbf{A}t_1)\exp(\mathbf{A}t_2)$
Identity	$\exp(\mathbf{0}) = \mathbf{I}$

where the derivative causes the lower limit of the sum to become $n = 1$ because $d\mathbf{I}/dt = \mathbf{0}$ (the zero matrix). The left-hand side is $\dot{\mathbf{y}}(t)$ and the right-hand side is $\mathbf{A}\mathbf{y}(t)$, thus verifying the solution in (6.220). Additional properties of the matrix exponential are summarized in Table 6.7.

Example 6.16 The second-order ODE in (6.181) can be written with $x(t) = 0$ (the homogeneous case) as follows:

$$\frac{d}{dt}\begin{bmatrix} y_0(t) \\ y_1(t) \end{bmatrix} = \begin{bmatrix} 0 & 1 \\ -0.5 & -2 \end{bmatrix}\begin{bmatrix} y_0(t) \\ y_1(t) \end{bmatrix}. \tag{6.225}$$

The matrix exponential is

$$\exp(\mathbf{A}t) = \mathbf{I} + \begin{bmatrix} 0 & 1 \\ -0.5 & -2 \end{bmatrix} t + \begin{bmatrix} -0.25 & -1 \\ 0.5 & 2 \end{bmatrix} t^2 + \cdots, \tag{6.226}$$

for which a closed-form solution is not easy to find. However, if the matrix has the eigendecomposition (see Chapter 3)

$$\mathbf{A} = \mathbf{Q}\mathbf{\Lambda}\mathbf{Q}^{-1}, \tag{6.227}$$

where $\mathbf{\Lambda}$ is a diagonal matrix containing the eigenvalues of $\mathbf{A}$, and the columns of $\mathbf{Q}$ are the corresponding normalized eigenvectors, then a closed-form expression is straightforward. Substituting (6.227) yields

$$\exp(\mathbf{A}t) = \exp(\mathbf{Q}\mathbf{\Lambda}\mathbf{Q}^{-1}t) = \mathbf{Q}\exp(\mathbf{\Lambda}t)\mathbf{Q}^{-1}. \tag{6.228}$$

The last expression where $\mathbf{Q}$ and $\mathbf{Q}^{-1}$ have factored from the exponent is easily shown by using the power series expansion in (6.224). Since $\mathbf{\Lambda}$ is diagonal of the form

$$\mathbf{\Lambda} = \begin{bmatrix} \lambda_1 & 0 & \cdots & 0 \\ 0 & \ddots & & \vdots \\ \vdots & & \ddots & 0 \\ 0 & \cdots & 0 & \lambda_N \end{bmatrix}, \tag{6.229}$$

it is clear that

$$\exp(\mathbf{\Lambda}t) = \begin{bmatrix} \exp(\lambda_1 t) & 0 & \cdots & 0 \\ 0 & \ddots & & \vdots \\ \vdots & & \ddots & 0 \\ 0 & \cdots & 0 & \exp(\lambda_N t) \end{bmatrix}. \tag{6.230}$$

This result follows because $\mathbf{\Lambda}^n$ is also diagonal for any $n \in \mathcal{N}$, and each diagonal term is derived from a power series expansion based on that eigenvalue. The original matrix exponential $\exp(\mathbf{A}t)$ is then generated by pre- and postmultiplying (6.230) by $\mathbf{Q}$ and $\mathbf{Q}^{-1}$, respectively. For this example, the eigenvalues are real: $\lambda_1, \lambda_2 = -1 \pm 1/\sqrt{2} \approx \{-0.2929, -1.7071\}$, and the eigenvector matrix is

$$\mathbf{Q} = \begin{bmatrix} 0.9597 & -0.5054 \\ -0.2811 & 0.8629 \end{bmatrix}. \tag{6.231}$$

The matrix exponential is then derived from the last expression in (6.228):

$$\exp(\mathbf{A}t) = \begin{bmatrix} 0.9597 & -0.5054 \\ -0.2811 & 0.8629 \end{bmatrix} \begin{bmatrix} \exp(-0.2929t) & 0 \\ 0 & \exp(-1.7071t) \end{bmatrix} \times \begin{bmatrix} 1.2578 & 0.7368 \\ 0.4097 & 1.3990 \end{bmatrix}, \tag{6.232}$$

and the components of the homogeneous solution in (6.220) are

$$\begin{aligned} y_0(t) &= [1.2071y(0) + 0.7071y'(0)]\exp(-0.2929t)u(t) \\ &\quad - [0.2071y(0) + 0.7071y'(0)]\exp(-1.7071t)u(t), \end{aligned} \tag{6.233}$$

$$\begin{aligned} y_1(t) &= -[0.3536y(0) + 0.2071y'(0)]\exp(-0.2929t)u(t) \\ &\quad + [0.3536y(0) + 1.2071y'(0)]\exp(-1.7071t)u(t), \end{aligned} \tag{6.234}$$

where $y_0(t) = y(t)$ is the output of the system. If $y_1(t) = y'(t)$ is integrated, the expression for $y(t)$ in (6.233) is derived (see Problem 6.33).

Another way to solve for $\exp(\mathbf{A}t)$ is to use the following identity for $\mathbf{A} \in \mathcal{R}^{N \times N}$:

$$\exp(\mathbf{A}t) = \sum_{n=0}^{N-1} \alpha_n(t)\mathbf{A}^n, \tag{6.235}$$

where $\mathbf{A}^0 = \mathbf{I}$. This result is due to the *Cayley–Hamilton theorem* where it can be shown that every $\mathbf{A}^M$ for $M \geq N$ can be written as a linear combination of lower powers of $\mathbf{A}$ (Kailath, 1980). Thus, the higher order terms in the power series expansion are combined with the lower order terms, resulting in the finite sum in (6.235). This requires an appropriate set of coefficients that are derived by solving

$$\begin{bmatrix} \alpha_0(t) \\ \vdots \\ \alpha_{N-1}(t) \end{bmatrix} = \begin{bmatrix} 1 & \lambda_1 & \lambda_1^2 & \cdots & \lambda_1^{N-1} \\ \vdots & \vdots & \vdots & \cdots & \vdots \\ 1 & \lambda_N & \lambda_N^2 & \cdots & \lambda_N^{N-1} \end{bmatrix}^{-1} \begin{bmatrix} \exp(\lambda_1 t) \\ \vdots \\ \exp(\lambda_N t) \end{bmatrix}. \tag{6.236}$$

Example 6.17 Continuing with the previous example, the coefficients in (6.236) are

$$\begin{aligned} \begin{bmatrix} \alpha_0(t) \\ \alpha_1(t) \end{bmatrix} &= \begin{bmatrix} 1 & -0.2929 \\ 1 & -1.7071 \end{bmatrix}^{-1} \begin{bmatrix} \exp(-0.2929t) \\ \exp(-1.7071t) \end{bmatrix} \\ &= \begin{bmatrix} 1.2071 & -0.2071 \\ 0.7071 & -0.7071 \end{bmatrix} \begin{bmatrix} \exp(-0.2929t) \\ \exp(-1.7071t) \end{bmatrix}, \end{aligned} \tag{6.237}$$

and the matrix exponential is

$$\begin{aligned} \exp(\mathbf{A}t) &= [1.2071\exp(-0.2929t) - 0.2071\exp(-1.7071t)]\mathbf{I} \\ &\quad + [0.7071\exp(-0.2929t) - 0.7071\exp(-1.7071t)] \begin{bmatrix} 0 & 1 \\ -0.5 & -2 \end{bmatrix}. \end{aligned} \tag{6.238}$$

Postmultiplying this expression by $[y(0), y'(0)]^T$ gives (6.233) and (6.234).

For the nonhomogeneous case with nonzero $x(t)$, the output is derived from a *matrix convolution*:

$$\mathbf{y}(t) = \exp(\mathbf{A}t)\mathbf{y}(0) + \int_0^t \exp(\mathbf{A}(t-\tau))\mathbf{b}x(\tau)d\tau. \tag{6.239}$$

The first term on the right-hand side is the homogeneous solution discussed earlier, and the second term is the *particular solution*. Of course, if the initial conditions are 0, then the right-hand side includes only the convolution between $x(t)$ and the *matrix impulse response function* $\mathbf{h}(t) = \exp(\mathbf{A}t)\mathbf{b}u(t)$. Since the input is a scalar in (6.210), the column vector $\mathbf{b}$ contains all zeros except for 1 in the last position. As a result, the convolution in (6.239) is actually a set of convolutions between the last column of $\exp(\mathbf{A}t)$ and $x(t)$.

In order to verify (6.239), we first multiply (6.214) by $\exp(-\mathbf{A}t)$:

$$\exp(-\mathbf{A}t)\dot{\mathbf{y}}(t) = \exp(-\mathbf{A}t)\mathbf{A}\mathbf{y}(t) + \exp(-\mathbf{A}t)\mathbf{b}x(t). \tag{6.240}$$

Bringing the first term on the right-hand side to the left-hand side, we recognize the product rule of differentiation:

$$\exp(-\mathbf{A}t)\dot{\mathbf{y}}(t) - \exp(-\mathbf{A}t)\mathbf{A}\mathbf{y}(t) = \frac{d}{dt}\exp(-\mathbf{A}t)\mathbf{y}(t), \tag{6.241}$$

where the derivative property in Table 6.7 has been used. Thus,

$$\frac{d}{dt}\exp(-\mathbf{A}t)\mathbf{y}(t) = \exp(-\mathbf{A}t)\mathbf{b}x(t), \tag{6.242}$$

and integrating both sides yields

$$\exp(-\mathbf{A}t)\mathbf{y}(t)|_0^t = \exp(-\mathbf{A}t)\mathbf{y}(t) - \mathbf{y}(0) = \int_0^t \exp(-\mathbf{A}\tau)\mathbf{b}x(\tau)d\tau. \tag{6.243}$$

Multiplying the last two expressions by $\exp(\mathbf{A}t)$ gives

$$\mathbf{y}(t) - \exp(\mathbf{A}t)\mathbf{y}(0) = \int_0^t \exp(\mathbf{A}(t-\tau))\mathbf{b}x(\tau)d\tau, \tag{6.244}$$

which completes the proof.

Example 6.18 For matrix $\mathbf{A}$ in Example 6.16, the matrix exponential is given in (6.232). Assuming zero initial conditions $\mathbf{y}(0) = \mathbf{0}$, we need to consider only the last column of $\exp(\mathbf{A}(t-\tau))$ in (6.239) because $\mathbf{b} = [0, 1]^T$. Thus,

$$\mathbf{y}(t) = \int_0^t \begin{bmatrix} 0.7071\exp(-0.2929(t-\tau)) - 0.7071\exp(-1.7071(t-\tau)) \\ -0.2071\exp(-0.2929(t-\tau)) - 1.2071\exp(-1.7071(t-\tau)) \end{bmatrix} x(\tau)d\tau. \tag{6.245}$$

For $x(t) = \delta(t)$, the first element of the vector in (6.245) is the impulse response function $h(t)$ from $x(t)$ to $y(t)$. The unit step response is also derived from this first element when $x(t) = u(t)$:

$$y(t) = \int_0^t [0.7071\exp(-0.2929\tau) - 0.7071\exp(-1.7071\tau)]u(t-\tau)d\tau, \tag{6.246}$$

where for convenience we have interchanged the arguments of the exponentials and $u(t)$. Since $u(t-\tau)$ gives the upper limit of integration, it can be dropped from the integrand, and the step response is produced by integrating the two terms:

$$\begin{aligned} y(t) &= -(0.7071/0.2929)\exp(-0.2929\tau)|_0^t + (0.7071/1.7071)\exp(-1.7071t)|_0^t \\ &= 2.4141[1-\exp(-0.2929t)]u(t) - 0.4142[1-\exp(-1.7071t)]u(t). \end{aligned} \tag{6.247}$$

The steady-state value as $t \to \infty$ is $y(\infty) = 2.4141 - 0.4142 \approx 2$.

PROBLEMS

Differential Equations

6.1 Determine which of the following ODEs are linear and give the order of each:

$$\text{(a)}\ \frac{d^2}{dt^2}y(t) + 2\frac{d}{dt}y(t) + 2y^2(t) = x(t), \quad \text{(b)}\ \frac{d^3}{dt^3}y(t) + 3y(t)\frac{d}{dt}y(t) + y(t) = x(t). \tag{6.248}$$

6.2 Repeat the previous problem for

$$\text{(a)}\ \frac{d^2}{dt^2}y(t) + 3\left(\frac{d}{dt}y(t)\right)^2 + 2y(t) = x(t), \quad \text{(b)}\ \frac{d}{dt}y(t) + 2x(t)y(t) + y(t) = x(t). \tag{6.249}$$

6.3 Verify that the following functions are solutions of the ODEs:

$$\text{(a)}\ \frac{d^2}{dt^2}y(t) - 4\frac{d}{dt}y(t) + 3y(t) = 0 \Rightarrow y(t) = [\exp(t) + \exp(3t)]u(t), \tag{6.250}$$

$$\text{(b)}\ \frac{d^2}{dt^2}y(t) + 4\frac{d}{dt}y(t) + 4y(t) = 0 \Rightarrow y(t) = t\exp(-2t)u(t). \tag{6.251}$$

First-Order Linear ODE

6.4 Derive the linear ODE that models the voltage across the resistor in Figure 6.1 and given in Table 6.1.

6.5 Repeat the previous problem for the current through the resistor.

6.6 Determine if the following first-order ODEs are separable:

$$\text{(a)}\ \frac{d}{dt}y(t) = t + (t-1)y(t) - 1, \qquad \text{(b)}\ \frac{d}{dt}y(t) + ty^2(t) = 2t^2. \tag{6.252}$$

6.7 If the coefficient a in (6.44) is a function of time $a(t)$, then the integrating factor in (6.47) is

$$|g(t)/g(t_o)| = \exp\left(\int_{t_o}^{t} a(t)dt\right). \tag{6.253}$$

Use this generalization to find the solution for

$$\frac{d}{dt}y(t) - ty(t) = u(t), \tag{6.254}$$

which is a nonhomogeneous ODE with input $u(t)$. Assume nonzero initial condition $y(0)$.

6.8 (a) Give the solution $y(t)$ for the first-order ODE with exponential input

$$\frac{d}{dt} + 2y(t) = \exp(-t)u(t), \tag{6.255}$$

which has initial condition $y(0) = 0$. Determine how the solution is modified for input (b) $x(t) = \exp(-(t-1))u(t-1)$ and (c) $x(t) = \exp(-(t-1))u(t)$.

6.9 A first-order system has impulse response function $h(t) = \exp(-3t)u(t)$. Use convolution to find the output $y(t)$ for input $x(t) = 2u(t-1)$.

6.10 Repeat the previous problem for the rectangular input $x(t) = u(t) - u(t-1)$.

Second-Order Linear ODE

6.11 Verify the three current results in Table 6.2 for the parallel RLC circuit by deriving expressions for the ODEs.

6.12 Show that (6.102) and (6.103) are the same underdamped solution for a second-order linear ODE.

6.13 Determine the type of homogeneous solution for the series RLC circuit for the following component values, and specify the damping ratio ζ and resonant frequency ω_o. (a) $R = 100\ \Omega$, $L = 1$ mH, and $C = 5\ \mu$F. (b) $R = 10\ \Omega$, $L = 2$ mH, and $C = 5\ \mu$F. (c) $R = 1000\ \Omega$, $L = 1$ mH, and $C = 20\ \mu$F.

6.14 Repeat the previous problem for the parallel RLC circuit using the same component values.

6.15 (a) For the device parameters $L = 1$ mH and $C = 10\ \mu$F of the series RLC circuit, specify the range of values for R for the three types of homogeneous solutions. (b) Give expressions for the three homogeneous solutions for the capacitor voltage $v_C(t)$ assuming initial conditions $v_C(0) = 5$ V and $v'_C(0) = 1$ V/s.

6.16 Repeat the previous problem for the parallel RLC circuit.

6.17 Derive the second-order linear ODE for the capacitor voltage of the RLC circuit in Figure 6.17.

6.18 Repeat the previous problem for the circuit in Figure 6.18.

Second-Order ODE Responses

6.19 Combine the results in (6.188)–(6.190) for the ODE in (6.187) to verify that $y(t)$ is its solution.

6.20 (a) Find the complete solution for the ODE

$$\frac{d^2}{dt^2}y(t) + 2\frac{d}{dt}y(t) + y(t) = 2u(t), \tag{6.256}$$

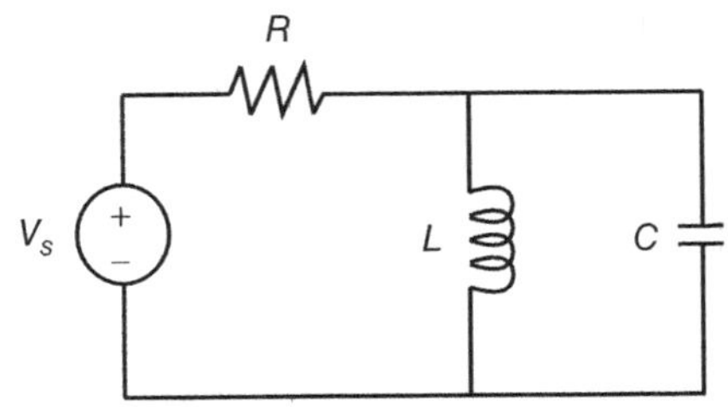

Figure 6.17 Second-order RLC circuit with voltage source V_s for Problem 6.17.

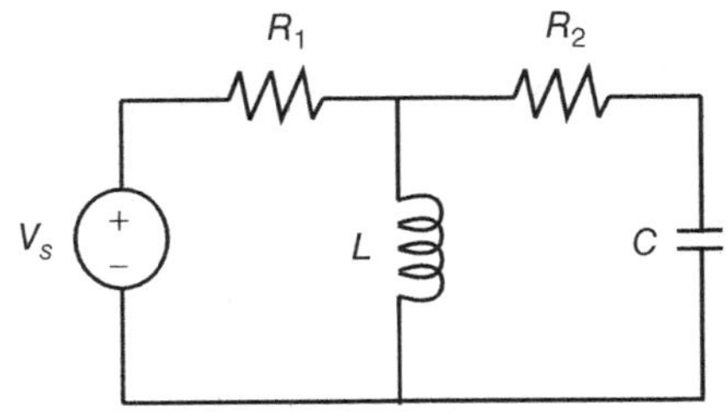

Figure 6.18 Second-order RLC circuit with voltage source V_s for Problem 6.18.

assuming zero initial conditions. (b) Give the impulse response function $h(t)$.

6.21 Repeat the previous problem for

$$\frac{d^2}{dt^2}y(t) + 3\frac{d}{dt}y(t) + y(t) = \exp(-t)u(t). \tag{6.257}$$

6.22 (a) Find the value of a_1 such that the output $y(t)$ is critically damped and write the homogeneous solution for

$$\frac{d^2}{dt^2}y(t) + a_1\frac{d}{dt}y(t) + 2 = x(t). \tag{6.258}$$

(b) Give the step response of this system assuming initial conditions $y(0) = y'(0) = 1$.

6.23 (a) Find the range of values for a_0 such that the output $y(t)$ is overdamped and write the homogeneous solution for

$$\frac{d^2}{dt^2}y(t) + 2\frac{d}{dt}y(t) + a_0 = x(t). \tag{6.259}$$

(b) Give the response of this system for $x(t) = \text{rect}(t - 1/2)$ assuming initial conditions $y(0) = y'(0) = 1$.

Convolution

6.24 Convolve the following two functions and include a sketch showing their overlap as t is varied:

$$f(t) = u(t-1), \quad g(t) = \text{rect}(t-1). \tag{6.260}$$

6.25 Repeat the previous problem for

$$f(t) = g(t) = \text{tri}(t-1). \tag{6.261}$$

6.26 Repeat Problem 6.24 for

$$f(t) = \exp(-t)u(t), \quad g(t) = [1 - \exp(-2t)]u(t). \tag{6.262}$$

6.27 For the second-order system in Problem 6.22, find the unit step response using convolution.

6.28 Cross-correlation is equivalent to convolution when one of the functions is reversed. Determine which of the following are equivalent to $f(t) * g(t)$: (a) $f(t) \star g(-t)$, (b) $f(-t) \star g(t)$, (c) $g(t) \star f(-t)$. Find expressions for all cases by using the functions in Problem 6.24.

6.29 Prove the associative property of convolution: $[f(t) * g(t)] * h(t) = f(t) * [g(t) * h(t)]$.

6.30 Prove the derivative property of convolution: $d[f(t) * g(t)]/dt = [df(t)/dt] * g(t) = [dg(t)/dt] * f(t)$.

System of ODE Equations

6.31 Specify matrix **A** for the ODE in Problem 6.21 and give the homogeneous solution for $\mathbf{y}(t)$ assuming initial conditions $y(0) = y'(0) = 1$.

6.32 (a) Find matrix **A** for the following third-order ODE:

$$\frac{d^3}{dt^3}y(t) + 6\frac{d^2}{dt^2}y(t) + 11\frac{d}{dt}y(t) + 6y(t) = x(t). \tag{6.263}$$

(b) Given that one eigenvalue is $\lambda_1 = -1$, find the other two eigenvalues.

6.33 Show that the integral for $y_1(t)$ with limits $\{0, t\}$ is the same expression as $y_0(t) = y(t)$ in (6.234).

6.34 Write an expression for the solution $\mathbf{y}(t)$ of the second-order ODE in (6.21) assuming zero initial conditions $\mathbf{y}(0) = \mathbf{0}$.

Computer Problems

6.35 Find the unit step response for the following system using the alternative initial condition method with coefficients $\{b_1, b_2\}$:

$$\frac{d^2}{dt^2}y(t) + 2\frac{d}{dt}y(t) + 5y(t) = u(t). \tag{6.264}$$

Plot the individual components $\{y_1(t), y_2(t)\}$ of the solution as well as the complete response using MATLAB.

6.36 Consider the following homogeneous third-order system:

$$\frac{d^3}{dt^3}y(t) + 6\frac{d^2}{dt^2}y(t) + 11\frac{d}{dt}y(t) + 6y(t) = 0. \tag{6.265}$$

(a) Give the system matrix $\mathbf{A}$ and use MATLAB to approximate $\exp(\mathbf{A}t)$ with five terms in its power series. (b) Find the eigendecomposition of $\mathbf{A}$ using eig and give an exact expression for the matrix exponential. (c) Derive the coefficients $\{\alpha_0(t), \alpha_1(t), \alpha_2(t)\}$ used in the finite form for $\exp(\mathbf{A}t)$ in (6.235).

6.37 (a) Use dsolve in MATLAB to solve the ODE in part (a) of Problem 6.6 with initial condition $y(0) = 2$ and plot the resulting function. (b) Repeat part (a) for the ODE in Problem 6.36 with $y(0) = y'(0) = y''(0) = 1$.

6.38 Repeat the previous problem using ode45 to numerically solve the ODEs. For the third-order system, it will be necessary to write it as a system of three equations in terms of matrix $\mathbf{A}$.

6.39 Convolve the two functions in Problem 6.24 using conv in MATLAB and plot the result. The functions can be implemented using heaviside and rectangularPulse.

7

LAPLACE TRANSFORMS AND LINEAR SYSTEMS

7.1 INTRODUCTION

In this chapter, we describe a complex-variable technique for solving linear ordinary differential equations (ODEs) more easily than using the time-domain methods of the previous chapter, especially for high-order systems (greater than second order). This technique also provides additional insight into the behavior of linear time-invariant (LTI) systems beyond that observed in the time domain. The *Laplace transform* is a particular integral transform of signal $x(t)$ that yields a function $X(s)$ of the complex variable $s \triangleq \sigma + j\omega$ in the *frequency domain*, which is also called the *s-domain*. This transformation is invertible such that for some $X(s)$ in the s-domain, it is possible to uniquely transform it to the time domain, yielding $x(t)$. The result for a system is identical to the solution that would be obtained entirely in the time domain using the ODE techniques from Chapter 6. The advantage of using the Laplace transform is that it converts an ODE into an *algebraic equation* of the same order that is simpler to solve, even though it is a function of a complex variable.

By way of analogy, this transformation is similar to that of logarithms, which are used to multiply two numbers or functions. For example, in order to compute the product xy, we can first transform x and y using logarithms and then *add* those results:

$$z = \log(x) + \log(y), \tag{7.1}$$

Mathematical Foundations for Linear Circuits and Systems in Engineering, First Edition. John J. Shynk.

Companion Website: http://www.wiley.com/go/linearcircuitsandsystems

where without loss of generality we have assumed that the base of the logarithm is 10. The final product in the original domain is then computed as

$$xy = 10^z. \tag{7.2}$$

The reason for using logarithms is that usually it is easier to perform additions instead of multiplications. The Laplace transform and the logarithm are *techniques* for converting uniquely from one domain to another domain, where it is easier to perform certain operations in the second domain.

7.2 SOLVING ODEs USING PHASORS

Before defining the Laplace transform, we describe how to solve ODEs using the phasor notation for sinusoidal signals that was covered in Chapter 5. The phasor approach has signal restrictions, and so, it is not as general as the Laplace transform. Since the phasor representation of a signal assumes that it is *sinusoidal* and extends for all time $t \in \mathcal{R}$, the results described in this section are not realizable solutions in practice because the signal duration is doubly infinite. However, phasors do provide insight into the properties of a system, and we see later that solving ODEs using the Laplace transform yields similar types of results for more general signals, including the generalized functions discussed in Chapter 5.

Consider the following nonhomogeneous first-order ODE with sinusoidal input:

$$\frac{d}{dt}y(t) + ay(t) = \cos(\omega_o t), \quad t \in \mathcal{R}, \tag{7.3}$$

where a is a constant and ω_o is angular frequency with units rad/s. This equation for the dependent variable can be solved by converting $y(t)$ into a phasor:

$$y(t) = A\cos(\omega_o t + \phi) \longrightarrow \mathbf{Y}\exp(j\omega_o t), \tag{7.4}$$

where $\mathbf{Y} \triangleq A\exp(j\phi)$ is the phasor, A is its amplitude, and ϕ is its phase. With this notation, the ODE becomes

$$\frac{d}{dt}\mathbf{Y}\exp(j\omega_o t) + a\mathbf{Y}\exp(j\omega_o t) = \exp(j\omega_o t), \tag{7.5}$$

where the cosine on the right-hand side of (7.3) has been replaced by the corresponding complex exponential. Since $\mathbf{Y}$ does not depend on t, the derivative is easily computed and $\exp(j\omega_o t)$ cancels from both sides of (7.5):

$$j\omega_o \mathbf{Y} + a\mathbf{Y} = 1, \tag{7.6}$$

which is now an algebraic equation. Solving this expression yields

$$\mathbf{Y} = \frac{1}{a + j\omega_o} = \frac{1}{a^2 + \omega_o^2}(a - j\omega_o), \tag{7.7}$$

where the numerator and denominator have been multiplied by the complex conjugate $a - j\omega_o$. Converting into polar form, the magnitude of **Y** is

$$|\mathbf{Y}| = \frac{1}{\sqrt{a^2 + \omega_o^2}}, \tag{7.8}$$

and its phase is

$$\arg(\mathbf{Y}) = \tan^{-1}(-\omega_o/a). \tag{7.9}$$

The final phasor can be written in two ways:

$$\mathbf{Y} = \frac{1}{\sqrt{a^2 + \omega_o^2}} \angle \tan^{-1}(-\omega_o/a) \tag{7.10}$$

$$= \frac{1}{\sqrt{a^2 + \omega_o^2}} \exp(\tan^{-1}(-\omega_o/a)). \tag{7.11}$$

The time-domain waveform $y(t)$ is derived by multiplying this expression with $\exp(j\omega_o t)$ and taking the real part, yielding

$$y(t) = \frac{1}{\sqrt{a^2 + \omega_o^2}} \cos(\omega_o t - \tan^{-1}(\omega_o/a)), \quad t \in \mathcal{R}, \tag{7.12}$$

where we have used the fact that tangent is an odd function to extract the minus sign. This is the same expression as the *steady-state* response in (6.72) as $t \to \infty$, which was derived for a first-order ODE with input $\cos(\omega_o t)u(t)$. Obviously, the phasor approach allowed for much simpler calculations compared with the ODE time-domain techniques developed in the previous chapter.

Example 7.1 For the ODE in (7.3), let $\omega_o = 4$ rad/s and $a = 2$ such that

$$|\mathbf{Y}| = \frac{1}{\sqrt{a^2 + \omega_o^2}} = 1/2\sqrt{5} \approx 0.2236, \qquad \arg(\mathbf{Y}) = \tan^{-1}(-\omega_o/a) \approx -1.1071, \tag{7.13}$$

which has the time-domain solution

$$y(t) = 0.2236\cos(4t - 1.1071), \quad t \in \mathcal{R}. \tag{7.14}$$

This waveform is plotted in Figure 7.1 along with $\cos(4t)$, which is the right-hand side of (7.3) (the input waveform of the ODE). The output voltage modeled by the ODE has the same frequency but a reduced amplitude, and it is shifted to the right (delayed) relative to the input cosine. As mentioned previously, this behavior is characteristic of an LTI system.

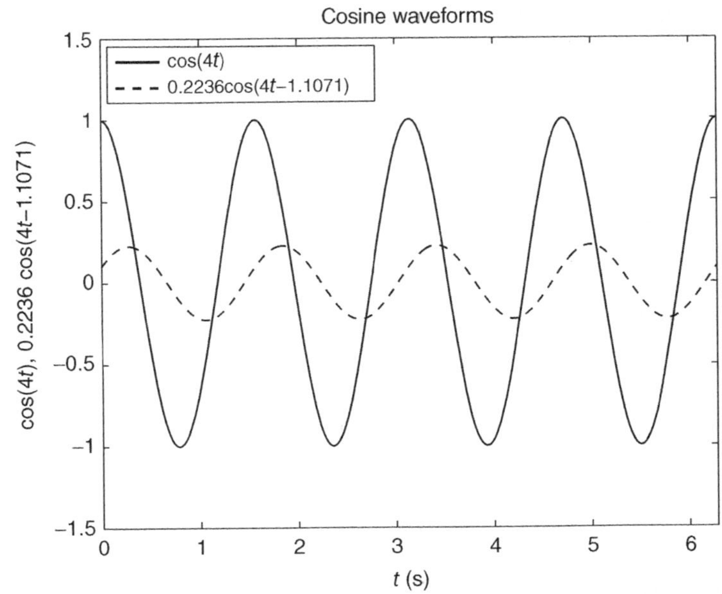

Figure 7.1 Cosine functions for the first-order ODE in Example 7.1.

For the second-order ODE in (6.81), we have using phasors

$$\frac{d^2}{dt^2}\mathbf{Y}\exp(j\omega_o t) + a_1\frac{d}{dt}\mathbf{Y}\exp(j\omega_o t) + a_0\mathbf{Y}\exp(j\omega_o t) = \exp(j\omega_o t), \tag{7.15}$$

which, after canceling the exponentials, yields the algebraic equation

$$-\omega_o^2\mathbf{Y} + j\omega_o a_1\mathbf{Y} + a_0\mathbf{Y} = 1 \implies \mathbf{Y} = \frac{a_0 - \omega_o^2 - j\omega_o a_1}{(a_0 - \omega_o^2)^2 + (\omega_o a_1)^2}. \tag{7.16}$$

The magnitude and phase are

$$|\mathbf{Y}| = \frac{1}{\sqrt{(a_0 - \omega_o^2)^2 + (\omega_o a_1)^2}}, \qquad \arg(\mathbf{Y}) = \tan^{-1}\left(\frac{\omega_o a_1}{\omega_o^2 - a_0}\right), \tag{7.17}$$

and the time-domain waveform is

$$y(t) = \frac{1}{\sqrt{(a_0 - \omega_o^2)^2 + (\omega_o a_1)^2}}\cos\left(\omega_o t + \tan^{-1}\left(\frac{\omega_o a_1}{\omega_o^2 - a_0}\right)\right), \quad t \in \mathcal{R}. \tag{7.18}$$

Although these expressions are more complicated than those for the first-order ODE, the output is still a cosine waveform with the magnitude and phase in (7.17). Since the input cosine has been active for all $t \in \mathcal{R}$, there is no transient response and the result in (7.18) is the steady-state response.

As mentioned after (7.12), a phasor solution also corresponds to the steady-state response of a system even if the input cosine is active starting at a finite time such as $\cos(\omega_o t)u(t)$. It is intuitive that as $t \to \infty$, the transient response tends to 0 (assuming a stable system), and the asymptotic solution is a cosine with the appropriate amplitude and phase. Phasors are useful for finding the steady-state response of a circuit only if the voltage and current sources are sinusoidal. If the input is a linear combination of sinusoidal signals with different frequencies, then *superposition* can be used to find the overall solution (see Chapter 5). For more general signals, the Laplace transform provides both the transient response and the steady-state response of an LTI system for signals starting at a finite time (such as $t = 0$).

7.3 EIGENFUNCTIONS

Consider again the first-order ODE in (7.3):

$$\frac{d}{dt}y(t) + ay(t) = 0, \tag{7.19}$$

where we have replaced the cosine function on the right-hand side with 0 so that the equation is homogeneous. If the initial condition $y(0)$ is nonzero, then we know from Chapter 6 that the form of the solution is exponential $y(t) = c\exp(\alpha t)u(t)$, where $\alpha < 0$ is a function of a, and the coefficient c depends on $y(0)$. Substituting $y(t)$ into (7.19) yields

$$\alpha c\exp(\alpha t) + ac\exp(\alpha t) = c\exp(\alpha t)(a + \alpha) = 0. \tag{7.20}$$

Assuming finite t such that the exponential is nonzero and finite, we find from the right-hand side that the exponent is $\alpha = -a$. The initial condition yields

$$y(0) = c\exp(-at)u(t)|_{t=0} = c, \tag{7.21}$$

and the solution is a decaying exponential for $a > 0$:

$$y(t) = y(0)\exp(-at)u(t). \tag{7.22}$$

A nonzero function that has the property in (7.20), where its substitution results in a scaled version of itself, is called an *eigenfunction* of that system. This is due to the fact that the derivative of the exponential function is another exponential function:

$$\frac{d}{dt}[c\exp(\alpha t)] = \alpha c\exp(\alpha t), \tag{7.23}$$

and it extends to derivatives of any order:

$$\frac{d^n}{dt^n}[c\exp(\alpha t)] = \alpha^n c\exp(\alpha t). \tag{7.24}$$

An eigenfunction is similar to an eigenvector of a matrix (see Chapter 3):

$$\mathbf{Ax} = \lambda\mathbf{x}, \tag{7.25}$$

where the eigenvector $\mathbf{x}$ (a column vector) appears on the right-hand side of the equation and is scaled by the eigenvalue λ. In (7.23) and (7.24), α and α^n are the eigenvalues of the differential *operators* d/dt and d^n/dt^n, respectively, and $\exp(\alpha t)$ is the eigenfunction.

Definition: Eigenfunction An *eigenfunction* of a linear operator L is a function such that L operating on it yields a scaled version of the same function.

In matrix algebra with $\mathbf{Ax} = \lambda\mathbf{x}$, the operator is the matrix $\mathbf{A}$, and for LTI systems modeled by an ODE, it is the derivative.

For the second-order homogeneous ODE:

$$\frac{d^2}{dt^2}y(t) + a_1\frac{d}{dt}y(t) + a_0y(t) = 0, \tag{7.26}$$

we know that a solution is of the form $y(t) = c\exp(st)$, where s may be complex even though the coefficients are real. Substituting $y(t)$ into (7.26) yields

$$s^2c\exp(st) + a_1sc\exp(st) + a_0c\exp(st) = (s^2 + a_1s + a_0)c\exp(st) = 0. \tag{7.27}$$

Assuming finite t, the term $c\exp(st)$ cancels from the equation, and then we can solve the *characteristic equation* $s^2 + a_1s + a_0 = 0$ to find its eigenvalues, which may be (i) real and distinct, (ii) real and repeated, or (iii) a complex conjugate pair. These three cases were examined in the time domain in Chapter 6. The usefulness of the Laplace transform for solving linear ODEs follows from the fact that the exponential function is an eigenfunction of LTI systems.

7.4 LAPLACE TRANSFORM

The Laplace transform is a specific type of integral transform.

Definition: Integral Transform An *integral transform* is an integral that maps a function of one variable to a different function of another variable:

$$X(p) \triangleq \int_{t_1}^{t_2} x(t)k(p,t)dt, \tag{7.28}$$

where $k(p,t)$ is called the *kernel function*. Uppercase letters are usually used to represent integral transforms.

TABLE 7.1 Integral Transforms

Kernel	Transform	Variable
$\exp(-st)$	Laplace transform	Complex $s = \sigma + j\omega$
$\exp(-j\omega t)$	Fourier transform	Imaginary $j\omega$
$1/\pi(p-t)$	Hilbert transform	Real p
$t^p - 1$	Mellin transform	Real p
$2t/\sqrt{t^2-p^2}$	Abel transform	Real p

Examples of different kernels are summarized in Table 7.1. These transforms are useful for a range of applications, though we focus on the Laplace transform and the Fourier transform.

Definition: Bilateral Laplace Transform The *bilateral Laplace transform* is an integral transform with kernel $k(s,t) = \exp(-st)$:

$$X(s) \triangleq \int_{-\infty}^{\infty} x(t)\exp(-st)dt, \tag{7.29}$$

where by convention a minus sign is included in the exponent, and $s \triangleq \sigma + j\omega$ is a complex variable. The following notation is used for the bilateral Laplace transform:

$$\mathcal{L}_b\{x(t)\} = X(s), \qquad x(t) \xrightarrow{\mathcal{L}_b} X(s). \tag{7.30}$$

In some mathematics books, real-valued variable p is used instead of s in the definition of the Laplace transform. However, since the roots associated with an ODE can be complex-valued as shown in Chapter 6, it is advantageous to use complex s in (7.29). The kernel $\exp(-st)$ of the Laplace transform is an eigenfunction of an LTI system that causes an ODE in the time domain to become an algebraic equation in the s-domain.

The values of s for which (7.29) yields a finite transform $X(s)$ is called the *region of convergence* (ROC). There are four ROCs depending on the type of function $x(t)$: (i) finite duration $t_1 \le t \le t_2$, (ii) right-sided $t \ge t_1$, (iii) left-sided $t \le t_2$, and (iv) two-sided $t \in \mathcal{R}$. Although we consider some functions for cases (iii) and (iv), we are concerned mainly with finite duration and right-sided functions, as is the case in most courses on circuits and systems.

If the function represents the impulse response $h(t)$ of a system, we are interested in whether or not that system is stable.

Definition: Stable System A system with impulse response function $h(t)$ is *stable* if it is absolutely integrable:

$$\int_{-\infty}^{\infty} |h(t)|dt < \infty. \tag{7.31}$$

This property is also known as bounded-input bounded-output (BIBO) stability.

Example 7.2 Examples of stable systems and their ROCs are given as follows and illustrated in Figure 7.2.

$$\textit{finite duration}\colon\ h_1(t) = u(t) - u(t-T)\colon \text{ entire } s\text{-plane}, \tag{7.32}$$

$$\textit{right-sided stable}\colon\ h_2(t) = \exp(-at)u(t)\colon \text{Re}(s) > a \text{ with } a > 0, \tag{7.33}$$

$$\textit{right-sided marginally stable}\colon\ h_3(t) = u(t)\colon \text{Re}(s) > 0, \tag{7.34}$$

$$\textit{right-sided unstable}\colon\ h_4(t) = \exp(at)u(t)\colon \text{Re}(s) > a \text{ with } a > 0. \tag{7.35}$$

When the ROC is the entire s-plane, we can also write $s \in C$. Because of the unit step function $u(t)$, these examples are nonzero only for $t \geq 0$. Such systems with impulse response function $h(t)$ starting at or after $t = 0$ are called causal.

Definition: Causal System An LTI system with impulse response function $h(t)$ is *causal* if $h(t)$ is nonzero only for $t \geq 0$.

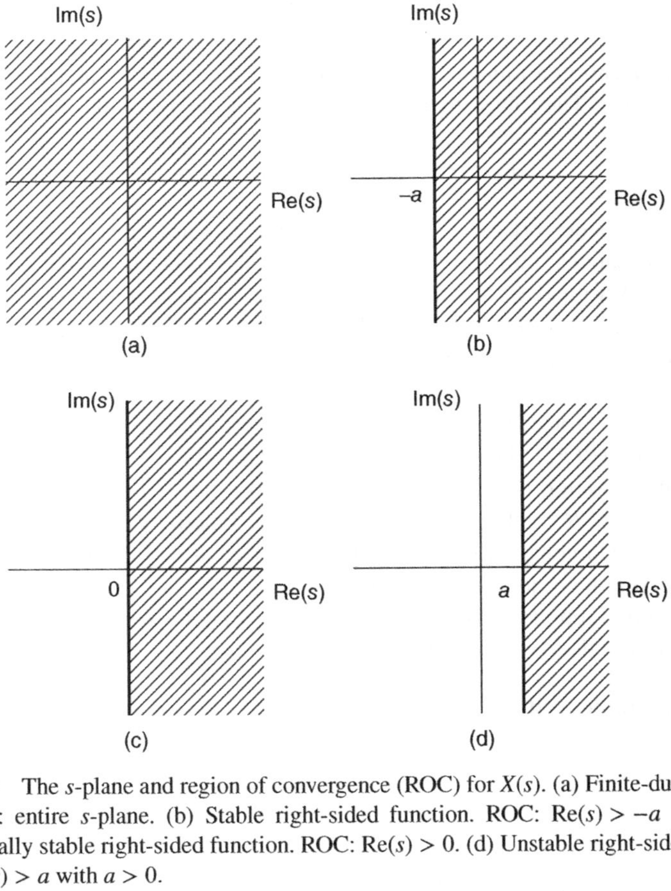

Figure 7.2 The s-plane and region of convergence (ROC) for $X(s)$. (a) Finite-duration function. ROC: entire s-plane. (b) Stable right-sided function. ROC: $\text{Re}(s) > -a$ with $a > 0$. (c) Marginally stable right-sided function. ROC: $\text{Re}(s) > 0$. (d) Unstable right-sided function. ROC: $\text{Re}(s) > a$ with $a > 0$.

For a linear system modeled by an ODE, we are generally interested in right-sided signals (or finite-duration signals) starting at the origin, and the unilateral (one-sided) Laplace transform is used to solve the ODE. In such cases, the dependent variable $y(t)$ and its derivatives may have nonzero initial values.

Definition: Unilateral Laplace Transform The *unilateral Laplace transform* is

$$X(s) \triangleq \int_{0^-}^{\infty} x(t)\exp(-st)dt, \tag{7.36}$$

where $t = 0^-$ is used in the event there is a singular function at the origin such as $\delta(t)$ or its derivatives. The following notation is used for the unilateral Laplace transform:

$$\mathcal{L}\{x(t)\} = X(s), \qquad x(t) \xrightarrow{\mathcal{L}} X(s). \tag{7.37}$$

Generally, the unilateral Laplace transform is known simply as the Laplace transform. For notational convenience, we often write the lower limit as 0 with the understanding that it is actually 0^-. Later it will be necessary to distinguish between 0^- and 0^+: "just before" $t = 0$ and "just after" $t = 0$, respectively. For example, the unit step function $u(t)$ is defined to be 0 at $t = 0^-$ and it is 1 at $t = 0^+$.

Definition: Initial State and Initial Condition The *initial state* of the output $y(t)$ of a system modeled by an ODE is $y(0^-)$ and its *initial condition* is $y(0^+)$.

In the initial value theorem (IVT) presented later, the initial condition is called the *initial value*. For example, the solution $y(t) = \exp(-\alpha t)u(t)$ has initial value $y(0^+) = 1$, but its initial state is $y(0^-) = 0$.

Example 7.3 In this example, we derive the Laplace transform for each type of system in (7.32)–(7.35) and illustrate how the ROC is determined. For (7.33):

$$\begin{aligned} H_2(s) &= \int_{-\infty}^{\infty} \exp(-\alpha t)u(t)\exp(-st)dt \\ &= \int_{0}^{\infty} \exp(-(s+\alpha)t)dt \\ &= -\frac{1}{s+b}\exp(-(s+\alpha)t)\bigg|_0^{\infty}, \end{aligned} \tag{7.38}$$

where $u(t)$ gives the lower limit of integration. This system is stable provided $\alpha > 0$ so that $\exp(-\alpha t)u(t)$ decays to 0. When evaluating the last expression at the limits of integration, the exponent must satisfy $\text{Re}(s + \alpha) > 0 \implies \text{Re}(s) > -\alpha$ in order for the transform to exist (be finite). Thus, the ROC is found by placing a bound on the exponent such that the exponential is 0 when evaluated at $t \to \infty$. The Laplace transform is

$$H_2(s) = \frac{1}{s+\alpha}, \qquad \text{ROC: Re}(s) > -\alpha. \tag{7.39}$$

This function is not defined at $s = -\alpha$, which is why this point on the s-plane is not included in the ROC; such a singularity is called a *pole* (see Chapter 5 and Appendix E). From this result, we expect for the finite duration signal in (7.32) that there will be no unboundedness issues because the limits of integration are finite:

$$\begin{aligned} H_1(s) &= \int_0^1 \exp(-st)dt = -\frac{1}{s}\exp(-st)\Big|_0^1 \\ &= \frac{1}{s}[1 - \exp(-s)], \qquad \text{ROC: } s \in C, \end{aligned} \tag{7.40}$$

where the step functions have provided both limits of integration. Since these limits are finite, there is no need to restrict s; the ROC is the entire s-plane. Note that $s = 0$ is not of concern because l'Hôpital's rule shows that the transform is bounded and well defined for that value:

$$H_1(0) = \frac{\frac{d}{ds}[1 - \exp(-s)]\Big|_{s=0}}{\frac{d}{ds}s\Big|_{s=0}} = 1. \tag{7.41}$$

Although there appears to be a pole at $s = 0$ in (7.40), it is actually a *removable singularity* as discussed in Chapter 5. For the third function in (7.34):

$$\begin{aligned} H_3(s) &= \int_0^\infty \exp(-st)dt \\ &= -\frac{1}{s}\exp(-st)\Big|_0^\infty = \frac{1}{s}, \qquad \text{ROC: Re}(s) > 0. \end{aligned} \tag{7.42}$$

The Laplace transform of this waveform has a pole on the imaginary axis where $s = j\omega$. Systems with one real pole or two complex conjugate poles on the imaginary axis are called *marginally stable*. Finally, for the unstable system in (7.35):

$$\begin{aligned} H_4(s) &= \int_0^\infty \exp(\alpha t)\exp(-st)dt \\ &= -\frac{1}{s-\alpha}\exp(-(s-\alpha)t)\Big|_0^\infty = \frac{1}{s-\alpha}, \qquad \text{ROC: Re}(s) > \alpha. \end{aligned} \tag{7.43}$$

Thus, the Laplace transform can be derived for an unstable system provided the ROC exists. From these examples, we find that a right-sided system with poles is stable if they all lie in the *left half* of the s-plane: $\text{Re}(s) < 0$. When there are poles in the right half of the s-plane for a right-sided system, then it is unstable. Since the ROC must lie to the right of all poles (again, for a right-sided system), we conclude that the system is stable if the ROC includes the $j\omega$ axis, as it does in Figure 7.2(a) and (b).

Example 7.4 Next, we illustrate how to compute the Laplace transform of the general exponential function $x(t) = \beta^{-\alpha t}u(t)$ with $\alpha > 0$ and $\beta \neq e$:

$$\int_0^\infty \beta^{-\alpha t} \exp(-st)dt = \int_0^\infty \exp(\ln(\beta^{-\alpha t}) - st)dt$$
$$= \int_0^\infty \exp(-(\alpha \ln(\beta) + s)t)dt, \tag{7.44}$$

where properties of $\exp(\cdot)$ and $\ln(\cdot)$ have been used. Thus,

$$\mathcal{L}\{\beta^{-\alpha t}u(t)\} = -\frac{\exp(-(\alpha \ln(\beta) + s)t)}{\alpha \ln(\beta) + s}\bigg|_0^\infty = \frac{1}{\alpha \ln(\beta) + s}, \tag{7.45}$$

which has ROC $\text{Re}(s) > -\alpha \ln(\beta)$. The pole is located at $s = -\alpha \ln(\beta)$, and the function is bounded only for $\beta \geq 1$ (and $\alpha > 0$). When $\beta = e$, the previous result in (7.39) for $\exp(-\alpha t)u(t)$ is obtained. This example illustrates that the "natural" choice for the exponential function is $\beta = e$, which avoids extra terms like $\ln(\beta)$ in (7.45).

Example 7.5 For a causal system represented by a linear ODE or an integro-differential equation, the input is usually assumed to be multiplied by the unit step function, which must be taken into account when finding the Laplace transform. This is illustrated for the following first-order ODE:

$$\frac{d}{dt}y(t) + ay(t) = bu(t), \tag{7.46}$$

whose Laplace transform is

$$sY(s) - y(0^-) + aY(s) = b/s \implies Y(s) = \frac{b}{s(s+a)} + \frac{y(0^-)}{s+a}, \tag{7.47}$$

which can be expanded as

$$Y(s) = \frac{b/a}{s} - \frac{b/a}{s+a} + \frac{y(0^-)}{s+a}. \tag{7.48}$$

The corresponding time-domain function is

$$y(t) = (b/a)u(t) + [y(0^-) - b/a] \exp(-at)u(t). \tag{7.49}$$

The steady-state response is $(b/a)u(t)$, and the last term is the transient response, which decays to 0. It is necessary that $1/s$ be included on the right-hand side of the first equation in (7.47) so that the correct causal solution is obtained. If $1/s$ is missing, then there would be no steady-state term on the right-hand side of (7.49), which is incorrect because the right-hand side of (7.46) is nonzero for $t \in \mathcal{R}^+$, implying a nonzero steady-state solution for $y(t)$.

Example 7.6 We briefly consider another integral transform. For the kernel function:

$$k(p,t) = \frac{1}{\pi(p-t)}, \tag{7.50}$$

with real-valued p, we have the *Hilbert transform*:

$$X(p) = \frac{1}{\pi}\int_{-\infty}^{\infty} \frac{x(t)}{p-t}dt. \tag{7.51}$$

This expression is actually a *convolution* between a system with impulse response function $h(t) = 1/\pi t$ and input $x(t)$:

$$X(p) = \frac{1}{\pi t} * x(t). \tag{7.52}$$

For $x(t) = \cos(\omega_o t)$:

$$X(p) = -(1/\pi)\int_{-\infty}^{\infty} \frac{\cos(\omega_o t)}{p-t}dt, \tag{7.53}$$

which can be evaluated by changing variables to $v \triangleq p - t$ and using a trigonometric identity for cosine:

$$\begin{aligned} X(p) &= -(1/\pi)\int_{-\infty}^{\infty} \frac{\cos(\omega_o(v+p))}{v}dv \\ &= -(1/\pi)\int_{-\infty}^{\infty} \left[\frac{\cos(\omega_o v)\cos(\omega_o p)}{v} - \frac{\sin(\omega_o v)\sin(\omega_o p)}{v}\right] dv. \end{aligned} \tag{7.54}$$

The first term is 0 because the ratio $\cos(\omega_o v)/v$ is an odd function. Thus

$$X(p) = (1/\pi)\sin(\omega_o p)\int_{-\infty}^{\infty} \frac{\sin(\omega_o v)}{v}dv = \sin(\omega_o p), \tag{7.55}$$

which simplifies because the improper integral is π. This result is easily verified because the following ratio known as the *sinc function* has unit area:

$$\text{sinc}(x) \triangleq \frac{\sin(\pi x)}{\pi x}, \tag{7.56}$$

where π is implicit on the left-hand side. As a result, changing variables to $\omega_o v = \pi u$ in (7.55) yields

$$\int_{-\infty}^{\infty} \frac{\sin(\omega_o v)}{v}dv = \int_{-\infty}^{\infty} \frac{\sin(\pi u)}{\pi u/\omega_o}(\pi/\omega_o)du = \pi\int_{-\infty}^{\infty} \text{sinc}(u)du = \pi, \tag{7.57}$$

TABLE 7.2 Laplace Transform Pairs: Impulsive, Step, and Ramp

Time-Domain $x(t)$	Laplace Transform $X(s)$	ROC ($\sigma = \text{Re}(s)$)
$\delta(t)$	1	$s \in C$
$\delta^{(n)}(t)$	s^n	$s \in C$ ($n \in \mathcal{Z}$)
$\text{rect}(t)$	$2\sinh(s/2)/s$	$s \in C$
$\text{tri}(t)$	$4\sinh^2(s/2)/s^2$	$s \in C$
$u(t)$	$1/s$	$\sigma > 0$
$u(-t)$	$-1/s$	$\sigma < 0$
$\text{sgn}(t)$	$2/s$	$\sigma = 0$ (Except $s = 0$)
$r(t)$	$1/s^2$	$\sigma > 0$
$t^n u(t)$	$n!/s^{n+1}$	$\sigma > 0$ ($n \in \mathcal{Z}$)
$\|t\|$	$2/s^2$	$\sigma = 0$ (Except $s = 0$)

where ω_o has canceled. It is a well-known result that the Hilbert transform of cosine is sine. The Hilbert transform is useful for studying amplitude modulated (AM) signals in analog communication systems. In particular, it can be used to model *single-sideband* (SSB) AM modulation as described later in Problem 8.25.

The Laplace transform is important for linear systems modeled by an ODE with constant coefficients because the ODE is transformed to an algebraic equation, as was demonstrated for the second-order system in (7.26), resulting in $s^2 + a_1 s + a_0 = 0$. This characteristic equation of the system is solved for s, and the s-domain solution is transformed to the time-domain solution $y(t)$ for the ODE.

Definition: Inverse Laplace Transform The *inverse Laplace transform* is

$$x(t) = \frac{1}{2\pi j}\int_{c-j\infty}^{c+j\infty} X(s)\exp(st)ds, \tag{7.58}$$

where the integration is performed along any vertical line on the s-plane as long as c is located in the ROC. The following notation is used:

$$\mathcal{L}^{-1}\{X(s)\} = x(t), \qquad X(s) \xrightarrow{\mathcal{L}^{-1}} x(t). \tag{7.59}$$

It turns out that for LTI systems, it is not necessary to compute this integral; instead, a *partial fraction expansion* (PFE) can be used to write a *rational* s-domain function as a sum of simpler functions called partial fractions for which the inverse transforms are readily found using a table of Laplace transform pairs such as Tables 7.2 and 7.3.

7.5 LAPLACE TRANSFORMS AND GENERALIZED FUNCTIONS

In this section, we derive Laplace transforms using the theory of generalized functions (distributions) that was discussed in Chapter 5. This requires that we define another

TABLE 7.3 Laplace Transform Pairs: Exponential and Sinusoidal ($\alpha > 0$ and $\omega_o > 0$)

Time-Domain $x(t)$	Laplace Transform $X(s)$	ROC ($\sigma = \text{Re}(s)$)
$\exp(-\alpha t)u(t)$	$1/(s+\alpha)$	$\sigma > -\alpha$
$[1-\exp(-\alpha t)]u(t)$	$\alpha/s(s+\alpha)$	$\sigma > 0$
$\exp(\alpha t)u(-t)$	$-1/(s-\alpha)$	$\sigma < \alpha$
$\exp(-\alpha\|t\|)$	$2\alpha/(\alpha^2 - s^2)$	$-\alpha < \sigma < \alpha$
$\exp(-\alpha t^2)$	$\sqrt{\pi/\alpha}\exp(s^2/4\alpha)$	$s \in C$
$t^n\exp(-\alpha t)u(t)$	$n!/(s+\alpha)^{n+1}$	$\sigma > -\alpha$ ($n \in \mathcal{Z}$)
$t^n\exp(\alpha t)u(-t)$	$-n!/(s+\alpha)^{n+1}$	$\sigma < \alpha$ ($n \in \mathcal{Z}$)
$\cosh(\beta t)u(t)$	$s/(s^2-\beta^2)$	$\sigma > \|\beta\|$
$\sinh(\beta t)u(t)$	$\alpha/(s^2-\beta^2)$	$\sigma > \|\beta\|$
$\cos(\omega_o t)u(t)$	$s/(s^2+\omega_o^2)$	$\sigma > 0$
$\sin(\omega_o t)u(t)$	$\omega_o/(s^2+\omega_o^2)$	$\sigma > 0$
$t\cos(\omega_o t)u(t)$	$(s^2-\omega_o^2)/(s^2+\omega_o^2)^2$	$\sigma > 0$
$t\sin(\omega_o t)u(t)$	$2\omega_o s/(s^2+\omega_o^2)^2$	$\sigma > 0$
$\exp(-\alpha t)\cos(\omega_o t)u(t)$	$(s+\alpha)/[(s+\alpha)^2+\omega_o^2]$	$\sigma > -\alpha$
$\exp(-\alpha t)\sin(\omega_o t)u(t)$	$\omega_o/[(s+\alpha)^2+\omega_o^2]$	$\sigma > -\alpha$
$t\exp(-\alpha t)\cos(\omega_o t)u(t)$	$[(s+\alpha)^2-\omega_o^2]/[(s+\alpha)^2+\omega_o^2]^2$	$\sigma > -\alpha$
$t\exp(-\alpha t)\sin(\omega_o t)u(t)$	$2\omega_o(s+\alpha)/[(s+\alpha)^2+\omega_o^2]^2$	$\sigma > -\alpha$

type of test function because the kernel $\exp(-st)$ does not have a compact support (Kanwal, 2004).

Definition: Test Function of Exponential Decay A *test function of exponential decay* has the following two properties: (i) $\phi(t)$ is infinitely differentiable (smooth) and (ii) all derivatives of $\phi(t)$ decrease to 0 more rapidly than the exponential function $\exp(\alpha t)$ as $t \to \infty$ for every $\alpha \in \mathcal{R}$. The second property can be written as

$$\left|\exp(\alpha t)\frac{d^n}{dt^n}\phi(t)\right| < c, \qquad |t| \to \infty, \tag{7.60}$$

for every $c > 0$ and $n \in \mathcal{Z}^+$.

We denote this class of test functions by $\mathcal{E}$.

Definition: Distribution of Exponential Growth A *distribution of exponential growth* $\langle x, \phi\rangle$ is a linear functional on the set $\mathcal{E}$ written as

$$\langle x, \phi\rangle \triangleq \int_{-\infty}^{\infty} x(t)\phi(t)dt, \qquad \phi(t) \in \mathcal{E}. \tag{7.61}$$

A function of exponential growth satisfies

$$\left|\frac{d^n}{dt^n}x(t)\right| \le c\exp(\alpha t), \tag{7.62}$$

as $|t| \to \infty$ for some $\alpha \in \mathcal{R}$ and $c > 0$. The dual space of distributions of exponential growth is denoted by $\mathcal{E}'$.

The linearity and continuity properties of functionals discussed in Chapter 5 also apply to $\mathcal{E}$. Since any test function with compact support satisfies (7.60), $\mathcal{D}$ has been expanded to $\mathcal{E}$ such that $\mathcal{D} \subset \mathcal{E}$. Some distributions in $\mathcal{D}'$ are not defined for (7.61), and as a result, $\mathcal{E}' \subset \mathcal{D}'$.

The definition in (7.60) also includes complex test functions, which are needed for the Laplace transform because of the complex exponential $\phi(t) = \exp(-st)$. For distribution $x(t)$, the Laplace transform can be written as

$$\langle x, \exp(-st)\rangle = \int_{-\infty}^{\infty} x(t)\exp(-st)dt. \tag{7.63}$$

The expression on the left-hand side applies to any generalized function defined on $\mathcal{E}$, including singular distributions.

Example 7.7 The Laplace transform of the Dirac delta function is

$$\langle \delta, \exp(-st)\rangle = \int_{-\infty}^{\infty} \delta(t)\exp(-st)dt = 1, \tag{7.64}$$

which follows from the sifting property of $\delta(t)$ (also from its definition). For its first derivative, the unit doublet $\delta'(t)$:

$$\begin{aligned}\langle \delta', \exp(-st)\rangle &= \int_{-\infty}^{\infty} \delta'(t)\exp(-st)dt \\ &= \delta(t)\exp(-st)|_{-\infty}^{\infty} - \int_{-\infty}^{\infty} \delta(t)\frac{d}{dt}\exp(-st)dt \\ &= -\left\langle \delta, \frac{d}{dt}\exp(-st)\right\rangle = \langle \delta, s\exp(-st)\rangle = s,\end{aligned} \tag{7.65}$$

which is the derivative property in (5.77). For the second derivative:

$$\langle \delta^{(2)}, \exp(-st)\rangle = -\left\langle \delta', \frac{d}{dt}\exp(-st)\right\rangle = \langle \delta', s\exp(-st)\rangle, \tag{7.66}$$

and repeating this operation yields

$$\langle \delta^{(2)}, \exp(-st)\rangle = -\left\langle \delta, s\frac{d}{dt}\exp(-st)\right\rangle = \langle \delta, s^2\exp(-st)\rangle = s^2. \tag{7.67}$$

It is clear from these cases that for the nth derivative:

$$\langle \delta^{(n)}, \exp(-st)\rangle = s^n. \tag{7.68}$$

The unilateral Laplace transform can also be defined for distributions of exponential growth by replacing (7.63) with

$$\langle x, \exp(-st)\rangle = \int_{0^-}^{\infty} x(t)\exp(-st). \tag{7.69}$$

Most of the properties of the bilateral Laplace transform are the same for the unilateral transform, except for its derivative properties. For the first derivative:

$$\begin{aligned}\langle x', \exp(-st)\rangle &= \int_{0^-}^{\infty} x'(t)\exp(-st)dt \\ &= x(t)\exp(-st)|_{0^-}^{\infty} - \int_{0^-}^{\infty} x(t)\frac{d}{dt}\exp(-st)dt \\ &= 0 - x(0^-) + s\langle x, \exp(-st)\rangle = sX(s) - x(0^-). \end{aligned}\tag{7.70}$$

The main difference in this expression is that the lower limit of integration is finite, in which case $\exp(0) = 1$, resulting in the term $x(0^-)$. Recall that the initial state $x(0^-)$ is not the initial condition of the time-domain waveform as used in Chapter 6 on ODEs. In this chapter, we denote the initial condition by $x(0^+)$, which is also called the initial value.

Example 7.8 For the unit step function $x(t) = u(t)$, these two initial quantities are $x(0^-) = 0$ and $x(0^+) = 1$, which are also the values for $x(t) = \exp(-\alpha t)u(t)$. Nonzero $x(0^-)$ can arise in a circuit that is in steady state before an input is applied at time 0. For example, the current through an inductor might be some nonzero $i_L(0^-)$ before a circuit switch is closed; this is its initial state. Since the current in an inductor cannot change instantaneously, it turns out that $i_L(0^-) = i_L(0^+)$. Similarly, since the voltage across a capacitor cannot change instantaneously, $v_C(0^-) = v_C(0^+)$. However, in general, we find that $x(0^-) \neq x(0^+)$, and in many problems $x(0^-) = 0$.

For the second distributional derivative:

$$\begin{aligned}\langle x^{(2)}, \exp(-st)\rangle &= x'(t)\exp(-st)|_{0^-}^{\infty} - \int_{0^-}^{\infty} x'(t)\frac{d}{dt}\exp(-st)dt \\ &= s\langle x', \exp(-st)\rangle - x'(0^-). \end{aligned}\tag{7.71}$$

Substituting the result from (7.70) yields

$$\begin{aligned}\langle x^{(2)}, \exp(-st)\rangle &= s^2\langle x, \exp(-st)\rangle - sx(0^-) - x'(0^-) \\ &= s^2X(s) - sx(0^-) - x'(0^-). \end{aligned}\tag{7.72}$$

For the nth derivative, a similar result is obtained:

$$\langle x^{(n)}, \exp(-st)\rangle = s^nX(s) - \sum_{m=0}^{n-1} s^{n-m-1}x^{(m)}(0^-). \tag{7.73}$$

Example 7.9 Consider the one-sided exponential function, which has a discontinuity at the origin: $x(t) = \exp(-\alpha t)u(t)$. Its derivative from the product rule is

$$\begin{aligned}\frac{d}{dt}\exp(-\alpha t)u(t) &= \exp(-\alpha t)\frac{d}{dt}u(t) + u(t)\frac{d}{dt}\exp(-\alpha t) \\ &= \delta(t)\exp(-\alpha t) - \alpha\exp(-\alpha t)u(t) \\ &= \delta(t) - \alpha\exp(-\alpha t)u(t), \qquad (7.74)\end{aligned}$$

where the sampling property of the Dirac delta function has been used for the first term of the third line. From the expression in (5.77) for distributional derivatives:

$$\begin{aligned}\left\langle\frac{d}{dt}\exp(-\alpha t)u(t), \phi(t)\right\rangle &= -\langle\exp(-\alpha t)u(t), \phi'(t)\rangle \\ &= -\int_0^\infty \exp(-\alpha t)\frac{d}{dt}\phi(t)dt \\ &= -\exp(-\alpha t)\phi(t)|_0^\infty - \alpha\int_0^\infty \exp(-\alpha t)\phi(t)dt \\ &= \phi(0) - \alpha\langle\exp(-\alpha t)u(t), \phi(t)\rangle. \qquad (7.75)\end{aligned}$$

The leading term $\phi(0)$ is the distribution of the Dirac delta function, and the result in (7.74) is verified. The Laplace transform is derived by substituting $\phi(t) = \exp(-st)$:

$$\begin{aligned}\left\langle\frac{d}{dt}\exp(-\alpha t)u(t), \exp(-st)\right\rangle &= \exp(0) - \alpha\langle\exp(-\alpha tu(t), \exp(-st)\rangle \\ &= 1 - \frac{\alpha}{s+\alpha} = \frac{s}{s+\alpha}, \qquad (7.76)\end{aligned}$$

which has ROC $\text{Re}(s) > -\alpha$. This is the same expression derived using (7.70):

$$\langle x', \exp(-st)\rangle = s\frac{1}{s+\alpha} - 0 = \frac{s}{s+\alpha}, \qquad (7.77)$$

with $x(0^-) = 0$.

We return to generalized functions in Chapter 8 on Fourier transforms where the transforms can yield singular distributions such as the Dirac delta function and its derivatives. Tables 7.2 and 7.3 provide a summary of several Laplace transform pairs, most of which are right-sided functions. The ROC for each case is specified in terms of σ, which is the real part of $s = \sigma + j\omega$. Appendix B has several *inverse* Laplace transform pairs where the transform is given first. The reader might find these useful because it is not necessary to perform a PFE on the transforms with multiple poles. Extensive summaries for several functions and their transforms are provided in Appendix A.

7.6 LAPLACE TRANSFORM PROPERTIES

In this section, we prove several properties of the Laplace transform that are useful for finding the transforms of nonstandard signals and impulse response functions. Most of the properties extend to the bilateral Laplace transform, which is defined for $t \in \mathcal{R}$; differences are mentioned for some cases.

- *Linearity:*

$$\mathcal{L}\{a_1 x_1(t) + a_2 x_2(t)\} = a_1 X_1(s) + a_2 X_2(s). \tag{7.78}$$

  Proof: Since integration is a linear operation:

$$\begin{aligned}\int_0^\infty [a_1 x_1(t) + a_2 x_2(t)] \exp(-st)dt &= a_1 \int_0^\infty x_1(t) \exp(-st)dt \\ &\quad + a_2 \int_0^\infty x_2(t) \exp(-st)dt \\ &= a_1 X_1(s) + a_2 X_2(s).\end{aligned} \tag{7.79}$$

  The ROC must hold for the sum of the two functions, which means it is the intersection of the individual ROCs: $\text{ROC}_1 \cap \text{ROC}_2$.

- *Time shift:*

$$\mathcal{L}\{x(t-t_o)u(t-t_o)\} = \exp(-t_o s)X(s), \qquad t_o > 0. \tag{7.80}$$

  The unit step function is also delayed to emphasize that the shifted function is 0 for $t < t_o$. *Proof*: The change of variables $\tau \triangleq t - t_o \implies t = \tau + t_o$ yields

$$\begin{aligned}\int_{t_o}^\infty x(t-t_o) \exp(-st)dt &= \exp(-st_o) \int_0^\infty x(\tau) \exp(-s\tau)d\tau \\ &= \exp(-st_o)X(s),\end{aligned} \tag{7.81}$$

  and the ROC is unchanged. We assume $t_o > 0$ so that the function shifts only to the right and remains causal. (For the bilateral Laplace transform, there is no restriction on t_o.)

- *Time scaling:*

$$\mathcal{L}\{x(\alpha t)\} = \frac{1}{\alpha} X(s/\alpha), \qquad \alpha > 0. \tag{7.82}$$

  Proof: The change of variables $\tau \triangleq \alpha t \implies t = \tau/\alpha$ yields

$$\begin{aligned}\int_0^\infty x(\alpha t) \exp(-st)dt &= (1/\alpha) \int_0^\infty x(\tau) \exp(-(s/\alpha)\tau)d\tau \\ &= (1/\alpha)X(s/\alpha).\end{aligned} \tag{7.83}$$

If the original ROC is $\text{Re}(s) > -a$ for $a > 0$, then the new ROC is $\text{Re}(s/\alpha) > -a \implies \text{Re}(s) > -\alpha a$. We assume $\alpha > 0$ so that time is not reversed: it is expanded for $\alpha > 1$ and contracted for $\alpha < 1$. (For the bilateral Laplace transform, α can be negative, in which case the scale factor in (7.83) is $1/|\alpha|$; the argument of the transform is still s/α.)

- *Frequency shift:*

$$\mathcal{L}\{\exp(s_o t)x(t)\} = X(s - s_o). \tag{7.84}$$

Proof: From the Laplace transform definition:

$$\int_0^\infty x(t)\exp(s_o t)\exp(-st)dt = \int_0^\infty x(t)\exp(-(s-s_o)t)dt$$
$$= X(s - s_o). \tag{7.85}$$

If the original ROC is $\text{Re}(s) > -a$ for $a > 0$, then the new ROC is $\text{Re}(s - s_o) > -a \implies \text{Re}(s) > \text{Re}(s_o) - a = s_o - a$. Since we consider only real signals in the time domain, s_o is real-valued so that the product $x(t)\exp(s_o t)$ is real.

- *Derivatives:*

$$\mathcal{L}\left\{\frac{d^n}{dt^n}x(t)\right\} = s^n X(s) - \sum_{m=0}^{n-1} s^{n-m-1} \left.\frac{d^m x(t)}{dt^m}\right|_{t=0^-}$$
$$= s^n X(s) - \sum_{m=0}^{n-1} s^{n-m-1} x^{(m)}(0^-), \tag{7.86}$$

where $x^{(m)}(t)$ is the mth ordinary derivative of $x(t)$, and $x^{(m)}(0^-)$ is the initial state "just before" $t = 0$. The ROC is unchanged unless s^n cancels all s terms in the denominator of $X(s)$, in which case the ROC is determined by the remaining poles in the denominator. (For the bilateral Laplace transform, the sum in (7.86) is 0.) *Proof*: This was derived earlier in (7.70)–(7.73) using the generalized function approach.

- *Integral:*

$$\mathcal{L}\left\{\int_0^t x(t)dt\right\} = \frac{1}{s}X(s). \tag{7.87}$$

Proof: This result is also proved using integration by parts. In order to avoid confusion with the notation, the variable of integration in (7.87) is replaced with τ:

$$\int_0^\infty \int_0^t x(\tau)d\tau \exp(-st)dt = -(1/s)\exp(-st)\int_0^t x(\tau)d\tau\Big|_{t=0}^{\infty}$$
$$+ (1/s)\int_0^t x(t)\exp(-st)dt$$
$$= \frac{1}{s}X(s). \tag{7.88}$$

The first term on the right-hand side of the first line is 0 as $t \to \infty$ because it is assumed that $\exp(-st) \to 0$ faster than $x(t)$ increases as $t \to \infty$ (the integral is finite). The ROC is $\text{Re}(s) > 0$ because of the $1/s$ term (which is a pole at $s = 0$) unless, of course, $X(s)$ already has one or more poles at $s = 0$, in which case the ROC is unchanged.

- *Convolution:*

$$\mathcal{L}\{x(t) * h(t)\} = X(s)H(s), \tag{7.89}$$

where

$$x(t) * h(t) \triangleq \int_0^t x(\tau)h(t-\tau)d\tau = \int_0^t x(t-\tau)h(\tau)d\tau. \tag{7.90}$$

This is a symmetric property: $\mathcal{L}\{h(t) * x(t)\} = \mathcal{L}\{x(t) * h(t)\} = H(s)X(s) = X(s)H(s)$. The limits of integration are determined by the fact that $x(t)$ and $h(t)$ are causal: in the first integral, the lower limit is 0 because $x(\tau)$ is nonzero for $\tau \geq 0$, and the upper limit is t because $h(t-\tau)$ is nonzero for $t - \tau \geq 0 \implies \tau \leq t$. *Proof*: From the identity $\exp(-st) = \exp(-s(t-\tau))\exp(-s\tau)$:

$$\begin{aligned}\mathcal{L}\{x(t) * h(t)\} &= \int_0^\infty \int_0^t x(\tau)h(t-\tau)\exp(-st)d\tau dt \\ &= \int_0^\infty \int_0^t x(\tau)\exp(-s\tau)h(t-\tau)\exp(-s(t-\tau))d\tau dt. \end{aligned} \tag{7.91}$$

This type of double integral is not straightforward to evaluate because the outer integral is defined over t, which is the upper limit of integration for the inner integral. This can be handled by recognizing that the integration is performed over the shaded region in Figure 7.3(a) defined by the line $t = \tau$ for $t \geq 0$: the inner integration is performed horizontally over $\tau \in [0, t]$ and the outer integration is performed vertically over $t \in [0, \infty)$. However, note from Figure 7.3(b) that the integration can be performed instead over $t \in [\tau, \infty)$ and then over $\tau \in [0, \infty)$. As a result, the integrals are interchanged and (7.91) is rewritten as

$$\mathcal{L}\{x(t) * h(t)\} = \int_0^\infty \int_\tau^\infty x(\tau)\exp(-s\tau)h(t-\tau)\exp(-s(t-\tau))dt d\tau. \tag{7.92}$$

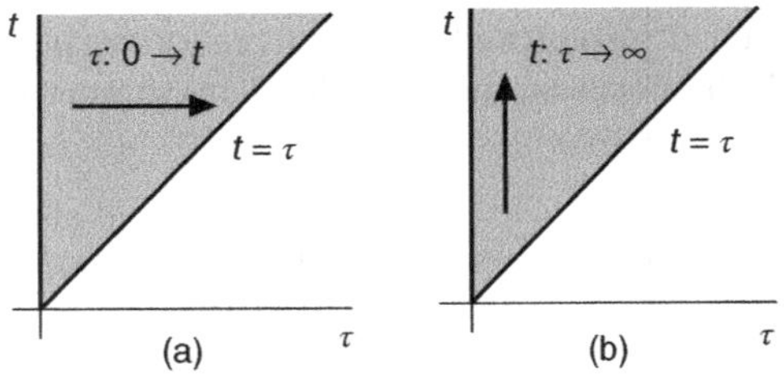

Figure 7.3 Region of integration for proving the convolution property of the Laplace transform. (a) Horizontally over $\tau \in [0, t]$ and vertically over $t \in [0, \infty)$. (b) Vertically over $t \in [\tau, \infty)$ and horizontally over $\tau \in [0, \infty)$.

Changing variables to $v \triangleq t - \tau$ yields 0 for the lower limit of the inner integral:

$$\begin{aligned}\mathcal{L}\{x(t) * h(t)\} &= \int_0^\infty \int_0^\infty x(\tau)\exp(-s\tau)h(v)\exp(-sv)dvd\tau \\ &= \int_0^\infty x(\tau)\exp(-s\tau)d\tau \int_0^\infty h(v)\exp(-sv)dv \\ &= X(s)H(s), \end{aligned} \tag{7.93}$$

which allows us to split the integrals into a product. The overall ROC is the intersection of the individual ROCs: $\text{ROC}_x \cap \text{ROC}_h$.

- *Cross-correlation:*

$$\mathcal{L}\{x(t) \star h(t)\} = X(-s)H(s), \tag{7.94}$$

where

$$x(t) \star h(t) \triangleq \int_{\max(0,-t)}^\infty x(\tau)h(t+\tau)d\tau \tag{7.95}$$

$$= \int_{\max(0,t)}^\infty x(\tau - t)h(\tau)d\tau. \tag{7.96}$$

This is *not* a symmetric property: $\mathcal{L}\{h(t) \star x(t)\} = H(-s)X(s) \neq X(-s)H(s)$. The limits of integration are determined by the fact that $x(t)$ and $h(t)$ are causal: the lower limit for the first integral is the maximum of 0 and $-t$ because $x(\tau)$ is nonzero for $\tau \geq 0$ and $h(t+\tau)$ is nonzero for $t + \tau \geq 0 \implies \tau \geq -t$. Similar reasoning leads to a different lower limit for the second integral. We use the following notation for cross-correlation functions (see Chapter 5):

$$c_{xh}(t) \triangleq x(t) \star h(t), \qquad c_{hx}(t) \triangleq h(t) \star x(t). \tag{7.97}$$

When $h(t) = x(t)$, $c_{xx}(t) = x(t) \star x(t)$ is the *autocorrelation function* of $x(t)$. *Proof*: The proof is similar to that used for convolution, requiring the slightly different identity $\exp(-st) = \exp(-s(t+\tau))\exp(s\tau)$:

$$\begin{aligned}\mathcal{L}\{x(t) \star h(t)\} &= \int_{-\infty}^\infty \int_{\max(0,-t)}^\infty x(\tau)\exp(s\tau)h(t+\tau)\exp(-s(t+\tau))d\tau dt \\ &= \int_{-\infty}^0 \int_{-t}^\infty x(\tau)\exp(s\tau)h(t+\tau)\exp(-s(t+\tau))d\tau dt \\ &\quad + \int_0^\infty \int_0^\infty x(\tau)\exp(s\tau)h(t+\tau)\exp(-s(t+\tau))d\tau dt, \end{aligned} \tag{7.98}$$

where the outer integral defined over t has been split into a sum so that the lower limit $\max(0, -t)$ of the inner integral can be evaluated. As was done for the convolution property, the order of integrations performed over the variables

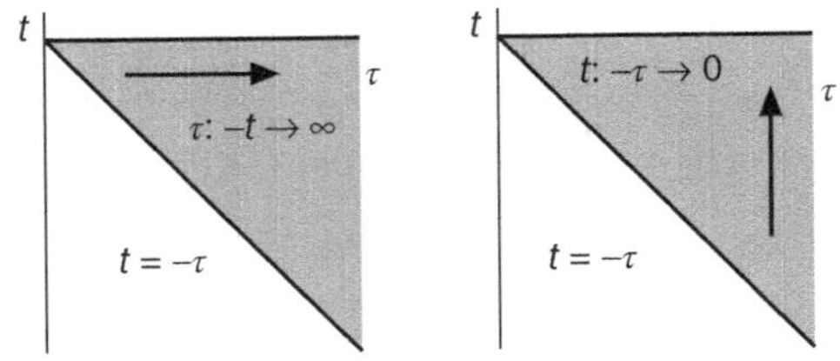

Figure 7.4 Region of integration for the first term in (7.98) for proving the cross-correlation property of the Laplace transform. (a) Horizontally over $\tau \in [-t, \infty)$ and vertically over $t \in (-\infty, 0]$. (b) Vertically over $t \in [-\tau, 0]$ and horizontally over $\tau \in [0, \infty)$.

is changed for the first term on the right-hand side to that of the shaded region in Figure 7.4(b):

$$\mathcal{L}(x(t) \star h(t)) = \int_0^\infty \int_{-\tau}^0 x(\tau)\exp(s\tau)h(t+\tau)\exp(-s(t+\tau))dtd\tau + \int_0^\infty \int_0^\infty x(\tau)\exp(s\tau)h(t+\tau)\exp(-s(t+\tau))dtd\tau. \quad (7.99)$$

Changing variables to $v \triangleq t + \tau$ yields

$$\mathcal{L}(x(t) \star h(t)) = \int_0^\infty \int_0^\tau x(\tau)\exp(s\tau)h(v)\exp(-sv)dvd\tau + \int_0^\infty \int_\tau^\infty x(\tau)\exp(s\tau)h(v)\exp(-sv)dvd\tau. \quad (7.100)$$

The inner integrals are combined and the double integral is split into a product:

$$\mathcal{L}(x(t) \star h(t)) = \int_0^\infty x(\tau)\exp(-(-s)\tau)d\tau \int_0^\infty h(v)\exp(-sv)dv = X(-s)H(s), \quad (7.101)$$

which completes the proof. The overall ROC is the intersection of the individual ROCs: $\text{ROC}_x \cap \text{ROC}_h$.

- *Product:*

$$\mathcal{L}\{x(t)h(t)\} = \frac{1}{2\pi j}\int_{c-j\infty}^{c+j\infty} X(v)H(s-v)dv, \quad (7.102)$$

where the integral is performed along a vertical line $\sigma = c$ in ROC_x (or in ROC_h if the integrand terms are interchanged). If ROC_x is $\text{Re}(s) > -\alpha_x$ and ROC_h is $\text{Re}(s) > -\alpha_h$ for positive $\{\alpha_x, \alpha_h\}$, then the ROC of the product is $\text{Re}(s) > -\alpha_x - \alpha_h$, which is *expanded* to the left on the s-plane when both terms are nonzero. The product property is the *dual* of the convolution property, where in this case the convolution is performed in the s-domain and the functions are multiplied in the time domain. *Proof*: Substituting the inverse Laplace transform in (7.58) for $x(t)$ and interchanging the integrals yields

$$\begin{aligned}\mathcal{L}\{x(t)h(t)\} &= \frac{1}{2\pi j}\int_0^\infty \int_{c-j\infty}^{c+j\infty} X(v)h(t)\exp(vt)\exp(-st)dvdt \\ &= \frac{1}{2\pi j}\int_{c-j\infty}^{c+j\infty} X(v)\int_0^\infty h(t)\exp(-(s-v)t)dtdv \\ &= \frac{1}{2\pi j}\int_{c-j\infty}^{c+j\infty} X(v)H(s-v)dv. \qquad (7.103)\end{aligned}$$

The last integration is performed in the ROC for $X(s)$ because the inverse transform for $x(t)$ is substituted here. The ROC for $H(s-v)$ is $\text{Re}(s-v) > -\alpha_h \implies \text{Re}(s) > -\alpha_h + v$, and since the integration performed over v can be done just to the right of $-\alpha_x$, we have an overall ROC of $\text{Re}(s) > -\alpha_h - \alpha_x$.

- *Time product:*

$$\mathcal{L}\{t^n x(t)\} = (-1)^n \frac{d^n}{ds^n}X(s). \qquad (7.104)$$

The ROC is unchanged because $\text{Re}(s) > 0$ for $t^n u(t)$ and we have used the ROC expression given earlier for the product property. *Proof*: Although this is a special case of the product property with $h(t) = t^n u(t)$, it is easier to start with the right-hand side of (7.104):

$$\begin{aligned}\frac{d^n}{ds^n}X(s) &= \frac{d^n}{ds^n}\int_0^\infty x(t)\exp(-st)dt \\ &= \int_0^\infty x(t)\frac{d^n}{ds^n}[\exp(-st)]dt \\ &= \int_0^\infty (-1)^n t^n x(t)\exp(-st)dt. \qquad (7.105)\end{aligned}$$

The derivative is simple to evaluate because only $\exp(-st)$ depends on s. Moving $(-1)^n$ to the left-hand side completes the proof.

- *Time division:*

$$\mathcal{L}\{x(t)/t\} = \int_s^\infty X(v)dv, \qquad (7.106)$$

assuming that $x(t)/t$ is defined as $t \to 0^+$. The ROC is unchanged. *Proof*: Starting with the right-hand side:

$$\begin{aligned}\int_s^\infty X(v)dv &= \int_s^\infty \int_0^\infty x(t)\exp(-vt)dtdv \\ &= \int_0^\infty x(t)\int_s^\infty \exp(-vt)dvdt. \qquad (7.107)\end{aligned}$$

TABLE 7.4 Properties of the Laplace Transform

Property	Function	Laplace Transform
Linearity	$a_1x_1(t) + a_2x_2(t)$	$a_1X_1(s) + a_2X_2(s)$
Time shift	$x(t - t_o)$	$\exp(-t_os)X(s)$
Time scaling	$x(\alpha t)$	$(1/\|\alpha\|)X(s/\alpha)$
Frequency shift	$\exp(s_ot)x(t)$	$X(s - s_o)$
Derivatives	$d^nx(t)/dt^n$	$s^nX(s) - \sum_{m=0}^{n-1} s^{n-m-1}x^{(m)}(0^-)$
Integral	$\int_0^t x(\tau)d\tau$	$(1/s)X(s)$
Double integral	$\int_0^{t_1}\int_0^{t_2} x(v)dvd\tau$	$(1/s^2)X(s)$
Convolution	$x(t) * h(t)$	$X(s)H(s)$
Cross-correlation	$x(t) \star h(t)$	$X(s)H(-s)$
Product	$x(t)h(t)$	$(1/2\pi j)\int_{\sigma-j\infty}^{\sigma+j\infty} X(v)H(s-v)dv$
Time product	$t^nx(t)$	$(-1)^nd^nX(s)/ds^n$
Time division	$x(t)/t$	$\int_s^\infty X(v)dv$

Performing the inner integration yields

$$\int_s^\infty X(v)dv = \int_0^\infty [x(t)/t]\exp(-st)dt, \tag{7.108}$$

which is the Laplace transform of $x(t)/t$ provided it exists.

These properties of the Laplace transform are summarized in Table 7.4.

Example 7.10 A simple example of the time division property is $x(t) = (t^n/t)u(t)$ for $n \geq 1$, which we know has Laplace transform $(n-1)!/s^n$:

$$\mathcal{L}\{(t^n/t)u(t)\} = \int_s^\infty (n!/v^{n+1})dv, \tag{7.109}$$

where $\mathcal{L}\{t^n\} = n!/s^{n+1}$ has been substituted. Thus,

$$\mathcal{L}\{(t^n/t)u(t)\} = -(n!/nv^n)|_s^\infty = (n-1)!/s^n, \tag{7.110}$$

of which the ramp function is a special case:

$$\mathcal{L}\{r(t)/t\} = \int_s^\infty (1/v^2)dv = -1/v|_s^\infty = 1/s. \tag{7.111}$$

This is the Laplace transform of the unit step function, as expected. For most of the functions in Tables 7.2 and 7.3, this property cannot be used because $x(t)/t$ is not defined at the origin. An exception is the sine function (see Problem 7.16).

Example 7.11 For the product of two exponential functions $\exp(-2t)u(t)$ and $\exp(-3t)u(t)$, it is easy to see that the ROC is $\text{Re}(s) > -2-3 = -5$ because we can just combine the exponents as $x(t) = \exp(-5t)u(t)$ and compute the transform. Suppose the product is $x(t) = \delta(t)\exp(-2t)u(t)$, which from the sampling property is $x(t) = \delta(t)$, and so, the ROC is the entire s-plane: $\text{Re}(s) > -\infty - 2 = -\infty \implies s \in C$.

Example 7.12 In this example, we verify the derivative property for the following function, which has a discontinuity at $t = 0$:

$$x(t) = \exp(-\alpha t)\cos(\omega_o t)u(t), \tag{7.112}$$

with $\alpha > 0$ and $\omega_o > 0$. The product rule of derivatives yields

$$\frac{d}{dt}x(t) = -\alpha\exp(-\alpha t)\cos(\omega_o t)u(t) - \omega_o\exp(-\alpha t)\sin(\omega_o t)u(t) \\ + \exp(-\alpha t)\cos(\omega_o t)\delta(t). \tag{7.113}$$

From the sampling property of the Dirac delta function, the last term is $\delta(t)$, and the final result is

$$\frac{d}{dt}x(t) = \delta(t) - \exp(-\alpha t)[\alpha\cos(\omega_o t) + \omega_o\sin(\omega_o t)]u(t). \tag{7.114}$$

The Laplace transform of (7.112) is (see Table 7.3)

$$X(s) = \frac{s+\alpha}{(s+\alpha)^2 + \omega_o^2}, \tag{7.115}$$

which has ROC $\text{Re}(s) > -\alpha$. Assuming $x(0^-) = 0$, the derivative property of the Laplace transform for $y(t) \triangleq dx(t)/dt$ gives

$$Y(s) = sX(s) = \frac{s(s+\alpha)}{(s+\alpha)^2 + \omega_o^2}. \tag{7.116}$$

This expression can be rearranged as follows:

$$Y(s) = 1 - \frac{\alpha s + \alpha^2 + \omega_o^2}{(s+\alpha)^2 + \omega_o^2} = 1 - \left[\alpha\frac{s+\alpha}{(s+\alpha)^2 + \omega_o^2} + \omega_o\frac{\omega_o}{(s+\alpha)^2 + \omega_o^2}\right], \tag{7.117}$$

which has the inverse Laplace transform in (7.114).

Because the Laplace transform is defined with lower limit 0^-, we must take into account any discontinuities at the origin, which means the Dirac delta function should appear in (7.114). However, for the exponentially weighted sine function, there is no Dirac delta function as shown in the next example.

Example 7.13 The exponentially weighted sine function is

$$x(t) = \exp(-\alpha t)\sin(\omega_o t)u(t), \tag{7.118}$$

and its derivative is

$$\frac{d}{dt}x(t) = -\alpha\exp(-\alpha t)\sin(\omega_o t)u(t) + \omega_o\exp(-\alpha t)\cos(\omega_o t)u(t) \\ + \exp(-\alpha t)\sin(\omega_o t)\delta(t). \tag{7.119}$$

Since $\sin(0) = 0$, the last term is 0 due to the sampling property of the Dirac delta function:

$$\frac{d}{dt}x(t) = \exp(-\alpha t)[\omega_o\cos(\omega_o t) - \alpha\sin(\omega_o t)]u(t). \tag{7.120}$$

The Laplace transform of $x(t)$ is (from Table 7.3)

$$X(s) = \frac{\omega_o}{(s+\alpha)^2 + \omega_o^2}, \tag{7.121}$$

with ROC $\text{Re}(s) > -\alpha$. The derivative property of the Laplace transform for $y(t) = dx(t)/dt$ yields

$$Y(s) = sX(s) = \frac{s\omega_o}{(s+\alpha)^2 + \omega_o^2}, \tag{7.122}$$

which does not have a leading 1, verifying that there is no Dirac delta function in the inverse Laplace transform in (7.120).

Example 7.14 In order to illustrate how the convolution and cross-correlation of two functions differ, suppose that $x(t) = u(t)$ and $h(t) = \exp(-t)u(t)$. Convolution yields the right-sided expression

$$\begin{aligned} y(t) &= \int_0^\infty u(\tau)\exp(-(t-\tau))u(t-\tau)d\tau \\ &= \exp(-t)u(t)\int_0^t \exp(\tau)d\tau = \exp(-t)u(t)\exp(\tau)|_0^t \\ &= \exp(-t)[\exp(t) - 1]u(t) = [1 - \exp(-t)]u(t), \end{aligned} \tag{7.123}$$

where $u(t)$ has been included because t must be nonnegative. Its Laplace transform is

$$Y(s) = \frac{1}{s} - \frac{1}{s+1} = \frac{1}{s(s+1)}, \tag{7.124}$$

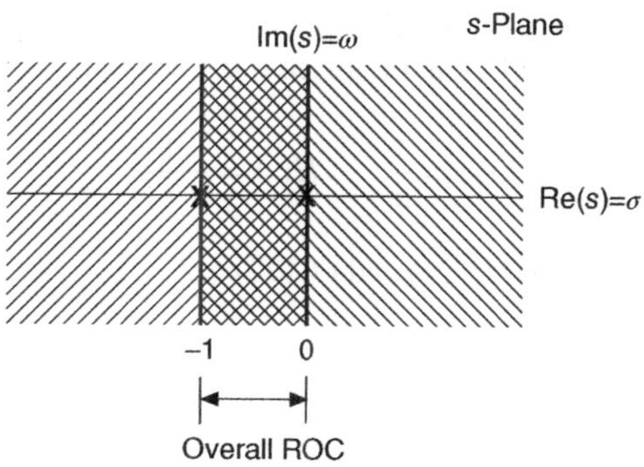

Figure 7.5 Region of convergence (ROC) for the Laplace transform in (7.130) of the cross-correlation function $c_{xh}(t)$ in Example 7.14, with poles at $s = \{0, -1\}$.

with ROC Re(s) > 0. The cross-correlation of these two functions is *two-sided*:

$$
\begin{aligned}
c_{xh}(t) &= \int_{\max(0,-t)}^{\infty} u(\tau)\exp(-(t+\tau))u(t+\tau)d\tau \\
&= \exp(-t)\int_{\max(0,-t)}^{\infty} \exp(-\tau)d\tau \\
&= \exp(-t)\exp(-\max(0,-t)),
\end{aligned} \tag{7.125}
$$

which can be written as

$$
\begin{aligned}
c_{xh}(t) &= \begin{cases} 1, & t < 0 \\ \exp(-t), & t \geq 0 \end{cases} \\
&= u(-t) + \exp(-t)u(t),
\end{aligned} \tag{7.126}
$$

and it is understood that the function is 1 at the origin (this is an example of a waveform where it is convenient to use $u(0) = 0.5$ because the two functions overlap at $t = 0$). Its *bilateral* Laplace transform is

$$
\begin{aligned}
C_{xh}(s) &= \int_{-\infty}^{0} \exp(-st)dt + \int_{0}^{\infty} \exp(-(s+1)t)dt \\
&= -\frac{1}{s} + \frac{1}{s+1} = \frac{-1}{s(s+1)},
\end{aligned} \tag{7.127}
$$

whose ROC is the intersection of Re(s) < 0 and Re(s) > −1, corresponding to the vertical strip on the s-plane shown in Figure 7.5: $-1 < \text{Re}(s) < 0$. The ROC is located between the poles at $s = \{0, 1\}$, which are denoted by X and are described further in a subsequent section. The individual Laplace transforms are

$$X(s) = \frac{1}{s}, \qquad H(s) = \frac{1}{s+1}, \tag{7.128}$$

which confirm the results in (7.124) and (7.127):

$$Y(s) = X(s)H(s) = \frac{1}{s(s+1)}, \tag{7.129}$$

$$C_{xh}(s) = X(-s)H(s) = -\frac{1}{s(s+1)}. \tag{7.130}$$

It is straightforward to show that the other cross-correlation function is also two-sided:

$$c_{hx}(t) = \begin{cases} \exp(t), & t < 0 \\ 1, & t \geq 0 \end{cases}$$

$$= \exp(t)u(-t) + u(t), \tag{7.131}$$

which has bilateral Laplace transform

$$C_{hx}(s) = \frac{1}{-s+1} + \frac{1}{s} = \frac{1}{s(-s+1)}, \tag{7.132}$$

and ROC $0 < \text{Re}(s) < 1$ (see Problem 7.13). The cross-correlation functions and $y(t)$ in (7.123) are plotted in Figure 7.6. The cross-correlation plots are reversed relative to each other, and the convolution result $y(t)$ increases to 1 (the solid line). Since $x(t) = u(t)$, the output $y(t)$ is the *step response* of the system with impulse response function $h(t)$. All three functions in the time domain, $y(t)$, $c_{xh}(t)$, and $c_{hx}(t)$, have an isolated step function due to s in the denominator of their transforms.

Example 7.15 For the following Laplace transform with ROC $\text{Re}(s) > -1$:

$$X(s) = \frac{2}{s+1} + \frac{2}{s+3}\exp(-4t) + \frac{s}{s+2}, \tag{7.133}$$

the time-domain waveform is

$$x(t) = 2\exp(-t)u(t) + 2\exp(-(t-4)t)u(t-4) + \frac{d}{dt}\exp(-2t)u(t)$$

$$= 2\exp(-t)u(t) + 2\exp(-(t-4)t)u(t-4) - 2\exp(-2t)u(t) + \delta(t), \tag{7.134}$$

where various properties of the Laplace transform have been used. The Dirac delta function appears because the last term in (7.133) is an improper rational function.

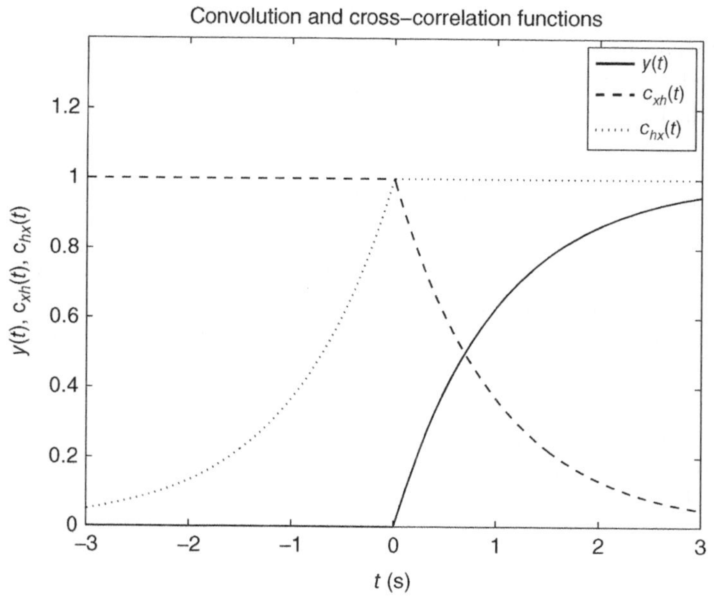

Figure 7.6 Cross-correlation functions and the convolution result in Example 7.14.

Thus, the last two terms of (7.134) can also be derived by applying long division to $s/(s+2)$ of (7.133):

$$\frac{s}{s+2} = 1 - \frac{2}{s+2} \xrightarrow{\mathcal{L}^{-1}} \delta(t) - 2\exp(-2t)u(t), \tag{7.135}$$

which confirms the results in (7.134).

Example 7.16 Finally, we give an example that illustrates a subtle difference between the derivative and integral properties of the Laplace transform. The I-V model for the inductor and its Laplace transform are given by

$$v(t) = L\frac{d}{dt}i(t) \xrightarrow{\mathcal{L}} V(s) = sLI(s) - Li(0^-), \tag{7.136}$$

where $i(0^-)$ is the current before any changes at $t = 0$, such as a switch opening or closing in the circuit; it is the initial state. Now consider the integral form for the inductor model and its Laplace transform:

$$i(t) = \frac{1}{L}\int_{0^-}^{t} v(t)dt \xrightarrow{\mathcal{L}} I(s) = V(s)/sL. \tag{7.137}$$

The integral property of the Laplace transform in (7.87) does not include $i(0^-)$ as does the derivative property. Solving (7.136) for $I(s)$ yields

$$I(s) = V(s)/sL + i(0^-)/s, \tag{7.138}$$

which is not the same as the result in (7.137). The reason for this discrepancy is due to the fact that the time-domain expression in (7.137) is *incorrect* when $i(0^-)$ is nonzero. If a dependent variable defined by an integral has a nonzero initial value, then it must be added as follows (the correct expression is also given in (2.23)):

$$i(t) = \frac{1}{L}\int_{0^-}^{t} v(t)dt + i(0^-). \tag{7.139}$$

Furthermore, this term acts as a step function because it appears in the expression for $t \in \mathcal{R}^+$. Thus, when taking the unilateral Laplace transform of (7.139), the transform of the constant $i(0^-)$ is $i(0^-)/s$, and the correct result in (7.138) is obtained.

The previous example illustrates that when taking the Laplace transform of an integral in an integro-differential equation, we must include any nonzero values of the function at $t = 0^-$ and consider them to be step functions. An example of this is given by the integro-differential equation in (6.82) for a series RLC circuit, repeated here

$$Ri(t) + L\frac{d}{dt}i(t) + (1/C)\int_{0}^{t} i(t)dt + v_C(0^-) = V_s u(t), \tag{7.140}$$

which includes $v_C(0^-)$ for the capacitor voltage. The Laplace transform of this equation is

$$RI(s) + sLI(s) - Li(0^-) + I(s)/sC + v_C(0^-)/s = V_s/s, \tag{7.141}$$

where the derivative property yields $Li(0^-)$. Since $v_C(0^-)$ and the voltage supply V_s are constants starting at $t = 0$, corresponding to step functions, they are divided by s in the transformation.

7.7 INITIAL AND FINAL VALUE THEOREMS

The two theorems in this section are useful for finding the initial value $x(0^+)$ and the final value $\lim_{t\to\infty} x(t)$ of the time-domain function $x(t)$ directly from the Laplace transform $X(s)$ without having to find its inverse Laplace transform.

- *Initial value theorem (IVT):*

$$\lim_{t\to 0^+} x(t) = \lim_{s\to\infty} sX(s). \tag{7.142}$$

Since $t = 0^+$, the value of the function is found just after any discontinuity at the origin; it is the initial condition. *Proof*: From the derivative property in (7.86):

$$\int_{0^-}^{\infty} \frac{d}{dt}x(t)\exp(-st)dt = sX(s) - x(0^-). \tag{7.143}$$

Splitting the integral about 0 to $[0^-, 0^+] \cup (0^+, \infty)$ gives

$$\int_{0^-}^{0^+} \left[\frac{d}{dt}x(t)\right] \exp(-st)dt + \int_{0^+}^{\infty} \left[\frac{d}{dt}x(t)\right] \exp(-st)dt = sX(s) - x(0^-). \tag{7.144}$$

In the first integral, $s = 0$ can be substituted because the exponential function is continuous:

$$\int_{0^-}^{0^+} \left[\frac{d}{dt}x(t)\right] \exp(-st)dt = \int_{0^-}^{0^+} \frac{d}{dt}x(t)dt = x(0^+) - x(0^-). \tag{7.145}$$

Substituting this result into (7.144) and rearranging the expression gives

$$x(0^+) + \int_{0^+}^{\infty} \left[\frac{d}{dt}x(t)\right] \exp(-st)dt = sX(s), \tag{7.146}$$

where $x(0^-)$ has canceled. The proof is completed by letting $s \to \infty$ on both sides of the last expression such that the integrand tends to 0.

There are some subtleties associated with the IVT. It is important to note that there is a difference between the function at $t = 0^-$ and the initial value at $t = 0^+$. Any nonzero values at $t = 0^-$ such as $v(0^-)$ in a circuit are needed when solving an ODE, and they must be included in the Laplace transforms. The initial voltage $v(0^+)$, on the other hand, is the initial value of the solution of the ODE. In some problems, these two quantities are the same, as is the case for the voltage across a capacitor, which cannot change instantaneously. However, in general, we cannot assume that $v(0^-)$ equals $v(0^+)$, and it is necessary that they be used properly when solving ODEs in the s-domain.

If there is a discontinuity at the origin due to a step, then

$$\left.\frac{d}{dt}x(t)\right|_{t=0} = [x(0^+) - x(0^-)]\delta(t) \tag{7.147}$$

should be substituted into (7.145):

$$\int_{0^-}^{0^+} [x(0^+) - x(0^-)]\delta(t)dt = x(0^+) - x(0^-), \tag{7.148}$$

where the sifting property of the Dirac delta function has been used. Thus, the IVT holds for step functions. For $x(t) = \delta(t)$, the derivative is the unit doublet $d\delta(t)/dt = \delta'(t)$ and

$$\int_{0^-}^{0^+} \delta'(t)\exp(-st)dt = -\left.\frac{d}{dt}\exp(-st)\right|_{t=0} = s, \tag{7.149}$$

where the sifting property of $\delta'(t)$ has been used for the right-hand side. Since the Dirac delta function is defined to be 0 at $t = 0^+$, the second integral in (7.144) is necessarily 0. The right-hand side of (7.144) is s because $X(s) = 1$ and $x(0^-) = 0$, so that the equation is valid for the Dirac delta function. However, note that (7.142) is infinite in this case, which demonstrates that the IVT is not useful for impulsive functions at the origin.

- *Final value theorem (FVT):*

$$\lim_{t\to\infty} x(t) = \lim_{s\to 0} sX(s). \tag{7.150}$$

Proof: The derivative property is also used to prove this theorem:

$$\lim_{s\to 0^-} \int_{0^-}^{\infty} \frac{d}{dt}x(t)\exp(-st)dt = \int_{0^-}^{\infty} \frac{d}{dt}x(t)dt = x(t)|_{t=0^-}^{\infty} = x(\infty) - x(0^-). \tag{7.151}$$

Equating this with (7.143) gives

$$\lim_{s\to 0^-} sX(s) - x(0^-) = x(\infty) - x(0^-), \tag{7.152}$$

and canceling $x(0^-)$ completes the proof. It is not necessary to distinguish between 0^- and 0^+ in the final value theorem (FVT) because we are interested in $x(t)$ as $t \to \infty$.

Observe that the variables of the two domains for these two properties have an inverse relationship: $s \to \infty$ for $x(0^+)$ and $s \to 0$ for $x(\infty)$, which are often mistakenly interchanged in practice.

Example 7.17 For the transform in (7.172), the IVT yields

$$\lim_{s\to\infty} \frac{s(s+\alpha)}{(s+\alpha)^2 + \omega_o^2} = 1, \tag{7.153}$$

which follows because both the numerator and the denominator are dominated by s^2 as s approaches ∞. The FVT gives

$$\lim_{s\to 0} \frac{s(s+\alpha)}{(s+\alpha)^2 + \omega_o^2} = 0. \tag{7.154}$$

Both of these results are consistent with the time-domain waveform given later in (7.172).

Example 7.18 Consider the unit step and ramp functions, which have transforms

$$x_1(t) = u(t) \xrightarrow{\mathcal{L}} X_1(s) = \frac{1}{s}, \tag{7.155}$$

$$x_2(t) = r(t) \xrightarrow{\mathcal{L}} X_2(s) = \frac{1}{s^2}, \tag{7.156}$$

both with ROC Re$(s) > 0$. The IVT yields

$$x_1(0^+) = \lim_{s\to\infty} \frac{s}{s} = 1, \qquad x_2(0^+) = \lim_{s\to\infty} \frac{s}{s^2} = 0, \tag{7.157}$$

and the FVT gives

$$x_1(\infty) = \lim_{s\to 0} \frac{s}{s} = 1, \qquad x_2(\infty) = \lim_{s\to 0} \frac{s}{s^2} = \infty. \tag{7.158}$$

The FVT is valid for the Dirac delta function and its derivative $\delta^{(n)}(t)$, which have Laplace transforms 1 and s^n, respectively:

$$\lim_{t\to\infty} \delta(t) = \lim_{s\to 0} s = 0, \qquad \lim_{t\to\infty} \delta^{(n)}(t) = \lim_{s\to 0} s^{n+1} = 0. \tag{7.159}$$

However, the IVT is not useful in either case as mentioned previously:

$$\lim_{t\to 0^+} \delta(t) \neq \lim_{s\to\infty} s = \infty, \qquad \lim_{t\to 0^+} \delta^{(n)}(t) \neq \lim_{s\to\infty} s^{n+1} = \infty, \tag{7.160}$$

whereas we know that these singular generalized functions are defined to be 0 at $t = 0^+$. The FVT does not hold for undamped or ramped sinusoidal functions, which is easily verified from the transform tables (also, see the discussion in Appendix B).

7.8 POLES AND ZEROS

The Laplace transform of a linear ODE with constant coefficients is a ratio of polynomials in s called a *rational function*. This follows from (6.10), which we repeat here:

$$\begin{aligned} a_N \frac{d^N}{dt^N} y(t) &+ a_{N-1} \frac{d^{N-1}}{dt^{N-1}} y(t) + \cdots + a_1 \frac{d}{dt} y(t) + a_0 y(t) \\ &= b_M \frac{d^M}{dt^M} x(t) + b_{M-1} \frac{d^{M-1}}{dt^{M-1}} x(t) + \cdots + b_1 \frac{d}{dt} x(t) + b_0 x(t). \end{aligned} \tag{7.161}$$

Assuming $y^{(n)}(0^-) = 0$, its Laplace transform is

$$\begin{aligned} a_N s^N Y(s) &+ a_{N-1} s^{N-1} Y(s) + \cdots + a_1 s Y(s) + a_0 Y(s) \\ &= b_M s^M X(s) + b_{M-1} s^{M-1} X(s) + \cdots + b_1 s X(s) + b_0 X(s), \end{aligned} \tag{7.162}$$

which can be rewritten as

$$\frac{Y(s)}{X(s)} = \frac{b_M s^M + b_{M-1} s^{M-1} + \cdots + b_1 s X(s) + b_0}{a_N s^N + a_{N-1} s^{N-1} + \cdots + a_1 s + a_0}, \tag{7.163}$$

where $X(s)$ and $Y(s)$ in (7.162) have been factored and written as a ratio.

Definition: Transfer Function The *transfer function* of an LTI system is the rational function

$$H(s) \triangleq \frac{Y(s)}{X(s)}, \tag{7.164}$$

where $X(s)$ and $Y(s)$ are the Laplace transforms of its input and output, respectively. The initial states are assumed to be 0.

A rational function can be categorized as one of two possible types.

Definition: Proper and Improper Rational Functions The rational function $H(s) = Y(s)/X(s)$ is *proper* if the numerator order M is less than N of the denominator. Otherwise, it is an *improper* rational function.

For improper rational functions, the IVT cannot be used because $h(t)$ would include some combination of the Dirac delta function and its derivatives. However, by using long division, it is possible to isolate those singular components and apply the IVT to the remaining part that is in proper form, for which we can determine its initial value at $t = 0^+$ by ignoring impulses at the origin. The significance of proper and improper rational functions will become evident later when we discuss the PFE technique for finding the inverse Laplace transform of $H(s)$.

Assume for now that $M = N$. Then from the polynomials in (7.163), it is clear that the numerator and denominator can be factored into a product as follows:

$$H(s) = \prod_{n=1}^{N} \frac{s - z_n}{s - p_n}, \tag{7.165}$$

where $\{z_n\}$ and $\{p_n\}$ are the roots of the two polynomials. The numerator is 0 for $s = z_n$ and the denominator is 0 for $s = p_n$. These roots may be complex-valued, and they may be repeated. If any root is complex of the form $c_n + jd_n$, then its complex conjugate $c_n - jd_n$ must also be a root because the coefficients $\{a_n, b_n\}$ of $H(s)$ are assumed to be real-valued.

Definition: Poles and Zeros The *poles* of transfer function $H(s)$ are $s = p_n$ such that $\lim_{s \to p_n} H(s) \to \pm\infty$, and the zeros are $s = z_n$ such that $\lim_{s \to z_n} H(s) \to 0$.

The transfer function $H(s)$ is undefined when s is evaluated at a pole (recall that the ROC excludes all poles). The locations of the poles of a transfer function yield information about the impulse response function $h(t)$ in the time domain.

Example 7.19 Consider the transfer function

$$H(s) = \frac{2s}{(s+1)(s+2)}, \tag{7.166}$$

which can be rewritten as follows:

$$H(s) = \frac{4}{s+2} - \frac{2}{s+1}. \tag{7.167}$$

This formulation is a PFE that can be verified by the reverse operation of combining terms over a common denominator:

$$H(s) = \frac{4(s+1) - 2(s+2)}{(s+1)(s+2)} = \frac{2s}{(s+1)(s+2)}. \tag{7.168}$$

This system has two real poles in the left half of the s-plane, and thus, we know from Table 7.3 that $H(s)$ is the Laplace transform of two exponential functions:

$$h(t) = 4\exp(-2t)u(t) - 2\exp(-t)u(t), \tag{7.169}$$

which have individual ROCs $\text{Re}(s) > -1$ and $\text{Re}(s) > -2$, respectively. The overall ROC is the intersection of these two regions: $(\text{Re}(s) > -1) \cap (\text{Re}(s) > -2) = \text{Re}(s) > -1$. This is an example of a general result: the ROC of a right-sided function for a *stable* system is the region on the s-plane just to the right of the pole with the smallest magnitude. A stable system has poles located only in the left half of the s-plane. The system in (7.168) also has a real zero at $s = 0$. Figure 7.7(a) shows a pole-zero plot for $H(s)$ where X denotes the pole locations and O indicates the zero location.

There are four types of poles with the following inverse transforms:

$$\textit{real pole}: \quad \frac{b}{s+a} \xrightarrow{\mathcal{L}^{-1}} b\exp(-at)u(t), \tag{7.170}$$

$$\textit{repeated real poles}: \quad \frac{b}{(s+a)^2} \xrightarrow{\mathcal{L}^{-1}} bt\exp(-at)u(t), \tag{7.171}$$

$$\textit{complex poles}: \quad \frac{s+a}{(s+a)^2 + \omega_o^2} \xrightarrow{\mathcal{L}^{-1}} \exp(-at)\cos(\omega_o t)u(t), \tag{7.172}$$

$$\textit{repeated complex poles}: \quad \frac{(s+a) - \omega_o^2}{[(s+a)^2 + \omega_o^2]^2} \xrightarrow{\mathcal{L}^{-1}} t\exp(-at)\cos(\omega_o t)u(t). \tag{7.173}$$

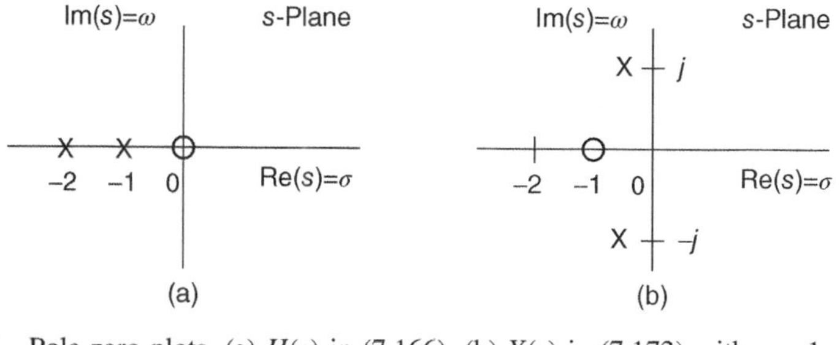

Figure 7.7 Pole-zero plots. (a) $H(s)$ in (7.166). (b) $X(s)$ in (7.172) with $a = 1$ and $\omega_o = 1$ rad/s.

Distinct real poles correspond to exponential functions in the time domain, and distinct complex conjugate poles correspond to decaying sinusoidal functions. Distinct poles are also called *simple poles*. For a second-order ODE, the first case in (7.173) (with real poles) is an overdamped system and the third case in (7.172) (with complex poles) is underdamped. If the complex poles lie on the imaginary axis ($a = 0$), then the sinusoidal function does not have any exponential weighting. This is the so-called *undamped* case because the cosine does not decay to 0. Repeated poles correspond to an exponentially weighted cosine function that is multiplied by t, which is a *ramped* cosine. Such systems are called *critically damped* (although (7.173) grows unbounded if $a = 0$).

Consider the right-sided cosine function $x(t) = \exp(-at)\cos(\omega_o t)u(t)$, which has the Laplace transform in (7.172). Its pole-zero plot on the s-plane is illustrated in Figure 7.7(b) for $a = 1$ and $\omega_o = 1$ rad/s. Since $X(s)$ is a complex-valued function, we can examine its magnitude after substituting $s = \sigma + j\omega$, with $a = 0$ in order to simplify the expression:

$$\begin{aligned}|X(s)| &= \frac{|\sigma + j\omega|}{|(\sigma + j\omega + j\omega_o)(\sigma + j\omega - j\omega_o)|} \\ &= \frac{\sqrt{(\sigma + j\omega)(\sigma - j\omega)}}{\sqrt{[\sigma + j(\omega + \omega_o)][\sigma - j(\omega + \omega_o)][(\sigma + j(\omega - \omega_o)][(\sigma - j(\omega - \omega_o)]}},\end{aligned} \quad (7.174)$$

where complex conjugate terms have been substituted into the numerator and the denominator in order to cancel all terms containing j. Multiplying all pairs of terms yields the final expression:

$$|X(s)| = \frac{\sqrt{\sigma^2 + \omega^2}}{\sqrt{[\sigma^2 + (\omega + \omega_o)^2][\sigma^2 + (\omega - \omega_o)^2]}}. \quad (7.175)$$

The logarithm of this function is plotted versus σ and ω in Figure 7.8 (the logarithm is used to show greater dynamic range). Observe that there is a zero at $s = 0$ where $20\log(|X(s|) \to -\infty$, and there are complex conjugate poles at $s = \pm j\omega_o$ where $20\log(|X(s|) \to \infty$. (Of course, the poles and zeros in the figure have finite values because of the finite resolution of the grid used in MATLAB to generate the three-dimensional plot.) The ROC is $\text{Re}(s) > 0$, which is denoted by the grid to the right of the solid line at $\sigma = 0$. Several more examples of $|X(s)|$ for important waveforms used in linear systems are provided in Appendix A.

It turns out that the dynamic characteristics of the time-domain function $x(t)$ can be determined from the pole locations on the s-plane. This is illustrated in Figure 7.9 for the transforms in (7.170) and (7.172), as well as

$$\textit{real pole at origin}: \ \frac{1}{s} \xrightarrow{\mathcal{L}^{-1}} u(t), \quad (7.176)$$

$$\textit{complex poles on imaginary axis}: \ \frac{s}{s^2 + \omega_o^2} \xrightarrow{\mathcal{L}^{-1}} \cos(\omega_o t)u(t). \quad (7.177)$$

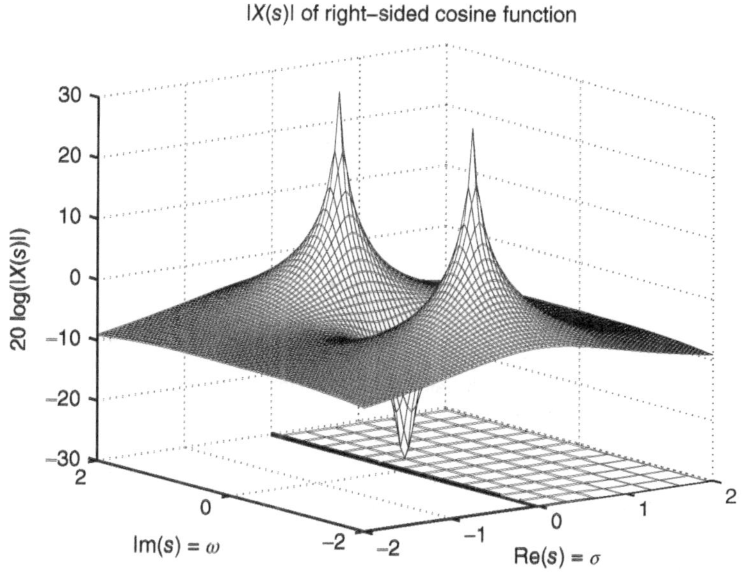

Figure 7.8 Truncated magnitude of the Laplace transform and its ROC for the right-sided cosine function in (7.172) with $a = 0$ and $\omega_o = 1$ rad/s.

If the poles are moved further to the left, $|a|$ increases and the exponential functions in (7.170) and (7.172) decay to 0 more quickly; they have a smaller time constant τ. Of course, these functions decay more slowly if the poles are moved to the right, and when they lie on the imaginary axis, we have the functions in (7.176) and (7.177), which do not decay to 0. The waveforms grow unbounded if any pole is located in the right half of the s-plane; generally, we are not concerned with such signals and systems in this book. If the complex conjugate poles are moved upward and downward away from the origin, then the frequency ω_o of the sinusoids in (7.172) and (7.177) increases.

The zeros of $X(s)$ do not change the basic shape of $x(t)$, except when the rational function is improper, in which case $x(t)$ would include a combination of the Dirac delta function and its derivatives. The zeros are related to time shifts (delays) in $x(t)$. For example, the Laplace transform of the right-sided sine function is

$$x(t) = \sin(\omega_o t)u(t) \xrightarrow{\mathcal{L}} X(s) = \frac{\omega_o}{s^2 + \omega_o^2}. \tag{7.178}$$

Comparing this with the cosine function, which is shifted by 90°, we find that the only difference between their Laplace transforms is the zero in (7.177). The zeros also affect the frequency characteristics of $x(t)$, which are determined by examining $X(s)$ on the imaginary axis where $s = j\omega$ ($\sigma = 0$). The resulting function $X(\omega)$ is the *Fourier transform* of $x(t)$, which is the topic of Chapter 8. Fourier transforms are

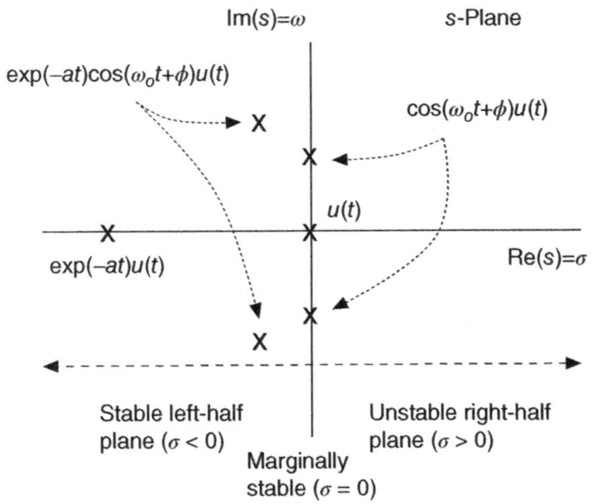

Figure 7.9 Four sets of pole locations. (i) $x(t) = u(t)$ with real pole at the origin ($s = 0$). (ii) $x(t) = \exp(-at)u(t)$ with real pole on the real axis ($s = -a$). (iii) $x(t) = \cos(\omega_o t + \phi)u(t)$ with complex conjugate poles on the imaginary axis ($s = \pm j\omega_o$). (iv) $x(t) = \exp(-at)\cos(\omega_o t + \phi)u(t)$ with complex conjugate poles on the left half of the s-plane ($s = a \pm j\omega_o$).

useful for describing the *frequency content* of a signal and the *frequency response* of an LTI system.

7.9 LAPLACE TRANSFORM PAIRS

The Laplace transforms for $\exp(-\alpha t)u(t)$, $u(t)$, and $\delta^{(n)}(t)$ were derived in (7.39), (7.42), and (7.68), respectively. In this section, we derive a few more transforms for some of the signals described in Chapter 5.

7.9.1 Constant Function

First, we point out a subtlety involving a constant function and the *unilateral* Laplace transform that can arise when analyzing linear circuits with nonzero initial conditions. Consider the following integral equation for the current in a parallel RL circuit (see Figure 2.29 with C and V_s removed):

$$i_L(t) = i_R(t) \implies (1/L)\int_{0^-}^{t} v(t)dt + i(0^-) = v(t)/R, \tag{7.179}$$

where $v(t)$ is the voltage across both elements and $i(0^-)$ is the initial current, which is treated as a constant. Although the *bilateral* Laplace transform of a constant does not exist, the unilateral Laplace transform exists because of the 0 lower limit:

$$\int_{0^-}^{\infty} i(0^-)\exp(-st)dt = i(0^-)/s, \tag{7.180}$$

with ROC Re(s) > 0. Thus, the one-sided Laplace transform of a constant gives the same result as that of a step function. This property is used later when we examine circuits that are modeled by integro-differential equations and solve them using Laplace transform techniques. Note, however, that $i(0^-)$ is *not* a step function in (7.179) even though the unilateral Laplace transform of the equation yields

$$V(s)/sL + i(0^-)/s = V(s)/R. \tag{7.181}$$

Since the current in an inductor cannot change instantaneously, which means $i(0^-) = i(0^+)$, the initial state $i(0^-)$ in (7.179) is not a step function. The s dividing $i(0^-)$ in (7.180) is due only to the finite lower limit of the unilateral Laplace transform.

7.9.2 Rectangle Function

For the standard rectangle function $x(t) = \text{rect}(t) = I_{[-1/2,1/2]}(t)$ where $I(t)$ is the indicator function, the bilateral Laplace transform is used:

$$\begin{aligned} X(s) &= \int_{-1/2}^{1/2} \exp(-st)dt = -\frac{\exp(-st)}{s}\bigg|_{-1/2}^{1/2} \\ &= \frac{\exp(s/2) - \exp(-s/2)}{s} = \frac{2\sinh(s/2)}{s}, \end{aligned} \tag{7.182}$$

whose ROC is the entire s-plane because the function has finite duration; the apparent pole at $s = 0$ is removable. For the shifted (causal) rectangle function $x(t) = u(t) - u(t-1)$:

$$\begin{aligned} X(s) &= \int_{0^-}^{\infty} [u(t) - u(t-1)] \exp(-st)dt \\ &= \int_{0}^{1} \exp(-st)dt = [1 - \exp(-s)]/s \\ &= \frac{2\exp(-s/2)\sinh(s/2)}{s}. \end{aligned} \tag{7.183}$$

This result can also be derived using the time-shift property of the Laplace transform in Table 7.4 with $t_o = 1/2$. Another derivation uses the Laplace transform of the unit step function:

$$u(t) \xrightarrow{\mathcal{L}} \frac{1}{s}, \qquad u(t-1) \xrightarrow{\mathcal{L}} \frac{\exp(-s)}{s}$$

$$\implies x(t) = u(t) - u(t-1) \xrightarrow{\mathcal{L}} X(s) = [1 - \exp(-s)]/s. \tag{7.184}$$

Although $u(t)$ and $u(t-1)$ individually have ROC Re(s) > 0 and there appears to be a pole at $s = 0$ (as mentioned earlier), note from l'Hôpital's rule that

$$\lim_{s\to 0}[1 - \exp(-s)]/s = \lim_{s\to 0} \frac{d}{ds}[1 - \exp(-s)] = 1, \tag{7.185}$$

since the derivative of the denominator is also 1. As a result, the Laplace transform of any finite rectangle function actually has no poles and the ROC is the entire s-plane. This is due to the fact that $u(t-1)$ exactly cancels $u(t)$ for $t \geq 1$, yielding a finite-duration function. In fact, for any finite-duration waveform written with unit step functions, l'Hôpital's rule can be used to show that the apparent poles are removable.

Example 7.20 Let the exponential function be weighted by unit step functions as follows:

$$x(t) = \exp(-\alpha t)[u(t-1) - u(t-2)], \tag{7.186}$$

which has finite duration. The Laplace transform is

$$\begin{aligned} X(s) &= \int_0^\infty \exp(-\alpha t)[u(t-1) - u(t-2)]\exp(-st)dt \\ &= \int_1^2 \exp(-(s+\alpha)t)dt \\ &= [\exp(-(s+\alpha)) - \exp(-2(s+\alpha))]/(s+\alpha), \end{aligned} \tag{7.187}$$

whose ROC is the entire s-plane because the apparent pole at $s = -\alpha$ is removable. The derivative of the numerator is

$$\begin{aligned} &\lim_{s\to-\alpha} \frac{d}{ds}[\exp(-(s+\alpha)) - \exp(-2(s+\alpha))] \\ &\quad = \lim_{s\to-\alpha}[-\exp(-2(s+\alpha)) + 2\exp(-(s+\alpha))] \\ &\quad = 2 - 1 = 1, \end{aligned} \tag{7.188}$$

and the derivative of the denominator is obviously 1.

7.9.3 Triangle Function

The standard triangle function is also centered about the origin: $x(t) \triangleq (1 - |t|)I_{[-1,1]}(t)$. Its Laplace transform is

$$\begin{aligned} X(s) &= \int_{-1}^1 (1-|t|)\exp(-st)dt \\ &= \int_{-1}^0 (1+t)\exp(-st)dt + \int_0^1 (1-t)\exp(-st)dt. \end{aligned} \tag{7.189}$$

Substituting the integral $\int t\exp(-st)dt = -\exp(-st)(st+1)/s^2$ from Appendix C yields

$$\begin{aligned}X(s) &= -[\exp(-st)/s + \exp(-st)(st+1)/s^2]|_{-1}^{0} \\ &\quad -[\exp(-st)/s - \exp(-st)(st+1)/s^2]|_{0}^{1} \\ &= -1/s - 1/s^2 + \exp(s)/s + \exp(s)(1-s)/s^2 \\ &\quad -\exp(-s)/s + \exp(-s)(s+1)/s^2 + 1/s - 1/s^2.\end{aligned} \tag{7.190}$$

Canceling and combining terms give the transform

$$X(s) = [\exp(s) + \exp(-s)]/s^2 - 2/s^2. \tag{7.191}$$

This can be factored as follows, resulting in the hyperbolic sine function:

$$X(s) = \frac{[\exp(s/2) - \exp(-s/2)]^2}{s^2} = \frac{4\sinh^2(s/2)}{s^2}, \tag{7.192}$$

whose ROC is the entire s-plane because the apparent double poles at the origin are removable. This Laplace transform is the square of that for the standard rectangle function, which follows from the fact that the triangle function is the convolution of two rectangle functions and $X(s)$ is the product of the two transforms in the s-domain.

Example 7.21 The time product in Table 7.4 can be used to verify the Laplace transform of the triangle function. From (7.189), we have three integrals:

$$X(s) = \int_{-1}^{1} \exp(-st)dt + \int_{-1}^{0} t\exp(-st)dt - \int_{0}^{1} t\exp(-st)dt. \tag{7.193}$$

The first term on the right-hand side is the Laplace transform of the rectangle function $\text{rect}(t/2)$ given by $[\exp(s) - \exp(-s)]/s$, where the time-scaling property has been applied to the second line of (7.182). Initially ignoring the multiplicative t, the second integral is the Laplace transform of the shifted rectangle function $\text{rect}(t+1/2)$ given by $\exp(s/2)[\exp(s/2) - \exp(-s/2)]/s$, and the third integral is the Laplace transform of $\text{rect}(t-1/2)$ given by $\exp(-s/2)[\exp(s/2) - \exp(-s/2)]/s$. Combining all the three terms, we have from the time product property:

$$\begin{aligned}X(s) &= \frac{\exp(s) - \exp(-s)}{s} - \frac{d}{ds}\frac{\exp(s) - 1}{s} + \frac{d}{ds}\frac{1 - \exp(-s)}{s} \\ &= \frac{\exp(s) - \exp(-s)}{s} - \frac{\exp(s)}{s} + \frac{\exp(s) - 1}{s^2} \\ &\quad + \frac{\exp(-s)}{s} - \frac{1 - \exp(-s)}{s^2},\end{aligned} \tag{7.194}$$

and canceling some terms yields the expression in (7.190).

7.9.4 Ramped Exponential Function

For the ramped exponential function

$$x(t) = t\exp(-\alpha t)u(t), \tag{7.195}$$

the Laplace transform is

$$\begin{aligned} X(s) &= \int_{0^-}^{\infty} t\exp(-\alpha t)\exp(-st)dt \\ &= -\frac{t\exp(-(s+\alpha)t}{s+\alpha}\bigg|_0^{\infty} + \frac{1}{s+\alpha}\int_0^{\infty} \exp(-(s+\alpha)t)dt, \end{aligned} \tag{7.196}$$

where integration by parts has been used to remove t from the integrand. The first term is 0 when evaluated at the two limits. The last integral is the Laplace transform of $\exp(-\alpha t)u(t)$:

$$X(s) = \frac{1}{(s+\alpha)^2}, \tag{7.197}$$

which has ROC $\text{Re}(s) > -\alpha$. Successive application of integration by parts is used to derive the Laplace transform of more general ramped exponential functions:

$$x(t) = t^n\exp(-\alpha t)u(t) \xrightarrow{\mathcal{L}} X(s) = \frac{n!}{(s+\alpha)^{n+1}}. \tag{7.198}$$

The case for $n = 2$ is included in Problem 7.11.

7.9.5 Sinusoidal Functions

The Laplace transforms of sinusoidal functions are easily handled by using Euler's formula and the previous result for an exponential function. For $x(t) = \cos(\omega_o t)u(t)$:

$$\begin{aligned} X(s) &= \frac{1}{2}\int_0^{\infty} [\exp(j\omega_o t) + \exp(-j\omega_o t)]\exp(-st)dt \\ &= -\frac{\exp(-(s-j\omega_o)t)}{s-j\omega_o}\bigg|_0^{\infty} - \frac{\exp(-(s+j\omega_o)t)}{s+j\omega_o}\bigg|_0^{\infty} \\ &= \frac{1}{2}\left(\frac{1}{s-j\omega_o} + \frac{1}{s+j\omega_o}\right), \end{aligned} \tag{7.199}$$

which has ROC $\text{Re}(s \pm j\omega_o) = \text{Re}(s) > 0$. This result holds even though the exponential functions are complex; the ROC is determined only by $\exp(-\sigma t)$, which weights the sinusoidal $\exp(\pm j\omega_o t)$. Rewriting (7.199) over a common denominator gives

$$X(s) = \frac{s}{s^2 + \omega_o^2}. \tag{7.200}$$

The Laplace transform for $x(t) = \sin(\omega_o t)u(t) = (1/2j)[\exp(j\omega_o t) - \exp(-j\omega_o t)]u(t)$ can be derived from (7.199) by subtracting the two terms and dividing by j:

$$X(s) = \frac{1}{2j}\left(\frac{1}{s - j\omega_o} - \frac{1}{s + j\omega_o}\right) = \frac{\omega_o}{s^2 + \omega_o^2}, \tag{7.201}$$

which also has ROC $\text{Re}(s) > 0$. Both of these functions have complex conjugate poles on the imaginary axis at $s = \pm j\omega_o$, and the cosine function has a zero $s = 0$ at the origin.

From the derivation leading to (7.199), it is straightforward to find the Laplace transform for the exponentially weighted cosine function $x(t) = \exp(-\alpha t)\cos(\omega_o t)u(t)$ by using $s + \alpha$ in place of s:

$$\begin{aligned} X(s) &= \frac{1}{2}\left(\frac{1}{s + \alpha - j\omega_o} + \frac{1}{s + \alpha + j\omega_o}\right) \\ &= \frac{s + \alpha}{(s + \alpha)^2 + \omega_o^2}, \end{aligned} \tag{7.202}$$

which has ROC $\text{Re}(s + \alpha) > 0 \implies \text{Re}(s) > -\alpha$, assuming $\alpha > 0$ for a bounded function. Similarly for the exponentially weighted sine function $x(t) = \exp(-\alpha t)\sin(\omega_o t)u(t)$:

$$X(s) = \frac{\omega_o}{(s + \alpha)^2 + \omega_o^2}, \tag{7.203}$$

which also has ROC $\text{Re}(s) > -\alpha$. This function has complex conjugate poles at $s = -\alpha \pm j\omega_o$, which are located on a vertical line defined by $s = -\alpha$ to the left of the imaginary axis for $\alpha > 0$. The exponentially weighted cosine function also has a zero at $s = -\alpha$. The poles are easily determined by the fact that $s^2 + \omega_o^2$ has roots at $s = \pm j\omega_o$, which means $(s + \alpha)^2 + \omega_o^2$ has roots at $s + \alpha = \pm j\omega_o \implies s = -\alpha \pm j\omega_o$.

Some additional Laplace transforms in Tables 7.2 and 7.3 are derived in the problems at the end of this chapter.

7.10 TRANSFORMS AND POLYNOMIALS

In this section, we provide some insight into *how* the Laplace transform converts ("transforms") time-domain functions into *rational polynomials* that are easier to manipulate. The most important input functions of a linear system are sinusoidal:

$$\exp(j\omega_o t), \qquad \cos(\omega_o t), \qquad \sin(\omega_o t), \tag{7.204}$$

where ω_o is angular frequency with units of rad/s. All three are *eigenfunctions* of an LTI system because the sinusoidal functions can be written in terms of the complex exponential using Euler's inverse formulas:

$$\cos(\omega_o t) = (1/2)[\exp(j\omega_o t) + \exp(-j\omega_o t)], \tag{7.205}$$

$$\sin(\omega_o t) = (1/2j)[\exp(j\omega_o t) - \exp(-j\omega_o t)], \tag{7.206}$$

and similarly, Euler's formula gives the complex exponential in terms of sine and cosine. This is significant because the kernel of the Laplace transform is also a complex exponential, and as a result, the exponents of its integrand *can be combined algebraically*. We demonstrate this with an example.

Example 7.22 Consider again the Laplace transform of $\exp(-\alpha t)u(t)$:

$$\int_0^\infty \exp(-\alpha t)\exp(-st)dt = \int_0^\infty \exp(-(s+\alpha)t)dt$$
$$= -\frac{1}{s+\alpha}\exp(-(s+\alpha)t)\bigg|_0^\infty . \quad (7.207)$$

The upper limit of infinity determines the ROC such that the last expression is 0 when evaluated at $t \to \infty$. The ROC for this right-sided function is $\text{Re}(s) > -\alpha$. The finite lower limit is important because when 0 is substituted into the last expression, the exponential function is 1 and the result is a rational polynomial in s:

$$\mathcal{L}\{\exp(-\alpha t)u(t)\} = \frac{1}{s+\alpha}. \quad (7.208)$$

We emphasize that this result is due to the fact that when exponential functions are multiplied, their exponents *add or subtract*, resulting in an algebraic expression when integrated and the limits of integration are substituted. This is the *mechanism* by which the Laplace integral converts functions and ODEs to algebraic equations in the complex variable s.

Polynomials are also obtained when generalized functions appear in the time-domain function:

$$u(t) \xrightarrow{\mathcal{L}} 1/s, \qquad \delta(t) \xrightarrow{\mathcal{L}} 1, \qquad \delta'(t) \xrightarrow{\mathcal{L}} s. \quad (7.209)$$

A polynomial can be weighted by exponential functions of the form $\exp(-st_o)$ when signals are delayed by t_o, and they also appear for finite duration waveforms such as the rectangle and triangle functions. This is not a problem, however, because the characteristic equation that determines the poles of a system is a polynomial in s; exponential terms that appear in the numerator of a transfer function are handled after performing a PFE of the rational part.

Example 7.23 For example, suppose the ODE for a system is

$$\frac{d^2}{dt^2}y(t) + 3\frac{d}{dt}y(t) + 2y(t) = u(t-1), \quad (7.210)$$

which has Laplace transform

$$(s^2+3s+2)Y(s) = \exp(-s)/s \implies Y(s) = \frac{\exp(-s)}{s(s+1)(s+2)}. \quad (7.211)$$

The PFE is performed by first factoring the exponential multiplier:

$$Y(s) = \exp(-s)\left[\frac{A_1}{s} + \frac{A_2}{s+1} + \frac{A_3}{s+2}\right], \tag{7.212}$$

where $A_1 = 1/2$, $A_2 = -1$, and $A_3 = -1/2$. As shown later in this chapter, these are obtained as follows:

$$A_1 = \lim_{s\to 0} s\frac{1}{s(s+1)(s+2)}, \qquad A_2 = \lim_{s\to -1}(s+1)\frac{1}{s(s+1)(s+2)}, \tag{7.213}$$

$$A_3 = \lim_{s\to -2}(s+2)\frac{1}{s(s+1)(s+2)}. \tag{7.214}$$

Thus, the inverse Laplace transform of the expression in brackets in (7.212) is

$$\left[\frac{A_1}{s} + \frac{A_2}{s+1} + \frac{A_3}{s+2}\right] \xrightarrow{\mathcal{L}^{-1}} [1/2 - \exp(-t) - (1/2)\exp(-2t)]u(t), \tag{7.215}$$

and from the time shift property of the Laplace transform, the leading $\exp(-s)$ delays the overall function by 1:

$$y(t) = [1/2 - \exp(-(t-1)) - (1/2)\exp(-2(t-1))]u(t-1). \tag{7.216}$$

This is the expected system output if the input unit step is delayed by $t_o = 1$ s.

Next, we elaborate on the fact that the exponent used in the kernel of the Laplace transform is complex: $s = \sigma + j\omega$. Suppose that $\omega = 0$ and instead we use the following integral as the Laplace transform:

$$\mathcal{L}\{x(t)\} = \int_0^\infty x(t)\exp(-\sigma t)dt. \tag{7.217}$$

As mentioned earlier, some books define the Laplace transform in this way with a real exponent, and this form is useful for many functions such as $\exp(-\alpha t)u(t)$, where from (7.208):

$$X(\sigma) = \frac{1}{\sigma + \alpha}, \tag{7.218}$$

with ROC $\sigma > -\alpha$. By using a real kernel, the s-domain function is restricted to the real axis on the s-plane. From the previous discussion on poles and zeros, this restriction obviously does not allow us to fully examine more general functions such as $x(t) = \exp(-\alpha t)\cos(\omega_o t)u(t)$, which has the Laplace transform in (7.202) with complex conjugate poles. Substituting $s = \sigma$ in that expression yields

$$X(\sigma) = \frac{\sigma + \alpha}{(\sigma + \alpha)^2 + \omega_o^2}, \tag{7.219}$$

which also has ROC $\sigma > -\alpha$. This transform has a zero at $\sigma = -\alpha$, and the poles are computed as follows:

$$(\sigma + \alpha)^2 + \omega_o^2 = 0 \implies \sigma = -\alpha \pm j\omega_o. \quad (7.220)$$

This is the same situation encountered in Chapter 4 when we attempted to solve the quadratic equation $x^2 + 1 = 0$, which has no solutions if the roots are restricted to be real numbers. The result in (7.219) is correct, but $X(\sigma)$ is the Laplace transform of $\exp(-\alpha t)\cos(\omega_o t)u(t)$ evaluated only on the real axis of the s-plane. In most engineering problems, it is preferable to utilize the entire s-plane and let s be a complex variable in the Laplace transform.

Suppose instead that $\sigma = 0$ in the Laplace transform such that

$$X(\omega) = \int_{-\infty}^{\infty} x(t)\exp(-j\omega t)dt. \quad (7.221)$$

This integral is the Fourier transform, and as mentioned earlier, it provides information about the frequency content of a signal or the frequency response of a system. It is the Laplace transform evaluated on the imaginary axis, provided the ROC includes this axis. If this is not the case, then in the limit there may be singular functions on the $j\omega$ axis, or else the Fourier transform does not exist. For the exponentially weighted cosine function, the Fourier transform is

$$X(\omega) = \frac{j\omega + \alpha}{(j\omega + \alpha)^2 + \omega_o^2} = \frac{j\omega + \alpha}{\alpha^2 + \omega_o^2 - \omega^2 + j2\alpha\omega}. \quad (7.222)$$

Unlike $X(\sigma)$, which is strictly a real function, $X(\omega)$ is complex in general and it can be expressed in polar form with magnitude $|X(\omega)|$ and phase $\theta(\omega)$. The Fourier transform is examined further in Chapter 8.

7.11 SOLVING LINEAR ODEs

Using the derivative property of the Laplace transform, it is straightforward to solve linear ODEs with nonzero initial states. Consider the second-order nonhomogeneous ODE:

$$\frac{d^2}{dt^2}y(t) + a_1\frac{d}{dt}y(t) + a_0 y(t) = x(t), \quad (7.223)$$

which has Laplace transform

$$s^2Y(s) - sy(0^-) - y'(0^-) + a_1 sY(s) - a_1 y(0^-) + a_0 Y(s) = X(s). \quad (7.224)$$

Solving this expression for $Y(s)$ yields two rational function components:

$$Y(s) = \frac{X(s)}{s^2 + a_1 s + a_0} + \frac{(s + a_1)y(0^-) + y'(0^-)}{s^2 + a_1 s + a_0}. \quad (7.225)$$

The first term on the right-hand side gives the transfer function of the system $H(s) \triangleq Y(s)/X(s)$ assuming zero initial states $\{y(0^-), y'(0^-)\}$. Both terms in (7.225) have the same set of poles, and so, they have the same type of response: (i) overdamped, (ii) underdamped, or (iii) critically damped.

Similarly, integro-differential equations can be solved using the derivative and integral properties of the Laplace transform. An example integro-differential equation is

$$\frac{d}{dt}y(t) + a_1 y(t) + a_0 \int_0^t y(t)dt + a_0 y(0^-) = \int_0^t x(t)dt, \tag{7.226}$$

where $y(0^-)$ is the initial state for the output of the system associated with the first integral. We assume that $x(0^-)$ associated with the integral on the right-hand side is zero. The Laplace transform yields

$$sY(s) - y(0^-) + a_1 Y(s) + a_0 Y(s)/s + a_0 y(0^-)/s = X(s)/s, \tag{7.227}$$

where the second $y(0^-)$ is divided by s because it appears as a step when using the one-sided Laplace transform. Factoring $Y(s)$ yields

$$Y(s)(s + a_1 + a_0/s) = X(s)/s + y(0^-) - a_0 y(0^-)/s. \tag{7.228}$$

Multiplying through by s and solving for $Y(s)$, we have

$$Y(s) = \frac{X(s)}{s^2 + a_1 s + a_0} + \frac{(s - a_0)y(0^-)}{s^2 + a_1 s + a_0}. \tag{7.229}$$

Observe the similarity between the s-domain results in (7.225) and (7.229). Differentiating the integro-differential equation in (7.226) yields the ODE in (7.223). We demonstrate in the next example that by applying the Laplace transform derivative property to (7.229), we generate the expression in (7.225).

Example 7.24 Define the left-hand side of (7.226) to be $g(t)$ and the right-hand side to be $f(t)$. Differentiating this equation gives

$$\frac{d}{dt}g(t) = \frac{d}{dt}f(t), \tag{7.230}$$

which has Laplace transform

$$sG(s) - g(0^-) = sF(s) - f(0^-) = X(s). \tag{7.231}$$

Since $F(s) = X(s)/s$ and we have assumed $x(0^-) = 0$ such that $f(0^-) = 0$, the right-hand side of (7.231) is $X(s)$. For the left-hand side of (7.231):

$$g(0^-) = \left.\frac{d}{dt}y(t)\right|_{t=0^-} + a_1 y(0^-) + 0 + a_0 y(0^-) = y'(0^-) + y(0^-)(a_1 + a_0), \tag{7.232}$$

where the third term in the middle expression is 0 because the integral is 0 when the upper limit is $t = 0^-$. Since $G(s)$ is the left-hand side of (7.227), we have

$$sG(s) = s^2Y(s) - sy(0^-) + a_1sY(s) + a_0Y(s) + a_0y(0^-). \tag{7.233}$$

Substituting this expression into (7.231) and using (7.232) yields

$$s^2Y(s) - sy(0^-) + a_1sY(s) + a_0Y(s) - y'(0^-) - a_1y(0^-) = X(s), \tag{7.234}$$

where $a_0y(0^-)$ has cancelled. This is identical to the equation in (7.224), and so we have the transform $Y(s)$ in (7.225).

The previous example demonstrates that although the ODE in (7.223) and the integro-differential equation in (7.226) are models for the same system, they lead to slightly different results in the way that initial states appear in the solution for $Y(s)$. This occurs because (7.223) is the derivative of (7.226), the derivative property of the Laplace transform introduces the initial states in a different way, and moreover, we have the derivative quantity $y'(0^-)$. Of course, when the initial states are all 0, the solution for $Y(s)$ is identical for both system models and it depends only on the input $X(s)$.

In general for linear ODEs with constant coefficients, the Laplace transform gives an expression for $Y(s)$ that can be written as a rational function. The Laplace transform of a third-order ODE is a cubic equation of the form $Y(s)(s^3 + a_2s^2 + a_1s + a_0) = X(s)/s$, and so on for higher-order ODEs. All of these algebraic equations can be factored into an expansion of terms with (i) distinct real poles, (ii) repeated real poles, (iii) complex poles, or (iv) repeated complex poles. A PFE factorization makes it is easy to find the inverse Laplace transform of the individual components using a table of transforms such as those in Tables 7.2 and 7.3.

7.12 IMPULSE RESPONSE AND TRANSFER FUNCTION

For a linear ODE with constant coefficients, the output is derived as a convolution of its impulse response function $h(t)$ and the input $x(t)$:

$$y(t) = h(t) * x(t) = x(t) * h(t), \tag{7.235}$$

assuming zero initial states. This was demonstrated in Chapter 6 for first- and second-order systems in the time domain, and it also holds for higher order LTI systems. From the convolution property of the Laplace transform, the output is the product of transforms:

$$Y(s) = H(s)X(s) = X(s)H(s), \tag{7.236}$$

where $H(s) = \mathcal{L}\{h(t)\}$ is the transfer function and $X(s) = \mathcal{L}\{x(t)\}$ is the input. From this expression, we find that the impulse response function of a system is generated

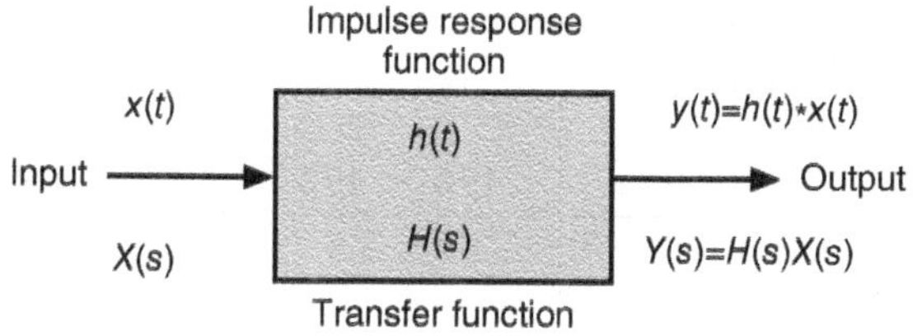

Figure 7.10 Time- and s-domain representations for an LTI system.

when $x(t) = \delta(t)$, which means $X(s) = 1 \implies Y(s) = H(s)$, and so the transfer function and impulse response function are a Laplace transform pair. These results are depicted in Figure 7.10. Once $H(s)$ is found for a particular ODE, the response of the system for any input is derived using (7.236); the inverse Laplace transform then yields the output $y(t)$ for that particular input.

Example 7.25 Consider the third-order ODE

$$\frac{d^3}{dt^3}y(t) + 6\frac{d^2}{dt^2}y(t) + 11\frac{d}{dt}y(t) + 6y(t) = x(t), \tag{7.237}$$

which has zero initial states. Its Laplace transform yields

$$s^3Y(s) + 6s^2Y(s) + 11sY(s) + 6Y(s) = X(s), \tag{7.238}$$

from which we have

$$H(s) = \frac{Y(s)}{X(s)} = \frac{1}{s^3 + 6s^2 + 11s + 6}. \tag{7.239}$$

The transfer function can be factored as

$$H(s) = \frac{1}{(s+1)(s+2)(s+3)}, \tag{7.240}$$

which reveals the poles of the system: $p_1 = -1, p_2 = -2$, and $p_3 = -3$, and it is now straightforward to perform a PFE:

$$H(s) = \frac{1/2}{s+1} - \frac{1}{s+2} + \frac{1/2}{s+3}, \tag{7.241}$$

which has the impulse response function

$$h(t) = [(1/2)\exp(-t) - \exp(-2t) + (1/2)\exp(-3t)]u(t). \tag{7.242}$$

The step response of the system is derived by multiplying $H(s)$ by the Laplace transform of $x(t) = u(t) \xrightarrow{\mathcal{L}} X(s) = 1/s$:

$$Y(s) = \frac{1}{s(s+1)(s+2)(s+3)} = \frac{1/6}{s} - \frac{1/2}{s+1} + \frac{1/2}{s+2} - \frac{1/6}{s+3}, \tag{7.243}$$

which has inverse transform

$$y(t) = [1/6 - (1/2)\exp(-t) + (1/2)\exp(-2t) - (1/6)\exp(-3t)]u(t) \tag{7.244}$$

and a steady-state value of 1/6. The responses for other inputs are derived using the same approach, which we find is easier than solving the original third-order ODE in the time domain.

The convolution between input signal $x(t)$ and impulse response function $h(t)$ is

$$y(t) = \int_{-\infty}^{\infty} x(\tau)h(t-\tau)d\tau = \int_{-\infty}^{\infty} h(\tau)x(t-\tau)d\tau, \tag{7.245}$$

where the support of each function determines the limits of integration. Depending on the specific functions under the integrals, one form may be easier to compute than the other. This occurs because the variable of integration is τ so that the function with argument $t - \tau$ is *reversed and shifted.* In the next two examples, we perform convolutions in the time domain and verify the results using the s-domain convolution property $Y(s) = H(s)X(s)$.

Example 7.26 Previously in Example 6.13, we demonstrated how to convolve two rectangular functions, which of course have finite duration, resulting in a finite-duration triangular function. In this example, one of the functions is rectangular $h(t) = u(t) - u(t - T)$, but the other function is the unit step $x(t) = u(t)$, such that their convolution has infinite duration. It is somewhat easier to use the second integral in (7.245) where the infinite-duration input $x(t)$ is reversed and shifted. This will be evident in the following because the finite-duration function determines the lower limit of integration, whereas the upper limit of integration depends on the specific time shift t. Figure 7.11(a) shows the reversed and shifted unit step function (note that the horizontal axis is the variable of integration τ). Figure 7.11(b) shows the rectangle function in terms of the variable τ. The goal when evaluating the convolution integral is to determine how these two functions overlap for different values of t, resulting in a nonzero integral. It is clear that for $t < 0$, there is no overlap and the convolution integral is 0. The shaded regions in Figure 7.11(a) and (b) indicate the amount of overlap between these two functions for $0 < t < T$. The resulting area from the integral for this particular t is illustrated by the point in Figure 7.11(c): $t \times 1 = t$. As t is increased toward T, the amount of function overlap increases, and so the output also increases. Initially, the limits of integration are $\{0, t\}$. When t exceeds T, the upper limit becomes T because the impulse response

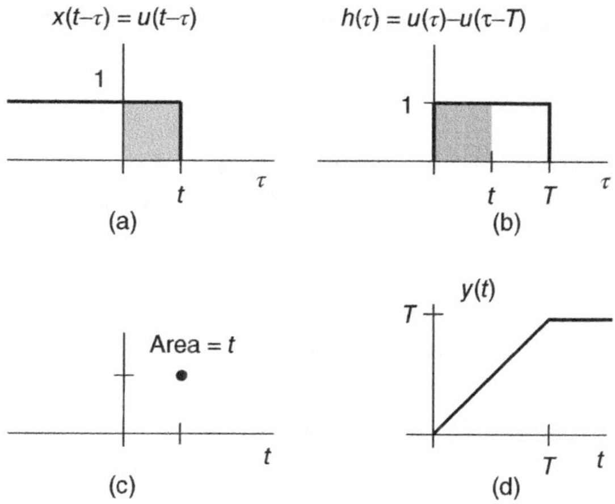

Figure 7.11 Convolution. (a) $x(t) = u(t)$ is reversed and shifted by t: $x(\tau - t) = u(\tau - t)$ with variable of integration τ. (b) $h(\tau) = u(\tau) - u(\tau - T)$ with variable of integration τ. (c) Area of overlapping functions (shaded regions) gives $y(t)$ for $0 < t < T$. (d) Convolution result $y(t)$.

function has finite duration, and from that point on the convolution gives a constant value. From (7.245), the convolution integral is written as

$$y(t) = \int_{-\infty}^{\infty} u(t-\tau)[u(\tau) - u(\tau - T)]d\tau$$
$$= \int_{0}^{\min(t,T)} d\tau = \min(t, T)u(t), \tag{7.246}$$

where the unit step functions have defined the limits of integration. The lower limit is derived from $u(\tau) \implies \tau \geq 0$, and the upper limit is derived using $u(t-\tau)u(t-T) \implies \min(t, T)$. The waveform for $y(t)$ is shown in Figure 7.11(d), which is linearly increasing until $t = T$, at which point it remains constant:

$$y(t) = \begin{cases} 0, & t < 0 \\ t, & 0 \leq t \leq T \\ T, & t > T. \end{cases} \tag{7.247}$$

For this example, we needed to consider three intervals for t, and we demonstrated that it is usually convenient to sketch plots as in Figure 7.11 to verify the limits of integration in (7.246) and determine the degree of overlap between the two functions. If we view $x(t)$ as the input of a system with impulse response function $h(t)$, then $y(t)$ in the figure is the system output. The s-domain output is

$$Y(s) = \frac{1}{s}\left[\frac{1}{s} - \frac{\exp(-sT)}{s}\right] = \frac{1}{s^2} - \frac{\exp(-sT)}{s^2}, \tag{7.248}$$

and the inverse Laplace transform yields two ramp functions, of which the second one is shifted by T:

$$y(t) = r(t) - r(t - T). \tag{7.249}$$

This is a more compact way of writing the result in (7.247), and it is straightforward to verify that $y(t)$ in (7.249) is fixed at T for $t \geq T$ by subtracting the two ramp functions.

Example 7.27 The convolution of $x(t) = u(t)$ and $h(t) = \exp(-t)u(t)$ was considered in Examples 6.6 and 7.14, where in the second case it was compared to cross-correlations of the two functions. The time-domain result is

$$\begin{aligned} y(t) &= \int_{-\infty}^{\infty} u(t-\tau)\exp(-\tau)u(\tau)d\tau \\ &= \int_{0}^{t} \exp(-\tau)d\tau = [1 - \exp(-t)]u(t), \end{aligned} \tag{7.250}$$

which is causal and increases exponentially to 1 in the limit as $t \to \infty$. The output in the s-domain is

$$Y(s) = \left(\frac{1}{s}\right)\left(\frac{1}{s+1}\right) = \frac{1}{s} - \frac{1}{s+1}, \tag{7.251}$$

which has inverse Laplace transform

$$y(t) = u(t) - \exp(-t)u(t), \tag{7.252}$$

and is the same result as in (7.250).

Since the output of an LTI system is a convolution between its input and impulse response function, we can determine the overall impulse response function for a *cascade* of systems $h_1(t)$ and $h_2(t)$. Let the intermediate signal be $s(t) = h_1(t) * x(t)$ and the overall output be $y(t) = h_2(t) * s(t)$. From the convolution integral:

$$\begin{aligned} y(t) &= \int_{-\infty}^{\infty} s(\tau)h_1(t-\tau)d\tau \\ &= \int_{-\infty}^{\infty}\int_{-\infty}^{\infty} x(v)h_1(\tau - v)h_2(t-\tau)dvd\tau, \end{aligned} \tag{7.253}$$

where the convolution for $s(t)$ has been substituted, and we have used a different integration variable v in order to avoid confusion with τ. Interchanging the two integrals yields

$$\begin{aligned} y(t) &= \int_{-\infty}^{\infty}\int_{-\infty}^{\infty} h_1(\tau - v)h_2(t-\tau)x(v)d\tau dv \\ &= \int_{-\infty}^{\infty}\int_{-\infty}^{\infty} h_1(t - v - w)h_2(w)dwx(v)dv, \end{aligned} \tag{7.254}$$

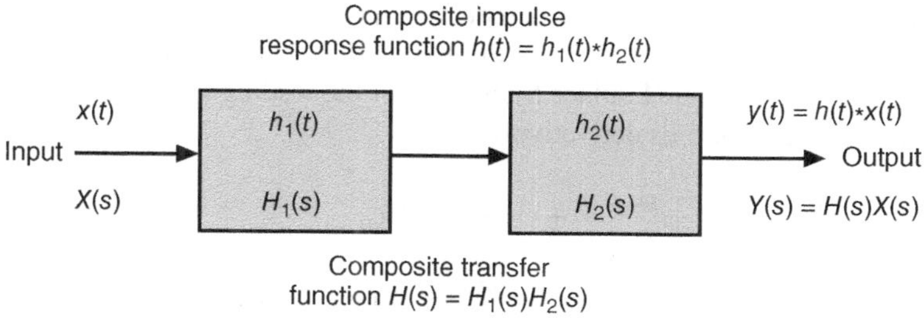

Figure 7.12 Cascaded LTI systems.

where we have changed variables to $w \triangleq t - \tau$ in the last expression. The inner integral is a convolution with argument $t - v$:

$$h(t-v) \triangleq \int_{-\infty}^{\infty} h_1(t-v-w)h_2(w)dw. \tag{7.255}$$

The overall output is the convolution of this composite impulse response function $h(t)$ and the input:

$$y(t) = \int_{-\infty}^{\infty} h(t-v)x(v)dv, \tag{7.256}$$

demonstrating that $y(t) = h(t) * x(t) = h_1(t) * h_2(t) * x(t)$ as depicted in Figure 7.12. From this result, we find in the s-domain that the composite transfer function is (see Problem 7.21)

$$H(s) = H_1(s)H_2(s). \tag{7.257}$$

7.13 PARTIAL FRACTION EXPANSION

The Laplace transform of a linear ODE is a rational function of polynomials in s with the following form:

$$X(s) = \frac{N(s)}{D(s)} \triangleq \frac{s^M + b_{M-1}s^{M-1} + \cdots + b_1 s + b_0}{s^N + a_{N-1}s^{N-1} + \cdots + a_1 s + a_0}, \tag{7.258}$$

where $N(s)$ is the numerator polynomial, which determines the zeros, and $D(s)$ is the denominator polynomial, which determines the poles. A rational function is in *proper form* if the numerator order is strictly less than the denominator order: $M < N$. In the event that $M \geq N$, long division is used to rewrite $X(s)$ as the sum of two terms where one term is a polynomial in s and the second term is a proper rational function:

$$X(s) = s^{M-N} + c_{M-1}s^{M-N-1} + \cdots + c_1 s + c_0 + \frac{d_{N-1}s^{N-1} + d_{N-2}s^{N-2} + \cdots + d_1 s + d_0}{s^N + a_{N-1}s^{N-1} + \cdots + a_1 s + a_0}. \tag{7.259}$$

The leading polynomial with coefficients $\{c_m\}$ is the *quotient*, and the numerator polynomial with coefficients $\{d_m\}$ is the *remainder*. This was illustrated earlier in Example 7.15. The inverse Laplace transforms of the leading terms of (7.259) are the Dirac delta function and its derivatives:

$$s^{M-N} + c_{M-1}s^{M-N-1} + \cdots + c_1 s + c_0$$
$$\xrightarrow{\mathcal{L}^{-1}} \delta^{(M-N)}(t) + c_{M-1}\delta^{(M-N-1)}(t) + \cdots + c_1\delta'(t) + c_0\delta(t). \quad (7.260)$$

Example 7.28 For the improper form with $M = N = 2$:

$$X(s) = \frac{s^2 + b_1 s + b_0}{s^2 + a_1 s + a_0}, \quad (7.261)$$

long division is performed by matching powers of s as follows:

$$\frac{s^2 + b_1 s + b_0}{s^2 + a_1 s + a_0} = 1 + \frac{(b_1 - a_1)s + b_0 - a_0}{s^2 + a_1 s + a_0}$$
$$= 1 + (b_1 - a_1)\frac{s + (b_0 - a_0)/(b_1 - a_1)}{s^2 + a_1 s + a_0}. \quad (7.262)$$

The inverse Laplace transform of 1 is $\delta(t)$, and the inverse Laplace transform of the second term on the right-hand side depends on the type of poles as discussed next.

The goal of a PFE is to write $N(s)/D(s)$ as a sum of simpler terms such that the order of the polynomial in each denominator is 1 or 2. A PFE can be viewed as the reverse operation of writing a sum of terms over a common denominator. For a linear ODE with constant coefficients, there are only four types of *partial fractions* to consider in the expansion:

- Distinct real poles of the form $s - p = 0$ such that $s = p$.
- Distinct complex conjugate poles of the form $(s + \alpha - j\beta)(s + \alpha + j\beta) = (s + \alpha)^2 + \beta^2 = 0$ such that $s_1, s_2 = -\alpha \pm j\beta$.
- Repeated real poles of the form $(s - p)^n = 0$ such that $s_1 = p, \ldots, s_n = p$.
- Repeated complex conjugate poles of the form $(s + \alpha - j\beta)^n(s + \alpha + j\beta)^n = [(s + \alpha)^2 + \beta^2]^n = 0$ such that $s_1, s_2 = -\alpha \pm j\beta, \ldots, s_{2n-1}, s_{2n} = -\alpha \pm j\beta$.

7.13.1 Distinct Real Poles

In order to proceed, it is necessary that the denominator be factored to explicitly show the poles. In general for distinct real poles, we have

$$X(s) = \frac{N(s)}{(s - p_1)\cdots(s - p_N)}, \quad (7.263)$$

for which the PFE is

$$X(s) = \sum_{k=1}^{N} \frac{A_k}{s - p_k}. \tag{7.264}$$

We assume that (7.263) is in proper form and it is stable, which means all $p_k \leq 0$. Although in some books $N(s)$ is factored into zeros, this is not necessary to find the *residues* $\{A_k\}$. The mth residue is

$$A_m = \lim_{s \to p_m} (s - p_m) X(s), \tag{7.265}$$

which can be verified from (7.264) as follows:

$$\begin{aligned} A_m &= \lim_{s \to p_m} (s - p_m) \sum_{k=1}^{N} \frac{A_k}{s - p_k} \\ &= A_m + \lim_{s \to p_m} \sum_{k=1, k \neq m}^{N} \frac{A_k (s - p_m)}{s - p_k} = A_m. \end{aligned} \tag{7.266}$$

Since the poles are distinct, there are no further cancellations in the sum such that when $s \to p_m$ all terms except A_m tend to 0. The resulting inverse Laplace transform is a sum of exponential functions:

$$x(t) = \sum_{k=1}^{N} A_k \exp(p_k t) u(t), \tag{7.267}$$

where the rate of decay for each term depends on its pole location on the s-plane. The time constant for the kth term is $\tau_k = -1/p_k$ (recall that p_k is negative for a stable system), and the exponential functions with poles closest to $s = 0$ decay more slowly to 0. Since the overall ROC of $X(s)$ is the intersection of the ROCs for the individual components in the PFE, it follows that the ROC lies just to the right of the pole with the smallest magnitude. This is depicted in Figure 7.13 for $X(s)$ with three distinct real poles.

Example 7.29 Consider a second-order system with transfer function in proper form:

$$H(s) = \frac{5s}{s^2 + 3s + 2} = \frac{A_1}{s + 1} + \frac{A_2}{s + 2}, \tag{7.268}$$

which has poles at $s_1, s_2 = \{-1, -2\}$. The two residues are

$$A_1 = \lim_{s \to -1} (s + 1) \frac{5s}{(s + 1)(s + 2)} = \lim_{s \to -1} \frac{5s}{s + 2} = -5, \tag{7.269}$$

$$A_2 = \lim_{s \to -2} (s + 2) \frac{5s}{(s + 1)(s + 2)} = \lim_{s \to -2} \frac{5s}{s + 1} = 10, \tag{7.270}$$

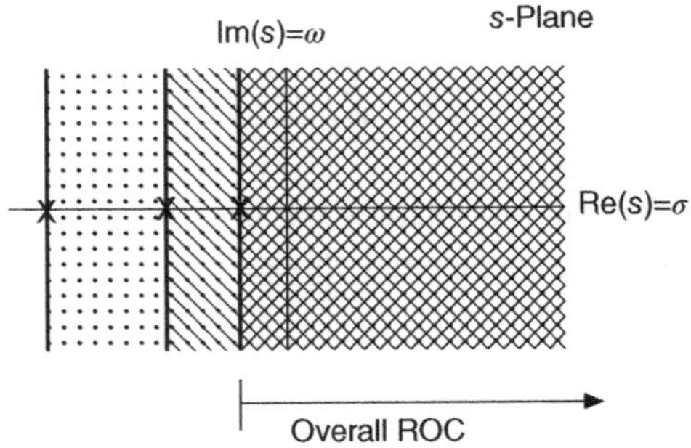

Figure 7.13 Overall region of convergence (ROC) for a transform with three distinct real poles.

and the PFE is

$$H(s) = -\frac{5}{s+1} + \frac{10}{s+2}. \tag{7.271}$$

The ROC is the intersection of the two individual ROCs: $\{\text{Re}(s) > -1\} \cap \{\text{Re}(s) > -2\} = \text{Re}(s) > -1$. It is straightforward to verify that the original numerator is derived by collecting terms over a common denominator:

$$H(s) = \frac{-5(s+2) + 10(s+1)}{(s+1)(s+2)} = \frac{5s}{(s+1)(s+2)}. \tag{7.272}$$

The time-domain function from Table 7.3 is

$$h(t) = [10\exp(-2t) - 5\exp(-t)]u(t). \tag{7.273}$$

Example 7.30 Since it is usually difficult to find the poles for high-order polynomials, we can resort to mathematics software such as MATLAB. The command for finding a PFE is

$$[\mathbf{r}, \mathbf{p}, \mathbf{k}] = \text{residue}(\mathbf{b}, \mathbf{a}), \tag{7.274}$$

where $\{\mathbf{b}, \mathbf{a}\}$ are vectors containing the coefficients of the numerator and denominator polynomials, respectively. (We have used bold notation to emphasize that the various quantities are vectors, though, of course, they are not bold in MATLAB.) The column vectors $\{\mathbf{r}, \mathbf{p}\}$ contain the residues and poles, respectively, and the row vector $\mathbf{k}$ contains the coefficients of the polynomial derived by long division in the event that the rational function is not in proper form. It is possible to generate the original rational function from the PFE by rearranging (7.274) as follows:

$$[\mathbf{b}, \mathbf{a}] = \text{residue}(\mathbf{r}, \mathbf{p}, \mathbf{k}). \tag{7.275}$$

For the rational function:

$$H(s) = \frac{5s}{s^3 + 6s^2 + 11s + 6}, \tag{7.276}$$

with $\mathbf{b} = [5, 0]^T$ and $\mathbf{a} = [1, 6, 11, 6]^T$, MATLAB gives

$$\mathbf{r} = \begin{bmatrix} -7.5 \\ 10 \\ -2.5 \end{bmatrix}, \quad \mathbf{p} = \begin{bmatrix} -3 \\ -2 \\ -1 \end{bmatrix}, \quad \mathbf{k} = [], \tag{7.277}$$

where the notation in the last term is an empty vector, meaning $H(s)$ is in proper form. Thus, the PFE is

$$H(s) = \frac{10}{s+2} - \frac{7.5}{s+3} - \frac{2.5}{s+1}, \tag{7.278}$$

with ROC $\text{Re}(s) > -1$, and the time-domain function is

$$h(t) = [10\exp(-2t) - 7.5\exp(-3t) - 2.5\exp(-t)]u(t). \tag{7.279}$$

7.13.2 Distinct Complex Poles

Similar results are obtained for rational functions with distinct complex conjugate poles. The PFE has the following form:

$$X(s) = \sum_{k=1}^{N/2} \frac{A_k}{s - p_k} + \sum_{k=1}^{N/2} \frac{A_k^*}{s - p_k^*}, \tag{7.280}$$

where the superscript $*$ denotes complex conjugation. Each sum extends only to $N/2$, and N is even because for each complex pole p_k, its complex conjugate p_k^* must also be present so that $X(s)$ has real coefficients. Of course, if $H(s)$ also has real poles, then N could be odd and the upper limit of the summations would need to be adjusted accordingly. Observe in (7.280) that we need only find $\{A_k\}$ for $N/2$ terms in the PFE because the other residues are derived via complex conjugation. Similar to (7.266), the mth residue is derived by multiplying $X(s)$ with $s - p_m$:

$$A_m = \lim_{s \to p_m} (s - p_m) \sum_{k=1}^{N/2} \frac{A_k}{s - p_k} + \lim_{s \to p_m} (s - p_m) \sum_{k=1}^{N/2} \frac{A_k^*}{s - p_k^*}, \tag{7.281}$$

and canceling that term from the denominator of the first sum:

$$A_m = A_m + \lim_{s \to p_m} \sum_{k=1, k \neq m}^{N/2} \frac{A_k(s - p_m)}{s - p_k} + \lim_{s \to p_m} \sum_{k=1}^{N/2} \frac{A_k^*(s - p_m)}{s - p_k^*} = A_m, \tag{7.282}$$

where $A_m/(s-p_m)$ has been factored in (7.281). Note that $s-p_m$ does not cancel $s-p_m^*$ in the second summation, and so, its limit is 0. Once the $N/2$ residues $\{A_m\}$ are found, it is a simple matter to conjugate them for the second sum in (7.280). Finally, we can rewrite (7.280) as

$$\begin{aligned} X(s) &= \sum_{k=1}^{N/2} \frac{A_k(s-p_k^*)+A_k^*(s-p_k)}{(s-p_k)(s-p_k^*)} \\ &= \sum_{k=1}^{N/2} \frac{(A_k+A_k^*)s-(A_kp_k^*+A_k^*p_k)}{s^2-(p_k+p_k^*)s+p_kp_k^*}, \end{aligned} \tag{7.283}$$

which is equivalent to

$$X(s) = 2\sum_{k=1}^{N/2} \frac{\text{Re}(A_k)s-\text{Re}(A_kp_k^*)}{s^2-2\text{Re}(p_k)+|p_k|^2}, \tag{7.284}$$

where $\text{Re}(A_k) \triangleq (A_k+A_k^*)/2$. This result confirms that the PFE has only real coefficients.

Example 7.31 Suppose (7.268) is modified so that it has complex conjugate poles:

$$H(s) = \frac{5s}{s^2+2s+2} = \frac{A_1}{s+1+j} + \frac{A_1^*}{s+1-j}. \tag{7.285}$$

The complex residue A_1 is derived in the same way as is done for distinct real poles, except, of course, the algebra for complex variables must be used:

$$\begin{aligned} A_1 &= \lim_{s\to-1-j}(s+1+j)\frac{5s}{(s+1+j)(s+1-j)} = \lim_{s\to-1-j}\frac{5s}{s+1-j} \\ &= \frac{-5-5j}{-2j} = (5/2)(1-j). \end{aligned} \tag{7.286}$$

The other residue is the complex conjugate of A_1, yielding the PFE

$$H(s) = \frac{(5/2)(1-j)}{s+1+j} + \frac{(5/2)(1+j)}{s+1-j}. \tag{7.287}$$

This result is verified by bringing the two terms over a common denominator:

$$\begin{aligned} H(s) &= \frac{(5/2)(1-j)(s+1-j)+(5/2)(1+j)(s+1+j)}{s^2+2s+2} \\ &= (5/2)\frac{s(1-j)-2j+s(1+j)+2j}{s^2+2s+2} = \frac{5s}{s^2+2s+2}. \end{aligned} \tag{7.288}$$

In order to prove that a PFE with residue A_1 must include its complex conjugate A_1^*, we examine one such pair where A_2 is used in place of A_1^*:

$$\frac{A_1}{s-p} + \frac{A_2}{s-p^*} = \frac{A_1(s-p^*) + A_2(s-p)}{s^2 - (p+p^*)s + pp^*}. \tag{7.289}$$

The denominator has real coefficients because $p + p^* = 2\text{Re}(p)$ and $pp^* = |p|^2$, as was shown in (7.284). The numerator is

$$N(s) = (A_1 + A_2)s - (A_1p^* + A_2p). \tag{7.290}$$

The coefficient of s is real only when $A_2 = A_1^*$ such that the imaginary parts cancel: $A_1 + A_2 = 2\text{Re}(A_1)$. From this choice of A_2, the other term of $N(s)$ is also real:

$$A_1p^* + A_1^*p = A_1p^* + (A_1p^*)^* = 2\text{Re}(A_1p^*), \tag{7.291}$$

which was illustrated in (7.284). Thus, we must have the form in (7.280) for complex conjugate poles and real polynomial coefficients.

Next, consider a second-order system with only two complex conjugate poles, which we rewrite by expressing the poles in terms of their real and imaginary parts $p = -\alpha + j\beta$ and $p^* = -\alpha - j\beta$ for $\alpha > 0$, yielding

$$H(s) = \frac{A}{s+\alpha-j\beta} + \frac{A^*}{s+\alpha+j\beta}. \tag{7.292}$$

The inverse Laplace transform is

$$\begin{aligned} h(t) &= [A\exp((-\alpha+j\beta)t) + A^*\exp((-\alpha-j\beta)t)]u(t) \\ &= \exp(-\alpha t)[A\exp(j\beta t) + A^*\exp(-j\beta t)]u(t), \end{aligned} \tag{7.293}$$

where the real exponential function common to both terms has factored. In order to continue, A is written in polar form $A = |A|\exp(j\theta)$ with phase component $\theta \triangleq \tan^{-1}(\text{Im}(A)/\text{Re}(A))$. Substituting this expression yields

$$\begin{aligned} h(t) &= |A|\exp(-\alpha t)[\exp(j(\beta t+\theta))\exp(-j(\beta t+\theta))]u(t) \\ &= 2|A|\exp(-\alpha t)\cos(\beta t+\theta)u(t). \end{aligned} \tag{7.294}$$

Using the trigonometric identity $\cos(x+y) = \cos(x)\cos(y) - \sin(x)\sin(y)$, we can also write this expression in sine/cosine form:

$$h(t) = 2|A|\exp(-\alpha t)[\cos(\theta)\cos(\beta t) - \sin(\theta)\sin(\beta t)]u(t). \tag{7.295}$$

As a result, (7.294) or (7.295) can be used directly for any pair of complex conjugate poles without having to repeat the previous derivations, though, of course, we must find the residue A.

The previous results imply some special cases for complex conjugate poles.

- $\text{Im}(A) = 0 \implies |A| = A$ and $\theta = 0$:

$$h(t) = 2A\exp(-\alpha t)\cos(\beta t)u(t). \tag{7.296}$$

In this case, the terms of (7.292) can be combined over a common denominator as follows:

$$\begin{aligned} H(s) &= \frac{A(s+\alpha+j\beta)+A(s+\alpha-j\beta)}{(s+\alpha)^2+\beta^2} \\ &= \frac{2A(s+\alpha)}{(s+\alpha)^2+\beta^2}. \end{aligned} \tag{7.297}$$

This is the same result as in (7.202) with $\omega_o = \beta$, except for the factor of $2A$.

- $\text{Re}(A) = 0 \implies \theta = 90°$:

$$h(t) = -2|A|\exp(-\alpha t)\sin(\beta t)u(t). \tag{7.298}$$

By combining (7.292) over a common denominator and using $A = jB$ such that $|A| = B$, we have

$$\begin{aligned} H(s) &= \frac{jB(s+\alpha+j\beta)-jB(s+\alpha-j\beta)}{(s+\alpha)^2+\beta^2} \\ &= \frac{-2B\beta}{(s+\alpha)^2+\beta^2}. \end{aligned} \tag{7.299}$$

This is the same result as in (7.203) with $\omega_o = \beta$, except for the factor of $-2B$.

Example 7.32 For $H(s)$ in (7.285) with the PFE in (7.287), we have $|A| = \sqrt{1^2+(-1)^2} = \sqrt{2}$ and $\theta = \tan^{-1}(-1) = -45°$ such that the inverse Laplace transform from (7.294) is

$$h(t) = 2\sqrt{2}\exp(-t)\cos(t-45°)u(t). \tag{7.300}$$

The form in (7.295) with $\cos(-45°) = \sqrt{2}/2$ and $\sin(-45°) = -\sqrt{2}/2$ yields

$$\begin{aligned} h(t) &= 2\sqrt{2}\exp(t)[(\sqrt{2}/2)\cos(t) - (\sqrt{2}/2)\sin(t)]u(t) \\ &= 2\exp(-t)[\cos(t) - \sin(t)]u(t). \end{aligned} \tag{7.301}$$

This sine/cosine form can be derived directly from $H(s)$ by rearranging it into the sum of two terms. For convenience, we repeat the Laplace transforms in Table 7.3 for exponentially weighted cosine and sine functions:

$$\exp(-\alpha t)\cos(\beta t)u(t) \xrightarrow{\mathcal{L}} \frac{s+\alpha}{(s+\alpha)^2+\beta^2}, \tag{7.302}$$

$$\exp(-\alpha t)\sin(\beta t)u(t) \xrightarrow{\mathcal{L}} \frac{\beta}{(s+\alpha)^2+\beta^2}. \tag{7.303}$$

The denominator for complex conjugate poles always has the form in these two expressions. Thus, it is only necessary that $H(s)$ be rewritten as the weighted sum of these two transforms, from which it is possible to write $h(t)$ using the sine/cosine form. There are two steps to this procedure. First, the denominator is written as earlier by completing the square (see Appendix C):

$$H(s) = \frac{5s}{s^2 + 2s + 2} = \frac{5s}{(s^2 + 2s + 1) + 1} = \frac{5s}{(s + 1)^2 + 1}, \tag{7.304}$$

which gives $\alpha = 1$ and $\beta = 1$. Second, the numerator is written as a weighted sum of $s + \alpha = s + 1$ and $\beta = 1$ with weights $\{a, b\}$:

$$H(s) = \frac{a(s + 1) + b \times 1}{(s + 1)^2 + 1}, \tag{7.305}$$

from which $a(s + 1) + b = 5s \implies a = 5$ and $a + b = 0 \implies b = -5$. These yield

$$H(s) = \frac{5(s + 1)}{(s^2 + 2s + 1) + 1} - \frac{5}{(s^2 + 2s + 1) + 1}, \tag{7.306}$$

which has inverse Laplace transform

$$h(t) = 5\exp(-t)\,[\cos(t) - \sin(t)]\,u(t). \tag{7.307}$$

We summarize the two methods of finding the inverse Laplace transform of $H(s)$ for a pair of complex conjugate poles:

$$H(s) = \frac{N(s)}{s^2 + a_1 s + a_0} = \frac{N(s)}{(s + \alpha - j\beta)(s + \alpha + j\beta)}, \tag{7.308}$$

where $H(s)$ is assumed to be in proper form. For the cosine form of $h(t)$, the residue is

$$A = \lim_{s \to -\alpha + j\beta} \frac{N(s)}{s + \alpha + j\beta} = \frac{N(-\alpha + j\beta)}{2j\beta}, \tag{7.309}$$

from which we have $|A|$, $\theta = \tan^{-1}(\text{Im}(A)/\text{Re}(A))$, and

$$h(t) = 2|A|\exp(-\alpha t)\cos(\beta t + \theta)u(t). \tag{7.310}$$

For the sine/cosine form, we complete the square in the denominator and then use the resulting quantities to rewrite $H(s)$ as the sum of two terms:

$$H(s) = \frac{N(s)}{(s + \alpha)^2 + \beta^2} = \frac{a(s + \alpha)}{(s + \alpha)^2 + \beta^2} + \frac{b\beta}{(s + \alpha)^2 + \beta^2}, \tag{7.311}$$

where $\{a, b\}$ are derived such that $a(s + \alpha) + b\beta = N(s)$. The inverse Laplace transform is

$$h(t) = \exp(-\alpha t)\,[a\cos(\beta t) + b\sin(\beta t)]\,u(t). \tag{7.312}$$

We provide another example of the second technique.

Example 7.33 The following transfer function has complex conjugate poles:

$$H(s) = \frac{s-3}{s^2+4s+13} = \frac{s-3}{(s+2)^2+9}, \tag{7.313}$$

where we have completed the square in the denominator, yielding $\alpha = 2$ and $\beta = 3$. In order to split this expression into the sum of two terms, the following equation is solved:

$$a(s+2)+3b = s-3 \implies a = 1 \text{ and } b = -5/3. \tag{7.314}$$

Thus

$$H(s) = \frac{s+2}{(s+2)^2+9} - (5/3)\frac{3}{(s+2)^2+9}, \tag{7.315}$$

and the inverse transform is

$$h(t) = \exp(-2t)\,[\cos(3t) - (5/3)\sin(3t)]\,u(t). \tag{7.316}$$

Finally, since complex conjugate roots can always be written as $(s+\alpha)^2+\beta^2$, this form can be used as the starting point to write polynomials in the standard form $s^2+a_1s+a_0 = s^2+2\alpha s+\alpha^2+\beta^2$. The fact that the roots are complex is verified again by using the quadratic formula:

$$s_1, s_2 = \frac{-2\alpha \pm \sqrt{4\alpha^2 - 4(\alpha^2+\beta^2)}}{2} = -\alpha \pm j\beta, \tag{7.317}$$

where $\alpha > 0$ is assumed for a stable system. This last result appears in (7.292).

7.13.3 Repeated Real Poles

Finding the PFE for repeated real poles requires more work than for distinct poles, which we illustrate with the following simple example:

$$\begin{aligned} X(s) &= \frac{s}{(s-p_1)(s-p_2)^2} \\ &= \frac{A_0}{s-p_1} + \frac{A_1}{(s-p_2)^2} + \frac{A_2}{s-p_2}. \end{aligned} \tag{7.318}$$

Using results from Table 7.3, the inverse Laplace transform is

$$x(t) = [A_0\exp(p_1t) + A_1t\exp(p_2t) + A_2\exp(p_2t)]\,u(t), \tag{7.319}$$

where the second term is a *ramped* exponential function. Observe that in addition to the term with denominator $(s-p_2)^2$, we also need to include in (7.318) a partial fraction with denominator $s - p_2$. The reason for this can be seen by combining all three terms over a common denominator:

$$X(s) = \frac{A_0(s-p_2)^2 + A_2(s-p_2)(s-p_1) + A_1(s-p_1)}{(s-p_1)(s-p_2)^2}$$
$$= \frac{(A_0+A_2)\,s^2 + (2A_0p_2 + A_2p_2 + A_2p_1 + A_1)\,s + (A_0\,p_2^2 + A_2p_1p_2 - A_1p_1)}{(s-p_1)(s-p_2)^2}, \tag{7.320}$$

where coefficients for the different powers of s have been collected together in the numerator. If the $s - p_2$ term is not included in (7.318), which corresponds to setting $A_2 = 0$ in (7.320), then

$$X(s) = \frac{A_0s^2 + (2A_0p_2 + A_1)\,s + (A_0\,p_2^2 - A_1p_1)}{(s-p_1)(s-p_2)^2}. \tag{7.321}$$

Comparing this expression with (7.318), we see that (7.321) does not have enough parameters to give the correct numerator. For the specific numerator in (7.318):

$$A_0 = 0, \qquad A_0p_2^2 - A_1p_1 = 0, \tag{7.322}$$

and so, it is not possible to have $2A_0p_2 + A_1 = 1$. There are three terms in the numerator of (7.321), but there are only two available parameters $\{A_0, A_1\}$ when $A_2 = 0$ (the poles $\{p_1, p_2\}$ are fixed and cannot be adjusted to give the correct numerator). By including the partial fraction with A_2 in (7.318), there is a sufficient number of parameters to produce the numerator in the first equation of (7.318), which can be seen from (7.320):

$$A_0 + A_2 = 0, \qquad 2A_0p_2 + A_2(p_1 + p_2) + A_1 = 1, \tag{7.323}$$
$$A_0p_2^2 + A_2p_1p_2 - A_1p_1 = 0. \tag{7.324}$$

These three equations with three unknowns can be written in matrix form:

$$\begin{bmatrix} 1 & 0 & 1 \\ 2p_2 & 1 & p_1 + p_2 \\ p_2^2 & -p_1 & p_1p_2 \end{bmatrix} \begin{bmatrix} A_0 \\ A_1 \\ A_2 \end{bmatrix} = \begin{bmatrix} 0 \\ 1 \\ 0 \end{bmatrix}. \tag{7.325}$$

Solving this matrix equation yields the residues for the PFE. The situation in (7.321) with only two parameters corresponds to an *underdetermined* system of linear equations that has no solution:

$$\begin{bmatrix} 1 & 0 \\ 2p_2 & 1 \\ p_2^2 & -p_1 \end{bmatrix} \begin{bmatrix} A_0 \\ A_1 \end{bmatrix} = \begin{bmatrix} 0 \\ 1 \\ 0 \end{bmatrix}. \tag{7.326}$$

Rearranging this in row-echelon form (see Chapter 3) yields

$$\begin{bmatrix} 1 & 0 \\ 0 & 1 \\ 0 & 0 \end{bmatrix} \begin{bmatrix} A_0 \\ A_1 \end{bmatrix} = \begin{bmatrix} 0 \\ 1 \\ p_1 \end{bmatrix}. \tag{7.327}$$

Since we assume that $p_1 \neq 0$ (otherwise there is no need to find a PFE for this Laplace transform), a solution does not exist; for the double pole $(s - p_2)^2$, the PFE must include a term with denominator $s - p_2$.

For the more general case with repeated poles $(s - p_k)^m$:

$$X(s) = \frac{N(s)}{(s - p_1)(s - p_2)^m}, \tag{7.328}$$

it is necessary that partial fractions with poles $\{(s - p_2)^{m-1}, \ldots, (s - p_2)^2, s - p_2\}$ be included. This follows from the fact that when the PFE terms are collected over a common denominator, the numerator will be of the form

$$N(s) = b_m s^m + \cdots + b_1 s + b_0. \tag{7.329}$$

In order to represent an arbitrary numerator $N(s)$, an equation is needed for each coefficient b_k. There must be $m + 1$ equations for the $m + 1$ coefficients, from which $m + 1$ residues $\{A_k\}$ are computed. Residue A_0 is associated with $s - p_1$, and $\{A_1, \ldots, A_m\}$ are associated with $\{(s - p_2)^m, \ldots, (s - p_2)^2, s - p_2\}$, respectively:

$$X(s) = \frac{A_0}{s - p_1} + \frac{A_1}{(s - p_2)^m} + \frac{A_2}{(s - p_2)^{m-1}} + \cdots + \frac{A_m}{s - p_2}. \tag{7.330}$$

Combining all terms over a common denominator leads to a complicated numerator, and so, the matrix representation used to solve for the parameters is also complicated. The inverse Laplace transform is

$$x(t) = A_0 \exp(p_1 t)\, u(t) + \left[\frac{A_1}{(m-1)!} t^{m-1} + \frac{A_2}{(m-2)!} t^{m-2} + \cdots + A_m \right] \exp(p_2 t)\, u(t), \tag{7.331}$$

which shows that m repeated poles yield an exponential function in the time domain that is weighted by a sum of terms containing t with exponents ranging from 0 to $m - 1$.

Example 7.34 Consider the following fourth-order system with repeated poles:

$$\begin{aligned} H(s) &= \frac{2s}{(s+1)(s+3)^3} \\ &= \frac{A_0}{s+1} + \frac{A_1}{(s+3)^3} + \frac{A_2}{(s+3)^2} + \frac{A_3}{s+3}. \end{aligned} \tag{7.332}$$

Although there are four residues in this example, A_0 can be found separately, so the matrix needed for solving the other three residues has rank 3. For the pole at $s = -1$:

$$A_0 = \lim_{s \to -1} \frac{2s}{(s+3)^3} = -1/4. \tag{7.333}$$

For the other residues, we examine the transfer function when combined over a common denominator:

$$H(s) = \frac{A_0(s+3)^3 + A_1(s+1) + A_2(s+1)(s+3) + A_3(s+1)(s+3)^2}{(s+1)(s+3)^3}, \tag{7.334}$$

whose numerator is

$$\begin{aligned} N(s) &= A_0(s^3 + 9s^2 + 27s + 27) + A_1(s+1) \\ &\quad + A_2(s^2 + 4s + 3) + A_3(s^3 + 7s^2 + 15s + 9) \\ &= (A_0 + A_3)s^3 + (9A_0 + A_2 + 7A_3)s^2 \\ &\quad + (27A_0 + A_1 + 4A_2 + 15A_3)s + (27A_0 + A_1 + 3A_2 + 9A_3). \end{aligned} \tag{7.335}$$

Setting this equation equal to the actual numerator $2s$ after substituting $A_0 = -1/4$ and equating coefficients of s^m gives three equations in three unknowns as follows:

$$\begin{bmatrix} 0 & 1 & 7 \\ 1 & 4 & 15 \\ 1 & 3 & 9 \end{bmatrix} \begin{bmatrix} A_1 \\ A_2 \\ A_3 \end{bmatrix} = \begin{bmatrix} 9/4 \\ 35/4 \\ 27/4 \end{bmatrix}. \tag{7.336}$$

Solving this matrix equation using MATLAB yields $A_1 = 3, A_2 = 1/2$, and $A_3 = 1/4$. Note that in this case, we can immediately find A_3 from the coefficient for s^3 in (7.335) because A_0 is already known: $A_0 + A_3 = 0 \implies A_3 = -A_0 = 1/4$. As a result, the other residues can be derived using a lower dimension matrix equation:

$$\begin{bmatrix} 1 & 4 \\ 1 & 3 \end{bmatrix} \begin{bmatrix} A_1 \\ A_2 \end{bmatrix} = \begin{bmatrix} 5 \\ 9/2 \end{bmatrix}, \tag{7.337}$$

which yields $A_1 = 3$ and $A_2 = 1/2$. We can simplify further by recognizing that A_1 is derived in a manner similar to that used to find A_0 as follows:

$$A_1 = \lim_{s \to -3} (s+3)^3 H(s) = \lim_{s \to -3} \frac{2s}{s+1} = 3. \tag{7.338}$$

Since $\{A_0, A_1, A_3\}$ are now known, $A_2 = 1/2$ can be found from any one of the three coefficients of s^m in (7.335) containing A_2:

$$9A_0 + A_2 + 7A_3 = 0, \quad 27A_0 + A_1 + 4A_2 + 15A_3 = 2, \quad 27A_0 + A_1 + 3A_2 + 9A_3 = 0. \tag{7.339}$$

From the previous discussion, we find that it is straightforward to compute the parameters of a PFE for the case of repeated real poles, but it may be tedious to solve the resulting matrix equation when there are many poles. However, as shown for the previous example, it is often possible to solve for some residues using the distinct pole method, which in turn can be substituted into the coefficient equations derived from the numerator in order to reduce the size of the matrix equation needed to solve for the remaining residues.

There is an alternative method for handling repeated poles that is based on derivatives with respect to s. We illustrate this procedure for the system in (7.318). The first two residues $\{A_0, A_1\}$ are computed in the usual manner:

$$A_0 = \lim_{s\to p_1} (s-p_1)\, X(s), \qquad A_1 = \lim_{s\to p_2} (s-p_2)^2 X(s). \tag{7.340}$$

The approach used for these two equations cannot be used for A_2, which is seen as follows:

$$\begin{aligned} A_2 &= \lim_{s\to p_2} (s-p_2) \left[\frac{A_0}{s-p_1} + \frac{A_1}{(s-p_2)^2} + \frac{A_2}{s-p_2} \right] \\ &= 0 + \lim_{s\to p_2} \frac{A_1}{s-p_2} + A_2, \end{aligned} \tag{7.341}$$

because the middle term on the right-hand side is undefined (infinite) in the limit. This problem is handled by multiplying $X(s)$ instead with $(s-p_2)^2$ and then *differentiating* the expression with respect to s before taking the limit:

$$\begin{aligned} A_2 &= \lim_{s\to p_2} \frac{d}{ds} \left[(s-p_2)^2 \frac{A_0}{s-p_1} + A_1 + A_2(s-p_2) \right] \\ &= \lim_{s\to p_2} \left[2(s-p_2)\frac{A_0}{s-p_1} - (s-p_2)^2 \frac{A_0}{(s-p_1)^2} + 0 + A_2 \right] = A_2, \end{aligned} \tag{7.342}$$

which is the desired result. Multiplying by $(s-p_2)^2$ eliminates the denominator of the A_1 term in (7.341), while ensuring that A_0 and A_2 are multiplied by either $s-p_2$ or $(s-p_2)^2$ so they tend to 0 as $s \to p_2$. For the more general case in (7.328), it is clear that for all terms with denominators of the form $(s-p_2)^k$, for $k = 1, \ldots, m$, the Laplace transform $X(s)$ is multiplied by $(s-p_2)^m$ (with the highest power m), and then we successively differentiate $(s-p_2)^m X(s)$ with respect to s and take the limit after each derivative to produce the residues $\{A_1, \ldots, A_m\}$, respectively. The only caveat is that each result must be scaled by $(m-1)!$, which we demonstrate for (7.328):

$$(s-p_2)^m X(s) = (s-p_2)^m \frac{A_0}{s-p_1} + A_1 + (s-p_2)\, A_2 + \cdots + (s-p_2)^{m-1} A_m. \tag{7.343}$$

Residue A_1 is generated when $s \to p_2$ because all other terms tend to 0. Differentiating once eliminates the A_1 term; residue A_2 then appears explicitly in the expression, and all other terms tend to 0 as $s \to p_2$ because they are multiplied by positive powers of $s - p_2$. The other residues are derived by continuing this process. In particular, we have after $m - 1$ derivatives:

$$\frac{d^{m-1}}{ds^{m-1}}(s-p_2)^m X(s) = \frac{d^{m-1}}{ds^{m-1}}(s-p_2)^m \frac{A_0}{s-p_1} + [(m-1) \times \cdots \times 2 \times 1]A_m. \tag{7.344}$$

For any partial fraction with poles other than p_2, we always obtain the form in the first term on the right-hand side of (7.344) where it is multiplied by $(s - p_2)^m$ with a power exceeding the order of the derivative, and there are no cancellations in the numerator and the denominator. Thus, in the limit as $s \to p_2$, such terms always tend to 0. The last term in (7.344) is premultiplied by $(m - 1)!$ because of the $m - 1$ derivatives, which in the limit yields the desired residue:

$$A_m = \frac{1}{(m-1)!} \lim_{s \to p_2} \frac{d^{m-1}}{ds^{m-1}}(s-p_2)^m X(s). \tag{7.345}$$

For all parameters $\{A_1, \ldots, A_m\}$ associated with $s = p_2$, the general expression for the kth residue is

$$A_k = \frac{1}{(k-1)!} \lim_{s \to p_2} \frac{d^{k-1}}{ds^{k-1}}(s-p_2)^m X(s), \tag{7.346}$$

for $k = 1, \ldots, m$ (with $0! \triangleq 1$). Note that the exponent of $s - p_2$ is always m for any value of k in this expression.

For the Laplace transform in (7.318), the inverse transforms associated with A_0 and A_2 are

$$\frac{A_0}{s-p_1} \xrightarrow{\mathcal{L}^{-1}} A_0 \exp(p_1 t)\, u(t), \tag{7.347}$$

$$\frac{A_2}{s-p_2} \xrightarrow{\mathcal{L}^{-1}} A_2 \exp(p_2 t)\, u(t). \tag{7.348}$$

For the partial fraction with residue A_1, the inverse Laplace transform of (7.197) is the ramped exponential function:

$$\frac{A_1}{(s-p_2)^2} \xrightarrow{\mathcal{L}^{-1}} A_1 t \exp(p_2)tu(t), \tag{7.349}$$

and so, the overall inverse Laplace transform is

$$x(t) = [A_0 \exp(p_1 t) + (A_1 t + A_2) \exp(p_2 t)]\, u(t). \tag{7.350}$$

Example 7.35 For the transfer function in Example 7.34, the residue for the pole at $s = -1$ is still derived using the approach described in (7.333), which yielded $A_0 = -1/4$. Likewise, A_1 is derived in the same manner as at the end of that example, which yielded $A_1 = 3$. The remaining two residues are generated using the derivative approach:

$$A_2 = \lim_{s\to-3} \frac{d}{ds}\frac{2s}{s+1} = \lim_{s\to-3}\left[\frac{2}{s+1} - \frac{2s}{(s+1)^2}\right] = 1/2, \tag{7.351}$$

and

$$\begin{aligned} A_3 &= (1/2)\lim_{s\to-3} \frac{d^2}{ds^2}\frac{2s}{s+1} = (1/2)\lim_{s\to-3}\frac{d}{ds}\left[\frac{2}{s+1} - \frac{2s}{(s+1)^2}\right] \\ &= (1/2)\lim_{s\to-3}\left[-\frac{2}{(s+1)^2} - \frac{2}{(s+1)^2} + \frac{4s}{(s+1)^3}\right] = 1/4, \end{aligned} \tag{7.352}$$

which are the results obtained by solving (7.336). Note that the denominator of the last term with $(s+1)^3$ is negative because of the odd exponent: $(-3+1)^3 = -8$. From (7.198), the inverse Laplace transform of (7.332) is

$$\begin{aligned} h(t) &= [A_0 \exp(-t) + [(A_1/2)t^2 + A_2 t + A_3]\exp(-3t)]\, u(t) \\ &= [(-1/4)\exp(-t) + [(3/2)t^2 + (1/2)t + (1/4)]\exp(-3t)]\, u(t). \end{aligned} \tag{7.353}$$

This impulse response function for the fourth-order system is plotted in Figure 7.14, which we see has a more complex behavior than second-order overdamped, underdamped, and critically damped systems. The components due to the simple real pole and the repeated pole are shown separately.

7.13.4 Repeated Complex Poles

Finally, we consider the case of repeated complex conjugate poles, which is illustrated by the following simple case:

$$X(s) = \frac{s}{(s-p_1)(s-p_2)^2(s-p_2^*)^2}. \tag{7.354}$$

This Laplace transform has a real pole at $s = p_1$ and repeated complex conjugate poles at $s = p_2$ and $s = p_2^*$, where in general $p_2 = -\alpha + j\beta$ and $p_2^* = -\alpha - j\beta$. Its PFE has the form

$$X(s) = \frac{A_0}{s-p_1} + \frac{A_1}{(s-p_2)^2} + \frac{A_1^*}{(s-p_2^*)^2} + \frac{A_2}{s-p_2} + \frac{A_2^*}{s-p_2^*}, \tag{7.355}$$

where from earlier results we know that for distinct complex poles, the residues must occur as complex conjugate pairs because the coefficients of $X(s)$ are real. Even though the residues are complex, they are computed in the same way as was done

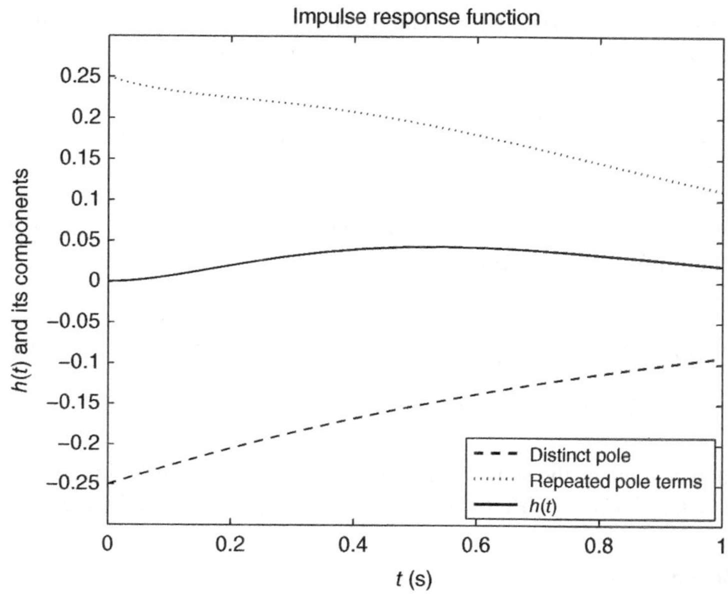

Figure 7.14 Impulse response function $h(t)$ and its components from Example 7.35.

for repeated real poles, such as the derivative approach discussed at the end of the previous section:

$$A_0 = \lim_{s \to p_1} (s - p_1)X(s), \tag{7.356}$$

$$A_1 = \lim_{s \to p_2} (s - p_2)^2 X(s), \tag{7.357}$$

$$A_2 = \lim_{s \to p_2} \frac{d}{ds}(s - p_2)^2 X(s). \tag{7.358}$$

The last two results also give the other two residues A_1^* and A_2^*. For m repeated complex poles, the formula in (7.346) is used, and again the results are copied over for the conjugated residues.

For the real pole in (7.354):

$$\frac{A_0}{s - p_1} \xrightarrow{\mathcal{L}^{-1}} A_0 \exp(p_1 t)\, u(t), \tag{7.359}$$

and for the terms associated with A_2, we have from (7.294):

$$\frac{A_2}{s - p_2} + \frac{A_2^*}{s - p_2^*} \xrightarrow{\mathcal{L}^{-1}} 2|A_2| \exp(-\alpha t) \cos(\beta t + \theta_2)\, u(t), \tag{7.360}$$

with $\theta_2 = \tan^{-1}(\text{Im }(A_2)/\text{Re }(A_2))$. The inverse Laplace transform for the $\{A_1, A_1^*\}$ terms is a ramped version of (7.360) (see Problem 7.29):

$$\frac{A_1}{(s-p_2)^2} + \frac{A_1^*}{(s-p_2^*)^2} \to 2|A_1| t \exp(-\alpha t)\cos(\beta t + \theta_1)\, u(t), \tag{7.361}$$

with $\theta_1 = \tan^{-1}(\text{Im }(A_1)/\text{ Re }(A_1))$. If the repeated pole in (7.354) is extended to order m, then the inverse Laplace transform is

$$x(t) = A_0 \exp(p_1 t)u(t) + 2\left[\frac{|A_1|}{(m-1)!}t^{m-1}\cos(\beta t + \theta_1) + \frac{|A_2|}{(m-2)!}t^{m-2}\cos(\beta t + \theta_2)\right.$$
$$\left. + \cdots + |A_m|\cos(\beta t + \theta_m)\right]\exp(-\alpha t)u(t), \tag{7.362}$$

where each cosine has the same frequency and exponential weighting, but a different phase $\theta_k = \tan^{-1}(\text{Im }(A_k)/\text{Re }(A_k))$, magnitude $|A_k|$, and factorial weighting.

Example 7.36 Suppose the Laplace transform is

$$H(s) = \frac{s}{(s+1)(s^2+4)^2}, \tag{7.363}$$

which has a real pole at $s = -1$ and repeated complex conjugate poles at $s = \pm j2$ (located on the imaginary axis). The residues of the PFE in (7.355) are

$$A_0 = \lim_{s\to-1} \frac{s}{s^2+4} = -1/5, \tag{7.364}$$

$$A_1 = \lim_{s\to-j2} \frac{s}{(s+1)(s-j2)^2} = \frac{-j2}{(-j2+1)(-j2-j2)^2} = (1/40)(j-2), \tag{7.365}$$

$$A_2 = \lim_{s\to-j2} \frac{d}{ds}\frac{s}{(s+1)(s-j2)^2}$$
$$= \lim_{s\to-j2}\left[\frac{1}{(s+1)(s-j2)^2} - \frac{s}{(s+1)^2(s-j2)^2} - \frac{2s}{(s+1)(s-j2)^3}\right]. \tag{7.366}$$

The three terms of A_2 are somewhat more complicated to evaluate than A_0 and A_1:

$$A_2 = \frac{1}{(-j2+1)(-j4)^2} + \frac{j2}{(-j2+1)^2(-j4)^2} + \frac{j4}{(-j2+1)(-j4)^3}$$
$$= -(1/80)(1+j2) + (j/200)(4+3j) + (1/80)(1+j2)$$
$$= (1/200)(4+3j). \tag{7.367}$$

It is not necessary to combine the partial fractions associated with residues $\{A_1, A_1^*\}$ and $\{A_2, A_2^*\}$; instead, we can immediately write the inverse Laplace transform using

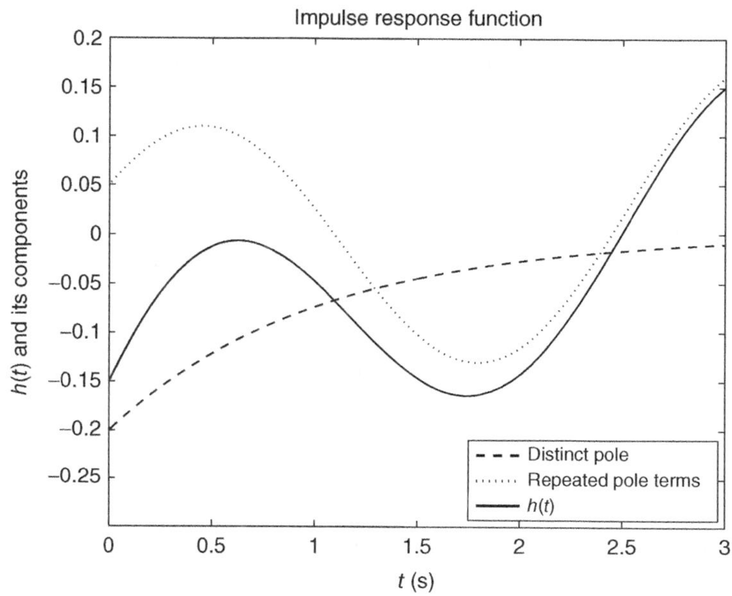

Figure 7.15 Impulse response function $h(t)$ and its components from Example 7.36.

the cosine forms in (7.294) and (7.361) with $\alpha = 0$ and $\beta = 2$:

$$h(t) = [-(1/5)\exp(-t) + (1/4\sqrt{5})\cos(2t - \tan^{-1}(2)) \\ +(1/20)t\cos(2t + \tan^{-1}(3/4))]\ u(t). \tag{7.368}$$

Since the poles are on the imaginary axis, these cosine terms are not exponentially weighted. The factor of t in the third term on the right-hand side causes the cosine to grow unbounded, which, of course, is due to the repeated complex conjugate poles. From this example, we find that the two terms associated with the repeated complex poles must have the same frequency, but their magnitudes and phase shifts generally differ because the residues A_1 and A_2 are not usually identical. The fifth-order impulse response function is plotted in Figure 7.15, along with its individual components due to the single real pole and the repeated complex conjugate poles. This function is quickly dominated by the cosine functions (the dotted line) because the exponential component (the dashed line) quickly tends to 0.

Table 7.5 provides a brief summary of the PFE residues and inverse Laplace transforms for the four different types of poles. For repeated poles, only the results for two poles are included; the residues for higher order repeated poles are computed using (7.346), and the inverse transforms for real and complex repeated poles are given in (7.331) and (7.362), respectively.

TABLE 7.5 Partial Fraction Expansion for $X(s)$

Type of Pole	Residue	Inverse Transform
Real p	$A = \lim_{s\to p}(s-p)X(s)$	$\exp(pt)u(t)$
Repeated real p	$A = \lim_{s\to p} d[(s-p)X(s)]/ds$	$t\exp(pt)(u(t)$
Complex $p = -\alpha + j\beta$	$A = \lim_{s\to p}(s-p)X(s)$	$2\lvert A\rvert \exp(-\alpha t)\cos(\beta t + \theta)u(t)$
Repeated complex p	$A = \lim_{s\to p} d[(s-p)X(s)]/ds$	$2\lvert A\rvert t\exp(-\alpha t)\cos(\beta t + \theta)u(t)$

Example 7.37 We conclude this section with a discussion of the three types of second-order system responses for the polynomial in (6.119), which we include in the following transfer function:

$$H(s) = \frac{1}{s^2 + 2\zeta\omega_o s + \omega_o^2}, \tag{7.369}$$

where ζ is the *damping ratio* and ω_o is the *resonant frequency*. The poles are given by

$$s_1, s_2 = -\zeta\omega_o \pm \omega_o\sqrt{\zeta^2 - 1}, \tag{7.370}$$

and the type of system depends on the value of ζ:

$$\textit{overdamped}: \ \zeta > 1 \implies s_1, s_2 = -\zeta\omega_o \pm \omega_o\zeta_d \tag{7.371}$$

$$\textit{underdamped}: \ \zeta < 1 \implies s_1, s_2 = -\zeta\omega_o \pm j\omega_d \tag{7.372}$$

$$\textit{critically damped}: \ \zeta = 1 \implies s_1 = s_2 = -\zeta\omega_o, \tag{7.373}$$

where $\omega_d \triangleq \omega_o\sqrt{1-\zeta^2}$, and we have defined $\zeta_d \triangleq \sqrt{\zeta^2 - 1}$ for convenience in the following derivation. For the underdamped system, the transfer function is

$$H(s) = \frac{1}{(s + \zeta\omega_o + j\omega_d)\,(s + \zeta\omega_o - j\omega_d)} = \frac{1}{(s+\zeta\omega_o)^2 + \omega_d^2}. \tag{7.374}$$

The Laplace transform pair for the exponentially weighted sine function is

$$\exp(-\alpha t)\sin(\beta t)\,u(t) \ \xrightarrow{\mathcal{L}} \ \frac{\beta}{(s+\alpha)^2 + \beta^2}, \tag{7.375}$$

from which we have the impulse response function:

$$h(t) = (1/\omega_d)\exp(-\zeta\omega_o t)\sin(\omega_d t)u(t). \tag{7.376}$$

The transfer function for the critically damped system is

$$H(s) = \frac{1}{(s+\zeta\omega_o)^2}, \tag{7.377}$$

which we find in Table 7.3 is an exponentially weighted ramp function:

$$h(t) = t\exp(-\omega_o t)\, u(t), \tag{7.378}$$

where $\zeta = 1$ has been substituted. For the overdamped system, we perform a PFE:

$$H(s) = \frac{A}{s + \omega_o(\zeta + \zeta_d)} + \frac{B}{s + \omega_o(\zeta - \zeta_d)}, \tag{7.379}$$

with residues

$$A = \left.\frac{1}{s + \omega_o(\zeta - \zeta_d)}\right|_{s=-\omega_o(\zeta+\zeta_d)} = -\frac{1}{2\omega_o\zeta_d} \tag{7.380}$$

$$B = \left.\frac{1}{s + \omega_o(\zeta + \zeta_d)}\right|_{s=-\omega_o(\zeta-\zeta_d)} = \frac{1}{2\omega_o\zeta_d}. \tag{7.381}$$

Thus, the impulse response function for the overdamped system is

$$\begin{aligned} h(t) = {} & (1/2\omega_o\sqrt{\zeta^2 - 1})[\exp(-\omega_o(\zeta - \sqrt{\zeta^2 - 1})t) \\ & - \exp(-\omega_o(\zeta + \sqrt{\zeta^2 - 1})t)]u(t). \end{aligned} \tag{7.382}$$

The purpose of this example is to investigate these three systems when ζ is close to 1, in order to determine how the waveforms change when transitioning from one type of second-order system to another. Toward this end, we examine the underdamped response in (7.376) with $\zeta = 1 - a$, and the overdamped response in (7.382) with $\zeta = 1 + a$, both for small $a > 0$ so they are close to being critically damped. The three impulse response functions are illustrated in Figure 7.16 for $\omega_o = 1$ rad/s and three values of a. Figure 7.16(a) shows the critically damped response, which we see is essentially the same as the other two responses for small $a = 0.1$ (the solid lines). With increasing a, Figure 7.16(b) shows that the underdamped response extends higher and lower than the critically damped response, whereas the overdamped response in Figure 7.16(c) does not have as much variation. This example demonstrates that for the same transfer function but different ζ, the critically damped response is the *transition waveform* ($\zeta = 1$) from a strictly exponential response ($\zeta > 1$) to one that is sinusoidal ($\zeta < 1$). Even though the three equations for $h(t)$ are quite different in general, they become equivalent as $\zeta \to 1$, and the transition across systems is continuous.

The previous results can also be verified by using power series expansions for the sine and exponential functions (see Appendix C). For the underdamped impulse response function in (7.376), ω_d is very small for $\zeta \approx 1$, which means $\sin(\omega_d t) \approx \omega_d t$ and

$$h(t) \approx (1/\omega_d)\exp(-\omega_o t)\omega_d t u(t) = t\exp(-\omega_o t)\, u(t), \tag{7.383}$$

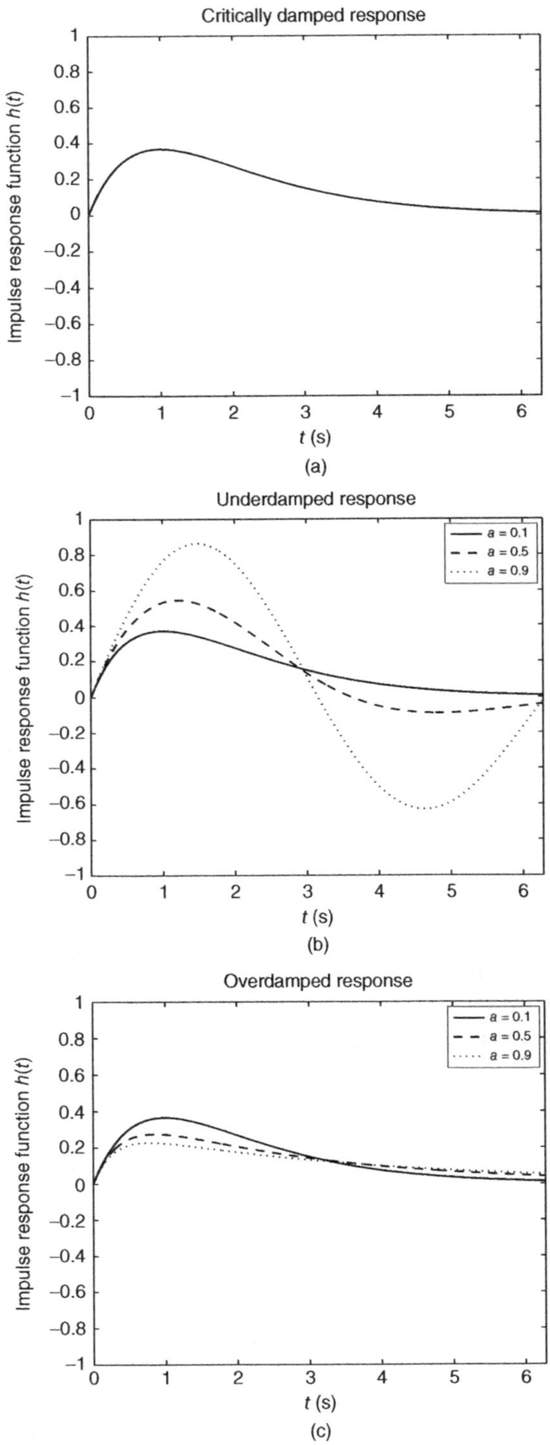

Figure 7.16 Impulse response functions for second-order systems with $\omega_o = 1$ rad/s. (a) Critically damped with $\zeta = 1$. (b) Underdamped with $\zeta = 1 - a$. (c) Overdamped with $\zeta = 1 + a$.

which is the critically damped response. For the overdamped response, we factor the common exponential function:

$$h(t) = (1/2\omega_o\sqrt{\zeta^2 - 1})\exp(-\omega_o t) \times \left[\exp(\omega_o t\sqrt{\zeta^2 - 1}) - \exp(-\omega_o t\sqrt{\zeta^2 - 1}))\right] u(t). \tag{7.384}$$

The exponential power series is approximated by $\exp(\pm x) \approx 1 \pm x$ for small x. Thus

$$h(t) \approx (1/2\omega_o\sqrt{\zeta^2 - 1})\exp(-\omega_o t) \times (2\omega_o t\sqrt{\zeta^2 - 1})\, u(t) = t\exp(-\omega_o t)\, u(t), \tag{7.385}$$

which again is the critically damped response.

7.14 LAPLACE TRANSFORMS AND LINEAR CIRCUITS

In this section, we demonstrate how to solve for circuit voltages and currents using the Laplace transform. The approach is similar to that described earlier using phasors for sinusoidal signals, though here the signals can be more general and start at a finite time.

The time-domain V-I and I-V models for the three passive circuit elements are

$$\textit{resistor}: v_R(t) = Ri_R(t), \qquad i_R(t) = v_R(t)/R, \tag{7.385}$$

$$\textit{inductor}: v_L(t) = L\frac{di_L(t)}{dt}, \qquad i_L(t) = (1/L)\int_0^t v_L(t)\, dt + i_L(0^-), \tag{7.387}$$

$$\textit{capacitor}: v_C(t) = (1/C)\int_0^t i_C(t)\, dt + v_C(0^-), \qquad i_C(t) = C\frac{dv_C(t)}{dt}. \tag{7.388}$$

Using properties of the Laplace transform, the corresponding s-domain expressions are provided in Table 7.6, which includes the time-domain initial states: $i(0^-)$ for the inductor and $v(0^-)$ for the capacitor. These follow from the derivative property of the Laplace transform:

$$v_L(t) = L\frac{di_L(t)}{dt} \xrightarrow{\mathcal{L}} V_L(s) = sLI_L(s) - Li_L(0^-), \tag{7.389}$$

$$i_C(t) = C\frac{dv_C(t)}{dt} \xrightarrow{\mathcal{L}} I_C(s) = sCV_C(s) - Cv_C(0^-), \tag{7.390}$$

which can be rearranged to give the other two expressions in the table:

$$I_L(s) = V_L(s)/sL + i_L(0^-)/s, \tag{7.391}$$

$$V_C(s) = I_C(s)/sC + v_C(0^-)/s. \tag{7.392}$$

These results can also be derived directly from the integral forms in (7.387) and (7.388), as long as the initial states are treated as step functions as shown in

TABLE 7.6 s-Domain Impedance of Linear Circuit Elements

Device	Impedance $Z(s)$	V-I Transform	I-V Transform
Resistor	R	$V(s) = RI(s)$	$I(s) = V(s)/R$
Inductor	sL	$V(s) = sLI(s) - Li(0^-)$	$I(s) = V(s)/sL + i(0^-)/s$
Capacitor	$1/sC$	$V(s) = I(s)/sC + v(0^-)/s$	$I(s) = sCV(s) - Cv(0^-)$

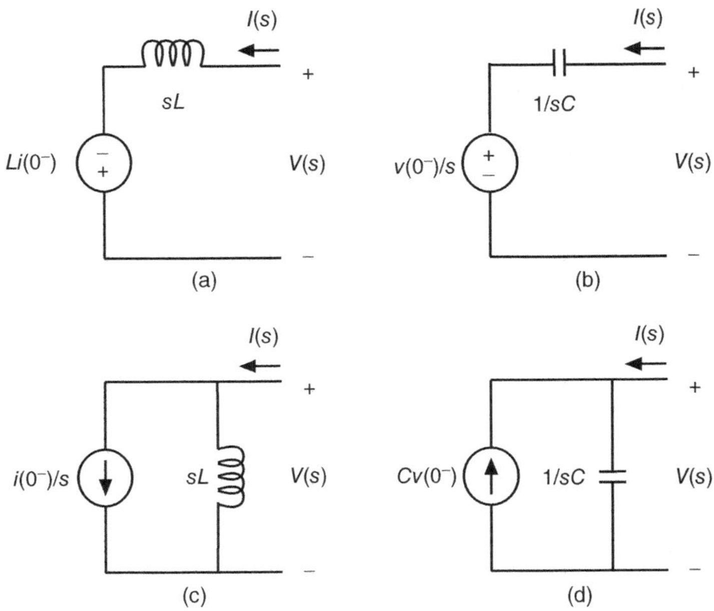

Figure 7.17 s-Domain circuit element models with initial states. (a) Inductor series model. (b) Capacitor series model. (c) Inductor parallel model. (d) Capacitor parallel model.

Example 7.16. Series and parallel circuit implementations of these equations are illustrated in Figure 7.17. Either configuration can be used, but one is usually more useful depending on the rest of the circuit. These are essentially Thévenin and Norton equivalent circuits (see Chapter 2) with impedance $Z(s)$ in place of resistance R. The s-domain impedance $Z(s) \triangleq V(s)/I(s)$ is calculated by assuming zero initial states (which, of course, was not an issue in the phasor definition of impedance $Z \triangleq V(\omega)/I(\omega)$ because the signals are assumed to be sinusoidal with doubly infinite duration). When analyzing a circuit, each element is replaced by its s-domain impedance, and the initial state is included as an independent voltage source or an independent current source.

Example 7.38 Consider again the first-order RC circuit in Figure 5.32, for which we want to find $v(t)$ for a general voltage input $v_s(t)$. The circuit is repeated in

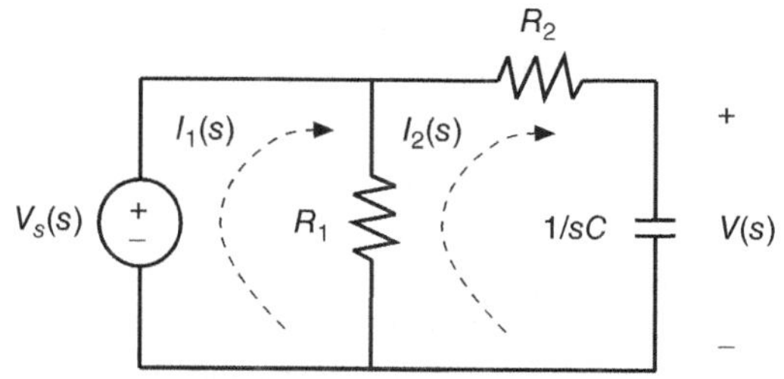

Figure 7.18 First-order circuit with capacitor C and voltage supply $V_s(s)$.

Figure 7.18, except that the phasor impedance $1/j\omega C$ has been replaced with the s-domain impedance $1/sC$. Since we are interested in the output voltage $v(t)$, the V-I transform for C in Table 7.6 is used, assuming a zero initial state. All quantities have been transformed to the s-domain, and the usual techniques for the analysis of resistive circuits can be used for this circuit. In this case, voltage division yields the desired result:

$$V(s) = \frac{1/sC}{1/sC + R} V_s(s) = \frac{1/RC}{s + 1/RC} V_s(s), \tag{7.393}$$

which has a pole at $p = -1/RC$. If $v_s(t) = \delta(t)$, then this ratio is the transfer function $H(s)$, and the corresponding impulse response function $h(t)$ is exponential:

$$h(t) = (1/RC)\exp(-t/RC)u(t). \tag{7.394}$$

If $v_s(t) = u(t)$, then a PFE can be used to find the component terms:

$$V(s) = \frac{1/RC}{s(s + 1/RC)} = \frac{A_1}{s} + \frac{A_2}{s + 1/RC}, \tag{7.395}$$

where

$$A_1 = \left.\frac{1/RC}{s + 1/RC}\right|_{s=0} = 1, \quad A_2 = \left.\frac{1/RC}{s}\right|_{s=-1/RC} = -1. \tag{7.396}$$

Thus, the output voltage (step response of the circuit) has steady-state and transient components:

$$v(t) = u(t) - \exp(-t/RC)\, u(t) = [1 - \exp(-t/RC)]\, u(t). \tag{7.397}$$

Example 7.39 Figure 7.19(a) shows the series RLC discussed in Chapter 2, but with a step function voltage source. Kirchoff's voltage law (KVL) in the time domain yields $V_s u(t) = v_R(t) + v_L(t) + v_C(t)$ and the following integro-differential equation:

$$L\frac{d}{dt}i(t) + Ri(t) + \frac{1}{C}\int_0^t i(t)\, dt + v_C(0^-) = V_s u(t), \tag{7.398}$$

which includes the initial capacitor voltage $v_C(0^-)$. Differentiating this expression gives a second-order ODE that models the circuit current:

$$L\frac{d^2}{dt^2}i(t) + R\frac{d}{dt}i(t) + \frac{1}{C}i(t) = V_s\delta(t), \tag{7.399}$$

which eliminates the constant $v_C(0^-)$. In this example, we show three methods of solving for the current $i(t)$ using s-domain techniques: (i) transform (7.398) to the s-domain, (ii) transform (7.399) to the s-domain, and (iii) find the current directly in the s-domain using the circuit in Figure 7.19(b), with the inductor and capacitor replaced by their s-domain models from Figure 7.17. (i) The Laplace transform of (7.398) is

$$sLI(s) - Li(0^-) + RI(s) + I(s)/sC + v_C(0^-)/s = V_s/s, \tag{7.400}$$

where the unilateral Laplace transform of the constant $v_C(0^-)$ is $v_C(0^-)/s$. Solving for $I(s)$ yields

$$I(s) = \frac{si(0^-) + V_s/L - v_C(0^-)/L}{s^2 + (R/L)s + 1/LC}. \tag{7.401}$$

The type of circuit (overdamped, underdamped, critically damped) depends on the specific parameter values for $\{R, L, C\}$. (ii) The Laplace transform of (7.399) is

$$Ls^2I(s) - sLi(0^-) - Li'(0^-) + RsI(s) - Ri(0^-) + I(s)/C = V_s, \tag{7.402}$$

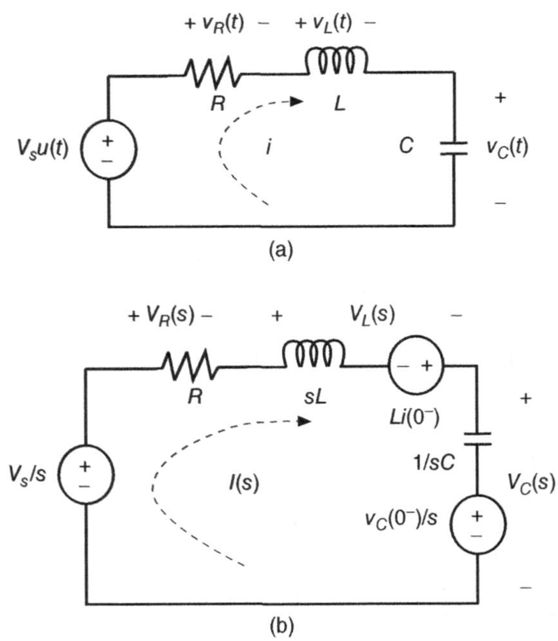

Figure 7.19 Second-order series circuit with resistor R, inductor L, and capacitor C. (a) Time-domain model. (b) s-Domain model.

and solving for $I(s)$ yields

$$I(s) = \frac{si(0^-) + Ri(0^-)/L + V_s/L - v_C(0^-)/L + i'(0^-)}{s^2 + (R/L)s + 1/LC}. \tag{7.403}$$

This expression is equivalent to that in (7.401) because the derivative $i'(0^-)$ can be written in terms of the circuit voltages using $v_L(t) + v_C(t) + v_R(t) = V_s u(t)$:

$$v_L(0^-) = L\frac{d}{dt}i(t)\bigg|_{t=0^-} \implies i'(0^-) = (1/L)\,[V_s u(0^-) - Ri(0^-) - v_C(0^-)]$$
$$= -(1/L)\,[Ri(0^-) + v_C(0^-)], \tag{7.404}$$

where $V_s u(0^-) = 0$ because of the unit step function. Substituting this result into (7.403) causes the $Ri(0^-)/L$ term to cancel, resulting in the same expression as in (7.401). This last step would normally be necessary in practice because derivative quantities such as $i'(0^-)$ are not typically given in a problem statement. (iii) Finally, from the s-domain model in Figure 7.19(b) and KVL, we have

$$V_R(s) + V_L(s) + V_C(s) = V_s/s$$
$$\implies RI(s) + sLI(s) - Li(0^-) + I(s)/sC + v_C(0^-)/s = V_s/s, \tag{7.405}$$

which is the same expression as in (7.400). This last result demonstrates that it is often simpler to work directly in the s-domain using the models in Figure 7.17, rather than first finding an integro-differential equation or an ODE in the time domain, and then transforming them to the s-domain. The corresponding time-domain current $i(t)$ is then derived via a PFE, which is straightforward because the rational function is in proper form.

Example 7.40 Suppose we want to find the voltages $v_R(t)$, $v_C(t)$, and $v_L(t)$ for the circuit in Figure 7.19(a). Since an expression has already been found in (7.401) for the s-domain current $I(s)$, and we have its time-domain waveform from a PFE, it is easy to find these quantities either in the time domain or the s-domain without writing another ODE. For the resistor:

$$v_R(t) = Ri(t) \implies V_R(s) = RI(s), \tag{7.406}$$

and for the inductor:

$$v_L(t) = L\frac{d}{dt}i(t) \implies V_L(s) = sLI(s) - Lv_L(0^-). \tag{7.407}$$

The initial voltage is derived from KVL: $v_R(0^-) + v_L(0^-) + v_C(0^-) = V_s u(0^-) = 0$, which yields

$$v_L(0^-) = -v_C(0^-) - Ri(0^-). \tag{7.408}$$

Finally, for the capacitor:

$$v_C(t) = \frac{1}{C}\int_{0^-}^{t} i(t)\,dt + v_C(0^-) \implies V_C(s) = I(s)/sC + v_C(0^-)/s. \tag{7.409}$$

It may be easier to differentiate or integrate the current in the time domain as in the first expressions of (7.407) and (7.409), rather than perform additional PFEs on the corresponding Laplace transforms.

Example 7.41 Finally, we derive the Thévenin equivalent circuit as seen by $V_s u(t)$ in Figure 7.19(a). The Thévenin impedance is derived by short circuiting the independent voltage sources due to the initial states:

$$Z_{\text{th}}(s) = R + sL + 1/sC. \tag{7.410}$$

The open circuit voltage depends only on the initial states because the current $I(s)$ is zero:

$$V_{\text{oc}}(s) = v_C(0^-)/s - Li(0^-). \tag{7.411}$$

The Norton equivalent circuit has the same impedance, and the short circuit current is derived by dividing (7.411) and (7.410):

$$I_{\text{sc}}(s) = \frac{v_C(0^-)/s - Li(0^-)}{sL + R + 1/sC} = \frac{v_C(0^-)/L - si(0^-)}{s^2 + (R/L)s + 1/LC}. \tag{7.412}$$

The two equivalent circuits in the s-domain are shown in Figure 7.20. These can be used to verify the circuit current derived in the previous example. From the Thévenin equivalent circuit:

$$I(s) = \frac{V_s/s - V_{\text{oc}}(s)}{Z_{\text{th}}(s)} = \frac{V_s/s - v_C(0^-)/s + Li(0^-)}{R + sL + 1/sC}, \tag{7.413}$$

which is the same as (7.401). From the Norton equivalent circuit and Kirchoff's current law (KCL):

$$I(s) = \frac{V_s/s}{Z_{\text{th}}(s)} - I_{\text{sc}}(s) = \frac{V_s/L}{s^2 + (R/L)s + 1/LC} - \frac{v_C(0^-)/L - si(0^-)}{s^2 + (R/L)s + 1/LC}, \tag{7.414}$$

which is also the same as (7.401).

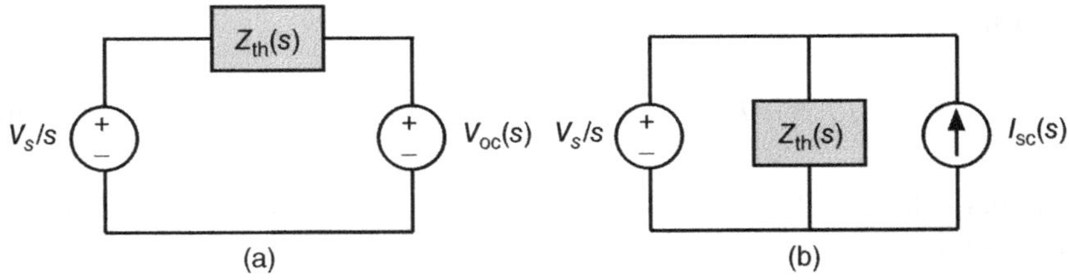

Figure 7.20 s-Domain equivalent circuits. (a) Thévenin. (b) Norton.

PROBLEMS

Solving ODEs Using Phasors

7.1 Solve the following first-order ODE using phasors:

$$\frac{d}{dt}y(t) + 2y(t) = \sin(t), \qquad t \in \mathcal{R}. \tag{7.415}$$

Give the magnitude and phase of **Y**, and then write an expression for $y(t)$.

7.2 Repeat the previous problem for the second-order system:

$$\frac{d^2}{dt^2}\, y(t) + 3\frac{d}{dt}\, y(t) + 2\, y(t) = \cos(2t + 45°), \qquad t \in \mathcal{R}. \tag{7.416}$$

7.3 Demonstrate how to solve the following integro-differential equation using phasors:

$$\frac{d}{dt}y(t) + \int_{-\infty}^{t} y(t)\, dt + y(t) = \cos(t - 30°), \qquad t \in \mathcal{R}. \tag{7.417}$$

Eigenfunctions

7.4 Determine if $y_1(t) = \sin(\omega_o t)$ or $y_2(t) = \exp(j\omega_o t)$ are eigenfunctions of the following ODEs:

$$\text{(a)}\ \frac{d^2}{dt^2}y(t) + 2y(t) = 0, \qquad \text{(b)}\ \frac{d}{dt}y(t) + 2y(t) = 0. \tag{7.418}$$

7.5 Find all eigenfunctions for the following third-order ODE by assuming $y(t) = \exp(st)$ with complex $s = \sigma + j\omega$:

$$\frac{d^3}{dt^3}y(t) + \frac{d^2}{dt^2}y(t) + 4\frac{d}{dt}y(t) + 4y(t) = 0. \tag{7.419}$$

Laplace Transform

7.6 (a) Show that the ROC for the bilateral Laplace transform of $x(t) = \exp(-2|t|)$ is a strip on the s-plane centered about the imaginary axis. (b) Repeat part (a) for $y(t) = \exp(-(t-1))u(t-1) + \exp(3t)u(-t)$.

7.7 (a) Demonstrate that for even function $x(t)$, the bilateral Laplace transform becomes

$$X(s) = \int_0^{\infty} x(t) \exp(-\sigma t) \cos(\omega t) dt, \tag{7.420}$$

and determine if $X(s)$ is also even. (b) Find a similar expression for odd $x(t)$ and determine if $X(s)$ is odd.

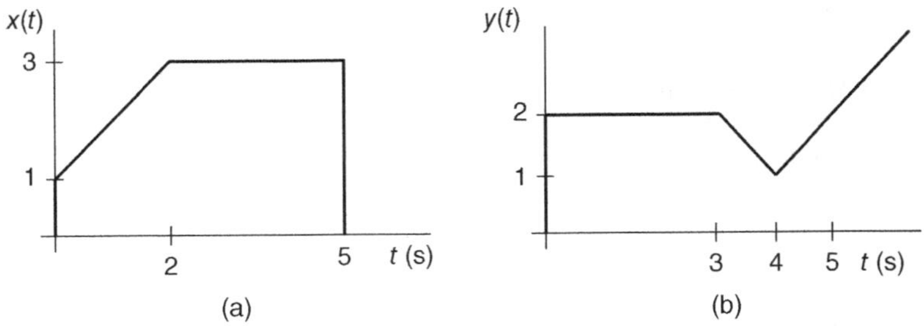

Figure 7.21 Time-domain functions for Problem 7.10.

7.8 Find the Laplace transform and specify the ROC for each of the following:

$$\text{(a) } x(t) = \sum_{n=0}^{\infty} \delta(t - nT_o), \qquad \text{(b) } y(t) = \sum_{n=0}^{\infty} f(t - nT_o), \tag{7.421}$$

with $T_o > 0$. The support of $f(t)$ is $[0, T_o/2]$ and its Laplace transform is $F(s)$ with ROC$_f$.

7.9 Repeat the previous problem for

$$\text{(a) } x(t) = \sin(\omega_o t + \phi)u(t), \qquad \text{(b) } y(t) = \text{rect}(t - 1/2)\cos(\omega_o t). \tag{7.422}$$

7.10 Find the Laplace transforms for $x(t)$ and $y(t)$ in Figure 7.21.

7.11 Use integration by parts to find the Laplace transform of $x(t) = t^2 \exp(-\alpha t)u(t)$.

7.12 Verify that the inverse Laplace transform equation in (7.58) is correct by direct substitution of $X(s)$ from (7.29).

Laplace Transform Properties

7.13 Derive the cross-correlation function in (7.131) and its bilateral Laplace transform in (7.132).

7.14 Find the Laplace transform for $d^2x(t)/dt^2$ using the derivative property:

$$x(t) = 2\exp(-t)u(t) + \exp(-2t)u(t-1). \tag{7.423}$$

7.15 For right-sided $h(t)$, use the time-shift property to find the inverse Laplace transform of

$$H(s) = \frac{s\exp(-s) + \exp(-2s)}{s^2 + 5s + 6}. \tag{7.424}$$

7.16 Use the time-division property to find the Laplace transform for $[\sin(\omega_o t)/t]u(t)$.

7.17 Find the initial and final values for the following Laplace transforms, and verify your results from the time-domain waveforms:

$$\text{(a) } Y_1(s) = \frac{s-2}{s^2+3s}, \qquad \text{(b) } Y_2(s) = \frac{s^2}{s^2+1}, \qquad \text{(c) } Y_3(s) = \frac{s}{s^2+7s+12}. \tag{7.425}$$

Solving Linear ODEs

7.18 Solve the following ODE by transforming it to the s-domain:

$$\frac{d^2}{dt^2}y(t) + 2\frac{d}{dt}y(t) + 2y(t) = u(t), \tag{7.426}$$

with initial states $y(0^-) = y'(0^-) = 1$.

7.19 Repeat the previous problem for

$$\frac{d^3}{dt^3}y(t) + 4\frac{d^2}{dt^2}y(t) + 9\frac{d}{dt}y(t) + 10y(t) = u(t), \tag{7.427}$$

which has one real pole at $s = -2$ and initial states $y(0^-) = y'(0^-) = y''(0^-) = 1$.

7.20 Assuming $y(0^-) = y'(0^-) = 0$, use Laplace transforms to solve for $y(t)$:

$$\frac{d^2}{dt^2}y(t) + 6\frac{d}{dt}y(t) + 9y(t) = 4\exp(-t)u(t). \tag{7.428}$$

Impulse Response and Transfer Function

7.21 Starting with the double integral in (7.253), take the Laplace transform and verify that the transfer function of two cascaded LTI systems with impulse response functions $h_1(t)$ and $h_2(t)$ is the product $H_1(s)H_2(s)$.

7.22 Repeat the derivation in (7.253) for the output of two cascaded systems assuming $\{x(t), h_1(t), h_2(t)\}$ are all causal so that the limits of the convolution integrals are $\{0, t\}$.

7.23 Find the transfer function $H(s)$ for the system represented by the ODE in Problem 7.5.

7.24 (a) Find the transfer function $H(s)$ and use it to derive (b) the step response and (c) the ramp response for the following impulse response function:

$$h(t) = [\exp(-2t) + (1/2)\exp(-t)]u(t). \tag{7.429}$$

7.25 For the transfer function

$$H(s) = \frac{s}{(s+1)(s+3)}, \tag{7.430}$$

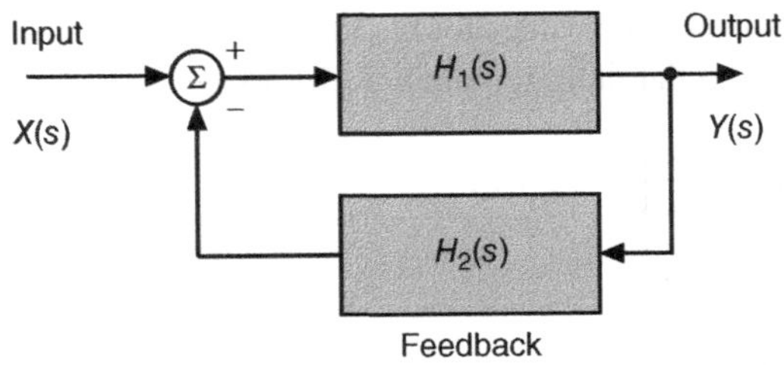

Figure 7.22 System with feedback for Problem 7.26.

substitute $s = j\omega$ (which gives the Fourier transform), and find expressions for (a) the magnitude $|H(\omega)|$ and (b) the phase $\angle H(\omega)$.

7.26 (a) Derive the transfer function $H(s)$ from $X(s)$ to $Y(s)$ for the feedback system in Figure 7.22. (b) Find the poles for $H(s)$ for the following system components:

$$H_1(s) = \frac{1}{s+2}, \qquad H_2(s) = \frac{2}{s+3}. \tag{7.431}$$

Convolution

7.27 Convolve the following functions and verify your results by finding the inverse Laplace transform of $H(s)X(s)$. (a) $x(t) = u(t-1)$ and $h(t) = u(t+1)$. (b) $x(t) = \exp(-t)u(t)$ and $h(t) = \exp(-2t)u(t)$.

7.28 Repeat the previous problem for (a) $x(t) = \text{tri}(t-1)$ and $h(t) = \text{rect}(t)$. (b) $x(t) = r(t)$ and $h(t) = \exp(-t)u(t)$.

Partial Fraction Expansion

7.29 Verify the Laplace transform pair for repeated complex poles in (7.361).

7.30 Find the inverse Laplace transform for the following transfer functions, assuming right-sided time-domain functions:

$$\text{(a) } H_1(s) = \frac{s+1}{s^2+5s+6}, \qquad \text{(b) } H_2(s) = \frac{s}{2s^2+4s+4}. \tag{7.432}$$

7.31 Repeat the previous problem for

$$\text{(a) } H_1(s) = \frac{s^2}{s^2+4s+3}, \qquad \text{(b) } H_2(s) = \frac{s}{s^2+4s+4}. \tag{7.433}$$

7.32 Repeat Problem 7.30 for the following transfer functions, assuming two-sided time-domain functions:

$$\text{(a) } H_1(s) = \frac{s+2}{s^2-s-2}, \qquad \text{(b) } H_2(s) = \frac{s^2+1}{s^2-s-6}. \tag{7.434}$$

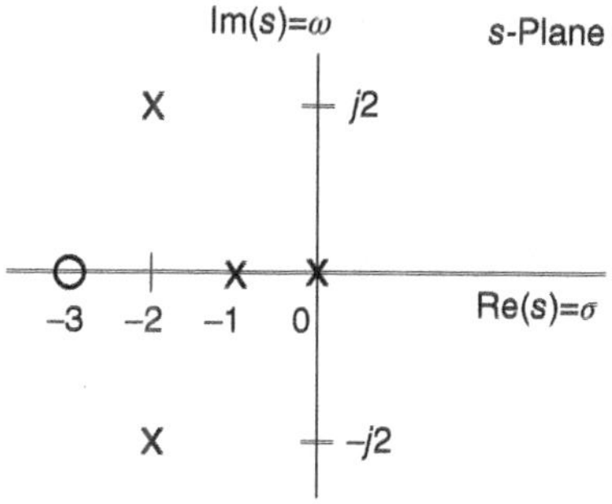

Figure 7.23 pole-zero plot for Problem 7.35.

7.33 Find the PFE for $H(s)$ using (a) the derivative approach for repeated poles and (b) the matrix approach described in Example 7.34:

$$H(s) = \frac{s^2}{(s+5)^2(s+2)}. \tag{7.435}$$

7.34 Repeat the previous problem for

$$H(s) = \frac{3s}{(s+1)(s^2+4)^2}. \tag{7.436}$$

7.35 Find the inverse Laplace transform for the system $H(s)$ represented by the pole-zero plot in Figure 7.23

7.36 Specify the system impulse response function $h(t)$ if its unit step response is

$$y(t) = [1/3 - (1/2)\exp(-t) + (1/6)\exp(-3t)]\, u(t). \tag{7.437}$$

Laplace Transform and Linear Circuits

7.37 For the parallel RLC circuit in Figure 7.24, use a nodal analysis in the s-domain to find the current through the inductor for $t \in \mathcal{R}^+$. Assume nonzero initial states $v_C(0^-)$ and $i_L(0^-)$.

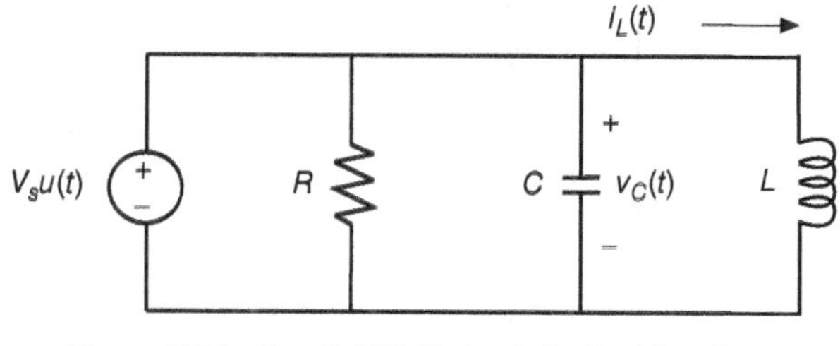

Figure 7.24 Parallel RLC circuit for Problem 7.37.

7.38 Repeat the previous problem for the RLC circuit in Figure 7.25

7.39 (a) Find an expression for the voltage $v(t)$ across the capacitor in Figure 7.26 using a nodal analysis in the s-domain, assuming $v_C(0^-) = 0$ and $i_l(0^-) = 2$ mA. (b) Find an equation for the capacitor current $i(t)$ starting with $V(s)$ in the s-domain. (c) Verify your result in part (b) by starting with $v(t)$ in the time domain.

Computer Problems

7.40 Use residue in MATLAB to find the poles and perform a PFE for the following transforms, and plot the time-domain functions $h_1(t)$ and $h_2(t)$:

$$\text{(a) } H_1(s) = \frac{5}{s^4 + 4s^3 + 4s^2 + 4s + 3}, \qquad \text{(b) } H_2(s) = \frac{s-1}{s^5 + 4s^4 + 9s^3 + 10s^2}. \tag{7.438}$$

7.41 Specify the transfer function for the following third-order ODE:

$$\frac{d^3}{dt^3}y(t) + 4\frac{d^2}{dt^2}y(t) + 5\frac{d}{dt}y(t) + 2 = u(t). \tag{7.439}$$

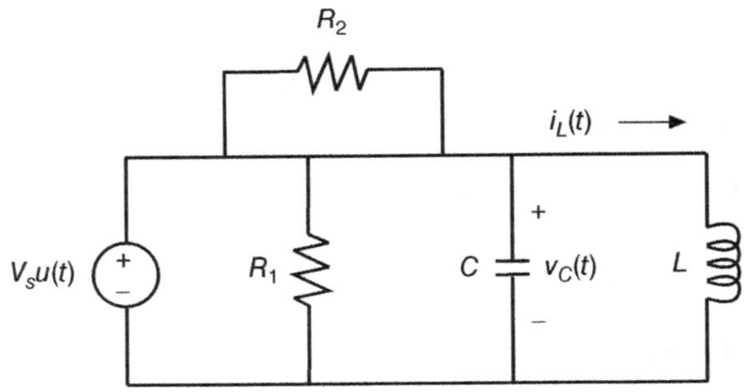

Figure 7.25 RLC circuit for Problem 7.38.

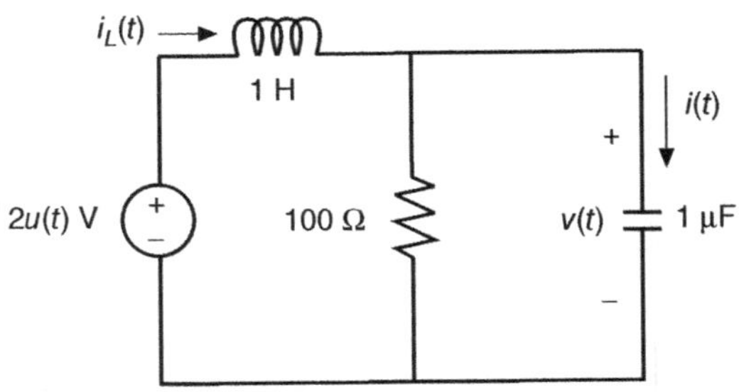

Figure 7.26 RLC circuit for Problem 7.39.

For a unit step input, use residue to find the PFE for $Y(s)$ and then obtain an expression for the time-domain function $y(t)$. The MATLAB function lsim can be used to numerically solve the ODE in (7.439):

$$\mathbf{y} = \text{lsim}(\mathbf{b}, \mathbf{a}, \mathbf{x}, \mathbf{t}), \tag{7.440}$$

where $\{\mathbf{b}, \mathbf{a}\}$ are the coefficient vectors defining the ODE, $\mathbf{t}$ is a vector of time instants, and $\mathbf{x}$ contains the corresponding input samples. Use lsim to generate $\mathbf{y}$, plot these values versus time, and compare the resulting curve to $y(t)$ derived earlier via the PFE. Although $\mathbf{a}$ for residue includes the pole at $s = 0$ due to the unit step input, that pole is excluded in $\mathbf{a}$ for lsim; instead samples of $\mathbf{x}$ should be generated using the function heaviside.

7.42 MATLAB generates Laplace transforms and inverse Laplace transforms by using syms to indicate that the variables t and s are symbolic. Once the functions $x(t)$ or $X(s)$ are defined, the transform commands are laplace and ilaplace. Use these to verify several of the transform pairs in Table 7.3. Also, transform some nonstandard functions such as $x(t) = t\exp(-t)\cos(t)\sin(t)u(t)$, and include delays in one or more arguments to see how the Laplace transforms change.

8

FOURIER TRANSFORMS AND FREQUENCY RESPONSES

8.1 INTRODUCTION

In this chapter, we describe another integral transform that can be viewed as a special case of the bilateral Laplace transform defined on the imaginary axis of the s-plane and generally for all $t \in \mathcal{R}$. In contrast to the unilateral Laplace transform whose lower limit of integration is zero, thus implying initial states and initial conditions, the Fourier transform is generally used to provide information about the frequency content of a signal or the frequency response of a linear time-invariant (LTI) system. As such, it is similar to a Fourier series except that the signals need not be periodic.

One important application of the Fourier transform is its description of an LTI system as a *filter* that enhances or removes certain frequency bands of a signal. For example, a *low-pass* filter emphasizes low frequencies, including the DC term, while rejecting high frequencies. Although DC usually refers to "direct current" in a circuit, it also corresponds to $f = 0$ Hz when describing the frequency content of a signal. The other major types of filters are *high-pass*, *band-pass*, and *band-reject* (also called band-stop). For sinusoidal waveforms, the angular frequency ω in radians/second (rad/s) is related to natural frequency f in hertz (Hz) (cycles/second) as follows:

$$\omega = 2\pi f = 2\pi/T, \tag{8.1}$$

where $T = 1/f$ is the period in seconds (s). Another parameter associated with a sinusoidal waveform is its wavelength, which takes into account the speed of light.

Mathematical Foundations for Linear Circuits and Systems in Engineering, First Edition. John J. Shynk.

Companion Website: http://www.wiley.com/go/linearcircuitsandsystems

Definition: Wavelength The *wavelength* of a waveform with frequency f is

$$\lambda \triangleq c/f, \tag{8.2}$$

where $c \triangleq 299{,}792{,}458$ m/s is the speed of light.

From this ratio, the units of wavelength are (m/s)/(cycles/s) = m/cycle, which means the wavelength is the distance in meters of one period T of the waveform. The speed of light is usually approximated by 300,000 km/s, which is the number used to derive the wavelengths of the different bands of the *electromagnetic spectrum* in Table 8.1; most of these are defined by the International Organization for Standardization (ISO, 2007). Of course, we are most familiar with the visible spectrum, but it is actually a very small band of frequencies. The other well-known band is the radio frequency band, which is exploited for various forms of communication. (Note that the audible and microwave frequency bands in the table overlap the radio band.) Table 8.2 summarizes the various nomenclature defined by the International Telecommunication Union (ITU, 2000) for radio frequency subbands. The United States government has allocated certain frequency bands for many different communications applications; some examples are mentioned in the table. Although frequencies in the ultra low

TABLE 8.1 Electromagnetic Spectrum

Frequency Band	Frequency Range f	Wavelength λ
Audible	20 Hz–20 kHz	15,000 km–15 km
Radio	300 Hz–300 GHz	1,000 km–1 mm
Microwave	300 MHz–300 GHz	1 m–1 mm
Infrared	300 GHz–395 THz	1 mm–760 nm
Visible	395 THz–789 THz	760 nm–380 nm
Ultraviolet	750 THz–30 PHz	400 nm–10 nm
X-rays	30 PHz–300 EHz	10 nm–1 pm
Gamma rays	300 EHz–30,000 EHz	1 pm–10 fm

TABLE 8.2 ITU Nomenclature for Radio Frequency Bands

Frequency Band	Frequency Range f	Example Usage
Ultra low frequency (ULF)	300–3,000 Hz	Seismic activity
Very low frequency (VLF)	3–30 kHz	Maritime mobile
Low frequency (LF)	30–300 kHz	Aeronautical mobile
Medium frequency (MF)	300–3000 kHz	AM radio
High frequency (HF)	3–30 MHz	Amateur radio
Very high frequency (VHF)	30–300 MHz	FM radio, VHF television
Ultra high frequency (UHF)	300–3000 MHz	UHF television, cellular
Super high frequency (SHF)	3–30 GHz	Satellite television
Extremely high frequency (EHF)	30–300 GHz	Radio astronomy

frequency (ULF) band have not been allocated for specific applications, they are often associated with seismic activity.

Radio frequency applications are implemented by combining a low-frequency message signal with a sinusoidal waveform called the *carrier*. It is the frequency of the carrier that determines the specific transmission band as summarized in the tables. The process of "combining" a message signal with a carrier is called *modulation*, as in amplitude modulation (AM) and frequency modulation (FM). These methods are easily examined in the *frequency domain* by using Fourier transform techniques. AM is covered later in this chapter, while FM is beyond the scope of this book (it is a nonlinear process).

8.2 FOURIER TRANSFORM

The Fourier transform is an integral transform with kernel $\exp(-j\omega t)$.

Definition: Fourier Transform The *Fourier transform* of $x(t)$ is

$$X(\omega) \triangleq \int_{-\infty}^{\infty} x(t)\exp(-j\omega t)dt, \tag{8.3}$$

which can be written in terms of natural frequency f by substituting $\omega = 2\pi f$:

$$X(f) = \int_{-\infty}^{\infty} x(t)\exp(-j2\pi ft)dt. \tag{8.4}$$

The following notation is used:

$$\mathcal{F}\{x(t)\} = X(\omega) \text{ or } X(f), \quad x(t) \xrightarrow{\mathcal{F}} X(\omega) \text{ or } X(f). \tag{8.5}$$

The Fourier transform of $x(t)$ exists if the following *Dirichlet conditions* hold:

- *Absolutely integrable*:

$$\int_{-\infty}^{\infty} |x(t)|dt < \infty. \tag{8.6}$$

- *Bounded discontinuities*: A finite number of bounded discontinuities in any finite-duration interval $[a, b] \subset \mathcal{R}$.
- *Bounded variation*: A finite number of minima and maxima in any finite-duration interval $[a, b] \subset \mathcal{R}$.

Observe from (8.3) that

$$\begin{aligned}|X(\omega)| &= \left|\int_{-\infty}^{\infty} x(t)\exp(-j\omega t)dt\right| \\ &\leq \int_{-\infty}^{\infty} |x(t)\exp(-j\omega t)|dt = \int_{-\infty}^{\infty} |x(t)|dt,\end{aligned} \tag{8.7}$$

where we have used the fact that the complex exponential has unit magnitude. Thus, if (8.6) holds, then the Fourier transform is at least bounded. These three conditions are sufficient but not necessary. For example, there are functions that are not absolutely integrable but have Fourier transforms; these include the unit step function, the signum function, the absolute value function, as well as periodic functions like sine and cosine. Such waveforms have Fourier transforms provided $X(\omega)$ is allowed to include *singular* functions: the Dirac delta function and its derivatives.

The Fourier transform always exists if the region of convergence (ROC) of the Laplace transform of $x(t)$ includes the imaginary axis. For these cases, $X(\omega)$ is generated from $X(s)$ by substituting $s = j\omega$. In fact, many books use the notation $X(j\omega)$ where j is explicitly shown in the argument because the Fourier transform is often derived as $X(j\omega) = X(s)|_{s=j\omega}$. However, for notational simplicity, we use $X(\omega)$ throughout this chapter and in the appendices. Moreover, it is possible for $X(\omega)$ to be real-valued, and so j may not actually appear in those transforms (although it does for most functions summarized in Tables 8.3 and 8.4).

The inverse Fourier transform is

$$x(t) = \frac{1}{2\pi} \int_{-\infty}^{\infty} X(\omega) \exp(j\omega t) d\omega, \tag{8.8}$$

TABLE 8.3 Fourier Transform Pairs: Impulsive, Step, and Ramp

Time Domain $x(t)$	Fourier Transform $X(\omega)$
1	$2\pi\delta(\omega)$
$\delta(t)$	1
$\sum_{m=-\infty}^{\infty} \delta(t - mT)$	$(2\pi/T) \sum_{m=-\infty}^{\infty} \delta(\omega - 2\pi m/T)$
$\delta(n)(t)$	$(j\omega)^n$
$\text{rect}(t)$	$\text{sinc}(\omega/2\pi)$
$\text{tri}(t)$	$\text{sinc}^2(\omega/2\pi)$
$u(t)$	$\pi\delta(\omega) + 1/j\omega$
$u(-t)$	$\pi\delta(\omega) - 1/j\omega$
$\text{sgn}(t)$	$2/j\omega$
$r(t)$	$j\pi\delta'(\omega) - 1/\omega^2$
$r(-t)$	$-j\pi\delta'(\omega) - 1/\omega^2$
$t^n u(t)$	$j^n \pi\delta^{(n)}(\omega) + n!/(j\omega)^{n+1}$
$\|t\|$	$-2/\omega^2$
$1/\sqrt{\|t\|}$	$\sqrt{2\pi/\|\omega\|}$
$1/t$	$-j\pi\text{sgn}(\omega)$
$1/t^2$	$\pi\|\omega\|$
$\text{sinc}(t)$	$\text{rect}(\omega/2\pi)$
$\text{sinc}^2(t)$	$\text{tri}(\omega/2\pi)$

TABLE 8.4 Fourier Transform Pairs: Exponential and Sinusoidal ($\alpha > 0$ and $\omega_o > 0$)

Time Domain $x(t)$	Fourier Transform $X(\omega)$
$\exp(-\alpha t)u(t)$	$1/(\alpha + j\omega)$
$[1 - \exp(-\alpha t)]u(t)$	$\pi\delta(\omega) + \alpha/(j\omega\alpha - \omega^2)$
$\exp(\alpha t)u(-t)$	$1/(\alpha - j\omega)$
$\exp(-\alpha\|t\|)$	$2\alpha/(\omega^2 + \alpha^2)$
$\exp(-\|t\|)\text{sgn}(t)$	$-j2\omega/(\alpha^2 + \omega^2)$
$\exp(-\alpha t^2)$	$\sqrt{\pi/\alpha}\exp(\omega^2/4\alpha)$
$t^n \exp(-\alpha t)u(t)$	$n!/(\alpha + j\omega)^{n+1}$
$t^n \exp(\alpha t)u(-t)$	$-n!/(\alpha + j\omega)^{n+1}$
$\cos(\omega_o t)u(t)$	$j\omega/(\omega_o^2 - \omega^2) + (\pi/2)[\delta(\omega + \omega_o) + \delta(\omega - \omega_o)]$
$\cos(\omega_o t)$	$\pi[\delta(\omega + \omega_o) + \delta(\omega - \omega_o)]$
$\cos^2(\omega_o t)$	$\pi^2[\delta(\omega + \omega_o) + \delta(\omega - \omega_o) + 2\delta(\omega)]$
$\exp(-\alpha t)\cos(\omega_o t)u(t)$	$(\alpha + j\omega)/[(\alpha + j\omega)^2 + \omega_o^2]$
$t\cos(\omega_o t)u(t)$	$(j\pi/2)[\delta'(\omega - \omega_o) + \delta'(\omega + \omega_o)] - (\omega^2 + \omega_o^2)/(\omega_o^2 - \omega^2)^2$
$t\exp(-\alpha t)\cos(\omega_o t)u(t)$	$[(\alpha + j\omega)^2 - \omega_o^2]/[(\alpha + j\omega)^2 + \omega_o^2]^2$
$\sin(\omega_o t)u(t)$	$\omega_o/(\omega_o^2 - \omega^2) + (j\pi/2)[\delta(\omega + \omega_o) - \delta(\omega - \omega_o)]$
$\sin(\omega_o t)$	$j\pi[\delta(\omega + \omega_o) - \delta(\omega - \omega_o)]$
$\sin^2(\omega_o t)$	$\pi^2[\delta(\omega + \omega_o) - \delta(\omega - \omega_o) + 2\delta(\omega)]$
$\exp(-\alpha t)\sin(\omega_o t)u(t)$	$\omega_o/[(\alpha + j\omega)^2 + \omega_o^2]$
$t\sin(\omega_o t)u(t)$	$(\pi/2)[\delta'(\omega - \omega_o) - \delta'(\omega + \omega_o)] + j2\omega_o\omega/(\omega_o^2 - \omega^2)^2$
$t\exp(-\alpha t)\sin(\omega_o t)u(t)$	$2\omega_o(\alpha + j\omega)/[(\alpha + j\omega)^2 + \omega_o^2]^2$

and in terms of f:

$$x(t) = \int_{-\infty}^{\infty} X(f)\exp(j2\pi ft)df. \tag{8.9}$$

The Fourier transform pairs in (8.4) and (8.9) are symmetric, whereas the Fourier transform pairs in (8.3) and (8.8) require the leading constant $1/2\pi$. Proof of the inverse transform in (8.8) is shown by substituting $X(\omega)$ and rearranging the two integrals:

$$\begin{aligned}\frac{1}{2\pi}\int_{-\infty}^{\infty} X(\omega)\exp(j\omega t)d\omega &= \frac{1}{2\pi}\int_{-\infty}^{\infty}\int_{-\infty}^{\infty} x(\tau)\exp(-j\omega\tau)\exp(j\omega t)d\omega d\tau \\ &= \frac{1}{2\pi}\int_{-\infty}^{\infty} x(\tau)\int_{-\infty}^{\infty}\exp(j(t-\tau)\omega)d\omega d\tau, \end{aligned} \tag{8.10}$$

where a different variable of integration τ has been used for the Fourier transform $X(\omega)$. The inner integral is the inverse Fourier transform of a constant, which is a shifted Dirac delta function $2\pi\delta(t-\tau)$ (shown later). Thus, from the sifting property of $\delta(t)$, we complete the proof:

$$\frac{1}{2\pi}\int_{-\infty}^{\infty} X(\omega)\exp(j\omega t)d\omega = \int_{-\infty}^{\infty} x(\tau)\delta(t-\tau)d\tau = x(t). \tag{8.11}$$

Recall that the inverse Laplace transform is an integral over the complex variable s, and it is preferable to use a partial fraction expansion (PFE). The inverse Fourier transform is just as easy to compute as the Fourier transform; the only difference is the positive argument of $\exp(j\omega t)$ and the multiplicative factor $1/2\pi$ in (8.8).

Example 8.1 From Table 7.2, the Laplace transform of $x(t) = \exp(-\alpha t)u(t)$ for $\alpha > 0$ is

$$X(s) = \frac{1}{s+\alpha}. \tag{8.12}$$

Since the ROC $\text{Re}(s) > -\alpha$ includes the $j\omega$ axis, we can immediately write

$$X(\omega) = \frac{1}{j\omega+\alpha} = \frac{\alpha - j\omega}{\alpha^2+\omega^2}. \tag{8.13}$$

This result is verified from the definition of the Fourier transform:

$$\begin{aligned} X(\omega) &= \int_{0}^{\infty} \exp(-\alpha t - j\omega t)dt \\ &= \frac{1}{\alpha + j\omega}\exp(-\alpha t - j\omega)\bigg|_{0}^{\infty}. \end{aligned} \tag{8.14}$$

Evaluating this expression at the upper limit gives 0 because α is positive, and we obtain the result in (8.13). Observe that $x(t)$ is absolutely integrable:

$$\begin{aligned} \int_{-\infty}^{\infty} |\exp(-\alpha t)u(t)|dt &= \int_{0}^{\infty} \exp(-\alpha t)dt \\ &= -\frac{1}{\alpha}\exp(-\alpha t)\bigg|_{0}^{\infty} = 1/\alpha < \infty. \end{aligned} \tag{8.15}$$

The magnitude of $X(s)$ is derived by substituting $s = \sigma + j\omega$ and separating the real and imaginary parts as follows:

$$|X(s)| = \frac{1}{|\sigma + j\omega + \alpha|} = \frac{1}{\sqrt{(\sigma+\alpha)^2+\omega^2}}. \tag{8.16}$$

This expression with $\alpha = 1$ is plotted in Figure A.13(d), which we repeat here in Figure 8.1(a). (The logarithm is used in the plot to show a greater dynamic range

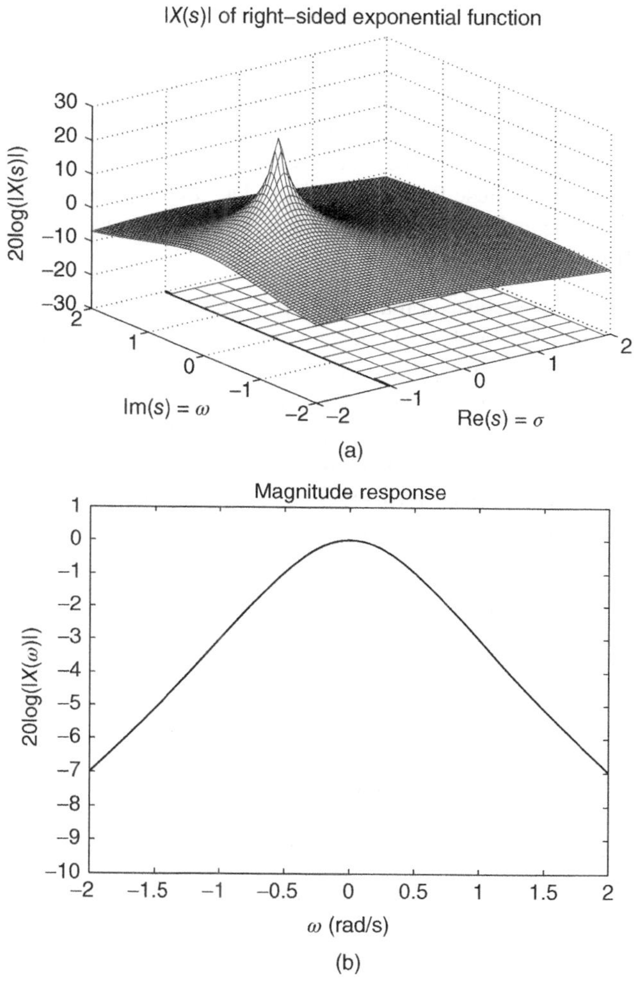

Figure 8.1 Laplace and Fourier transforms of the right-sided exponential function with $\alpha = 1$. (a) Truncated $20\log(|X(s)|)$ with ROC (lower horizontal grid). (b) $20\log(|X(\omega)|)$, corresponding to $20\log(|X(s)|)$ viewed along the $\sigma = 0$ axis.

for ease of viewing.) The magnitude of the Fourier transform corresponds to $|X(s)|$ evaluated at $s = j\omega$, which means $\sigma = 0$ in (8.16):

$$|H(\omega)| = \frac{1}{\sqrt{\alpha^2 + \omega^2}}. \tag{8.17}$$

This result is shown in Figure 8.1(b), and is valid because the imaginary axis is located within the ROC. The Laplace transform provides useful information about the time-domain function from the locations of its poles and zeros (see Figure 7.9). Since (8.12) has a pole on the real axis, we know that $x(t)$ is an exponential function. The time constant of $x(t)$ decreases if the pole is moved further to the left of the imaginary axis, which means the exponential function decays to 0 more rapidly.

The Fourier transform $X(\omega)$ provides information about the frequency content of $x(t)$. If $h(t)$ is the impulse response function of a system, then $H(\omega)$ is its frequency response. For this example, we see that $X(\omega)$ in Figure 8.1(b) has the characteristic of a low-pass signal because frequencies about DC are emphasized. Of course, this is due to the pole located on the real axis at $\sigma = -1$, which causes the shape of $|H(s)|$ in Figure 8.1(a).

Example 8.2 Suppose instead that $\alpha < 0$ in the previous example such that $x(t)$ is not absolutely integrable and $X(\omega)$ does not exist. From Chapter 7, we know that the Laplace transform exists: it is also given by (8.12) but with ROC $\text{Re}(s) = \sigma > -\alpha$, which does not include the imaginary axis for $-\alpha > 0$. If $s = j\omega$ is substituted into (8.12), then the same expression in (8.13) is obtained. However, this is *incorrect* because (8.12) holds only for $\text{Re}(s) > -\alpha > 0$. Since $\text{Re}(s) = 0$ for the Fourier transform, we cannot use the result in (8.13) for this function; the Fourier transform does not exist for $x(t) = \exp(-\alpha t)u(t)$ when $\alpha < 0$ because the function grows unbounded exponentially.

Example 8.3 The Fourier transform of the Dirac delta function $x(t) = \delta(t)$ is a constant for all frequencies:

$$X(f) = \int_{-\infty}^{\infty} \delta(t)\exp(-j2\pi ft)dt = 1, \quad f \in \mathcal{R}, \tag{8.18}$$

where the sifting property of $\delta(t)$ has been used (see Chapter 5). All frequencies appear equally for the Fourier transform of $\delta(t)$, which is the only "function" whose spectrum is flat over $f \in \mathcal{R}$. From the duality property discussed later, we conclude that the Fourier transform of a constant $x(t) = 1$ is the Dirac delta function:

$$X(f) = \int_{-\infty}^{\infty} \exp(-j2\pi ft)dt = \delta(f), \tag{8.19}$$

which is derived in Example 8.5 starting with the rectangle function. The Fourier transform written in angular frequency is the same:

$$X(\omega) = \int_{-\infty}^{\infty} \delta(t)\exp(j\omega t)dt = 1, \quad \omega \in \mathcal{R}, \tag{8.20}$$

but the inverse transform is slightly different:

$$\frac{1}{2\pi}\int_{-\infty}^{\infty} (1)\exp(j\omega t)d\omega = \delta(t) \implies \int_{-\infty}^{\infty} (1)\exp(j\omega t)d\omega = 2\pi\delta(t), \tag{8.21}$$

which has the factor 2π. Thus, the Fourier transform in angular frequency of a constant is

$$\mathcal{F}\{1\} = \int_{-\infty}^{\infty} (1)\exp(-j\omega t)dt = 2\pi\delta(\omega). \tag{8.22}$$

The Fourier transform can also be viewed as a decomposition of a waveform into its frequency components, as is the case for the Fourier series of a periodic function. The difference here is that $\omega \in \mathcal{R}$ is a continuous variable, whereas only the fundamental frequency ω_o and its harmonics $n\omega_o$ for $n \in \mathcal{Z}$ appear in the Fourier series expansion.

Definition: Spectrum The *spectrum* of signal $x(t)$ is its Fourier transform $X(\omega)$. It is a frequency-domain representation of the time-domain signal that indicates the relative strength of its frequency components for the continuous variable $\omega \in \mathcal{R}$.

The electromagnetic spectrum in Table 8.1 summarizes various frequency bands. The band in which a signal is located determines its wavelength and energy. Additional properties of a signal depend on the actual shape of the spectrum; for example, it might have a "notch" (low magnitude/high attenuation) for a narrow band of frequencies.

Example 8.4 The Fourier transform of the rectangle function $x(t) = \text{rect}(t)$ is

$$\begin{aligned} X(f) &= \int_{-1/2}^{1/2} \exp(-j2\pi ft)dt \\ &= \frac{1}{j2\pi f}\left[\exp(j\pi f) - \exp(-j\pi f)\right] \\ &= \frac{\sin(\pi f)}{\pi f} \triangleq \text{sinc}(f), \end{aligned} \tag{8.23}$$

where π is suppressed in the definition of the sinc function. The Fourier transform in angular frequency is

$$\begin{aligned} X(\omega) &= \frac{1}{j\omega}[\exp(j\omega/2) - \exp(-j\omega/2)] \\ &= (2/\omega)\sin(\omega/2) = (2\pi/\omega)\text{sinc}(\omega/2\pi). \end{aligned} \tag{8.24}$$

Observe from Euler's formula that the integral in (8.23) can be written as

$$X(f) = \int_{-1/2}^{1/2} \cos(2\pi ft)dt - j\int_{-1/2}^{1/2} \sin(2\pi ft)dt = \int_{-1/2}^{1/2} \cos(2\pi ft)dt. \tag{8.25}$$

The second integral is 0 because the rectangle function is even, sine is an odd function, and their product is an odd function that integrates to 0 for these symmetric limits of integration. For an even function, the Fourier transform reduces to the *cosine transform*. Figure 8.2(a) shows a plot of the rectangle function and its product with $\cos(2\pi ft)$ for $f = 2$ Hz and 6 Hz. Since the argument of the cosine function is an integer multiple of π, the integrals (areas) of these products are 0. This is evident from the figure where we see exactly two periods of the cosine function for $f = 2$ Hz and

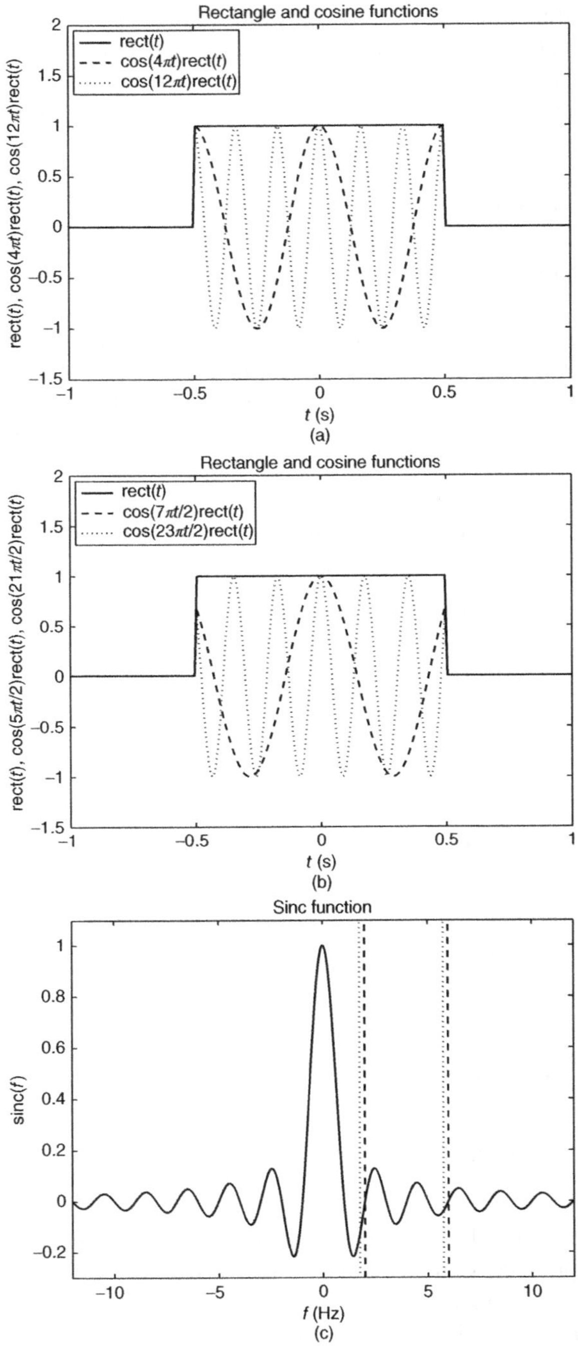

Figure 8.2 Product of rectangle function and two cosine functions. (a) f is an integer multiple of π. (b) f is a noninteger multiple of π. (c) Fourier transform of rectangle function: sinc(f). Vertical dotted lines denote frequencies $f = 7/4 = 1.75$ Hz and $23/4 = 5.75$ Hz, giving sinc(f) values of -0.0909 and -0.0277, respectively. The sinc function is 0 at the vertical dashed lines where $f = 2$ Hz and $f = 6$ Hz.

exactly 6 periods for $f = 6$ Hz. These integrals correspond to zero-crossings of the sinc function in Figure 8.2(c) (denoted by the vertical dashed lines). The frequencies of the cosine functions in Figure 8.2(b) are noninteger multiples of 2π, so that the sinc function is nonzero at $f = 7/4$ Hz and $f = 23/4$ Hz (denoted by the vertical dotted lines in Figure 8.2(c)). The corresponding values of sinc(f) are -0.0909 and -0.0277, respectively.

The curves in Figure 8.2(b) and (c) also illustrate the *cross-correlation* property of the Fourier transform. The lower frequency $\cos(5\pi t/2)\text{rect}(t)$ resembles the rectangle function more than $\cos(21\pi t/2)\text{rect}(t)$, which explains why the magnitude of the sinc function decreases with increasing frequency.

Example 8.5 Next, we examine the Fourier transform of a function that is not absolutely integrable. For the constant function $x(t) = c$ for $t \in \mathcal{R}$, we start with the rectangle function and extend it at both ends by scaling its argument:

$$c = \lim_{\alpha\to\infty} c\ \text{rect}(x/\alpha). \tag{8.26}$$

From the previous example and the time-scaling property of the Fourier transform given later:

$$\mathcal{F}\{\text{rect}(x/\alpha)\} = \text{sinc}(\alpha f), \tag{8.27}$$

which yields

$$\mathcal{F}\{c\} = \lim_{\alpha\to\infty} c\ \text{sinc}(\alpha f) = c\delta(f). \tag{8.28}$$

This result is to be expected because the Fourier transform of the Dirac delta function is a constant. This Fourier transform does not exist in the usual sense of the definition, but instead *exists in the limit*.

Example 8.6 Consider the signum function

$$\text{sgn}(t) = \begin{cases} -1, & t < 0 \\ 0, & t = 0 \\ 1, & t > 0. \end{cases} \tag{8.29}$$

It is not straightforward to derive its Fourier transform from the definition:

$$\int_{-\infty}^{\infty} \text{sgn}(t)\exp(-j2\pi ft)dt = \int_{-\infty}^{0} (-1)\exp(-j2\pi ft)dt + \int_{0}^{\infty} (1)\exp(-j2\pi ft)dt, \tag{8.30}$$

from which we find that neither integral is finite. Instead, we compute the Fourier transform in the limit by first writing the signum function as the limit of two exponential functions:

$$\text{sgn}(t) = \lim_{a\to 0}[\exp(-at)u(t) - \exp(at)u(-t)] = u(t) - u(-t). \tag{8.31}$$

The Fourier transform of the function in brackets is

$$\int_{-\infty}^{\infty}[\exp(-at)u(t)-\exp(at)u(-t)]\exp(-j2\pi ft)dt=\int_{0}^{\infty}\exp(-at)\exp(-j2\pi ft)dt$$
$$-\int_{-\infty}^{0}\exp(at)\exp(-j2\pi ft)dt. \tag{8.32}$$

The first integral on the right-hand side was computed in Example 8.1 with $a=\alpha$, and by a change of variables in the second integral, we obtain a similar result:

$$\int_{-\infty}^{\infty}[\exp(-at)u(t)-\exp(at)u(-t)]\exp(-j2\pi ft)dt=\frac{1}{j2\pi f+a}-\frac{1}{-j2\pi f+a}$$
$$=\frac{-j4\pi f}{4\pi^2f^2+a^2}. \tag{8.33}$$

Taking the limit as $a\rightarrow 0$ gives the Fourier transform of the signum function:

$$\int_{-\infty}^{\infty}\text{sgn}(t)\exp(-j2\pi ft)dt=\frac{2}{j2\pi f}=\frac{1}{j\pi f}, \tag{8.34}$$

which is strictly imaginary.

Example 8.7 The Fourier transform of the unit step function is derived by writing it in terms of the signum function and a constant:

$$u(t)=(1/2)[\text{sgn}(t)+1]. \tag{8.35}$$

Thus

$$\int_{-\infty}^{\infty}u(t)\exp(-j2\pi ft)dt=(1/2)\left[\frac{1}{j\pi f}+\delta(f)\right]=\frac{1}{j2\pi f}+(1/2)\delta(f), \tag{8.36}$$

and in terms of angular frequency:

$$\int_{-\infty}^{\infty}u(t)\exp(-j\omega t)dt=\frac{1}{j\omega}+\pi\delta(\omega). \tag{8.37}$$

Recall from Chapter 7 that the Laplace transform of the unit step function is

$$\mathcal{L}\{u(t)\}=\frac{1}{s}, \tag{8.38}$$

which has ROC $\text{Re}(s)>0$. Thus, the Fourier transform for this case is not derived by substituting $s=j\omega$, and this is because the ROC does not include the imaginary axis. Instead, we must derive the Fourier transform in the limit, which was achieved by starting with the signum function and a constant, or by using the generalized function methods described later. The spectrum of the unit step function has a DC component because of the Dirac delta function at the origin.

8.3 MAGNITUDE AND PHASE

For fixed ω, the Fourier transform is a point on the imaginary axis of the complex plane. Since the spectrum $X(\omega)$ is generally complex-valued, it can be written as the product of two *functions* of ω as follows:

$$X(\omega) = |X(\omega)| \exp(j\theta(\omega)), \tag{8.39}$$

where $X(\omega)$ is its *magnitude* and $\theta(\omega)$ is its *phase*. There are two methods for deriving these functions as illustrated in the next example.

Example 8.8 In the first method, $X(\omega)$ is written in rectangular complex form $c = a + jb$. The only difference compared with simple complex numbers is that $\{a, b, c\}$ here are functions of ω. For the Fourier transform in Example 8.1:

$$X(\omega) = \frac{1}{\alpha + j\omega} = \frac{\alpha}{\alpha^2 + \omega^2} + j\frac{-\omega}{\alpha^2 + \omega^2}. \tag{8.40}$$

Since $|c| = \sqrt{a^2 + b^2}$ from Chapter 4, we have in this case

$$|X(\omega)| = \left[\frac{\alpha^2}{(\alpha^2 + \omega^2)^2} + \frac{\omega^2}{(\alpha^2 + \omega^2)^2}\right]^{1/2} = \frac{1}{\sqrt{\alpha^2 + \omega^2}}. \tag{8.41}$$

The phase component is $\tan^{-1}(b/a)$, which for $X(\omega)$ is

$$\theta(\omega) = \tan^{-1}(-\omega/\alpha) = -\tan^{-1}(\omega/\alpha). \tag{8.42}$$

Combining these results gives an expression that is equivalent to (8.40):

$$X(\omega) = \frac{1}{\sqrt{\alpha^2 + \omega^2}} \exp(-\tan^{-1}(\omega/\alpha)). \tag{8.43}$$

This form is useful because now it is possible to plot the two terms separately versus ω: they are real-valued functions as depicted in Figure 8.3 for two values of α. If the ω-axis is extended, it is clear from (8.41) and (8.42) that $|X(\omega)| \to 0$ and $\theta(\omega) \to 90°$. In the second method, the magnitude and phase for the numerator and denominator are found separately and then combined:

$$X(\omega) = \frac{1}{\alpha + j\omega} = \frac{1}{\sqrt{\alpha^2 + \omega^2} \exp(\tan^{-1}(\omega/\alpha))}. \tag{8.44}$$

The overall magnitude is derived by dividing the numerator and denominator magnitudes. Subtracting the numerator and denominator phases gives the overall phase, and so again we have the result in (8.43).

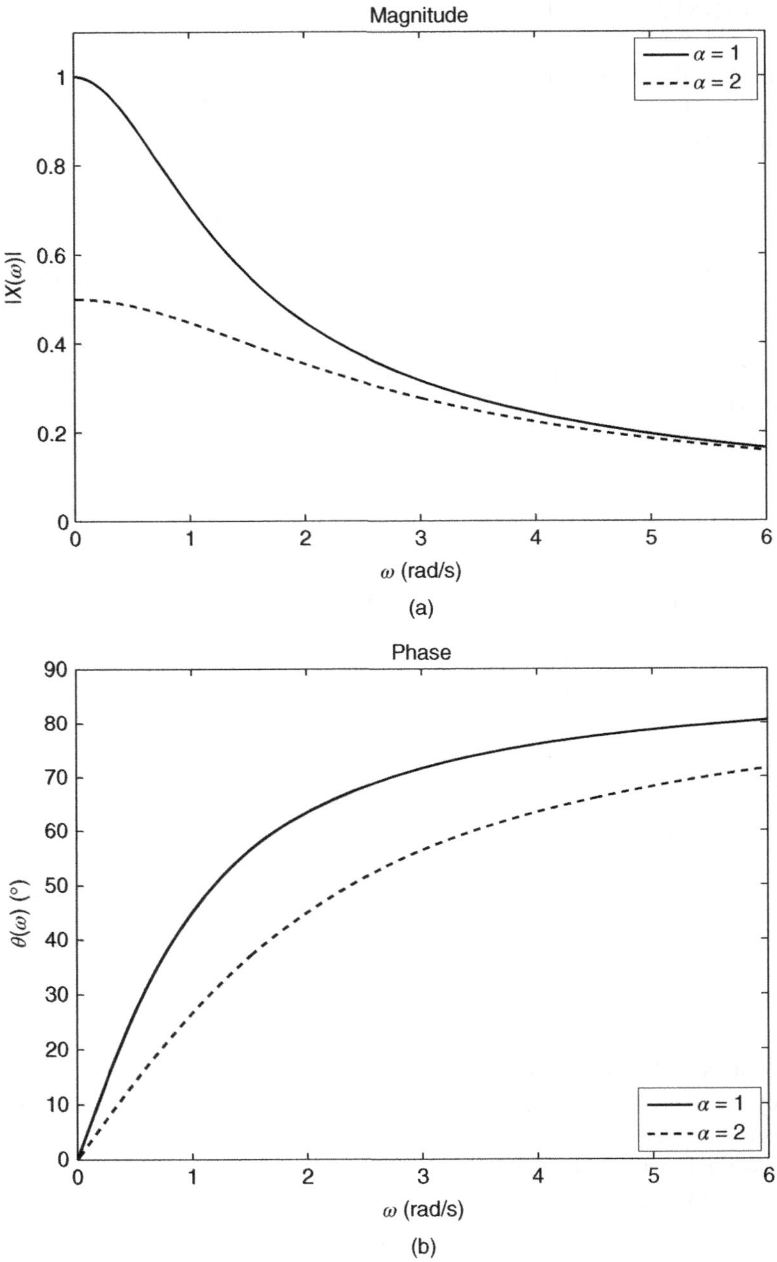

Figure 8.3 Magnitude and phase of first-order $X(\omega)$ in Example 8.8. (a) Magnitude. (b) Phase.

The magnitude gives the strength of $X(\omega)$ for a particular frequency ω, and the phase determines the delay (time shift) of $x(t)$ in the time domain. The magnitude and phase are also important when $H(\omega)$ is derived from the transfer function $H(s)$ of a system, in which case they describe the frequency response of a *filter* operating on the input signal.

The magnitude and phase have the following two basic properties for real-valued $x(t)$.

- *Even magnitude* $|X(\omega)|$:

$$|X(\omega)| = |X(-\omega)|. \tag{8.45}$$

Proof: Since $x(t)$ is real:

$$X(-\omega) = \int_{-\infty}^{\infty} x(t)\exp(j\omega t)dt = X^*(\omega). \tag{8.46}$$

Taking the absolute value of both sides completes the proof: $|X(-\omega)| = |X^*(\omega)| = |X(\omega)|$.

- *Odd phase* $\theta(\omega)$:

$$\theta(\omega) = -\theta(\omega). \tag{8.47}$$

Proof: This also follows from (8.46):

$$\arg(X(-\omega)) = \arg(X^*(\omega)) = -\arg(X(\omega)). \tag{8.48}$$

The last step is due to the definition of the phase: $\theta(\omega) = \tan^{-1}(\text{Im}(X(\omega))/\text{Re}(X(\omega))$. Conjugating $X(\omega)$ changes the sign of the imaginary part, and we use the fact that arctangent is an odd function.

8.4 FOURIER TRANSFORMS AND GENERALIZED FUNCTIONS

The function $x(t)$ is a mapping of each $t \in \mathcal{R}$ to the number represented by $x(t)$, which we can write as the ordered pair $\{t, x(t)\}$. The *functional* $X(\phi)$ is a mapping of the function $\phi(t)$ to the number $X(\phi)$ via the integral

$$X(\phi) = \int_{-\infty}^{\infty} x(t)\phi(t)dt, \tag{8.49}$$

which we write as $X(\phi) = \langle x, \phi \rangle$ for notational convenience. A distribution is a functional as defined earlier with the additional properties discussed in Chapter 5. Recall that the set $\mathcal{D}$ of test functions $\{\phi(t)\}$ have compact support, and the dual space of distributions defined on $\mathcal{D}$ is denoted by $\mathcal{D}'$. Since $\exp(-j\omega t)$ of the Fourier integral does not have compact support, the Fourier transform of the distribution $x(t)$ is not defined. This situation requires that we expand $\mathcal{D}$ to a new set of test functions $\mathcal{S}$, called test functions of rapid decay, which are also known as *Schwartz functions* (Kanwal, 2004).

Consider Parseval's theorem for two functions, which is discussed later in this chapter:

$$\int_{-\infty}^{\infty} x(t)\phi^*(t)dt = \int_{-\infty}^{\infty} X(f)\Phi^*(f)df, \tag{8.50}$$

and note that the integrand on the left-hand side can be written as

$$x(t)\phi^*(t) = \mathcal{F}^{-1}\{X(f)\}\phi^*(t), \tag{8.51}$$

where the inverse Fourier transform has been substituted for $x(t)$. Similarly for the integrand on the right-hand side of (8.50)

$$X(f)\Phi^*(f) = X(f)[\mathcal{F}\{\phi(t)\}]^* = X(f)\mathcal{F}^{-1}\{\phi(t)\}, \tag{8.52}$$

where we have assumed that $x(t)$ and $\phi(t)$ are real-valued. The last expression is derived as follows:

$$[\mathcal{F}\{\phi(t)\}]^* = \int_{-\infty}^{\infty} \phi(t)[\exp(-j2\pi f)]^* dt = \int_{-\infty}^{\infty} \phi(t)\exp(j2\pi f)]dt, \tag{8.53}$$

which is the inverse Fourier transform $\mathcal{F}^{-1}\{\phi(t)\}$, with the variables t and f interchanged. Combining (8.51) and (8.52) according to (8.50) yields

$$\int_{-\infty}^{\infty} \mathcal{F}^{-1}\{X(f)\}\phi(t)dt = \int_{-\infty}^{\infty} X(f)\mathcal{F}^{-1}\{\phi(t)\}df, \tag{8.54}$$

which we can write using distribution notation:

$$\langle \mathcal{F}^{-1}\{X(f)\}, \phi(t)\rangle = \langle X(f), \mathcal{F}^{-1}\{\phi(t)\}\rangle. \tag{8.55}$$

As we have seen previously with the derivative property for distributions, the operation on the distribution is "transferred" to the test function, which is smooth and well-defined. Instead of the derivative, in this case it is the inverse Fourier transform. This result illustrates why test functions with compact support cannot be used with Fourier transforms because even if $\phi(t)$ has compact support, this is generally not the case for its inverse transform $\mathcal{F}^{-1}\{\phi(t)\}$ on the right-hand side of (8.55). Thus, the test functions of D with compact support must be extended to include rapidly decreasing test functions, and this leads to the set of Schwartz functions.

Definition: Rapidly Decreasing Test Function A *rapidly decreasing test function* $\phi(t)$ has the following two properties: (i) $\phi(t)$ is smooth and (ii) all derivatives of $\phi(t)$ decrease to 0 more rapidly than the inverse of a polynomial:

$$\left| t^p \frac{d^n}{dt^n}\phi(t) \right| < c_{n,p}, \quad \text{as } |t| \to \infty, \tag{8.56}$$

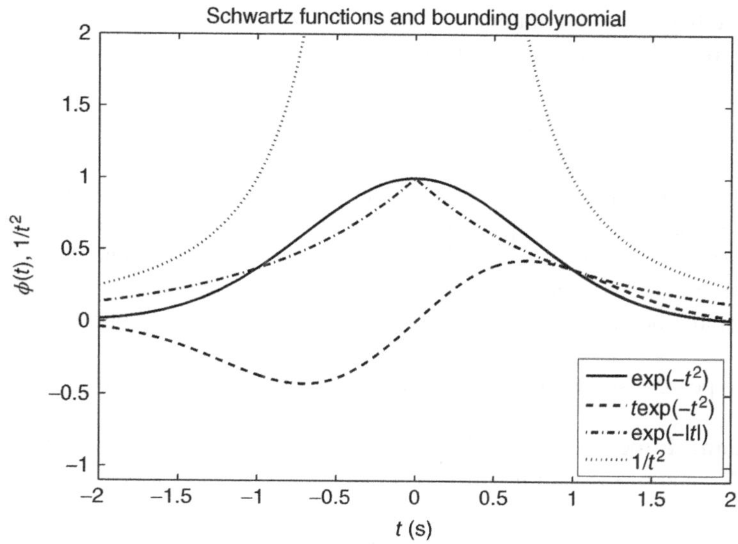

Figure 8.4 Example Schwartz functions and a bounding inverse polynomial $1/t^2$.

where $c_{n,p} \in \mathcal{R}^+$ is a coefficient that may vary with $\{n, p, \phi\}$ such that the inequality holds for every $n \in \mathcal{Z}^+$ and $p \in \mathcal{Z}^+$.

A test function $\phi(t)$ of $\mathcal{D}$ is also in $\mathcal{S}$ because it is 0 outside its compact support. Example Schwartz functions that are not elements of $\mathcal{D}$ include $\phi_1(t) = \exp(-\alpha|t|)$, $\phi_2(t) = \exp(-\alpha t^2)$, and $\phi_3(t) = t^q \exp(-\alpha t^2)$ for $\alpha > 0$ and $q \in \mathcal{Z}^+$. These functions are plotted in Figure 8.4 for $\alpha = 1$ and $q = 1$, all of which are bounded by $1/t^2$ so that (8.56) is satisfied for $n = 0$ (the nondifferentiated $\phi(t)$) with $c_{n,p} = c_{0,2} = 1$. It is clear that an upper bound can be found for these functions for any n by an appropriate choice for $c_{n,p}$.

Next, we define distributions based on the Schwartz test functions.

Definition: Tempered Distribution A *tempered distribution* $\langle x, \phi \rangle$ is a linear functional on the set $\mathcal{S}$ written as

$$\langle x, \phi \rangle \triangleq \int_{-\infty}^{\infty} x(t)\phi(t)dt, \quad \phi(t) \in \mathcal{S}. \tag{8.57}$$

This definition is essentially the same as that for classical distributions, except that $\mathcal{S}$ has replaced $\mathcal{D}$. Likewise, the dual space $\mathcal{D}'$ of all distributions is replaced by $\mathcal{S}'$, which is the set of all tempered distributions. Since the test functions of $\mathcal{S}$ are not as "strict" as those in $\mathcal{D}$ (which have compact support), the number of functions $x(t)$ for which (8.57) holds is less than the number when using $\mathcal{D}$. As a result, $\mathcal{S}' \subset \mathcal{D}'$: every tempered distribution in $\mathcal{S}'$ must also be in $\mathcal{D}'$. By expanding the set of test functions to $\mathcal{S}$, the Fourier integral is well defined for tempered distributions in $\mathcal{S}'$.

Example 8.9 The distribution for $u(t)$ is in $\mathcal{D}'$ and $\mathcal{S}'$ because the following is defined for both classes of test functions:

$$\langle u, \phi \rangle = \int_{-\infty}^{\infty} u(t)\phi(t)dt = \int_{0}^{\infty} \phi(t)dt. \tag{8.58}$$

This is obvious when $\phi(t)$ has compact support (for $\mathcal{D}$), and it is also the case when $\phi(t) \in \mathcal{S}$ because of the upper bound in (8.56). Observe, however, that a function like $\exp(t^2)$ is a distribution in $\mathcal{D}'$ because it is locally integrable and the $\{\phi(t)\}$ have compact support, whereas it is not a distribution in $\mathcal{S}'$ because it grows too fast relative to the rapidly decreasing test functions in $\mathcal{S}$. It does not have a Fourier transform.

A tempered distribution is also called a *distribution of slow growth.*

Definition: Function of Slow Growth For a *function of slow growth*, there exist $c, \alpha \in \mathcal{R}^+$ and $p \in \mathcal{Z}^+$ for $n \in \mathcal{Z}^+$ such that

$$\left| \frac{d^n}{dt^n} x(t) \right| \leq c|t|^p, \quad \text{as } |t| > \alpha. \tag{8.59}$$

Observe that (8.59) is essentially the "dual" of (8.56), and so all functions of slow growth have tempered distributions (though slow growth is not a requirement).

The reason that tempered distributions are important in this chapter is that all elements of $\mathcal{S}'$ have a Fourier transform, which is not the case for every distribution in $\mathcal{D}'$. The Fourier transform $X(\omega)$ of tempered distribution $x(t)$ is

$$\langle \mathcal{F}\{x\}, \phi \rangle = \langle x, \mathcal{F}\{\phi\} \rangle, \tag{8.60}$$

The left-hand side of this expression is

$$\langle \mathcal{F}\{x\}, \phi \rangle = \int_{-\infty}^{\infty} X(\omega)\phi(\omega)d\omega = \int_{-\infty}^{\infty} \int_{-\infty}^{\infty} x(t)\exp(-j\omega t)\phi(\omega)dtd\omega, \tag{8.61}$$

where ω is the usual independent variable of $X(\omega)$. If $x(t)$ is a regular function, then the inner integral on the right-hand side of (8.61) is the standard Fourier transform; otherwise, for singular functions such as the Dirac delta function, it is symbolic. Interchanging the integrals yields the right-hand side of (8.60):

$$\begin{aligned} \langle \mathcal{F}\{x\}, \phi \rangle &= \int_{-\infty}^{\infty} x(t) \int_{-\infty}^{\infty} \phi(\omega)\exp(-j\omega t)d\omega dt \\ &= \int_{-\infty}^{\infty} x(t)\Phi(t)dt = \langle x, \Phi \rangle = \langle x, \mathcal{F}\{\phi\} \rangle, \end{aligned} \tag{8.62}$$

where

$$\Phi(t) = \int_{-\infty}^{\infty} \phi(\omega)\exp(-j\omega t)d\omega. \tag{8.63}$$

The notation in the last expression may be somewhat confusing because t and ω are interchanged from the usual definition of the Fourier transform, which occurs because the integrals are interchanged in (8.62). (We saw the same type of interchange earlier when discussing Parseval's theorem.) However, $\Phi(t)$ is still the Fourier transform of a test function with argument ω replaced by t. The next example illustrates how to interpret (8.60) where the Fourier transform integral operates on ϕ of the right-hand side of (8.61).

Example 8.10 Consider the singular distribution $\delta(t-t_o)$. From (8.60), we have

$$\langle \mathcal{F}\{\delta(t-t_o)\},\phi\rangle = \langle \delta(t-t_o),\Phi\rangle = \int_{-\infty}^{\infty}\delta(t-t_o)\int_{-\infty}^{\infty}\phi(\omega)\exp(-j\omega t)d\omega dt$$
$$= \int_{-\infty}^{\infty}\phi(\omega)\exp(-j\omega t_o)d\omega = \langle \exp(-j\omega t_o),\phi\rangle, \tag{8.64}$$

where we have used the sifting property of the Dirac delta function to give t_o in the exponent of the exponential. The Fourier transform of the Dirac delta function has been "transferred" to the Fourier transform of the test function. Thus, from the first entry in each of the angle brackets of the first and second lines, we have $\mathcal{F}\{\delta(t-t_o)\} = \exp(-j\omega t_o)$, and in particular for $t_o = 0$, the Fourier transform of $\delta(t)$ is 1.

The Fourier transform of the derivative of a tempered distribution is easily found using (8.60):

$$\langle \mathcal{F}\{x'\},\phi\rangle = \langle x',\Phi\rangle = -\langle x,\Phi'\rangle, \tag{8.65}$$

where the derivative property of distributions has been used for the last result.

Example 8.11 The Fourier transform of the unit doublet $\delta'(t)$ is derived as follows:

$$\langle \mathcal{F}\{\delta'\},\phi\rangle = \langle \delta',\Phi\rangle = -\langle \delta,\Phi'\rangle. \tag{8.66}$$

Since $\phi(t)$ is a smooth function, the derivative property of the Fourier transform (shown later) yields

$$\Phi'(t) = -j\omega\mathcal{F}\{\phi\}. \tag{8.67}$$

Combining (8.66) and (8.67) gives

$$\langle \mathcal{F}\{\delta'\},\phi\rangle = -\langle \delta,-j\omega\mathcal{F}\{\phi(\omega)\}\rangle$$
$$= \langle \mathcal{F}\{\delta\},j\omega\phi\rangle = \langle j\omega\mathcal{F}\{\delta\},\phi\rangle. \tag{8.68}$$

From the first element of the first and last set of angle brackets, we have $\mathcal{F}\{\delta'\} = j\omega\mathcal{F}\{\delta\} = j\omega$ because $\mathcal{F}\{\delta\} = 1$.

8.5 FOURIER TRANSFORM PROPERTIES

Several Fourier transform properties in terms of angular frequency ω and ordinary frequency f are summarized in Tables 8.5 and 8.6, respectively. Two tables are included because there are subtle differences for some properties; for example, those involving integrals of Fourier transforms have the multiplicative factor $1/2\pi$ when using ω, but not when using f. Also, the term $\pi X(0)\delta(\omega)$ for the integral property becomes $(1/2)X(0)\delta(f)$ because of the scaling property of the Dirac delta function, $\delta(\alpha\omega) = \delta(\omega)/|\alpha|$, such that

$$\delta(\omega) = \delta(2\pi f) = \delta(f)/2\pi. \tag{8.69}$$

Most of the properties of the Laplace transform carry over to the Fourier transform; differences for some cases are mentioned.

- *Time scaling*: The Fourier transform of a time-scaled waveform is

$$\mathcal{F}\{x(\alpha t)\} = \frac{1}{|\alpha|}X(\omega/\alpha). \tag{8.70}$$

TABLE 8.5 Properties of the Fourier Transform $X(\omega)$

Property	Function	Fourier Transform
Linearity	$c_1x_1(t) + c_2x_2(t)$	$c_1X_1(\omega) + c_2X_2(\omega)$
Time shift	$x(t - t_o)$	$\exp(-j\omega t_o)X(\omega)$
Time scaling	$x(\alpha t)$	$(1/\lvert\alpha\rvert)X(\omega/\alpha)$
Frequency shift	$\exp(j\omega_o t)x(t)$	$X(\omega - \omega_o)$
Derivatives	$d^nx(t)/dt^n$	$(j\omega)^nX(\omega)\quad (n \in \mathcal{N})$
Integral	$\int_{-\infty}^{t} x(\tau)d\tau$	$(1/j\omega)X(\omega) + \pi X(0)\delta(\omega)$
Convolution	$x(t) * h(t)$	$X(\omega)H(\omega)$
Cross-correlation	$x(t) \star h(t)$	$X(\omega)H(-\omega)$
Autocorrelation	$x(t) \star x(t)$	$\lvert X(\omega)\rvert^2$
Product	$x(t)h(t)$	$(1/2\pi)\int_{-\infty}^{\infty} X(v)H(\omega - v)dv$
Cosine modulation	$x(t)\cos(\omega_o t)$	$(1/2)[X(\omega - \omega_o) + X(\omega + \omega_o)]$
Sine modulation	$x(t)\sin(\omega_o t)$	$(1/2j)[X(\omega - \omega_o) - X(\omega + \omega_o)]$
Time product	$t^nx(t)$	$j^nd^nX(\omega)/d\omega^n$
Time area	$\int_{-\infty}^{\infty} x(t)dt$	$X(0)$
Frequency area	$x(0)$	$(1/2\pi)\int_{-\infty}^{\infty} X(\omega)d\omega$
Duality	$X(t)$	$2\pi x(-\omega)$
Energy	$\int_{-\infty}^{\infty} x^2(t)dt$	$(1/2\pi)\int_{-\infty}^{\infty} \lvert X(\omega)\rvert^2d\omega$
Even/odd components	$x(t) = x_E(t) + x_O(t)$	$X(\omega) = X_E(\omega) - jX_O(\omega)$
Even function	Real and even $x(t)$	Real and even $X(\omega)$
Odd function	Real and odd $x(t)$	Imaginary and odd $X(\omega)$

TABLE 8.6 Properties of the Fourier Transform $X(f)$

Property	Function	Fourier Transform
Linearity	$c_1x_1(t) + c_2x_2(t)$	$c_1X_1(f) + c_2X_2(f)$
Time shift	$x(t - t_o)$	$\exp(-j2\pi f t_o)X(f)$
Time scaling	$x(\alpha t)$	$(1/\lvert\alpha\rvert)X(f/\alpha)$
Frequency shift	$\exp(j2\pi f_o t)x(t)$	$X(f - f_o)$
Derivatives	$d^n x(t)/dt^n$	$(j2\pi f)^n X(f)$ $(n \in \mathcal{N})$
Integral	$\int_{-\infty}^{t} x(\tau)d\tau$	$(1/j2\pi f)X(f) + (1/2)X(0)\delta(f)$
Convolution	$x(t) * h(t)$	$X(f)H(f)$
Cross-correlation	$x(t) \star h(t)$	$X(f)H(-f)$
Autocorrelation	$x(t) \star x(t)$	$\lvert X(f)\rvert^2$
Product	$x(t)h(t)$	$\int_{-\infty}^{\infty} X(v)H(f - v)dv$
Cosine modulation	$x(t)\cos(f_o t)$	$\pi[X(f - f_o) + X(f + f_o)]$
Sine modulation	$x(t)\sin(f_o t)$	$(\pi/j)[X(f - f_o) - X(f + f_o)]$
Time product	$t^n x(t)$	$j^n d^n X(f)/df^n$
Time area	$\int_{-\infty}^{\infty} x(t)dt$	$X(0)$
Frequency area	$x(0)$	$\int_{-\infty}^{\infty} X(f)df$
Duality	$X(t)$	$x(-f)$
Energy	$\int_{-\infty}^{\infty} x^2(t)dt$	$\int_{-\infty}^{\infty} \lvert X(f)\rvert^2 df$
Even/odd components	$x(t) = x_E(t) + x_O(t)$	$X(f) = X_E(f) - jX_O(f)$
Even function	Real and even $x(t)$	Real and even $X(f)$
Odd function	Real and odd $x(t)$	Imaginary and odd $X(f)$

Unlike the unilateral Laplace transform, α can be negative because the Fourier transform is a two-sided integral, which causes a time reversal in addition to time scaling. *Proof*: Changing variables to $\tau \triangleq \alpha t \Rightarrow t = \tau/\alpha$ for $\alpha > 0$ yields

$$\int_{-\infty}^{\infty} x(\alpha t)\exp(-j\omega t)dt = \frac{1}{\alpha}\int_{-\infty}^{\infty} x(\tau)\exp(-j(\omega/\alpha)\tau)d\tau, \qquad (8.71)$$

and for $\alpha < 0$, the integration limits must be interchanged:

$$\int_{-\infty}^{\infty} x(\alpha t)\exp(-j\omega t)dt = \frac{1}{\alpha}\int_{\infty}^{-\infty} x(\tau)\exp(-j(\omega/\alpha)\tau)d\tau$$

$$= -\frac{1}{\alpha}\int_{-\infty}^{\infty} x(\tau)\exp(-j(\omega/\alpha)\tau)d\tau, \qquad (8.72)$$

Equations (8.71) and (8.72) together give (8.70).

- *Time shift*: A time-shifted waveform has the following Fourier transform:

$$\mathcal{F}\{x(t - t_o)\} = \exp(-j\omega t_o)X(\omega), \qquad (8.73)$$

where $t_o \in \mathcal{R}$. For $t_o > 0$, the waveform is shifted to the right, and for $t_o < 0$, it is shifted to the left. Recall that only $t_o > 0$ was allowed for the unilateral Laplace transform, meaning the function could only be delayed. *Proof*: From the transformation of variables $\tau = t - t_o$:

$$\mathcal{F}\{x(t-t_o)\} = \int_{-\infty}^{\infty} x(t-t_o)\exp(-j\omega t)dt$$
$$= \int_{-\infty}^{\infty} x(\tau)\exp(-j\omega(\tau+t_o))dt. \quad (8.74)$$

Factoring $\exp(-j\omega t_o)$ completes the proof.

- *Frequency shift*: This property is the dual of a time shift:

$$\mathcal{F}\{\exp(j\omega_o t)x(t)\} = X(\omega - \omega_o). \quad (8.75)$$

Proof: Similar to the expression in (8.74):

$$\int_{-\infty}^{\infty} x(t)\exp(j\omega_o t)\exp(-j\omega t)dt = \int_{-\infty}^{\infty} x(t)\exp(-j(\omega-\omega_o)t)dt. \quad (8.76)$$

We recognize that the last result is the Fourier transform of $x(t)$ with ω replaced by $\omega - \omega_o$.

- *Duality*: The duality property is straightforward for the Fourier transform expressed in natural frequency f:

$$\mathcal{F}\{x(t)\} = X(f) \implies \mathcal{F}\{X(t)\} = x(-f). \quad (8.77)$$

For the Fourier transform based on angular frequency ω, the duality property is

$$\mathcal{F}\{x(t)\} = X(\omega) \implies \mathcal{F}\{X(t)\} = 2\pi x(-\omega), \quad (8.78)$$

which includes the factor 2π. *Proof*: Starting with the inverse Fourier transform

$$x(t) = \int_{-\infty}^{\infty} X(f)\exp(j2\pi ft)df, \quad (8.79)$$

we let $t \to -t$:

$$x(-t) = \int_{-\infty}^{\infty} X(p)\exp(-j2\pi pt)dp, \quad (8.80)$$

where the variable of integration has been replaced with p. Replacing t with f on both the sides completes the proof:

$$x(-f) = \int_{-\infty}^{\infty} X(p)\exp(-j2\pi pf)dp = \mathcal{F}\{X(p)\}. \quad (8.81)$$

The proof for the Fourier transform with angular frequency is considered in Problem 8.13.

- *Area*: The area of a function is derived from $X(\omega)$ as follows:

$$\int_{-\infty}^{\infty} x(t)dt = X(0). \tag{8.82}$$

The DC component $X(0)$ indicates whether or not the function has zero area. *Proof*: This property follows directly from the definition of the Fourier transform:

$$X(f)|_{f=0} = \int_{-\infty}^{\infty} x(t)\exp(-j2\pi f)dt\bigg|_{f=0} = \int_{-\infty}^{\infty} x(t)dt. \tag{8.83}$$

- *Derivatives*: The Fourier transform of the nth derivative of a function is related to that of the original function:

$$\mathcal{F}\left\{\frac{d^n}{dt^n}x(t)\right\} = (j\omega)^n X(\omega). \tag{8.84}$$

Proof: The inverse Fourier transform yields

$$\begin{aligned}\frac{d^n}{dt^n}x(t) &= \frac{d^n}{dt^n}\left[\frac{1}{2\pi}\int_{-\infty}^{\infty} X(\omega)\exp(j\omega t)d\omega\right] \\ &= \frac{(j\omega)^n}{2\pi}\int_{-\infty}^{\infty} X(\omega)\exp(j\omega t)d\omega = (j\omega)^n x(t).\end{aligned} \tag{8.85}$$

Taking the Fourier transform of both sides completes the proof.

- *Convolution*: In Chapters 6 and 7, we demonstrated that for an LTI system with zero initial states, the output $y(t)$ is derived from the input $x(t)$ by a convolution:

$$y(t) = \int_{-\infty}^{\infty} h(\tau)x(t-\tau)d\tau = \int_{-\infty}^{\infty} x(\tau)h(t-\tau)d\tau, \tag{8.86}$$

where $h(t)$ is the system impulse response function. The corresponding operation in the s-domain is

$$Y(s) = H(s)X(s) = X(s)H(s), \tag{8.87}$$

and so in the frequency domain, we have

$$Y(\omega) = H(\omega)X(\omega) = X(\omega)H(\omega). \tag{8.88}$$

Proof: Taking the Fourier transform of $y(t)$ yields

$$\int_{-\infty}^{\infty} y(t)\exp(-j\omega t)dt = \int_{-\infty}^{\infty}\int_{-\infty}^{\infty} x(\tau)h(t-\tau)\exp(-j\omega t)d\tau dt, \tag{8.89}$$

where one of the convolution integrals has been substituted. Changing variables to $v \triangleq t-\tau$ gives

$$\int_{-\infty}^{\infty} y(t)\exp(-j\omega t)dt = \int_{-\infty}^{\infty}\int_{-\infty}^{\infty} x(\tau)h(v)\exp(-j\omega(v+\tau))d\tau dv, \tag{8.90}$$

which splits into a product

$$\int_{-\infty}^{\infty} y(t)\exp(-j\omega t)dt = \int_{-\infty}^{\infty} x(\tau)\exp(-j\tau)d\tau \int_{-\infty}^{\infty} h(v)\exp(-j\omega v)dv. \quad (8.91)$$

Each of the three integrals is a Fourier transform, proving the result in (8.88).

- *Integral*: The Fourier transform of the integral of a function is somewhat more complicated:

$$\mathcal{F}\left\{\int_{-\infty}^{t} x(t)dt\right\} = \frac{X(\omega)}{j\omega} + \pi X(0)\delta(\omega), \quad (8.92)$$

where from a previous property $X(0)$ is the area of $x(t)$. If the signal has no DC component, then the term containing the Dirac delta function is dropped (many textbooks assume this condition for simplicity). *Proof*: This result is proved by rewriting the integral as a convolution with $u(t)$:

$$\int_{-\infty}^{t} x(t)dt = \int_{-\infty}^{\infty} x(\tau)u(t-\tau)d\tau = x(t) * u(t). \quad (8.93)$$

The upper limit on the left-hand side is due to the fact that $u(t-\tau) = 1$ for $t - \tau \geq 0 \implies \tau \leq t$. From the convolution property:

$$\mathcal{F}\{x(t) * u(t)\} = X(\omega)\left[\frac{1}{j\omega} + \pi\delta(\omega)\right], \quad (8.94)$$

where the term in brackets is the Fourier transform of the unit step function. Thus

$$\mathcal{F}\left\{\int_{-\infty}^{t} x(t)dt\right\} = \frac{X(\omega)}{j\omega} + \pi X(\omega)\delta(\omega). \quad (8.95)$$

The sampling property of the Dirac delta function yields the final result because $X(\omega)\delta(\omega) = X(0)\delta(\omega)$.

- *Parseval's theorem*: This theorem provides an identity for finding the energy of a waveform from its Fourier transform:

$$\int_{-\infty}^{\infty} x^2(t)dt = \frac{1}{2\pi}\int_{-\infty}^{\infty} |X(\omega)|^2 d\omega, \quad (8.96)$$

where we note the factor of $1/2\pi$ on the right-hand side. (Earlier, we used a different form of this theorem involving two functions when discussing the Fourier transform and generalized functions.) It is used to determine the amount of energy contributed by different frequency bands to the overall energy of a signal (see Problem 8.21). *Proof*: Substituting the inverse Fourier transform on the left-hand side and using different variables under the integrals yield

$$\begin{aligned}\int_{-\infty}^{\infty} |x(t)|^2 dt &= \frac{1}{4\pi^2} \int_{-\infty}^{\infty} \int_{-\infty}^{\infty} X(\omega_1) \exp(j\omega_1 t) d\omega_1 \int_{-\infty}^{\infty} X^*(\omega_2) \exp(-j\omega_2 t) d\omega_2 dt \\ &= \frac{1}{4\pi^2} \int_{-\infty}^{\infty} \int_{-\infty}^{\infty} \int_{-\infty}^{\infty} \exp(-j(\omega_2 - \omega_1)t) dt X(\omega_1) X^*(\omega_2) d\omega_1 d\omega_2.\end{aligned} \tag{8.97}$$

In order to continue, $|x(t)|^2 = x(t)x^*(t)$ is used even though $x(t)$ is assumed to be real so that the second exponential on the right-hand side has the correct sign. In the second equation, the innermost integral is the Fourier transform of a constant with frequency $\omega_2 - \omega_1$, which we know is $2\pi\delta(\omega_2 - \omega_1)$. Thus

$$\begin{aligned}\int_{-\infty}^{\infty} x^2(t) dt &= \frac{1}{2\pi} \int_{-\infty}^{\infty} \int_{-\infty}^{\infty} \delta(\omega_2 - \omega_1) X(\omega_1) X^*(\omega_2) d\omega_1 d\omega_2 \\ &= \frac{1}{2\pi} \int_{-\infty}^{\infty} X(\omega_2) X^*(\omega_2) d\omega_2 = \frac{1}{2\pi} \int_{-\infty}^{\infty} |X(\omega)|^2 d\omega,\end{aligned} \tag{8.98}$$

where the sifting property of the Dirac delta function has been used to evaluate the inner integral and complete the proof.

- *Even and odd symmetry*: Since any function can be expressed as the sum of even and odd components $x(t) = x_E(t) + x_O(t)$, we find that (8.3) can also be written as

$$\begin{aligned}X(\omega) &= \int_{-\infty}^{\infty} [x_E(t) + x_O(t)] \cos(\omega t) dt - j \int_{-\infty}^{\infty} [x_E(t) + x_O(t)] \sin(\omega t) dt \\ &= \int_{-\infty}^{\infty} x_E(t) \cos(\omega t) dt - j \int_{-\infty}^{\infty} x_O(t) \sin(\omega t) dt \\ &\triangleq X_E(\omega) - jX_O(\omega),\end{aligned} \tag{8.99}$$

where $X_E(\omega)$ and $-X_O(\omega)$ are the even/real and odd/imaginary parts, respectively, of $X(\omega)$. The following properties are also concluded in the event that $x(t)$ is strictly even or odd:

$$\text{even } x(t) \implies X(\omega) \text{ is real and even.} \tag{8.100}$$

$$\text{odd } x(t) \implies X(\omega) \text{ is imaginary and odd.} \tag{8.101}$$

Proof: For the second line in (8.99), the symmetric integral of the product of odd $x_O(t)$ and even $\cos(\omega t)$ is 0, and likewise for the product of even $x_E(t)$ and odd $\sin(\omega t)$. The even and odd properties of the components in the last line of (8.99) are verified as follows:

$$X_E(-\omega) = \int_{-\infty}^{\infty} x_E(t)\cos(-\omega t)dt = \int_{-\infty}^{\infty} x_E(t)\cos(\omega t)dt = X_E(\omega), \tag{8.102}$$

$$X_O(-\omega) = \int_{-\infty}^{\infty} x_O(t)\sin(-\omega t)dt = -\int_{-\infty}^{\infty} x_O(t)\sin(\omega t)dt = -X_O(\omega). \tag{8.103}$$

From these results, we find that using $\exp(-j\omega t)$ as the kernel in the Fourier transform integral (instead of sine or cosine alone) allows it to handle functions with even and odd parts, yielding a transform that has even and odd parts.

Example 8.12 The Fourier transform of the ramp function $r(t) = tu(t)$ can be derived from the time product in Table 8.5 with $n = 1$:

$$\begin{aligned} \mathcal{F}\{r(t)\} &= j\frac{d}{d\omega}\mathcal{F}\{u(\omega)\} = j\frac{d}{d\omega}[\pi\delta(\omega) + 1/j\omega] \\ &= j\pi\delta'(\omega) - 1/\omega^2, \end{aligned} \tag{8.104}$$

where $\delta'(\omega)$ is a generalized derivative. Using the time scaling property with $\alpha = -1$ in Table 8.5, the Fourier transform of the left-sided ramp function $r(-t) = -tu(-t)$ is

$$\mathcal{F}\{r(-t)\} = j\pi\delta'(-\omega) - 1/\omega^2 = -j\pi\delta'(\omega) - 1/\omega^2, \tag{8.105}$$

where we have used the fact that the unit doublet is an odd generalized function.

Finally, we describe how the Fourier transform integral can be interpreted as the cross-correlation of waveform $x(t)$ with the complex exponential function $\exp(j\omega t)$ (this is not the cross-correlation property given in the tables). From Euler's formula

$$X(\omega) = \int_{-\infty}^{\infty} x(t)\cos(\omega t)dt - j\int_{-\infty}^{\infty} x(t)\sin(\omega t)dt, \tag{8.106}$$

we find that $X(\omega)$ is the degree to which $x(t)$ is similar to a cosine waveform and a sine waveform, both having the same frequency ω. Since j can be viewed as a marker for the imaginary component of a complex number (see Chapter 4), the Fourier transform simultaneously performs two cross-correlations. Thus, using the notation for cross-correlation in Chapter 5, (8.106) can be written as

$$X(\omega) \triangleq c_{xc}(\omega) - jc_{xs}(\omega), \tag{8.107}$$

where $c_{xc}(\omega)$ is the cross-correlation function of $x(t)$ with $\cos(\omega t)$, and $c_{xs}(\omega)$ is the cross-correlation function of $x(t)$ with $\sin(\omega t)$. Note that the argument is ω to indicate the sinusoidal frequency, instead of the lag τ, which is 0 because the functions in (8.106) are not shifted. This correlation interpretation is similar to that used for the Fourier series in Chapter 5. The difference here is that $x(t)$ need not be periodic and the domain is $\omega \in \mathcal{R}$, whereas for the Fourier series, $x(t)$ must be periodic with period

T_o and only integer multiples of the fundamental frequency $\omega_o = 2\pi/T_o$ are used to generate the Fourier series coefficients.

The product property (also called *modulation*) is considered next in the context of a communication system based on AM.

8.6 AMPLITUDE MODULATION

Consider the following sinusoidal signal with angular frequency ω_o:

$$c(t) = A\cos(\omega_o t), \tag{8.108}$$

which has Fourier transform

$$C(\omega) = A\pi\delta(\omega - \omega_o) + A\pi\delta(\omega + \omega_o). \tag{8.109}$$

In a communication system, such a waveform is called the carrier because signal information is "carried" across the channel at this frequency. Let $x(t)$ be an arbitrary signal with Fourier transform (spectrum) $X(\omega)$. Modulation is defined to be the product of these two waveforms in the time domain:

$$y(t) = x(t)c(t) = c(t)x(t), \tag{8.110}$$

which, of course, is a symmetric operation like convolution. However, since ω_o in a communication system is usually much greater than the highest frequency component of the *message signal* $x(t)$, we say that $x(t)$ modulates $c(t)$ (Haykin, 2001).

The transform of this product is

$$Y(\omega) = \int_{-\infty}^{\infty} c(t)x(t)\exp(-j\omega t)dt. \tag{8.111}$$

Substituting the inverse Fourier transform for each signal yields

$$Y(\omega) = \frac{1}{(2\pi)^2}\int_{-\infty}^{\infty}\int_{-\infty}^{\infty} C(u)\exp(jut)du \int_{-\infty}^{\infty} X(v)\exp(jvt)dv\exp(-j\omega t)dt, \tag{8.112}$$

where different variables of integration have been used to avoid confusion across terms. Rearranging this expression yields

$$Y(\omega) = \frac{1}{(2\pi)^2}\int_{-\infty}^{\infty}\int_{-\infty}^{\infty} C(u)X(v)\int_{-\infty}^{\infty} \exp(-j(\omega - u - v)t)dtdudv. \tag{8.113}$$

The innermost integral with respect to t is the Fourier transform of a constant, which is the Dirac delta function $2\pi\delta(\omega - u - v)$. Thus,

$$Y(\omega) = \frac{1}{2\pi}\int_{-\infty}^{\infty}\int_{-\infty}^{\infty} C(u)X(v)\delta(\omega - u - v)dudv. \tag{8.114}$$

Integrating over v, the sifting property of the Dirac delta function yields the final result:

$$Y(\omega) = \frac{1}{2\pi}\int_{-\infty}^{\infty} C(u)X(\omega - u)du, \tag{8.115}$$

which is a convolution in the frequency domain (scaled by $1/2\pi$). Integrating instead over u would give the symmetric result:

$$Y(\omega) = \frac{1}{2\pi}\int_{-\infty}^{\infty} X(v)C(\omega - v)dv. \tag{8.116}$$

This result is not surprising because of the duality property of the Fourier transform: convolution in the time domain gives a product in the frequency domain, and so we would expect that a product in the time domain yields a convolution of their Fourier transforms. The only difference is the $1/2\pi$ scaling factor, which appears because we have written the convolution using angular frequency ω. This term is not present when using ordinary frequency f (see Problem 8.23):

$$Y(f) = \int_{-\infty}^{\infty} C(u)X(f-u)du = \int_{-\infty}^{\infty} X(u)C(f-u)du. \tag{8.117}$$

For the cosine carrier $c(t)$ and arbitrary $x(t)$, the output in the frequency domain is

$$\begin{aligned} Y(\omega) &= \frac{A}{2\pi}\int_{-\infty}^{\infty} \pi[\delta(\omega - \omega_o - v) + \delta(\omega - \omega_o - v)]X(v)dv \\ &= (A/2)X(\omega - \omega_o) + (A/2)X(\omega + \omega_o). \end{aligned} \tag{8.118}$$

Thus, modulation in the time domain causes the spectrum of $x(t)$ to be shifted both right and left in the frequency domain, centered at $\pm\omega_o$ and scaled by $A/2$. This type of modulation is called AM with *suppressed carrier*, or *double-sideband* AM with suppressed carrier. The carrier is suppressed because only the signal spectrum $X(\omega)$ appears at $\pm\omega_o$; there are no Dirac delta functions in $Y(\omega)$ as there are in $C(\omega)$. Of course, the delta functions are not present in the expression because of the sifting property of the Dirac delta function used to derive (8.118).

Example spectra associated with AM are illustrated in Figure 8.5 for a signal with the following (ideal) rectangular spectrum:

$$X(\omega) = \begin{cases} 2, & |\omega| \le \omega_c \\ 0, & \text{else} \end{cases} = 2\text{rect}(\omega/2\omega_c), \tag{8.119}$$

where ω_c is the cutoff frequency for this low-pass response. Observe that the spectrum has been replicated at $\pm\omega_o$ and scaled by a factor of $1/2$ (we assume $A = 1$ for the carrier). The corresponding waveform $y(t)$ is called a *narrowband passband signal* because $\omega_c \ll \omega_o$ and its positive and negative components are centered about $\pm\omega_o$. In this communications application, the low-pass waveform $x(t)$ is called a *baseband signal*. The double-sideband description refers to the fact that the components of $Y(\omega)$ are even functions about ω_o, which occurs because $X(\omega)$ is even about the origin and $x(t)$ is a real waveform. Because of this symmetry, the modulated signal has redundancy, and so it is possible to remove either the upper or the lower sideband at $\pm\omega_o$ without losing information about the message. Such a modulated signal, which is more complex to implement and demodulate, is called *single-sideband* AM, and is considered in Problem 8.25.

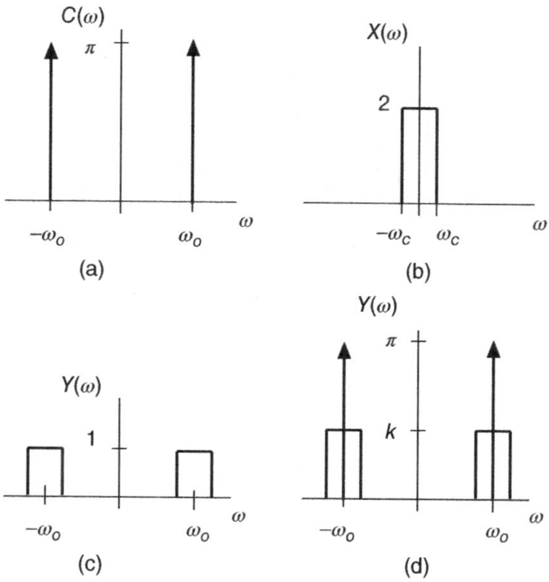

Figure 8.5 Amplitude modulation. (a) Carrier spectrum $C(\omega)$ with $A = 1$. (b) Baseband signal spectrum $X(\omega)$. (c) AM with suppressed carrier: $Y(\omega) = X(\omega) * C(\omega)$. (d) Conventional AM: $Y(\omega) = C(\omega) + kX(\omega) * C(\omega)$.

Example 8.13 Let the (artificial) message signal be

$$x(t) = \cos(\pi t/4), \tag{8.120}$$

which has frequency $f_m = 1/8$ Hz ($\omega_m = \pi/4$ rad/s). The carrier $c(t)$ has $A = 1$ and $\omega_o = 2\pi$ rad/s, which corresponds to $f_o = 1$ Hz such that the message signal has a lower frequency. Figure 8.6 shows these signals along with the modulated waveform $y(t)$ for a duration of 10 s. We have also illustrated the *envelope* of the modulated signal in Figure 8.6(c) (the dotted curves), which are plus and minus replicas of the message waveform $x(t)$. The information/message $x(t)$ of the modulated signal $y(t)$ is contained in this envelope. The composite signal $y(t)$ is transmitted across a communication channel, and a *receiver* is designed to extract $x(t)$ from $y(t)$ and thus obtain the original message. Of course, the channel introduces impairments such as noise so that the *detected* signal $\hat{x}(t)$ is only an estimate of $x(t)$.

The receiver for AM with suppressed carrier is somewhat complicated due to the fact that the plus and minus envelopes usually intersect each other as illustrated in Figure 8.6(c). The details of various detection methods are beyond the scope of this book, but we can provide some intuition by examining the waveform for *conventional* AM where the transmitted waveform $y(t)$ includes the carrier signal $c(t)$. This is done by modifying (8.110) as follows:

$$y(t) = [1 + kx(t)]c(t) = A[1 + kx(t)]\cos(\omega_o t), \tag{8.121}$$

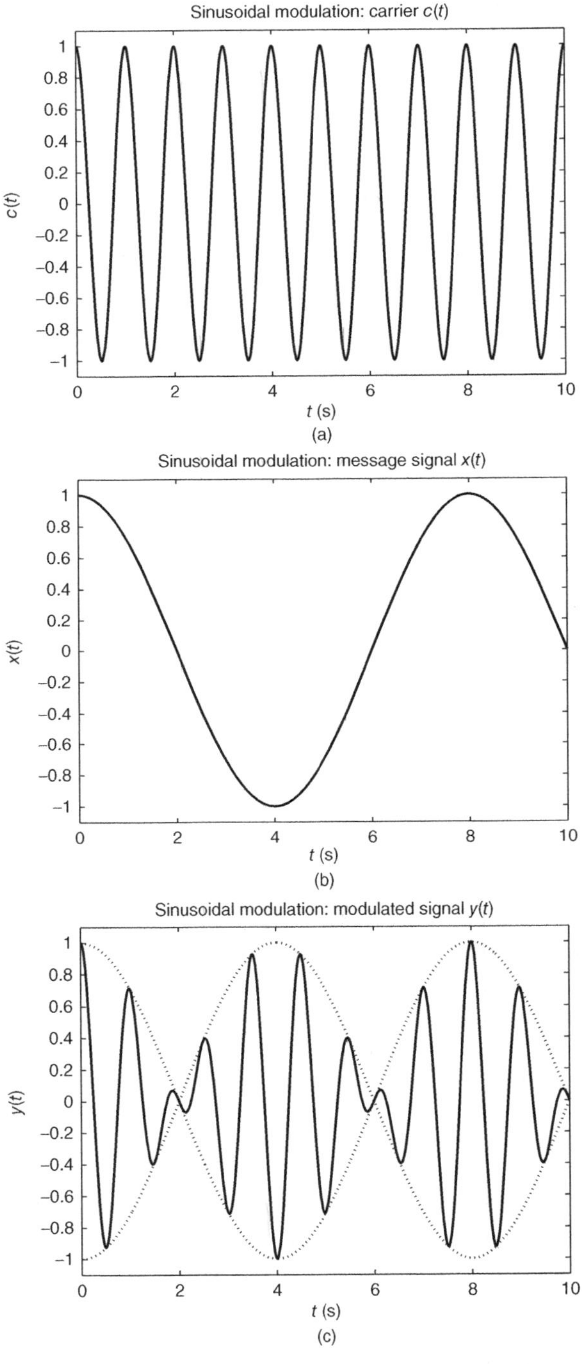

Figure 8.6 Sinusoidal modulation of Example 8.13. (a) Carrier $c(t)$ with $A = 1$. (b) Modulating signal $x(t)$. (c) AM with suppressed carrier $y(t)$. The dotted lines show the envelope of the waveform, corresponding to overlapping $\pm x(t)$ waveforms.

whose frequency domain representation is

$$\begin{aligned}Y(\omega) &= A[1 + kX(\omega)] * [\pi\delta(\omega - \omega_o) + \pi\delta(\omega - \omega_o)] \\ &= A\pi\delta(\omega - \omega_o) + A\pi\delta(\omega - \omega_o) + (Ak/2)X(\omega - \omega_o) + (Ak/2)X(\omega + \omega_o) \\ &= C(\omega) + (Ak/2)X(\omega - \omega_o) + (Ak/2)X(\omega + \omega_o), \end{aligned} \qquad (8.122)$$

which includes the transform $C(\omega)$ of the carrier. The advantage of this form is that by proper choice of k, it is possible to separate the plus and minus envelopes shown in Figure 8.6(c) so they no longer overlap. This is illustrated in Figure 8.7 for two values of the *amplitude sensitivity* k. Since these envelopes do not intersect each other, a simple *envelope detector* can be used to recover the top (positive) envelope, which is exactly $x(t)$ (assuming no channel impairments). If an envelope detector is applied to the waveform in Figure 8.6(c), it will recover only the positive dotted waveform, which we know is not correct for the message signal in Figure 8.6(b). By including k in the modulation process and scaling $x(t)$, Figure 8.7 shows a positive envelope that is the message $x(t)$.

In order to avoid the *overmodulation* that can occur in AM with suppressed carrier, we need to ensure that the term in brackets in (8.121) does not change sign. Since communication signals tend to have an average (mean) near zero, the sign of $x(t)$ usually changes often. By scaling $x(t)$ with k and adding 1 to $kx(t)$, the positive and negative envelopes of the modulated signals will not have any zero crossings. Thus, we require that for all t:

$$|kx(t)| < 1. \qquad (8.123)$$

It is also assumed that $\omega_c \gg \omega_o$ so there is no confusion as to which signal is the carrier (with much higher frequency) and which is the message. The actual magnitude of the envelope is not important because it can always be scaled after detection, and in fact, the received signal is usually *amplified* at some point in the receiver circuit. The variations and *relative* amplitudes of the waveform over time determine the information content.

Although conventional AM allows for a simple receiver, such as an RC circuit with postfiltering and buffering, the disadvantage is that power is wasted by not suppressing the carrier. Power is also wasted because both sidebands are transmitted; this can be reduced using single-sideband (SSB) modulation as mentioned in Problem 8.25.

8.7 FREQUENCY RESPONSE

In Chapter 7, we examined linear systems in the s-domain and defined the transfer function of a system to be the ratio of the output signal transform $Y(s)$ and the input signal transform $X(s)$:

$$H(s) \triangleq Y(s)/X(s), \qquad (8.124)$$

which follows from the convolution of $x(t)$ and $h(t)$ in the time domain. This definition for $H(s)$ assumes that all initial states are zero, such as the initial voltage across a

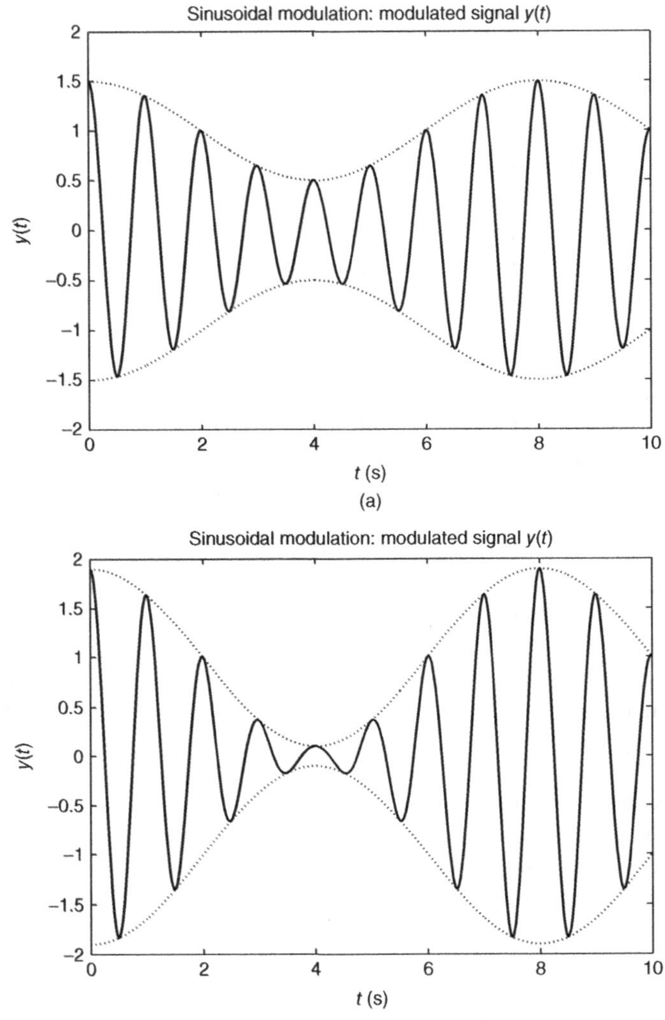

(a)

(b)

Figure 8.7 Conventional AM $y(t)$ of Example 8.13 with amplitude sensitivity k for the signal waveform $x(t)$ and carrier $c(t)$ in Figure 8.6. (a) $k = 0.5$. (b) $k = 0.9$.

capacitor in an RC circuit. The transfer function provides insight into the properties of $h(t)$. In particular, we found that the locations of the *poles* on the s-plane indicate the degree to which the system has an exponential or sinusoidal response:

- Real poles in the left half of the s-plane $\Longrightarrow$ decaying exponential.
- Complex conjugate poles in the left half of the s-plane $\Longrightarrow$ damped sinusoid.
- Complex conjugate poles on the imaginary axis $\Longrightarrow$ undamped sinusoid.

Next, we demonstrate that the pole locations determine another feature of a system called its *frequency response* $H(\omega)$, which is derived from $H(s)$ by substituting $s = j\omega$, assuming that the ROC includes the imaginary axis. For the linear circuits and systems covered in this book, $H(\omega)$ is a rational function: it is the ratio of two polynomials of the single variable ω (because $\text{Re}(s) = \sigma = 0$ in the substitution $s = j\omega$).

8.7.1 First-Order Low-Pass Filter

We begin with definitions of the three different frequency bands for a low-pass filter, which are summarized in Figure 8.8.

Definition: Passband, Stopband, and Transition Band The *passband* of a low-pass filter is the frequency range $[0, \omega_c]$ where $|H(\omega)|$ decreases from its maximum $H_{\max}$ at $\omega = 0$ to $H_{\max}/\sqrt{2}$ at the cutoff frequency ω_c. The *transition band* is the frequency range $(\omega_c, \omega_{\min}]$ where $\omega_{\min}$ is the frequency corresponding to $H_{\min} \triangleq |H(\omega_{\min})|$. The *stopband* is the frequency range $(\omega_{\min}, \infty)$ where $|H(\omega)| < H_{\min}$.

The maximum $H_{\max}$ is usually determined by the gain at $\omega = 0$, whereas $\omega_{\min}$ and $H_{\min}$ are given as *specifications* for the desired width and depth of the transition band of the filter. Thus, a narrow transition band depends on the following: (i) how close $\omega_{\min}$ is to ω_c, (ii) how close $H_{\min}$ is to 0, and (iii) the order of the denominator of the transfer function (the number of poles).

Consider the first-order system:

$$H(s) = \frac{a}{s + a} \implies H(\omega) = \frac{a}{j\omega + a}, \tag{8.125}$$

with real parameter $a > 0$, ROC $\text{Re}(s) > -a$, and impulse response function

$$h(t) = a \exp(-at)u(t). \tag{8.126}$$

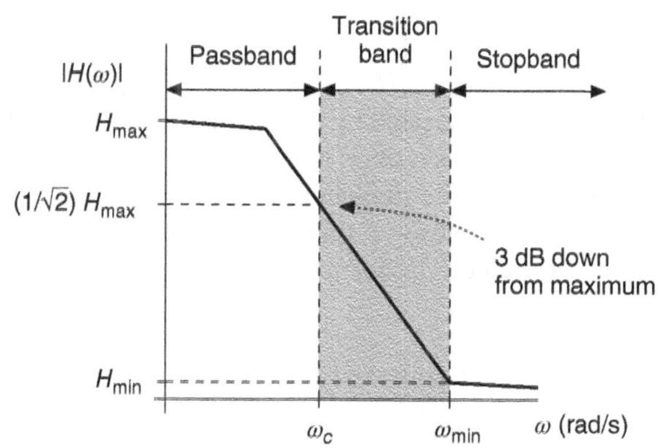

Figure 8.8 Magnitude response of a low-pass filter showing the passband, transition band, and stopband.

The magnitude $|H(\omega)|$ is derived from

$$|H(\omega)|^2 = H(\omega)H^*(\omega) = \frac{a^2}{(j\omega + a)(-j\omega + a)}$$
$$= \frac{a^2}{\omega^2 + a^2}, \tag{8.127}$$

which is necessarily real. Thus,

$$|H(\omega)| = \frac{|a|}{\sqrt{\omega^2 + a^2}}. \tag{8.128}$$

The phase is derived by writing $H(\omega)$ in rectangular complex variable form:

$$H(\omega) = \frac{a(-j\omega + a)}{(j\omega + a)(-j\omega + a)} = \frac{a^2}{\omega^2 + a^2} - j\frac{a\omega}{\omega^2 + a^2}, \tag{8.129}$$

and then taking the ratio of the imaginary and real parts as follows:

$$\theta(\omega) = \tan^{-1}(-\omega/a). \tag{8.130}$$

This system has the characteristic of a *low-pass filter* because it passes low frequencies and rejects high frequencies:

$$\lim_{\omega \to 0} |H(\omega)| = 1, \quad \lim_{\omega \to \infty} |H(\omega)| = 0. \tag{8.131}$$

The *bandwidth* is defined to be the cutoff frequency ω_c where $|H(\omega)|^2$ is one-half its maximum value, which is $H_{\max} = 1$ for the low-pass filter in (8.125). Thus, the following expression is solved for ω_c:

$$|H(\omega_c)|^2 = (1/2)|H(0)|^2 \implies \frac{a^2}{\omega_c^2 + a^2} = 1/2 \implies \omega_c = a. \tag{8.132}$$

At this frequency, the magnitude is $|H(\omega_c)| = 1/\sqrt{2} \approx 0.7071$ and the phase is $\theta = \tan^{-1}(-1) = -45°$. These are indicated by the dotted lines in Figure 8.9 for $a = 1$.

Since this is only a first-order filter with a single pole, it turns out that the bandwidth ω_c and the depth of the stopband defined by $H_{\min}$ at $\omega_{\min}$ are competing specifications. If we want a narrower transition band for the same depth of the stopband, then a smaller cutoff frequency is required as illustrated in the next example.

Example 8.14 For the filter response in Figure 8.9 with $\omega_c = a = 1$, let the original specification be $\omega_{\min} = 4$ rad/s, corresponding to

$$H_{\min} = \frac{1}{\sqrt{4^2 + 1^2}} \approx 0.2425. \tag{8.133}$$

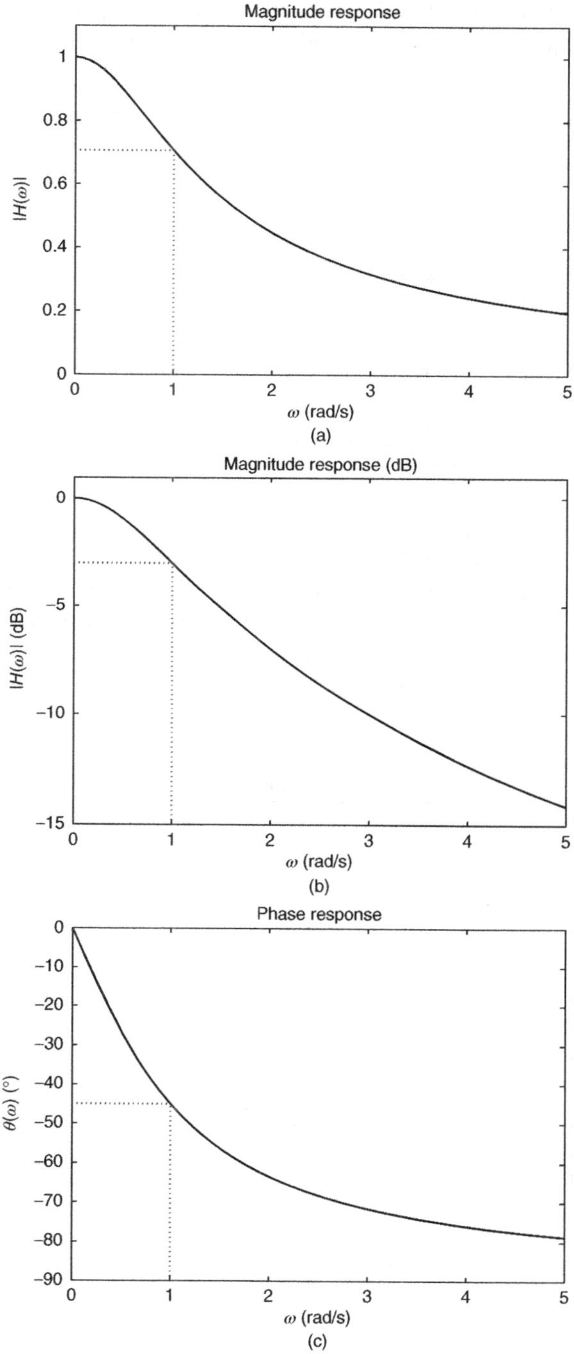

Figure 8.9 Low-pass filter. (a) Magnitude response. (b) Magnitude response in dB. (c) Phase response. The vertical dotted lines denote the cutoff frequency $\omega_c = 1$ rad/s.

Suppose we want this value for H_{min} to occur instead at $\omega_{\text{min}} = 3$ rad/s, which means a new cutoff frequency ω_c must be found. This is the same as finding a new value for a as follows:

$$H_{\text{min}} = 0.2425 = \frac{a}{\sqrt{\omega_{\text{min}}^2 + a^2}} = \frac{a}{\sqrt{3^2 + a^2}}. \tag{8.134}$$

Solving this expression for a yields

$$a^2 = \frac{9(0.2425)^2}{1 - (0.2425)^2} \approx 0.5623 \implies a = \omega_c \approx 0.7499. \tag{8.135}$$

Rounding this value to 0.75 rad/s, the new transfer function is

$$H(s) = \frac{0.75}{s + 0.75} \implies |H(\omega)| = \frac{0.75}{\sqrt{\omega^2 + 0.5625}}. \tag{8.136}$$

Plots of $|H(\omega)|$ for this new cutoff frequency and the previous one at $\omega_c = 1$ rad/s are shown in Figure 8.10. The bandwidth of the filter has been reduced to $\omega_c = 0.75$ rad/s, but the width of the transition band is narrower: $3 - 0.75 = 2.25$ rad/s versus the previous $4 - 1 = 3$ rad/s. If we want to keep the same cutoff frequency, then higher order filters with more poles are needed, such as that provided by the Butterworth filter discussed at the end of this chapter.

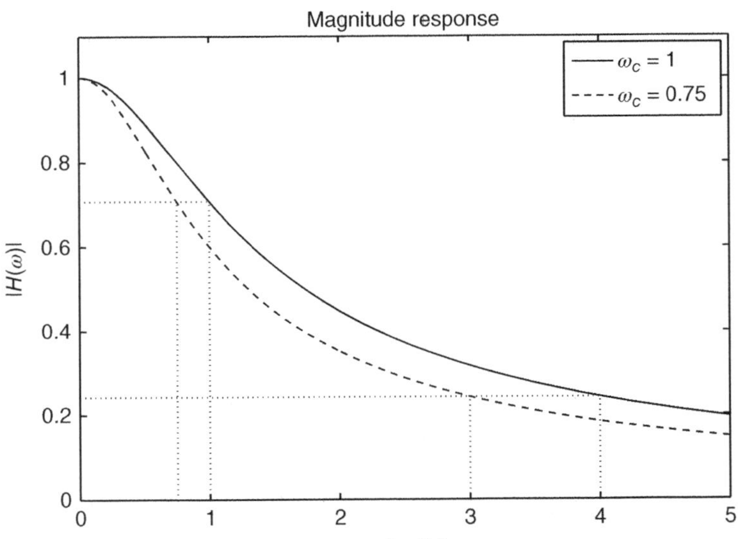

Figure 8.10 Magnitude response of a low-pass filter with different cutoff frequencies denoted by the two vertical dotted lines on the left. The two vertical dotted lines on the right show the upper bound of the transition band for each case.

Figure 8.9(b) shows the magnitude response in dB given by

$$10\log(|H(\omega)|^2) = 20\log(|H(\omega)|), \tag{8.137}$$

which is done to provide a greater dynamic range than what is observable in a linear plot. In particular, the logarithmic plot allows us to easily view very small values of the magnitude ($\ll 1$), which is important when examining the depth of the stopband. This advantage is not so obvious for this first-order low-pass filter because it has a wide transition band; it is not a sharp filter. The logarithmic plot is more advantageous for filters with a narrow transition band such as a high-order Butterworth filter. Observe in Figure 8.9(b) that the magnitude at the cutoff frequency ω_c is approximately 3 dB down from its maximum of 0 dB at $\omega = 0$. This, of course, follows from the definition of ω_c:

$$10\log(|H(\omega_c)|^2) = 10\log(1/2) = -10\log(2) \approx -3.0103 \text{ dB}. \tag{8.138}$$

The magnitude and phase can also be derived by computing them separately for the numerator (N) and denominator (D):

$$H(\omega) = \frac{N(\omega)}{D(\omega)} \implies |H(\omega)| = \frac{|N(\omega)|}{|D(\omega)|}, \quad \theta(\omega) = \theta_N(\omega) - \theta_D(\omega). \tag{8.139}$$

The magnitude components divide and the phase components subtract because they appear in the exponent of the exponential functions in polar form. Using this approach, it is not necessary to rewrite $H(\omega)$ in rectangular complex variable form as demonstrated in the next section.

8.7.2 First-Order High-Pass Filter

The following modified first-order transfer function has a zero at the origin:

$$H(s) = \frac{s}{s+a} \implies H(\omega) = \frac{j\omega}{j\omega + a}, \tag{8.140}$$

with impulse response function:

$$H(s) = 1 - \frac{a}{s+a} \implies h(t) = \delta(t) - a\exp(-at)u(t). \tag{8.141}$$

The last term is the impulse response function of the previous low-pass filter, and so we find that the output $y(t)$ for the high-pass filter is generated by subtracting the low-pass response from the input $x(t)$:

$$y(t) = h(t) * x(t) = x(t) - ax(t) * \exp(-at)u(t). \tag{8.142}$$

The magnitude response is

$$|H(\omega)| = \frac{|j\omega|}{|j\omega + a|} = \frac{|\omega|}{\omega^2 + a^2}, \tag{8.143}$$

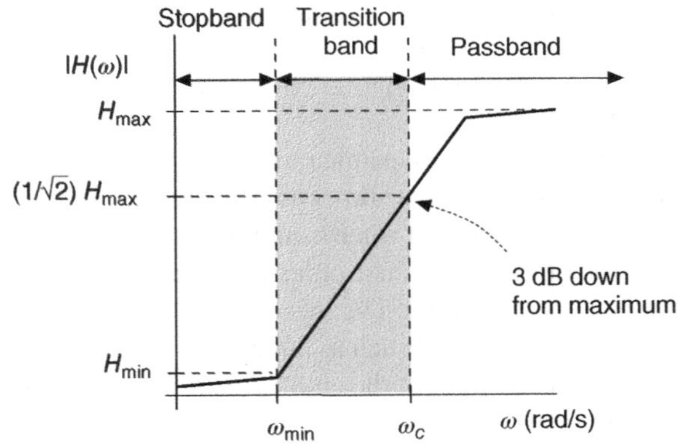

Figure 8.11 Magnitude response of high-pass filter showing the passband, transition band, and stopband.

and the phase is

$$\theta(\omega) = \tan^{-1}(\omega/0) - \tan^{-1}(\omega/a) = 90^\circ - \tan^{-1}(\omega/a). \tag{8.144}$$

This transfer function has the characteristic of a *high-pass filter*:

$$\lim_{\omega\to 0} |H(\omega)| = 0, \quad \lim_{\omega\to\infty} |H(\omega)| = 1. \tag{8.145}$$

The typical magnitude response of a high-pass filter is shown in Figure 8.11. The cutoff frequency is found by solving

$$|H(\omega_c)|^2 = \frac{a^2\omega_c^2}{\omega_c^2 + a^2} = 1/2 \implies \omega_c = a, \tag{8.146}$$

which is the same as the previous low-pass filter. The magnitude and phase responses of this filter for $a = 1$ are shown in Figure 8.12, where the vertical dotted lines denote the cutoff frequency $\omega_c = 1$ rad/s. At this frequency, the magnitude is $|H(\omega_c)| = 1/\sqrt{2} \approx 0.7071$ and the phase is $\theta = 90^\circ - \tan^{-1}(1) = 45^\circ$. As is the case for the low-pass filter, ω_{min} and H_{min} are the filter specifications: the desired width of the transition band and the depth of the stopband.

There are two other standard filter frequency responses: *band-pass* and *band-reject* (also called band-stop). Both of these require at least a second-order polynomial in the denominator of the transfer function $H(s)$.

8.7.3 Second-Order Band-Pass Filter

The typical magnitude response for a band-pass filter is shown in Figure 8.13. The following second-order transfer function has complex conjugate poles at $s = -\alpha \pm j\beta$ with $\alpha, \beta > 0$:

$$H(s) = \frac{a_1 s}{s^2 + a_1 s + a_0} = \frac{2\alpha s}{(s + \alpha + j\beta)(s + \alpha - j\beta)}. \tag{8.147}$$

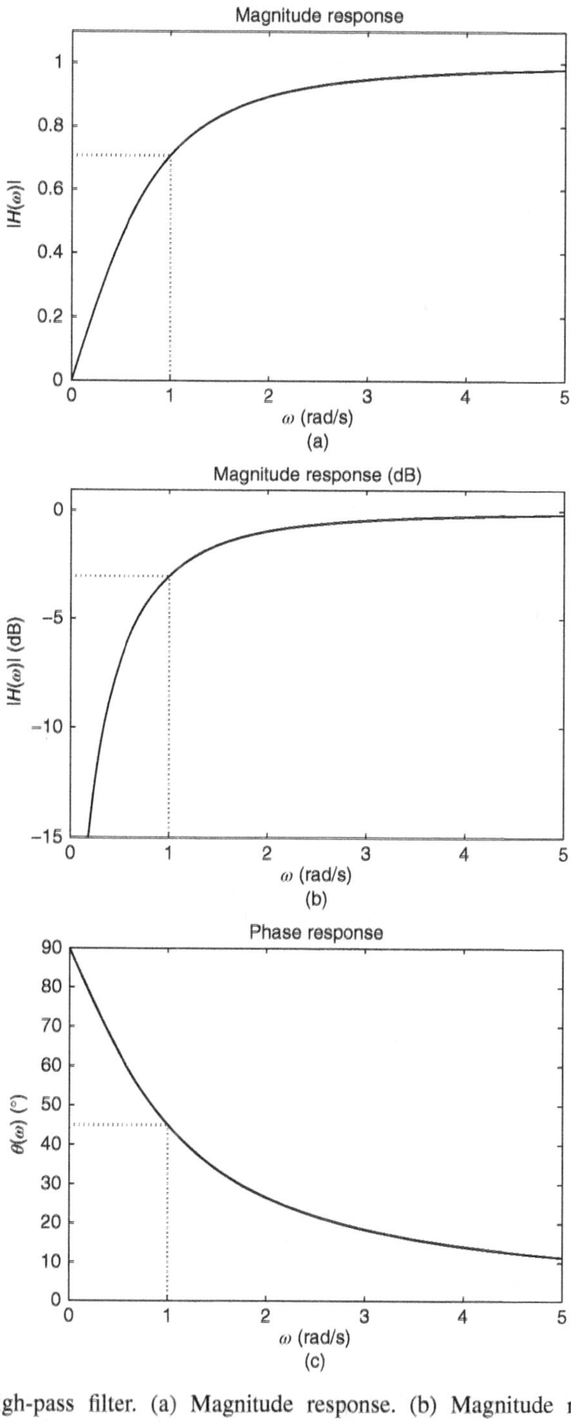

Figure 8.12 High-pass filter. (a) Magnitude response. (b) Magnitude response in dB. (c) Phase response. The vertical dotted lines denote the cutoff frequency $\omega_c = 1$ rad/s.

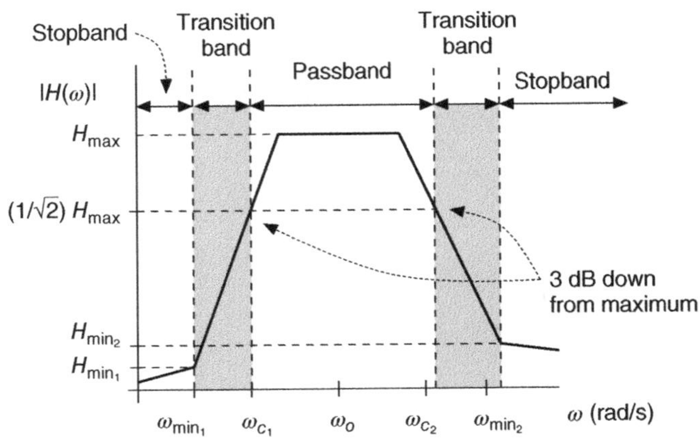

Figure 8.13 Magnitude response of band-pass filter showing the passband, two transition bands, and two stopbands.

This is not the most general form for a second-order band-pass filter because it does not allow for distinct real poles (corresponding to an overdamped system). This system is underdamped in general or critically damped in the event that $\beta = 0$. A more general transfer function for a band-pass filter is considered later. The frequency response of (8.147) is

$$H(\omega) = \frac{j2\alpha\omega}{[\alpha + j(\omega + \beta)][\alpha + j(\omega - \beta)]}, \tag{8.148}$$

its magnitude response is

$$|H(\omega)| = \frac{2\alpha|\omega|}{\sqrt{[\alpha^2 + (\omega + \beta)^2][\alpha^2 + (\omega - \beta)^2]}}, \tag{8.149}$$

and the phase is

$$\theta(\omega) = 90^\circ - \tan^{-1}((\omega + \beta)/\alpha) - \tan^{-1}((\omega - \beta)/\alpha). \tag{8.150}$$

A band-pass filter has five parameters:

- *Center frequency* ω_o where $|H(\omega)|$ is maximum.
- *Lower cutoff frequency* ω_{c_1} where $|H(\omega_{c_1})|^2 = (1/2)|H(\omega_o)|^2$.
- *Upper cutoff frequency* ω_{c_2} where $|H(\omega_{c_2})|^2 = (1/2)|H(\omega_o)|^2$.
- *Bandwidth* $BW \triangleq \omega_{c_2} - \omega_{c_1}$.
- *Quality factor* $Q \triangleq \omega_o/B$.

The quality factor Q is a dimensionless quantity that is a measure of the width (sharpness) of the filter transition band relative to its center frequency.

It is straightforward to show that the center frequency for this filter is (see Problem 8.28)

$$\omega_o = \sqrt{\alpha^2 + \beta^2}, \tag{8.151}$$

with $|H(\omega_o)| = 1$. In order to find the two cutoff frequencies $\{\omega_{c_1}, \omega_{c_2}\}$, the following expression is solved for ω_c:

$$\frac{4\alpha^2\omega_c^2}{[\alpha^2 + (\omega_c + \beta)^2][\alpha^2 + (\omega_c - \beta)^2]} = 1/2, \tag{8.152}$$

which becomes

$$\omega_c^4 - (6\alpha^2 + 2\beta^2)\omega_c^2 + (\alpha^2 + \beta^2)^2 = 0. \tag{8.153}$$

This is a quadratic equation in ω_c^2 with solution (see Problem 8.28)

$$\omega_c^2 = 3\alpha^2 + \beta^2 \pm \alpha\sqrt{9 + 4\beta^2 - \alpha^2}. \tag{8.154}$$

The square root of this equation yields four frequencies; however, two of those frequencies correspond to negative ω_c (which occur because the impulse response function is real); the positive cutoff frequencies are

$$\omega_{c_1} = \sqrt{3\alpha^2 + \beta^2 - \alpha\sqrt{9 + 4\beta^2 - \alpha^2}}, \tag{8.155}$$

$$\omega_{c_2} = \sqrt{3\alpha^2 + \beta^2 + \alpha\sqrt{9 + 4\beta^2 - \alpha^2}}, \tag{8.156}$$

with $\omega_{c_2} > \omega_{c_1}$. The bandwidth is the difference of these two quantities. The magnitude and phase characteristics of this filter for $\alpha = 1$ and $\beta = 3$ are shown in Figure 8.14, where $\omega_o \approx 3.1623$ rad/s, $\omega_{c_1} \approx 2.3166$ rad/s, $\omega_{c_2} \approx 4.3166$ rad/s, and $B = 2$ rad/s. The phase response extends for 180° over $[-90°, 90°]$, and it is exactly 0 at the center frequency ω_o. (The range of angles is only 90° for a single pole as shown previously for the low-pass and high-pass filters.) The magnitude plots are asymmetric because the frequency ranges about ω_o differ: $[0, \omega_o)$ versus (ω_o, ∞). For higher order filters and a larger center frequency, it is possible to design band-pass filters with a more symmetric response about ω_o.

8.7.4 Second-Order Band-Reject Filter

In order to implement a filter with a band-reject frequency characteristic, it is necessary that complex conjugate *zeros* be included in the transfer function:

$$H(s) = \frac{s^2 + a_0}{s^2 + a_1 s + a_0} = \frac{(s + j\gamma)(s - j\gamma)}{(s + \alpha + j\beta)(s + \alpha - j\beta)}, \tag{8.157}$$

which has the same denominator as the band-pass filter. We have chosen a numerator with zeros exactly on the imaginary axis at $s = \pm j\gamma = \pm j\sqrt{\alpha^2 + \beta^2}$ so that

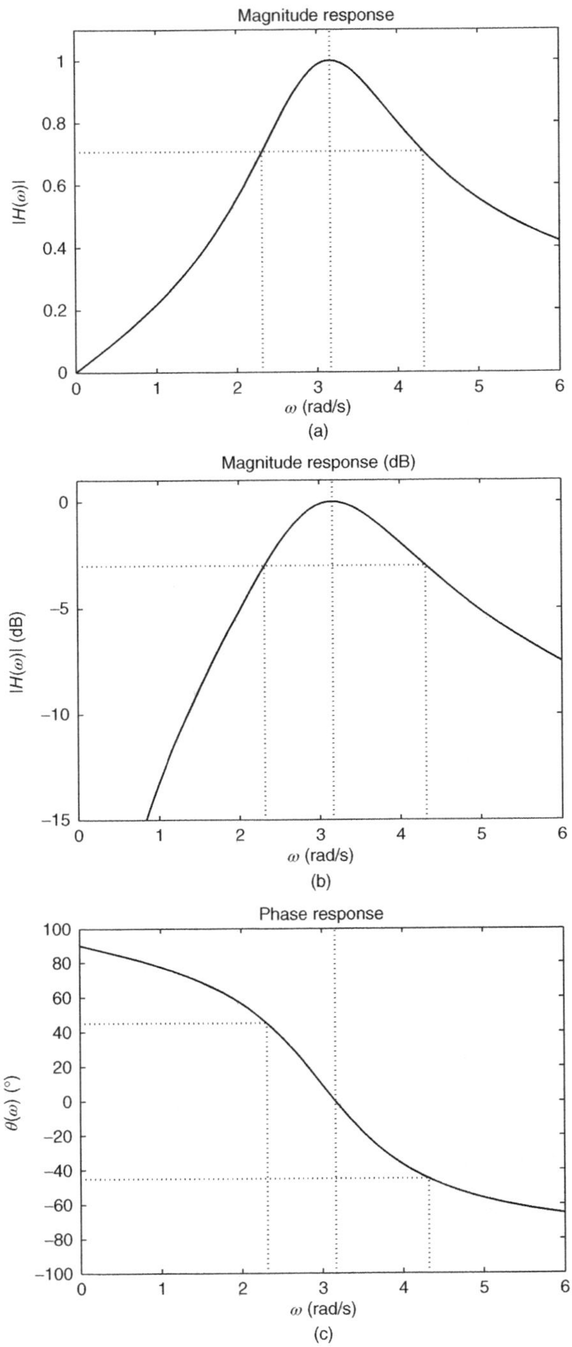

Figure 8.14 Band-pass filter. (a) Magnitude response. (b) Magnitude response in dB. (c) Phase response. The vertical dotted lines denote the center frequency $\omega_o \approx 3.1623$ rad/s and the lower and upper cutoff frequencies $\{\omega_{c_1}, \omega_{c_2} \approx 2.3166, 4.3166\}$ rad/s with a bandwidth of $BW = 2$ rad/s.

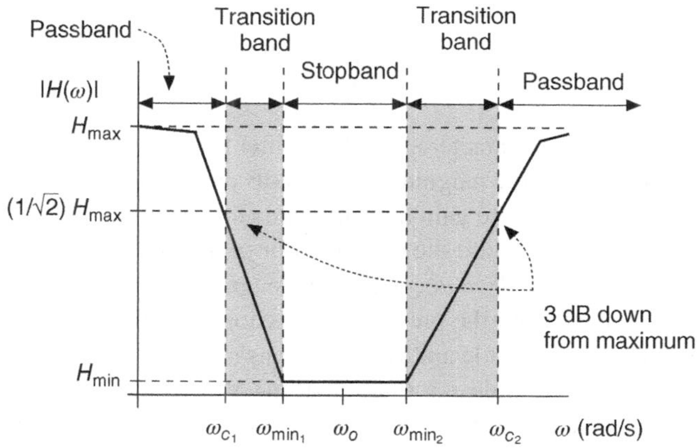

Figure 8.15 Magnitude response of band-reject filter showing two passbands, two transition bands, and the stopband.

$H(\omega) = 0$ at $\omega = \pm\gamma$. The typical magnitude response for a band-reject filter is shown in Figure 8.15.

Using results from the previous band-pass filter, we find that the squared magnitude is

$$|H(\omega)|^2 = \frac{(\omega^2 - \alpha^2 - \beta^2)^2}{[\alpha^2 + (\omega + \beta)^2][\alpha^2 + (\omega - \beta)^2]}, \tag{8.158}$$

and

$$|H(\omega)| = \frac{|\omega^2 - \alpha^2 - \beta^2|}{\sqrt{[\alpha^2 + (\omega + \beta)^2][\alpha^2 + (\omega - \beta)^2]}}. \tag{8.159}$$

The center frequency is obtained when the numerator is 0, which yields

$$\omega_o = \sqrt{\alpha^2 + \beta^2}, \tag{8.160}$$

and is the same as ω_o for the band-pass filter, by design for this particular numerator. The two cutoff frequencies $\{\omega_{c_1}, \omega_{c_2}\}$ are found by solving

$$\frac{(\omega_c^2 - \alpha^2 - \beta^2)^2}{[\alpha^2 + (\omega_c + \beta)^2][\alpha^2 + (\omega_c - \beta)^2]} = 1/2. \tag{8.161}$$

Rearranging this expression yields the quartic equation in (8.153), and so this band-reject filter has the same cutoff frequencies as the previous band-pass filter.

The phase is derived from $H(s)$ with $s = j\omega$:

$$H(\omega) = \frac{\alpha^2 + \beta^2 - \omega^2}{\alpha^2 + \beta^2 - \omega^2 + j2\alpha\omega}, \tag{8.162}$$

which gives

$$\theta(\omega) = -\tan^{-1}\left(\frac{2\alpha\omega}{\alpha^2+\beta^2-\omega^2}\right). \tag{8.163}$$

The magnitude and phase characteristics of this filter for $\alpha = 1$ and $\beta = 3$ are shown in Figure 8.16. Observe that the magnitude plot in dB clearly illustrates the band-reject nature of the filter because the gain actually tends to $-\infty$ at ω_o. There is a discontinuity in the phase at ω_o because the denominator of the phase expression in (8.163) changes sign at $\omega^2 = \alpha^2 + \beta^2 = \omega_o^2$, where the magnitude response is 0 dB.

It should be evident from the previous discussions that the type and quality of a filter are determined by the pole and zero locations relative to the real and imaginary axes. Thus, it is possible to design filters that meet the desired frequency response specifications by judiciously placing a sufficient number of poles and zeros on the s-plane. Since the filter should have real coefficients, a transfer function with complex poles and zeros must include their complex conjugates.

8.8 FREQUENCY RESPONSE OF SECOND-ORDER FILTERS

In this section, we describe the standard transfer functions for different types of second-order filters. Although band-pass and band-reject filters require at least second-order denominator polynomials, we show that low-pass and high-pass filters can also be implemented using second-order polynomials by an appropriate choice of the transfer function numerator.

- *Low-pass filter*:

$$H_{\mathrm{LP}}(s) = \frac{\omega_o^2}{s^2+2\zeta\omega_o s+\omega_o^2}, \quad |H_{\mathrm{LP}}(\omega)| = \frac{\omega_o^2}{\sqrt{(\omega_o^2-\omega^2)^2+(2\zeta\omega_o\omega)^2}}. \tag{8.164}$$

  $|H_{\mathrm{LP}}(\omega)| = 1$ for $\omega = 0$ and $|H_{\mathrm{LP}}(\omega)| = 0$ as $\omega \to \infty$.

- *High-pass filter*:

$$H_{\mathrm{HP}}(s) = \frac{s^2}{s^2+2\zeta\omega_o s+\omega_o^2}, \quad |H_{\mathrm{HP}}(\omega)| = \frac{\omega^2}{\sqrt{(\omega_o^2-\omega^2)^2+(2\zeta\omega_o\omega)^2}}. \tag{8.165}$$

  $|H_{\mathrm{HP}}(\omega)| = 0$ for $\omega = 0$ and $|H_{\mathrm{HP}}(\omega)| = 1$ as $\omega \to \infty$.

- *Band-pass filter*:

$$H_{\mathrm{BP}}(s) = \frac{2\zeta\omega_o s}{s^2+2\zeta\omega_o s+\omega_o^2}, \quad |H_{\mathrm{BP}}(\omega)| = \frac{|2\zeta\omega_o\omega|}{\sqrt{(\omega_o^2-\omega^2)^2+(2\zeta\omega_o\omega)^2}}. \tag{8.166}$$

  $|H_{\mathrm{BP}}(\omega)| = 0$ for $\omega = 0$ and as $\omega \to \infty$, $|H_{\mathrm{BP}}(\omega)| = 1$ for $\omega = \omega_o$.

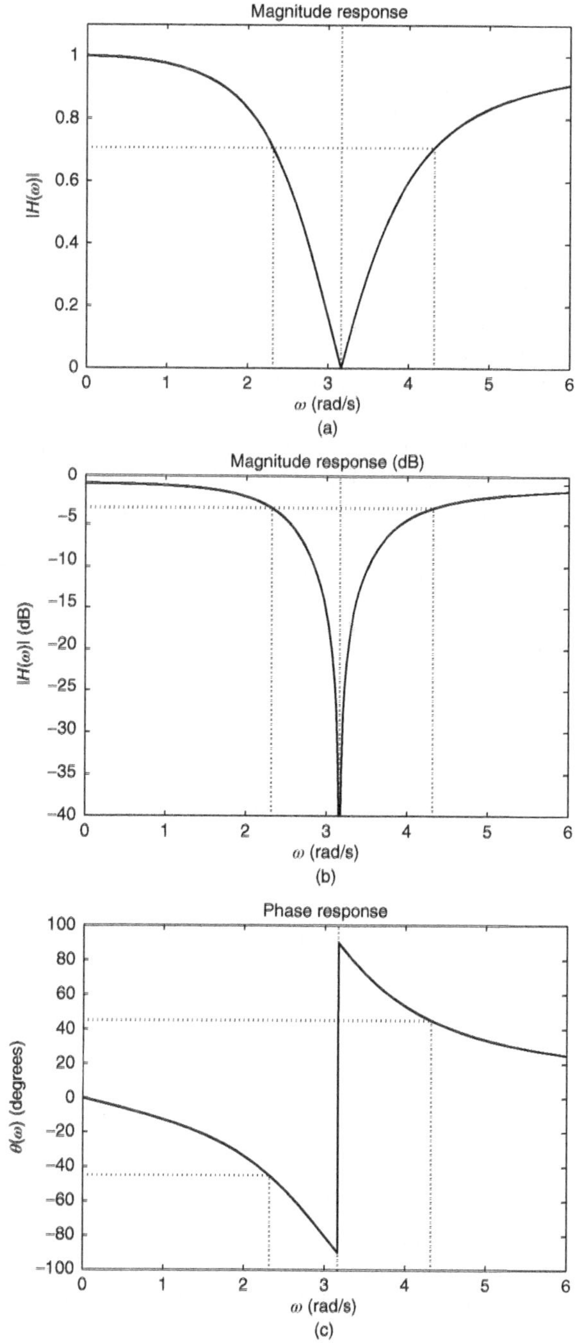

Figure 8.16 Band-reject filter. (a) Magnitude response. (b) Magnitude response in dB. (c) Phase response. The center frequency ω_o, cutoff frequencies $\{\omega_{c_1}, \omega_{c_2}\}$, and bandwidth BW are the same as those in Figure 8.14 for the band-pass filter.

- *Band-reject filter*:

$$H_{\text{BR}}(s) = \frac{s^2 + \omega_o^2}{s^2 + 2\zeta\omega_o s + \omega_o^2}, \quad |H_{\text{BR}}(\omega)| = \frac{|\omega_o^2 - \omega^2|}{\sqrt{(\omega_o^2 - \omega^2)^2 + (2\zeta\omega_o\omega)^2}}. \tag{8.167}$$

$|H_{\text{BR}}(\omega)| = 1$ for $\omega = 0$ and as $\omega \to \infty$, $|H_{\text{BR}}(\omega)| = 0$ for $\omega = \omega_o$.

The damping ratio ζ and resonant frequency ω_o were mentioned in Chapter 7. The numerator for each $H(s)$ has been chosen so that $|H(\omega)| = 1$ at either $\omega = 0$, $\omega = \omega_o$, or as $\omega \to \infty$ depending on the type of filter. With these specific transfer functions, we show later that the cutoff frequencies are all proportional to ω_o, and the proportionality constant varies with the damping ratio ζ.

When describing the type of filter, its transfer function should be evaluated at $s = 0$ and as $s \to \infty$. Observe that $|H(s)|$ for the low-pass filter and the band-pass filter both approach 0 as $s \to \infty$ because the order of the denominator exceeds that of the numerator. The numerator and denominator for the high-pass and band-reject filters, on the other hand, have the same order which is why $|H(s)|$ is nonzero as $s \to \infty$.

The roots of the denominator polynomial for each filter are the poles

$$p_1, p_2 = -\zeta\omega_o \pm \sqrt{\zeta^2\omega_o^2 - \omega_o^2} = -\omega_o\zeta \pm \omega_o\sqrt{\zeta^2 - 1}, \tag{8.168}$$

and so there are three different cases, as covered in Chapter 7, which we repeat here for convenience:

- *Distinct real poles* ($\zeta > 1$, overdamped):

$$p_1, p_2 = -\omega_o\zeta \pm \omega_o\sqrt{\zeta^2 - 1}. \tag{8.169}$$

- *Complex conjugate poles* ($\zeta < 1$, underdamped):

$$p_1, p_2 = -\omega_o\zeta \pm j\omega_o\sqrt{1 - \zeta^2}. \tag{8.170}$$

- *Repeated real poles* ($\zeta = 1$, critically damped):

$$p_1 = p_2 = -\omega_o\zeta. \tag{8.171}$$

For the underdamped case, the quantity $\omega_d \triangleq \omega_o\sqrt{1 - \zeta^2}$ is called the *damped resonant frequency*; it is the frequency of the sinusoidal waveform in the time domain obtained via an inverse Laplace transform.

The pole–zero plots for these four filters are illustrated in Figure 8.17 for the underdamped case ($\zeta < 1$) with complex conjugate poles. Figure 8.18(a) shows the magnitude of the two poles for $\omega_o = 2$ rad/s as ζ is varied from 0 to 2. For $\zeta > 1$, the poles are distinct, as mentioned earlier, whereas for $\zeta \leq 1$, the magnitude of each pole is a constant 2 because they form a complex conjugate pair. The pole locations

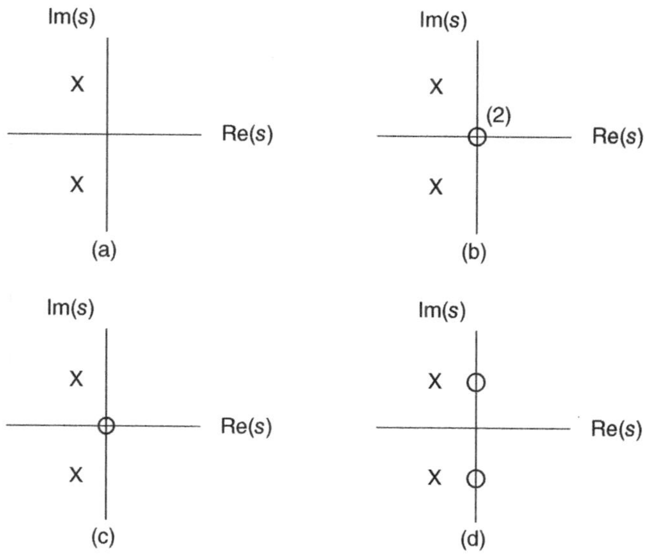

Figure 8.17 Pole–zero plots for underdamped second-order filters. (a) Low-pass. (b) High-pass. (c) Band-pass. (d) Band-reject.

on the s-plane are shown in Figure 8.18(b) for the same variation in ζ. For $\zeta > 1$, one pole moves to the left and the other pole moves to the right on the real axis. When $\zeta < 1$, the pole that moved right now traces a circle of radius 2 in the clockwise direction (the solid line), while the other pole moves counterclockwise on the same circle (the dashed line). Of course, these traces are mirror images of each other because the poles must form a complex conjugate pair for a second-order polynomial with real coefficients.

The magnitude response for each of the four types of filters with $\omega_o = 1$ rad/s and variable ζ is shown in Figures 8.19 and 8.20. It is clear that ω_o is the "center" frequency of the band-pass and band-reject filters where $|H(\omega)|$ is maximum (= 1) and 0, respectively. Observe from Figure 8.19 that the low-pass and high-pass responses always intersect each other at $\omega_o = 1$ rad/s as ζ and the bandwidth are varied. The magnitude response of the low-pass and high-pass filters at this frequency is

$$|H_{\rm LP}(\omega_o)| = 1/2\zeta = |H_{\rm HP}(\omega_o)|. \tag{8.172}$$

The cutoff frequency for the low-pass filter is derived by solving

$$|H_{\rm LP}(\omega_c)| = \frac{\omega_o^2}{\sqrt{(\omega_o^2 - \omega_c^2)^2 + (2\zeta\omega_o\omega_c)^2}} = 1/\sqrt{2}, \tag{8.173}$$

from which we have a quadratic equation in ω_c^2:

$$\omega_c^4 + 2\omega_o^2(2\zeta^2 - 1)\omega_c^2 - \omega_o^4 = 0. \tag{8.174}$$

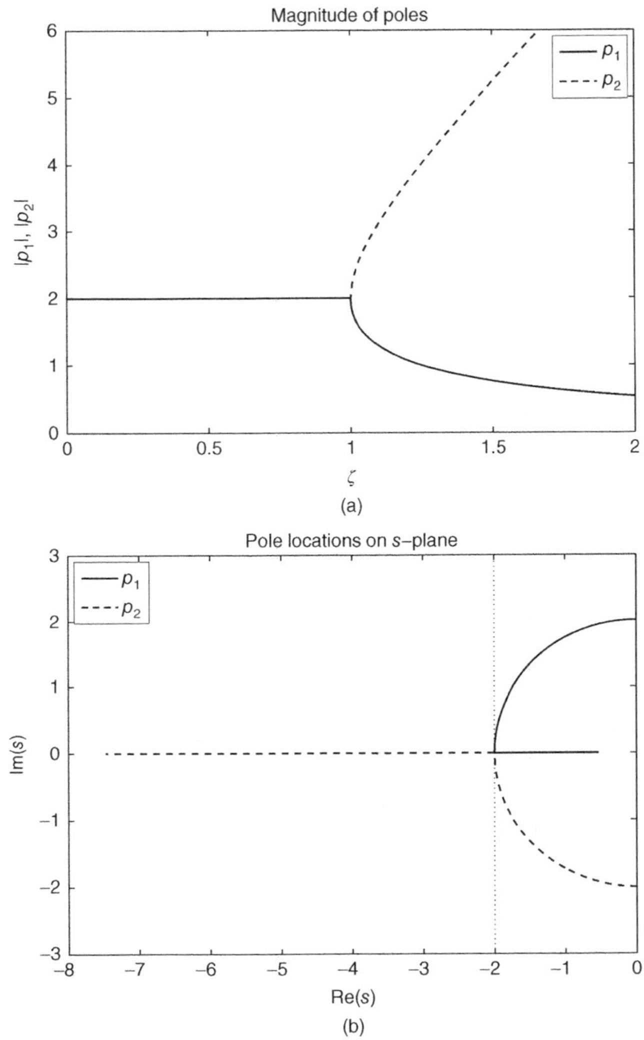

Figure 8.18 Poles for a second-order system with $\omega_o = 2$ rad/s as the damping ratio ζ is varied from 0 to 2. (a) Magnitude of poles versus ζ. (The solid and dashed lines merge at the horizontal line for $\zeta < 1$, corresponding to complex conjugate poles, which of course have the same magnitude.) (b) Poles on the s-plane. The vertical dotted line is the boundary where the two real poles for $\zeta > 1$ move to the left and right on the real axis.

The solution for ω_c^2 is

$$\omega_c^2 = -(2\zeta^2 - 1)\omega_o^2 + 2\omega_o^2\sqrt{\zeta^4 - \zeta^2 + 1/2}, \tag{8.175}$$

where only the positive square root is allowed so that the overall right-hand side is nonnegative:

$$\omega_c = \omega_o\sqrt{-(2\zeta^2 - 1) + 2\sqrt{\zeta^4 - \zeta^2 + 1/2}}. \tag{8.176}$$

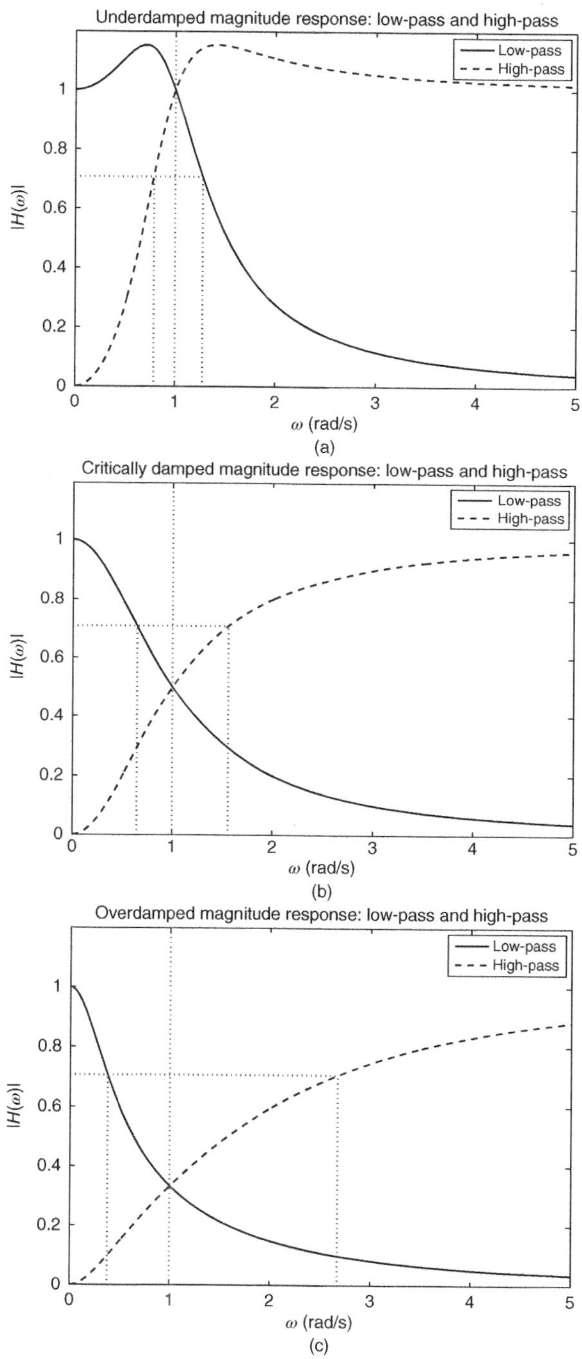

Figure 8.19 Magnitude response for second-order systems: low-pass and high-pass filters with $\omega_o = 1$ rad/s. (a) Underdamped: $\zeta = 1/2$. (b) Critically damped: $\zeta = 1$. (c) Overdamped: $\zeta = 3/2$. The vertical dotted lines show ω_o and the cutoff frequency ω_c for both filter types.

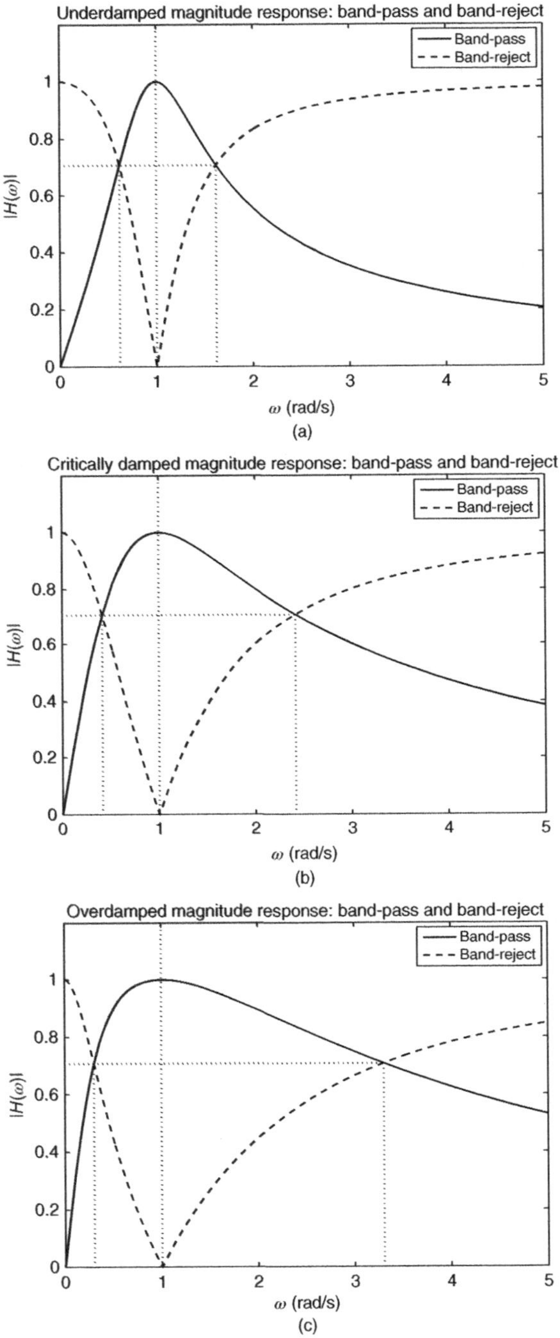

Figure 8.20 Magnitude response for second-order systems: band-pass and band-reject filters with $\omega_o = 1$ rad/s. (a) Underdamped: $\zeta = 1/2$. (b) Critically damped: $\zeta = 1$. (c) Overdamped: $\zeta = 3/2$. The vertical dotted lines show ω_o and the cutoff frequencies $\{\omega_{c_1}, \omega_{c_2}\}$, which are the same for both filter types.

For $\omega_o = 1$ rad/s and $\zeta = \{1/2, 1, 3/2\}$, we obtain the following set of positive cutoff frequencies: $\{1.2720, 0.6436, 0.3742\}$ rad/s, respectively, as shown by the vertical dotted lines in Figure 8.19. A similar equation is obtained for the high-pass filter:

$$|H_{\text{HP}}(\omega_c)| = \frac{\omega_c^2}{\sqrt{(\omega_o^2 - \omega_c^2)^2 + (2\zeta\omega_o\omega_c)^2}} = 1/\sqrt{2}, \tag{8.177}$$

which gives

$$\omega_c^2 = (2\zeta^2 - 1)\omega_o^2 + 2\omega_o^2\sqrt{\zeta^4 - \zeta^2 + 1/2}, \tag{8.178}$$

where again only the positive square root is retained. Thus,

$$\omega_c = \omega_o\sqrt{(2\zeta^2 - 1) + 2\sqrt{\zeta^4 - \zeta^2 + 1/2}}, \tag{8.179}$$

which differs from (8.176) only by the negative leading term under the outer square root (a similar result was obtained in the previous section). For $\omega_o = 1$ rad/s and $\zeta = \{1/2, 1, 3/2\}$, the positive cutoff frequencies are $\{0.7682, 1.5538, 2.6721\}$ rad/s, respectively. For the low-pass filter, the bandwidth is given by the cutoff frequency ω_c relative to $\omega = 0$. The bandwidth for a high-pass filter is not as easily defined because the dominant magnitude response extends from ω_c to $\omega \to \infty$.

In order to find the cutoff frequencies that define the bandwidth BW for the band-pass filter, we examine

$$|H_{\text{BP}}(\omega_c)| = \frac{2\zeta\omega_o|\omega_c|}{\sqrt{(\omega_o^2 - \omega_c^2)^2 + (2\zeta\omega_o\omega_c^2)^2}} = 1/\sqrt{2}. \tag{8.180}$$

Squaring and rearranging this expression yield

$$(2\zeta\omega_o\omega_c)^2 = (\omega_o^2 - \omega_c^2)^2. \tag{8.181}$$

Taking the square root of both sides, we have

$$\omega_o^2 - \omega_c^2 = \pm 2\zeta\omega_o\omega_c \implies \omega_c^2 \pm 2\zeta\omega_o\omega_c - \omega_o^2 = 0, \tag{8.182}$$

which is a quadratic equation in ω_c. Thus, four cutoff frequencies are obtained:

$$\omega_c = \pm\zeta\omega_o \pm \sqrt{(\zeta\omega_o)^2 + \omega_o^2}, \tag{8.183}$$

which we label as follows:

$$\pm\,\omega_{c_1} = \pm\omega_o(-\zeta + \sqrt{\zeta^2 + 1}), \quad \pm\omega_{c_2} = \pm\omega_o(\zeta + \sqrt{\zeta^2 + 1}), \tag{8.184}$$

with $\omega_{c_1} < \omega_{c_2}$. Although negative $-\zeta$ appears in ω_{c_1}, the square-root term exceeds ζ, and so when they are added together, a positive cutoff frequency is obtained. For $\omega_o =$

1 rad/s and $\zeta = \{1/2, 1, 3/2\}$, the positive cutoff frequencies are $\{0.6180, 1.6180\}$, $\{0.4142, 2.4142\}$, and $\{0.3028, 3.3028\}$ rad/s, respectively. These results are denoted by the vertical dotted lines in Figure 8.20. The original quartic equation in (8.181) also yields the negative cutoff frequencies $\{-\omega_{c_1}, -\omega_{c_2}\}$, which occur because the magnitude is an even function (which, of course, is due to the fact that the second-order system has real coefficients). The difference of each pair of numbers with the same sign yields the bandwidth in each case:

$$BW \triangleq \omega_{c_2} - \omega_{c_1} = \zeta\omega_o - (-\zeta\omega_o) = 2\zeta\omega_o, \tag{8.185}$$

which is the coefficient of s in the denominator of the transfer function. The same equations for the cutoff frequencies of the band-reject filter are obtained by solving

$$|H_{\text{BR}}(\omega_c)| = \frac{|\omega_o^2 - \omega_c^2|}{\sqrt{(\omega_o^2 - \omega_c^2)^2 + (2\zeta\omega_o\omega_c)^2}} = 1/\sqrt{2}. \tag{8.186}$$

Rearranging this expression yields

$$2(\omega_o^2 - \omega_c^2)^2 = (\omega_o^2 - \omega_c^2)^2 + (2\zeta\omega_o\omega_c)^2, \tag{8.187}$$

which is identical to (8.181), and so the cutoff frequencies are the same as those in (8.184).

The band-pass and band-reject results derived in terms of ω_o and ζ are similar to those in the previous section, which were expressed in terms of the poles $p_1 = -\alpha + j\beta$ and $p_2 = -\alpha - j\beta$. Substituting $\omega_o = \sqrt{\alpha^2 + \beta^2}$ and $\zeta = \alpha/\sqrt{\alpha^2 + \beta^2}$ into (8.184) yields the same expressions as in the previous section for the band-pass and band-reject filters. This equation for ζ is derived by equating the numerators of the two representations of the second-order band-pass filter such that $2\alpha s = 2\zeta\omega_o s$, solving for ζ, and substituting ω_o. However, the transfer functions used in this section are more general because they allow for all three types of systems: underdamped, overdamped, and critically damped. The results in the previous section are not completely general because they assume complex conjugate poles: the same α is used for the two poles. Thus, it is not possible to implement an overdamped system with distinct poles using (8.148) and (8.162), nor is it possible to implement an undamped system with $\alpha = 0$ because the transfer functions would be either 0 or fixed at 1. A critically damped system is possible when $\beta = 0$, resulting in a transfer function with double poles at $p_1 = p_2 = -\alpha$.

Summarizing the second-order transfer functions with identical poles given at the beginning of this section, these filters operate as low-pass, high-pass, band-pass, or band-reject depending on the type of *numerator*. The pole locations of the *denominator* determine the type of damping. The transfer functions for the high-pass and band-reject filters are actually improper, and in order to derive the corresponding

TABLE 8.7 Transfer Function Limits for Series RLC Circuit with Denominator $s^2 + (R/L)s + 1/LC$

Output $y(t)$	$H(0)$	$H(\infty)$	$H(s)$ Numerator	Filter Type
$y_1(t) = v_R(t)$	0	0	$(R/L)s$	Band-pass
$y_2(t) = v_L(t)$	0	1	s^2	High-pass
$y_3(t) = v_C(t)$	1	0	$1/LC$	Low-pass
$y_4(t) = v_L(t) + v_C(t)$	1	1	$s^2 + 1/LC$	Band-reject
$y_5(t) = v_R(t) + v_L(t)$	0	1	$s^2 + (R/L)s$	High-pass
$y_6(t) = v_R(t) + v_C(t)$	1	0	$(R/L)s + 1/LC$	Low-pass

impulse response functions, long division must be performed before writing a PFE. For the high-pass filter:

$$H_{\text{HP}}(s) = 1 - \frac{2\zeta\omega_o s + \omega_o^2}{s^2 + 2\zeta\omega_o s + \omega_o^2}, \tag{8.188}$$

and for the band-reject filter:

$$H_{\text{BR}}(s) = 1 - \frac{2\zeta\omega_o s}{s^2 + 2\zeta\omega_o s + \omega_o^2}. \tag{8.189}$$

Both of these filters have a Dirac delta function in the time domain; the low-pass and band-pass filters do not. It is interesting to note from the expressions in (8.188) and (8.189), compared with those in (8.164)–(8.167), that the transfer functions of the four second-order filters are related as follows:

$$H_{\text{BR}}(s) = H_{\text{LP}}(s) + H_{\text{HP}}(s) = 1 - H_{\text{BP}}(s), \tag{8.190}$$

$$H_{\text{BP}}(s) = 1 - H_{\text{LP}}(s) - H_{\text{HP}}(s) = 1 - H_{\text{BR}}(s). \tag{8.191}$$

These results are consistent with the series RLC circuit results shown in Table 8.7 (which are discussed in the next section), where we find that a band-reject response is produced across L and C together, and a band-pass response is produced across R. Similarly, a high-pass response is derived across L, whereas low-pass and band-pass responses are derived across C and R, respectively. A different low-pass response is derived across C and R together, but note that this overlaps with the band-pass response across R. Likewise, a different high-pass response is derived from L and R together, but this also overlaps with the band-pass response across R.

8.9 FREQUENCY RESPONSE OF SERIES RLC CIRCUIT

Next, we consider the second-order series RLC circuit shown in Figure 8.21 and demonstrate that depending on where the output is selected, the four second-order filters described in the previous section are all possible. For input $x(t) = V_s\delta(t)$, we

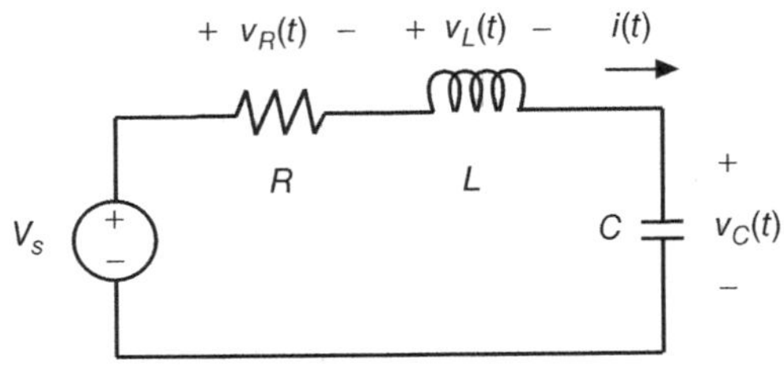

Figure 8.21 Second-order series RLC circuit with resistor R, inductor L, and capacitor C.

consider the following output voltages: $y_1(t) = v_R(t)$, $y_2(t) = v_L(t)$, $y_3(t) = v_C(t)$, $y_4(t) = v_L(t) + v_C(t)$, $y_5(t) = v_R(t) + v_L(t)$, $y_6(t) = v_R(t) + v_C(t)$, and $y_7(t) = v_R(t) + v_L(t) + v_C(t)$. From Kirchoff's voltage law (KVL), the last case is identical to V_s, and so it need not be considered any further: its transfer function is 1. The other six cases are summarized in Table 8.7. In order to physically generate $y_6(t)$, R and L should be interchanged in the circuit. Using the Laplace transform techniques in Chapter 7, voltage division yields the output in the s-domain for $y_1(t)$:

$$Y_1(s) = \frac{R}{R + sL + 1/sC} V_s. \tag{8.192}$$

The transfer function is

$$H_1(s) = \frac{(R/L)s}{s^2 + (R/L)s + 1/LC}, \tag{8.193}$$

and the magnitude of its frequency response is

$$|H_1(\omega)| = \frac{(R/L)|\omega|}{\sqrt{(1/LC - \omega^2)^2 + (\omega R/L)^2}}. \tag{8.194}$$

It is clear from this voltage division that the same denominator appears in the magnitude response for every case; only the numerator varies as shown in Table 8.7. The other five cases are provided as follows:

$$H_2(s) = \frac{s^2}{s^2 + (R/L)s + 1/LC}, \quad |H_2(\omega)| = \frac{\omega^2}{\sqrt{(1/LC - \omega^2)^2 + (\omega R/L)^2}}, \tag{8.195}$$

$$H_3(s) = \frac{1/LC}{s^2 + (R/L)s + 1/LC}, \quad |H_3(\omega)| = \frac{1/LC}{\sqrt{(1/LC - \omega^2)^2 + (\omega R/L)^2}}, \tag{8.196}$$

$$H_4(s) = \frac{s^2 + 1/LC}{s^2 + (R/L)s + 1/LC}, \quad |H_4(\omega)| = \frac{|1/LC - \omega^2|}{\sqrt{(1/LC - \omega^2)^2 + (\omega R/L)^2}}, \tag{8.197}$$

$$H_5(s) = \frac{s^2 + (R/L)s}{s^2 + (R/L)s + 1/LC}, \quad |H_5(\omega)| = \frac{\sqrt{(\omega R/L)^2 + \omega^4}}{\sqrt{(1/LC - \omega^2)^2 + (\omega R/L)^2}}, \tag{8.198}$$

$$H_6(s) = \frac{(R/L)s + 1/LC}{s^2 + (R/L)s + 1/LC}, \quad |H_6(\omega)| = \frac{\sqrt{(\omega R/L)^2 + (1/LC)^2}}{\sqrt{(1/LC - \omega^2)^2 + (\omega R/L)^2}}. \quad (8.199)$$

Comparing with the standard second-order denominator in (8.164)–(8.167), we find that $\omega_o = 1/\sqrt{LC}$ rad/s and $2\zeta\omega_o = R/L \Rightarrow \zeta = (R/2)\sqrt{C/L}$. The resonant frequency ω_o is the frequency where the inductor and capacitor impedances cancel each other:

$$j\omega L + \frac{1}{j\omega C} = j(\omega L - 1/\omega C) = 0. \quad (8.200)$$

Solving this expression yields $\omega_o^2 = 1/LC$, in which case the circuit appears to be purely resistive with resistance R, a condition known as *resonance*.

The type of filter can generally be determined by substituting $\omega = 0$ and $\omega \to \infty$. These results are also summarized in Table 8.7, where we see that the numerators of the first four cases correspond exactly to the standard second-order transfer functions in (8.164)–(8.167). Thus, we can use the expressions for the cutoff frequencies derived in the previous section. The second set of low-pass and high-pass filters, $H_5(s)$ and $H_6(s)$, do not have the standard forms as in (8.164) and (8.165). The numerator of the low-pass filter $H_6(s)$ has a zero at $s = -1/RC$, whereas $H_3(s)$ does not have any (finite) zeros. The high-pass filter $H_5(s)$ has zeros at $s = 0$ and $s = -R/L$, whereas both zeros of $H_2(s)$ are located at the origin. It is interesting that for this simple RLC circuit, 18 different types of frequency responses are possible because each transfer function can realize any of the three types of damping: underdamped, overdamped, and critically damped.

The cutoff frequency for the second high-pass filter $H_5(s)$ is determined by solving the following equation for ω_c:

$$\frac{(\omega_c R/L)^2 + \omega_c^4}{(1/LC - \omega_c^2)^2 + (\omega_c R/L)^2} = 1/2. \quad (8.201)$$

Rearranging this expression yields a quadratic equation in ω_c^2:

$$\omega_c^4 + (R/L^2 + 2/LC)\omega_c^2 + 1/(LC)^2 = 0, \quad (8.202)$$

of which the only valid positive solution is

$$\omega_c = \sqrt{-(R^2/2L^2 + 1/LC) + (1/2L)\sqrt{R^4/L^2 + 4R^2/LC + 8/C^2}}. \quad (8.203)$$

The cutoff frequency for the second low-pass filter $H_6(s)$ is (see Problem 8.31)

$$\omega_c = \sqrt{(R^2/2L^2 + 1/LC) + (1/2L)\sqrt{R^4/L^2 + 4R^2/LC + 8/C^2}}. \quad (8.204)$$

Observe that these two cutoff frequencies have similar expressions, except that the leading term in (8.203) is negative. Although the equations are quartic in ω_c for both cases, the inner square root is not subtracted because that would yield a complex number.

Example 8.15 In this example, we show the magnitude response for each of the different types of filters summarized in Table 8.7. Let $R = 2500\ \Omega$, $L = 1$ H, and $C = 1\ \mu$F such that $\omega_o^2 = 1/(1 \times 10^{-6}) \Rightarrow \omega_o = 1000$ rad/s and $2\zeta\omega_o = 2500 \Rightarrow \zeta = 2500/2000 = 1.25$. Thus, the denominator polynomial is overdamped with real poles

$$p_1, p_2 = -1000(1.25) \pm 1000\sqrt{(1.25)^2 - 1} = -500, 2000. \tag{8.205}$$

Figure 8.22 shows the frequency response for each of the six filters. Observe again that the band-pass and band-reject filters intersect at the two cutoff frequencies $\{\omega_{c_1}, \omega_{c_2}\}$. They both have bandwidth $BW = 2500$ rad/s and quality factor $Q = 0.4$. The second high-pass filter has a sharper transition band due to $(R/L)s$ in the numerator, causing $|H_5(\omega)|$ to increase more rapidly for small ω. Likewise, the magnitude response of the low-pass filter $H_6(s)$ increases initially because of the zero in the numerator; the other low-pass filter has only a constant in the numerator. Figure 8.23 shows the results when the resistor value is decreased to $R = 1000\ \Omega$, resulting in an underdamped circuit with $\zeta = 0.5$ and complex conjugate poles

$$p_1, p_2 = -1000(0.5) \pm j1000\sqrt{1 - (0.5)^2} \approx -500 \pm j866. \tag{8.206}$$

These plots have sharper transition bands than with the larger resistor, and the quality factor for the band-pass and band-reject filters is now $Q = 1$ with smaller bandwidth $BW = 1000$ rad/s.

8.10 BUTTERWORTH FILTERS

In the previous sections, we investigated first- and second-order transfer functions and their frequency responses. It turns out that for such low-order systems, the transition from passband to stopband is relatively gradual. In order to have a faster transition and more precise frequency filtering, corresponding to a narrow transition band (a sharp filter), it is necessary that high-order polynomials be used in the denominator of the transfer function.

Although many different high-order filters have a narrow transition band, there are three well-known filters that offer different frequency characteristics:

- *Butterworth filter:* Maximally flat response in the passband.
- *Chebyshev filter:* Narrower transition band than the Butterworth filter, but at the expense of ripple in either the passband or the stopband.
- *Elliptic filter:* Narrower transition band than either the Butterworth or Chebyshev filters, but at the expense of ripple in both the passband and the stopband.

For the rest of this chapter, we consider only the Butterworth filter.

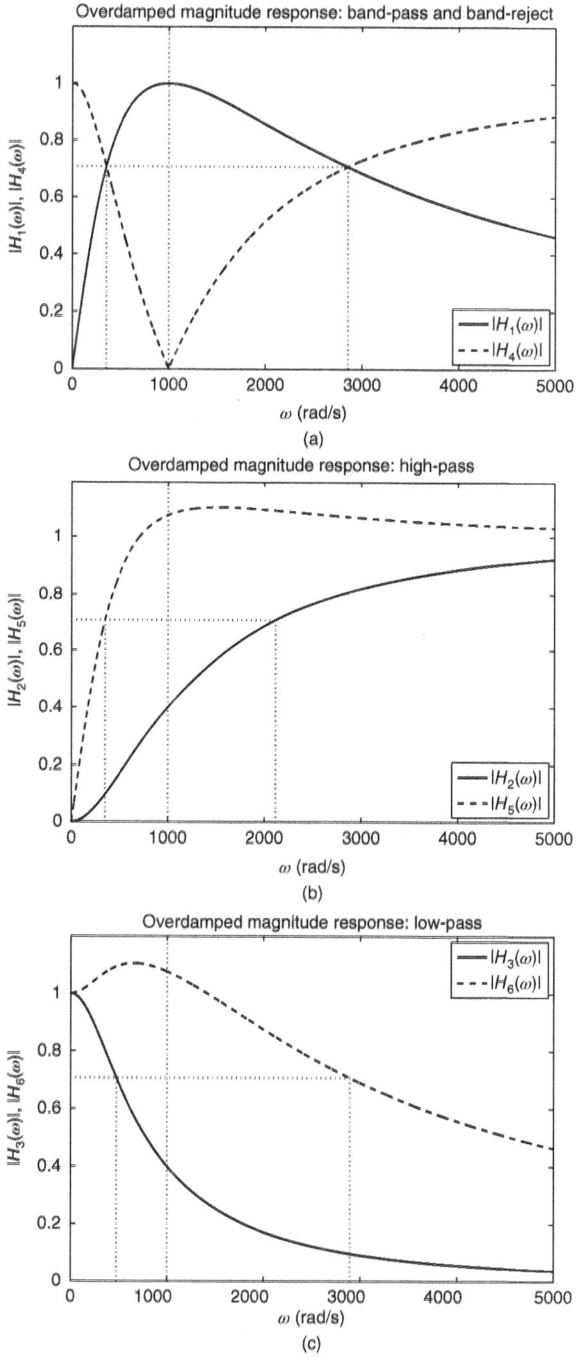

Figure 8.22 Frequency responses of overdamped series RLC circuit in Example 8.15 with $\omega_o = 1000$ rad/s and $\zeta = 1.25$. (a) Band-pass and band-reject filter responses. (b) Two high-pass filter responses. (c) Two low-pass filter responses. The vertical dotted lines show ω_o and the cutoff frequencies, $\{\omega_{c_1}, \omega_{c_2}\}$ or ω_c, for each type of filter.

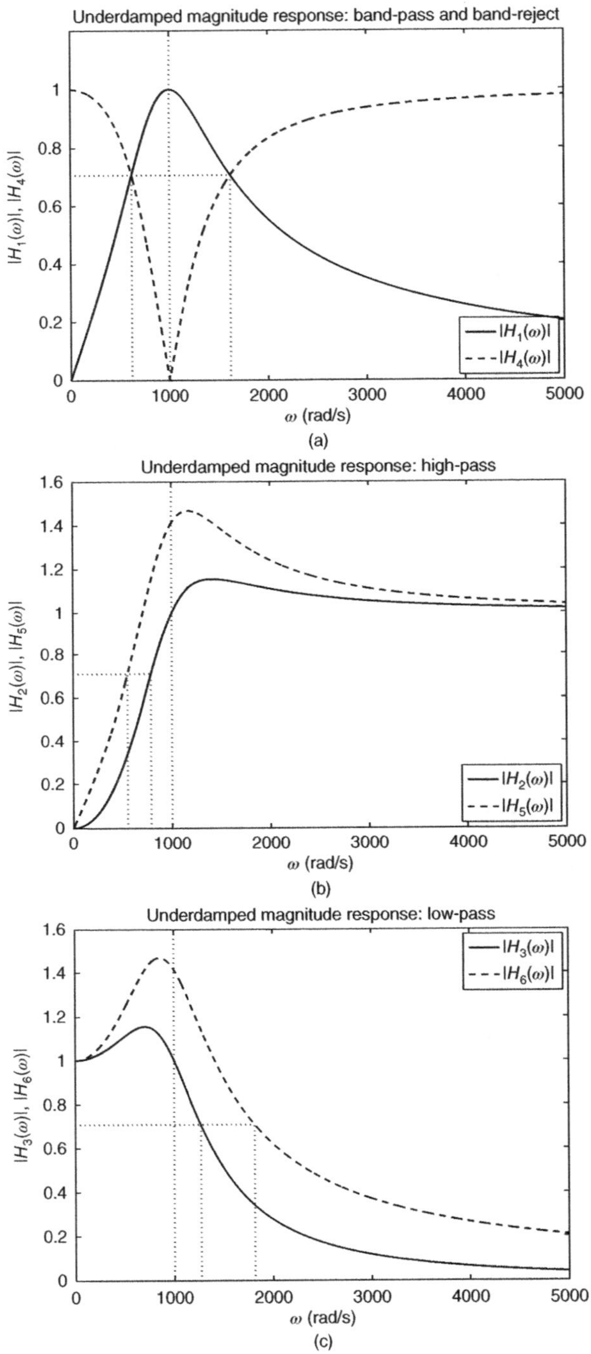

Figure 8.23 Frequency responses of underdamped series RLC circuit in Example 8.15 with $\omega_o = 1000$ rad/s and $\zeta = 0.5$. (a) Band-pass and band-reject filter responses. (b) Two high-pass filter responses. (c) Two low-pass filter responses. The vertical dotted lines show ω_o and the cutoff frequencies, $\{\omega_{c_1}, \omega_{c_2}\}$ or ω_c, for each type of filter.

8.10.1 Low-Pass Filter

Definition: Butterworth Low-Pass Filter A *Butterworth low-pass filter* has the following magnitude response in the frequency domain:

$$|H(\omega)| = \frac{|K|}{\sqrt{1+(\omega/\omega_c)^{2n}}}, \tag{8.207}$$

where K is the DC gain, ω_c is the cutoff frequency, and $n \in \mathcal{N}$ (a natural number).

The cutoff frequency is defined in the usual manner; it is the frequency where the squared magnitude is one-half its maximum value:

$$|H(\omega_c)|^2 = \frac{K^2}{1+(\omega_c/\omega_c)^{2n}} = K^2/2. \tag{8.208}$$

The magnitude response in (8.207) is plotted (in dB) in Figure 8.24 for $K = 1$, $\omega_c = \pi$ rad/s, and three values of n. Observe that all three curves intersect each other at the cutoff frequency where the magnitude is ≈ -3 dB, as expected because for any $n \in \mathcal{N}$:

$$20\log(|H(\omega_c)|) = 20\log(K/\sqrt{2}) = 10\log(2) \approx -3 \text{ dB}, \tag{8.209}$$

with $K = 1$. These plots illustrate that the transition band becomes narrower with increasing n, corresponding to more poles in the denominator of $H(s)$. In fact, in the limit as $n \to \infty$, the squared magnitude response is rectangular:

$$\lim_{n\to\infty} |H(\omega)|^2 = \begin{cases} K^2, & |\omega| < \omega_c \\ K^2/2, & |\omega| = \omega_c \\ 0, & |\omega| > \omega_c. \end{cases} \tag{8.210}$$

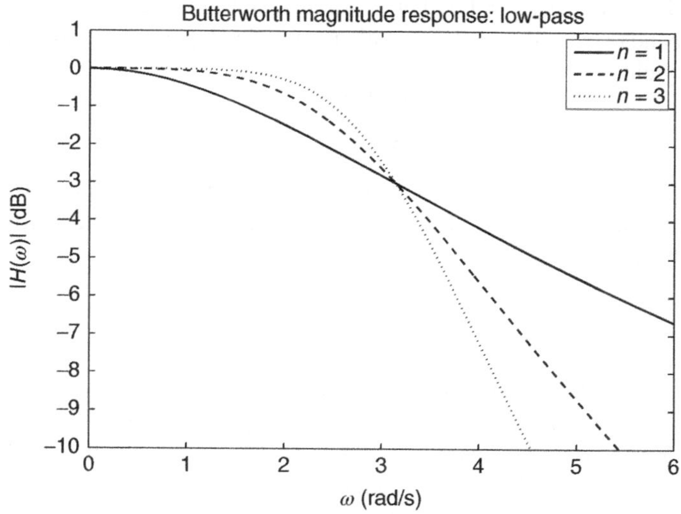

Figure 8.24 Magnitude response of Butterworth low-pass filter.

This result follows by examining the denominator of the square of (8.207): for $|\omega| < \omega_c$, the ratio is $\omega/\omega_c < 1$ and $(\omega/\omega_c)^{2n} \to 0$. Similarly, for $|\omega| > \omega_c$, the ratio is $|\omega/\omega_c| > 1$ and $(\omega/\omega_c)^{2n} \to \infty$. For $|\omega| = \omega_c$, the ratio is a constant $(\omega/\omega_c)^{2n} = 1$ for all n.

It is possible to determine the size of n needed to achieve a specific transition band using the following relationship in the stopband:

$$|H(\omega_{\text{min}})| = \frac{H_{\text{max}}}{\sqrt{1 + (\omega_{\text{min}}/\omega_c)^{2n}}} \leq H_{\text{min}}. \tag{8.211}$$

The two ratios $H_{\text{max}}/H_{\text{min}}$ and $\omega_{\text{min}}/\omega_c$ together specify n, which is derived by squaring (8.211) and taking logarithms:

$$\log((H_{\text{max}}/H_{\text{min}})^2 - 1) \leq n \log((\omega_{\text{min}}/\omega_c)^2). \tag{8.212}$$

The smallest integer value of n satisfying the inequality is chosen:

$$n \geq \frac{\log((H_{\text{max}}/H_{\text{min}})^2 - 1)}{\log((\omega_{\text{min}}/\omega_c)^2)}. \tag{8.213}$$

The base of the logarithm does not matter in this calculation because of the ratio of logarithms. Since H_{max} and ω_c are usually known for a particular problem, we find from this expression that there are two degrees of freedom for choosing n: H_{min} and the corresponding angular frequency ω_{min}. This is evident from Figure 8.24. Consider the curve for $n = 3$ with $H_{\text{max}} = 1$ and $\omega_c = \pi$ rad/s. The transition band could be defined by any pair of values $\{\omega_{\text{min}}, H_{\text{min}}\}$ along the dotted curve; these values depend on the problem specifications as illustrated by the next example.

Example 8.16 Let $H_{\text{max}} = 1$ and $\omega_c = \pi$ rad/s, and suppose we want the end of the transition band to be at $\omega_{\text{min}} = 1.5\omega_c$ with magnitude $H_{\text{min}} = H_{\text{max}}/20$. The condition in (8.213) yields

$$n \geq \frac{\log(400 - 1)}{\log(2.25)} \approx 7.3853, \tag{8.214}$$

from which we choose $n = 8$. The resulting magnitude response is shown in Figure 8.25 (the solid curve). The dotted lines intersect at the specified end of the transition band:

$$\omega_{\text{min}} = 1.5\pi \approx 4.7124 \text{ rad/s}, \quad H_{\text{min}} = 1/20 \Rightarrow \approx -26.02 \text{ dB}. \tag{8.215}$$

The magnitude response curve lies below this point, which, of course, is due to the fact that n must be an integer in (8.213). The inequality in (8.213) ensures that the magnitude response will *meet or exceed* the transition band specification. For convenience, we have also included the frequency response curves for $n = 10$ and $n = 12$ to illustrate how steep the transition band becomes with increasing n.

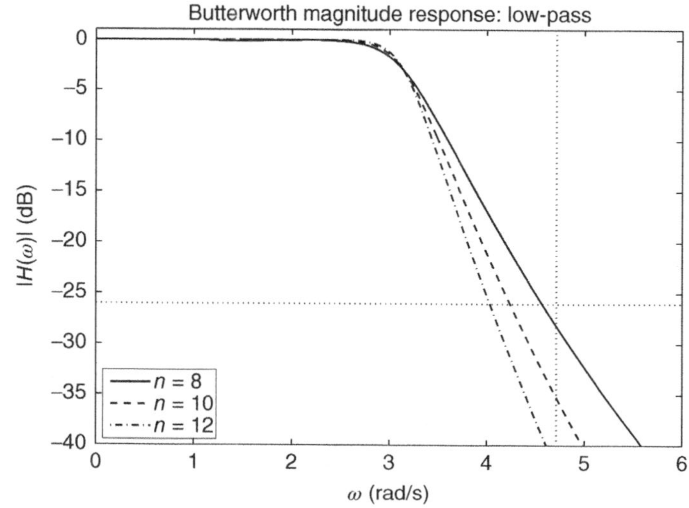

Figure 8.25 Magnitude response of Butterworth low-pass filter. The dotted lines intersect at the desired transition band specification, showing that $n = 8$ is sufficient.

Converting the squared magnitude response $|H(\omega)|^2 = H(\omega)H(-\omega)$ to its s-domain equivalent by substituting $s = j\omega \implies \omega = s/j$, we have the transfer function product

$$H(s)H(-s) = \frac{K^2}{1 + (s/j\omega_c)^{2n}}. \tag{8.216}$$

(Note that $H(s)H(-s)$ is *not* the same as $|H(s)|^2$ used in the summaries of Appendix A.) Furthermore, only $H(s)$ is the transfer function of the system with poles in the left half of the s-plane. The poles of $H(-s)$ are the mirror image of those of $H(s)$ about the imaginary axis, and they are located in the right half of the s-plane. The reason for the form in (8.216) is due to the squared magnitude $|H(\omega)|^2 = H(\omega)H(-\omega)$ with $j\omega$ replaced by s, yielding $H(s)H(-s)$. However, we emphasize that the physical filter is derived only from $H(s)$. The poles of (8.216) are found by solving

$$1 + (s/j\omega_c)^{2n} = 0 \implies s^{2n} = -(j\omega_c)^{2n}. \tag{8.217}$$

In order to continue, we use the fact that $j = \exp(j\pi/2)$ and $-1 = \exp(jm\pi)$ for odd positive integer m. The last expression can be written as $-1 = \exp(j(2k-1)\pi)$ for $k \in \mathcal{N}$. Thus,

$$s^{2n} = \omega_c^{2n} \exp(j(2k-1)\pi)\exp(j\pi 2n/2) = \omega_c^{2n} \exp(j(2k+n-1)\pi), \tag{8.218}$$

and so the $2n$ poles of (8.216) are

$$p_k = \omega_c \exp(j(2k+n-1)\pi/2n), \quad k = 1, \ldots, 2n. \tag{8.219}$$

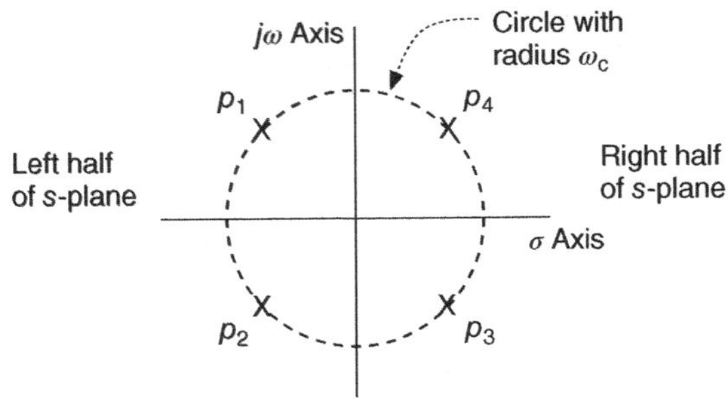

Figure 8.26 Butterworth poles on the s-plane for $H(s)H(-s)$ and $n = 2$.

These are equally spaced on a circle with radius ω_c on the complex plane. For example, when $n = 2$, the poles are

$$p_1 = \omega_c \exp(j3\pi/4) = \omega_c(-1+j)/\sqrt{2}, \quad p_2 = \omega_c \exp(j5\pi/4) = \omega_c(-1-j)/\sqrt{2}, \tag{8.220}$$

$$p_3 = \omega_c \exp(j7\pi/4) = \omega_c(1-j)/\sqrt{2}, \quad p_4 = \omega_c \exp(j9\pi/4) = \omega_c(1+j)/\sqrt{2}, \tag{8.221}$$

which are depicted in Figure 8.26. Poles $\{p_1, p_2\}$ form a complex conjugate pair associated with $H(s)$, and poles $\{p_3, p_4\}$ form a complex conjugate pair for $H(-s)$, which are the mirror image of the other two poles about the imaginary axis. The poles of stable $H(s)$ are necessarily located in the left half of the s-plane, which correspond to $k = 1, \ldots, n$, for the general case in (8.219). Table 8.8 summarizes the denominator polynomials and gives the poles in the left half of the s-plane for orders up to $n = 8$ and with $\omega_c = 1$ (along the unit circle). Thus, $H(s)$ is given by

$$H(s) = \frac{\tilde{K}}{\prod_{k=1}^{n}(s - p_k)}, \tag{8.222}$$

where the denominator is the nth-order polynomial in Table 8.8, which we have written as a product of the poles in the left half of the s-plane. The constant $\tilde{K}$ in the numerator is determined by the desired gain at some frequency, usually $\omega = 0$ for a low-pass filter. For example, if we want unity DC gain, then substituting $s = 0$ in (8.222) yields $\tilde{K} = \prod_{k=1}^{n} |p_k|$, where we have used magnitude because the $\{p_k\}$ are generally complex.

8.10.2 High-Pass Filter

Definition: Butterworth High-Pass Filter A *Butterworth high-pass filter* has the following magnitude response in the frequency domain:

$$|H(\omega)| = \frac{|K|}{\sqrt{1 + (\omega_c/\omega)^{2n}}}, \tag{8.223}$$

where K is the DC gain, ω_c is the cutoff frequency, and $n \in \mathcal{N}$.

TABLE 8.8 Butterworth Low-Pass Filter Poles ($\omega_c = 1$)

Order n	Denominator Polynomial	Poles $\{p_k\}$
1	$s + 1$	-1
2	$s^2 + \sqrt{2}s + 1$	$-0.7071 \pm 0.7071j$
3	$(s + 1)(s^2 + s + 1)$	$-1, -0.5 \pm 0.8660\,j$
4	$(s^2 + 0.7654s + 1)$	$-0.3827 \pm 0.9239j$
	$(s^2 + 1.8478s + 1)$	$-0.9239 \pm 0.3826j$
5	$(s + 1)(s^2 + 0.6180s + 1)$	$-1, -0.3090 \pm 0.9511j$
	$(s^2 + 1.6180s + 1)$	$-0.8090 \pm 0.5878j$
6	$(s^2 + 0.5176s + 1)$	$-0.2588 \pm 0.9659j$
	$(s^2 + \sqrt{2}s + 1)$	$-0.7071 \pm 0.7071j$
	$(s^2 + 1.9319s + 1)$	$-0.9659 \pm 0.2587j$
7	$(s + 1)(s^2 + 0.4450s + 1)$	$-1, -0.2225 \pm 0.9749j$
	$(s^2 + 1.2470 + 1)$	$-0.6235 \pm 0.7818j$
	$(s^2 + 1.8019s + 1)$	$-0.9010 \pm 0.4339j$
8	$(s^2 + 0.3902s + 1)$	$-0.1951 \pm 0.9808j$
	$(s^2 + 1.1111s + 1)$	$-0.5555 \pm 0.8315j$
	$(s^2 + 1.6629s + 1)$	$-0.8314 \pm 0.5556j$
	$(s^2 + 1.9616s + 1)$	$-0.9808 \pm 0.1950j$

This expression has a form identical to (8.207) of the Butterworth low-pass filter, except that ω and ω_c have been interchanged. It is straightforward to show that the bound on n is (see Problem 8.33)

$$n \geq \frac{\log((H_{\max}/H_{\min})^2 - 1)}{\log((\omega_c/\omega_{\min})^2)}, \tag{8.224}$$

where ω_c and $\omega_{\min}$ in (8.213) have been interchanged.

In order to derive the s-domain expression for $H(\omega)H(-\omega)$, we substitute $\omega = s/j$, yielding

$$H(s)H(-s) = \frac{K^2}{1 + (j\omega_c/s)^{2n}}. \tag{8.225}$$

This result is also derived from the low-pass equation in (8.216) via the *transformation* $s/\omega_c \rightarrow \omega_c/s$. Factoring the $(j\omega_c/s)^{2n}$ component, we find that the Butterworth high-pass filter actually has multiple zeros at the origin:

$$H(s)H(-s) = \frac{K^2(s/j\omega_c)^{2n}}{1 + (s/j\omega_c)^{2n}}, \tag{8.226}$$

where the denominator now matches that of the Butterworth low-pass filter. Thus, (8.226) has $2n$ poles equally spaced about a circle with radius ω_c on the s-plane, just like the Butterworth low-pass filter. But it also has $2n$ zeros at the origin, which convert the low-pass response to a high-pass response. The transfer function of the Butterworth high-pass filter is obtained from (8.222) by including s^n in the numerator:

$$H(s) = \frac{\tilde{K}s^n}{\prod_{k=1}^{n}(s - p_k)}, \tag{8.227}$$

where $\tilde{K}$ is chosen to have some desired gain at a particular frequency, usually at $\omega \to \infty$ for a high-pass filter. We can verify that this transfer function has the response of a high-pass filter by noting that $H(0) = 0$ and $\lim_{s\to\infty} H(s) = \tilde{K}$, similar to the results found for the second-order transfer function in (8.165).

Example 8.17 Suppose we want to design a Butterworth high-pass filter with the same magnitude specifications used for the Butterworth low-pass filter in Example 8.16: $H_{\max} = 1$ and $H_{\min} = H_{\max}/20$, but with $\omega_c = 1.5\pi$ rad/s and $\omega_{\min} = \pi$ rad/s (these are reversed compared with the low-pass specifications). The order of the high-pass filter from (8.224) is

$$n \geq \frac{\log(400 - 1)}{\log(2.25)} \approx 7.3853 \implies n = 8, \tag{8.228}$$

which is necessarily the same result as that of the low-pass filter because the width of the transition band is the same $\omega_c - \omega_{\min} = 0.5\pi$. Figure 8.27 shows the resulting magnitude response (the solid curve). The dotted lines intersect at the end of the transition band with specifications

$$\omega_{\min} = \pi \approx 3.1416 \text{ rad/s}, \qquad H_{\min} = 1/20 \approx -26.02 \text{ dB}. \tag{8.229}$$

The results for $n = 10$ and 12 are also shown.

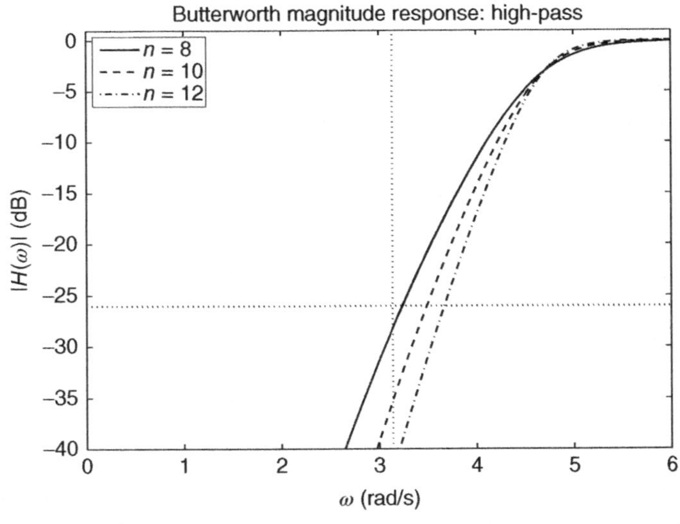

Figure 8.27 Magnitude response of Butterworth high-pass filter. The dotted lines intersect at the desired transition band specification, showing that $n = 8$ is sufficient.

8.10.3 Band-Pass Filter

A Butterworth low-pass filter can be transformed into a band-pass filter by substituting $\omega_c = 1$ and replacing s in (8.216) with

$$s \longrightarrow \frac{s^2 + \omega_{c_1}\omega_{c_2}}{s(\omega_{c_2} - \omega_{c_1})}, \tag{8.230}$$

where $\{\omega_{c_1}, \omega_{c_2}\}$ are the lower and upper cutoff frequencies of the band-pass filter. Because of the specific pole structure of the Butterworth low-pass filter, only these two frequencies need to be specified. The center frequency and other features of the frequency response are determined from the resulting denominator polynomial. In this section, however, we do not consider this nonlinear mapping any further and instead focus on a combination of the previous low-pass and high-pass filters. Problem 8.36 considers an example of the transformation in (8.230) starting with the low-pass filter $H_{\text{LP}}(s) = 1/(s+1)$.

A band-pass filter with the Butterworth filter characteristic (maximally flat in the passbands) is also achieved by placing a low-pass filter in *cascade* with a high-pass filter, as depicted in Figure 8.28. Since the intermediate output is $Y_{\text{LP}}(\omega) = H_{\text{LP}}(\omega)X(\omega)$ and the overall output is $Y(\omega) = H_{\text{HP}}(\omega)Y_{\text{LP}}(\omega)$, it is clear that the band-pass transfer function is the product:

$$H_{\text{BP}}(\omega) = H_{\text{LP}}(\omega)H_{\text{HP}}(\omega) = H_{\text{HP}}(\omega)H_{\text{LP}}(\omega), \tag{8.231}$$

which, of course, is commutative. In order for the product to function properly as a band-pass filter, we see from Figures 8.25 and 8.27 that the magnitude responses must *overlap* to some extent in the two transition bands. If the cutoff frequencies of the low-pass and high-pass filters are denoted by ω_{c_L} and ω_{c_H}, respectively, then we must have $\omega_{c_L} > \omega_{c_H}$ for overlapping transition bands. Otherwise, the stopband of the low-pass filter will reject frequencies passed by the high-pass filter, and similarly, the stopband of the high-pass filter will reject frequencies passed by the low-pass filter. In order to prevent this, the two cascaded Butterworth filters should have cutoff frequencies that satisfy

$$\omega_{c_L} \geq 2\omega_{c_H}, \tag{8.232}$$

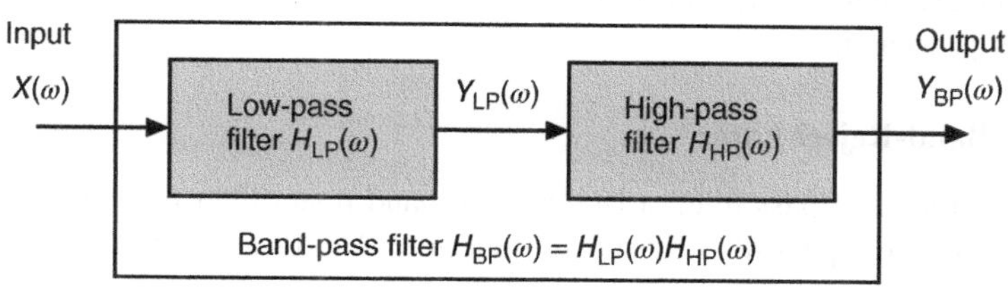

Figure 8.28 Band-pass filter implemented as a cascade combination of low-pass and high-pass filters.

as demonstrated in the next example. A cascade band-pass filter with the property in (8.232) is called a *broadband* filter. It is shown later for a specific example that if this condition is not satisfied then the low-pass and high-pass filter responses do not have much overlap and the overall passband is relatively narrow. Moreover, the gain at the center frequency ω_o is no longer unity, though this could be modified by a follow-on gain circuit.

Example 8.18 In this example, we implement a band-pass filter with center frequency $\omega_o = 4$ rad/s and a bandwidth of $BW = \omega_{c_2} - \omega_{c_1} = 4$ rad/s. This is achieved by choosing $\omega_{c_H} = 2$ rad/s and $\omega_{c_L} = 6$ rad/s, which satisfy the condition in (8.232) because $\omega_{c_L} = 3\omega_{c_H}$. Using the low-pass formula in (8.213) for n, let $H_{\max} = 1$, $H_{\min} = 0.1$, and $\omega_{\min,L} = \omega_{c_L} + 1 = 7$ rad/s. Thus, $20\log(1/0.1) = -20$ dB and

$$n \geq \log(100 - 1)/\log((7/6)^2) \approx 14.9046 \implies n = 15. \tag{8.233}$$

For the high-pass filter, we choose similar parameters: $H_{\max} = 1$, $H_{\min} = 0.1$, and $\omega_{\min,H} = \omega_{c_H} - 1 = 1$ rad/s such that $20\log(1/0.1) = -20$ dB and (8.224) gives

$$n \geq \log(100 - 1)/\log((2/1)^2) \approx 3.3147 \implies n = 4. \tag{8.234}$$

The magnitude responses for the low-pass and high-pass filters are shown in Figure 8.29(a) and (b), respectively, where we see they meet their individual specifications. The overall band-pass response generated as the product of the low-pass and high-pass responses is provided in Figure 8.29(c).

Example 8.19 Suppose now that we modify the cutoff frequencies to be $\omega_{c_L} = 4.5$ rad/s and $\omega_{c_H} = 3.5$ rad/s, which do not satisfy (8.232). Let the center frequency and values for $\{H_{\min}, H_{\max}\}$ remain unchanged. Assuming that the two values for $\omega_{\min}$ are again 1 rad/s away from the cutoff frequencies, giving $4.5 + 1 = 5.5$ rad/s and $3.5 - 1 = 2.5$ rad/s, the order n is 12 and 7, respectively, for the low-pass and high-pass filters. The resulting frequency response of the band-pass filter is shown in Figure 8.30, which we see is not as broadband as the response in Figure 8.29(c), and its magnitude is slightly lower at the center frequency. This occurs because the low-pass and high-pass frequency responses have less overlap, which for the cascade structure reduces the gain of the frequency components around ω_o, creating a narrower passband.

8.10.4 Band-Reject Filter

Similar to the band-pass filter, a Butterworth band-reject filter can be derived from a Butterworth low-pass filter by substituting $\omega_c = 1$ rad/s and replacing s in (8.216) with

$$s \longrightarrow \frac{s(\omega_{c_2} - \omega_{c_1})}{s^2 + \omega_{c_1}\omega_{c_2}}, \tag{8.235}$$

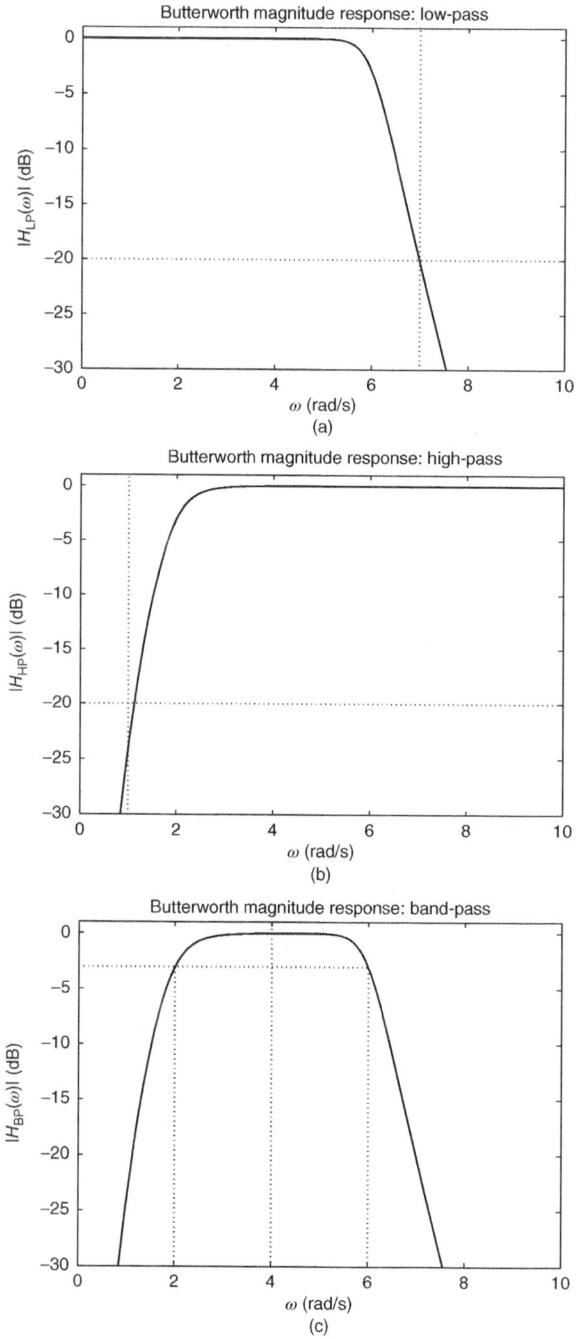

Figure 8.29 Band-pass filter implemented as a cascade combination of Butterworth low-pass and high-pass filters. (a) Low-pass response with $\omega_{c_L} = 6$ rad/s ($n = 15$). (b) High-pass response with $\omega_{c_H} = 2$ rad/s ($n = 4$). The dotted lines in (a) and (b) intersect at the specifications for $H_{\min}$ and $\omega_{\min}$. (c) Band-pass response with $\omega_o = 4$ rad/s and bandwidth $BW = 4$ rad/s. The vertical dotted lines in (c) denote ω_o and $\{\omega_{c_1}, \omega_{c_2}\}$.

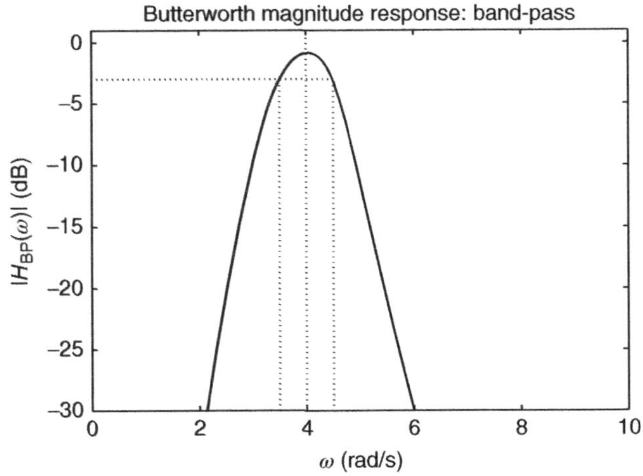

Figure 8.30 Band-pass filter frequency response of Example 8.19. The vertical dotted lines denote ω_o and $\{\omega_{c_1}, \omega_{c_2}\}$.

where $\{\omega_{c_1}, \omega_{c_2}\}$ are the lower and upper cutoff frequencies of the band-reject filter. This is the inverse of the transformation in (8.230) used to generate a band-pass Butterworth filter. As in the previous section, we do not consider this mapping approach any further, and instead focus on another combination of the low-pass and high-pass filters. Problem 8.37 examines the band-reject transformation starting with the low-pass filter $H_{\text{LP}}(s) = 1/(s+1)$.

For a band-reject filter, the goal is to attenuate a narrow band of frequencies while retaining relatively flat passbands above and below the rejected frequencies. This cannot be achieved using the cascade structure in Figure 8.29 because the stopband of the low-pass filter removes high frequencies, and the stopband of the high-pass filter removes low frequencies. Instead, we use the *parallel* implementation shown in Figure 8.31 where the filter outputs are added together:

$$Y_{\text{BR}}(\omega) = Y_{\text{LP}}(\omega) + Y_{\text{HP}}(\omega), \tag{8.236}$$

which has transfer function

$$H_{\text{BR}}(\omega) = H_{\text{LP}}(\omega) + H_{\text{HP}}(\omega). \tag{8.237}$$

Since the low-pass filter allows low frequencies to pass, and the high-pass filter allows high frequencies to pass, it is possible to attenuate a band of frequencies between these two passbands by judiciously aligning their stopbands. This is illustrated in the next example.

Example 8.20 As in Example 8.18, let the center frequency of the band-reject filter be $\omega_o = 4$ rad/s and the bandwidth be $BW = 4$ rad/s. This means the cutoff frequency of the low-pass filter is $\omega_{c_L} = 2$ rad/s and that of the high-pass filter is $\omega_{c_H} = 6$ rad/s.

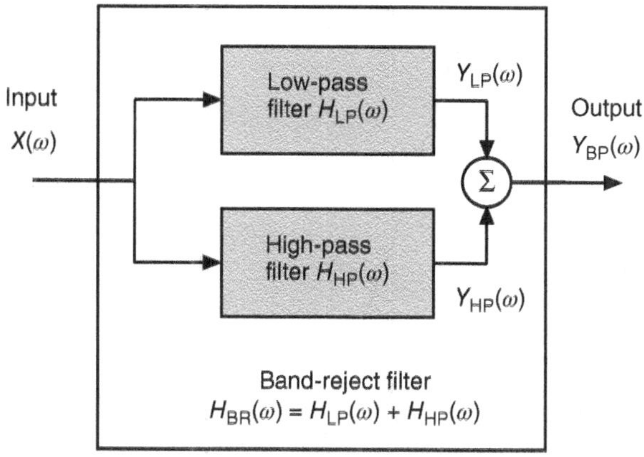

Figure 8.31 Band-reject filter implemented as a parallel combination of low-pass and high-pass filters.

Also as in Example 8.18, let the frequency $\omega_{\min}$ for each component filter be 1 rad/s from the cutoff frequency with $H_{\min} = 0.1$. Thus, (8.213) for the low-pass filter yields

$$n \geq \log(100-1)/\log((3/2)^2) \approx 5.6665 \implies n = 6, \tag{8.238}$$

and (8.224) for the high-pass filter gives

$$n \geq \log(100-1)/\log((6/5)^2) \approx 12.6017 \implies n = 13. \tag{8.239}$$

The results are shown in Figure 8.32. Steeper transition bands are achieved by using higher order low-pass and high-pass filters in (8.237) (see Problem 8.40).

Example 8.21 Next, we modify the cutoff frequencies to be $\{\omega_{c_L} = 3, \omega_{c_H} = 5\}$ rad/s, which just barely satisfy the broadband condition in (8.232). The center frequency and $\{H_{\min}, H_{\max}\}$ remain unchanged for the component low-pass and high-pass filters. If the values for $\omega_{\min}$ are 1 rad/s away from the cutoff frequencies, we have $3 + 1 = 4$ rad/s and $5 - 1 = 4$ rad/s, which yield $n = 8$ and 11, respectively, for the low-pass and high-pass filters. The resulting frequency response of the band-reject filter is shown in Figure 8.33, which is narrower and not as deep as the response in Figure 8.32(c). This occurs because the low-pass and high-pass frequency responses have greater overlap, which for the parallel structure allows more frequency components to have a higher gain, and so the degree of rejection is less.

Although we used only Butterworth filters to demonstrate how to implement band-pass and band-reject filters with cascade and parallel architectures, respectively, these implementations can be used for any type of filter such as the Chebyshev and elliptic filters mentioned earlier.

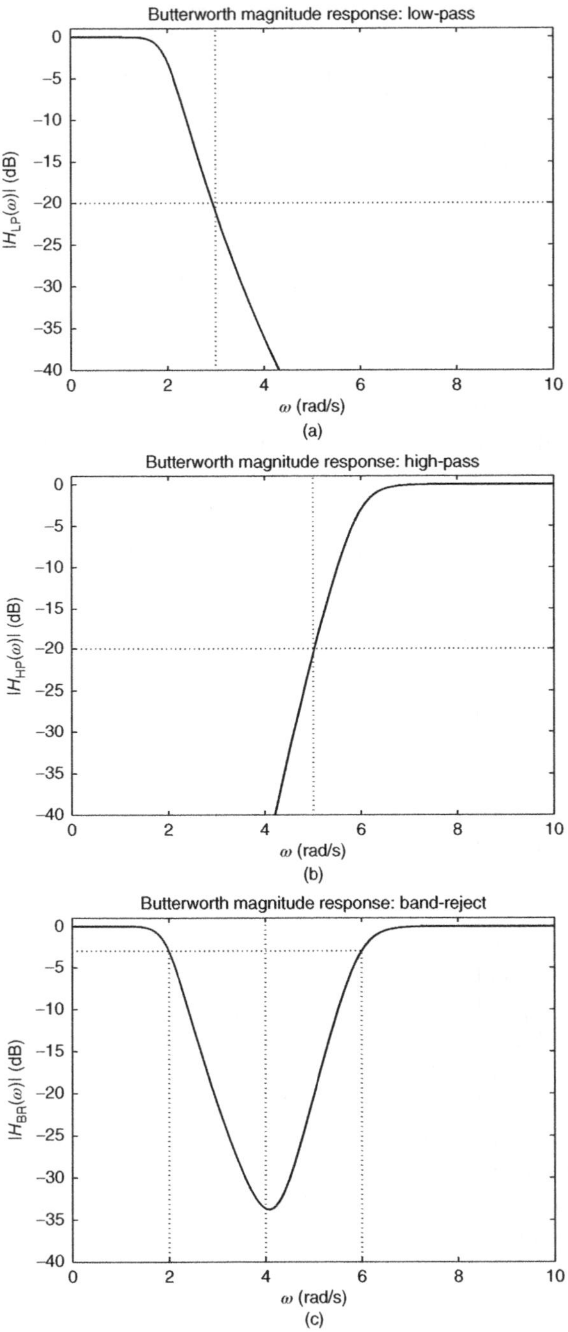

Figure 8.32 Band-reject filter implemented as a parallel combination of Butterworth low-pass and high-pass filters. (a) Low-pass response with $\omega_{c_L} = 2$ rad/s ($n = 6$). (b) High-pass response with $\omega_{c_H} = 6$ rad/s ($n = 13$). The dotted lines in (a) and (b) intersect at the specifications for $H_{\min}$ and $\omega_{\min}$. (c) Band-reject response with $\omega_o = 4$ rad/s and bandwidth $BW = 4$ rad/s. The vertical dotted lines in (c) denote ω_o and $\{\omega_{c_1}, \omega_{c_2}\}$.

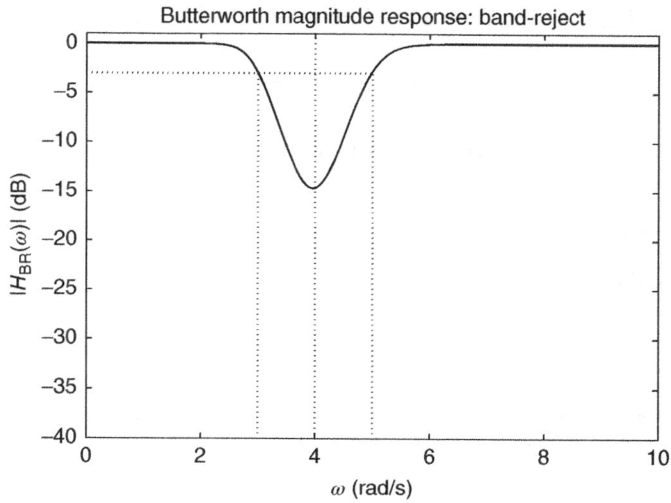

Figure 8.33 Band-reject filter frequency response of Example 8.21. The vertical dotted lines denote ω_o and $\{\omega_{c_1}, \omega_{c_2}\}$.

PROBLEMS

Fourier Transform

8.1 Determine which of the following functions are absolutely integrable. (a) $x_1(t) = \exp(t^2)u(-t)$. (b) $x_2(t) = \sin(\omega_o/t)[u(t-1) - u(t-2)]$. (c) $x_3(t) = 1/(1+t^2)$.

8.2 Determine if any of the functions in the previous problem are square integrable:

$$\int_{-\infty}^{\infty} |x(t)|^2 dt < \infty. \tag{8.240}$$

8.3 Find the Fourier transform of $u(t_o - t)$ for any $t_o \in \mathcal{R}$.

8.4 The Fourier transform of $x(t) = \exp(\alpha t)$ does not exist for $\alpha > 0$. Find the Fourier transform for the time-limited function $y(t) = x(t)[u(t+T) - u(t-T)]$ with $T > 0$.

8.5 Suppose the Fourier transform is defined as

$$X(\omega) \triangleq \frac{1}{\sqrt{2\pi}} \int_{-\infty}^{\infty} x(t) \exp(-j\omega t) dt. \tag{8.241}$$

Derive the corresponding Fourier transform inversion formula.

8.6 Derive the inversion formula for the Fourier cosine transform:

$$X_c(\omega) = \int_{-\infty}^{\infty} x(t) \cos(\omega t) dt. \tag{8.242}$$

8.7 Use the Laplace transform and a PFE to find the inverse Fourier transform of

$$X(\omega) = \frac{5}{2 - \omega^2 + 2j\omega}, \tag{8.243}$$

by first rewriting the denominator in terms of $s = j\omega$.

Magnitude and Phase

8.8 Derive the magnitude and phase for

$$X(\omega) = \frac{\alpha + j\omega}{(\alpha + j\omega)^2 + \beta^2}. \tag{8.244}$$

8.9 Find the magnitude and phase for (a) $Y_1(\omega) = X_1(\omega)X_2(\omega)$ and (b) $Y_2(\omega) = X_1(\omega) + X_2(\omega)$ where

$$X_1(\omega) = \frac{2}{j\omega}, \qquad X_2(\omega) = \exp(-j\alpha\omega). \tag{8.245}$$

8.10 The Hilbert transform of $x(t)$ in the Fourier transform domain is

$$Y(\omega) = -j\text{sgn}(\omega)X(\omega). \tag{8.246}$$

Show how the magnitude and phase for $X_1(\omega)$ and $X_2(\omega)$ in the previous problem are altered by the Hilbert transform.

8.11 Derive the following phase of $X(\omega)$ for the rectangle function in Appendix A:

$$\theta(\omega) = \pi\text{sgn}(\omega) \sum_{n=1}^{\infty} \text{rect}([|\omega| - (4n-1)\pi]/2\pi). \tag{8.247}$$

8.12 (a) Derive the following magnitude of the Laplace transform for the rectangle function in Appendix A:

$$|X(s)| = \frac{2\sqrt{\cosh^2(\sigma/2)\cos^2(\omega/2) + \sinh^2(\sigma/2)\sin^2(\omega/2)}}{\sqrt{\sigma^2 + \omega^2}}, \tag{8.248}$$

and (b) show that it reduces to $|X(\omega)| = |\text{sinc}(\omega/2\pi)|$ on the imaginary axis.

Fourier Transforms and Properties

8.13 Prove the duality property in (8.78).

8.14 Derive the Fourier transform for $x(t) = 1/t^2$ given in Table 8.3.

8.15 Repeat the previous problem for $x(t) = t^n u(t)$.

8.16 Find the Fourier transform for each of the following functions:

$$\text{(a) } x_1(t) = \frac{2}{\alpha^2 + t^2}, \quad \text{(b) } x_2(t) = \sin(t)/t. \qquad (8.249)$$

8.17 Repeat the previous problem for

$$\text{(a) } x_1(t) = \exp(t-1)u(-t-1), \quad \text{(b) } x_2(t) = \exp(-|t|)\text{rect}(t/2). \qquad (8.250)$$

8.18 Derive the Fourier transforms in Table 8.4 for (a) $\cos^2(\omega_o t)$ and (b) $\sin^2(\omega_o t)$ using the product property.

8.19 (a) Find $|X(\omega)|$ and $\theta(\omega)$ for $X(s) = 3/s(s+2)$. (b) Find $H(s)$ from $H(\omega) = 4/(1+j\omega)(2-\omega^2)$.

8.20 Assuming $X(\omega) = \exp(-\omega^2)$, (a) find the Fourier transform of

$$y(t) = 2x(t-1) + 4\frac{d}{dt}x(t) - 3tx(t-2), \qquad (8.251)$$

and (b) verify your result by finding $x(t)$.

8.21 Find the energy in the frequency band $\omega \in [-2\pi, 2\pi]$ for the standard rectangle function in Appendix A.

Amplitude Modulation

8.22 Suppose the carrier waveform $c(t) = \sin(\omega_o t)$ is modulated by a message signal $x(t)$ with the rectangular spectrum in Figure 8.5(b). Give an expression for the modulator output $Y(\omega)$ for (a) AM with suppressed carrier and (b) conventional AM. Sketch plots similar to those in Figure 8.5.

8.23 Derive the modulation property in (8.117) for the Fourier transform based on the natural frequency f.

8.24 At a receiver, the transmitted signal $x(t)$ in (8.110) with a cosine carrier is multiplied by $r(t) = \cos(\omega_o t)$. (a) Show how it is possible to recover the message signal $x(t)$ using this approach followed by a low-pass filter. (b) Suppose instead that $r(t) = \cos(\omega_o t + \phi)$ where ϕ is a nonzero fixed phase shift. Determine if the message signal can be recovered using the approach in part (a).

8.25 Quadrature amplitude modulation (QAM) has the following transmitted signal $y(t) = x_1(t)\cos(\omega_o t) + x_2(t)\sin(\omega_o t)$ where $\{x_1(t), x_2(t)\}$ are two message signals that may or may not be independent. Let $x_1(t)$ have a rectangular spectrum and suppose $x_2(t)$ is generated by filtering $x_1(t)$ with the Hilbert transform filter $H(\omega) = -j\text{sgn}(\omega)$. Find and sketch the resulting spectrum $Y(\omega)$ for this SSB modulation.

Frequency Response

8.26 For the first-order RC circuit in Figure 8.34, find transfer functions from the voltage source to the voltage across (a) the horizontal resistor R and then across (b) the capacitor C. Describe the type of frequency response for each case and find the cutoff frequencies.

8.27 Repeat the previous problem with the capacitor C replaced by inductor L.

8.28 Derive the expressions in (8.151) and (8.154) for the resonant frequency ω_o and the cutoff frequency ω_c of the second-order band-pass filter in (8.148).

8.29 Derive the range of values for the proportionality constants weighting ω_o in (8.176), (8.179), and (8.184) that specify the filter cutoff frequencies for (a) underdamped and (b) overdamped systems.

Frequency Response of RLC Circuit

8.30 For the series RLC circuit, let $R = 1000\ \Omega$ and $L = 1$ H. Determine the range of values for C to have (a) an underdamped circuit and (b) an overdamped circuit. In each case, give the range of values for the resonant frequency ω_o.

8.31 Derive the cutoff frequency in (8.204) for the low-pass filter $H_5(s)$ in (8.198) of the series RLC circuit.

8.32 (a) Find the transfer function from the voltage source V_s to the voltage across the inductor for the RLC circuit in Figure 8.35. (b) Derive an expression for the cutoff frequencies and specify the type of frequency response.

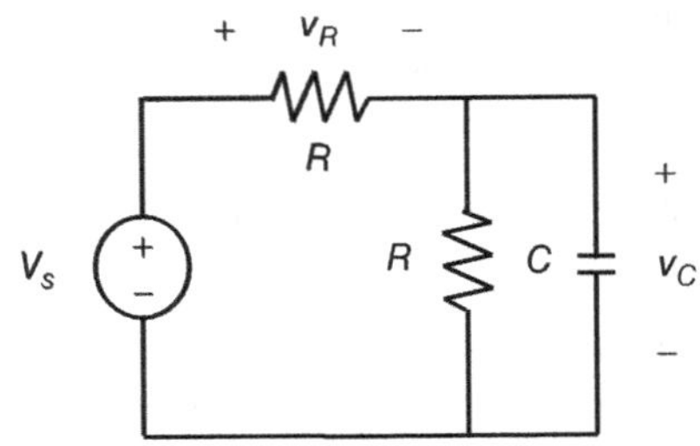

Figure 8.34 First-order RC circuit with resistor R and capacitor C.

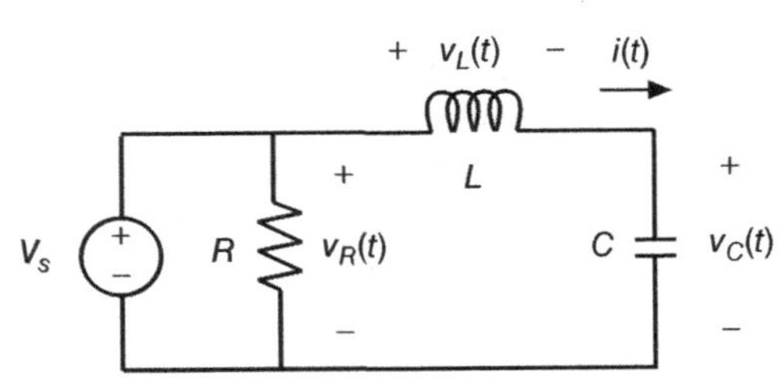

Figure 8.35 Second-order RLC circuit with resistor R, inductor L, and capacitor C.

Butterworth Filters

8.33 Derive the bound on n in (8.224) for the Butterworth high-pass filter.

8.34 Determine the size n of a low-pass Butterworth filter with $H_{\max} = 1$ and cutoff frequency $\omega_c = \pi$ rad/s for each of the following specifications. (a) $\omega_{\min} = 1.2\pi$ rad/s and $H_{\min} = H_{\max}/20$. (b) $\omega_{\min} = 1.4\pi$ rad/s and $H_{\min} = H_{\max}/30$.

8.35 Determine the size n of a high-pass Butterworth filter with $H_{\max} = 1$ and cutoff frequency $\omega_c = 3\pi$ rad/s for each of the following specifications. (a) $\omega_{\min} = 2\pi$ rad/s and $H_{\min} = H_{\max}/20$. (b) $\omega_{\min} = \pi$ rad/s and $H_{\min} = H_{\max}/30$.

8.36 Design a Butterworth band-pass filter using the transformation in (8.230), starting with the first-order Butterworth low-pass filter $H_{\text{LP}}(s) = 1/(s+1)$. The cut-off frequencies are $\omega_{c_1} = 800$ rad/s and $\omega_{c_2} = 1200$ rad/s. Specify the resonant frequency ω_o and the type of damping.

8.37 Repeat the previous problem using the transformation in (8.235) for the Butterworth band-reject filter.

8.38 Design a Butterworth band-pass filter using a cascade of low-pass and high-pass filters with the following specifications: $\omega_o = 2000$ rad/s, $\omega_{c_L} = 2200$ rad/s, $\omega_{c_H} = 1800$ rad/s, $\omega_{\min,L} = 2300$ rad/s, and $\omega_{\min,H} = 1700$ rad/s. Let $H_{\max} = 1$ and $H_{\min} = 0.1$ for the low-pass and high-pass filters.

Computer Problems

8.39 The MATLAB command freqs(**b**, **a**) plots the magnitude and phase of a system given its transfer function coefficients:

$$H(s) = \frac{b_M s^M + b_{M-1} s^{M-1} + \cdots + b_1 s + b_0}{a_N s^N + a_{N-1} s^{N-1} + \cdots + a_1 s + a_0}. \tag{8.252}$$

The vectors contain the coefficients in reverse order: $\mathbf{b} = [b_M, \ldots, b_0]^T$ and $\mathbf{a} = [a_N, \ldots, a_0]^T$. The angular frequency and the magnitude axes are logarithmic, and the phase axis is in degrees. Use freqs to plot the frequency response for the following second-order systems:

$$\text{(a) } H_1(s) = \frac{4}{s^2 + 5s + 4}, \quad \text{(b) } H_2(s) = \frac{s^2 + 4}{s^2 + 2s + 2}. \tag{8.253}$$

8.40 The MATLAB command $[\mathbf{z}, \mathbf{p}, k] =$ butter($n, 2\pi f$, 'ftype','s') provides the zeros **z**, poles **p**, and gain k for a Butterworth filter given the order n and the cutoff frequency f in Hz. The argument "ftype" specifies the type of filter: "low," "high," "band-pass," or "stop." The command $[\mathbf{b}, \mathbf{a}] =$ zp2tf($\mathbf{z}, \mathbf{p}, k$) converts the zeros and poles into the transfer function coefficients in reverse order (as defined in the vectors following (8.252)). Repeat the band-reject filter design in Example 8.20 that is a parallel combination of low-pass and high-pass Butterworth filters. Use butter to design higher order filters so that the transition bands are steeper than those in Figure 8.32(c). The command sys $=$ tf($\mathbf{b}, \mathbf{a}$) creates a transfer function representation based on the numerator and denominator coefficients. Once these are generated for the low-pass and high-pass filters, denoted by sysL and sysH, their parallel combination is produced as sysP $=$ parallel(sysL, sysH). The numerator and denominator coefficients are derived from $[\mathbf{b}, \mathbf{a}] =$ tfdata(sysP,'v'), and these are used in freqs to generate plots of the magnitude and phase for the band-reject filter. The argument 'v' returns the numerator and denominator coefficients as vectors (instead of as cell arrays).

APPENDICES

INTRODUCTION TO APPENDICES

In the following appendices, some background material is included to supplement the topics covered in the chapters.

- **Appendix A**: Additional properties of the Laplace transform and the Fourier transform are discussed. Extensive summaries of several functions and their transforms are provided for ease of reference. The summaries are organized as follows: impulsive functions, piecewise linear functions (such as the unit step and ramp functions), exponential functions, and sinusoidal functions. One-sided and two-sided functions are included, and some of the exponential and sinusoidal functions are weighted by the ramp function.
- **Appendix B**: Two tables of inverse Laplace transforms are provided where the transforms are given first, some with multiple poles, so that the time-domain function can be found without performing a partial fraction expansion. There are also discussions of an improper rational Laplace transform, an unbounded system, and a double integrator with feedback.
- **Appendix C**: Several identities, derivatives, and integrals are summarized. Additional topics include completing the square, quadratic and cubic formulas, and closed-form expressions for infinite and finite summations.
- **Appendix D**: This appendix gives a brief review of set theory. Properties of set operations are summarized, and Venn diagrams are included to describe some of the properties.
- **Appendix E**: Series expansions and different types of singularities are covered. These include the Taylor series, the Maclaurin series, and the Laurent series for complex functions.
- **Appendix F**: The final appendix discusses the Lambert W-function, which can be used to write explicit expressions for the solutions of nonlinear equations. It includes examples of a nonlinear diode circuit and a nonlinear system of equations.

APPENDIX A

EXTENDED SUMMARIES OF FUNCTIONS AND TRANSFORMS

In this appendix, we summarize several functions used in the book and provide expressions for their Fourier transforms and Laplace transforms.

A.1 FUNCTIONS AND NOTATION

The following notations are used for time-domain functions and frequency-domain transforms:

$x(t)$ general function of time,
$X(s)$ Laplace transform of $x(t)$,
$X(\omega)$ Fourier transform of $x(t)$ (angular frequency),
$X(f)$ Fourier transform of $x(t)$ (natural frequency).

Independent variables:

t continuous time (s),
f natural frequency (Hz),
s complex variable (of Laplace transform),
σ real part of s,
ω imaginary part of s, angular frequency (rad/s).

Mathematical Foundations for Linear Circuits and Systems in Engineering, First Edition. John J. Shynk.

Companion Website: http://www.wiley.com/go/linearcircuitsandsystems

Basic functions:

$\delta(t)$	Dirac delta,
$u(t)$	unit step,
$\cos(\omega_o t)$	cosine,
$\sin(\omega_o t)$	sine,
$\exp(-\alpha t)u(t)$	exponential,
$r(t) = tu(t)$	ramp.

Parameters:

ω_o	specific angular frequency used in sine and cosine,
f_o	specific natural frequency used in sine and cosine,
T_o	period of sine and cosine,
α	exponent of decaying exponential,
E	energy,
P	power.

Some combinations of these functions are the solutions of linear ODEs with constant coefficients. Examples include the exponentially weighted cosine function $\exp(-\alpha t)\cos(\omega_o t)u(t)$ and the ramped and exponentially weighted sine function $t\exp(-\alpha t)\sin(\omega_o t)u(t)$. We also consider some two-sided functions such as the Gaussian function $\exp(-\alpha t^2)$.

A.2 LAPLACE TRANSFORM

The Laplace transform is derived from the following improper integral:

$$X(s) = \int_{-\infty}^{\infty} x(t)\exp(-st)dt, \tag{A.1}$$

which has a region of convergence (ROC) on the complex plane of the form $\text{Re}(s) = \sigma > a$ for right-sided functions and $a < \sigma < b$ for two-sided functions (we do not explicitly consider left-sided functions with ROC $\sigma < b$, though of course they are part of two-sided functions). If the ROC includes the imaginary axis ($s = j\omega$), then the function is bounded (stable); otherwise it may be unbounded. For example, the ROC of the ramp function $r(t) = tu(t)$ is $\sigma > 0$, and clearly the function grows without bound. For the signals and systems considered in this book, the Laplace transform generally is the ratio of two polynomials (a rational function):

$$X(s) = \frac{N(s)}{D(s)} = \frac{\prod_{m=0}^{M-1}(s - z_m)}{\prod_{n=0}^{N-1}(s - p_n)}. \tag{A.2}$$

The roots $\{z_m\}$ of the numerator polynomial $N(s)$ are called zeros, and $\{p_n\}$ of the denominator polynomial $D(s)$ are called poles. The poles largely determine the time-domain properties of a function; the reader will observe the following trends in the summaries:

- Functions with $\sin(\omega_o t)$ or $\cos(\omega_o t)$ have complex conjugate poles with imaginary parts $\pm j\omega_o$.
- Functions with $\exp(-\alpha t)u(t)$ have poles with real part $-\alpha$.
- Functions with $tu(t)$ have repeated poles.

The summaries specify the s-plane locations for finite poles and zeros; poles at infinity are not considered. For example, the Laplace transform $X(s) = s$ has a zero at $s = 0$, which could be interpreted as a pole at $s = \infty$. Similarly, $X(s) = 1/s$ has a pole at $s = 0$, which could also be viewed as a zero at $s = \infty$.

Plots of $|X(s)|$ are shown on the s-plane for the various functions. The magnitude is derived by substituting $s = \sigma + j\omega$ and finding the real and imaginary parts of the complex-valued function. For example, the Laplace transform of $x(t) = \cos(\omega_o t)u(t)$ is

$$X(s) = \frac{s}{s^2 + \omega_o^2} = \frac{\sigma + j\omega}{(\sigma + j\omega)^2 + \omega_o^2}, \tag{A.3}$$

from which we have

$$|X(s)| = \frac{\sqrt{\sigma^2 + \omega^2}}{\sqrt{[\sigma^2 + (\omega + \omega_o)^2][\sigma^2 + (\omega - \omega_o)^2]}}. \tag{A.4}$$

Note that $X(s)$ exists only in the ROC, which for (A.3) is $\sigma > 0$. However, the magnitude $|X(s)|$ is plotted on the entire s-plane so that the poles and zeros can be seen, even though the ROC does not include any poles.

There are some Laplace transforms in this appendix whose ROC is the line defined by $\sigma = 0$, but excluding $s = 0$ (the origin on the s-plane). This is demonstrated for the signum function, which we model using two exponential functions:

$$\text{sgn}(t) = \lim_{\alpha \to 0}[\exp(-\alpha t)u(t) - \exp(\alpha t)u(-t)]. \tag{A.5}$$

The function in brackets is shown in Figure A.1 for two nonzero values of α. The Laplace transform of (A.5) is

$$\begin{aligned} X(s) &= -\int_{-\infty}^{0} \exp(\alpha t)\exp(-st)dt + \int_{0}^{\infty} \exp(-\alpha t)\exp(-st)dt \\ &= \int_{\infty}^{0} \exp((s-\alpha)t)dt + \int_{0}^{\infty} \exp(-(s+\alpha)t)dt \\ &= \frac{1}{s-\alpha} + \frac{1}{s+\alpha} = \frac{2s}{s^2 - \alpha^2}, \end{aligned} \tag{A.6}$$

with ROC given by the strip $-\alpha < \sigma < \alpha$. In the limit as $\alpha \longrightarrow 0$, the Laplace transform is

$$\mathcal{L}(\text{sgn}(t)) = \frac{2}{s}, \tag{A.7}$$

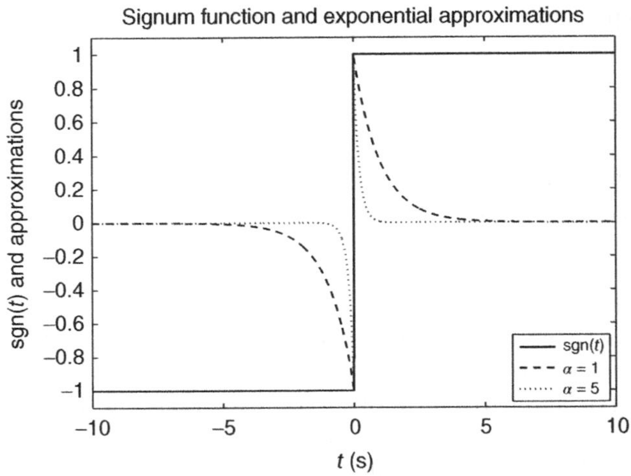

Figure A.1 Signum function and exponential function approximation in (A.5).

whose ROC is $\sigma = 0$. However, note that $s = 0$ must be excluded because a single pole is located there. This is also evident from the last expression in (A.6), which is zero for $s = 0$ before taking the limit. The Laplace transform of the signum function is essentially equivalent to its Fourier transform because the ROC forces $\sigma = 0$ in $\exp(-st)$, yielding $\exp(-j\omega t)$.

A.3 FOURIER TRANSFORM

In the summaries, Fourier transforms are given as a function of angular frequency ω. The corresponding expressions in terms of natural frequency f are generated by substituting $\omega = 2\pi f$. An exception to this rule is the Dirac delta function whose scaling property yields

$$\delta(\omega - \omega_o) \longrightarrow \delta(2\pi f - 2\pi f_o) = \frac{1}{2\pi}\delta(f - f_o). \tag{A.8}$$

The factor of $1/2\pi$ must be included when converting delta functions of ω to natural frequency f. For its derivative the unit doublet, the scale factor is $1/4\pi^2$:

$$\delta'(\omega - \omega_o) \longrightarrow \delta'(2\pi f - 2\pi f_o) = \frac{1}{4\pi^2}\delta'(f - f_o). \tag{A.9}$$

As mentioned in Chapter 7, the Laplace transform is more general than the Fourier transform because of the complex variable $s = \sigma + j\omega$ of $\exp(-st)$, which results in an ROC where $X(s)$ is defined. Given that we have an expression for the Laplace transform $X(s)$, the corresponding Fourier transform $X(\omega)$ can be derived from $X(s)$ depending on the type of ROC:

- The ROC includes the $j\omega$ axis and has the form $a < \sigma < b$ with $a < 0$ and $b > 0$:

$$X(\omega) = X(s)|_{s=j\omega}. \tag{A.10}$$

  This holds for finite-duration functions, right-sided functions with $b = \infty$, and left-sided functions with $a = -\infty$.

- Either $a = 0$ or $b = 0$ of $a < \sigma < b$. This means that one or more singular generalized functions are located on the $j\omega$ axis, and these must be included in the Fourier transform:

$$X(\omega) = X(s)|_{s=j\omega} + \text{singular generalized functions}. \tag{A.11}$$

- Neither of these cases: $X(\omega)$ does not exist.

All of the functions summarized in this appendix have a Fourier transform, but they may not have a Laplace transform as described later.

The first case in (A.10) is obviously straightforward. For the second case in (A.11) with singular generalized functions on the imaginary axis, the Fourier transform exists in the limit. Consider the Laplace transform $X(s) = 1/s$ of the unit step function $x(t) = u(t)$, which has ROC $\sigma > 0$. Clearly, the following improper integral is not defined:

$$\int_{-\infty}^{\infty} u(t)\exp(-j\omega t)dt = \int_{0}^{\infty} \exp(-j\omega t)dt. \tag{A.12}$$

However, suppose we approximate $u(t)$ by the exponential function $x(t) = \exp(-\alpha t)u(t)$ and let $\alpha \longrightarrow 0$ after computing its Fourier transform. Since the Laplace transform of $x(t)$ is

$$X(s) = \frac{1}{s+\alpha}, \tag{A.13}$$

its Fourier transform is

$$X(\omega) = \frac{1}{j\omega+\alpha} = \frac{\alpha}{\omega^2+\alpha^2} - \frac{j\omega}{\omega^2+\alpha^2}. \tag{A.14}$$

In the limit as $\alpha \longrightarrow 0$, the second term in the final expression is $-j/\omega$, and the first term becomes the Dirac delta function $\pi\delta(\omega)$. This last result is derived by recognizing that the first term is the Fourier transform of $(1/2)\exp(-\alpha|t|)$. From the area property of Fourier transforms:

$$\frac{1}{2\pi}\int_{-\infty}^{\infty} \frac{\alpha}{\omega^2+\alpha^2}d\omega = (1/2)\exp(-\alpha|t|)|_{t=0} = 1/2, \tag{A.15}$$

which demonstrates that the area of $\alpha/(\omega^2+\alpha^2)$ is π for any α. In the limit as $\alpha \longrightarrow 0$, the first term in (A.14) is zero for $\omega \neq 0$, and it is $1/\alpha \longrightarrow \infty$ for $\omega = 0$. (A similar model of the Dirac delta function was presented in Chapter 5 as the limit of rectangle functions.) Thus, the Fourier transform of the unit step function is

$$X(\omega) = \frac{1}{j\omega} + \pi\delta(\omega). \tag{A.16}$$

In general for N distinct poles on the $j\omega$ axis at $\omega = \omega_n$, the Fourier transform is

$$X(\omega) = X(s)|_{s=j\omega} + \pi \sum_{n=1}^{N} \delta(\omega - \omega_n). \tag{A.17}$$

For $X(s) = 1/s^2$, we have

$$X(\omega) = -\frac{1}{\omega^2} + j\pi\delta'(\omega), \tag{A.18}$$

and for a repeated pole at $\omega = \omega_o$ of order m, the Fourier transform includes derivatives of the Dirac delta function:

$$X(\omega) = X(s)|_{s=j\omega} + \pi \frac{j^{m-1}}{(m-1)!} \delta^{(m-1)}(\omega - \omega_o), \tag{A.19}$$

which holds for $m \geq 1$ with $0! \triangleq 1$.

Since $s = \sigma + j\omega$ is used in the Laplace transform with $\sigma \neq 0$, it is generally true that $X(s)$ exists for functions $x(t)$ that do not have a Fourier transform. This result follows because, in effect, $x(t)$ is multiplied by $\exp(-\sigma t)$, and so the product $x(t)\exp(-\sigma t)$ might be absolutely integrable for some range of values for σ, which of course defines the ROC. However, it turns out that there are some functions that have a Fourier transform (in the limit), but do not have a bilateral Laplace transform. In this appendix, they are the following two-sided functions: the constant 1, $\cos(\omega_o t)$, and $\sin(\omega_o t)$. The Laplace transforms of these functions do not exist for any $s \neq 0$, and for the line defined by $\sigma = 0$ (excluding $s = 0$), the Laplace transform is zero as shown later.

A.4 MAGNITUDE AND PHASE

The spectrum of a signal and the frequency response of a system can be written in terms of their magnitude and phase as follows:

$$X(\omega) = |X(\omega)| \exp(j\theta(\omega)). \tag{A.20}$$

If $X(\omega)$ is written in rectangular form $X(\omega) = X_R(\omega) + jX_I(\omega)$ where $\{X_R(\omega), X_I(\omega)\}$ are the real and imaginary parts, respectively, then

$$|X(\omega)| = \sqrt{X_R^2(\omega) + X_I^2(\omega)}, \quad \theta(\omega) = \tan^{-1}(X_I(\omega)/X_R(\omega)). \tag{A.21}$$

When $X_I(\omega) = 0$, this does not mean $|X(\omega)| = X_R(\omega)$ because $X_R(\omega)$ could be negative. Instead, when the imaginary part is 0, we have $|X(\omega)| = |X_R(\omega)|$. If $|X_R(\omega)|$

removes sign information about $X(\omega)$, then the phase component will be nonzero. This is illustrated for the rectangle function $\text{rect}(t)$ whose Fourier transform is

$$X(\omega) = \text{sinc}(\omega/2\pi), \tag{A.22}$$

which is negative for specific intervals of ω. Obviously this expression is real, and so we have

$$X_R(\omega) = \text{sinc}(\omega/2\pi), \quad X_I(\omega) = 0, \quad |X(\omega)| = |\text{sinc}(\omega/2\pi)|. \tag{A.23}$$

The phase is strictly zero for all ω where $X(\omega)$ is nonnegative. When $X(\omega)$ is negative for some ω, we must take into account a nonzero phase by multiplying $|X(\omega)|$ with $\exp(\pm j\pi) = -1$ for those particular regions of ω. The phase is π for positive ω and $-\pi$ for negative ω. For the sinc function in (A.22), this leads to the rectangular phase shown later in Figure A.11(c).

Figure A.2(a) shows a plot of $\tan(\omega)$ where ω is in radians, which we see repeats every π radians. The inverse tangent function $\tan^{-1}(x)$ shown in Figure A.2(b) asymptotically approaches $\pm\pi/2$ as $x \longrightarrow \pm\infty$. The radian units can be changed to degrees by multiplying the result by $180°/\pi$, giving the equivalent range $[-90°, 90°]$. The composite functions $\tan^{-1}(\tan(\omega))$ and $\tan(\tan^{-1}(x))$ are shown in Figure A.3. Due to the periodic nature of the waveform in Figure A.2(a), we find that $\tan^{-1}(\tan(\omega))$ is also periodic, but with the ramp (sawtooth) waveform in Figure A.3(a). For the other case $\tan(\tan^{-1}(x))$ with nonperiodic $\tan^{-1}(x)$ in Figure A.2(b), the exact inverse is obtained: $\tan(\tan^{-1}(x)) = x$ as shown in Figure A.3(b).

We illustrate the sawtooth behavior of the phase for the shifted Dirac delta function whose Fourier transform is $X(\omega) = \exp(-j\omega_o t)$. From Euler's formula:

$$\exp(-j\omega t_o) = \cos(\omega t_o) - j\sin(\omega t_o), \tag{A.24}$$

with phase

$$\theta(\omega) = \tan^{-1}(-\sin(\omega t_o)/\cos(\omega t_o)) = -\tan^{-1}(\tan(\omega t_o)), \tag{A.25}$$

which is the negative of the waveform in Figure A.3(a) (with x replaced by ωt_o). This phase is plotted later (with units of degrees) in Figure A.4(c). A similar derivation is used for the phase of the Fourier transform for the unit doublet shown later in Figure A.5(c).

Finally, we comment on the magnitude and phase of a Fourier transform involving the Dirac delta function $\delta(\omega)$ or its derivative the unit doublet $\delta'(\omega)$. Since these generalized functions are defined by their properties under an integral, the meaning of $|\delta(\omega)|$ and $|\delta'(\omega)|$ is not clear. By definition, we let $|\delta(\omega)| = \delta(\omega)$, and so the phase is zero. This approach is consistent when the Dirac delta function is viewed as the limit of increasingly narrow rectangle functions. The magnitude of the doublet is less clear. Recall that it is represented graphically by two pulses at the origin with opposite directions: upward for $\omega = 0^-$ and downward for $\omega = 0^+$ (it is zero in between

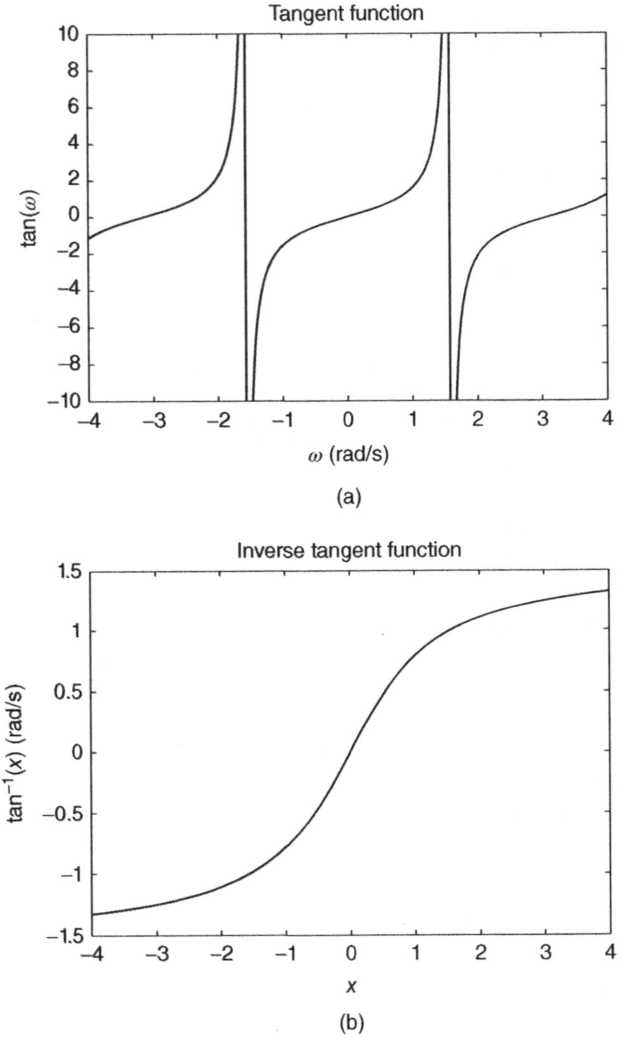

Figure A.2 Tangent functions. (a) $\tan(\omega)$. (b) $\tan^{-1}(x)$.

at $\omega = 0$). Thus, we represent $|\delta'(\omega)|$ on a plot by two closely spaced upward arrows, with the understanding that these pulses must be kept together as one symbol (as are the up and down impulses of the unit doublet). In order to represent the phase on a plot, we use the analogy of the signum function $X(\omega) = -j\text{sgn}(\omega)$, which has unit magnitude and phase

$$\angle X(\omega) = -(\pi/2)\text{sgn}(\omega) = \begin{cases} \pi/2, & \omega < 0 \\ 0, & \omega = 0 \\ -\pi/2, & \omega > 0. \end{cases} \tag{A.26}$$

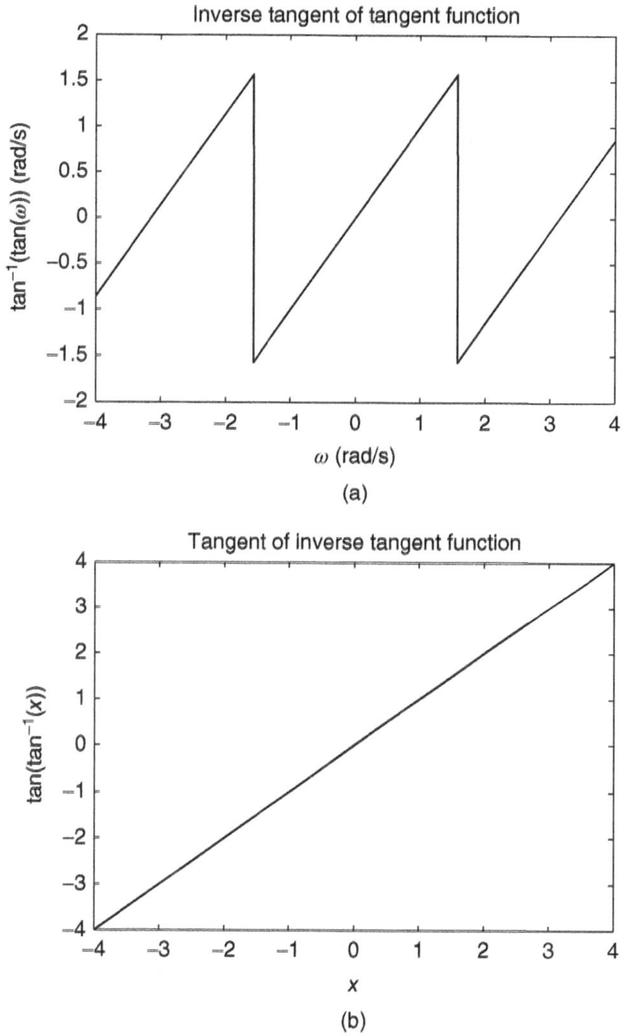

Figure A.3 Composite tangent functions. (a) $\tan^{-1}(\tan(\omega))$. (b) $\tan(\tan^{-1}(x))$.

For the unit doublet $j\delta'(\omega)$, the Kronecker delta function is used to symbolically represent the phase at two points with opposite sign about the origin (like the signum function):

$$\angle j\delta'(\omega) = (\pi/2)(\delta[\omega - 0^-] - \delta[\omega - 0^+]), \tag{A.27}$$

where

$$\delta[\omega] \triangleq \begin{cases} 1, & x = 0 \\ 0, & x \neq 0. \end{cases} \tag{A.28}$$

A.5 IMPULSIVE FUNCTIONS

A.5.1 Dirac Delta Function (Shifted)

Parameters: $t_o > 0$. Support: $t = t_o$. Range: singular generalized function.

$$x(t) = \delta(t - t_o) \triangleq \begin{cases} \text{undefined}, & t = t_o \\ 0, & t \neq t_o, \end{cases} \qquad \int_{-\infty}^{\infty} \delta(t - t_o)dt = 1,$$

$X(s) = \exp(-st_o)$, ROC: entire s-plane, poles: none, zeros: none,

$$|X(s)| = \exp(-\sigma t_o), \quad X(\omega) = \exp(-j\omega t_o),$$

$$|X(\omega)| = 1, \quad \theta(\omega) = -\tan^{-1}(\tan(\omega t_o)).$$

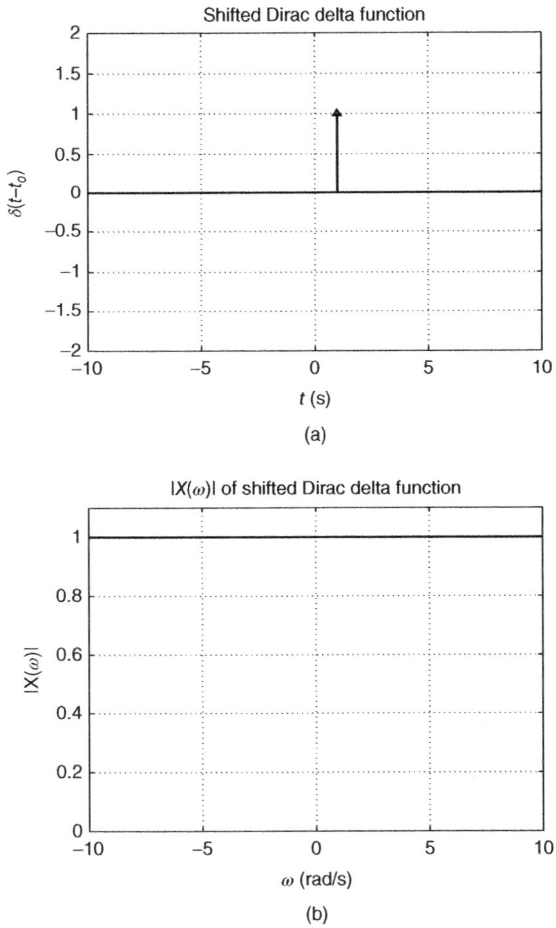

Figure A.4 Shifted Dirac delta function with $t_o = 1$ s. (a) $x(t) = \delta(t - t_o)$. The Dirac delta function has area 1. (b) $|X(\omega)|$.

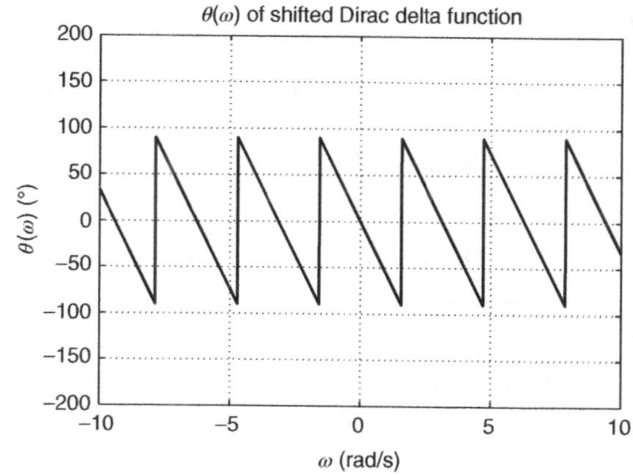

(c)

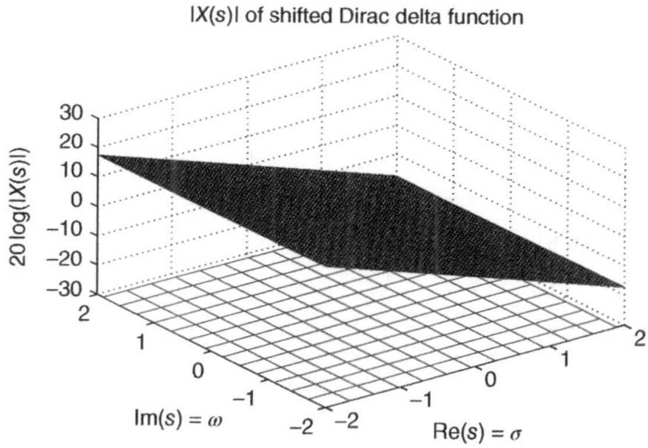

(d)

Figure A.4 Shifted Dirac delta function (continued). (c) $\theta(\omega)$. (d) $20\log(|X(s)|)$ and ROC: entire s-plane (lower grid).

- Phase from $X(\omega) = \cos(\omega t_o) - j\sin(\omega t_o)$:

$$\theta(\omega) = \tan^{-1}(-\sin(\omega t_o)/\cos(\omega t_o)) = -\tan^{-1}(\tan(\omega t_o)). \tag{A.29}$$

$$(\theta(\omega) = 0 \text{ for } t_o = 0.)$$

- Identities:

$$\delta(t - t_o) = \frac{d}{dt}u(t - t_o), \quad \delta(t - t_o) = \frac{d^2}{dt^2}r(t - t_o), \tag{A.30}$$

$$f(t)\delta(t - t_o) = f(t_o)\delta(t), \quad \int_{-\infty}^{\infty} f(t)\delta(t - t_o)dt = f(t_o). \tag{A.31}$$

(Assumes $f(t)$ is continuous at $t = t_o$.)

A.5.2 Unit Doublet (Shifted)

Parameters: $t_o > 0$. Support: $t = t_o$. Range: singular generalized function.

$$x(t) = \delta'(t - t_o) = \begin{cases} \text{undefined}, & t = t_o \\ 0, & t \neq t_o, \end{cases} \qquad \int_{-\infty}^{\infty} \delta'(t - t_o) dt = 0,$$

$$X(s) = s\exp(-st_o), \quad \text{ROC: entire } s\text{-plane}, \quad \text{poles: none}, \quad \text{zeros: } s = 0,$$

$$|X(s)| = \sqrt{\sigma^2 + \omega^2}\exp(-\sigma t_o), \quad X(\omega) = j\omega\exp(-j\omega t_o),$$

$$|X(\omega)| = |\omega|, \quad \theta(\omega) = \tan^{-1}(\cot(\omega t_o)).$$

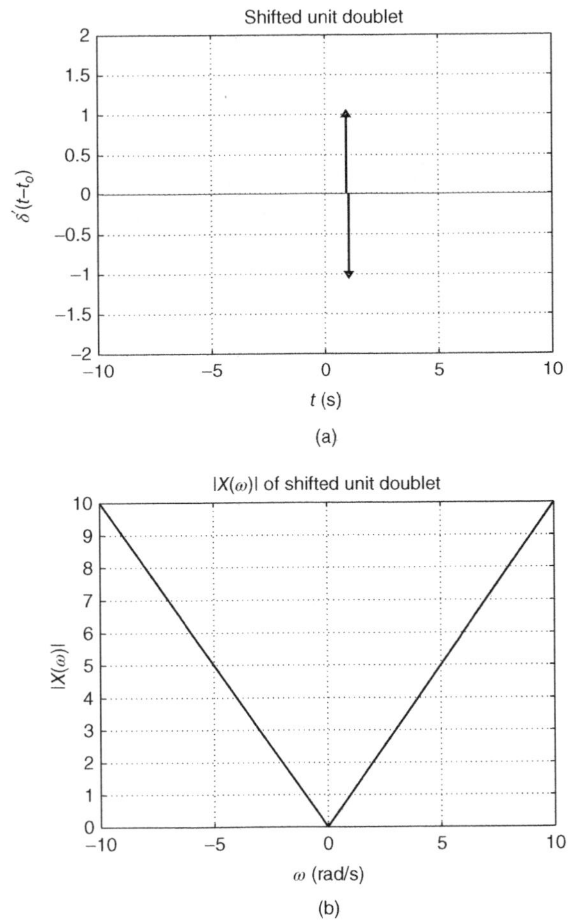

Figure A.5 Shifted unit doublet with $t_o = 1$ s. (a) $x(t) = \delta(t - t_o)$. Each component of the coupled impulses has infinite area. (b) $|X(\omega)|$.

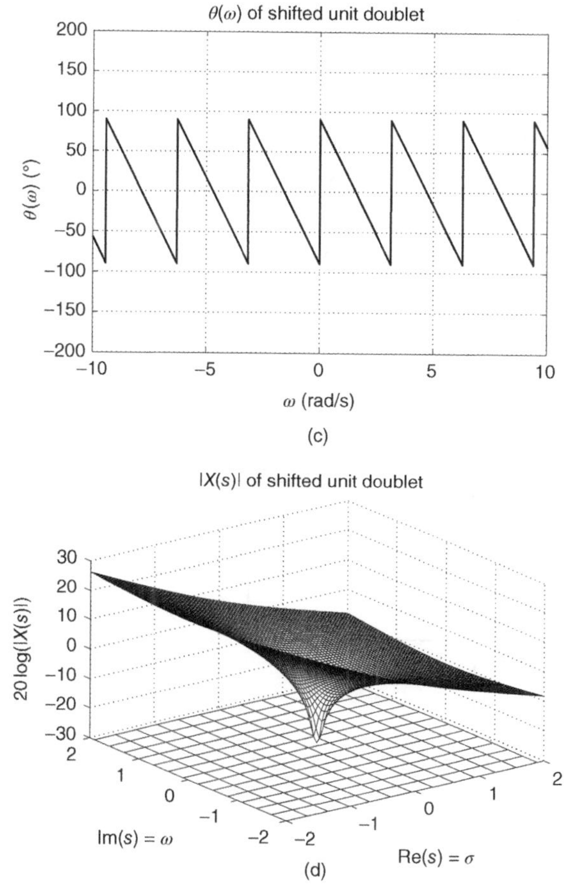

Figure A.5 Shifted unit doublet (continued). (c) $\theta(\omega)$. (d) $20\log(|X(s)|)$ and ROC: entire s-plane (lower grid).

- Phase from $X(\omega) = j\omega[\cos(\omega t_o) - j\sin(\omega t_o)] = j\omega\cos(\omega_o t) + \omega\sin(\omega t_o)$:

$$\theta(\omega) = \tan^{-1}(\cos(\omega t_o)/\sin(\omega t_o)) = \tan^{-1}(\cot(\omega t_o)). \tag{A.32}$$

$$(\text{For } t_o = 0,\ \theta(\omega) = \lim_{a\to 0}\tan^{-1}(\omega/a) = (\pi/2)\text{sgn}(\omega).)$$

- Identities:

$$\delta'(t-t_o) = \frac{d}{dt}\delta(t-t_o), \quad \delta'(t-t_o) = \frac{d^2}{dt^2}u(t-t_o), \tag{A.33}$$

$$f(t)\delta'(t-t_o) = f(t_o)\delta'(t-t_o) - f'(t_o)\delta(t-t_o), \quad f(t) * \delta'(t_o - t) = f'(t_o). \tag{A.34}$$

(Assumes $f(t)$ is continuous at $t = t_o$.)

A.6 PIECEWISE LINEAR FUNCTIONS

A.6.1 Unit Step Function

Parameters: none. Support: $t \in \mathcal{R}^+$. Range: $x(t) \in \{0, 1\}$.

$$x(t) = u(t) \triangleq I_{\mathcal{R}^+}(t), \quad X(s) = \frac{1}{s},$$

$$\text{ROC: } \sigma > 0, \quad \text{poles: } s = 0, \quad \text{zeros: none,}$$

$$|X(s)| = \frac{1}{\sqrt{\sigma^2 + \omega^2}}, \quad X(\omega) = \frac{1}{j\omega} + \pi\delta(\omega) \qquad \text{(exists in the limit),}$$

$$|X(\omega)| = \frac{1}{|\omega|} + \pi\delta(\omega), \quad \theta(\omega) = -(\pi/2)\text{sgn}(\omega).$$

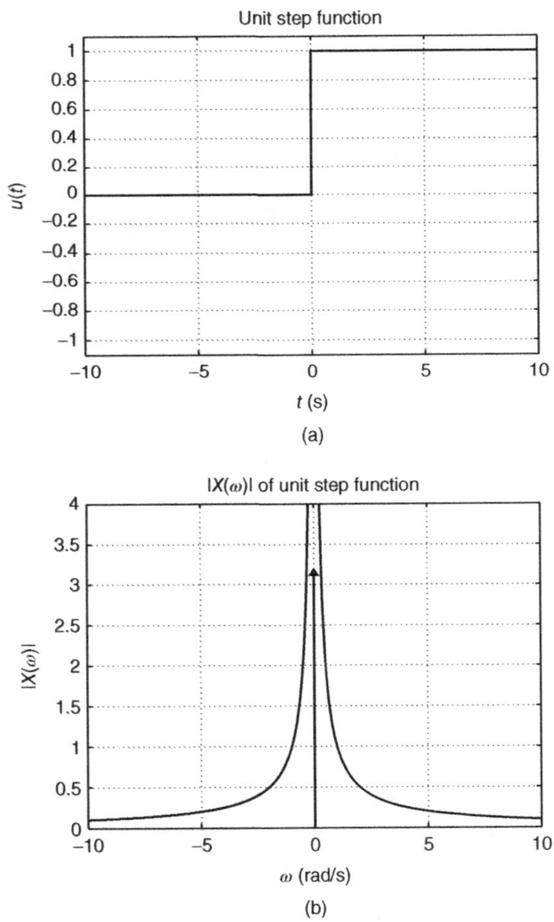

Figure A.6 Unit step function. (a) $x(t) = u(t)$. (b) Truncated $|X(\omega)|$. The Dirac delta function has area π.

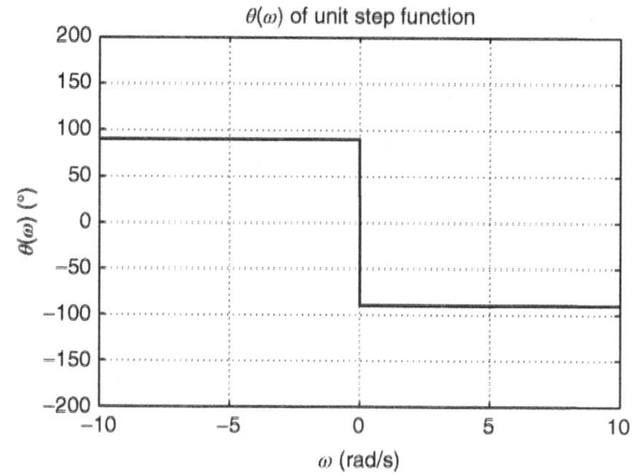

(c)

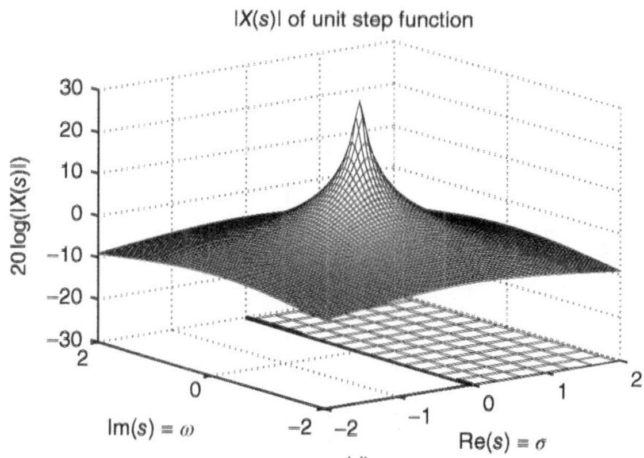

(d)

Figure A.6 Unit step function (continued). (c) $\theta(\omega)$. (d) Truncated $20\log(|X(s)|)$ and ROC: $\sigma > 0$ (lower grid excluding line at $\sigma = 0$).

- Power signal:

$$P = \lim_{T\to\infty} \frac{1}{T} \int_0^{T/2} dt = 1/2. \tag{A.35}$$

- Phase from $X(\omega) = -j/\omega$:

$$\theta(\omega) = \lim_{a\to 0} \tan^{-1}(-1/a\omega) = -(\pi/2)\text{sgn}(\omega). \tag{A.36}$$

- Identities:

$$\frac{d}{dt}u(t) = \delta(t), \quad \frac{d}{dt}r(t) = u(t), \tag{A.37}$$

$$u(t) = \frac{1}{2\pi} \int_{-\infty}^{\infty} \left(\frac{1}{j\omega} + \pi\delta(\omega)\right) \exp(j\omega t)d\omega = (1/2)\text{sgn}(t) + 1/2. \tag{A.38}$$

A.6.2 Signum Function

Parameters: none. Support: $t \in \mathcal{R}$. Range: $x(t) \in \{-1, 0, 1\}$.

$$x(t) = \text{sgn}(t) \triangleq \begin{cases} 1, & t > 0 \\ 0, & t = 0 \\ -1, & t < 0, \end{cases} \quad X(s) = \frac{2}{s},$$

ROC: $\sigma = 0$ (except $s = 0$), poles: $s = 0$, zeros: none,

$$|X(s)| = \frac{2}{\sqrt{\sigma^2 + \omega^2}}, \quad X(\omega) = \frac{2}{j\omega} \quad \text{(exists in the limit)},$$

$$|X(\omega)| = \frac{2}{|\omega|}, \quad \theta(\omega) = -(\pi/2)\text{sgn}(\omega).$$

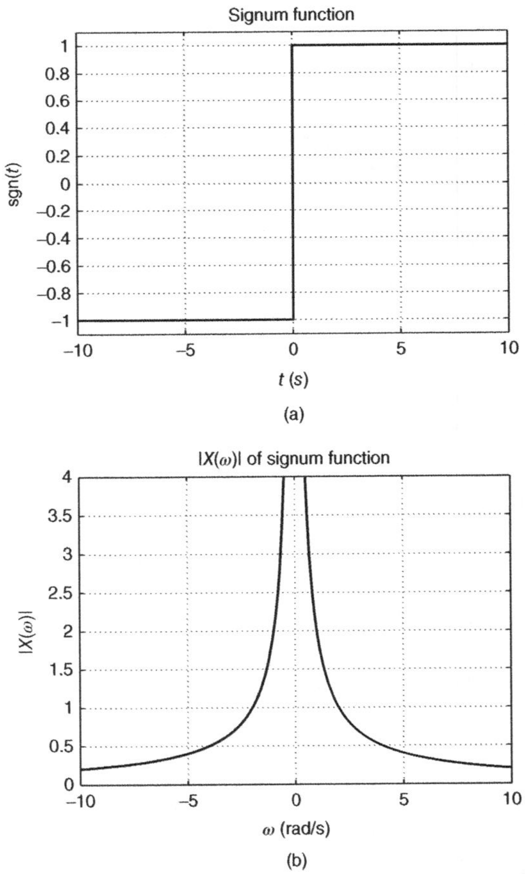

Figure A.7 Signum function. (a) $x(t) = \text{sgn}(t)$. (b) Truncated $|X(\omega)|$.

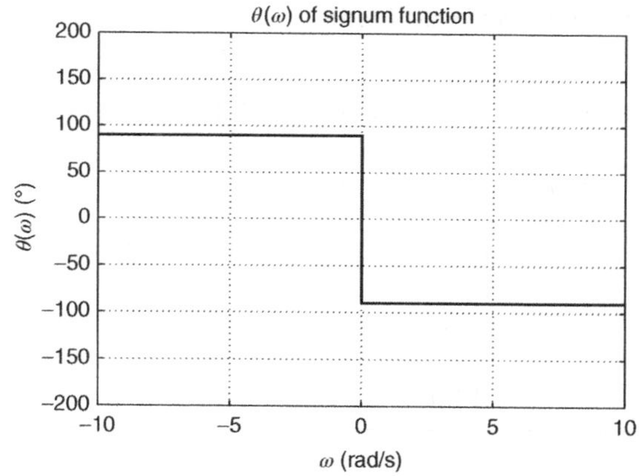

(c)

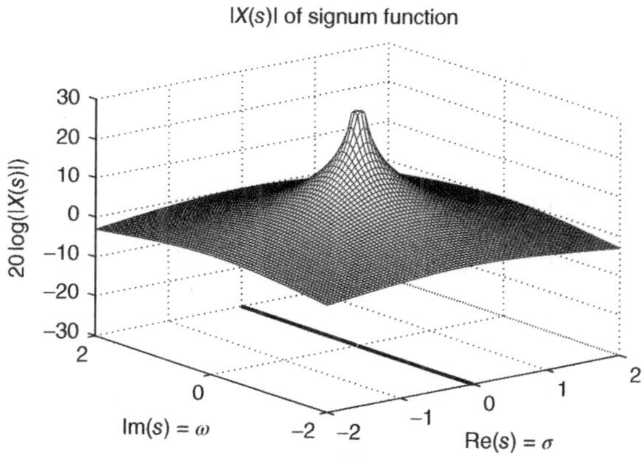

(d)

Figure A.7 Signum function (continued). (c) $\theta(\omega)$. (d) Truncated $20\log(|X(s)|)$ and ROC: $\sigma = 0$ (solid line excluding $s = 0$).

- Power signal:

$$P = \lim_{T\to\infty} \frac{1}{T}\int_{-T/2}^{T/2} dt = 1. \tag{A.39}$$

- Phase from $X(\omega) = -j2/\omega$:

$$\theta(\omega) = \lim_{a\to 0} \tan^{-1}(-2/a\omega) = -(\pi/2)\text{sgn}(\omega). \tag{A.40}$$

- Identities:

$$\frac{d}{dt}\text{sgn}(t) = 2\delta(t), \quad \text{sgn}(t) = 2u(t) - 1, \tag{A.41}$$

$$\text{sgn}(t) = \frac{d}{dt}|t| = t/|t| = |t|/t \quad \text{(excluding } t = 0\text{)}. \tag{A.42}$$

A.6.3 Constant Function (Two-Sided)

Parameters: none. Support: $t \in \mathcal{R}$. Range: $x(t) \in \{1\}$.

$$x(t) = 1, \quad X(s) = \text{does not exist (bilateral)},$$

$$\text{ROC: none}, \quad \text{poles: none}, \quad \text{zeros: none},$$

$$X(\omega) = 2\pi\delta(\omega) \quad \text{(exists in the limit)},$$

$$|X(\omega)| = 2\pi\delta(\omega), \quad \theta(\omega) = 0.$$

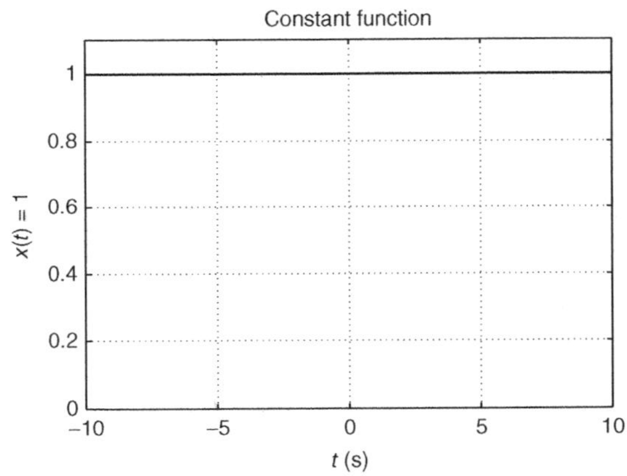

(a)

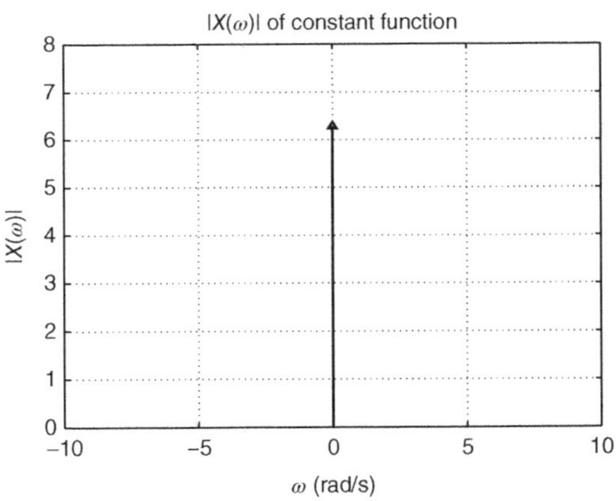

(b)

Figure A.8 Two-sided constant function. (a) $x(t) = 1$. (b) $|X(\omega)|$. The Dirac delta function has area 2π.

- Power signal:

$$P = \lim_{T\to\infty} \frac{1}{T} \int_{-T/2}^{T/2} dt = 1. \tag{A.43}$$

- Phase: $\theta(\omega) = 0$ because $X(\omega)$ is real and nonnegative.
- Identity:

$$x(t) = \frac{1}{2\pi} \int_{-\infty}^{\infty} 2\pi\delta(\omega) \exp(j\omega t) d\omega = \int_{-\infty}^{\infty} \delta(\omega) d\omega. \tag{A.44}$$

- Unlike the absolute value function, the constant function does not have a bilateral Laplace transform even though the two functions have some similarity:

$$\begin{aligned}\mathcal{L}_b\{1\} &= \int_{-\infty}^{0} \exp(-st)dt + \int_{0}^{\infty} \exp(-st)dt \\ &= (-1/s)\exp(-st)|_{-\infty}^{0} + (-1/s)\exp(-st)|_{0}^{\infty} = -\frac{1}{s} + \frac{1}{s}.\end{aligned} \tag{A.45}$$

Although the two individual ROCs match those of the absolute value function, $\sigma = 0$ (excluding $s = 0$) is not the ROC for (A.45) because the two terms cancel each other. This result is also derived from the Laplace transform of the two-sided exponential function:

$$\mathcal{L}_b\{1\} = \lim_{\alpha\to 0} \frac{2\alpha}{\alpha^2 - s^2} = 0. \tag{A.46}$$

- Unilateral Laplace transform:

$$\mathcal{L}\{1\} = \int_{0}^{\infty} \exp(-st)dt = \frac{1}{s}, \tag{A.47}$$

with ROC $\sigma > 0$. This result is identical to the Laplace transform of the unit step function. It arises when solving an integro-differential equation where the integral term has a nonzero initial state. For example, the voltage across a capacitor may be nonzero $v_C(0^-)$, and so it is treated as a constant (not a step function because this voltage cannot change instantaneously). Its unilateral Laplace transform is $v_C(0^-)/s$, which is similar to a step function, but occurs only because the lower limit of the transform is $t = 0^-$.

A.6.4 Ramp Function

Parameters: none. Support: $t \in \mathcal{R}^+$. Range: $x(t) \in \mathcal{R}^+$.

$$x(t) = tu(t), \quad X(s) = \frac{1}{s^2},$$

$$\text{ROC: } \sigma > 0, \quad \text{poles: } s = 0 \text{ (double)}, \quad \text{zeros: none},$$

$$|X(s)| = \frac{1}{\sigma^2 + \omega^2}, \quad X(\omega) = -\frac{1}{\omega^2} + j\pi\delta'(\omega) \quad \text{(exists in the limit)},$$

$$|X(\omega)| = \frac{1}{\omega^2} + \pi|\delta'(\omega)|, \quad \theta(\omega) = \pi\text{sgn}(\omega) + (\pi/2)(\delta[\omega - 0^-] - \delta[\omega - 0^+]).$$

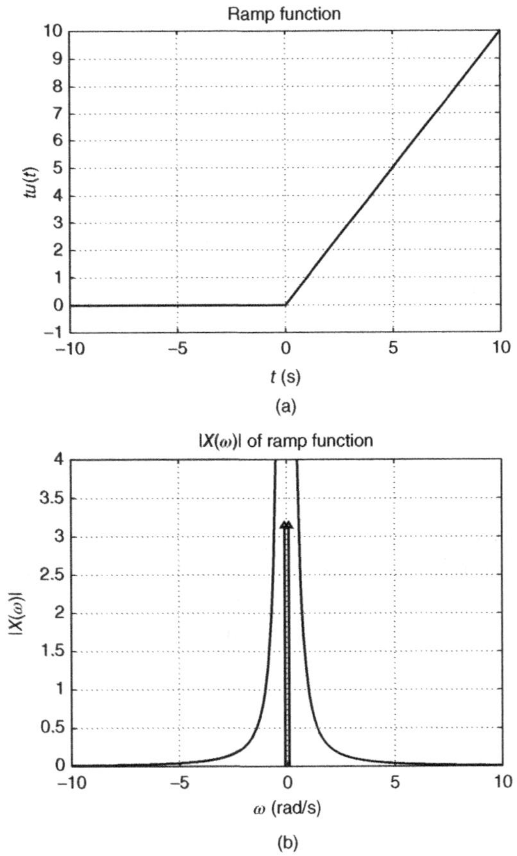

Figure A.9 Ramp function. (a) $x(t) = tu(t)$. (b) Truncated $|X(\omega)|$. The coupled upward arrows represent $\pi|\delta'(\omega)|$.

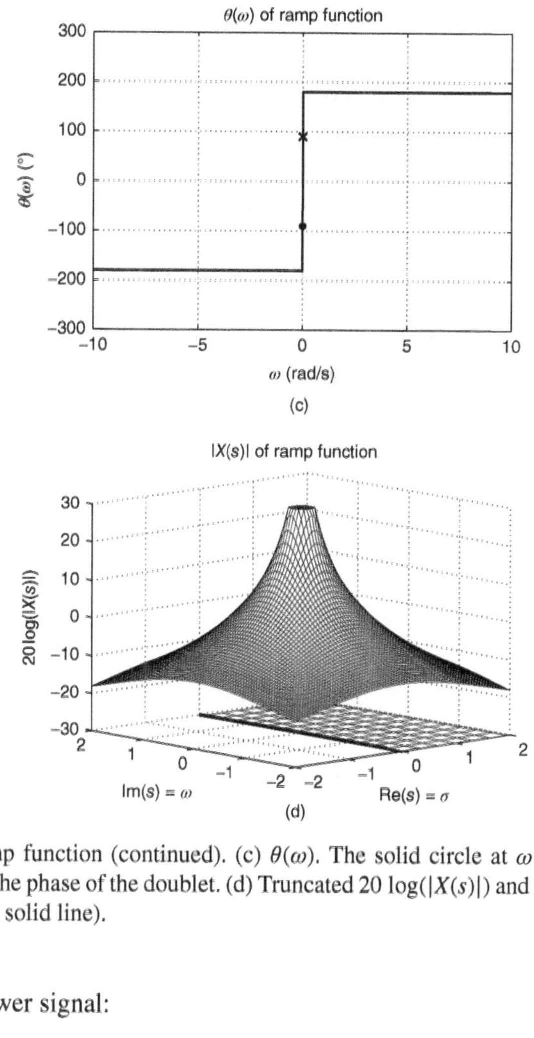

Figure A.9 Ramp function (continued). (c) $\theta(\omega)$. The solid circle at $\omega = 0^+$ and the $\times$ at $\omega = 0^-$ represent the phase of the doublet. (d) Truncated $20\log(|X(s)|)$ and ROC: $\sigma > 0$ (lower grid excluding the solid line).

- Infinite power signal:

$$P = \lim_{T\longrightarrow\infty} \frac{1}{T}\int_0^{T/2} t^2 dt = \lim_{T\rightarrow\infty} \frac{1}{T}(T^3/24) \longrightarrow \infty. \tag{A.48}$$

- Identities:

$$r(t) = (t + |t|)/2, \quad r(t) = u(t) * u(t), \quad u(t) = \frac{d}{dt} r(t), \tag{A.49}$$

$$r(t) = -\frac{1}{2\pi}\int_{-\infty}^{\infty} (1/\omega^2)\exp(j\omega t)d\omega + \frac{1}{2\pi}\int_{-\infty}^{\infty} j\pi\delta'(\omega)\exp(j\omega t)d\omega$$
$$= -\frac{1}{2\pi}\int_{-\infty}^{\infty} (1/\omega^2)\exp(j\omega t)d\omega + t/2 = (t/2)\text{sgn}(t) + t/2. \tag{A.50}$$

A.6.5 Absolute Value Function (Two-Sided Ramp)

Parameters: none. Support: $t \in \mathcal{R}$. Range: $x(t) \in \mathcal{R}^+$.

$$x(t) = |t|, \quad X(s) = \frac{2}{s^2},$$

ROC: $\sigma = 0$ (except $s = 0$), poles: $s = 0$ (double), zeros: none,

$$|X(s)| = \frac{2}{\sigma^2 + \omega^2}, \quad X(\omega) = -\frac{2}{\omega^2} \quad \text{(exists in the limit)},$$

$$|X(\omega)| = \frac{2}{\omega^2}, \quad \theta(\omega) = \pi \text{sgn}(\omega).$$

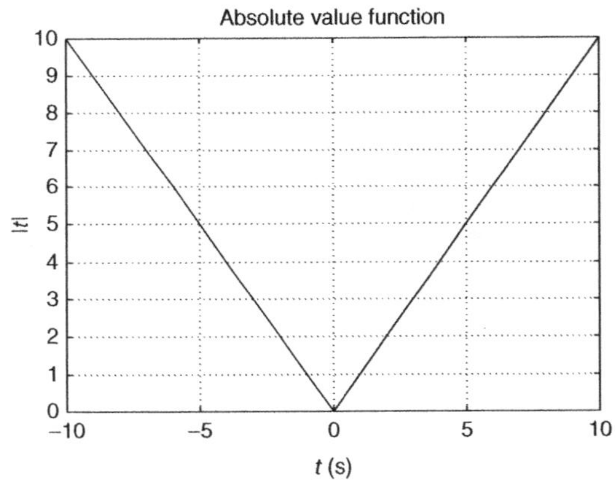

(a)

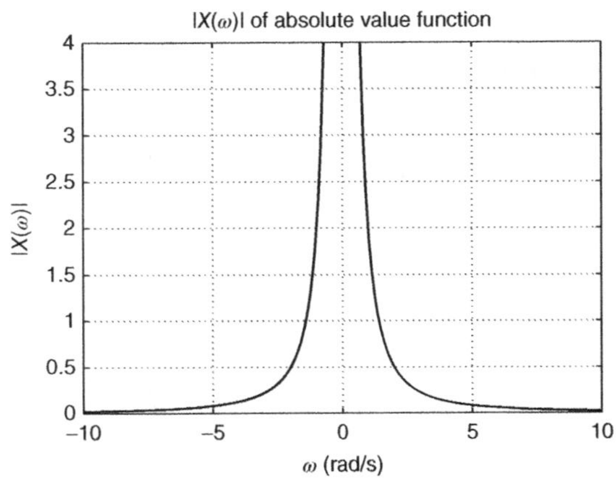

(b)

Figure A.10 Absolute value function. (a) $x(t) = |t|$. (b) Truncated $|X(\omega)|$.

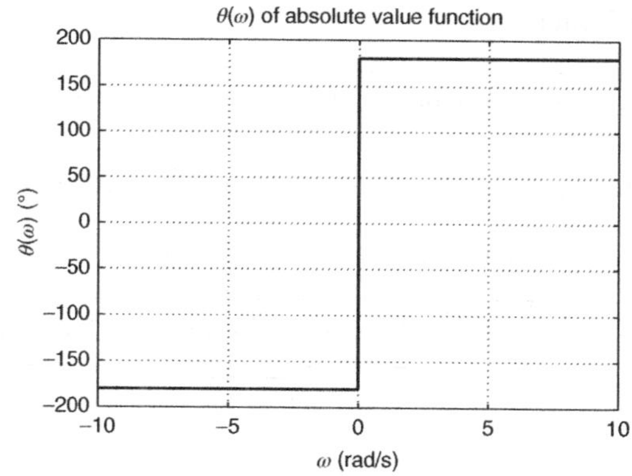

(c)

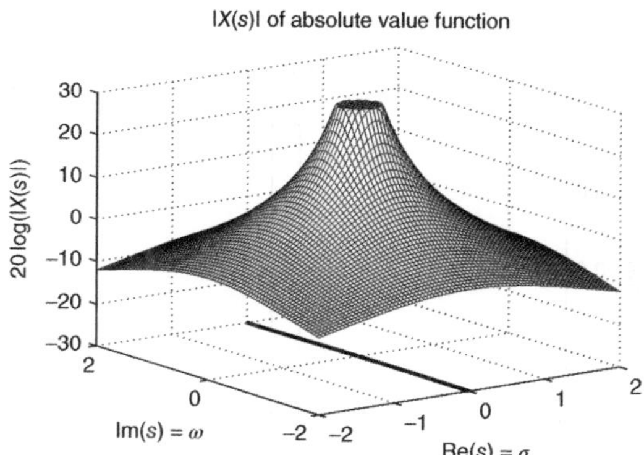

(d)

Figure A.10 Absolute value function (continued). (c) $\theta(\omega)$. (d) Truncated $20\log(|X(s)|)$ and ROC: $\sigma = 0$ (solid line excluding $s = 0$).

- Infinite power signal:

$$P = \lim_{T\to\infty} \frac{2}{T} \int_0^{T/2} t^2 dt = \lim_{T\to\infty} \frac{1}{T}(T^3/12) \longrightarrow \infty. \tag{A.51}$$

- Fourier transform from ramp functions (using (8.104) and (8.105)):

$$\begin{aligned} \mathcal{F}\{|t|\} &= \mathcal{F}\{r(t)\} + \mathcal{F}\{r(-t)\} \\ &= [j\pi\delta'(\omega) - 1/\omega^2] + [-j\pi\delta'(\omega) - 1/\omega^2] = -2/\omega^2, \end{aligned} \tag{A.52}$$

- Identities:

$$|t| = t\,\mathrm{sgn}(t), \quad |t| = r(t) + r(-t), \tag{A.53}$$

$$\mathrm{sgn}(t) = |t|/t = t/|t|, \quad \mathrm{sgn}(t) = \frac{d}{dt}|t| \text{ (excluding } t = 0). \tag{A.54}$$

A.6.6 Rectangle Function

Parameters: none. Support: $t \in [-1/2, 1/2]$. Range: $x(t) \in \{0, 1\}$.

$$x(t) = \text{rect}(t) \triangleq I_{[-1/2,1/2]}(t), \quad X(s) = \frac{2\sinh(s/2)}{s}, \quad X(\omega) = \text{sinc}(\omega/2\pi),$$

ROC: entire s-plane, poles: none (removable), zeros: none,

$$|X(s)| = \frac{2\sqrt{\cosh^2(\sigma/2)\cos^2(\omega/2) + \sinh^2(\sigma/2)\sin^2(\omega/2)}}{\sqrt{\sigma^2 + \omega^2}},$$

$$|X(\omega)| = |\text{sinc}(\omega/2\pi)|, \quad \theta(\omega) = \pi\text{sgn}(\omega) \sum_{n=1}^{\infty} \text{rect}([|\omega| - (4n-1)\pi]/2\pi).$$

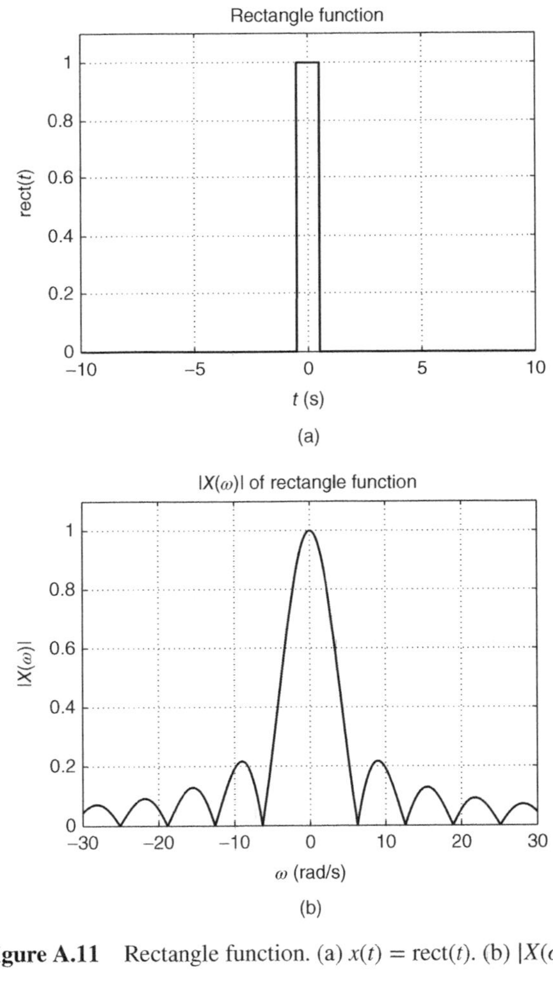

Figure A.11 Rectangle function. (a) $x(t) = \text{rect}(t)$. (b) $|X(\omega)|$.

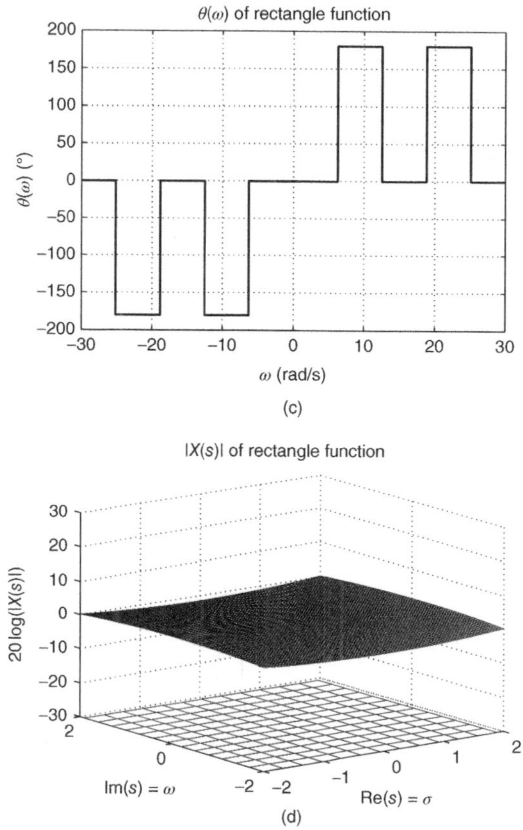

Figure A.11 Rectangle function (continued). (c) $\theta(\omega)$. (d) $20\log(|X(s)|)$ and ROC: entire s-plane (lower grid).

- Energy signal:

$$E = \int_{-1/2}^{1/2} dt = 1. \tag{A.55}$$

- Phase: Since $X(\omega)$ changes sign periodically, the phase function is a square waveform. The phase is negative during intervals of duration 2π given by $\omega \in [k2\pi, (k+1)2\pi]$ for $k = \pm 1, \pm 3, \cdots$ (odd integer values). These regions can be represented by the rectangle function $\text{rect}([|\omega| - (4n-1)\pi]/2\pi)$ for $n \in \mathcal{N}$. Scaling the sum of the shifted rectangles by $\pi\text{sgn}(\omega)$ gives $\theta(\omega)$.
- Identities:

$$\text{rect}(t) = u(t+1/2) - u(t-1/2), \quad \frac{d}{dt}\text{rect}(t) = \delta(t+1/2) - \delta(t-1/2). \tag{A.56}$$

A.6.7 Triangle Function

Parameters: none. Support: $t \in [-1, 1]$. Range: $x(t) \in [0, 1]$.

$$x(t) = \text{tri}(t) \triangleq (1 - |t|)I_{[-1,1]}(t), \quad X(s) = \frac{4\sinh^2(s/2)}{s^2},$$

ROC: entire s-plane, poles: none (removable), zeros: none,

$$|X(s)| = \frac{4[\cosh^2(\sigma/2)\cos^2(\omega/2) + \sinh^2(\sigma/2)\sin^2(\omega/2)]}{\sigma^2 + \omega^2},$$

$$X(\omega) = |X(\omega)| = \text{sinc}^2(\omega/2\pi), \quad \theta(\omega) = 0.$$

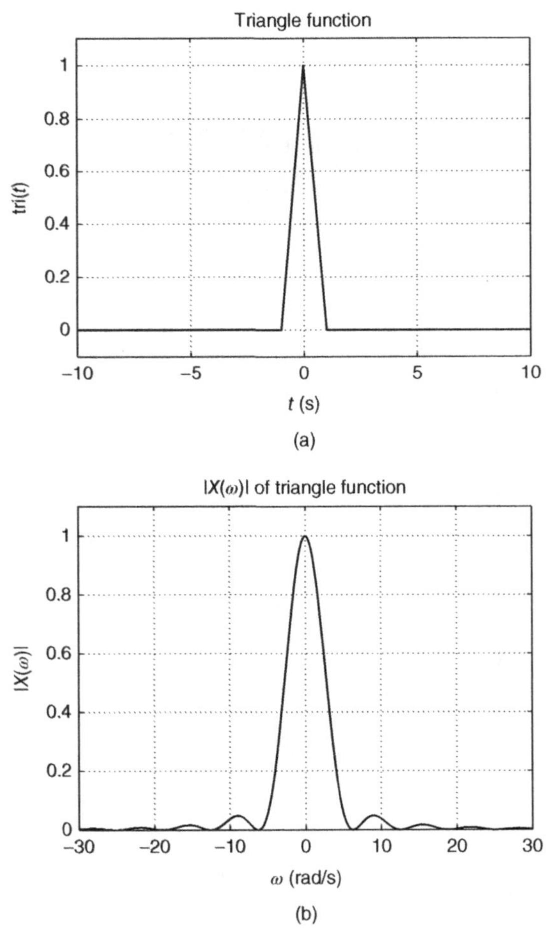

Figure A.12 Triangle function. (a) $x(t) = \text{tri}(t)$. (b) $|X(\omega)|$.

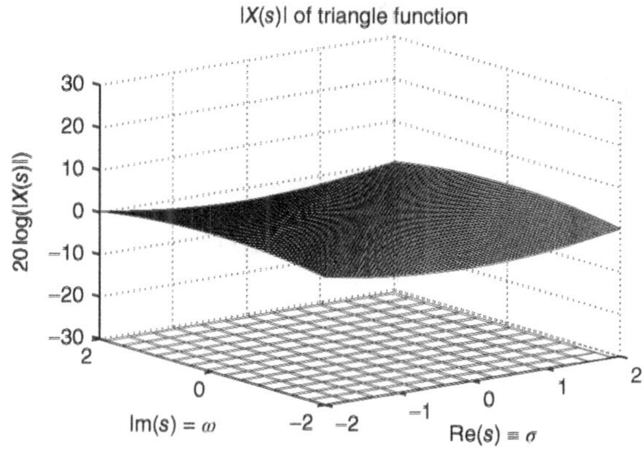

Figure A.12 Triangle function (continued). (c) $20 \log(|X(s)|)$ and ROC: entire s-plane (lower grid).

- Energy signal:

$$E = \int_{-1}^{1} (1 - |t|)^2 dt = 2 \int_{0}^{1} (1 - 2t + t^2) dt = 2/3. \qquad \text{(A.57)}$$

- Phase: $\theta(\omega) = 0$ because $X(\omega)$ is real and nonnegative.
- Fourier transform from Laplace transform:

$$\begin{aligned} X(\omega) &= \frac{4\sinh^2(j\omega/2)}{(j\omega)^2} = \frac{[\exp(j\omega/2) - \exp(-j\omega/2)]^2}{(j\omega)^2} \\ &= \frac{\sin^2(\pi\omega/2\pi)}{(\pi\omega/2\pi)^2} = \text{sinc}^2(\omega/2\pi). \end{aligned} \qquad \text{(A.58)}$$

(A similar approach is used for the Fourier transform of the rectangle function.)

- Identities:

$$\text{tri}(t) = \text{rect}(t) * \text{rect}(t), \quad \frac{d}{dt}\text{tri}(t) = -\text{sgn}(t) I_{[-1,1]}(t). \qquad \text{(A.59)}$$

A.7 EXPONENTIAL FUNCTIONS

A.7.1 Exponential Function (Right-Sided)

Parameters: $\alpha > 0$. Support: $t \in \mathcal{R}^+$. Range: $x(t) \in [0, 1]$.

$$x(t) = \exp(-\alpha t)u(t), \quad X(s) = \frac{1}{s+\alpha},$$

$$\text{ROC: } \sigma > -\alpha, \quad \text{poles: } s = -\alpha, \quad \text{zeros: none,}$$

$$|X(s)| = \frac{1}{\sqrt{(\sigma+\alpha)^2+\omega^2}}, \quad X(\omega) = \frac{1}{\alpha + j\omega},$$

$$|X(\omega)| = \frac{1}{\sqrt{\alpha^2+\omega^2}}, \quad \theta(\omega) = -\tan^{-1}(\omega/\alpha).$$

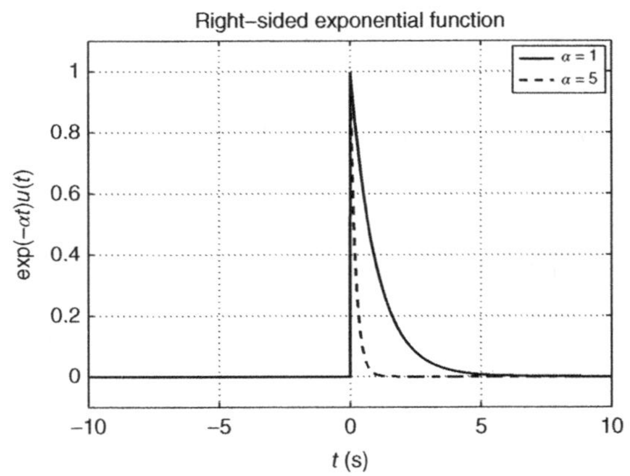

(a)

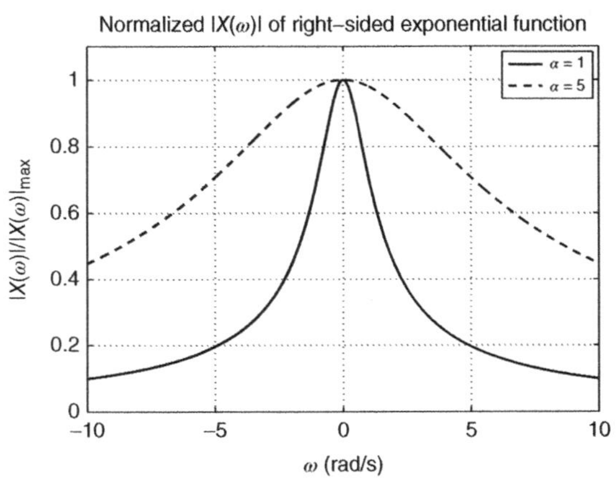

(b)

Figure A.13 Right-sided exponential function. (a) $x(t) = \exp(-\alpha t)u(t)$. (b) Normalized $|X(\omega)|$.

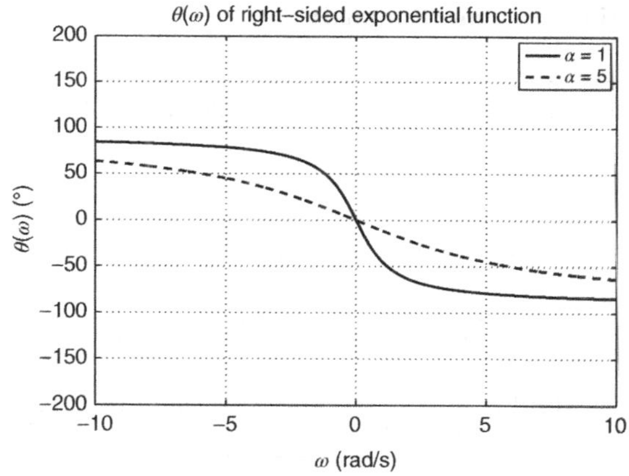

(c)

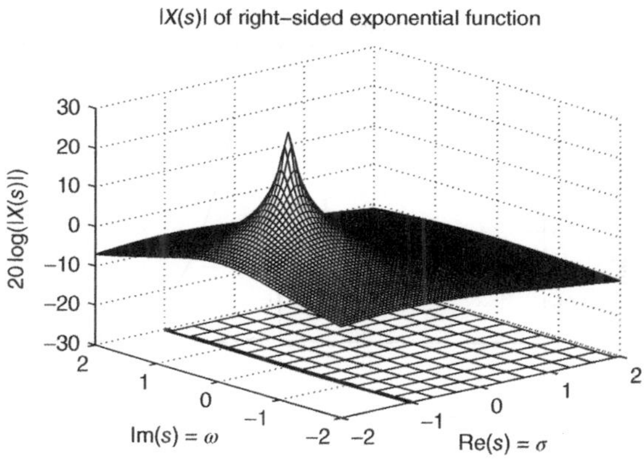

(d)

Figure A.13 Right-sided exponential function (continued). (c) $\theta(\omega)$. (d) Truncated $20\log(|X(s)|)$ with $\alpha = 1$ and ROC: $\sigma > -1$ (lower grid excluding the solid line).

- Energy signal:

$$E = \int_0^{\infty} \exp(-2\alpha t)dt = 1/2\alpha. \tag{A.60}$$

- Identities:

$$\frac{d}{dt}\exp(-\alpha t)u(t) = \delta(t) - \alpha\exp(-\alpha t)u(t), \tag{A.61}$$

$$\frac{d}{dt}[1 - \exp(-\alpha t)]u(t) = \alpha\exp(-\alpha t)u(t). \tag{A.62}$$

A.7.2 Exponential Function (Ramped)

Parameters: $\alpha > 0$. Support: $t \in \mathcal{R}^+$. Range: $x(t) \in [0, 1/\alpha e]$.

$$x(t) = t\exp(-\alpha t)u(t), \quad X(s) = \frac{1}{(s+\alpha)^2},$$

ROC: $\sigma > -\alpha$, poles: $s = -\alpha$ (double), zeros: none,

$$|X(s)| = \frac{1}{(\sigma+\alpha)^2+\omega^2}, \quad X(\omega) = \frac{1}{(\alpha+j\omega)^2},$$

$$|X(\omega)| = \frac{1}{\alpha^2+\omega^2}, \quad \theta(\omega) = -2\tan^{-1}(\omega/\alpha).$$

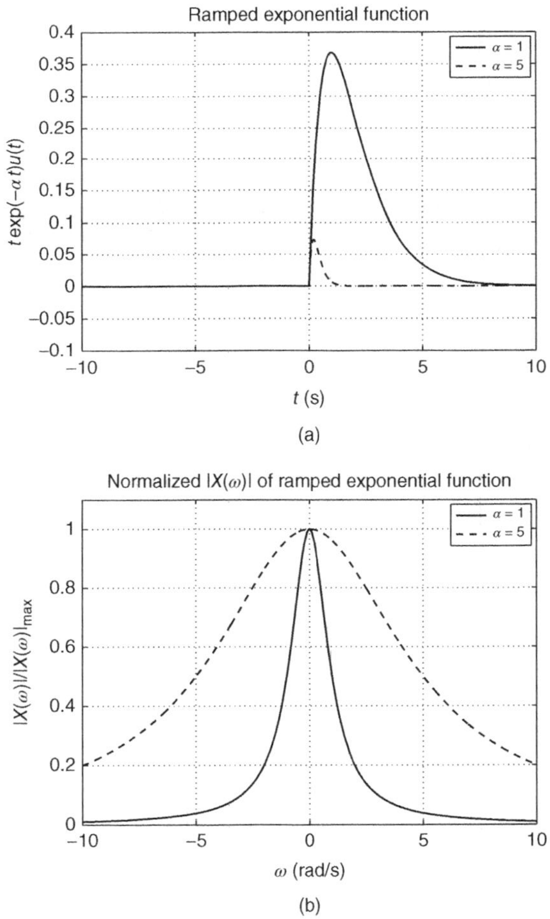

Figure A.14 Ramped exponential function. (a) $x(t) = t\exp(-\alpha t)u(t)$. (b) Normalized $|X(\omega)|$.

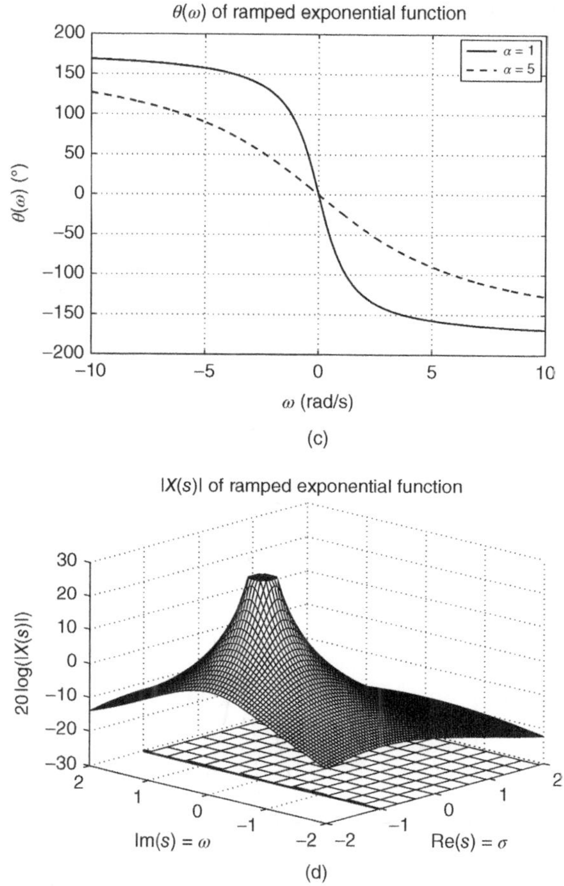

Figure A.14 Ramped exponential function (continued). (c) $\theta(\omega)$. (d) Truncated $20\log(|X(s)|)$ with $\alpha = 1$ and ROC: $\sigma > -1$ (lower grid excluding the solid line).

- Energy signal:

$$\begin{aligned} E &= \int_0^{\infty} t^2 \exp(-2\alpha t)dt \\ &= -t^2 \exp(-2\alpha t)/2\alpha - \exp(-2\alpha t)(2\alpha t + 1)/4\alpha^3 |_0^{\infty} \\ &= 1/4\alpha^3. \end{aligned} \tag{A.63}$$

- Identities:

$$\frac{d}{dt} t \exp(-\alpha t)u(t) = (1 - \alpha t)\exp(-\alpha t)u(t), \tag{A.64}$$

$$\frac{d^2}{dt^2} t \exp(-\alpha t)u(t) = \delta(t) - (2\alpha - \alpha^2 t)\exp(-\alpha t)u(t). \tag{A.65}$$

A.7.3 Exponential Function (Two-Sided)

Parameters: $\alpha > 0$. Support: $t \in \mathcal{R}$. Range: $x(t) \in [0, 1]$.

$$x(t) = \exp(-\alpha|t|), \quad X(s) = \frac{-2\alpha}{(s-\alpha)(s+\alpha)},$$

$$\text{ROC: } -\alpha < \sigma < \alpha, \quad \text{poles: } s = \pm\alpha, \quad \text{zeros: none},$$

$$|X(s)| = \frac{2\alpha}{\sqrt{(\alpha^2 - \sigma^2 + \omega^2)^2 + 4\sigma^2\omega^2}}, \quad X(\omega) = \frac{2\alpha}{\alpha^2 + \omega^2},$$

$$|X(\omega)| = \frac{2\alpha}{\alpha^2 + \omega^2}, \quad \theta(\omega) = 0.$$

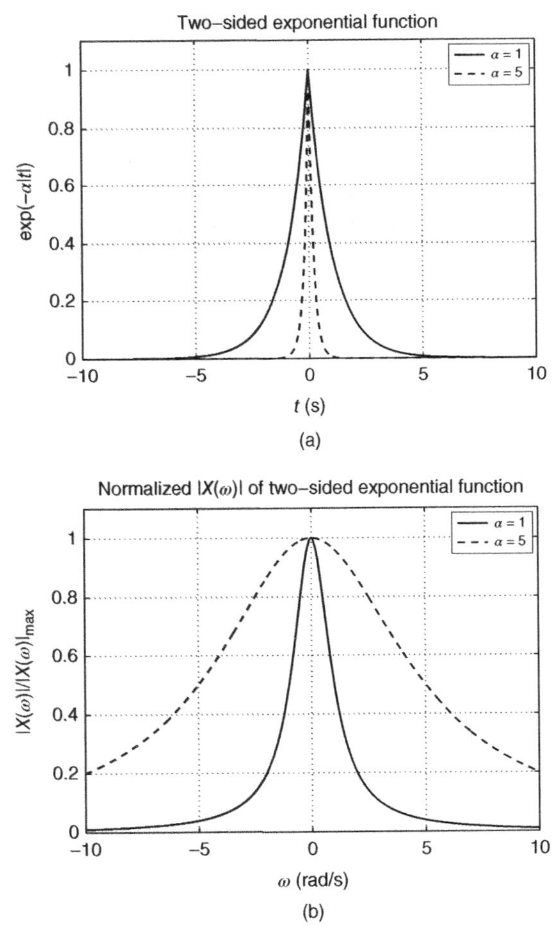

Figure A.15 Two-sided exponential function. (a) $x(t) = \exp(-\alpha|t|)$. (b) Normalized $|X(\omega)|$.

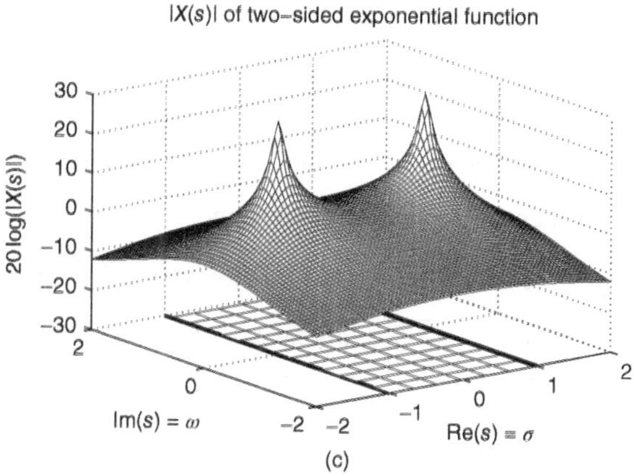

Figure A.15 Two-sided exponential function (continued). (c) Truncated $20\log(|X(s)|)$ with $\alpha = 1$ and ROC: $-1 < \sigma < 1$ (lower grid excluding the two solid lines).

- Energy signal:

$$E = \int_{-\infty}^{\infty} \exp(-2\alpha|t|)dt = 2\int_{0}^{\infty} \exp(-2\alpha t)dt = 1/\alpha. \tag{A.66}$$

- Phase: $\theta(\omega) = 0$ because $X(\omega)$ is real and nonnegative.
- Laplace transform from one-sided exponential functions:

$$\begin{aligned}\mathcal{L}_b\{\exp(-\alpha|t|)\} &= \mathcal{L}\{\exp(-\alpha t)u(t)\} + \mathcal{L}_b\{\exp(\alpha t)u(-t)\} \\ &= \frac{1}{s+\alpha} + \frac{1}{-s+\alpha} = \frac{-2\alpha}{s^2-\alpha^2}.\end{aligned} \tag{A.67}$$

- Identity:

$$\frac{d}{dt}\exp(-\alpha|t|) = -\alpha\exp(-\alpha|t|)\text{sgn}(t) \text{ (excluding } t = 0). \tag{A.68}$$

The scaled function $(\alpha/2)\exp(-\alpha|t|)$ is the Laplace probability density function with unit area, zero mean, and variance $2/\alpha^2$. The energy result in (A.66) follows from the unit area property, but with variance $\sigma^2 = 1/2\alpha^2$ because of the factor of 2 in the exponent:

$$\int_{-\infty}^{\infty} \alpha\exp(-2\alpha|t|)dt = 1 \implies E = 1/\alpha. \tag{A.69}$$

A.7.4 Gaussian Function

Parameters: $\alpha > 0$. Support: $t \in \mathcal{R}$. Range: $x(t) \in [0, 1]$.

$$x(t) = \exp(-\alpha t^2), \quad X(s) = \sqrt{\pi/\alpha}\exp(s^2/4\alpha),$$

ROC: entire s-plane, poles: none, zeros: none,

$$|X(s)| = \sqrt{\pi/\alpha}\exp((\sigma^2 - \omega^2)/4\alpha), \quad X(\omega) = \sqrt{\pi/\alpha}\exp(-\omega^2/4\alpha),$$

$$|X(\omega)| = \sqrt{\pi/\alpha}\exp(-\omega^2/4\alpha), \quad \theta(\omega) = 0.$$

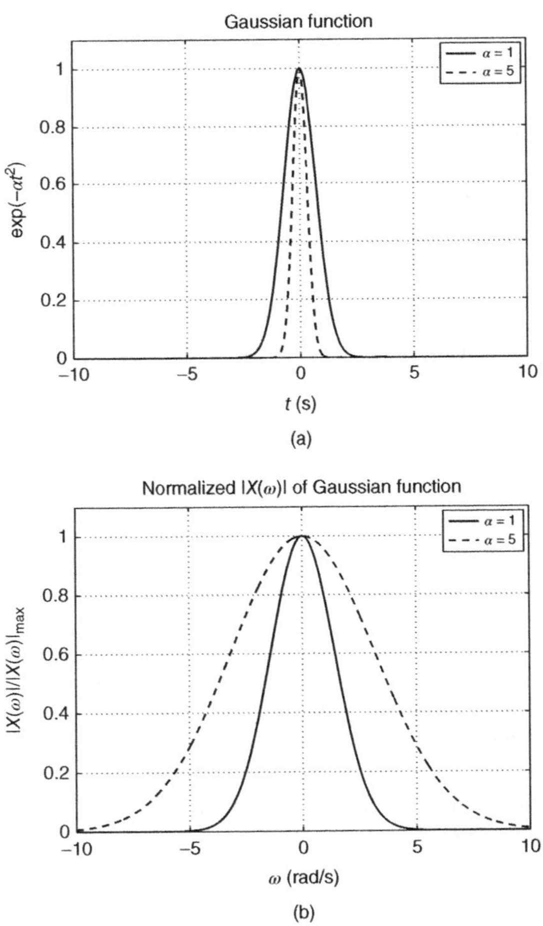

Figure A.16 Gaussian function. (a) $x(t) = \exp(-\alpha t^2)$. (b) Normalized $|X(\omega)|$.

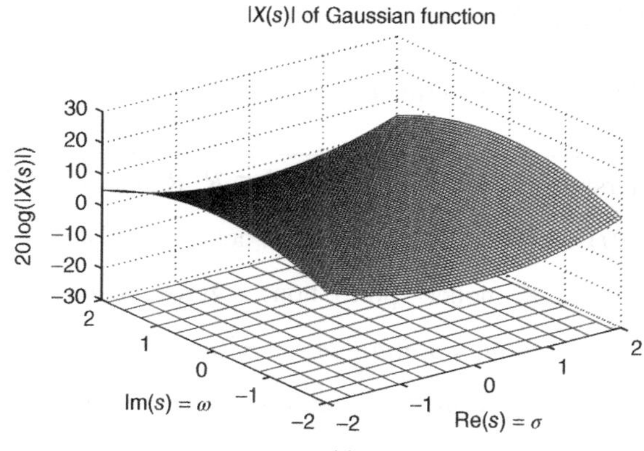

(c)

Figure A.16 Gaussian function (continued). (c) $20\log(|X(s)|)$ with $\alpha = 1$ and ROC: entire s-plane (lower grid).

- Energy signal:

$$E = \int_{-\infty}^{\infty} \exp(-2\alpha t^2)dt = \sqrt{\pi/2\alpha}. \tag{A.70}$$

- Phase: $\theta(\omega) = 0$ because $X(\omega)$ is real and nonnegative.
- Identity:

$$\frac{d}{dt}\exp(-\alpha t^2) = -2\alpha t\exp(-\alpha t^2). \tag{A.71}$$

The scaled function $\sqrt{\alpha/\pi}\exp(-\alpha t^2)$ is the Gaussian probability density function with unit area, zero mean, and variance $\sigma^2 = 1/2\alpha$. The energy result in (A.70) follows from the unit area property, but with variance $\sigma^2 = 1/4\alpha$ because of the factor of 2 in the exponent:

$$\int_{-\infty}^{\infty} \sqrt{2\alpha/\pi}\exp(-2\alpha t^2)dt = 1 \implies E = \sqrt{\pi/2\alpha}. \tag{A.72}$$

A.8 SINUSOIDAL FUNCTIONS

A.8.1 Cosine Function (Two-Sided)

Parameters: $\omega_o = 2\pi f_o > 0$. Support: $t \in \mathcal{R}$. Range: $x(t) \in [-1, 1]$.

$$x(t) = \cos(\omega_o t), \quad X(s) = \text{does not exist (bilateral)},$$

$$\text{ROC: none}, \quad \text{poles: none}, \quad \text{zeros: none},$$

$$X(\omega) = \pi\delta(\omega + \omega_o) + \pi\delta(\omega - \omega_o),$$

$$|X(\omega)| = \pi\delta(\omega + \omega_o) + \pi\delta(\omega - \omega_o), \quad \theta(\omega) = 0.$$

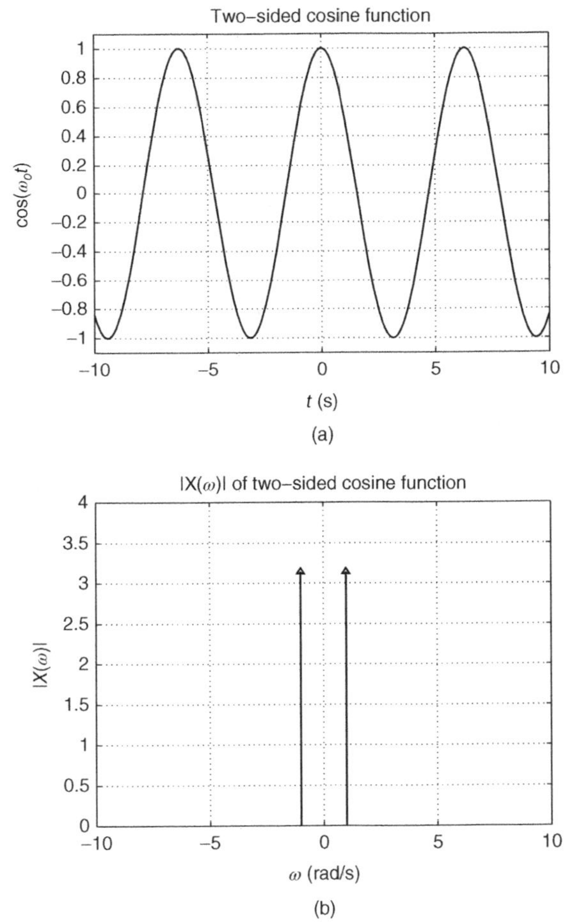

Figure A.17 Two-sided cosine function with $\omega_o = 1$ rad/s ($T_o = 2\pi$). (a) $x(t) = \cos(\omega_o t)$. (b) $|X(\omega)|$. Each Dirac delta function has area π.

- Power signal:

$$P = \lim_{T\to\infty} \frac{1}{T} \int_{-T/2}^{T/2} \cos^2(\omega_o t)dt$$
$$= \lim_{T\to\infty} \frac{1}{T} \int_{-T/2}^{T/2} (1/2)[1 + \cos(2\omega_o t)]dt = 1/2. \quad \text{(A.73)}$$

- Phase: $\theta(\omega) = 0$ because $X(\omega)$ is real and nonnegative.
- The bilateral Laplace transform does not exist because all terms cancel (as was the case for the constant function in (A.45)):

$$\mathcal{L}_b(\cos(\omega_o t)) = \int_{-\infty}^{0} \cos(\omega_o t)\exp(-st)dt + \int_{0}^{\infty} \cos(\omega_o t)\exp(-st)dt$$
$$= \frac{1}{2}\int_{-\infty}^{0} [\exp(-(s - j\omega_o)t) + \exp(-(s + j\omega_o)t)]dt$$
$$+\frac{1}{2}\int_{0}^{\infty} [\exp(-(s - j\omega_o)t) + \exp(-(s + j\omega_o)t)]dt$$
$$= \frac{-1}{2(s - j\omega_o)} + \frac{-1}{2(s + j\omega_o)} + \frac{1}{2(s - j\omega_o)} + \frac{1}{2(s + j\omega_o)}. \quad \text{(A.74)}$$

- Identity (Euler's inverse formula):

$$\cos(\omega_o t) = \frac{1}{2\pi}\int_{-\infty}^{\infty} [\pi\delta(\omega + \omega_o) + \pi\delta(\omega - \omega_o)]\exp(j\omega t)d\omega$$
$$= (1/2)[\exp(-j\omega_o t) + \exp(j\omega_o t)]. \quad \text{(A.75)}$$

A.8.2 Cosine Function (Right-Sided)

Parameters: $\omega_o = 2\pi f_o > 0$. Support: $t \in \mathcal{R}^+$. Range: $x(t) \in [-1, 1]$.

$$x(t) = \cos(\omega_o t)u(t), \quad X(s) = \frac{s}{s^2 + \omega_o^2},$$

$$\text{ROC: } \sigma > 0, \quad \text{poles: } s = \pm j\omega_o, \quad \text{zeros: } s = 0,$$

$$|X(s)| = \frac{\sqrt{\sigma^2 + \omega^2}}{\sqrt{([\sigma^2 + (\omega + \omega_o)^2][\sigma^2 + (\omega - \omega_o)^2]}},$$

$$X(\omega) = \frac{j\omega}{\omega_o^2 - \omega^2} + \frac{\pi}{2}\delta(\omega + \omega_o) + \frac{\pi}{2}\delta(\omega - \omega_o) \quad \text{(exists in the limit)},$$

$$|X(\omega)| = \frac{|\omega|}{|\omega_o^2 - \omega^2|} + \frac{\pi}{2}\delta(\omega + \omega_o) + \frac{\pi}{2}\delta(\omega - \omega_o),$$

$$\theta(\omega) = (\pi/2)\mathrm{sgn}(\omega/(\omega_o^2 - \omega^2)).$$

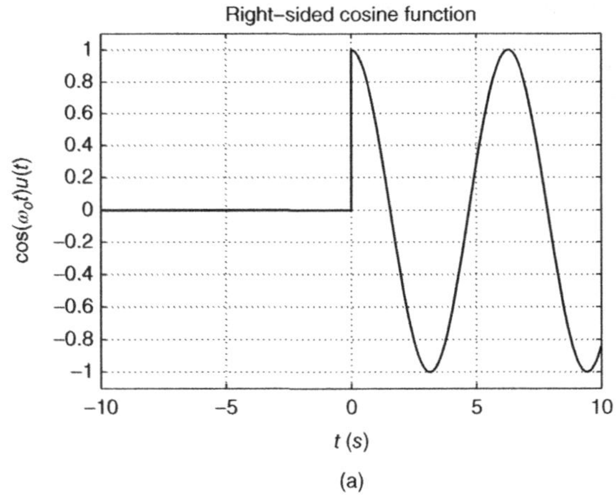

Figure A.18 Right-sided cosine function with $\omega_o = 1$ rad/s ($T_o = 2\pi$). (a) $x(t) = \cos(\omega_o t)u(t)$.

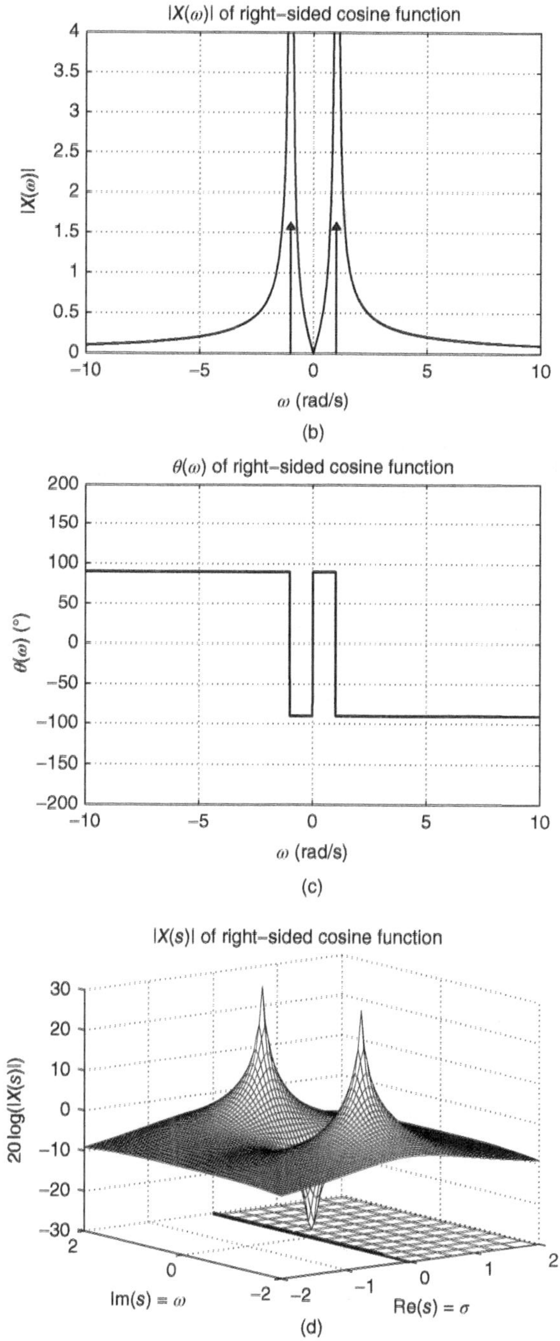

Figure A.18 Right-sided cosine function (continued). (b) Truncated $|X(\omega)|$. Each Dirac delta function has area $\pi/2$. (c) $\theta(\omega)$. (d) Truncated $20\log(|X(s)|)$ and ROC: $\sigma > 0$ (lower grid excluding the solid line).

- Power signal:

$$P = \lim_{T\to\infty} \frac{1}{T} \int_0^{T/2} \cos^2(\omega_o t)dt$$
$$= \lim_{T\to\infty} \frac{1}{T} \int_0^{T/2} (1/2)[1+\cos(2\omega_o t)]dt = 1/4. \tag{A.76}$$

- Phase:

$$\theta(\omega) = \lim_{a\to 0} \tan^{-1}(\omega/a(\omega_o^2-\omega^2)) = (\pi/2)\text{sgn}(\omega/(\omega_o^2-\omega^2)). \tag{A.77}$$

- Identity:

$$\cos(\omega_o t)u(t) = \frac{1}{2\pi}\int_{-\infty}^{\infty}\left[\frac{j\omega}{\omega_o^2-\omega^2} + \frac{\pi}{2}\delta(\omega+\omega_o) + \frac{\pi}{2}\delta(\omega-\omega_o)\right]\exp(j\omega t)d\omega$$
$$= \frac{j}{2\pi}\int_{-\infty}^{\infty}\frac{\omega}{\omega_o^2-\omega^2}\exp(j\omega t)d\omega + (1/2)\cos(\omega_o t)$$
$$= (1/2)\cos(\omega_o t)\text{sgn}(t) + (1/2)\cos(\omega_o t), \tag{A.78}$$

where the signum function causes the two terms to cancel for $t < 0$.

A.8.3 Cosine Function (Exponentially Weighted)

Parameters: $\alpha > 0$, $\omega_o = 2\pi f_o > 0$. Support: $t \in \mathcal{R}^+$.
Range: $x(t) \in [-\exp(-\alpha\pi/\omega_o), 1]$.

$$x(t) = \exp(-\alpha t)\cos(\omega_o t)u(t), \quad X(s) = \frac{s+\alpha}{(s+\alpha)^2+\omega_o^2},$$

ROC: $\sigma > -\alpha$, poles: $s = -\alpha \pm j\omega_o$, zeros: $s = -\alpha$,

$$|X(s)| = \frac{\sqrt{(\sigma+\alpha)^2+\omega^2}}{\sqrt{[(\sigma+\alpha)^2+(\omega+\omega_o)^2][(\sigma+\alpha)^2+(\omega-\omega_o)^2]}},$$

$$X(\omega) = \frac{\alpha+j\omega}{(\alpha+j\omega)^2+\omega_o^2},$$

$$|X(\omega)| = \frac{\sqrt{\alpha^2+\omega^2}}{\sqrt{[\alpha^2+(\omega+\omega_o)^2][\alpha^2+(\omega-\omega_o)^2]}},$$

$$\theta(\omega) = \tan^{-1}(\omega/\alpha) - \tan^{-1}((\omega+\omega_o)/\alpha) - \tan^{-1}((\omega-\omega_o)/\alpha).$$

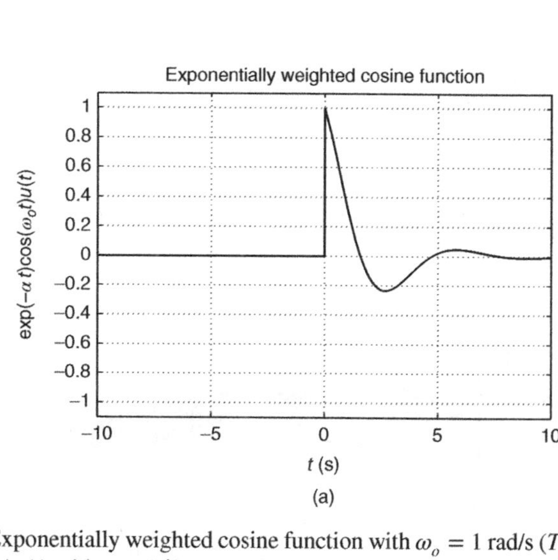

Figure A.19 Exponentially weighted cosine function with $\omega_o = 1$ rad/s ($T_o = 2\pi$). (a) $x(t) = \exp(-\alpha t)\cos(\omega_o t)u(t)$ with $\alpha = 1/2$.

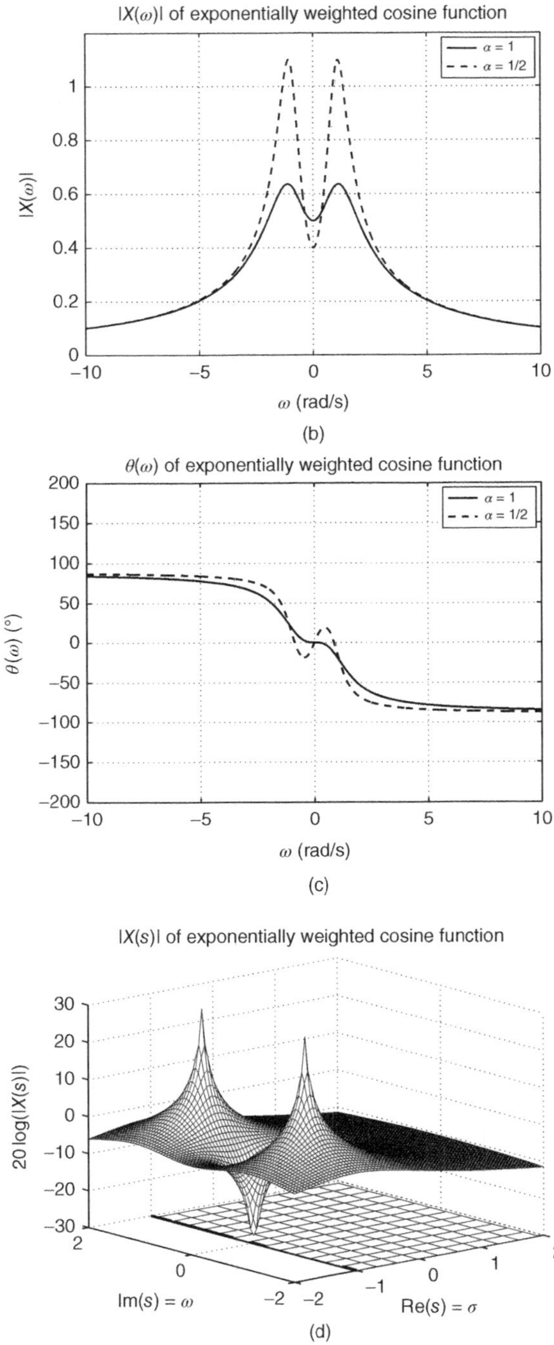

Figure A.19 Exponentially weighted cosine function (continued). (b) $|X(\omega)|$. (c) $\theta(\omega)$. (d) Truncated $20\log(|X(s)|)$ with $\alpha = 1$ and ROC: $\sigma > -1$ (lower grid excluding the solid line).

- Energy signal:

$$\begin{aligned}E &= \int_0^\infty \exp(-2\alpha t)\cos^2(\omega_o t)dt \\ &= (1/2)\int_0^\infty \exp(-2\alpha t)[1+\cos(2\omega_o t)]dt \\ &= 1/4\alpha + \alpha/4(\alpha^2+\omega_o^2) = (2\alpha^2+\omega_o^2)/4\alpha(\alpha^2+\omega_o^2). \qquad \text{(A 79)}\end{aligned}$$

- Phase: The fluctuations on each side of the origin are due to $\tan^{-1}((\omega \pm \omega_o)/\alpha)$.
- Identities:

$$\frac{d}{dt}\exp(-\alpha t)\cos(\omega_o t)u(t) = \delta(t) - [\alpha\cos(\omega_o t) + \omega_o\sin(\omega_o t)]\exp(-\alpha t)u(t), \qquad \text{(A.80)}$$

$$\begin{aligned}\int_0^t \exp(-\alpha t)\cos(\omega_o t)u(t)dt &= \frac{\exp(-\alpha t)}{\alpha^2+\omega_o^2}[\omega_o\sin(\omega_o t) - \alpha\cos(\omega_o t)]u(t) \\ &\quad + \frac{\alpha}{\alpha^2+\omega_o^2}u(t). \qquad \text{(A.81)}\end{aligned}$$

A.8.4 Cosine Function (Exponentially Weighted and Ramped)

Parameters: $\alpha > 0$, $\omega_o = 2\pi f_o > 0$. Support: $t \in \mathcal{R}^+$. Range: complicated.

$$x(t) = t\exp(-\alpha t)\cos(\omega_o t)u(t), \quad X(s) = \frac{(s+\alpha)^2 - \omega_o^2}{[(s+\alpha)^2 + \omega_o^2]^2},$$

$$\text{ROC}: \sigma > -\alpha, \quad \text{poles: } s = -\alpha \pm j\omega_o \text{ (double pair)}, \quad \text{zeros: } s = -\alpha \pm \omega_o,$$

$$|X(s)| = \frac{\sqrt{[(\sigma+\alpha)^2 - \omega^2 - \omega_o^2]^2 + 4(\sigma+\alpha)^2\omega^2}}{[(\sigma+\alpha)^2 - \omega^2 + \omega_o^2]^2 + 4(\sigma+\alpha)^2\omega^2},$$

$$X(\omega) = \frac{(\alpha + j\omega)^2 - \omega_o^2}{[(\alpha + j\omega)^2 + \omega_o^2]^2},$$

$$|X(\omega)| = \frac{\sqrt{(\alpha^2 - \omega^2 - \omega_o^2)^2 + 4\alpha^2\omega^2}}{(\alpha^2 - \omega^2 + \omega_o^2)^2 + 4\alpha^2\omega^2},$$

$$\theta(\omega) = \tan^{-1}(\omega/(\alpha+\omega_o)) + \tan^{-1}(\omega/(\alpha-\omega_o))$$
$$-2\tan^{-1}((\omega+\omega_o)/\alpha) - 2\tan^{-1}((\omega-\omega_o)/\alpha).$$

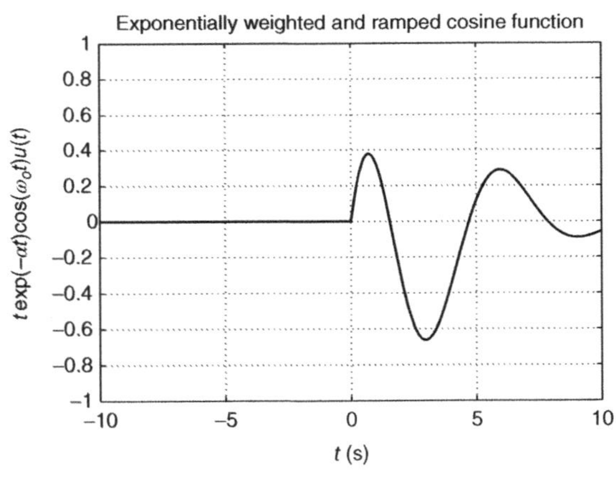

Figure A.20 Exponentially weighted and ramped cosine function with $\omega_o = 1$ rad/s ($T_o = 2\pi$). (a) $x(t) = t\exp(-\alpha t)\cos(\omega_o t)u(t)$ with $\alpha = 1/2$.

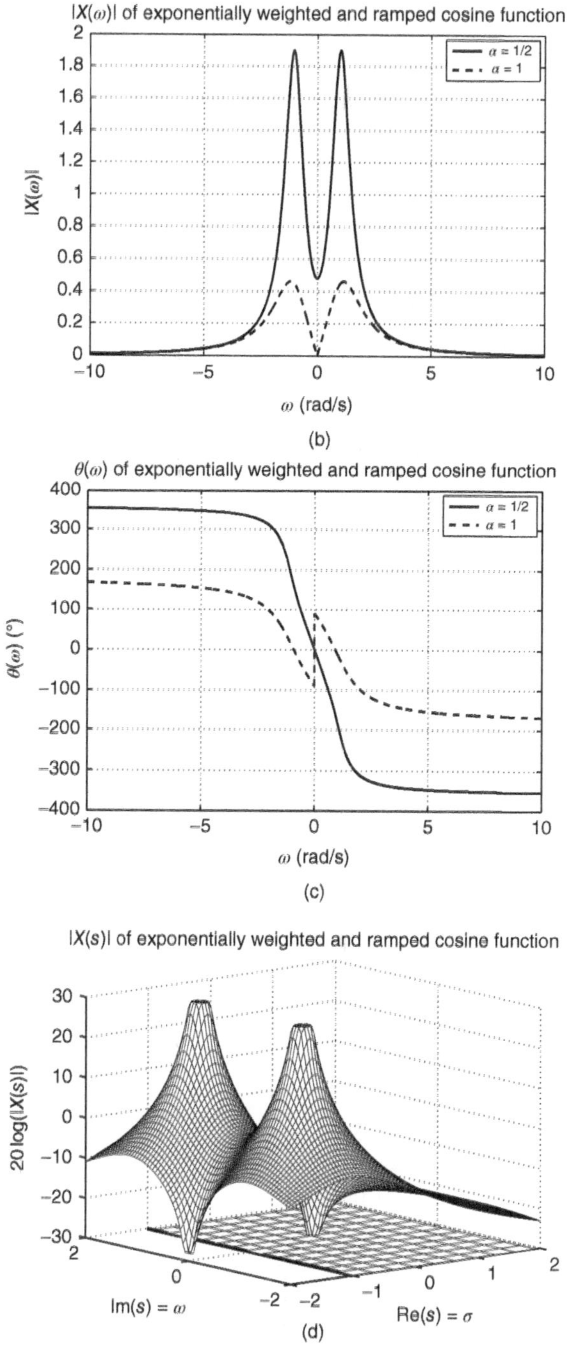

Figure A.20 Exponentially weighted and ramped cosine function (continued). (b) $|X(\omega)|$. (c) $\theta(\omega)$. (d) Truncated $20\log(|X(s)|)$ with $\alpha = 1$ and ROC: $\sigma > -1$ (lower grid excluding the solid line).

- Energy signal:

$$E = \int_0^\infty t^2 \exp(-2\alpha t)\cos^2(\omega_o t)dt$$
$$= (1/2)\int_0^\infty t^2 \exp(-2\alpha t)[1 + \cos(2\omega_o t)]dt$$
$$= 1/8\alpha^3 + (\alpha^3 - 3\alpha\omega_o^2)/8(\alpha^2 + \omega_o^2)^3. \quad \text{(A.82)}$$

- Phase: The fluctuations on each side of the origin are due to $\tan^{-1}((\omega \pm \omega_o)/\alpha)$.
- Identity:

$$\frac{d}{dt} t\exp(-\alpha t)\cos(\omega_o t)u(t) = (1 - \alpha t)\exp(-\alpha t)\cos(\omega_o t)u(t)$$
$$+ \omega_o t\exp(-\alpha t)\sin(\omega_o t)u(t). \quad \text{(A.83)}$$

A.8.5 Sine Function (Two-Sided)

Parameters: $\omega_o = 2\pi f_o > 0$. Support: $t \in \mathcal{R}$. Range: $x(t) \in [-1, 1]$.

$$x(t) = \sin(\omega_o t), \quad X(s) = \text{does not exist (bilateral)},$$

$$\text{ROC: none}, \quad \text{poles: none}, \quad \text{zeros: none},$$

$$X(\omega) = j\pi\delta(\omega + \omega_o) - j\pi\delta(\omega - \omega_o) \quad \text{(exists in the limit)},$$

$$|X(\omega)| = \pi\delta(\omega + \omega_o) + \pi\delta(\omega - \omega_o), \quad \theta(\omega) = (\pi/2)(\delta[\omega + \omega_o] - \delta[\omega - \omega_o]).$$

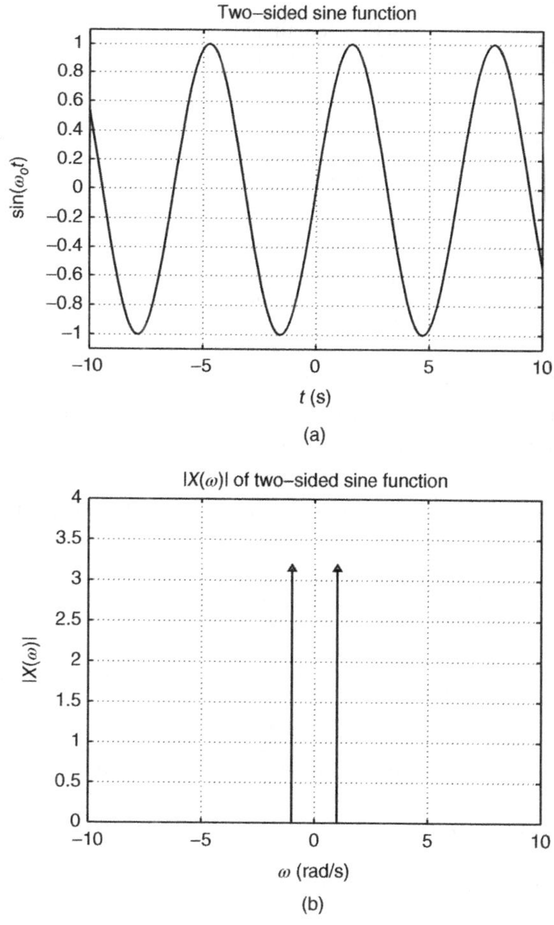

Figure A.21 Two-sided sine function with $\omega_o = 1$ rad/s ($T_o = 2\pi$). (a) $x(t) = \sin(\omega_o t)$. (b) $|X(\omega)|$. Each Dirac delta function has area π.

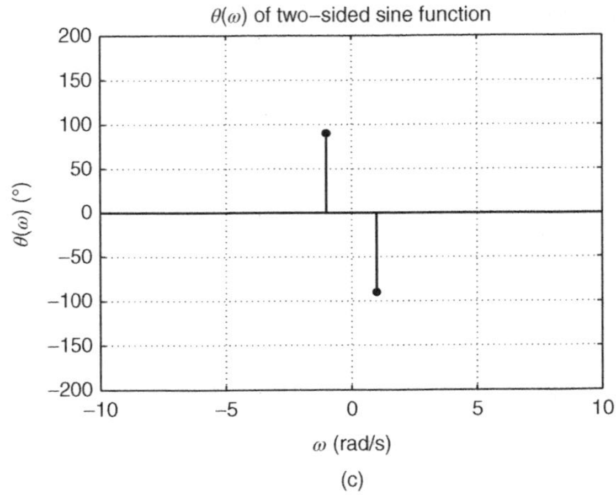

Figure A.21 Two-sided sine function (continued). (c) $\theta(\omega)$. The solid circles at $\omega = \pm\omega_o$ represent the phase of the Dirac delta component.

- Power signal:

$$P = \lim_{T\to\infty} \frac{1}{T} \int_{-T/2}^{T/2} \sin^2(\omega_o t)dt$$

$$= \lim_{T\to\infty} \frac{1}{T} \int_{-T/2}^{T/2} (1/2)[1 - \cos(2\omega_o t)]dt = 1/2. \tag{A.84}$$

- Phase: $\theta(\omega)$ is nonzero only at $\pm\omega_o$.
- The bilateral Laplace transform does not exist because all terms cancel (as was the case for the two-sided cosine function):

$$\mathcal{L}_b(\sin(\omega_o t)) = \int_{-\infty}^{0} \sin(\omega_o t)\exp(-st)dt + \int_{0}^{\infty} \sin(\omega_o t)\exp(-st)dt$$

$$= \frac{1}{2j}\int_{-\infty}^{0} [\exp(-(s - j\omega_o)t) - \exp(-(s + j\omega_o)t)]dt$$

$$+\frac{1}{2j}\int_{0}^{\infty} [\exp(-(s - j\omega_o)t) - \exp(-(s + j\omega_o)t)]dt$$

$$= \frac{-1}{2j(s - j\omega_o)} - \frac{-1}{2j(s + j\omega_o)} + \frac{1}{2j(s - j\omega_o)} - \frac{1}{2j(s + j\omega_o)}. \tag{A.85}$$

- Identity (Euler's inverse formula):

$$\sin(\omega_o t) = \frac{1}{2\pi}\int_{-\infty}^{\infty} [\pi j\delta(\omega + \omega_o) - \pi j\delta(\omega - \omega_o)\exp(j\omega t)]d\omega$$

$$= (1/2j)[\exp(j\omega_o t) - \exp(-j\omega_o t)]. \tag{A.86}$$

A.8.6 Sine Function (Right-Sided)

Parameters: $\omega_o = 2\pi f_o > 0$. Support: $t \in \mathcal{R}^+$. Range: $x(t) \in [-1, 1]$.

$$x(t) = \sin(\omega_o t)u(t),\ X(s) = \frac{\omega_o}{s^2 + \omega_o^2},$$

$$\text{ROC: } \sigma > 0, \quad \text{poles: } s = \pm j\omega_o, \quad \text{zeros: none},$$

$$|X(s)| = \frac{\omega_o}{\sqrt{([\sigma^2 + (\omega + \omega_o)^2][\sigma^2 + (\omega - \omega_o)^2]}},$$

$$X(\omega) = \frac{\omega_o}{\omega_o^2 - \omega^2} + \frac{j\pi}{2}\delta(\omega + \omega_o) - \frac{j\pi}{2}\delta(\omega - \omega_o) \quad \text{(exists in the limit)},$$

$$|X(\omega)| = \frac{\omega_o}{|\omega_o^2 - \omega^2|} + \frac{\pi}{2}\delta(\omega + \omega_o) + \frac{\pi}{2}\delta(\omega - \omega_o),$$

$$\theta(\omega) = \pi\text{sgn}(\omega/(\omega_o^2 - \omega^2)) + (\pi/2)(\delta[\omega + \omega_o] - \delta[\omega - \omega_o]).$$

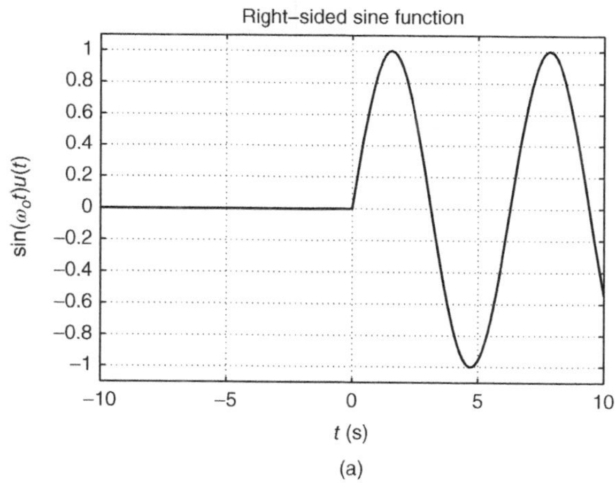

Figure A.22 Right-sided sine function with $\omega_o = 1$ rad/s ($T_o = 2\pi$). (a) $x(t) = \sin(\omega_o t)u(t)$.

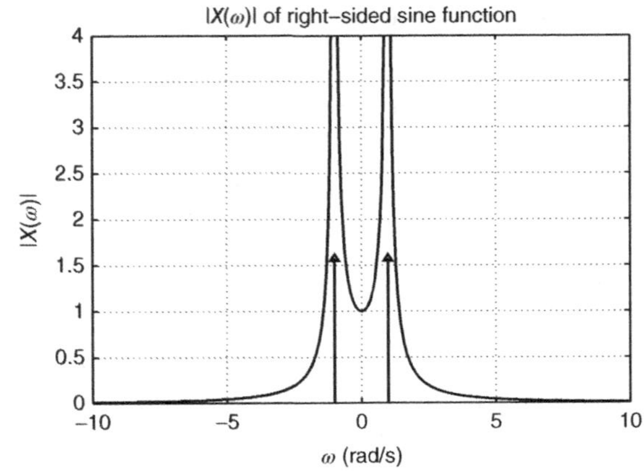

(b)

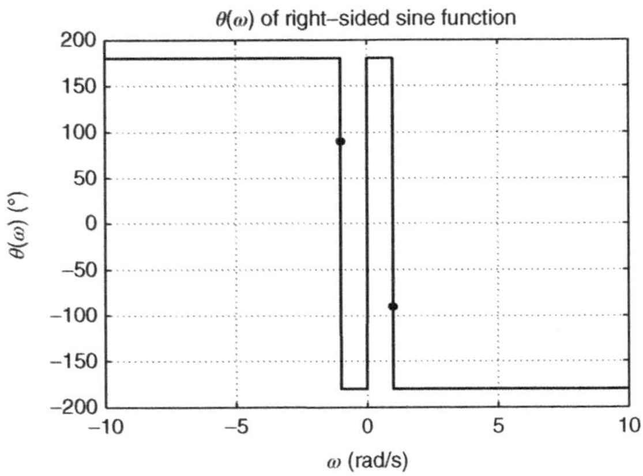

(c)

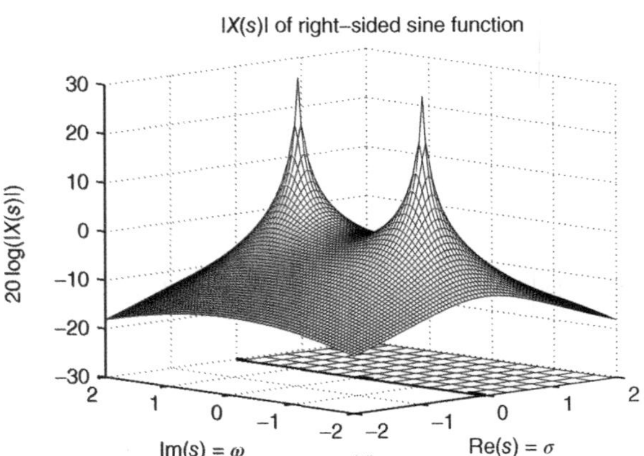

(d)

Figure A.22 Right-sided sine function (continued). (b) Truncated $|X(\omega)|$. Each Dirac delta function has area $\pi/2$. (c) $\theta(\omega)$. The solid circles at $\omega = \pm\omega_o$ represent the phase of the Dirac delta component. (d) Truncated $20 \log(|X(s)|)$ and ROC: $\sigma > 0$ (lower grid excluding the solid line).

- Power signal:

$$P = \lim_{T\to\infty} \frac{1}{T} \int_0^{T/2} \sin^2(\omega_o t)dt$$
$$= \lim_{T\to\infty} \frac{1}{T} \int_0^{T/2} (1/2)[1 - \cos(2\omega_o t)]dt = 1/4. \quad \text{(A.87)}$$

- Identity:

$$\sin(\omega_o t)u(t) = \frac{1}{2\pi} \int_{-\infty}^{\infty} \left[\frac{j\omega_o}{\omega_o^2 - \omega^2} + \frac{j\pi}{2}\delta(\omega + \omega_o) - \frac{j\pi}{2}\delta(\omega - \omega_o) \right] \exp(j\omega t)d\omega$$
$$= \frac{j}{2\pi} \int_{-\infty}^{\infty} \frac{\omega_o}{\omega_o^2 - \omega^2} \exp(j\omega t)d\omega + (1/2)\sin(\omega_o t)$$
$$= (1/2)\sin(\omega_o t)\mathrm{sgn}(t) + (1/2)\sin(\omega_o t), \quad \text{(A.88)}$$

where the signum function causes the two terms to cancel for $t < 0$.

A.8.7 Sine Function (Exponentially Weighted)

Parameters: $\alpha > 0$, $\omega_o = 2\pi f_o > 0$. Support: $t \in \mathcal{R}^+$. Range: complicated.

$$x(t) = \exp(-\alpha t)\sin(\omega_o t)u(t), \quad X(s) = \frac{\omega_o}{(s+\alpha)^2 + \omega_o^2},$$

$$\text{ROC: } \sigma > -\alpha, \quad \text{poles: } s = -\alpha \pm j\omega_o, \quad \text{zeros: none},$$

$$|X(s)| = \frac{\omega_o}{\sqrt{[(\sigma+\alpha)^2 + (\omega+\omega_o)^2][(\sigma+\alpha)^2 + (\omega-\omega_o)^2]}},$$

$$X(\omega) = \frac{\omega_o}{(\alpha + j\omega)^2 + \omega_o^2},$$

$$|X(\omega)| = \frac{\omega_o}{\sqrt{[\alpha^2 + (\omega+\omega_o)^2][\alpha^2 + (\omega-\omega_o)^2]}},$$

$$\theta(\omega) = -\tan^{-1}((\omega+\omega_o)/\alpha) - \tan^{-1}((\omega-\omega_o)/\alpha).$$

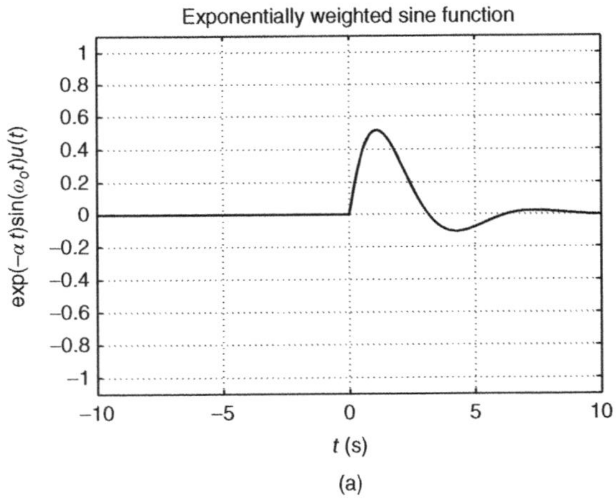

Figure A.23 Exponentially weighted sine function with $\omega_o = 1$ rad/s ($T_o = 2\pi$). (a) $x(t) = \exp(-\alpha t)\sin(\omega_o t)u(t)$ with $\alpha = 1/2$.

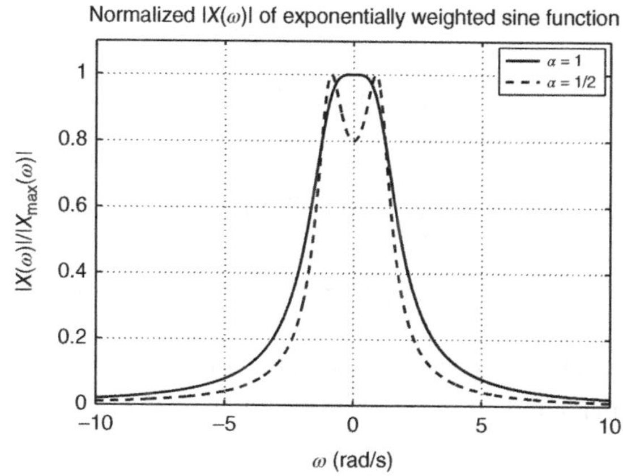

(b)

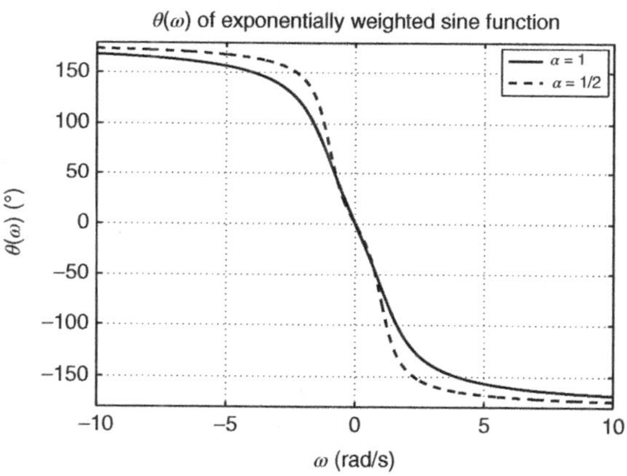

(c)

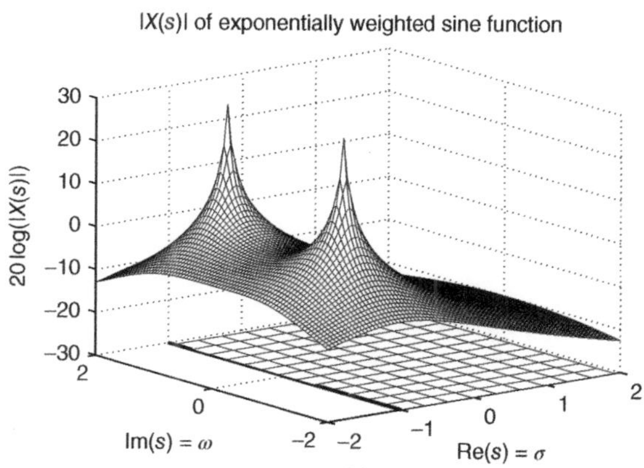

(d)

Figure A.23 Exponentially weighted sine function (continued). (b) Normalized $|X(\omega)|$. (c) $\theta(\omega)$. (d) Truncated $20\log(|X(s)|)$ with $\alpha = 1$ and ROC: $\sigma > -1$ (lower grid excluding the solid line).

- Energy signal:

$$
\begin{aligned}
E &= \int_0^\infty \exp(-2\alpha t)\sin^2(\omega_o t)dt \\
&= (1/2)\int_0^\infty \exp(-2\alpha t)[1-\cos(2\omega_o t)]dt \\
&= 1/4\alpha - \alpha/4(\alpha^2+\omega_o^2) = \omega_o^2/4\alpha(\alpha^2+\omega_o^2). \qquad \text{(A.89)}
\end{aligned}
$$

- Phase: The fluctuations on each side of the origin are due to $\tan^{-1}((\omega \pm \omega_o)/\alpha)$.
- Maximum magnitude:

$$
|X(\omega)|_{\max} = \begin{cases} \omega_o/(\alpha^2+\omega_o^2) \text{ at } \omega = 0, & \alpha \geq \omega_o \\ 1/2\alpha \text{ at } \omega = \pm\sqrt{\omega_o^2-\alpha^2}, & \alpha < \omega_o. \end{cases} \qquad \text{(A.90)}
$$

- Identities:

$$
\frac{d}{dt}\exp(-\alpha t)\sin(\omega_o t)u(t) = [\omega_o\cos(\omega_o t) - \alpha\sin(\omega_o t)]\exp(-\alpha t)u(t), \qquad \text{(A.91)}
$$

$$
\begin{aligned}
\int_0^t \exp(-\alpha t)\sin(\omega_o t)u(t)dt &= -\frac{\exp(-\alpha t)}{\alpha^2+\omega_o^2}[\omega_o\cos(\omega_o t) + \alpha\sin(\omega_o t)]u(t) \\
&+ \frac{\omega_o}{\alpha^2+\omega_o^2}u(t). \qquad \text{(A.92)}
\end{aligned}
$$

A.8.8 Sine Function (Exponentially Weighted and Ramped)

Parameters: $\alpha > 0$, $\omega_o = 2\pi f_o > 0$. Support: $t \in \mathcal{R}^+$. Range: complicated.

$$x(t) = t\exp(-\alpha t)\sin(\omega_o t)u(t), \quad X(s) = \frac{2\omega_o(s+\alpha)}{[(s+\alpha)^2+\omega_o^2]^2},$$

ROC: $\sigma > -\alpha$, poles: $s = -\alpha \pm j\omega_o$ (double pair), zeros: $s = -\alpha$,

$$|X(s)| = \frac{2\omega_o\sqrt{(\sigma+\alpha)^2+\omega^2}}{[(\sigma+\alpha)^2-\omega^2+\omega_o^2]^2+4(\sigma+\alpha)^2\omega^2},$$

$$X(\omega) = \frac{2\omega_o(\alpha+j\omega)}{[(\alpha+j\omega)^2+\omega_o^2]^2},$$

$$|X(\omega)| = \frac{2\omega_o\sqrt{\alpha^2+\omega^2}}{(\alpha^2-\omega^2+\omega_o^2)^2+4\alpha^2\omega^2},$$

$$\theta(\omega) = \tan^{-1}(\omega/\alpha) - 2\tan^{-1}((\omega+\omega_o)/\alpha) - 2\tan^{-1}((\omega-\omega_o/\alpha).$$

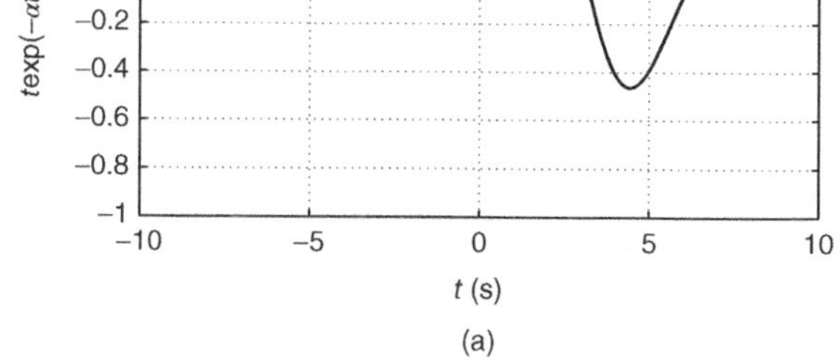

Figure A.24 Exponentially weighted and ramped sine function with $\omega_o = 1$ rad/s ($T_o = 2\pi$). (a) $x(t) = t\exp(-\alpha t)\sin(\omega_o t)u(t)$ with $\alpha = 1/2$.

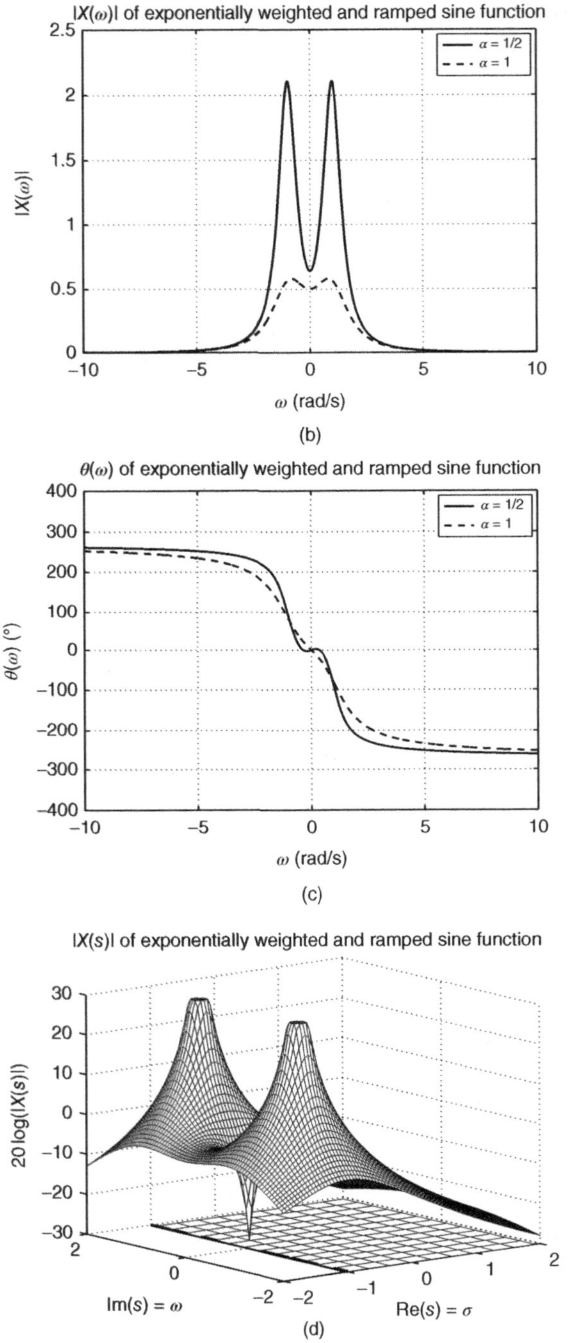

Figure A.24 Exponentially weighted and ramped sine function (continued). (b) $|X(\omega)|$. (c) $\theta(\omega)$. (d) Truncated $20\log(|X(s)|)$ with $\alpha = 1$ and ROC: $\sigma > -1$ (lower grid excluding the solid line).

- Energy signal:

$$\begin{aligned}E &= \int_0^\infty t^2 \exp(-2\alpha t)\sin^2(\omega_o t)dt \\ &= (1/2)\int_0^\infty t^2 \exp(-2\alpha t)[1 - \cos(2\omega_o t)]dt \\ &= 1/8\alpha^3 - (\alpha^3 - 3\alpha\omega_o^2)/8(\alpha^2 + \omega_o^2)^3.\end{aligned} \tag{A.93}$$

- Phase: The fluctuations on each side of the origin are due to $\tan^{-1}((\omega \pm \omega_o)/\alpha)$.
- Identity:

$$\begin{aligned}\frac{d}{dt} t \exp(-\alpha t) \sin(\omega_o t)u(t) = (1 - \alpha t) \exp(-\alpha t) \sin(\omega_o t)u(t) \\ + \omega_o t \exp(-\alpha t) \cos(\omega_o t)u(t).\end{aligned} \tag{A.94}$$

APPENDIX B

INVERSE LAPLACE TRANSFORMS

In this appendix, we provide additional *unilateral* Laplace transform pairs in Table B.1 and B.2, giving the s-domain expression first. These tables are useful because they include results with multiple poles, and so a partial fraction expansion (PFE) is avoided (though the reader should be familiar with that approach for finding inverse Laplace transforms of rational functions). All functions in this table are right-sided, which means the region of convergence (ROC) lies to the right of the pole with the smallest magnitude on the left-half of the s-plane, including the imaginary axis. For a transform with three nonzero poles, they have been arranged as $-c < -b < -a$ such that the ROC is $\sigma = \text{Re}(s) > -a$.

In the following sections, we consider three Laplace transform pairs, describe the corresponding ordinary differential equations (ODEs), and give integrator implementations for the systems, one of which is a double integrator modified by feedback.

B.1 IMPROPER RATIONAL FUNCTION

Consider the improper Laplace transform in Table B.1:

$$H(s) = \frac{s+d}{s+a}, \tag{B.1}$$

Mathematical Foundations for Linear Circuits and Systems in Engineering, First Edition. John J. Shynk.

Companion Website: http://www.wiley.com/go/linearcircuitsandsystems

TABLE B.1 Inverse Laplace Transforms: Step, Ramp, and Exponential

Laplace Transform $X(s)$	Time-Domain $x(t)$	ROC
1	$\delta(t)$	$s \in C$
s	$\delta'(t)$	$s \in C$
$1/s$	$u(t)$	$\sigma > 0$
$1/s^2$	$tu(t)$	$\sigma > 0$
$1/s^n$	$[t^{n-1}/(n-1)!]u(t)\ (n \in \mathcal{N})$	$\sigma > 0$
$1/(s+a)$	$\exp(-at)u(t)$	$\sigma > -a$
$1/(s+a)^2$	$t\exp(-at)u(t)$	$\sigma > -a$
$1/(s+a)^n$	$[t^{n-1}/(n-1)!]\exp(-at)u(t)\ (n \in \mathcal{N})$	$\sigma > -a$
$(s+d)/(s+a)$	$\delta(t) + (d-a)\exp(-at)u(t)$	$\sigma > -a$
$1/s(s+a)$	$(1/a)[1-\exp(-at)]u(t)$	$\sigma > 0$
$(s+d)/s(s+a)$	$(1/a^2)[d - d\exp(-at)$ $+ (a^2 - ad)t\exp(-at)]u(t)$	$\sigma > 0$
$1/s^2(s+a)$	$(1/a^2)[\exp(-at) + at - 1]u(t)$	$\sigma > 0$
$1/s(s+a)^2$	$(1/a^2)[1 - \exp(-at) - at\exp(-at)]u(t)$	$\sigma > 0$
$1/(s+a)(s+b)$	$[1/(b-a)][\exp(-at) - \exp(-bt)]u(t)$	$\sigma > -a$
$(s+d)/(s+a)(s+b)$	$[1/(b-a)][(d-a)\exp(-at)$ $-(d-b)\exp(-bt)]u(t)$	$\sigma > -a$
$1/s(s+a)(s+b)$	$(1/ab)[1 - b\exp(-at)/(b-a)$ $+ a\exp(-bt)/(b-a)]u(t)$	$\sigma > 0$
$(s+d)/s(s+a)(s+b)$	$(1/ab)[d - b(d-a)\exp(-at)/(b-a)$ $+ a(d-b)\exp(-bt)/(b-a)]u(t)$	$\sigma > 0$
$1/(s+a)(s+b)(s+c)$	$[\exp(-at)/(c-a)(b-a)$ $+ \exp(-bt)/(c-b)(a-b)$ $+ \exp(-ct)/(b-c)(a-c)]u(t)$	$\sigma > -a$
$(s+d)/(s+a)(s+b)(s+c)$	$[(d-a)\exp(-at)/(c-a)(b-a)$ $+ (d-b)\exp(-bt)/(c-b)(a-b)$ $+ (d-c)\exp(-ct)/(b-c)(a-c)]u(t)$	$\sigma > -a$

which has a real pole at $s = -a$ and a real zero at $s = -d$. Long division yields

$$H(s) = 1 + \frac{d-a}{s+a}, \tag{B.2}$$

and so the inverse Laplace transform is

$$h(t) = \delta(t) - (d-a)\exp(-at)u(t). \tag{B.3}$$

This impulse response function includes a direct path from the input $x(t)$ to the output $y(t)$ of the system. An integrator implementation of the system is shown in Figure B.1,

TABLE B.2 Inverse Laplace Transforms: Sinusoidal and Hyperbolic

Laplace Transform $X(s)$	Time-Domain $x(t)$	ROC
$\omega_o/(s^2+\omega_o^2)$	$\sin(\omega_o t)u(t)$	$\sigma > 0$
$s/(s^2+\omega_o^2)$	$\cos(\omega_o t)u(t)$	$\sigma > 0$
$a/(s^2-b^2)$	$\sinh(bt)u(t)$	$\sigma > \lvert b\rvert$
$s/(s^2-b^2)$	$\cosh(bt)u(t)$	$\sigma > \lvert b\rvert$
$(s+d)\omega_o/(s^2+\omega_o^2)$	$\sqrt{d^2+\omega_o^2}\sin(\omega_o t+\theta)u(t)$	
	with $\theta = \tan^{-1}(\omega_o/d)$	$\sigma > 0$
$(s-\omega_o^2/d)d/(s^2+\omega_o^2)$	$\sqrt{d^2+\omega_o^2}\cos(\omega_o t+\theta)u(t)$	
	with $\theta = \tan^{-1}(\omega_o/d)$	$\sigma > 0$
$[s\sin(\theta)+\omega_o\cos(\theta)]/(s^2+\omega_o^2)$	$\sin(\omega_o t+\theta)u(t)$	$\sigma > 0$
$[s\cos(\theta)-\omega_o\sin(\theta)]/(s^2+\omega_o^2)$	$\cos(\omega_o t+\theta)u(t)$	$\sigma > 0$
$\omega_o^2/s(s^2+\omega_o^2)$	$[1-\cos(\omega_o t)]u(t)$	$\sigma > 0$
$2\omega_o s/(s^2+\omega_o^2)^2$	$t\sin(\omega_o t)u(t)$	$\sigma > 0$
$(s^2-\omega_o^2)/(s^2+\omega_o^2)^2$	$t\cos(\omega_o t)u(t)$	$\sigma > 0$
$\omega_o^3/s^2(s^2+\omega_o^2)$	$[\omega_o t-\sin(\omega_o t)]u(t)$	$\sigma > 0$
$2\omega_o^3/(s^2+\omega_o^2)^2$	$[\sin(\omega_o t)-\omega_o t\cos(\omega_o t)]u(t)$	$\sigma > 0$
$(s^2-\omega_o^2)s/(s^2+\omega_o^2)^2$	$[\cos(\omega_o t)-\omega_o t\sin(\omega_o t)]u(t)$	$\sigma > 0$
$2\omega_o s^2/(s^2+\omega_o^2)^2$	$[\sin(\omega_o t)+\omega_o t\cos(\omega_o t)]u(t)$	$\sigma > 0$
$(s^2+3\omega_o^2)s/(s^2+\omega_o^2)^2$	$[\cos(\omega_o t)+\omega_o t\sin(\omega_o t)]u(t)$	$\sigma > 0$
$(\omega_{o1}^2-\omega_{o2}^2)/(s^2+\omega_{o1}^2)(s^2+\omega_{o2}^2)$	$[(1/\omega_{o2})\sin(\omega_{o2}t)$	
	$-\ (1/\omega_{o1})\sin(\omega_{o1}t)]u(t)\ (\omega_{o1}\neq\omega_{o2})$	$\sigma > 0$
$\omega_o/[(s+a)^2+\omega_o^2]$	$\exp(-at)\sin(\omega_o t)u(t)$	$\sigma > -a$
$(s+a)/[(s+a)^2+\omega_o^2]$	$\exp(-at)\cos(\omega_o t)u(t)$	$\sigma > -a$
$b/[(s-a)^2+b^2]$	$\exp(-at)\sinh(bt)u(t)$	$\sigma > -a+\lvert b\rvert$
$(s-a)/[(s-a)^2+b^2]$	$\exp(-at)\cosh(bt)u(t)$	$\sigma > -a+\lvert b\rvert$
$(s+d)\omega_o/[(s+a)^2+\omega_o^2]$	$[\omega_o\cos(\omega_o t)+(d-a)\sin(\omega_o t)]$	
	$\times\exp(-at)u(t)$	$\sigma > -a$
$[(s+a)^2-\omega_o^2]/[(s+a)^2+\omega_o^2]^2$	$t\exp(-at)\cos(\omega_o t)u(t)$	$\sigma > -a$
$2\omega_o(s+a)/[(s+a)^2+\omega_o^2]^2$	$t\exp(-at)\sin(\omega_o t)u(t)$	$\sigma > -a$

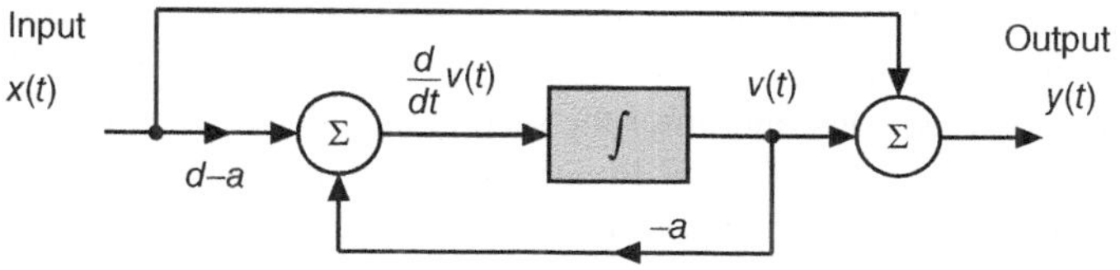

Figure B.1 Integrator implementation of an improper first-order transfer function.

which is similar to the one given earlier in Figure 6.2 except for the direct input/output path (and we have assumed zero initial conditions). The corresponding ODE for the ratio in (B.2) is

$$\frac{d}{dt}v(t) + av(t) = (d-a)x(t), \tag{B.4}$$

where $v(t)$ is the output of the integrator.

B.2 UNBOUNDED SYSTEM

Next, we examine the following Laplace transform in Table B.2:

$$H(s) = \frac{2\omega_o s}{(s^2+\omega_o^2)^2}, \tag{B.5}$$

with impulse response function

$$h(t) = t\sin(\omega_o t)u(t). \tag{B.6}$$

This system grows unbounded because of the double poles, which yield the ramp t. Since the poles are located on the imaginary axis, there is no exponential damping. Note that the final value theorem does not hold for this system; it gives a value of 0, which is obviously incorrect. This is due to the undamped sinusoidal nature of $h(t)$, which has an average value of 0 over one period.

The ODE for this system is derived by rewriting $H(s) = Y(s)/X(s)$ as

$$(s^4 + 2\omega_o^2 s^2 + \omega_o^4)Y(s) = 2\omega_o sX(s), \tag{B.7}$$

which yields

$$\frac{d^4}{dt^4}y(t) + 2\omega_o^2\frac{d^2}{dt^2}y(t) + \omega_o^4 y(t) = 2\omega_o\frac{d}{dt}x(t). \tag{B.8}$$

An integrator implementation of this system is shown in Figure B.2, which also includes a differentiator for the input. It is the repeated nature of the two sets of double integrators along with the specific feedback coefficients that cause the system to be unstable. This is due to the fact that two cascaded integrators without feedback have Laplace transform $1/s^2$, which corresponds to the ramp function $r(t)$.

We mention that it is possible to remove the differentiator so that only integrators are used in the implementation. This is easily done by noting in Figure B.2 that the input signal is not fed back until after the second integrator. Thus, we can move $x(t)$ to the right of the first integrator and drop the derivative as shown in Figure B.3. Note, however, that the first two integrator output labels are no longer the same as those in Figure B.2 because the derivative of $x(t)$ is no longer present in the first summation. The second set of integrator labels is unchanged because the signals in that section of the implementation are the same as before.

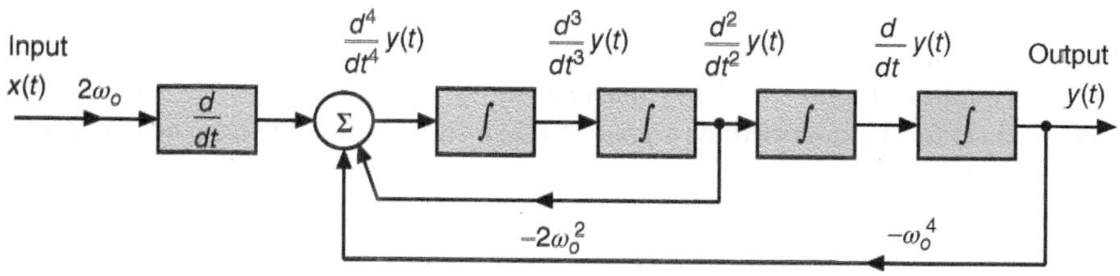

Figure B.2 Integrator/differentiator implementation of an unstable system with repeated poles.

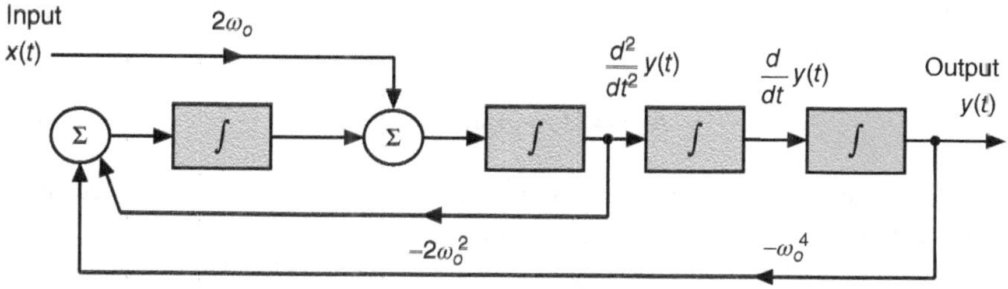

Figure B.3 Integrator-only implementation of an unstable system with repeated poles.

B.3 DOUBLE INTEGRATOR AND FEEDBACK

As mentioned in the previous section, the inverse Laplace transform of $1/s^2$ is the ramp function $r(t)$, which obviously grows without bound. Here, we demonstrate how to modify the pole locations with *feedback* as illustrated in Figure B.4. Feedback is

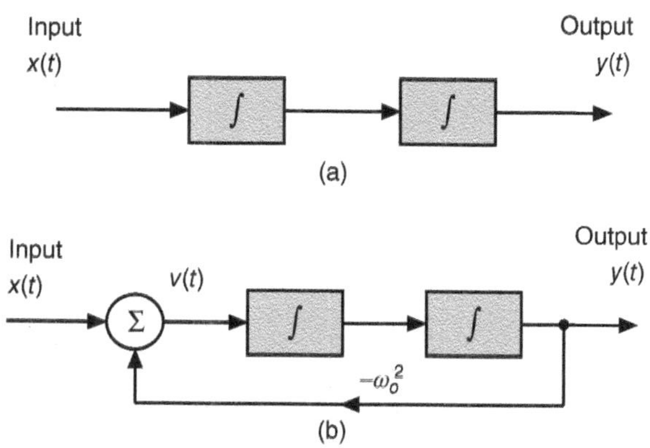

Figure B.4 (a) Double integrator implementation of the ramp function $h(t) = r(t)$ and (b) using feedback to modify the pole locations.

important for stability in control systems; this topic was not covered in Chapter 7, though an example is considered in Problem 7.26.

The system in Figure B.4(b) is analyzed in the s-domain by first writing $V(s) = X(s) - aY(s)$ for the intermediate signal and then substituting this into the expression for the output:

$$Y(s) = V(s)/s^2 = [X(s) - \omega_o^2 Y(s)]/s^2. \tag{B.9}$$

Solving for $Y(s)$ yields

$$Y(s)(1 + \omega_o^2/s^2) = X(s)/s^2 \implies H(s) = Y(s)/X(s) = 1/(s^2 + \omega_o^2), \tag{B.10}$$

whose poles are located at $s = \pm\sqrt{-\omega_o^2} \implies s_1, s_2 = \pm j\omega_o$, and the impulse response function is now sinusoidal:

$$h(t) = (1/\omega_o)\sin(\omega_o t)u(t), \tag{B.11}$$

which is a marginally stable system. Observe that *negative* feedback is used in this implementation. If $+\omega_o^2$ is used instead, then the poles are located at $s_1, s_2 = \pm\omega_o$, corresponding to an unstable system because one of them is located on the right-half of the s-plane. This simple example illustrates how feedback can be used to modify a system and the importance of using negative feedback for proper pole placement.

APPENDIX C

IDENTITIES, DERIVATIVES, AND INTEGRALS

C.1 TRIGONOMETRIC IDENTITIES

- *Basic identities*:

$$\cos(x)\cos(y) = (1/2)[\cos(x-y) + \cos(x+y)], \tag{C.1}$$

$$\sin(x)\sin(y) = (1/2)[\cos(x-y) - \cos(x+y)], \tag{C.2}$$

$$\sin(x)\cos(y) = (1/2)[\sin(x-y) + \sin(x+y)], \tag{C.3}$$

$$\cos(x \pm y) = \cos(x)\cos(y) \mp \sin(x)\cos(y), \tag{C.4}$$

$$\sin(x \pm y) = \sin(x)\cos(y) \pm \cos(x)\sin(y), \tag{C.5}$$

$$\cos(x \pm \pi/2) = \mp \sin(x), \quad \sin(x \pm \pi/2) = \pm \cos(x), \tag{C.6}$$

$$\sin(\tan^{-1}(x)) = \frac{x}{\sqrt{1+x^2}}, \quad \cos(\tan^{-1}(x)) = \frac{1}{\sqrt{1+x^2}}, \tag{C.7}$$

Mathematical Foundations for Linear Circuits and Systems in Engineering, First Edition. John J. Shynk.

Companion Website: http://www.wiley.com/go/linearcircuitsandsystems

$$\tan(x) = \frac{\sin(x)}{\cos(x)}, \quad \tan^{-1}(x) - x = -\tan^{-1}(x), \tag{C.8}$$

$$\cos^2(x) = (1/2)[1 + \cos(2x)], \quad \sin^2(x) = (1/2)[1 - \cos(2x)]. \tag{C.9}$$

- *Rectangular and polar forms*:

$$r\cos(x + \phi) = r\cos(\phi)\cos(x) - r\sin(\phi)\sin(x), \tag{C.10}$$

$$\triangleq a\cos(x) - b\sin(x), \tag{C.11}$$

$$a = r\cos(\phi), \quad b = r\sin(\phi), \quad r = \sqrt{a^2 + b^2}, \quad \phi = \tan^{-1}(b/a). \tag{C.12}$$

- *Euler's formulas*:

$$\cos(x) = (1/2)[\exp(jx) + \exp(-jx)], \tag{C.13}$$

$$\sin(x) = (1/2j)[\exp(jx) - \exp(-jx)], \tag{C.14}$$

$$\exp(\pm jx) = \cos(x) \pm j\sin(x), \quad \exp(j\pi) = -1. \tag{C.15}$$

- *Hyperbolic functions*:

$$\sinh(x) = (1/2)[\exp(x) - \exp(-x)], \tag{C.16}$$

$$\cosh(x) = (1/2)[\exp(x) + \exp(-x)], \tag{C.17}$$

$$\tanh(x) = \frac{\sinh(x)}{\cosh(x)} = \frac{\exp(2x) - 1}{\exp(2x) + 1}, \tag{C.18}$$

$$\cosh^2(x) - \sinh^2(x) = 1, \quad \cosh(x) \pm \sinh(x) = \exp(\pm x), \tag{C.19}$$

$$\cosh(x + y) = \cosh(x)\cosh(y) + \sinh(x)\sinh(y), \tag{C.20}$$

$$\sinh(x + y) = \sinh(x)\cosh(y) + \cosh(x)\sinh(y), \tag{C.21}$$

$$\cos(x + jy) = \cos(x)\cosh(y) - j\sin(x)\sinh(y), \tag{C.22}$$

$$\sin(x + jy) = \sin(x)\cosh(y) + j\cos(x)\sinh(y). \tag{C.23}$$

C.2 SUMMATIONS

- *Infinite sums*:

$$\sum_{n=0}^{\infty} x^n = \frac{1}{1 - x}, \quad |x| < 1, \tag{C.24}$$

$$\sum_{n=m}^{\infty} x^n = \frac{x^m}{1 - x}, \quad |x| < 1, \tag{C.25}$$

$$\sum_{n=1}^{\infty} nx^n = \frac{x}{(1-x)^2}, \quad |x| < 1, \tag{C.26}$$

$$\sum_{n=1}^{\infty} n^2 x^n = \frac{x(1+x)}{(1-x)^3}, \quad |x| < 1. \tag{C.27}$$

- *Finite sums*:

$$\sum_{n=0}^{N} x^n = \begin{cases} \left(1 - x^{N+1}\right)/(1-x), & x \neq 1 \\ N+1, & x = 1, \end{cases} \tag{C 28}$$

$$\sum_{n=1}^{N} nx^n = \begin{cases} x\left[1 - (N+1)x^N + Nx^{N+1}\right]/(1-x)^2, & x \neq 1 \\ (1/2)N(N+1), & x = 1. \end{cases} \tag{C.29}$$

C.3 MISCELLANEOUS

- *Minimum*:

$$\min(x, y) \triangleq (1/2)(x + y - |x - y|), \tag{C.30}$$

- *Maximum*:

$$\max(x, y) \triangleq (1/2)(x + y + |x - y|), \tag{C.31}$$

- *Factorial*:

$$n! \triangleq n \times (n-1) \times \cdots \times 2 \times 1, \tag{C.32}$$

- *Binomial coefficient*:

$$\binom{n}{m} \triangleq \frac{n!}{m!(n-m)!}. \tag{C.33}$$

C.4 COMPLETING THE SQUARE

The quadratic equation

$$f(x) = ax^2 + bx + c, \tag{C.34}$$

can be rewritten in the form

$$f(x) = a(x + d_1)^2 + d_2, \tag{C.35}$$

with

$$d_1 = b/2a, \quad d_2 = c - b^2/4a. \tag{C.36}$$

This result is verified by factoring a in (C.34), adding and subtracting $b^2/4a^2$, and rearranging the expression as follows:

$$\begin{aligned} f(x) &= a\left[x^2 + (b/a)x\right] + c \\ &= a\left[x^2 + (b/a)x + b^2/4a^2 - b^2/4a^2\right] + c \\ &= a\left[x^2 + (b/a)x + b^2/4a^2\right] + c - b^2/4a, \end{aligned} \tag{C.37}$$

which becomes

$$f(x) = a(x + b/2a)^2 + c - b^2/4a \tag{C.38}$$

and matches (C.35).

C.5 QUADRATIC AND CUBIC FORMULAS

The two roots of the quadratic equation in (C.34) are given by the quadratic formula:

$$x = \frac{-b \pm \sqrt{b^2 - 4ac}}{2a}. \tag{C.39}$$

The types of roots are determined by examining the *discriminant*

$$\Delta \triangleq b^2 - 4ac, \tag{C.40}$$

resulting in three different cases:

$$\Delta > 0 \implies \text{two distinct real roots,} \tag{C.41}$$

$$\Delta = 0 \implies \text{two repeated real roots,} \tag{C.42}$$

$$\Delta < 0 \implies \text{two complex conjugate roots.} \tag{C.43}$$

These are illustrated in Figure C.1(a), where we find that the function $f(x)$ crosses the horizontal axis ($f(x) = 0$) twice (the solid line, distinct real roots) or not at all (the dash-dot line, complex roots). For repeated real roots (the dashed line), the function touches the horizontal axis at one point. Since $c = 1$ for all three cases, the three curves intersect each other at $x = 0$ with value $f(0) = 1$.

The general form for a cubic equation is

$$f(x) = ax^3 + bx^2 + cx + d = 0, \tag{C.44}$$

which has three roots. The discriminant is

$$\Delta \triangleq 18abcd + b^2c^2 - 4b^3d - 4ac^3 - 27a^2d^2, \tag{C.45}$$

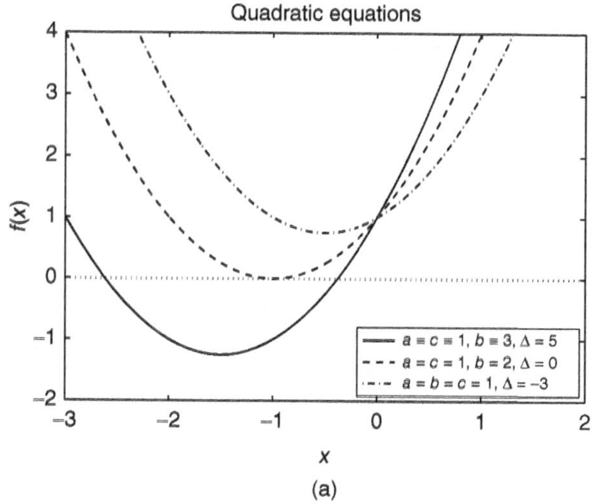

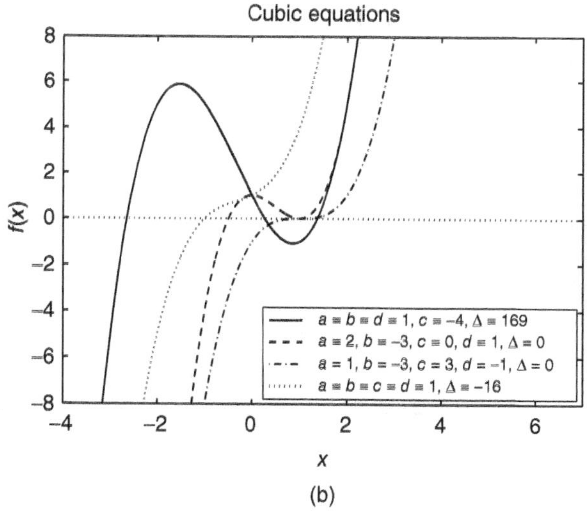

Figure C.1 (a) Quadratic equations. (b) Cubic equations.

and the three different cases are

$$\Delta > 0 \implies \text{three distinct real roots,} \tag{C.46}$$

$$\Delta = 0 \implies \text{one real root and two repeated real roots,} \tag{C.47}$$

$$\text{or three repeated real roots,} \tag{C.48}$$

$$\Delta < 0 \implies \text{one real root and two complex conjugate roots.} \tag{C.49}$$

Two types of repeated roots can occur when $\Delta = 0$ as illustrated in Figure C.1(b), where the function $f(x)$ crosses the horizontal axis three times (the solid line, distinct real roots) or only once (the dotted line, one real root and two complex conjugate roots). When there are one real root and two repeated real roots, the function crosses the horizontal axis once and touches it at another value of x (the dashed line). For three repeated real roots, the function touches the horizontal axis at a one point (the dash-dotted line). Since $d = 1$ for three of the cases, those curves intersect each other at $x = 0$ with value $f(0) = 1$.

Example C.1 The three types of roots for a quadratic equation are easily verified by examples. (i) Two distinct real roots: $f(x) = (x-1)(x-2) = x^2 - 3x + 2$. (ii) Two repeated real roots: $f(x) = (x-1)^2 = x^2 - 2x + 1$. (iii) Two complex conjugate roots: $f(x) = (x-j)(x+j) = x^2 + 1$. These are the only cases; it is not possible to have a single complex root if the coefficients $\{a, b, c\}$ are real-valued. The discriminants are $\Delta = \{1, 0, -4\}$, respectively. The four types of roots for a cubic equation are also verified by examples. (i) Three distinct real roots: $f(x) = (x-1)(x-2)(x-3) = x^3 - 6x^2 + 11x - 6$. (ii) One real root and two repeated real roots: $f(x) = (x-1)(x-2)^2 = x^3 - 5x^2 + 8x - 4$. (iii) Three repeated real roots: $f(x) = (x-1)^3 = x^3 - 3x^2 + 3x - 1$. (iv) One real root and two complex conjugate roots: $f(x) = (x-1)(x-j)(x+j) = x^3 - x^2 + x - 1$. The discriminants are $\Delta = \{4, 0, 0, -16\}$, respectively.

The three roots of a cubic equation can be derived using different methods; we present one approach known as Cardan's solution assuming $a = 1$. By first defining

$$p \triangleq c - b^2/3, \quad q \triangleq 2b^3/27 - bc/3 + d, \quad r \triangleq -1/2 + j\sqrt{3}/2, \tag{C.50}$$

and

$$s_1 = \sqrt[3]{-q/2 + \sqrt{q^2/4 + p^3/27}}, \quad s_2 = \sqrt[3]{-q/2 - \sqrt{q^2/4 + p^3/27}}, \tag{C.51}$$

the three roots are

$$x_1 = -b/3 + s_1 + s_2, \tag{C.52}$$

$$x_2 = -b/3 + rs_1 + r^* s_2, \tag{C.53}$$

$$x_3 = -b/3 + r^* s_1 + rs_2, \tag{C.54}$$

where r^* is the complex conjugate of r.

Example C.2 We verify the formulas in (C.52)–(C.54) for two of the cases in the previous example:

$$f(x) = (x-1)^3 = x^3 - 3x^2 + 3x - 1 \implies p = q = 0, \tag{C.55}$$

which yield $s_1 = s_2 = 0$ and $x_1 = x_2 = x_3 = -(-3)/3 = 1$. For

$$f(x) = (x-1)(x-2)^2 = x^2 - 5x^2 + 8x - 4 \implies p = -1/3, \quad q \approx 0.0741, \tag{C.56}$$

we have $s_1 = s_2 = -1/3$ (the square roots in (C.51) are 0). Since $s_1 = s_2$:

$$x_1 = 5/3 - 2/3 = 1, \tag{C.57}$$

$$x_2 = 5/3 - (1/3)(r + r^*) = 5/3 - (2/3)\mathrm{Re}(r)$$
$$= 5/3 + 1/3 = 2 = x_3. \tag{C.58}$$

This example illustrates additional properties of the cubic equation regarding repeated roots. For $\Delta = 0$, there are three repeated roots when $p = q = 0$ because $s_1 = s_2 = 0$. These are determined completely by the b coefficient: $x_1 = x_2 = x_3 = -b/3$. When the square roots in (C.51) are 0, we have the case of one real root and two repeated roots with $s_1 = s_2 = \sqrt[3]{q/2}$ such that $rs_1 + r^*s_2 = s_1(r + r^*) = 2s_1\mathrm{Re}(r)$. Thus, $x_1 = -b/3 + 2s_1$ and $x_2 = x_3 = -b/3 - s_1$.

C.6 DERIVATIVES

- *Product rules*:

$$\frac{d}{dx} f(x)g(x) = f'(x)g(x) + f(x)g'(x), \tag{C.59}$$

$$\frac{d}{dx} f(x)g(x)h(x) = f'(x)g(x)h(x) + f(x)g'(x)h(x) + f(x)g(x)h'(x), \tag{C.60}$$

$$\frac{d^2}{dx^2} f(x)g(x) = 2f'(x)g'(x) + f(x)\frac{d^2}{dx^2}g(x) + g(x)\frac{d^2}{dx^2}f(x), \tag{C.61}$$

$$\frac{d}{dx} f^m(x)g^n(x) = f^{m-1}(x)g^{n-1}(x)\left[mg(x)f'(x) + nf(x)g'(x)\right], \tag{C.62}$$

$$\frac{d^n}{dx^n} f(x)g(x) = \sum_{m=0}^{n} \binom{n}{m} \frac{d^{n-m}}{dx^{n-m}} f(x) \frac{d^m}{dx^m} g(x), \tag{C.63}$$

  with $d^0 f(x)/dx^0 \triangleq f(x)$.

- *Quotient rules*:

$$\frac{d}{dx}\frac{f(x)}{g(x)} = \frac{g'(x)}{f(x)} - \frac{f(x)g'(x)}{g^2(x)}, \tag{C.64}$$

$$\frac{d}{dx}\frac{f^m(x)}{g^n(x)} = \frac{f^{m-1}(x)}{g^{n-1}(x)}\left[mg(x)f'(x) - nf(x)g'(x)\right]. \tag{C.65}$$

- *Exponent rules*:

$$\frac{d}{dx}b^{f(x)} = \ln(b)b^{f(x)}\frac{d}{dx}f(x), \tag{C.66}$$

$$\frac{d}{dx}f^{g(x)}(x) = g(x)f^{g(x)-1}(x)\frac{d}{dx}f(x) + \ln(f(x))f^{g(x)}(x)\frac{d}{dx}g(x). \tag{C.67}$$

- *Chain rules*:

$$\frac{d}{dx}f(g(x)) = g'(x)\frac{d}{dg(x)}f(g(x)), \tag{C.68}$$

$$\frac{d^2}{dx^2}f(g(x)) = \frac{d^2}{dx}g(x)\frac{d}{dg(x)}f(g(x)) + \left[\frac{d}{dx}g(x)\right]^2\frac{d^2}{dg(x)^2}f(g(x)). \tag{C.69}$$

- *Leibniz's integral rules*:

$$\frac{\partial}{\partial v}\int_{a(v)}^{b(v)} f(u,v)du = \int_{a(v)}^{b(v)} \frac{\partial}{\partial v}f(u,v)du + f(b(v),v)\frac{\partial}{\partial v}b(v) - f(a(v),v)\frac{\partial}{\partial v}a(v), \tag{C.70}$$

$$\frac{d}{dx_2}\int_{x_1}^{x_2} f(x)dx = f(x_2), \tag{C.71}$$

$$\frac{d}{dx_1}\int_{x_1}^{x_2} f(x)dx = -f(x_1). \tag{C.72}$$

- *Basic derivatives*:

$$\frac{d}{dx}x^n = nx^{n-1}, \tag{C.73}$$

$$\frac{d}{dx}\sqrt{x} = 1/2\sqrt{x}, \tag{C.74}$$

$$\frac{d}{dx}\exp(\alpha x) = \alpha\exp(\alpha x), \tag{C.75}$$

$$\frac{d}{dx}\ln(x) = 1/x, \tag{C.76}$$

$$\frac{d}{dx}\log_b(x) = \log_b(e)/x. \tag{C.77}$$

- *Trigonometric*:

$$\frac{d}{dx}\cos(x) = -\sin(x), \quad \frac{d}{dx}\sin(x) = \cos(x), \tag{C.78}$$

$$\frac{d}{dx}\cos^{-1}(x) = -\frac{1}{\sqrt{1-x^2}}, \quad \frac{d}{dx}\sin^{-1}(x) = \frac{1}{\sqrt{1-x^2}}, \tag{C.79}$$

$$\frac{d}{dx}\tan(x) = \sec^2(x), \quad \frac{d}{dx}\tan^{-1}(x) = \frac{1}{1+x^2}, \tag{C.80}$$

$$\frac{d}{dx}\cosh(x) = \sinh(x), \quad \frac{d}{dx}\sinh(x) = \cosh(x). \tag{C.81}$$

C.7 INDEFINITE INTEGRALS

- *Polynomial*:

$$\int \frac{dx}{a+bx} = (1/b)\ln(a+bx), \tag{C.82}$$

$$\int \frac{x}{a+bx}dx = x/b - (a/b^2)\ln(a+bx), \tag{C.83}$$

$$\int \frac{dx}{(a+bx)^2} = -1/(a+bx)b, \tag{C.84}$$

$$\int \frac{x}{(a+bx)^2}dx = (1/b^2)\left[\ln(a+bx) + a/(a+bx)\right], \tag{C.85}$$

$$\int \frac{dx}{a^2+x^2} = (1/a)\tan^{-1}(x/a). \tag{C.86}$$

- *Logarithmic*:

$$\int \frac{dx}{x} = \ln(x), \tag{C.87}$$

$$\int \frac{f'(x)}{f(x)}dx = \ln(f(x)), \tag{C.88}$$

$$\int \ln(x)dx = x\ln(x) - x, \tag{C.89}$$

$$\int x\ln(x)dx = (x^2/2)\ln(x) - x^2/4. \tag{C.90}$$

- *Exponential*:

$$\int \exp(\alpha x)dx = \exp(\alpha x)/\alpha, \tag{C.91}$$

$$\int x\exp(\alpha x)dx = \left[(\alpha x - 1)/\alpha^2\right]\exp(\alpha x), \tag{C.92}$$

$$\int b^{\alpha x}dx = b^{\alpha x}/\alpha\ln(b). \tag{C.93}$$

- *Trigonometric*:

$$\int \cos(ax)dx = (1/a)\sin(x), \tag{C.94}$$

$$\int \sin(ax)dx = -(1/a)\cos(x), \tag{C.95}$$

$$\int x\cos(ax)dx = (1/a^2)\cos(ax) + (x/a)\sin(ax), \tag{C.96}$$

$$\int x\sin(ax)dx = (1/a^2)\sin(ax) - (x/a)\cos(ax), \tag{C.97}$$

$$\int x^2\cos(ax)dx = (2x/a^2)\cos(ax) + \left[(a^2x^2-2)/a^3\right]\sin(ax), \tag{C.98}$$

$$\int x^2\sin(ax)dx = (2x/a^2)\sin(ax) - \left[(a^2x^2-2)/a^3\right]\cos(ax), \tag{C.99}$$

$$\int \exp(\alpha x)\cos(ax)dx = \exp(\alpha x)\left[\alpha\cos(ax) + a\sin(ax)\right]/(\alpha^2+a^2), \tag{C.100}$$

$$\int \exp(\alpha x)\sin(ax)dx = \exp(\alpha x)\left[\alpha\sin(ax) - a\cos(ax)\right]/(\alpha^2+a^2), \tag{C.101}$$

$$\begin{aligned}\int x\exp(\alpha x)\cos(ax)dx &= x\exp(\alpha x)\left[\alpha\cos(ax) + a\sin(ax)\right]/(\alpha^2+a^2)\\ &- \exp(\alpha x)(\alpha^2-a^2)\cos(ax)/(\alpha^2+a^2)^2\\ &- 2a\alpha\exp(\alpha x)\sin(ax)/(\alpha^2+a^2)^2,\end{aligned} \tag{C.102}$$

$$\begin{aligned}\int x\exp(\alpha x)\sin(ax)dx &= x\exp(\alpha x)\left[\alpha\sin(ax) - a\cos(ax)\right]/(\alpha^2+a^2)\\ &- \exp(\alpha x)(\alpha^2-a^2)\sin(ax)/(\alpha^2+a^2)^2\\ &+ 2a\alpha\exp(\alpha x)\cos(ax)/(\alpha^2+a^2)^2.\end{aligned} \tag{C.103}$$

C.8 DEFINITE INTEGRALS

- *Integration by parts*:

$$\int_{x_1}^{x_2} f(x)g'(x)dx = f(x_2)g(x_2) - f(x_1)g(x_1) - \int_{x_1}^{x_2} g(x)f'(x)dx. \tag{C.104}$$

- *Exponential* ($\alpha > 0$):

$$\int_0^\infty x^n \exp(-\alpha x)dx = n!/\alpha^{n+1}, \quad n \in \mathcal{N}, \tag{C.105}$$

$$\int_0^\infty \exp(-\alpha x)\cos(bx)dx = \alpha/(\alpha^2 + b^2), \tag{C.106}$$

$$\int_0^\infty \exp(-\alpha x)\cos(bx)dx = \alpha/(\alpha^2 + b^2), \tag{C.107}$$

$$\int_0^\infty x\exp(-\alpha x)\sin(bx)dx = 2\alpha/(\alpha^2 + b^2)^2, \tag{C.108}$$

$$\int_0^\infty x\exp(-\alpha x)\cos(bx)dx = \alpha/(\alpha^2 + b^2), \tag{C.109}$$

$$\int_0^\infty \exp(-\alpha x)\sin(bx)dx = (\alpha^2 - b^2)/(\alpha^2 + b^2)^2, \tag{C.110}$$

$$\int_0^\infty \exp(-\alpha x^2)dx = (1/2)\sqrt{\pi/\alpha}, \tag{C.111}$$

$$\int_0^\infty x\exp(-\alpha x^2)dx = 1/2\alpha. \tag{C.112}$$

- *Trigonometric*:

$$\int_0^\pi \sin(ax)\sin(bx)dx = \int_0^\pi \cos(ax)\cos(bx)dx = 0, \quad a, b \in \mathcal{Z}, a \neq b, \tag{C.113}$$

$$\int_0^\pi \sin(ax)\cos(ax)dx = \int_0^{\pi/a} \sin(ax)\cos(ax)dx = 0, \tag{C.114}$$

$$\int_0^\pi \sin(ax)\cos(bx)dx = \begin{cases} 2a/(a^2 - b^2), & a - b \text{ odd} \\ 0, & a - b \text{ even}, \end{cases} \tag{C.115}$$

$$\int_0^\pi \sin^2(ax)dx = \int_0^\pi \cos^2(ax)dx = \pi/2, \tag{C.116}$$

$$\int_{-\infty}^\infty \frac{\sin(x)}{x}dx = \pi, \quad \int_{-\infty}^\infty \frac{\sin(\pi x)}{\pi x}dx = 1. \tag{C.117}$$

- *Polynomial*:

$$\int_0^\infty \frac{a}{a^2 + x^2} dx = \pi/2, \tag{C.118}$$

$$\int_0^\infty \frac{dx}{\sqrt{a^2 - x^2}} = \pi/2. \tag{C.119}$$

APPENDIX D

SET THEORY

This appendix provides a brief review of set theory.

D.1 SETS AND SUBSETS

Some basic definitions and examples are covered in this section.

Definition: Set A *set* is a collection of objects or numbers that represent those objects. The components of a set are called its *elements* or *points*.

In this book, we consider only sets with numerical elements.

Example D.1 Set $A = \{ \dots, -1, 0, 1, \dots \}$ consists of all integers, which are denoted by $\mathcal{Z}$. The elements of set $B = (-\infty, \infty)$ are the real numbers $\mathcal{R}$. Note that $\pm\infty$ do not correspond to real numbers; they are *symbols* frequently used in mathematics, such as when taking limits of the form $n \to \infty$. Additional examples of sets include the closed interval $C = [0, 1]$ of real numbers, the natural numbers $D = \mathcal{N}$, and so on. Sets A and D are *discrete*, while sets B and C are *continuous*.

Sets of numbers can also be described by equations.

Example D.2 Set $A = \{x + 1 : x \geq 1\} = [2, \infty)$ is continuous; in this context, the colon means "such that," and the statement defines the set of all $x + 1$ such that $x \geq 1$.

Mathematical Foundations for Linear Circuits and Systems in Engineering, First Edition. John J. Shynk.

Companion Website: http://www.wiley.com/go/linearcircuitsandsystems

The set $B = \{x^2 - 1 : x = 0, 1, 2\} = \{-1, 0, 3\}$ is discrete. The values of x to the right of the colon give the *support* of the function that describes the set.

The next definition involves specific relationships between two sets.

Definition: Subset $A \subset B$ and Equality $A = B$ Set A is a *subset* of B if all elements of A are also in B. The notation $A \subseteq B$ allows for situations where A and B might be equal. Sets A and B are *equal* when $A \subset B$ and $B \subset A$ such that they have exactly the same elements and $A = B$.

Example D.3 Example subsets include $\mathcal{R}^+ \subset \mathcal{R}$, $\mathcal{N} \subset \mathcal{Z}$, and $[0, 1) \subset [0, 1] \subset \mathcal{R}^+$. Additional examples include $\mathcal{Z} \subset \mathcal{R}$, $\mathcal{Q} \subset \mathcal{R}$, and $\{0, 1, 4, 9\} \subset \{x^2 : x \in \mathcal{Z}^+\}$. An example of set equality is $\mathcal{Z}^+ \backslash \{0\} = \mathcal{N}$, where the backslash operator (defined in the next section) removes element 0 from the set of nonnegative integers $\mathcal{Z}^+$, yielding the natural numbers $\mathcal{N}$.

In order to define set operations, especially set complement, we need to specify the set of all elements.

Definition: Universal Set Ω The *universal set* Ω is the set of all elements. It is also called the *universe*, and in probability, it is known as the *sample space*.

Example D.4 When we are interested in functions of continuous x, the universal set could be the real line $\Omega = \mathcal{R}$, the nonnegative real line $\Omega = \mathcal{R}^+$ (which includes zero), or even an interval such as $\Omega = [0, 10]$. For discrete x, the universal set might be the entire set of integers $\Omega = \mathcal{Z}$ or a subset such as the natural numbers $\Omega = \mathcal{N}$.

Definition: Set Complement A^c The *complement* A^c contains all elements of Ω that are not in A. It can be written as $A^c = \{x : x \notin A\}$, and the notation $\overline{A}$ is often used.

The *Venn diagram* in Figure D.1 is a useful graphic for visualizing the relationships of various sets. The universal set of all elements is represented by the rectangle, and subsets of Ω are represented by the circles. When sets have common elements, their circles overlap in a Venn diagram, and when a set is a subset of another set, one circle lies entirely within the other circle as depicted in Figure D.1(b). Unless otherwise specified, we assume that all elements of Ω are contained within the circles.

Example D.5 Suppose the universal set is the open interval $\Omega = (0, 2)$ and we are interested in the set $A = (0, 1]$. Then $A^c = (1, 2)$. If the universal set is extended to $\Omega = (0, \infty)$, then $A^c = (1, \infty)$. Since $\pm\infty$ are symbols, we always use open or semi-open intervals of the form $\mathcal{R} = (-\infty, \infty)$ and $\mathcal{R}^+ = [0, \infty)$.

Definition: Empty Set ϕ The *empty set* ϕ is the set without any elements. It is also called the *null set*, and it is the complement of the universal set: $\phi = \Omega^c$.

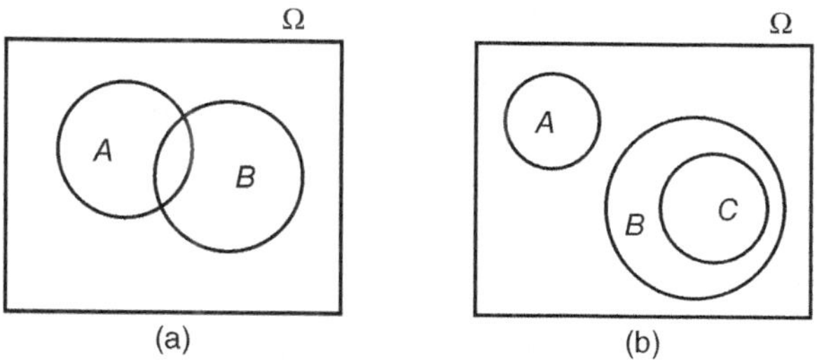

Figure D.1 Venn diagrams. (a) Overlapping sets A and B. (b) Subset $C \subset B$ and complement $A = B^c$.

D.2 SET OPERATIONS

The two basic set operations are union and intersection.

Definition: Union $A \cup B$ The *union* of sets A and B consists of all elements in A, B, or both. It can be written as

$$A \cup B = \{x : x \in A \text{ or } x \in B\}, \tag{D.1}$$

and is easily extended to multiple sets such as $A \cup B \cup C$.

Example D.6 For continuous $\Omega = \mathcal{R}$, let $A = [0, 1]$, $B = (0, 2)$, and $C = [1, 2]$. Then $A \cup B = [0, 2)$, $A \cup C = [0, 2]$, and $B \cup C = (0, 2]$. For discrete $\Omega = \mathcal{Z}^+$, let $D = \{0, 2, 4, 5\}$, $E = \{1, 2, 5\}$, and $F = \mathcal{N}$. Then $D \cup E = \{0, 1, 2, 5\}$, $D \cup F = \mathcal{Z}^+ = \Omega$, and $E \cup F = \mathcal{N}$.

Definition: Intersection $A \cap B$ The *intersection* of sets A and B consists of all elements common to both. It can be written as

$$A \cap B = \{x : x \in A \text{ and } x \in B\}, \tag{D.2}$$

and is easily extended to multiple sets such as $A \cap B \cap C$. Notationally, it is more convenient to write AB and ABC.

Example D.7 For the continuous Ω in Example D.6, $AB = (0, 1]$, $AC = \{1\}$, and $BC = [1, 2)$. For the discrete Ω in that example, $DE = \{2, 5\}$, $DF = \{2, 4, 5\}$, and $EF = \{1, 2, 5\}$.

The commutative, associative, and distributive properties of union and intersection are summarized in Table D.1.

Definition: Mutually Exclusive Sets A and B are *mutually exclusive* if $AB = \phi$. Such sets are also called *disjoint*.

TABLE D.1 Properties of Set Operations

Properties	Expressions
Commutative	$A \cup B = B \cup A, AB = BA$
Associative	$(A \cup B) \cup C = A \cup (B \cup C), (AB)C = A(BC)$
Distributive	$A \cup (BC) = (A \cup B)(A \cup C), A(B \cup C) = (AB) \cup (AC)$
Mutually exclusive	$AB = \phi$
Difference	$A - B = A \backslash B = AB^c$
Exclusive or	$A \oplus B = (A - B) \cup (B - A) = A \cup B - AB$
De Morgan's laws	$(A \cup B)^c = A^c B^c, (AB)^c = A^c \cup B^c$

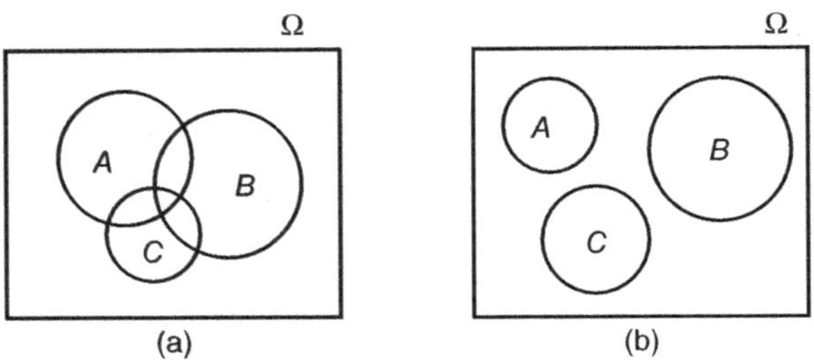

Figure D.2 Venn diagrams. (a) Collectively exhaustive sets. (b) Partition of Ω.

Obviously, A and A^c are mutually exclusive for any A: $AA^c = \phi$. Mutually exclusive sets can be used to partition the universal set.

Definition: Collectively Exhaustive and Partition Sets $\{A_n\}$ are *collectively exhaustive* when $\cup_n A_n = \Omega$. They cover every element in the universal set. If all $\{A_n\}$ are mutually disjoint, then they form a *partition* of Ω.

Figure D.2 shows examples of collectively exhaustive sets and a partition for three sets $\{A, B, C\}$. Such sets are not unique; the universal set can be partitioned in different ways. The simplest type of partition is some set A and its complement A^c: $A \cup A^c = \Omega$.

Example D.8 For $\Omega = \mathcal{R}$, sets $A = (-\infty, 0)$, $B = [0, 1]$, and $C = (1, \infty)$ form a partition, whereas $C = (-\infty, 1]$ and $D = [0, \infty)$ are collectively exhaustive.

Table D.1 summarizes three additional set operations, which can be written in terms of union, intersection, and complement. The *difference* $A - B = A \backslash B$ consists of all elements in A except those in common with B. From the Venn diagram in Figure D.3(a), it easy to verify that $A - B = AB^c$. The *exclusive or* operation is known as the symmetric difference:

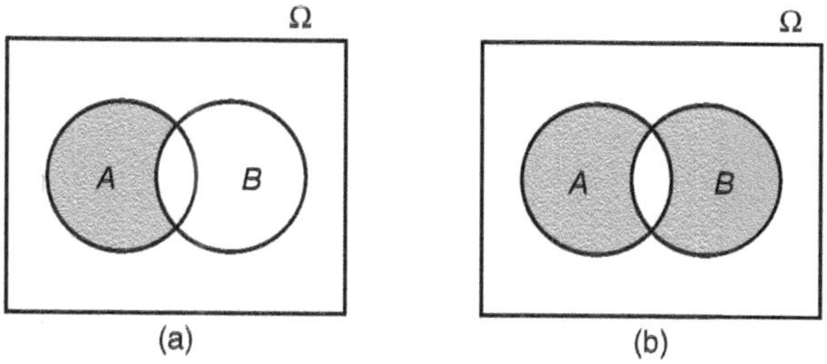

Figure D.3 Set operations (results are shaded). (a) Difference $A - B = AB^c$. (b) Exclusive or (symmetric difference) $A \oplus B = (A - B) \cup (B - A) = AB^c \cup A^cB = A \cup B - AB$.

$$\begin{aligned} A \oplus B &= (A - B) \cup (B - A) \\ &= AB^c \cup A^cB = A \cup B - AB. \end{aligned} \tag{D.3}$$

It removes the overlapping regions of two sets as illustrated in Figure D.3(b). Observe that every expression in (D.3) is symmetric: interchanging A and B gives the same results. Finally, De Morgan's laws in the table are two expressions derived by complementing the union and intersection of two sets. These can be proved by examining an individual element: if $x \in (A \cup B)^c$, then $x \notin A$ and $x \notin B$. Thus, $x \in A^c$ and $x \in B^c$, yielding

$$(A \cup B)^c = A^cB^c. \tag{D.4}$$

The proof for the other form of De Morgan's law is similar.

APPENDIX E

SERIES EXPANSIONS

In this appendix, we describe *power series* expansions for function $f(z)$ with complex argument z. A power series is a sum of powers of $z - z_o$ given by $(z - z_o)^n$ for nonnegative integers $n \in \mathcal{Z}^+$. We also describe the Laurent series expansion for which n can also be negative. The corresponding expansions for function $f(x)$ with real argument x are derived from $f(z)$ by replacing z with x.

E.1 TAYLOR SERIES

The *Taylor series* expansion of smooth function $f(z)$, which is infinitely differentiable at $z = z_o$ on the complex plane, is

$$f(z) = \sum_{n=0}^{\infty} \frac{f^{(n)}(z_o)}{n!}(z - z_o)^n = \sum_{n=0}^{\infty} c_n (z - z_o)^n, \tag{E.1}$$

where the derivative notation means

$$f^{(n)}(z_o) \triangleq \left. \frac{d^n}{dz^n} f(z) \right|_{z=z_o}. \tag{E.2}$$

The *coefficients* of the expansion are

$$c_n \triangleq \frac{f^{(n)}(z_o)}{n!}, \tag{E.3}$$

Mathematical Foundations for Linear Circuits and Systems in Engineering, First Edition. John J. Shynk.

Companion Website: http://www.wiley.com/go/linearcircuitsandsystems

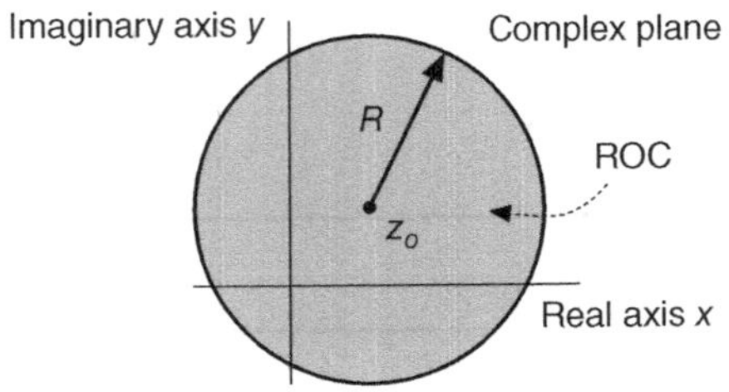

Figure E.1 Radius R defining a circle of points about z_o for which a Taylor series is convergent.

which we note includes the factorial term in the denominator. Generally, there is a circle with radius R about z_o that defines a region for z on the complex plane for which the series is *convergent* to a finite value (usually a different value for each z). Outside of this radius, the series is divergent. This region of convergence (ROC) is depicted in Figure E.1, and it is shown next that the circle boundary (the solid line) is *not* included in the ROC. The ROC in this context involving a circle is also called the *radius of convergence*. (The ROC for the Laplace transform in Chapter 7 is a vertical *strip* on the complex plane, and so a radius does not apply in that case.) For real x, the circular ROC reduces to an *open interval* of the form (a, b) on the real line with $a, b \in \mathcal{R}$.

Example E.1 Consider the function

$$\begin{aligned} f(z) &= \frac{1}{1-2z} = \sum_{n=0}^{\infty} (2z)^n \\ &= 1 + 2z + 4z^2 + 8z^3 + \cdots , \end{aligned} \tag{E.4}$$

which has been expanded as a Taylor series about $z_o = 0$. It is convergent provided $|2z| < 1$, and so the ROC on the complex plane is the circle defined by $|z| < 1/2$ (the strict inequality excludes the circle boundary). The coefficients are $\{c_n = 2^n\}$, and are derived using (E.2):

$$f'(z) = \frac{2}{(1-2z)^2}, \qquad f^{(2)}(z) = \frac{8}{(1-2z)^3}, \qquad f^{(3)}(z) = \frac{48}{(1-2z)^4}, \tag{E.5}$$

and so on. Thus, the general form is

$$f^{(n)}(z_o) = \frac{2^n n!}{(1-2z_o)^{n+1}} \implies c_n = \frac{2^n}{(1-2z_o)^{n+1}}. \tag{E.6}$$

Substituting $z_o = 0$ into (E.6) yields

$$f^{(n)}(z)|_{z=0} = 2^n n! \implies c_n = 2^n. \tag{E.7}$$

For the expansion about $z_o = 1$, we still use (E.6), but $z_o = 1$ is substituted:

$$f^{(n)}(z)|_{z=1} = (-1)^{n+1}2^n n! \implies c_n = (-1)^{n+1}2^n. \tag{E.8}$$

This yields

$$\begin{aligned} f(z) &= \sum_{n=0}^{\infty} (-1)^{n+1}2^n(z-1)^n \\ &= -1 + 2(z-1) - 4(z-1)^2 + 8(z-1)^3 - \cdots, \end{aligned} \tag{E.9}$$

which converges for $|2(z-1)| < 1 \implies |z-1| < 1/2$. The expansions in (E.4) and (E.9) converge in nonoverlapping regions (circles) on the complex plane.

From the previous example, we see that a Taylor series for function $f(z)$ can be derived about different points on the complex plane. Of course, it is the same function, but different series expansions will have different ROCs. Thus, when using series representations of $f(z)$, it is important to choose an appropriate expansion point z_o depending on the application.

E.2 MACLAURIN SERIES

When a Taylor series expansion is defined about $z_o = 0$, it is called a *Maclaurin series*:

$$f(z) = \sum_{n=0}^{\infty} \frac{f^{(n)}(0)z^n}{n!} = \sum_{n=0}^{\infty} c^n z^n. \tag{E.10}$$

The expansion in (E.4) of Example E.1 is actually a Maclaurin series.

Example E.2 The Maclaurin series of $f(z) = \cos(z)$ is derived by finding the derivatives:

$$f'(z) = -\sin(z), \qquad f^{(2)}(z) = -\cos(z), \qquad f^{(3)}(z) = \sin(z), \tag{E.11}$$

and so the general expression after substituting $z = 0$ is

$$f^{(n)}(z)|_{z=0} = \begin{cases} (-1)^n, & n \text{ even} \\ 0, & n \text{ odd}. \end{cases} \tag{E.12}$$

Thus, the Maclaurin series expansion is

$$\begin{aligned} \cos(z) &= \sum_{n=0}^{\infty} \frac{(-1)^n z^{2n}}{(2n)!} \\ &= 1 - z^2/2 + z^4/24 - z^6/720 + \cdots, \end{aligned} \tag{E.13}$$

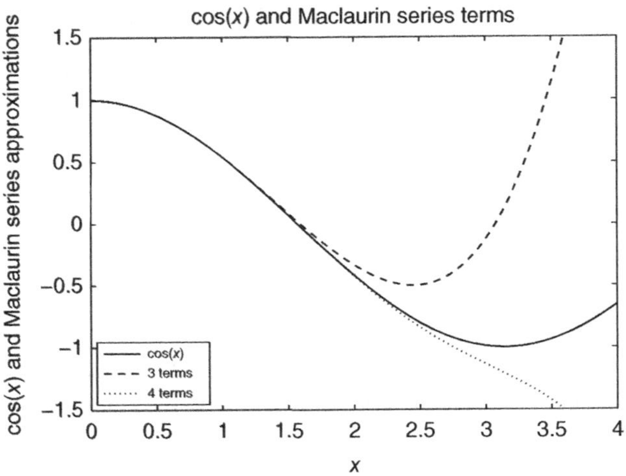

Figure E.2 Maclaurin series approximation of cos(x) for real-valued x with three and four nonzero expansion terms added together.

where we have used $2n$ instead of n in the sum to ensure that only the z terms with even exponents are nonzero. It can be shown that the ROC is the entire complex plane $z \in C$. The cosine function for real-valued x and a few nonzero terms from the Maclaurin series expansion are shown in Figure E.2. Observe that the series approximation is relatively accurate for small x (< 2); for larger x, increasingly more nonzero terms from the expansion are needed for an accurate approximation. The Maclaurin series expansion for the sine function has a form similar to (E.13), except that only the terms with odd exponents are nonzero:

$$\begin{aligned} \sin(z) &= \sum_{n=0}^{\infty} \frac{(-1)^n z^{2n+1}}{(2n+1)!} \\ &= z - z^3/6 + z^5/120 - z^7/5040 + \cdots, \end{aligned} \tag{E.14}$$

whose ROC is also the entire complex plane.

Several Maclaurin series expansions are summarized in Table E.1. The last entry

$$(1+z)^{\upsilon} = \sum_{n=0}^{\infty} \binom{\upsilon}{n} z^n \tag{E.15}$$

is a special series known as the *binomial series* expansion that holds for any complex exponent $\upsilon \in C$. It is based on the *generalized* binomial coefficient:

$$\binom{\upsilon}{n} \triangleq \frac{\upsilon(\upsilon-1)\cdots(\upsilon-n+1)}{n!}, \tag{E.16}$$

TABLE E.1 Maclaurin Series Expansions

Function	Series	ROC
$\sin(z)$	$\sum_{n=0}^{\infty} (-1)^n z^{2n+1}/(2n+1)!$	$z \in C$
$\cos(z)$	$\sum_{n=0}^{\infty} (-1)^n z^{2n}/(2n)!$	$z \in C$
$\sinh(z)$	$\sum_{n=0}^{\infty} z^{2n+1}/(2n+1)!$	$z \in C$
$\cosh(z)$	$\sum_{n=0}^{\infty} z^{2n}/(2n)!$	$z \in C$
$\tan^{-1}(z)$	$\sum_{n=0}^{\infty} (-1)^n z^{2n+1}/(2n+1)$	$\|z\| < 1$
$\exp(z)$	$\sum_{n=0}^{\infty} z^n/n!$	$z \in C$
$\ln(1+z)$	$\sum_{n=0}^{\infty} (-1)^{n+1} z^n/n$	$\|z\| < 1$
$1/(1-z)$	$\sum_{n=0}^{\infty} z^n$	$\|z\| < 1$
$1/(1-z)^2$	$\sum_{n=0}^{\infty} nz^{n-1}$	$\|z\| < 1$
$(1+z)^{\upsilon}$	$\sum_{n=0}^{\infty} \binom{\upsilon}{n} z^n$	$\|z\| < 1$

where $n \in \mathcal{Z}^+$. This binomial coefficient is 0 for integer $n < 0$, and it is equal to 1 for $n = 0$. Example binomial coefficients for real-valued noninteger $\upsilon = 3.5$ are summarized in Table E.2, where we see that the coefficients can be negative for $n > \upsilon + 1 = 4.5$. If $\upsilon = m \in \mathcal{Z}^+$ is a nonnegative integer, then one of the terms in the numerator of (E.16) is 0 for $m > n$. As a result, (E.16) for this case reduces to the standard binomial coefficient:

$$\binom{m}{n} \triangleq \frac{m!}{(n-m)!n!}, \tag{E.17}$$

and there is a finite number of terms in the series expansion of (E.15):

$$(1+z)^m = \sum_{n=0}^{m} \binom{m}{n} z^n. \tag{E.18}$$

Since the sum is finite, the ROC is the entire complex plane $z \in C$. This last expression is a special case of the *binomial formula* with $x = 1$ (also called the *binomial theorem*):

$$(x+y)^m = \sum_{n=0}^{m} \binom{m}{n} x^{m-n} y^n. \tag{E.19}$$

Example E.3 In this example, we examine the binomial series expansion in Table E.1 for integer and noninteger υ. For integer $\upsilon = m = 3$, the binomial coefficients are summarized in Table E.2, where we see that only four terms are nonzero. Figure E.3(a) shows a plot of $(1+x)^3$ for real-valued $0 \leq x < 1$, along with the sum in (E.18) having only two and three terms of the binomial series expansion (of course, including all four terms in the sum yields exactly the original function). Observe that the plot for three terms in the sum is reasonably close to the actual

TABLE E.2 Example Binomial Coefficients $\binom{v}{n}$

n	$v = 3.5$	$v = 3$
0	1	1
1	3.5	3
2	$(3.5)(2.5)/2 = 4.375$	3
3	$(3.5)(2.5)(1.5)/6 = 2.1875$	1
4	$(3.5)(2.5)(1.5)(0.5)/24 = 0.2734375$	0
5	$(3.5)(2.5)(1.5)(0.5)(-0.5)/120 = -0.02734375$	0
6	$(3.5)(2.5)(1.5)(0.5)(-0.5)(-1.5)/720 = 0.0068359375$	0

function. Figure E.3(b) shows a plot of $(1+x)^{3.5}$ along with the sum in (E.15) containing only two, three, and four terms. In this case, including four terms in the sum yields a close approximation of the original function, which is expected because the binomial coefficients in Table E.2 become small rather quickly with increasing n. Although we could extend the horizontal axis beyond $x = 1$ in both plots, this should not be done in Figure E.3(b) if the number of terms in the binomial expansion approaches infinity because the series is divergent for $x \geq 1$.

E.3 LAURENT SERIES

The Taylor series in (E.1) does not allow for expansions around *singular points* on the complex plane. For example, the function $f(z) = 1/(1-z)$ in Table E.1 has a singularity at $z = 1$ where it is unbounded, which is why the ROC is $|z| < 1$ for the expansion about $z_o = 0$. The *Laurent series* expansion allows for expansions around singular points, and it can be viewed as an extension of the Taylor series to include negative integers n:

$$f(z) = \sum_{n=-\infty}^{\infty} c_n (z - z_o)^n, \tag{E.20}$$

where the sum is now doubly infinite. This expression can be rewritten as

$$f(z) = \sum_{n=0}^{\infty} c_n (z - z_o)^n + \sum_{m=1}^{\infty} \frac{c_m}{(z - z_o)^m}, \tag{E.21}$$

where we have split the sum into two parts and changed variables in the second sum to $m \triangleq -n$ in order to emphasize that $(z - z_o)^m$ actually appears in the denominator for negative n. The coefficients $\{c_n\}$ in (E.20) are computed from $f(z)$ as follows:

$$c_n = \frac{1}{2\pi j} \oint_C \frac{f(z)}{(z - z_o)^{n+1}} dz. \tag{E.22}$$

The integration is performed counterclockwise along a closed contour C within the ROC that encloses $z = z_o$ and where $f(z)$ is *analytic* (see the definition in Chapter 5).

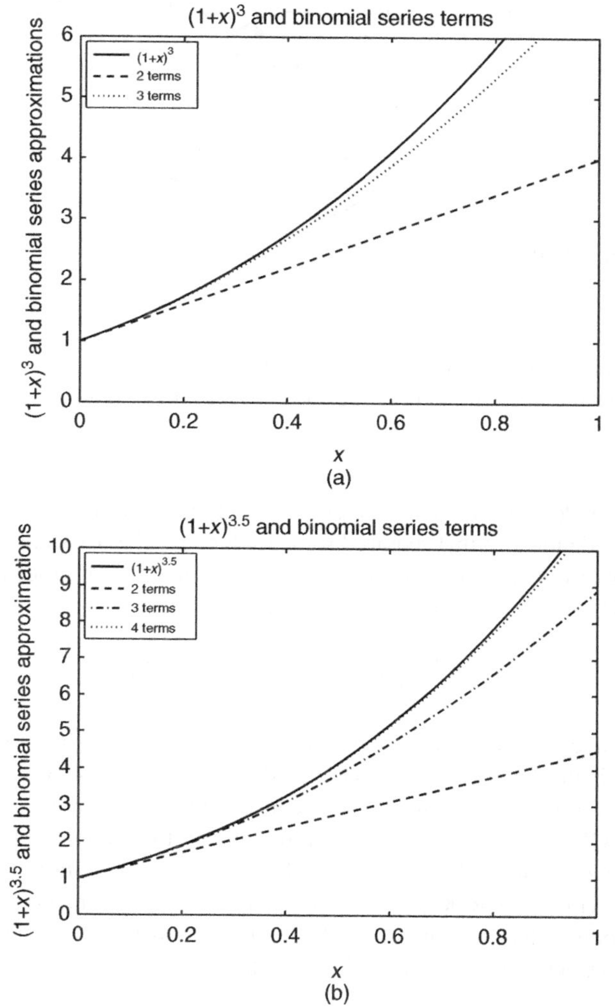

Figure E.3 Binomial series expansions for $(1 + x)^{\upsilon}$. (a) Integer $\upsilon = m = 3$. (b) Noninteger $\upsilon = 3.5$.

This is evident from the second sum in (E.21) where $f(z)$ may not be defined (is infinite) at $z = z_o$. There are three basic types of singularities (also called *singular points*):

- *Poles*: The second sum in (E.21) has a finite number of terms.
- *Essential singular points*: The second sum in (E.21) has an infinite number of terms.
- *Removable singular points*: The second sum in (E.21) has no terms, so the expansion reduces to a Taylor series.

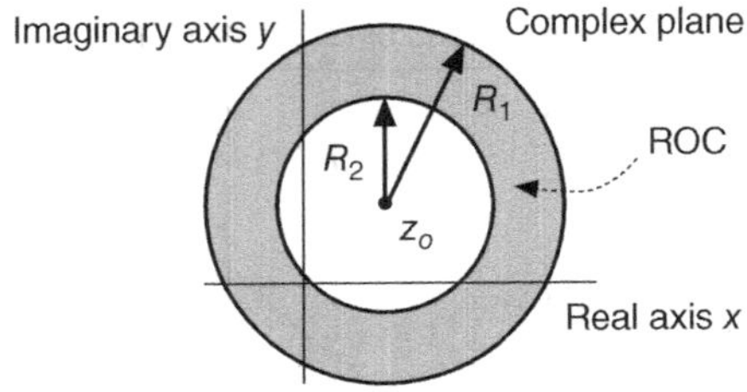

Figure E.4 Radii $\{R_1, R_2\}$ defining annulus of points about z_o for which a Laurent series is convergent.

These singularities are discussed further in Chapter 5.

In order for the first sum in (E.21) to be convergent, we require

$$|z - z_o| < R_1 \tag{E.23}$$

for some radius $R_1 > 0$ that defines a circle centered at z_o as depicted in Figure E.4. This result is identical to that required for a Taylor series, as expected because (E.1) is the same expression as the first sum in (E.21). Similarly, in order for the second sum in (E.21) to be convergent, we must have

$$\frac{1}{|z - z_o|} < R_2 \implies |z - z_o| > R_2, \tag{E.24}$$

which yields a region extending beyond a circle of radius $R_2 > 0$ because the $(z - z_o)^m$ terms appear in the denominator. This is also shown in Figure E.4, where it is clear that for both sums of the Laurent series to be convergent the radii must satisfy $R_2 < R_1$, and so the ROC is the *intersection* of the regions on the complex plane defined by (E.23) and (E.24). The shaded region in the figure is called an *annulus* (which is a type of ring), and within this ROC $f(z)$ is analytic. The contour of integration in (E.22) is performed counterclockwise in the shaded region enclosing $z = z_o$; the contour need not be circular, but it should form a closed path.

Example E.4 Consider again the function $f(z) = 1/(1 - z)$ which we would like to expand again about $z_o = 0$, as was done in Table E.1 using a Taylor series, but in this case let the ROC be $|z| > 1$. This function has a real pole at $z = 1$. In order to have the type of ROC in (E.24), the summation in (E.20) is performed over negative n or, equivalently, over positive m with $z - z_o$ in the denominator as in (E.21). Observe that

$$\sum_{n=-\infty}^{0} z^n = \sum_{m=0}^{\infty} \frac{1}{z^m} = \frac{1}{1 - 1/z}, \tag{E.25}$$

where we have changed variables to $m = -n$ and used the closed-form expression for a geometric series (see Appendix C). This result can be rearranged as follows:

$$\frac{1}{1 - 1/z} = \frac{z}{z - 1} = 1 + \frac{1}{z - 1} = 1 - f(z). \tag{E.26}$$

Thus,

$$1 - f(z) = \sum_{m=0}^{\infty} \frac{1}{z^m} \Rightarrow f(z) = 1 - \sum_{m=0}^{\infty} \frac{1}{z^m} = -\sum_{m=1}^{\infty} \frac{1}{z^m}, \tag{E.27}$$

where the leading 1 has canceled the $m = 0$ term in the sum. The last expression is the Laurent series expansion of $f(z)$ about $z_o = 0$ with ROC $|z| > 1$, which is the region outside the singularity. In this example, all coefficients have the same value $c_m = -1$.

When the second sum of (E.21) is 0, the coefficients are computed using (E.6), and when the first sum is 0, they are derived using the *theory of residues*. The residue technique is widely used to evaluate the inverse z-transform for discrete-time signals and systems. The z-transform is closely related to the Laplace transform in Chapter 7, but it assumes that the function is nonzero only for discrete values of time, whereas the Laplace transform assumes continuous time t.

APPENDIX F

LAMBERT W-FUNCTION

In this appendix, we give a brief overview of the Lambert W-function and illustrate how it is used to write an explicit expression for the nonlinear diode circuit in Chapter 2.

F.1 LAMBERT W-FUNCTION

The Lambert W-function (Corless *et al.*, 1996) for real-valued x is the solution w of the following equation:

$$x = w\exp(w). \tag{F.1}$$

It is not possible to write w explicitly as a function of x in terms of the ordinary functions described in this book. For example, if we take the logarithm of both sides:

$$\ln(x) = \ln(w) + w, \tag{F.2}$$

it is still not possible to solve explicitly for w. Figure F.1(a) shows a plot of (F.1), from which we see there is a region where two values of w map to x. It is straightforward to show that this region is the open interval $w \in (-\infty, 0)$. The dotted line at $x = -1/e$ is the minimum value of x (where $w = -1$), and the dashed line at $x = -1/2e$ is an example where two values of w map to a single x. The solution of (F.1) is

$$w = W(x), \tag{F.3}$$

Mathematical Foundations for Linear Circuits and Systems in Engineering, First Edition. John J. Shynk.

Companion Website: http://www.wiley.com/go/linearcircuitsandsystems

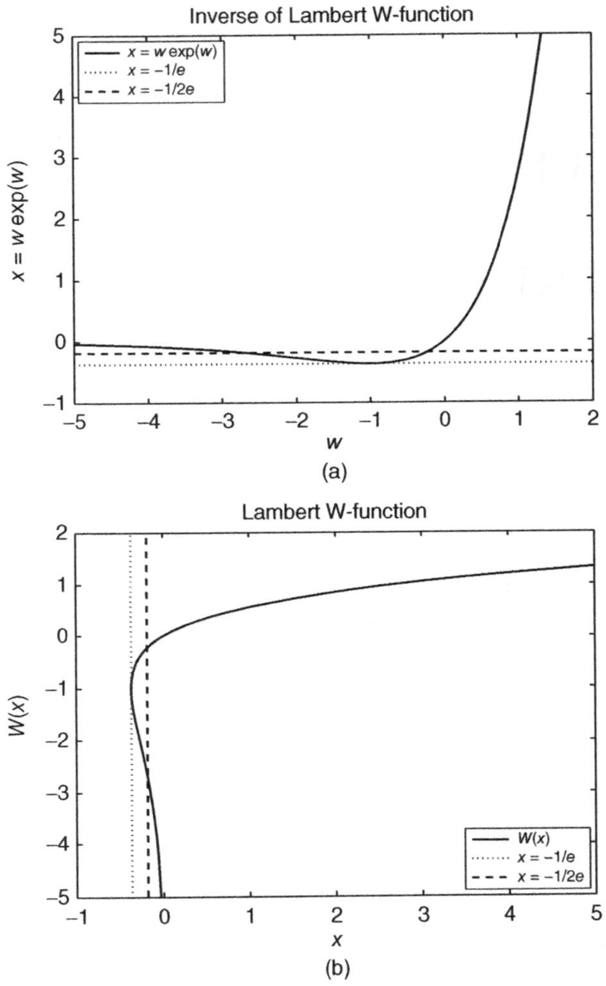

Figure F.1 (a) Inverse of Lambert W-function. (b) Lambert W-function (inverse image of the function in (a)).

where $W(x)$ is the notation for the Lambert W-function and its argument is the left-hand side of (F.1). Since the solution $W(x)$ is a function of x, (F.1) is often written as

$$x = W(x)\exp(W(x)). \tag{F.4}$$

There are only a few results for $W(x)$ that are obvious from the form in (F.4), such as

$$x = 0 \implies 0 = 0\exp(0) = 0 \implies W(0) = 0, \tag{F.5}$$

$$x = e \implies e = W(e)\exp(W(e)) \implies W(e) = 1, \tag{F.6}$$

$$x = -1/e \implies -1/e = W(-1/e)\exp(W(-1/e)) \implies W(-1/e) = -1. \tag{F.7}$$

In general, numerical methods are needed to evaluate $W(x)$ for other values of x. The derivative of $W(x)$ is derived by rearranging (F.4) as

$$W(x) = x\exp(-W(x)), \tag{F.8}$$

and using the product and chain rules:

$$W'(x) = \exp(-W(x)) + x\exp(-W(x))W'(x), \tag{F.9}$$

which yields

$$W'(x) = \frac{\exp(-W(x))}{1 + x\exp(-W(x))}. \tag{F.10}$$

Multiplying and dividing the last expression by x, it simplifies as follows when using (F.8):

$$W'(x) = \frac{W(x)}{x + xW(x)}. \tag{F.11}$$

Observe that $W'(0) = 1$ from (F.5) and (F.10), $W'(e) = 1/2e$ from (F.6) and (F.11), and $W'(-1/e) \longrightarrow \infty$ from (F.6) and (F.11).

The Lambert W-function can be examined by rearranging the plot in Figure F.1(a) so that the horizontal axis is x and the vertical axis is $W(x)$. The result shown in Figure F.1(b) is a curve representing the Lambert W-function for $x \in [-1/e, 5]$. The horizontal lines are now vertical lines in the new plot, located at the same values of x. Figure F.1(b) demonstrates that $W(x)$ is actually *multivalued*: every $x \in (-1/e, 0)$ maps to two values of $W(x)$. Thus, $W(x)$ is not the inverse function of the plot in Figure F.1(a); it is called the *inverse image* (see the definition in Chapter 1). Note from the figure that $W(0) = 0$, $W(-1/e) = -1$, $W'(0) = 1$, and $W'(-1/e) \longrightarrow \infty$, as shown earlier. The Lambert W-function is complex-valued for $x < -1/e$.

From Figure F.1(b), we find that $W(x) = -1$ is the dividing point between the two sets of values of $W(x)$ derived from the same x. The *branch* of the function for $W(x) \geq -1$ is denoted by $W_0(x)$, and that for $W(x) \leq -1$ is $W_{-1}(x)$. Although an explicit expression cannot be derived for $W(x)$, the plot in Figure F.1(b) can be used to find $W(x)$ given a specific value for x. The particular application determines if $W_0(x)$ or $W_{-1}(x)$ should be used when $x \in (-1/e, 0)$.

An equation is solved in terms of the Lambert W-function if it can be rearranged in the form of (F.4), as demonstrated by the next example.

Example F.1 Consider the following nonlinear equation for which there is no explicit solution for x:

$$ax + b^x = 0. \tag{F.12}$$

In order to write this in the form of (F.1), we use the identity $\exp(\ln(b)) = b$ assuming $b > 0$:

$$ax + [\exp(\ln(b))]^x = 0 \implies ax + \exp(x\ln(b)) = 0. \tag{F.13}$$

Solving for $1/a$ yields

$$1/a = -x\exp(-x\ln(b)), \tag{F.14}$$

which almost has the form in (F.1). Multiplying both sides by $\ln(b)$ gives the desired right-hand side:

$$\ln(b)/a = -x\ln(b)\exp(-x\ln(b)), \tag{F.15}$$

where in the notation of (F.4), the left-hand side is x and the term multiplying the exponential function is $W(x)$. Thus,

$$W(\ln(b)/a) = -x\ln(b) \implies x = -W(\ln(b)/a)/\ln(b), \tag{F.16}$$

which is the desired solution for x. Suppose that $a = b = 3$. Then using the MATLAB function lambertw, we find that $x = -W(\ln(3)/3)/\ln(3) \approx -0.2526$ with $W(\ln(3)/3) \approx 0.2775$. This is verified by the original equation in (F.12): $3(-0.2526) + 3^{-0.2526} = 0$.

Once the form in (F.4) is derived, the steps for writing x in terms of $W(x)$ are as follows, using the result in (F.15) as an example.

- The left-hand side of (F.4) is the argument of $W(\cdot)$, which is $\ln(b)/a$ in the previous example, yielding $W(\ln(b)/a)$.
- The quantity multiplying the exponential function in (F.4) equals $W(\cdot)$, which for the previous example is $-x\ln(b)$.
- Equating these two terms gives the desired equation $W(\ln(b)/a) = -x\ln(b)$.
- In the final step, we solve for x, which yields (F.16) for the previous example.

Example F.2 The following equation also has no explicit solution in terms of ordinary functions:

$$a + bx^x = 0. \tag{F.17}$$

Rearranging this expression and taking logarithms yields (assuming $-a/b > 0$):

$$\ln(-a/b) = x\ln(x), \tag{F.18}$$

and

$$\ln(-a/b) = \ln(x)\exp(\ln(x)). \tag{F.19}$$

Thus,

$$W(\ln(-a/b)) = \ln(x) \implies x = \exp(W(\ln(-a/b)). \tag{F.20}$$

The equation in (F.17) is plotted in Figure F.2 for $a = 3$ and $b = -3$ (the solid line). Observe that it has a real solution because it crosses the horizontal dotted line at 0. The MATLAB function lambertw gives $x = 1$, which is easily verified in (F.17) (actually,

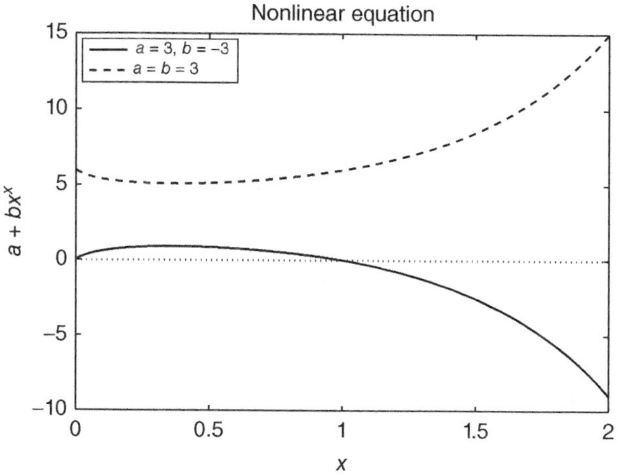

Figure F.2 Nonlinear equation of Example F.2.

the solution is obvious for these values of a and b). Observe that when $b = 3$, the dashed line does not cross the horizontal line at 0, which means it does not have a real solution. However, as we know from Chapter 4, it has a *complex* solution. Thus, it is not necessary to restrict $-a/b > 0$ as earlier; the logarithm yields a complex number for a negative argument, and $W(x)$ is also complex. The result for $a = b = 3$ is $1.6904 + j1.8699$, which can be verified by substitution into (F.17). (Likewise, we need not restrict b in Example F.1: allowing $b < 0$ causes $W(x)$ to be complex.)

F.2 NONLINEAR DIODE CIRCUIT

The Lambert W-function can be used to write a solution for the diode circuit described in Chapter 2 (Banwell and Jayakumar, 2000; Ortiz-Conde *et al.*, 2000), without having to use the iterative techniques discussed there. Recall that the relevant I-V equations for the diode in series with resistor R and voltage source V_s are

$$i = (V_s - v)/R, \qquad i = I_s \exp(v/V_T), \tag{F.21}$$

with $I_s = 10^{-15}$ A and $V_T = 0.026$ V. Substituting the first equation $v = V_s - iR$ into the second equation yields

$$i = I_s \exp((V_s - iR)/V_T), \tag{F.22}$$

which has only one independent variable i. In order to proceed, the function is rearranged as follows:

$$i = I_s \exp(V_s/V_T)\exp(-iR/V_T) \implies I_s \exp(V_s/V_T) = i\exp(iR/V_T). \tag{F.23}$$

Multiplying both sides by R/V_T gives the desired form in (F.4):

$$(I_sR/V_T)\exp(V_s/V_T) = (iR/V_T)\exp(iR/V_T), \tag{F.24}$$

which yields

$$W((I_sR/V_T)\exp(V_s/V_T)) = iR/V_T, \tag{F.25}$$

and the following equation for the current:

$$i = (V_T/R)W((I_sR/V_T)\exp(V_s/V_T)). \tag{F.26}$$

Example F.3 For $V_s = 1.2$ V and $R = 100\ \Omega$, we find from lambertw that $i \approx 0.0044$ A, and the voltage is $v \approx 0.7571$ V from $v = V_s - iR$. Substituting this voltage into the second equation of (F.21) verifies i given by lambertw, and these two values match those generated by the iterative method in Example 2.11.

F.3 SYSTEM OF NONLINEAR EQUATIONS

Finally, we show how to solve the nonlinear equations in (1.13) and (1.14), which we repeat as follows:

$$a_{11}y_1(t) + a_{12}\exp(\alpha y_2(t)) = x_1(t), \tag{F.27}$$

$$a_{21}y_1(t) + a_{22}y_2(t) = x_2(t). \tag{F.28}$$

Solving the second equation for $y_1(t)$ and substituting it into the first equation yields

$$a_{11}[x_2(t) - a_{22}y_2(t)]/a_{21} + a_{12}\exp(\alpha y_2(t)) = x_1(t). \tag{F.29}$$

For notational convenience, we drop the time argument and define the following quantities: $c_1 \triangleq -a_{11}x_2/a_{21}$ and $c_2 \triangleq -a_{11}a_{22}/a_{21}$ such that

$$c_2y_2 + a_{12}\exp(\alpha y_2) = x_1 + c_1. \tag{F.30}$$

Multiplying by $\alpha\exp(-\alpha y_2)/c_2$ gives

$$\alpha y_2\exp(-\alpha y_2) + \alpha a_{12}/c_2 = \alpha(x_1 + c_1)\exp(-\alpha y_2)/c_2, \tag{F.31}$$

so that y_2 now multiplies the exponential function on the left-hand side (recall that we need to rearrange this expression to have the form in (F.4) in order to solve for y_2). Next, we bring the first exponential to the right-hand side and factor it from the two terms:

$$\alpha a_{12}/c_2 = \alpha(-y_2 + x_1/c_2 + c_1/c_2)\exp(-\alpha y_2). \tag{F.32}$$

In the last step, we multiply by $\exp(\alpha(x_1 + c_1)/c_2)$ to obtain the final form:

$$(\alpha a_{12}/c_2)\exp(\alpha(x_1 + c_1)/c_2)$$
$$= \alpha(-y_2 + x_1/c_2 + c_1/c_2)\exp(\alpha(-y_2 + x_1/c_2 + c_1/c_2)), \tag{F.33}$$

yielding

$$y_2 = (x_1 + c_1)/c_2 - (1/\alpha)W((\alpha a_{12}/c_2)\exp(\alpha(x_1 + c_1)/c_2))$$
$$= x_2/a_{22} - a_{21}x_1/a_{11}a_{22}$$
$$- (1/\alpha)W((-\alpha a_{12}a_{21}/a_{11}a_{22})\exp(-\alpha a_{21}(x_1 - a_{11}x_2/a_{21})/a_{11}a_{22})), \tag{F.34}$$

where $\{c_1, c_2\}$ have been substituted so that the expression is written in terms of the original parameters.

Example F.4 The parameters in Example 1.4 are $\alpha = 4$, $a_{11} = a_{21} = a_{22} = 1$, $a_{12} = -0.1$, $x_1 = 0$, and $x_2 = 1$, such that (F.34) becomes

$$y_2 = 1 - (1/4)W((0.4)\exp(4)). \tag{F.35}$$

The lambertw function gives $y_2 \approx 0.4336$, and from (F.28) we have

$$y_1 + y_2 = 1 \implies y_1 \approx 0.5664, \tag{F.36}$$

which are the same values generated by the iterative technique in Example 1.4.

GLOSSARY

SUMMARY OF NOTATION

0^-: lower limit of integral includes singular functions at 0

0^+: lower limit of integral excludes singular functions at 0

$\mathbf{0}$: zero matrix

a: acceleration (m/s^2)

a: real part of complex c

$\tilde{a}$, $\tilde{\tilde{a}}$: modified elements of matrix $\mathbf{A}$ in row-echelon form

arg(c): argument (angle) of complex number c

adj($\mathbf{A}$): adjugate matrix of $\mathbf{A}$

$\mathbf{a}$: column of matrix $\mathbf{A}$

$\bar{\mathbf{a}}^T$: row of matrix $\mathbf{A}$

b: base of logarithm or imaginary part of complex c

A: ampere (C/s)

A_{mn}: cofactor of matrix $\mathbf{A}$

$\mathbf{A}$: matrix

$\mathbf{A}^T$: matrix transpose

$\mathbf{A}^H$: matrix transpose and complex conjugation

Mathematical Foundations for Linear Circuits and Systems in Engineering, First Edition. John J. Shynk.

Companion Website: http://www.wiley.com/go/linearcircuitsandsystems

$\mathbf{A}^{-1}$: matrix inverse

B: damping constant (N s/m)

BW: bandwidth

c: complex number or speed of light

$c(t)$: carrier waveform in amplitude modulation

$c_{fg}(\tau)$: cross-correlation function

C: coulomb

C: capacitor, capacitance (F), or contour of integration

$C(\mathbf{A})$: column space of matrix **A**

$\mathbf{C}$: matrix representation for complex numbers

d: distance (m)

det(**A**): determinant of matrix **A**

dB: decibel

D: diode symbol

e: Napier's constant 2.718281828459 … or energy (J)

$\mathbf{e}$: unit vector

exp(**A***t*): matrix exponential

E: energy (J)

$\mathbf{E}$: elementary matrix

f: natural frequency (Hz)

$f_E(t)$: even function

$f_O(t)$: odd function

F: farad

F: force (N)

$F(x)$: antiderivative of $f(x)$

$\mathbf{F}$: phasor of function $f(t)$

g: gram

g: acceleration due to gravity (m/s^2)

$g(t)$: integrating factor

h: height (m) or quaternion

$h(t)$: impulse response function

H: henry

$H(s)$: transfer function

$H(\omega)$: frequency response

$\mathbf{H}$: matrix representation for quaternions

i: current (A)

$\{i, j, k\}$: quaternion markers

I: constant current (A) or moment of inertia

$I(t)$: indicator function

$\mathbf{I}$: identity matrix or phasor current

$\tilde{\mathbf{I}}$: exchange matrix

j: $\sqrt{-1}$, imaginary marker of complex number

J: joule

$\mathbf{J}$: exchange matrix

$k(p, t)$: kernel of integral transform

K: spring constant (N/m)

$\log_b(\cdot)$: logarithm with base b

$\ln(\cdot)$: natural logarithm with base e

L: inductor, inductance (H), or length (m)

$L(\mathbf{A})$: left null space of matrix $\mathbf{A}$

$\mathbf{L}$: lower triangular matrix

m: meter

M: mass (g)

$\mathbf{M}_{mn}$: minor matrix

$\max(\cdot)$: maximum

$\min(\cdot)$: minimum

$n!$: factorial

$\binom{n}{m}$: binomial coefficient

N: newton

$N(\mathbf{A})$: null space of matrix $\mathbf{A}$

p: instantaneous power (W) or matrix pivots

p_n: transfer function pole

P: average power (W)

$\mathbf{P}$: permutation matrix

q: charge (C)

Q: total charge (C) or circuit quality factor

$\mathbf{q}$: normalized eigenvector

$\mathbf{Q}$: matrix of normalized eigenvectors

r: radius

$r(t)$: ramp function

$\text{rect}(t)$: rectangle function

R: resistor, resistance (Ω), matrix rank, or radius

$R(\mathbf{A})$: row space of matrix $\mathbf{A}$

$\mathbf{R}$: rotation matrix

s: second

s: complex variable $s = \sigma + j\omega$

sgn(t): signum function
sinc(t): sinc function
t: time (s)
tr(**A**): trace of matrix **A**
T: period (s)
tri(t): triangle function
$u(t)$: unit step function
$u_n(t)$: compact notation for ramp, step, Dirac delta, and doublet
U: upper triangular matrix
v: voltage (V) or velocity (m/s)
V: constant voltage (V)
V: phasor voltage
w: work (J)
W: watt or Lambert W
$W(t)$: Wronskian
$W(x)$: Lambert W-function
$\tilde{x}, \tilde{\tilde{x}}$: modified elements of vector **x** in row-echelon form
X: reactance
$X(f)$: Fourier transform of $x(t)$ (natural frequency)
$X(\omega)$: Fourier transform of $x(t)$ (angular frequency)
$X(s)$: Laplace transform of $x(t)$
$y_h(t)$: homogeneous solution of ODE
$y_p(t)$: particular solution of ODE
y_s: steady-state step response
$y_t(t)$: transient step response
$\mathbf{y}_h$: homogeneous solution vector of matrix equation
$\mathbf{y}_p$: particular solution vector of matrix equation
z: complex variable of series expansions
z_n: transfer function zero
Z: impedance

GREEK SYMBOLS

α: Neper frequency (rad/s)
$\delta[\omega]$: Kronecker delta function
$\delta(t)$: Dirac delta function
$\delta'(t)$: unit doublet
$\delta''(t)$: unit triplet
$\delta^{(n)}(t)$: nth derivative of Dirac delta function

Δ: discriminant or a small interval

Δx: small interval on x

λ: eigenvalue or wavelength

Λ: diagonal matrix of eigenvalues

ω: angular frequency (rad/s) or imaginary part of complex variable s

ω_c: center/cutoff frequency or carrier frequency

ω_d: damped ω_o

ω_o: specific angular frequency or resonant frequency

Ω: ohm or universal set

π: 3.14159265358979323846...

ϕ: angle (radians or degrees) or empty set

$\phi(t)$: test function

$\Phi(\omega)$: Fourier transform of test function

σ: real part of complex variable s

Σ: summation

τ: time constant, delay (s), or torque

θ: angle (radians or degrees)

ζ: damping ratio

CALLIGRAPHIC SYMBOLS

$\mathcal{C}$: complex numbers

$\mathcal{D}$: set of test functions with compact support

$\mathcal{D}'$: dual space for $\mathcal{D}$

$\mathcal{E}$: set of test functions of exponential decay

$\mathcal{E}'$: dual space for $\mathcal{E}$

$\mathcal{F}$: field

$\mathcal{F}\{\cdot\}$: Fourier transform

$\mathcal{F}^{-1}\{\cdot\}$: inverse Fourier transform

$\mathcal{H}$: quaternions

$\mathcal{I}$: imaginary numbers

$\mathcal{L}\{\cdot\}$: unilateral Laplace transform

$\mathcal{L}_b\{\cdot\}$: bilateral Laplace transform

$\mathcal{L}^{-1}\{\cdot\}$: inverse Laplace transform

$\mathcal{N}$: natural numbers $\{1, 2, \cdots\}$

$\mathcal{Q}$: rational numbers

$\mathcal{P}(f(t))$: use Cauchy principal value

$\mathcal{R}$: real numbers $(-\infty, \infty)$

$\mathcal{R}^+$: nonnegative real numbers $[0, \infty)$

$\mathcal{S}$: subspace or set of test functions of rapid decay
$\mathcal{S}^{\perp}$: orthogonal complement of subspace $\mathcal{S}$
$\mathcal{S}'$: dual space of $\mathcal{S}$
$\mathcal{V}$: vector space
$\mathcal{Z}$: integers $\{\dots, -2, -1, 0, 1, 2, \dots\}$
$\mathcal{Z}^{+}$: nonnegative integers $\{0, 1, 2, \dots\}$

MATHEMATICAL NOTATION

$*$: convolution or complex conjugation superscript
$\star$: correlation
$\longrightarrow$: to next step
$\underset{\longrightarrow}{\mathcal{F}}$: Fourier transform
$\underset{\longrightarrow}{\mathcal{F}^{-1}}$: inverse Fourier transform
$\underset{\longrightarrow}{\mathcal{L}}$: Laplace transform
$\underset{\longrightarrow}{\mathcal{L}^{-1}}$: inverse Laplace transform
$\Longrightarrow$: implies
$\triangleq$: defined as
$\equiv$: equivalent to
$(a, \mathbf{b})$: quaternion
$\angle\theta$: angle θ of polar form
$\langle f, \phi \rangle$: generalized function f with test function ϕ
$<$: less than
$\ll$: much less than
$\leq$: less than or equal to
$>$: greater than
$\gg$: much greater than
$\geq$: greater than or equal to
$\in$: element of
$\notin$: not an element of
$|t|$: absolute value of t
$\|\mathbf{v}\|$: vector norm
$\|\mathbf{v}\|^2$: vector squared norm
$f^{-1}(y)$: inverse image of function $y = f(x)$
$\dot{x}$: derivative with respect to time
x': ordinary derivative
x'': second ordinary derivative
$x^{(n)}$: nth ordinary derivative

$|A|$: cardinality of set A

$A \subset B$: A is a subset of B

$A^c, \overline{A}$: complement of set A

$A \cup B$: union of sets

$A \cap B, AB$: intersection of sets

$A - B, A \backslash B$: difference of sets

$A \oplus B$: exclusive or of sets

PHYSICAL PARAMETERS

g: acceleration due to gravity (9.80665 m/s^2)

q_e: elementary charge (1.6021×10^{-19} C)

I_s: saturation current (10^{-15} A)

V_T: thermal voltage (0.026 V)

ABBREVIATIONS

a: acceleration

arg: argument

A: ampere

AM: amplitude modulation

BP: band-pass

BR: band-reject

C: coulomb

CPV: Cauchy principal value

dB: decibel

DC: direct current

DE: differential equation

EHF: extremely high frequency

F: farad

FM: frequency modulation

FVT: final value theorem

g: gram

GE: Gaussian elimination

H: henry

HF: high frequency

HP: high-pass

Hz: Hertz

ISO: International Organization for Standardization
ITU: International Telecommunication Union
I–V: current–voltage characteristic
IVT: initial value theorem
J: joule
KCL: Kirchoff's current law
KVL: Kirchoff's voltage law
LDU: lower triangular/diagonal/upper triangular matrix decomposition
LF: low frequency
LP: low-pass
LTI: linear and time-invariant
LU: lower/upper triangular matrix decomposition
m: meter
MF: medium frequency
MIMO: multiple-input multiple-output
N: newton
NM: Newton's method
oc: open circuit subscript
ODE: ordinary differential equation
PDE: partial differential equation
PFE: partial fraction expansion
QAM: quadrature amplitude modulation
rad: radian
RC: resistor/capacitor circuit
RL: resistor/inductor circuit
RLC: resistor/inductor/capacitor circuit
ROC: region of convergence or radius of convergence
s: second
sc: short circuit subscript
SHF: super high frequency
SISO: single-input single-output
SSB: single sideband
th: Thévenin subscript
ULF: ultra low frequency
UHF: ultra high frequency
V: volt
V–I: voltage–current characteristic
VLF: very low frequency
VHF: very high frequency

BIBLIOGRAPHY

A. Agarwal and J. H. Lang, *Foundations of Analog and Digital Electronic Circuits*, Morgan Kaufmann Publishers, Amsterdam, 2005.

L. C. Andrews and R. L. Phillips, *Mathematical Techniques for Engineers and Scientists*, SPIE Press, Bellingham, WA, 2003.

T. C. Banwell and A. Jayakumar, "Exact analytical solution for current flow through diode with series resistance," *IEE Electronics Letters*, vol. 36, no. 4, pp. 291–292, 2000.

H. Bateman, *Tables of Integral Transforms*, vols. I and II, McGraw-Hill, New York, 1954.

M. Beck, G. Marchesi, D. Paxton, and L. Sabalka, A First Course in Complex Analysis, version 1.4, 2012, http://math.sfsu.edu/beck/complex.html.

F. P. Beer and E. R. Johnston, Jr., *Vector Mechanics for Engineers: Statics and Dynamics*, second edition, McGraw-Hill, New York, 1972.

W. H. Beyer, *Standard Mathematical Tables*, twenty-fourth edition, CRC Press, Cleveland, OH, 1976.

R. N. Bracewell, *The Fourier Transform and Its Applications*, second edition, McGraw-Hill, New York, 1978.

J. W. Brown and R. V. Churchill, *Complex Variables and Applications*, eighth edition, McGraw-Hill, New York, 2009.

J. R. Buck, M. M. Daniel, and A. C. Singer, *Computer Explorations in Signals and Systems Using MATLAB*, second edition, Prentice-Hall, Upper Saddle River, NJ, 2002.

F. J. Bueche, *Introduction to Physics for Scientists and Engineers*, second edition, McGraw-Hill, New York, 1975.

Mathematical Foundations for Linear Circuits and Systems in Engineering, First Edition. John J. Shynk.

Companion Website: http://www.wiley.com/go/linearcircuitsandsystems

B. L. Burrows and D. J. Colwell, "The Fourier transform of the unit step function," *International Journal of Mathematical Education in Science and Technology*, vol. 21, no. 4, pp. 629–635, 2011.

J. A. Cadzow and H. F. Van Landingham, *Signals, Systems, and Transforms*, Prentice-Hall, Englewood Cliffs, NJ, 1985.

G. E. Carlson, *Signal and Linear System Analysis*, Houghton Mifflin, Boston, MA, 1992.

G. F. Carrier and C. E. Pearson, *Partial Differential Equations: Theory and Technique*, Academic Press, New York, 1976.

R. M. Corless, G. H. Gonnet, D. E. G. Hare, D. J. Jeffrey, and D. E. Knuth, "On the Lambert *W* function," *Advances in Computational Mathematics*, vol. 5, no. 1, pp. 329–359, 1996.

J. J. D'Azzo, C. H. Houpis, and S. N. Sheldon, *Linear Control System Analysis and Design with MATLAB*, fifth edition, CRC Press, Boca Raton, FL, 2003.

J. J. DiStefano III, A. R. Stubberud, and I. J. Williams, *Schaum's Outline of Theory and Problems of Feedback and Control Systems*, second edition, McGraw-Hill, New York, 1990.

R. C. Dorf, *Modern Control Systems*, second edition, Addison-Wesley, Reading, MA, 1974.

H. Eves, *Elementary Matrix Theory*, Dover Publications, Mineola, NY, 1980.

F. Farassat, "Introduction to generalized functions with applications in aerodynamics and aeroacoustics," NASA Technical Paper 3428, April 1996.

R. A. Gabel and R. A. Roberts, *Signals and Linear Systems*, third edition, John Wiley & Sons, Inc., New York, 1987.

Z. Gajić, *Linear Dynamical Systems and Signals*, Prentice-Hall, Upper Saddle River, NJ, 2003.

F. R. Gantmacher, *The Theory of Matrices*, vol. 1, Chelsea, New York, 1959.

W. J. Gilbert, *Modern Algebra with Applications*, John Wiley & Sons, Inc., New York, 1976.

R. Goldman, *Rethinking Quaternions: Theory and Computation*, Morgan & Claypool, San Rafael, CA, 2010.

G. H. Golub and C. F. Van Loan, *Matrix Computations*, The Johns Hopkins University Press, Baltimore, MD, 1983.

I. S. Gradshteyn and I. M. Ryzhik, *Table of Integrals, Series, and Products*, Academic Press, New York, 1980.

U. Graf, *Applied Laplace Transforms and z-Transforms for Scientists and Engineers: A Computational Approach Using a Mathematica Package*, Birkhäuser, Boston, MA, 2004.

R. M. Gray and J. W. Goodman, *Fourier Transforms: An Introduction for Engineers*, Springer Science, New York, 1995.

V. S. Groza and S. Shelley, *Precalculus Mathematics*, Holt, Rinehart, and Winston, New York, 1972.

A. J. Hanson, *Visualizing Quarternions*, Morgan Kaufmann Publishers, Amsterdam, 2006.

W. W. Harman and D. W. Lytle, *Electrical and Mechanical Networks: An Introduction to Their Analysis*, McGraw-Hill, New York, 1962.

S. Haykin, *Communication Systems*, fourth edition, John Wiley & Sons, Inc., Hoboken, NJ, 2001.

S. Haykin and B. Van Veen, *Signals and Systems*, John Wiley & Sons, Inc., New York, 1999.

W. H. Hayt Jr., J. E. Kemmerly, and S. M. Durbin, *Engineering Circuit Analysis*, eighth edition, McGraw-Hill, New York, 2007.

R. F. Hoskins, *Delta Functions: An Introduction to Generalised Functions*, Horwood Publishing, Chichester, UK, 1999.

H. P. Hsu, *Schaum's Outline of Theory and Problems of Signals and Systems*, McGraw-Hill, New York, 1995.

J. D. Irwin, *Basic Engineering Circuit Analysis*, fourth edition, Macmillan, New York, 1993.

J. D. Irwin and R. M. Nelms, *Basic Engineering Circuit Analysis*, eighth edition, John Wiley & Sons, Inc., Hoboken, NJ, 2005.

ISO (International Organization for Standardization), ISO 21348: Space Environment (Natural and Artificial)—Process for Determining Solar Irradiances, first edition, Geneva, Switzerland, 2007.

ITU (International Telecommunication Union), Nomenclature of the frequency and wavelength bands used in telecommunications, Recommendation ITU-R V.431-7, 2000.

D. E. Johnson, J. R. Johnson, and J. L. Hilburn, *Electric Circuit Analysis*, second edition, Prentice-Hall, Englewood Cliffs, NJ, 1992.

R. E. Johnson and F. L. Kiokemeister, *Calculus with Analytic Geometry*, second edition, Allyn & Bacon, Boston, MA, 1960.

D. S. Jones, *The Theory of Generalised Functions*, second edition, Cambridge University Press, Cambridge, UK, 1982.

T. Kailath, *Linear Systems*, Prentice-Hall, Englewood Cliffs, NJ, 1980.

E. W. Kamen and B. S. Heck, *Fundamentals of Signals and Systems Using MATLAB®*, Prentice-Hall, Upper Saddle River, NJ, 1997.

R. P. Kanwal, *Generalized Functions: Theory and Applications*, third edition, Birkhäuser, Boston, MA, 2004.

B. Kolman, *Elementary Linear Algebra*, second edition, Macmillan, New York, 1977.

E. Kreyszig, *Advanced Engineering Mathematics*, fourth edition, John Wiley & Sons, Inc., New York, 1979.

B. R. Kusse and E. A. Westwig, *Mathematical Physics: Applied Mathematics for Scientists and Engineers*, Wiley-VCH, Weinheim, Germany, 2006.

B. P. Lathi, *Signals, Systems, and Communication*, John Wiley & Sons, Inc., New York, 1965.

B. P. Lathi, *Linear Systems and Signals*, Berkeley-Cambridge, Carmichael, CA, 1992.

A. J. Laub, *Matrix Analysis for Scientists and Engineers*, Society for Industrial and Applied Mathematics, Philadelphia, PA, 2005.

S. J. Leon, *Linear Algebra with Applications*, third edition, Macmillan, New York, 1990.

K. H. Lundberg, H. R. Miller, and D. L. Trumper, "Initial conditions, generalized functions, and the Laplace transform: troubles at the origin," *IEEE Control Systems Magazine*, vol. 27, no. 1, pp. 22–35, 2007.

E. Maor, *e: The Story of a Number*, Princeton University Press, Princeton, NJ, 1994.

J. H. McClellan, R. W. Schafer, and M. A. Yoder, *Signal Processing First*, Prentice-Hall, Upper Saddle River, NJ, 2003.

J. McLeish, *The Story of Numbers: How Mathematics Has Shaped Civilization*, Fawcett Columbine, New York, 1991.

C. D. Meyer, *Matrix Analysis and Applied Linear Algebra*, Society for Industrial and Applied Mathematics, Philadelphia, PA, 2000.

I. Miller and J. E. Freund, *Probability and Statistics for Engineers*, second edition, Prentice-Hall, Englewood Cliffs, NJ, 1977.

I. Miller and S. Green, *Algebra and Trigonometry*, second edition, Prentice-Hall, Englewood Cliffs, NJ, 1970.

H. R. Miller, D. L. Trumper, and K. H. Lundberg, "A brief treatment of generalized functions for use in teaching the Laplace transform," in Proceedings of the IEEE Conference on Decision and Control, San Diego, CA, pp. 3885–3889, Dec. 2006.

D. A. Neamen, *Microelectronics: Circuit Analysis and Design*, fourth edition, McGraw-Hill, New York, 2010.

T. Needham, *Visual Complex Analysis*, Oxford University Press, Oxford, UK, 1999.

J. W. Nilsson and S. A. Riedel, *Electric Circuits*, seventh edition, Pearson Prentice-Hall, Upper Saddle River, NJ, 2005.

M. O'Flynn and E. Moriarty, *Linear Systems: Time Domain and Transform Analysis*, Harper & Row, New York, 1987.

A. V. Oppenheim, A. S. Willsky, and I. T. Young, *Signals and Systems*, Prentice-Hall, Englewood Cliffs, NJ, 1983.

A. Ortiz-Conde, F. J. García Sánchez, and J. Muci, "Exact analytical solutions of the forward non-ideal diode equation with series and shunt parasitic resistances," *Solid-State Electronics*, vol. 44, no. 10, pp. 1861–1864, 2000.

B. Osgood, *Lecture Notes for EE 261: The Fourier Transform and Its Applications*, Electrical Engineering Department, Stanford University, 2007.

A. Papoulis, *The Fourier Integral and Its Applications*, McGraw-Hill, New York, 1962.

C. L. Phillips and J. M. Parr, *Signals, Systems, and Transforms*, Prentice-Hall, Englewood Cliffs, NJ, 1995.

A. D. Poularikas, Ed., *The Transforms and Applications Handbook*, second edition, CRC Press, Boca Raton, FL, 2000.

M. J. Roberts, *Signals and Systems: Analysis Using Transform Methods and MATLAB®*, McGraw-Hill, New York, 2004.

M. J. Roberts, *Fundamentals of Signals & Systems*, McGraw-Hill, New York, 2008.

S. Ross, *A First Course in Probability*, Macmillan, New York, 1976.

S. L. Ross, *Introduction to Ordinary Differential Equations*, second edition, John Wiley & Sons, Inc., New York, 1977.

K. A. Ross, *Elementary Analysis: The Theory of Calculus*, Springer-Verlag, New York, 1980.

R. J. Schilling and H. Lee, *Engineering Analysis: A Vector Space Approach*, John Wiley & Sons, Inc., New York, 1988.

A. S. Sedra and K. C. Smith, *Microelectronic Circuits*, fifth edition, Oxford University Press, New York, 2004.

I. S. Sokolnikoff and R. M. Redheffer, *Mathematics of Physics and Modern Engineering*, second edition, McGraw-Hill, New York, 1966.

G. Strang, *Linear Algebra and Its Applications*, second edition, Academic Press, New York, 1980.

R. S. Strichartz, *A Guide to Distribution Theory and Fourier Transforms*, World Scientific, River Edge, NJ, 1994.

G. B. Thomas Jr., *Calculus and Analytic Geometry, Part One: Functions of One Variable and Analytic Geometry*, fourth edition, Addison-Wesley, Reading, MA, 1968a.

G. B. Thomas Jr., *Calculus and Analytic Geometry, Part Two: Vectors and Functions of Several Variables*, fourth edition, Addison-Wesley, Reading, MA, 1968b.

R. E. Thomas, A. J. Rosa, and G. J. Toussaint, *The Analysis and Design of Linear Circuits*, seventh edition, John Wiley & Sons, Inc., Hoboken, NJ, 2012.

S. A. Tretter, *Introduction to Discrete-Time Signal Processing*, John Wiley & Sons, Inc., New York, 1976.

F. T. Ulaby and M. M. Maharbiz, *Circuits*, second edition, National Technology & Science Press, Allendale, NJ, 2013.

U.S. Department of Commerce, National Telecommunications and Information Administration, Office of Spectrum Management, United States Frequency Allocations: The Radio Spectrum, Oct. 2003.

V. S. Vladimirov, *Methods of the Theory of Generalized Functions*, Taylor & Francis, New York, 2002.

R. H. Williams, *Electrical Engineering Probability*, West Publishing, St. Paul, MN, 1991.

R. E. Ziemer, W. H. Tranter, and D. R. Fannin, *Signals and Systems: Continuous and Discrete*, third edition, Macmillan, New York, 1993.

INDEX

Mathematical Foundations for Linear Circuits and Systems in Engineering, First Edition. John J. Shynk.

Companion Website: http://www.wiley.com/go/linearcircuitsandsystems

Printed and bound by CPI Group (UK) Ltd, Croydon, CR0 4YY
17/09/2026
14964286-0008